# Surface Engineering of Biomaterials

Surface engineering provides one of the most important means of engineering product differentiation in terms of quality, performance, and life cycle cost. It is essential to achieve predetermined functional properties of materials such as mechanical strength, biocompatibility, corrosion resistance, wear resistance, and heat and oxidation resistance. *Surface Engineering of Biomaterials* addresses this topic across a diverse range of process technologies and healthcare applications.

- Introduces biomaterial surface science and surface engineering and includes criteria for biomaterial surface selection.
- Focuses on a broad array of materials including metals, ceramics, polymers, alloys, and composites.
- Discusses corrosion, degradation, and material release issues in implant materials.
- Covers various processing routes to develop biomaterial surfaces, including for smart and energy applications.
- Details techniques for post-modification of biomaterial surfaces.

This reference work helps researchers working at the intersection of materials science and biotechnology to engineer functional biomaterials for a variety of applications.

# Emerging Materials and Technologies

*Series Editor: Boris I. Kharissov*

The *Emerging Materials and Technologies* series is devoted to highlighting publications centered on emerging advanced materials and novel technologies. Attention is paid to those newly discovered or applied materials with potential to solve pressing societal problems and improve quality of life, corresponding to environmental protection, medicine, communications, energy, transportation, advanced manufacturing, and related areas.

The series takes into account that, under present strong demands for energy, material, and cost savings, as well as heavy contamination problems and worldwide pandemic conditions, the area of emerging materials and related scalable technologies is a highly interdisciplinary field, with the need for researchers, professionals, and academics across the spectrum of engineering and technological disciplines. The main objective of this book series is to attract more attention to these materials and technologies and invite conversation among the international R&D community.

*Nanomaterials for Energy Applications*
Edited by L. Syam Sundar, Shaik Feroz, and Faramarz Djavanroodi

*Wastewater Treatment with the Fenton Process: Principles and Applications*
Dominika Bury, Piotr Marcinowski, Jan Bogacki, Michal Jakubczak, and Agnieszka Jastrzebska

*Mechanical Behavior of Advanced Materials: Modeling and Simulation*
Edited by Jia Li and Qihong Fang

*Shape Memory Polymer Composites: Characterization and Modeling*
Nilesh Tiwari and Kanif M. Markad

*Impedance Spectroscopy and its Application in Biological Detection*
Edited by Geeta Bhatt, Manoj Bhatt and Shantanu Bhattacharya

*Nanofillers for Sustainable Applications*
Edited by N.M Nurazzi, E. Bayraktar, M.N.F. Norrrahim, H.A. Aisyah, N. Abdullah, and M.R.M. Asyraf

*Chemistry of Dehydrogenation Reactions and its Applications*
Edited by Syed Shahabuddin, Rama Gaur and Nandini Mukherjee

*Biosorbents: Diversity, Bioprocessing, and Applications*
Edited by Pramod Kumar Mahish, Dakeshwar Kumar Verma and Shailesh Kumar Jadhav

For more information about this series, please visit:
www.routledge.com/Emerging-Materials-and-Technologies/book-series/CRCEMT

# Surface Engineering of Biomaterials
## Synthesis and Processing Techniques

Edited by
Ajit Behera
Debasis Nayak
Biswajit Kumar Swain

CRC Press
Taylor & Francis Group
Boca Raton  London  New York

CRC Press is an imprint of the
Taylor & Francis Group, an **informa** business

First edition published 2024
by CRC Press
2385 Executive Center Drive, Suite 320, Boca Raton, FL 33431

and by CRC Press
4 Park Square, Milton Park, Abingdon, Oxon, OX14 4RN

*CRC Press is an imprint of Taylor & Francis Group, LLC*

ISBN: 978-1-032-55282-8 (hbk)
ISBN: 978-1-032-55285-9 (pbk)
ISBN: 978-1-003-42992-0 (ebk)

DOI: 10.1201/9781003429920

Typeset in Times New Roman
by MPS Limited, Dehradun

# Contents

# SECTION 2  Surface Synthesis and Engineering of Biomaterials

**Chapter 15** Energy Biomaterial Surface ............................................. 306

*Debasis Nayak and Ajit Behera*

# SECTION 3    Post-Modification of Biomaterial Surface

**Chapter 16** Surface Treatment of Polymeric, Ceramic, Metallic, and
Composite Biomaterials for Bioimplants and Medical Device
Applications ................................................................ 319

*Garima Mittal and Shiladitya Paul*

**Chapter 17**    Surface Functionalization for Biomaterials....................................344

*Mona M. Agwa, Hesham S.M. Soliman, Heba Elmotasem, and Sally Sabra*

**Chapter 18**    Surface Modification Technologies and Methods of
                      Biomaterials.................................................................................362

*Mojtaba Najafizadeh, Payam Sarir, Sahar Yazdi,
Ehsan Marzban Shirkharkolaei, Mansoor Bozorg,
Morteza Hosseinzadeh, and Pasquale Cavaliere*

**Chapter 23** LASER-based Surface Modification Techniques for Fatigue
Life Improvement of Biomaterials .................................................503

*T. Rajesh Kumar Dora, Karthik Dhandapani,*
*Sarada Prasanna Mallick, and Pratik Shukla*

# Editor Biographies

**Ajit Behera** is an assistant professor in the Metallurgical and Materials Department at the National Institute of Technology, Rourkela, India. He completed his PhD at IIT Kharagpur.

**Debasis Nayak** is a research associate in the Department of Material Science and Metallurgy, University of Cambridge, UK. He completed his PhD at IIT Kharagpur.

**Biswajit Kumar Swain** is a research scientist at Sahajanand Medical Technologies, India. He received his PhD in surface engineering of smart materials from the Department of Metallurgical & Materials Engineering, National Institute of Technology Rourkela, India.

# Contributors

**Hani Nasser Abdelhamid**
Advanced Multifunctional Materials
  Laboratory
Department of Chemistry
Faculty of Science
Assiut University
Assiut, Egypt
and
Nanotechnology Research Centre
  (NTRC)
The British University in Egypt
Cairo, Egypt

**Mona M. Agwa**
Department of Chemistry of Natural
  and Microbial Products
Pharmaceutical and Drug Industries
  Research Institute
National Research Centre
Giza, Egypt

**Shampa Aich**
Department of Metallurgical and
  Materials Engineering
Indian Institute of Technology
  Kharagpur
West Bengal, India

**Mohammad Azadi**
Faculty of Mechanical Engineering
Semnan University
Semnan, Iran

**Mahboobeh Azadi**
Faculty of Materials and Metallurgical
  Engineering
Semnan University
Semnan, Iran

**Ajit Behera**
Department of Metallurgical & Materials
  Engineering
National Institute of Technology
Rourkela, Odisha, India

**Asit Behera**
School of Mechanical Engineering
KIIT
Bhubaneswar, India

**Amir Hossein Beyzavi**
Faculty of Materials and Metallurgical
  Engineering
Semnan University
Semnan, Iran

**Samapika Bhuyan**
Biomaterials and Tissue
  Regeneration Lab
CETMS, Institute of Technical
  Education and Research
Siksha 'O' Anusandhan (Deemed to be
  University)
Bhubaneswar, Odisha, India

**Mansoor Bozorg**
Faculty of Chemical and Materials
  Engineering
Shahrood University of Technology
Shahrood, Iran

**Pasquale Cavaliere**
Department of Innovation Engineering
University of Salento
Lecce, Italy

**Shivani Chaudhary**
Department of Chemistry
Institute of Basic Sciences
Dr. Bhimrao Ambedkar University
Agra, India

**Modupeola Dada**
Chemical, Metallurgical and Materials
    Engineering
Tshwane University of Technology
Pretoria, South Africa

**Shokouh Dezianian**
Faculty of Mechanical Engineering
Semnan University
Semnan, Iran

**Karthik Dhandapani**
Department of Physics
School of Engineering
Presidency University
Bangalore, India

**T. Rajesh Kumar Dora**
Gitam School of Technology (GST)
GITAM
Visakhapatnam, AP, India

**Heba Elmotasem**
Pharmaceutical Technology
    Department
Pharmaceutical and Drug Industries
    Research Institute
National Research Centre
Giza, Egypt

**Santosh G**
Department of Mechanical Engineering
NMAM Institute of Technology
Karnataka, India

**Mansoureh Ganjali**
NourZoha Materials Engineering
    Research Group
Tehran, Iran

**Monireh Ganjali**
Biomaterials Group, Department of
    Nanotechnology & Advanced
    Materials
Materials and Energy Research Center
Karaj, Iran

**Arash Ghalandarzadeh**
School of Metallurgy and Materials
    Engineering
Iran University of Science and
    Technology
Tehran, Iran

**Morteza Hosseinzadeh**
Department of Engineering
Ayatollah Amoli branch
Islamic Azad University
Amol, Iran

**Gautam Jaiswar**
Department of Chemistry
Institute of Basic Sciences
Dr. Bhimrao Ambedkar University
Agra, India

**V. John Kennedy**
National Isotope Center
GNS Science Limited
Lowerhutt, New Zealand

**Tae Yub Kwon**
Dept. of Dental Biomaterials
Kyungpook National University
Daegu, Republic of Korea

**Rojaleen Lenka**
Biomaterials and Tissue
    Regeneration Lab
CETMS, Institute of Technical
    Education and Research
Siksha 'O' Anusandhan (Deemed to be
    University)
Bhubaneswar, Odisha, India

**Huiyan Li**
School of Engineering
University of Guelph
Guelph, Ontario, Canada

**Sibani Mahapatra**
Department of Metallurgical and
  Materials Engineering
Indian Institute of Technology
  Kharagpur
West Bengal, India

**Priyabrata Mallick**
Department of Metallurgical &
  Materials Engineering
National Institute of Technology
Rourkela, Odisha, India

**Sarada Prasanna Mallick**
Department of Biotechnology
Koneru Lakshmaiah Education
  Foundation
Guntur, AP, India

**Sangita Mangaraj**
Biomaterials and Tissue
  Regeneration Lab
CETMS, Institute of Technical
  Education and Research
Siksha 'O'Anusandhan (Deemed to be
  University)
Bhubaneswar, Odisha, India

**Neeraj Mehta**
Department of Physics
Institute of Science
Banaras Hindu University
Varanasi, India

**Manjusri Misra**
School of Engineering
University of Guelph
Guelph, Ontario, Canada
and
Bioproducts Discovery and
  Development Centre
Department of Plant Agriculture
University of Guelph
Guelph, Ontario, Canada

**Akash Mishra**
Department of Metallurgical &
  Materials Engineering
National Institute of Technology
Rourkela, Odisha, India

**Sapna Mishra**
Biomaterials and Tissue
  Regeneration Lab
Institute of Technical Education and
  Research
Siksha 'O'Anusandhan (Deemed to be
  University)
Bhubaneswar, Odisha, India

**Saswati Mishra**
Department of Biotechnology
School of Allied Health Sciences
Malla Reddy University
Hyderabad, Telangana, India

**Wrootchit Mishra**
Biological Sciences
Carnegie Mellon University
Pittsburgh, Pennsylvania, USA

**Garima Mittal**
Independent researcher
Omaha, Nebraska, USA

**Amar K. Mohanty**
School of Engineering
University of Guelph
Guelph, Ontario, Canada
and
Bioproducts Discovery and
    Development Centre
Department of Plant Agriculture
University of Guelph
Guelph, Ontario, Canada

**Bijayinee Mohapatra**
PG Department of Physics
Government Autonomous College
Angul, Odisha, India

**Mojtaba Najafizadeh**
Faculty of Chemical and Materials
    Engineering
Shahrood University of Technology
Shahrood, Iran
and
Department of Innovation Engineering
University of Salento
Lecce, Italy

**Hassan Namazi**
Polymer Research Laboratory
Department of Organic and
    Biochemistry
Faculty of Chemistry
University of Tabriz
Tabriz, Iran
and
Research Center for Pharmaceutical
    Nanotechnology
Biomedicine Institute
Tabriz University of Medical Science
Tabriz, Iran

**Debasis Nayak**
Department of Materials Science and
    Metallurgy
University of Cambridge
Cambridge, UK

**Shiladitya Paul**
Materials Innovation Centre
School of Engineering
University of Leicester, UK
and
Materials Performance and Integrity
    Group
TWI
Cambridge, UK

**Malihe Pooresmaeil**
Polymer Research Laboratory
Department of Organic and
    Biochemistry
Faculty of Chemistry
University of Tabriz
Tabriz, Iran

**Patricia Popoola**
Chemical, Metallurgical and Materials
    Engineering
Tshwane University of Technology
Pretoria, South Africa

**Tapash R. Rautray**
Biomaterials and Tissue
    Regeneration Lab
CETMS, Institute of Technical
    Education and Research
Siksha 'O' Anusandhan (Deemed to be
    University)
Bhubaneswar, Odisha, India

**Sally Sabra**
Department of Biotechnology
Institute of Graduate Studies and
  Research
Alexandria University
Alexandria, Egypt

**Priyatosh Sahoo**
Institute of Materials Science
Technische Universität Darmstadt
Darmstadt, Germany

**Amlan Prabhujyoti Sahu**
Department of Metallurgical &
  Materials Engineering
National Institute of Technology
Rourkela, Odisha, India

**Payam Sarir**
College of Civil Engineering
Tongji University
Shanghai, China

**Negar Sarrafan**
School of Dentistry
Urmia University of Medical Sciences
Urmia, Iran

**Sanjay Sharma**
Department of Mechanical Engineering
NMAM Institute of Technology
Karnataka, India

**Ehsane Marzban Shirkharkolaei**
Departement of Mechanical
  Engineering
Isfahan University of Technology
Isfahan, Iran

**Pratik Shukla**
The Manufacturing Technology
  Centre (MTC)
Coventry, UK

**Hesham S.M. Soliman**
Department of Pharmacognosy
Faculty of Pharmacy
Helwan University
Ain Helwan, Cairo, Egypt
and
Pharm D Program
Egypt-Japan University of Science and
  Technology
New Borg El-Arab City, Egypt

**Gisela Strohle**
School of Engineering
University of Guelph
Guelph, Ontario, Canada

**Priyabrata Swain**
Biomaterials and Tissue
  Regeneration Lab
CETMS, Institute of Technical
  Education and Research
Siksha 'O'Anusandhan (Deemed to be
  University)
Bhubaneswar, Odisha, India

**Subhasmita Swain**
Biomaterials and Tissue
  Regeneration Lab
CETMS, Institute of Technical
  Education and Research
Siksha 'O'Anusandhan (Deemed to be
  University)
Bhubaneswar, Odisha, India

**Sushree Sangita Swain**
Department of Materia Medica
Dr. Abhin Chandra Homoeopathic
  Medical College & Hospital
Bhubaneswar, India

**Valeh Talebsafa**
Faculty of Materials and Metallurgical
  Engineering
Semnan University
Semnan, Iran

**Chloe Tan**
School of Engineering
University of Guelph
Guelph, Ontario, Canada

**Sampara Ila Grace Victoria**
Department of Metallurgical and
   Materials Engineering
Indian Institute of Technology
   Kharagpur
West Bengal, India

**Sahar Yazdi**
School of Medical Sciences
Shahrood Branch
Islamic Azad University
Shahrood, Iran

# Section 1

## Introduction to Biomaterials Surface Engineering

# 1 Introduction to Biomaterial Surface Engineering

*Asit Behera*
School of Mechanical Engineering, KIIT, Bhubaneswar, India

*Sushree Sangita Swain*
Department of Materia Medica, Dr. Abhin Chandra
Homoeopathic Medical College & Hospital, Bhubaneswar,
India

*Debasis Nayak*
Department of Materials Science & Metallurgy, University of
Cambridge, Cambridge, UK

*Ajit Behera*
Department of Metallurgical & Materials Engineering,
National Institute of Technology, Rourkela, Odisha, India

## 1.1 INTRODUCTION

Biomaterials are a class of materials that are designed and engineered to interact with biological systems, ranging from molecules and cells to whole organisms as a support structure or as a replacement. They are used in various fields, including implant, medicine, biotechnology, and tissue engineering, with the aim of improving healthcare and advancing scientific research. The field of biomaterials combines principles from materials science, biology, chemistry, and engineering to develop materials that are compatible with living tissues and can perform specific functions within the body. These materials can be natural, synthetic, or a combination of both, and they are carefully selected and designed to have desirable properties for specific applications. Biomaterials can serve different functions depending on their intended use. Some common applications include **implant materials, drug delivery system, tissue engineering, diagnostic tools,** etc. (Figure 1.1). Biomaterials are utilized to manufacture a wide range of medical implants, such as joint replacements, dental implants, and cardiovascular stents [1]. These materials must be biocompatible; that is, they do not involve to create any harmful immune responses and can integrate with surrounding tissues. Biomaterials can be used to hold and release drugs to a

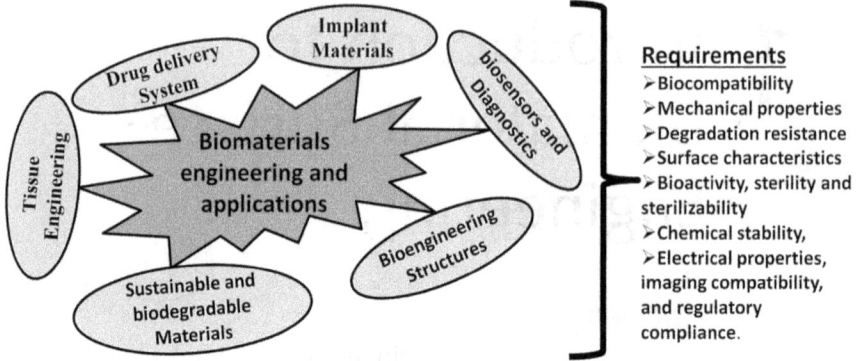

**Requirements**
➢ Biocompatibility
➢ Mechanical properties
➢ Degradation resistance
➢ Surface characteristics
➢ Bioactivity, sterility and sterilizability
➢ Chemical stability,
➢ Electrical properties, imaging compatibility, and regulatory compliance.

**FIGURE 1.1**  Biomaterials engineering and applications.

predetermined target sites in the required part of the body. They can control the release rate of drugs, protect them from degradation, and enhance their therapeutic efficacy. Biomaterials play a crucial role as scaffold materials [2] to support the development and regeneration of new tissues. These scaffolds provide a temporary framework that guides cells to form new tissue structures. Biomaterials are utilized for making diagnostic tools, such as biosensors and lab-on-a-chip devices. These materials can interact with biological molecules to detect and analyze specific biomarkers for disease diagnosis and monitoring. When designing biomaterials, researchers consider various factors, including biocompatibility, mechanical properties, degradation rate, surface characteristics, and interactions with cells and tissues. They also take into account the specific requirements of the intended application, such as the need for antimicrobial properties or the ability to support cellular adhesion and proliferation [3]. Overall, biomaterials have revolutionized the field of medicine and have the potential to improve patient outcomes, enhance medical treatments, and advance our understanding of biological systems. Ongoing research and advancements in biomaterials continue to expand their applications and impact various areas of healthcare.

## 1.2  ESSENTIAL REQUIREMENTS ESTABLISHED BY BIOMATERIALS

Biomaterials play a crucial role in today's scenario across various fields due to their significant importance and applications. Some key requirements for the biomaterials have for which those are included in **medical and healthcare advancements, implants and prosthetics, drug delivery systems, bioengineering and tissue engineering, biosensors and diagnostics, environmental applications, and sustainable and biodegradable materials** [4]. Biomaterials have revolutionized medical and healthcare industries by enabling advancements in areas such as tissue engineering, regenerative medicine, medical implants, and drug delivery structures. They are used to develop biocompatible and biodegradable materials that can interact with living tissues, promoting healing and tissue regeneration. Biomaterials are essential for the development of implants and prosthetics [5,6], such as artificial joints, dental implants, cardiovascular stents, and bone scaffolds. These materials need to possess high biocompatibility, mechanical strength, and durability to integrate seamlessly with the human body and restore function. Biomaterials are used to create drug delivery structures that enhance the efficacy and safety of therapeutic associates. They can encapsulate drugs and release them in a controlled manner, ensuring targeted delivery, sustained release, and reduced side effects. This enables more effective treatment for various diseases such as cancer, diabetes, and chronic conditions. Biomaterials provide a framework for the growth of cell and tissue formation, facilitating the regeneration of damaged or diseased tissues. Biomaterials are used in the development of biosensors and diagnostic tools. They enable the detection and monitoring of various biological markers, pathogens, and analytes. Biomaterial-based sensors offer high sensitivity, specificity, and selectivity, leading to improved disease diagnosis and monitoring [7]. Biomaterials are also significant in addressing environmental challenges. They are used in wastewater treatment, pollutant adsorption, and environmental remediation. Biomaterial-based solutions help in removing contaminants from water, soil, and air, contributing to a cleaner and healthier environment. With growing concerns about environmental sustainability, biomaterials offer a promising solution [8]. They can be derived from renewable sources and designed to be biodegradable, reducing the environmental impact associated with traditional materials.

## 1.3 SCIENTIFIC VALUES FOR THE CONSIDERATION OF BIOMATERIAL

When considering materials for biomedical applications, there are several scientific values that are typically taken into account. These values help assess the relevance of a material for the specific biomedical implementation and ensure its safety and effectiveness. The important scientific values to consider for a material include **biocompatibility, mechanical properties, degradation rate, surface characteristics, bioactivity, sterility and sterilizability, chemical stability, electrical properties, imaging compatibility, and regulatory compliance** [9,10]. Biocompatibility can be defined as the ability of a material to execute its required action within a specific application without extracting harmful reactions in the body. It involves assessing the material's interaction with living tissues, cells, and the immune system. The mechanical behavior of a material, such as materials strength, stiffness, and elasticity, is crucial for biomedical applications. For example, orthopedic implants require materials with sufficient strength to withstand loads and stresses experienced in the body [11]. For temporary biomedical implants or drug delivery systems, the degradation rate of the material is important. It should degrade at an appropriate rate, aligning with the healing process or drug release requirements. The surface properties of a material, including roughness, porosity, and surface chemistry, play a vital role in interactions with biological entities, such as cells and proteins [12]. Various surface modification techniques can be employed to improve the biocompatibility, cell adhesion, or drug delivery. Certain biomedical applications, such as bone regeneration, benefit from materials with inherent bioactivity. Bioactive materials can stimulate cell adhesion, proliferation, and differentiation, promoting tissue growth and integration. Materials used in biomedical applications must be sterilized to ensure they are free from bacteria, viruses, and other contaminants. The material's ability to withstand sterilization methods, such as autoclaving or ethylene oxide sterilization, is important. Biomedical materials should exhibit stability in the presence of bodily fluids, such as blood or interstitial fluid [13]. They should not undergo significant chemical changes that may cause adverse effects or compromise their functionality. In some biomedical applications, electrical properties of materials are essential. For instance, in neural interfaces or bioelectrodes, materials with appropriate electrical conductivity and charge transfer properties are required. Materials used in biomedical devices or implants should not interfere with imaging techniques such as MRI, X-rays, and ultrasound [14]. They should allow clear imaging of the surrounding tissues. Compliance with relevant regulatory standards and guidelines, such as the U.S. Food and Drug Administration (FDA) regulations, is critical for the approval and commercialization of biomedical materials. These scientific values help guide the selection, development, and evaluation of materials for biomedical applications, ensuring their safety, efficacy, and suitability for specific uses in healthcare [15].

## 1.4  GLOBAL DEMAND ON BIOMATERIALS

The global biomaterials market has been experiencing significant growth, and it is projected to continue growing at a compound annual growth rate (CAGR) of 15.52% from 2023 to 2030 [16]. The increasing prevalence of musculoskeletal and chronic skeletal medical conditions is driving the demand for biomaterial-based implants, especially in the field of orthopedics. With a growing geriatric population, there is a higher risk of conditions like osteoarthritis and osteoporosis, further fueling the demand for orthopedic implants. Additionally, the COVID-19 pandemic has created a demand for biomaterials and biomedical devices for detecting and treating the virus. Early detection of COVID-19 is crucial in limiting its transmission, and various biomaterial and biomedical devices are being developed for this purpose. Existing biomedical devices based on polymerase chain reaction (PCR) or non-PCR methods can be used for detecting infections from the virus [17]. These advancements in biomaterials and biomedical devices play a vital role in both improving healthcare outcomes and addressing the challenges posed by the COVID-19 pandemic. Technological advancements have indeed made biomaterials more versatile and have greatly expanded their applications in various fields of healthcare, including bioengineering and tissue engineering. The development of smart biomaterials that can interact with biological systems has opened up new possibilities for biomedical applications. Smart biomaterials have the ability to respond to specific stimuli or conditions in their environment, making them highly adaptable and useful in a range of applications [18]. They can be designed to release bioactive molecules in a controlled manner, enabling targeted drug delivery and improving therapeutic outcomes. This controlled release of drugs is particularly beneficial as it enhances the efficacy and safety of treatments while reducing side effects. The growing interest in smart biomaterials and their potential in drug delivery has driven revenue generation in the market. Companies are actively exploring novel drug delivery approaches, including the use of biomaterials, to develop controlled-release systems. This has led to collaborations and partnerships between different companies with complementary expertise. One such example is the partnership between DSM Biomedical and ProMed Pharma [19]. DSM Biomedical, a leading provider of biomedical materials, teamed up with ProMed Pharma, a company specializing in micro molding and extrusion capabilities. By combining their respective technologies and expertise, they aimed to create novel controlled-release drug implants and combination devices [20]. This partnership demonstrates the increasing focus on leveraging biomaterials for advanced drug delivery systems. Overall, the ongoing developments in smart biomaterials and their applications in drug delivery are fueling the growth of the biomaterials market in the healthcare industry. These advancements hold significant promise for improving patient care and advancing medical treatments.

Porous overlays and fully or partially porous orthopedic implants have gained enormous popularity in recent years. This is because the porous structure reduces the modulus of elasticity and stimulates bone growth around the implant. Powder metallurgy, 3D printing, and additive manufacturing are some possible technologies for the fabrication of porous metal and ceramic implants [21]. The increasing demand for smart biomaterials that can generate and transmit bioelectrical signals similar to natural tissues for precise physiological functions is expected to drive the market growth. Piezoelectric scaffolds are smart materials that play an important role in tissue engineering. They stimulate signaling pathways that improve tissue regeneration in damaged areas. In terms of revenue, the polymer products segment will dominate the market with a share of 28.47% in 2022 and is expected to continue its leadership throughout the forecast period due to its wide range of applications [22]. Widespread adoption of biopolymers and advanced polymers for bioresorbable tissue fixation and other orthopedic applications is also expected to accelerate revenue generation in this segment. Polymeric biomaterials are one of the cornerstones of tissue engineering. Continuous advances in technologies such as microfabrication, surface modification, drug delivery, nanotechnology, and high-throughput control play an integral role in expanding the use of polymeric materials in the field of tissue engineering.

Natural biomaterials are expected to grow at a CAGR of 17.51% from 2023 to 2030 due to their multiple advantages over synthetic biomaterials in terms of biodegradability, biocompatibility, and reconstitution [23]. Because of these advantages, they are increasingly used to replace or restore the structure and function of damaged organs or tissues. For example, a biosensor is an analytical device used to detect analytes such as biomolecules or bioelements produced by microorganisms such as tissues, enzymes, organelles, etc. Sensors mainly consist of three parts: Receptors (enzymes, antibodies, nucleic acids), sensors (electrochemical, electrical, optical, thermal, magnetic resonance), and electronic systems including signal amplifiers [24].

The orthopedic applications segment dominates the biomaterials market in terms of revenue, accounting for a share of 23.15% in 2022 [25]. The increasing adoption of metal biomaterials in orthopedic applications due to their high load-bearing capacity is one of the factors driving the growth of this segment. Additionally, the

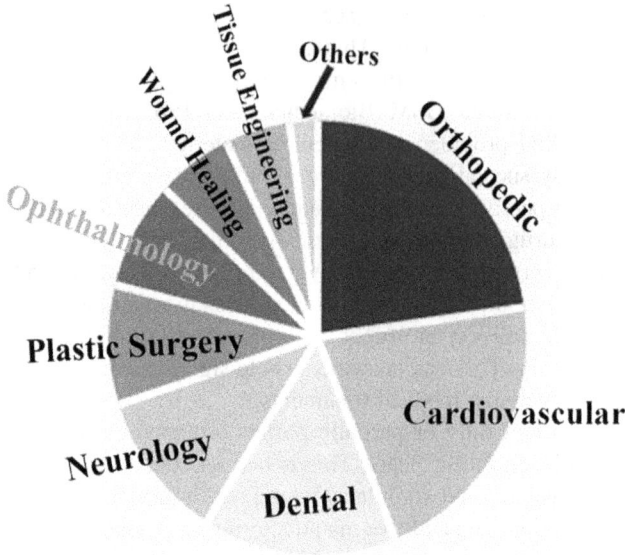

**FIGURE 1.2** Global biomaterials market for various products.

continued growth of market vendors launching advanced orthopedic implants is also expected to drive revenue generation (Figure 1.2). For example, in November 2019, DiFusion Inc. received FDA approval for the Xiphos-ZF spinal intervertebral device, which is based on a different biomaterial Zfuze (combination of ketone polyether ether and titanium) [26]. This novel biomaterial significantly reduced inflammatory cytokine markers associated with fibrogenesis [27].

The plastic surgery segment is projected to grow at the fastest CAGR of 16.99% during 2023–2030 due to the increase in the number of cosmetic procedures and the increasing use of biomaterials in these procedures [28]. According to a 2019 publication by the International Society of Aesthetic Plastic Surgery, approximately 4.3 million cosmetic procedures were performed in the United States in 2018, which positively affected the growth of the segment [29].

## 1.5   SURFACE PROPERTIES FOR THE BIOMATERIALS

Surface properties play a crucial role in determining the performance and interactions of biomaterials. Major surface properties of biomaterials are roughness, topography, wettability, charge, chemical composition, biodegradability, coating, and surface treatments. Surface roughness affects cell adhesion, protein adsorption, and bacterial colonization. Rough surfaces can enhance cell adhesion and promote osseointegration in bone implants [30]. However, excessive roughness may lead to increased wear and tissue damage. Surface topography refers to the microscale and nanoscale features present on the biomaterial surface. Topographical cues can influence cell behavior, including adhesion, migration, proliferation, and differentiation. For example, nanostructured surfaces can promote cell alignment and guide tissue regeneration. Wettability describes the ability of a surface to attract or repel liquids. It is characterized by contact angle measurements. Hydrophilic surfaces have low contact angles and promote the spreading of aqueous solutions. Hydrophobic surfaces have high contact angles and tend to repel water. Wettability affects cell attachment and protein adsorption [31]. The surface charge of biomaterials can influence protein adsorption, cell adhesion, and cellular signaling. Biomaterials can be positively charged (cationic), negatively charged (anionic), or neutral. The surface charge can be modulated to control interactions with specific biomolecules or cells. The chemical composition of the surface determines its reactivity and the types of interactions it can form. Biomaterials can be modified with various functional groups to control surface chemistry, such as introducing carboxyl, amino, or hydroxyl groups. Surface functionalization can be used to enhance biocompatibility, control protein adsorption, or enable specific biological interactions [32]. For biodegradable biomaterials, the surface properties can affect the degradation rate and degradation products. Surface modifications can be employed to control the degradation behavior and release of bioactive molecules, such as drugs or growth factors. Biomaterial surfaces can be modified through coatings or surface treatments to enhance specific properties [33]. Examples include applying thin films, plasma treatment, or chemical functionalization to improve biocompatibility, antimicrobial properties, or reduce friction.

## 1.6   BIOCOMPATIBILITY OF A MATERIALS

Biocompatibility refers to the ability of a material to interact with biological systems without causing any harmful effects. When developing materials for use in medical or biological applications, biocompatibility is a crucial consideration to ensure the safety and effectiveness of the material. The key factors that contribute to the biocompatibility of materials are **immunogenicity, cytotoxicity, hemocompatibility, tissue compatibility, biodegradability, corrosion resistance, mechanical properties, and surface characteristics**. Materials should have minimal immunogenic responses, meaning they should not trigger significant immune reactions such as inflammation or immune system activation. Immunogenicity tests assess the material's interaction with the immune system. A biocompatible material should not be toxic to cells. Cytotoxicity tests evaluate the effect of the material on cell viability and determine if it causes any adverse reactions or cell death [34]. If a material comes into contact with blood, it should not induce blood clotting or cause hemolysis (rupturing of red blood cells). Hemocompatibility tests evaluate the material's effects on blood components and coagulation. A biocompatible material should not cause significant irritation, inflammation, or damage to surrounding tissues [35]. Tissue compatibility tests assess the material's interaction with different types of tissues. For materials intended for temporary use, such as sutures or drug delivery systems, biodegradability is important. Biodegradable materials can be broken down and eliminated from the body over time, reducing the need for additional surgical procedures. If a material is used in long-term implantable devices, it should be corrosion-resistant to prevent the release of toxic ions or degradation products into the body [36]. The mechanical properties of a material should be compatible with the intended application. For example, if a material is used in bone implants, it should have sufficient strength and stiffness to support the load-bearing requirements. The surface of a material can influence its bio-compatibility [37]. Surface properties such as roughness, hydrophobicity, and surface charge can affect cell adhesion, protein adsorption, and tissue response. These factors are assessed through various in vitro and in vivo tests, including cell culture studies, animal studies, and clinical trials. It's important to note that biocompatibility is a complex topic, and regulatory bodies, such as the U.S. Food and Drug Administration (FDA), have specific guidelines and standards for evaluating the biocompatibility of materials intended for medical use.

## 1.7   SURFACE ENGINEERING FOR A BIOMATERIALS

Surface engineering plays a crucial role in enhancing the performance and functionality of biomaterials. It involves modifying the surface properties of materials to improve their biocompatibility, bioactivity, mechanical properties, and interactions with biological systems. Before any surface treatment, it is essential to clean the biomaterial to remove contaminants such as dust, oils, or residues from manufacturing processes. Cleaning can be done using solvents, detergents, ultrasonic baths, or other appropriate methods. Cleaning can be followed by etching. Etching is a process that involves the use of chemicals or physical methods to remove a thin layer of material from the surface. It can help improve the surface roughness, increase surface area, and create microstructures. Different etching techniques include chemical etching, plasma etching, or laser ablation [38].

Here are some common surface engineering techniques used in biomaterials:

**Surface Modification:** Various surface modification techniques can be em-ployed to alter the surface properties of biomaterials. This includes physical methods like sandblasting, polishing, or etching, as well as chemical methods such as plasma treatment, chemical vapor deposition (CVD), or wet chemical treatments. Surface modification techniques can change the surface roughness, morphology, chemistry, and wettability, thereby influencing cell adhesion, protein adsorption, and other biological interactions.

**Coating Deposition:** Coatings can be applied onto biomaterial surfaces to improve their functionality. For example, biocompatible and bioactive coatings like hydroxyapatite (HA), titanium nitride (TiN), or bioactive glasses can be deposited onto metallic implants to promote osseointegration [39]. Coating techniques include physical vapor deposition (PVD), chemical vapor deposition (CVD), electroche-mical deposition, and sol-gel methods.

**Biomolecule Immobilization:** Biomolecules, such as proteins, peptides, growth factors, or enzymes, can be immobilized onto biomaterial surfaces to enhance specific biological functions. This can be achieved through techniques like physical adsorption, covalent bonding, or layer-by-layer assembly. Biomolecule immobilization can regulate cell adhesion, differentiation, and tissue regeneration processes [40].

**Surface Patterning:** Patterning biomaterial surfaces at the micro- and nano-scale can control cellular behavior and tissue development. Techniques like photolithography, soft lithography, or laser ablation can create well-defined patterns and structures on biomaterial surfaces [41]. Surface patterning can guide cell alignment, migration, and organization, which is particularly useful in tissue engineering and regenerative medicine applications.

**Surface Functionalization:** Biomaterial surfaces can be functionalized with specific chemical groups or functional molecules to impart desired properties. For instance, the introduction of hydrophilic or hydrophobic moieties can influence surface wettability and protein adsorption. Functionalization techniques include grafting, self-assembled monolayers (SAMs), or click chemistry.

**Surface Characterization:** Surface engineering of biomaterials requires thorough characterization to assess the efficacy of modifications. Techniques such as scanning electron microscopy (SEM), atomic force microscopy (AFM), X-ray photoelectron spectroscopy (XPS), and contact angle measurements can provide valuable information about surface topography, chemistry, and wettability.

These surface engineering approaches can be combined or tailored according to the specific requirements of the biomaterial and intended application. The ultimate goal is to create biomaterial surfaces that interact favorably with biological systems, promoting improved biocompatibility, integration, and therapeutic outcomes. Creating a biocompatible surface involves designing a material or modifying an existing material to be compatible with living tissues and organisms. Biocompatible surfaces are crucial in various fields such as medicine, biomedical engineering, and biotechnology. Some general steps to make a biocompatible surface are **material selection, surface modification techniques (surface coatings, plasma treatment, chemical functionalization, physical techniques), surface roughening, sterilization, surface characterization, biocompatibility testing, and long-term assessment**.

Choose a material that is known to be biocompatible or has the potential for modification to enhance biocompatibility. Common biocompatible materials include metals (such as titanium and stainless steel), ceramics (such as hydroxyapatite), polymers (such as polyethylene and silicone), and biodegradable materials. Several techniques can be employed to modify the surface of the material to enhance its biocompatibility. Surface coating is a common treatment method to modify the properties of biomaterials [42]. Coatings can provide functionalities such as improved biocompatibility, reduced friction, antimicrobial properties, or enhanced cell adhesion. Coating techniques include physical vapor deposition (PVD), chemical vapor deposition (CVD), electroplating, dip coating, or spray coating. Surface coatings generally apply a biocompatible coating onto the material's surface; for example, depositing a thin layer of hydroxyapatite on a metal surface can improve its bioactivity [43]. Surface modification techniques alter the chemistry of the biomaterial surface to enhance its performance. Methods like plasma treatment, ion implantation, or chemical functionalization can introduce specific functional groups or change the surface energy of the material. Plasma treatment is a process to expose the material to a plasma environment to modify its surface properties. Plasma treatment can introduce functional groups or change the surface energy, which can influence the material's interaction with biological systems. The chemical functionalization is to attach biocompatible molecules or functional groups onto the material's surface. This can be done using various chemical reactions, such as silanization or grafting of biomolecules. Physical techniques utilize physical methods like laser ablation, ion beam implantation, or electrospinning to modify the material surface at a micro- or nano-scale level. These techniques can alter surface roughness, topography, or create nanostructures that promote cell adhesion and growth. Increasing the surface roughness can promote cell adhesion and improve osseointegration for implantable biomaterials [44]. Techniques such as sandblasting, acid etching, or microabrasion can be employed to create controlled surface roughness. Biomaterials intended for medical applications must undergo sterilization to eliminate microorganisms and ensure biocompatibility. Common sterilization methods include steam autoclaving, ethylene oxide (EtO) sterilization, gamma irradiation, or hydrogen peroxide plasma sterilization.

Surface characterization is to analyze the modified surface to ensure the desired modifications have been achieved. Characterization techniques like scanning electron microscopy (SEM), atomic force microscopy (AFM), X-ray photoelectron spectroscopy (XPS), or contact angle measurements can be used to assess surface roughness, topography, chemistry, and wettability, whereas, biocompatibility testing is used to evaluate the biocompatibility of the modified surface using appropriate in vitro and/or in vivo tests. Common tests include cell viability assays, cytotoxicity tests, adhesion studies, and assessments of the material's interaction with biological systems. These tests help determine if the surface modifications have successfully rendered the material biocompatible. Long-term assessment of biomaterials considers the long-term stability and durability of the biocompatible surface. Factors such as material degradation, wear resistance, and compatibility with physiological environments should be taken into account. It's important to note that the specific techniques and modifications required for creating a biocompatible surface depend on the material, intended application, and the desired level of biocompatibility. Consultation with experts in the field of biomaterials or surface modification can provide valuable guidance and insights tailored to your specific needs.

## 1.8 IMPORTANT AREAS FOR THE FUTURE RESEARCH ON BIOMATERIALS?

The following three exciting technologies point the way to biomaterials in the fields of immunomodulation, injectable biomaterials, and supramolecular biomaterials. Immunomodulation is accustoming the immune action to a required limit. Immunomodulatory biomaterials can assist to treat widespread chronic diseases such as type 1 diabetes, an autoimmune disease in which the body's defenses destroy insulin-producing cells in the pancreas [45]. Researchers have recently developed an injectable synthetic biomaterial that can reverse type 1 diabetes in non-obese diabetic mice, an important step toward developing a biodegradable platform to manage the effects of the disease. Injectable biomaterials are increasingly used to deliver therapeutic agents such as drugs, genetic material, and proteins. They offer the potential to treat a variety of conditions by providing targeted delivery while avoiding uptake by the immune system [46]. Ongoing research using synthetic and natural injectable biomaterials could one day be used to treat bone defects, cancer, and heart disease. Supramolecular biomaterials are complexes of molecules that have the ability to sense and respond beyond their own boundaries, making them ideal materials for treating injury or disease. Researchers are investigating the development of supramolecular biomaterials that can be turned on or off in response to physiological signals or mimic natural biological signals. Apart from these areas, there are various research gaps in smart biomaterials, bioactive materials, 3D bioprinting, self-healing biomaterials, sustainable biomaterials, nanotechnology in nanomaterials, biomimetic materials, and advanced coatings. Advanced materials such as hydrogels, shape memory polymers, biodegradable polymers, and graphene and carbon nanotubes also require rigorous research in the field of next-generation biomaterials.

## 1.9 SUMMARY

Biomaterials are of paramount importance in today's scenario, influencing medical advancements, diagnostics, environmental sustainability, and fostering innovation. Their unique properties make them indispensable in addressing critical challenges and improving human health and well-being. Ongoing advancements in material science, nanotechnology, and biotechnology contribute to the production of new biomaterials with improved functionalities. These materials pave the way for novel applications and possibilities in diverse fields. It's important to note that different biomaterials and applications may require specific surface properties tailored to their intended use. Researchers and engineers often employ various surface modification techniques to optimize the performance and biocompatibility of biomaterials in specific biomedical applications. The choice of surface treatment procedures is influenced by the specific requirements of the biomaterial, its intended use, and the desired surface properties. Manufacturers and researchers often employ a combination of these techniques to achieve the desired surface characteristics for biomaterials.

## REFERENCES

1. Dhandayuthapani B, Yoshida Y, Maekawa T, Sakthi Kumar D Polymeric Scaffolds in Tissue Engineering Application: A Review. Int J Polymer Sci. 2011; 2011: Article ID 290602, 19 pages. 10.1155/2011/290602
2. Fenton OS, Olafson KN, Pillai PS, Mitchell MJ, Langer R Advances in Biomaterials for Drug Delivery. Adv Mater. 2018 May 7:e1705328. 10.1002/adma.201705328. Epub ahead of print. PMID: 29736981; PMCID: PMC6261797.
3. Williams DF Biocompatibility Pathways and Mechanisms for Bioactive Materials: The Bioactivity Zone. Bioact Mater. 2021 Aug 26; 10:306–322. 10.1016/j.bioactmat.2021.08.014. PMID: 34901548; PMCID: PMC8636667.
4. Bhat S, Kumar A Biomaterials and Bioengineering Tomorrow's Healthcare. Biomatter. 2013 Jul-Sep; 3(3):e24717. 10.4161/biom.24717. Epub 2013 Apr 1. PMID: 23628868; PMCID: PMC3749281.
5. Al-Shalawi FD, Mohamed Ariff AH, Jung D-W, MohdAriffin MKA, Seng Kim CL, Brabazon D, Al-Osaimi MO Biomaterials as Implants in the Orthopedic Field for Regenerative Medicine: Metal versus Synthetic Polymers. Polymers. 2023; 15(12):2601. 10.3390/polym15122601

6. Todros S, Todesco M, Bagno A Biomaterials and Their Biomedical Applications: From Replacement to Regeneration. Processes. 2021; 9(11):1949. 10.3390/pr9111949

7. Andryukov BG, Lyapun IN, Matosova EV, Somova LM Biosensor Technologies in Medicine: From Detection of Biochemical Markers to Research into Molecular Targets (Review). Sovrem Tekhnologii Med. 2021; 12(6):70–83. 10.17691/stm2020.12.6.09. Epub 2020 Dec 28. PMID: 34796021; PMCID: PMC8596237.

8. Elbasiouny H, Darwesh M, Elbeltagy H, Abo-Alhamd FG, Amer AA, Elsegaiy MA, Khattab IA, Elsharawy EA, Ebehiry F, El-Ramady H, Brevik EC Ecofriendly Remediation Technologies for Wastewater Contaminated with Heavy Metals with Special Focus on Using Water Hyacinth and Black Tea Wastes: A Review. Environ Monit Assess. 2021 Jun 26; 193(7):449. 10.1007/s10661-021-09236-2. Erratum in: Environ Monit Assess. 2021 Jul 31;193(8):542. PMID: 34173877; PMCID: PMC8233605.

9. Huzum B, Puha B, Necoara RM, Gheorghevici S, Puha G, Filip A, Sirbu PD, Alexa O Biocompatibility Assessment of Biomaterials Used in Orthopedic Devices: An Overview (Review). Exp Ther Med. 2021 Nov; 22(5):1315. 10.3892/etm.2021.10750. Epub 2021 Sep 17. PMID: 34630669; PMCID: PMC8461597.

10. Sidambe AT Biocompatibility of Advanced Manufactured Titanium Implants-A Review. Materials (Basel). 2014 Dec 19; 7(12):8168–8188. 10.3390/ma7128168. PMID: 28788296; PMCID: PMC5456424.

11. Moghadasi K, Isa MSM, Ariffin MA, ZulhiqmiMohdjamil M, Raja S, Wu B, Yamani M, Muhamad MRB, Yusof F, Jamaludin MF, MohdSayuti bin Ab Karim, Bushroabinti Abdul Razak, Nukman bin Yusoff A Review on Biomedical Implant Materials and the Effect of Friction Stir Based Techniques on Their Mechanical and Tribological Properties. J Mater Res Technol. 2022; 17:1054–1121, ISSN 2238-7854. 10.1016/j.jmrt.2022.01.050

12. Song R, Murphy M, Li C, Ting K, Soo C, Zheng Z Current Development of Biodegradable Polymeric Materials for Biomedical Applications. Drug Des Devel Ther. 2018 Sep 24; 12:3117–3145. 10.2147/DDDT.S165440. PMID: 30288019; PMCID: PMC6161720.

13. Pérez Davila S, González Rodríguez L, Chiussi S, Serra J, González P How to Sterilize Polylactic Acid Based Medical Devices? Polymers (Basel). 2021 Jun 28; 13(13):2115. 10.3390/polym13132115. PMID: 34203204; PMCID: PMC8271615.

14. Ravichandran R, Sundarrajan S, Venugopal JR, Mukherjee S, Ramakrishna S Applications of Conducting Polymers and Their Issues in Biomedical Engineering. J R Soc Interface. 2010 Oct 6; 7 Suppl 5(Suppl 5):S559–S579. 10.1098/rsif.2010.0120.focus. Epub 2010 Jul 7. PMID: 20610422; PMCID: PMC2952180.

15. Marchetti S, Schellens JH The Impact of FDA and EMEA Guidelines on Drug Development in Relation to Phase 0 Trials. Br J Cancer. 2007 Sep 3; 97(5):577–581. 10.1038/sj.bjc.6603925. Epub 2007 Aug 28. PMID: 17726450; PMCID: PMC2360360.

16. https://www.globenewswire.com/en/news-release/2023/03/29/2636463/0/en/Biomaterials-Market-Size-Growth-Analysis-and-Forecast-2023-2030.html

17. Ertas YN, Mahmoodi M, Shahabipour F, Jahed V, Diltemiz SE, Tutar R, Ashammakhi N Role of Biomaterials in the Diagnosis, Prevention, Treatment, and Study of Coronavirus Disease 2019 (COVID-19). Emergent Mater. 2021; 4(1):35–55. 10.1007/s42247-021-00165-x. Epub 2021 Mar 16. PMID: 33748672; PMCID: PMC7962632.

18. Victor SP, Selvam S, Sharma CP Recent Advances in Biomaterials Science and Engineering Research in India: A Minireview. ACS Biomater Sci Eng. 2019 Jan 14; 5(1):3–18. 10.1021/acsbiomaterials.8b00233. Epub 2018 Jun 7. PMID: 33405853.

19. Martínez-Ballesta M, Gil-Izquierdo Á, García-Viguera C, Domínguez-Perles R Nanoparticles and Controlled Delivery for Bioactive Compounds: Outlining Challenges

for New "Smart-Foods" for Health. Foods. 2018 May 7; 7(5):72. 10.3390/foods7050072. PMID: 29735897; PMCID: PMC5977092.

20. Murua A, Herran E, Orive G, Igartua M, Blanco FJ, Pedraz JL, Hernández RM Design of a Composite Drug Delivery System to Prolong Functionality of Cell-Based Scaffolds. Int J Pharm. 2011 Apr 4; 407(1-2):142–150. 10.1016/j.ijpharm.2010.11. 022. Epub 2010 Nov 19. PMID: 21094235.

21. Li Z, Wang Q, Liu G A Review of 3D Printed Bone Implants. Micromachines (Basel). 2022 Mar 27; 13(4):528. 10.3390/mi13040528. PMID: 35457833; PMCID: PMC9025296.

22. Najjari A, MehdinavazAghdam R, Ebrahimi SAS, Suresh KS, Krishnan S, Shanthi C, Ramalingam M Smart Piezoelectric Biomaterials for Tissue Engineering and Regenerative Medicine: A Review. Biomed Tech (Berl). 2022 Mar 22; 67(2):71–88. 10.1515/bmt-2021-0265. PMID: 35313098.

23. Brovold M, Almeida JI, Pla-Palacín I, Sainz-Arnal P, Sánchez-Romero N, Rivas JJ, Almeida H, Dachary PR, Serrano-Aulló T, Soker S, Baptista PM Naturally-Derived Biomaterials for Tissue Engineering Applications. Adv Exp Med Biol. 2018; 1077:421–449. 10.1007/978-981-13-0947-2_23. PMID: 30357702; PMCID: PMC7526297.

24. Bhalla N, Jolly P, Formisano N, Estrela P Introduction to Biosensors. Essays Biochem. 2016 Jun 30; 60(1):1–8. 10.1042/EBC20150001. PMID: 27365030; PMCID: PMC4986445.

25. https://www.grandviewresearch.com/industry-analysis/biomaterials-industry# :~:text=The%20orthopedic%20application%20segment%20dominated.factors %20driving%20the%20segment%20growth.

26. Prasad K, Bazaka O, Chua M, Rochford M, Fedrick L, Spoor J, Symes R, Tieppo M, Collins C, Cao A, Markwell D, Ostrikov KK, Bazaka K Metallic Biomaterials: Current Challenges and Opportunities. Materials (Basel). 2017 Jul 31; 10(8):884. 10.3390/ma10080884. PMID: 28773240; PMCID: PMC5578250.

27. de la Torre BG, Albericio F The Pharmaceutical Industry in 2021. An Analysis of FDA Drug Approvals from the Perspective of Molecules. Molecules. 2022; 27(3):1075. 10.3390/molecules27031075

28. https://www.grandviewresearch.com/industry-analysis/cosmetic-surgery-procedure-market

29. Al Ghadeer HA, AlAlwan MA, AlAmer MA, Alali FJ, Alkhars GA, Alabdrabulrida SA, Al Shabaan HR, Buhlaigah AM, AlHewishel MA, Alabdrabalnabi HA Impact of Self-Esteem and Self-Perceived Body Image on the Acceptance of Cosmetic Surgery. Cureus. 2021 Oct 16; 13(10):e18825. 10.7759/cureus.18825. PMID: 34804682; PMCID: PMC8592260.

30. Cui L, Yao Y, Yim EKF The Effects of Surface Topography Modification on Hydrogel Properties. APL Bioeng. 2021 Jul 27; 5(3):031509. 10.1063/5.0046076. PMID: 34368603; PMCID: PMC8318605.

31. Liu F, Xu J, Wu L, Zheng T, Han Q, Liang Y, Zhang L, Li G, Yang Y The Influence of the Surface Topographical Cues of Biomaterials on Nerve Cells in Peripheral Nerve Regeneration: A Review. Stem Cells Int. 2021 Jul 24; 2021:8124444. 10.1155/2021/8124444. PMID: 34349803; PMCID: PMC8328695.

32. Bose S, Robertson SF, Bandyopadhyay A Surface Modification of Biomaterials and Biomedical Devices Using Additive Manufacturing. ActaBiomater. 2018 Jan 15; 66:6–22. 10.1016/j.actbio.2017.11.003. Epub 2017 Nov 3. PMID: 29109027; PMCID: PMC5785782.

33. Visan AI, Popescu-Pelin G, Socol G Degradation Behavior of Polymers Used as Coating Materials for Drug Delivery—A Basic Review. Polymers. 2021; 13(8):1272. 10.3390/polym13081272

34. Andorko JI, Jewell CM Designing Biomaterials with Immunomodulatory Properties for Tissue Engineering and Regenerative Medicine. Bioeng Transl Med. 2017 May 16; 2(2):139–155. 10.1002/btm2.10063. PMID: 28932817; PMCID: PMC5579731.

35. Weber M, Steinle H, Golombek S, Hann L, Schlensak C, Wendel HP, Avci-Adali M Blood-Contacting Biomaterials: In Vitro Evaluation of the Hemocompatibility. Front Bioeng Biotechnol. 2018 Jul 16; 6:99. 10.3389/fbioe.2018.00099. PMID: 30062094; PMCID: PMC6054932.

36. Sheikh Z, Najeeb S, Khurshid Z, Verma V, Rashid H, Glogauer M Biodegradable Materials for Bone Repair and Tissue Engineering Applications. Materials (Basel). 2015 Aug 31; 8(9):5744–5794. 10.3390/ma8095273. PMID: 28793533; PMCID: PMC5512653.

37. Prakasam M, Locs J, Salma-Ancane K, Loca D, Largeteau A, Berzina-Cimdina L Biodegradable Materials and Metallic Implants-A Review. J Funct Biomater. 2017 Sep 26; 8(4):44. 10.3390/jfb8040044. PMID: 28954399; PMCID: PMC5748551.

38. Sultana A, Zare M, Luo H, Ramakrishna S Surface Engineering Strategies to Enhance the In Situ Performance of Medical Devices Including Atomic Scale Engineering. Int J Mol Sci. 2021; 22(21):11788. 10.3390/ijms222111788

39. AmirtharajMosas KK, Chandrasekar AR, Dasan A, Pakseresht A, Galusek D Recent Advancements in Materials and Coatings for Biomedical Implants. Gels. 2022 May 21; 8(5):323. 10.3390/gels8050323. PMID: 35621621; PMCID: PMC9140433.

40. Nguyen HH, Lee SH, Lee UJ, Fermin CD, Kim M Immobilized Enzymes in Biosensor Applications. Materials (Basel). 2019 Jan 2; 12(1):121. 10.3390/ma1201 0121. PMID: 30609693; PMCID: PMC6337536.

41. Wang S, Li J, Zhou Z, Zhou S, Hu Z Micro-/Nano-Scales Direct Cell Behavior on Biomaterial Surfaces. Molecules. 2018 Dec 26; 24(1):75. 10.3390/molecules2401 0075. PMID: 30587800; PMCID: PMC6337445.

42. Spałek J, Ociepa P, Deptuła P, Piktel E, Daniluk T, Król G, Góźdź S, Bucki R, Okła S Biocompatible Materials in Otorhinolaryngology and Their Antibacterial Properties. Int J Mol Sci. 2022 Feb 25; 23(5):2575. 10.3390/ijms23052575. PMID: 35269718; PMCID: PMC8910137.

43. Fotovvati B, Namdari N, Dehghanghadikolaei A On Coating Techniques for Surface Protection: A Review. J Manuf Materials Process. 2019; 3(1):28. 10.3390/jmmp3010028

44. Small M, Faglie A, Craig AJ, Pieper M, Fernand Narcisse VE, Neuenschwander PF, Chou S-F Nanostructure-Enabled and Macromolecule-Grafted Surfaces for Biomedical Applications. Micromachines. 2018; 9(5):243. 10.3390/mi9050243

45. Rathod S Novel Insights into the Immunotherapy-Based Treatment Strategy for Autoimmune Type 1 Diabetes. Diabetology. 2022; 3(1):79–96. 10.3390/diabetology3 010007

46. Han X, Alu A, Liu H, Shi Y, Wei X, Cai L, Wei Y Biomaterial-Assisted Biotherapy: A Brief Review of Biomaterials Used in Drug Delivery, Vaccine Development, Gene Therapy, and Stem Cell Therapy. Bioactive Materials. 2022; 17:29–48, ISSN 2452-199X. 10.1016/j.bioactmat.2022.01.011

# 2 Criteria for Biomaterial Surface Selection

*Samapika Bhuyan, Subhasmita Swain, and Tapash R. Rautray*
Biomaterials and Tissue Regeneration Lab, CETMS, Institute of Technical Education and Research, Siksha 'O' Anusandhan (Deemed to be University), Bhubaneswar, Odisha, India

## 2.1 HISTORY OF BIOMATERIALS

Biomaterials are extensively used in diverse fields of tissue engineering, biotechnology, dentistry, and medicine in the modern era. Almost 70 years ago, the word "biomaterials" as we are familiar with it today, did not even come into existence [1]. In the last 200 years, evidence of the use of biomaterials as implants and prostheses on numerous Roman, Egyptian, Greek, and Etruscan human body parts, such as bones or skulls, has been found, convincing that biomaterials have been applied in the human body since ancient times [2]. The use of biomaterials significantly expanded after World War II. The latest advances in medicine have developed a number of devices made of high-performance metal, ceramic, and specifically polymeric materials to repair or replace tissues or parts of the human body that are damaged [1].

They are categorized into three generations in accordance with the development history of biomaterial for bone regeneration based on their features and functions. The first generation of biomaterials for implants was created in the 1960s and 1970s, The "bioinertness" of these implants was a prominent quality. Materials that don't provoke a biological reaction or interact with living things are termed bioinert biomaterials. Bioactive materials were first introduced as second-generation biomaterials in the mid-1980s, when biomaterials were developed to be either bioactive or bioresorbable. A substance is considered bioresorbable if it can progressively replace the original substance with new tissues (like bone) after being ingested by a person. Materials that are considered to be bioactive are those that allow for a direct chemical interaction to take place between an implant and the bone of the recipient. New generations of biomaterials that interact with host tissues have been developed with greater efficiency, utilizing biofunctions. The third-generation biomaterials include a variety of biological processes with the aim of creating substances that, when implanted, could facilitate the body's healing process or trigger genes and cells that promote tissue regeneration [3,4].

DOI: 10.1201/9781003429920-3

The terminology associated with biomaterials was updated in 2018 at the Chengdu Conference due to the fact that certain phrases were still relevant while others were no longer required, and most crucially because new phrases had to be introduced because of the rapid advancements in biomaterials. New definitions were released in 2019, and these are the ones that are currently being used [5].

## 2.2  INTRODUCTION

The term *biomaterials* refers to any substance, natural or manufactured, that may be utilized to repair damaged areas of the body by interacting with human cells, tissues, or organs. This engagement permits them to regain their functions, which is beneficial to the implant as well as to the recipient's bone [6]. There are multiple accepted definitions of biomaterials. The National Institutes of Health (NIH) describes a biomaterial as "any substance (not a drug) or combination of substances that can be used as a whole or as a component of a system to treat, replace, or enhance organs, tissues or functions of the body for any period of time." Due to advancements in tissue engineering and regenerative medicine, the definition of this term has been broadened to include "all materials used in medical devices that interact with biological systems." Thus, structures and combination devices that interact with the human body can now be included in this category [7].

Various kinds of biomaterials are available for use in medical applications. These include (1) implants like intraocular lenses, vascular grafts, heart valves, ligaments, and dental implants; (2) scaffolds that provide the necessary structural support for a particular tissue type; and (3) during the last several decades, it has been obvious that the requirement for biomaterials has expanded considerably because the elderly population that happens in almost every country could be at greater risk of hard tissue (tissues with a high hardness and elastic modulus (E), such as human, animal bones, and teeth) deficiency. As a consequence, biomaterials are employed in device-based treatments, regenerative medicine, tissue engineering, drug delivery, and imaging for medical purposes [8]. In 2000, the United States spent around $14 billion on healthcare overall, but the U.S. biomaterials industry was worth $9 billion when the worldwide incidence of biomaterials demand was taken into account. By 2015, it is anticipated to reach $904 billion. Additionally, keep in mind that annual prices in some poor countries might be two to three times higher than in the United States [9]. According to a distinct investigation, more than 13 million medical gadgets and prostheses are implanted annually in the United States alone [6].

Biomaterials provide locations for cell attachment and the start of signal transduction pathways produced by the matrix. For instance, tissue culture substrates have traditionally been made of natural substances like collagen and polysaccharides. Additionally, cells may be enclosed in a three-dimensional (3D) environment using biomaterials like hydrogels. Additionally, biomaterials control the development of stem cells by modulating matrix elasticity and biomechanical signals [10,11].

In the context of tissue engineering and medical fields, biomaterials are used as scaffolds having various applications. The structure of scaffolds should ideally resemble that of the tissue they are meant to replace. The best scaffold designs for a certain application must be identified in order to create an optimal scaffold. This may be done by testing a variety of different scaffold designs. To create better biomaterials, it is essential to comprehend these aspects at both the macro- and nano-scales. Surface topography, chemistry, charge, structural heterogeneity, and culture conditions all influence how cells react to the surfaces of biomaterials [11]. Diverse nanostructures made of the same material induce various changes in cell behavior, such as changes in cell adhesion, orientation, and motility, as well as the modification of intracellular signaling pathways that control transcriptional activity and gene expression [12]. It is common for biomaterial engineers to change the surface of the biomaterials in order to elicit a specific reaction or to completely quiet all responses since the biomaterial surface plays a significant role in these complicated interactions. Different parts of the body may have different reactions to the same biomaterial [13], depending on factors such as the biomaterial's location and composition. We'll talk more about each of these attributes later.

## 2.3 CLASSIFICATIONS OF BIOMATERIALS

Depending on their abundance in nature, macrostructures, biocompatibility or host reaction, and chemical bonding, materials may be divided into multiple categories, as displayed in Figure 2.1. On the basis of their availability in nature, materials may also be categorized as natural or synthetic. Materials that occur in nature are known as "natural materials," and they include materials like bone, coral, rock, and wood. The most prominent natural materials are ceramics, polymers, and their composites. On the other hand, materials that are manufactured by humans are termed synthetic materials and they serve a certain function. These materials include titanium and its alloys for implants, ceramics for bone tissue engineering, polymers for ocular lenses, and metallic ones like steels for fracture management devices [14]. Biomaterials are further classified into two categories, i.e., (i) resorbable materials: Materials that degrade after being inserted into a human body and are then absorbed by the body's tissues; (ii) non-resorbable materials: Biomaterials that do not undergo disintegration and are not reabsorbed by the tissues of the human body after being implanted in the body [15].

Materials can also be classified based on their macrostructures, such as dense or porous. Most natural materials, such as tissue and wood, are porous materials. Porosity in these materials can serve various purposes which will be described in the second half of this book paper. When materials do not show any porosity, those are called dense materials. Most metallic materials are dense in nature with residual porosity <1%. The majority of the time, dense materials are isotropic and readily formable using a variety of forming processes [14].

Biomaterials can further be categorized into three main types based on biocompatibility and host response:

a. *Bioactive:* The presence of a bioactive substance in the bone tissue environment may facilitate osteogenesis by forming chemical interactions with bone tissues. Bioactive materials again are classified into osteoconductive and osteoinductive materials. Bone may develop over the surface of the bioactive materials thanks to the osteoconductive materials. These

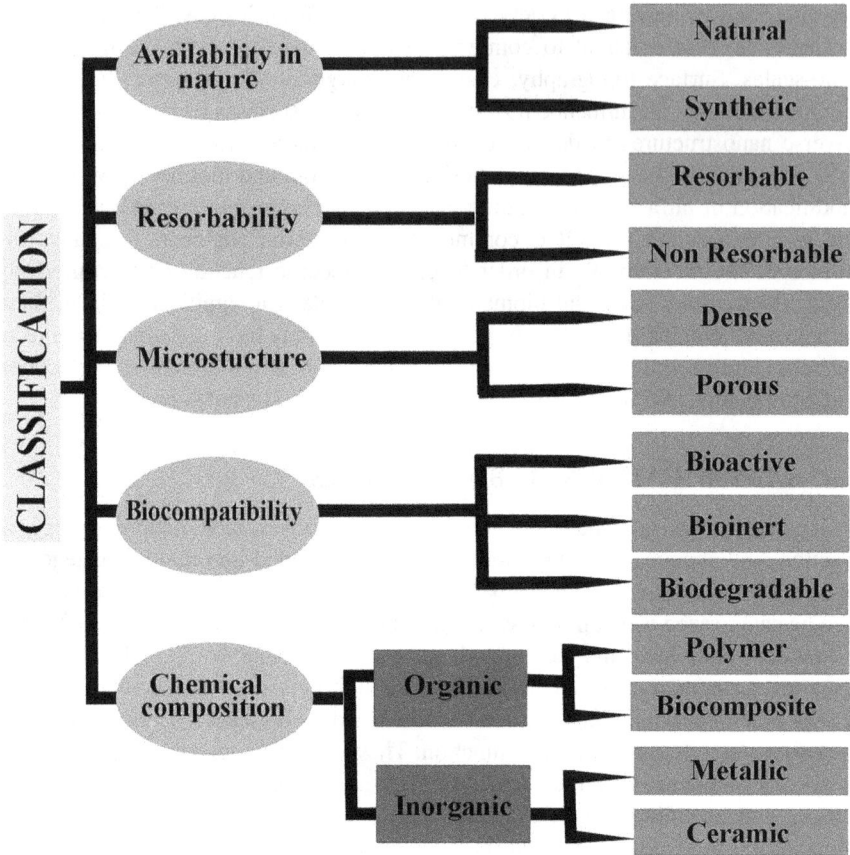

**FIGURE 2.1**  Classification of biomaterials.

osteoconductive materials include, for example, the ceramics hydroxy-apatite and tricalcium phosphate. Materials that are osteoinductive may encourage the formation of new bone. Because they may encourage bone formation distant from the implant site, certain osteoinductive materials are also known as osteoproductive materials. Examples of such osteopro-ductive materials are bioactive glasses [16–18].

  b. *Bioinert and biotolerant:* Implants and bone cannot form a biological interfacial connection when made of bioinert materials. Stainless steel, titanium, alumina, zirconia that has partly been stabilized, and ultra high molecular weight polyethylene are a few examples of these materials. The biofunctionality of bioinert implants often depends on tissue integration via the implant since a fibrous capsule may grow around them [19].

Although biotolerant materials are absorbed by the host, a fibrous tissue (scar tissue) forms to keep them apart from the host tissue. Ions, corrosion products, and chemical substances released by the implant cause the development of this fibrous tissue layer [20,21]. This group includes the majority of metals and almost all synthetic polymers.

    c. ***Bioresorbable:*** Polymers like polyglycolic and polylactic acids, as well as their co-polymers, are used as biodegradable materials. Calcium phosphates are examples of biodegradable ceramics, while magnesium is an example of a biodegradable metal [19,22].

Biomaterials are often categorized into four different groups based on their chemical bonding properties: Metals, ceramics, and polymers. Materials bound together by metallic bonding are known as metals. Due to the presence of free electrons in metallic bonds, metals have mechanical qualities that are malleable and are both electrically and thermally conductive. Materials that are mostly ionic or covalently linked are called ceramics. Because covalent and ionic bonding do not produce any free electrons, usually ceramics are non-conductive materials, both thermally and electrically. However, because of the migration of defects at high temperatures, certain ceramics show conductivity. Polymers, on the other hand, are substances that are built on a long carbon chain and are covalently joined together with some secondary bonding. Covalent bonding causes polymers to be non-conductive materials. A new class of materials called composites will arise when any three of the aforementioned components are mixed without compromising their fundamental qualities [23].

## 2.3.1 POLYMERS

Although their ability to exchange is restricted, polymers are biological molecules that form vast chains of repeating units and have a broader variety of uses than metallic implants. Good biocompatibility, flexibility, low weight, resistance to biochemical assault, availability in a range of compositions, and ease of production into desired forms are their particular qualities that make them frequently employed [24]. Both natural and synthesized polymers, which are either biodegradable or not, are employed in biomaterials. Natural polymers like starch, collagen, and chitin are often used as biomaterials since they are readily manufactured and biodegradable. The majority of non-biodegradable synthetic polymers were initially developed for non-medical uses, but due to their exceptional physical-mechanical properties, they are now widely used as biomedical materials in bandages, dental products, disposable medical supplies, prosthetics, grafts, and specialty tissue products. In comparison to bioceramics or metallic biomaterials, synthetic biopolymers offer a number of benefits, such as the capacity to produce a wide range of designs (such as sheet, fiber, film, and latex), a cheap cost, and the ease with which particular mechanical properties may be attained. Polypropylene, polyethylene, polyethylene terephthalate, polymethyl methacrylate, and polyurethane are a few examples [25,26].

    Polymers are further categorized into two types: Thermoplastics and Thermosetting. When heated above their glass transition temperature (Tg), thermoplastic polymers may be shaped in different ways. The temperature at which a fluid solidifies and experiences the minimum amount of molecular mobility is known as the glass transition temperature (Tg). These polymers include nylon and polyethylene, which are employed in implant coatings and catheters, respectively. Whereas thermosets cannot be reshaped, once they have been crosslinked, raising the temperature of the material will cause chemical decomposition without resulting in remelting. Examples include the use of vulcanized rubber in certain cardiovascular catheters and orthopedic prosthetics [27]. Another category of polymers is hydrogels, which expand when in contact with water due to their hydrophilic properties. In the pharmaceutical industry as well as in the medical profession, hydrogels have been developed for usage in products including contact lenses, membranes, drug delivery, vocal cord reconstruction, wound healing, and an extracellular matrix for tissue engineering [28].

Polymeric biomaterials have a variety of disadvantages in addition to their numerous benefits, such as the fact that they absorb water and protein into the human body, are easily polluted and difficult to sterilize, are leachable, biodegrade, and are subject to wear and disintegration. It is also crucial to remember that the increasing usage of non-biodegradable polymers generates issues with rubbish disposal and environmental contamination [27].

### 2.3.2 METALS

In the past, the exceptional mechanical qualities of these materials have made them ideal for load-bearing body components in orthopedics, such as bones, knees, and dental implants [25]. The most commonly employed metals include titanium and titanium-base alloys (TiAl4VELI, Ti-6Al-4V, Ti6Al17Nb), stainless steels (316 L), and cobalt alloys (Co-Cr-Mo). Titanium and its alloys are notable for their extraordinary resistance to corrosion, good biocompatibility, and long-term stability among these materials. Furthermore, Co-Cr alloys have exceptional mechanical qualities, particularly excellent wear resistance, although they are not very bio-compatible [24,29]. Bone resorption may occur due to stress shielding when a metallic implant is placed in a body that has a lower stiffness than the implant itself. Wherever they are transplanted, these previously mentioned traits progressively impact the host living thing. The effects of a transplant may be terminated by a combination of these variables, weakening the grafts and lowering their compatibility with live organisms (R4). Thus, metal coating with bioactive ceramics and chemical modification of metal surfaces by polymer and biomolecule binding enables the biofunctionalization of metallic materials as well as control over biodegradability and biocompatibility. Due to their enhanced mechanical and biological characteristics and low corrosion resistance, magnesium alloys (Mg-Zn, Mg-Ca), and magnesium matrix composites (magnesium hydroxyapatite, magnesium tricalcium phosphate, Mg-calcium phosphate particle) are among the metallic materials that have nowadays attracted an increasing amount of interest [24,29,30].

### 2.3.3 CERAMICS

Ceramics are non-metallic, polycrystalline, inorganic materials that have corrosion resistance, wear resistance, superior strength, and stiffness. They are distinguished by their hardness, brittleness, and low density. Other desirable characteristics of these materials include their poor electrical conductivity. Because of these properties, they are commonly employed in dental and orthopedic applications such as bone replacement in hip and knee surgery. Ceramics have evolved into a broad class of biomaterials based on the body's reaction, with three fundamental types now available: Bioinert, bioresorbable, and bioactive ceramics [7,24].

Bioinert ceramics are biocompatible materials that retain mechanical and physical properties after implantation while causing no immunologic host responses. Bionert bioceramics are widely employed in structural support applications and are prominent in orthopedics. They are also recognized for their high wear resistance that makes them suitable for gliding functions. Examples of bioinert ceramics include alumina, zirconia, silicon nitrides, and pyrolitic carbon [31].

Bioactive ceramics make direct chemical interactions with the bone or soft tissues in the body environment. Bioglass and ceravital are examples of commonly used bioactive ceramics. These materials are often used as cement for bone fillers, implant interfacial coatings, and restorative composites [32,33].

Bioresorbable/biodegradable ceramics have limited interactions with body tissues. This kind of ceramic, when implanted, eventually gets replaced by bone tissues, showing the highest level of response out of all the categories described. The ability to promote osteoblast integration or osteoconductive ability leads to osteoid development, which is an important characteristic. Calcium sulfate dihydrate (plaster of Paris), calcium phosphates, and hydroxyapatite are examples of resorbable bioceramics. For tissue engineering applications, the use of resorbable scaffolds implanted with cells has the potential to function as a synthetic extracellular matrix [31].

There are many advantages of using bioceramics in orthopedics. They have osteoconductive properties comparable to hard tissues without the risk of disease transmission or immunogenicity. They also have high resistance to corrosion, low toxicity, compressive forces, and promote materialization of new hard tissue. Due to their proportional increase in strength and comparable chemical properties, bioceramics with a higher calcium phosphate (Ca/P) ratio are more desirable. Scientists and clinicians have utilized them increasingly for bioimplant applications [34].

Due to their low mechanical strength values (between 10.5 and 16.1 MPa) and their stress-shielding effects on certain implant locations with high elastic modulus values, bioceramics have limited applications in tissue engineering. Shielding effect occurs when the bone shows high fracture toughness to the implanted ceramic, in spite of its high compressive strength. One of the main disadvantages of using ceramics and glasses as implants is their brittleness and low tensile characteristics [34].

## 2.3.4 COMPOSITES

The composite material is composed of a matrix material and filler (reinforcement) material with distinctive chemical, physical, and morphological properties that are widely used in structural applications. Filler enhances the biocompatibility, bioactivity, and structural properties of the matrix. It has been created in a variety of ways, depending on the component characteristics, to produce materials with physical, chemical, and mechanical properties tailored to particular applications, as well as exhibit better mechanical characteristics than the component materials. These biomaterials are categorized as ceramic matrix composites (HA stainless/ steel), metal matrix composites (Ti/HA, Ti6Al4V/HA, Mg/HA), and polymer matrix composites (carbon fiber/polyether ether ketone, high-density polyethylene/ HA) [35].

Internal prosthesis composites are usually composed of a polymeric substance that has been reinforced with hammered carbon fiber. They have superior tensile properties to the component elements. Carbon fiber biocomposites are currently not used for larger devices. The reasons for this are that they discharge carbon fibers into the surrounding tissues, and forecasting polymer resorption in larger load-bearing devices is difficult. Besides these, the major biocompatibility problems for composites are lipid absorption, satisfying tissue biomechanics, hypersensitivity, and combined particle surface exposure [36].

## 2.4  CRITERIA FOR BIOMATERIAL SURFACE SELECTION

Physical, chemical, biological, and mechanical aspects linked to their surface and bulk properties are the most important material characteristics of biomaterials. The chemistry of a substance, in particular, its bonding, composition, and atomic structures, is connected with the possession of chemical qualities. Phases, microstructures, density, and many types of porosity are all affected by the physical qualities, in contrast. Material toughness and strength, as well as hardness and other forms of failure, are all related to mechanical characteristics. However, the behavior of materials in a living context depends on biological features. In vivo properties relate to characteristics that are measured inside of an animal or a person; in vitro properties refer to characteristics that are created in a petri dish. When all bonds have been released, a material's surface properties are its outermost layer's features. The dangling bonds in materials cause surface features to often differ greatly from bulk values. The chemistry, structure, and processing of the materials are all relevant to these material characteristics [9,36]. Important attributes of an ideal scaffold selection for implantation are depicted in Figure 2.5.

### 2.4.1  COMPOSITION OF ECM

The extracellular matrix (ECM) and the cells are the primary building blocks of tissue architecture, and the ECM plays a key role in the regulation and control of many tissue activities. The extracellular matrix (ECM) is made of different proteins that help cells stick together, grow, and divide. The proteoglycans and fibrous proteins play a significant role in the ECM, which is mostly composed of minerals, water, proteoglycans, specialized proteins, and structural proteins. Figure 2.1 depicts the composition of ECM. The cellular constituents that the ECM supports are continually synthesizing, modifying, and remodeling the structural matrix that makes up the ECM. The relationship between the cell nucleus and ECM regulates all aspects of cell behavior, including adhesion, proliferation, differentiation, and death. As a result, it has been thought that the most important phase in the process

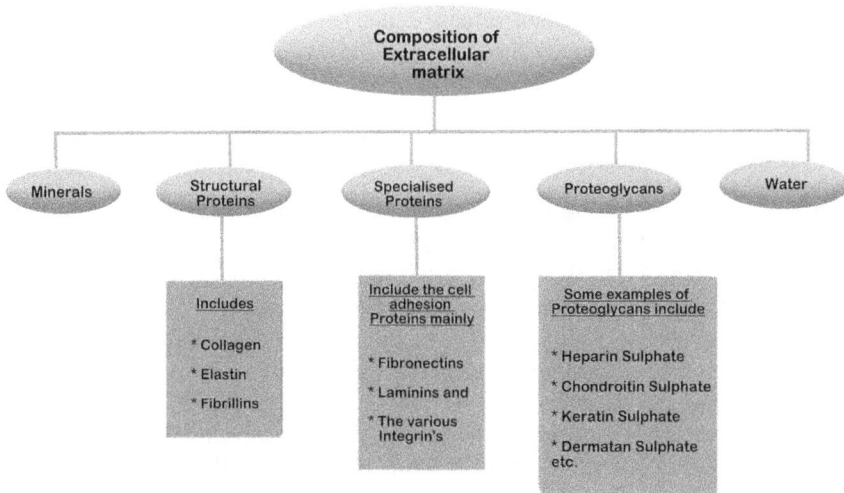

FIGURE 2.2    Composition of extracellular matrix.

of tissue regeneration is to re-create the artificial microenvironments and synthetic ECM structure. Stem cells are able to either directly or indirectly adhere to components of the extracellular matrix. The ECM may have a significant impact on the behavior of mesenchymal stem cells via interactions with other cells, such as those with specialist sticky ligands, or more generally through its physicochemical properties. Regeneration of bone tissue is facilitated by MSC by their differentiation into osteoblasts and osteocytes. Numerous polymers have been used in an effort to re-create the structural diversity of the ECM. Poly(ethylene glycol) (PEG) hydrogels, for instance, are a common biomaterial for stem cell encapsulation. They may be designed to have mechanical characteristics that are comparable to different natural tissue types and imitate the natural ECM in their capacity to absorb vast quantities of water. When a biomaterial is implanted, it should be able to provide the cells with the necessary matrices that resemble their natural environment [37,38]. Figure 2.2 depicts the composition of ECM.

## 2.4.2   HOST TISSUE RESPONSE

When a biomaterial is implanted into a living organism, it harms the host. A series of processes that define the biomaterial's destiny within the host system are triggered by this damage, which directly exposes the biomaterial to blood, and proteins quickly bind to the surface of the biomaterials. In this process, the proteins adsorb onto the biomaterial surface and the process begins, which is known as hemostasis. To provide the mechanical and physiological conditions necessary for osseointegration, the blood clot that forms on the implant surface is of greatest importance. As a result, the "contact osteogenesis" process is enabled, resulting in the production of juvenile bone with irregularly distributed, interwoven collagenous fibers and, as a final step, mature lamellar bone via bone remodeling. As a result of this chain of events, cells do not contact directly with the surface of the implant. Instead, the interface is tightly mediated and regulated by the proteins that have been adsorbed to the surface [38,39].

### 2.4.3  PROTEIN ABSORPTION AND SURFACE INTERACTIONS

Proteins are known to undergo conformational shift upon adsorption to the majority of materials, and the type of this conformational change will considerably influence cellular interactions, making it a crucial characteristic of biomaterial that should be taken into account. Not all alterations will be advantageous for cell attachment; for instance, surface denatured fibronectin won't help cells adhere and proliferate [39].

Adsorbed proteins could act as binding ligands for integrin-family receptor molecules, which in turn alter how cells react to surfaces. Proteins that have been adsorbed onto the surface of the substance may improve the activity of cells because they have specialized binding sites for those proteins in cells. Adhesive proteins include, for instance, vitronectin (VN) and fibronectin (FN). The binding domain arginine-glycine-aspartic acid (RGD), which is present in VN as well, allows FN to activate cell $\alpha_5\beta_1$integrins [40,41].

Determining the kind of bioactive sites that the adsorbed layer of proteins presents is thus crucial for regulating cellular response at the most basic level. The number and kind of proteins adsorbed, as well as their orientation, conformation, and packing arrangement on the biomaterial surface, may all be manipulated to achieve this result [39].

The complex dynamic process of protein adsorption includes non-covalent interactions such as hydrogen bonds, electrostatic forces, van der Waals forces, and hydrophobic interactions. It can be influenced by various factors, including protein properties (structure, surface charge, hydrophilicity), surface characteristics (charge, wettability, roughness), and solution parameters (pH, composition, temperature), which are listed in Table 2.1 [42,43].

### 2.4.4  CELLULAR INTERACTION OF BIOMATERIALS

Cellular behavior is governed by extrinsic signals from the microenvironment, such as cell-cell contact, cell-ECM interactions, and cell-soluble factor interactions when a biomaterial comes into touch with host tissues/cells/blood, with a few of nanoseconds, the surface will be completely covered with water molecules, shown in Figure 2.3 [42]. This first phase establishes the structural arrangement of the water molecules, primes the surface for protein reaction, and heavily relies on the surface characteristics of the material. The second process, which happens after the surface has been moistened, entails coating the surface with a layer of proteins drawn from bodily fluids such as lymphatic fluid, blood, extracellular fluid, and tears. In seconds to hours that follow implantation, an adsorbed protein layer is formed, with the low molecular weight (MW) proteins that reach the surface initially being replaced by comparatively greater MW proteins that arrive later. The extracellular matrix (ECM) proteins, such as laminin, vitronectin, fibronectin, and collagen, are examples of adhesive factors that help proteins adhere to one another.

**TABLE 2.1**

**Principal Parameters Influencing Protein Adsorption on Biomaterial Surface**

| Parameters | General Rules of Thumb |
|---|---|
| Topography/roughness | Higher surface roughness ≥ higher amount of adsorbed proteins |
| Hydrophobicity (non-polar surfaces) Hydrophilicity (polar surfaces, with a net surface charge) | Higher hydrophobicity ≥ higher amount of adsorbed proteins and denaturation degree; hydrophobic interaction as adsorption mechanism. Different mechanisms of adsorption on hydrophilic surfaces: Electrostatic, van der Waals, dipole-dipole; adsorbed water must be removed for adsorption |
| Chemistry (functional groups, metal ions) | Influence on the surface charge |
| Amino acid chain | Affects structural stability |
| Hydrophilicity/hydrophobicity | Surface charges and non-polar residues are always present; they can be differently arranged according to the environment; hydrophobic residues interact with hydrophobic surfaces |
| Charge | Higher amount of adsorbed proteins at IEP |
| Molecular weight | Small proteins adsorb quicker; large proteins replace the smaller ones and make stronger bonds with the hydrophobic surfaces |
| Structural stability | Soft proteins change easier configuration and adsorb larger on hydrophilic surfaces; denaturation can enhance or reduce biological activity |
| pH | Affects surface charge of both proteins and surfaces |
| Ionic strength | Adsorbed ions reduce repulsive effects among proteins; some ions compete with proteins for adsorption |
| Protein concentration | Higher protein concentration; higher amount of adsorption |
| Protein mixture(single, binary, or more complex) | Vroman effect |
| Temperature | Higher temperature ≥ faster kinetics of adsorption |

Blood proteins, also known as plasma proteins, are often the first proteins to be adsorbed. This includes proteins like fibronectin, immunoglobulin G (IgG), fibrinogen, and albumin. These are soon replaced by the other proteins, like Hageman factor (factor XII) and high MW kininogen. This process of protein deposition in a hierarchical manner is known as the Veroman effect. Initiating the blood coagulation is primarily carried out by the enzyme activated factor XII. The adsorbed proteins may improve the development of the temporary matrix on and surrounding the biomaterial, where the fibrin network and associated adhesive proteins serve as a natural scaffold for cell adhesion and proliferation. Within minutes after the implantation of a biomaterial, several events start to happen. The

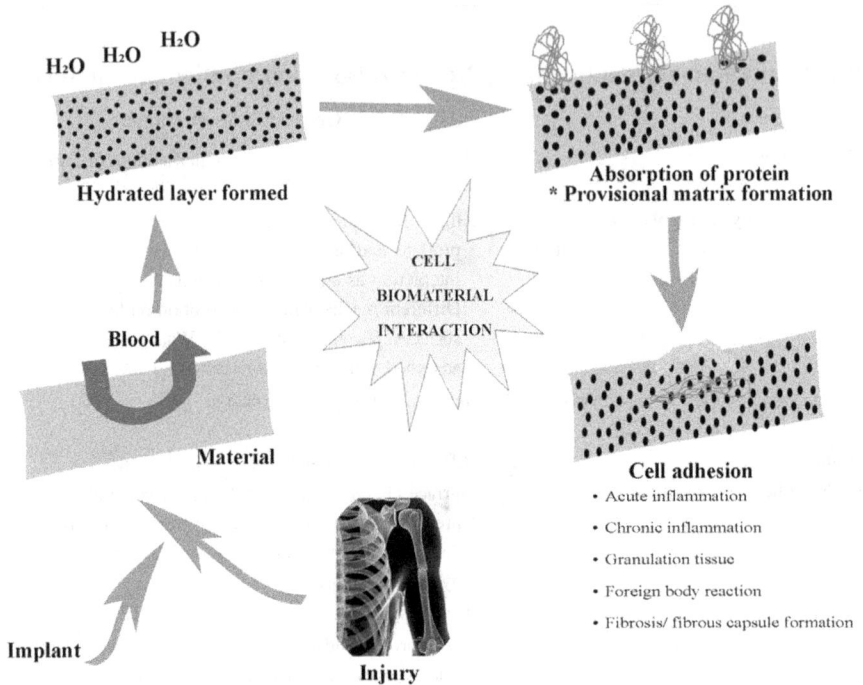

**FIGURE 2.3** The series of events that take place at the material's surface when it comes into contact with its host biological environment.

healing process and tissue repair are both triggered by the signal molecules released by the connected cells, which include matrix metalloproteinases (MMPs), cytokines, chemokines, and inhibitors of tissue matrix metalloproteinases (TIMPs) [44].

### 2.4.5 OSSEOINTEGRATION

According to definitions of osseointegration, it refers to the direct anchoring of an implant via the development of bone tissue around the implant without the creation of fibrous tissue at the bone-implant interface. The stable and effective union of the bone and an implanted surface is known as osseointegration. This occurrence happens after the device is implanted into the bone and the bone cells have moved to the bone's surface [45]. Osteoblastic cell adherence with biomaterial surface is shown in Figure 2.4(A-D).

Surface topography, surface chemistry, and surface roughness all have a significant impact on how well osseointegration develops. Surface roughness has a significant impact on how cells cling to surfaces and differentiate. A rougher surface may enhance the surface area where host proteins and cells can interact with implants. In vitro, rough surfaces encourage differentiation and mineralization, whereas smooth surfaces encourage osteoblast cell adhesion and proliferation [17,46].

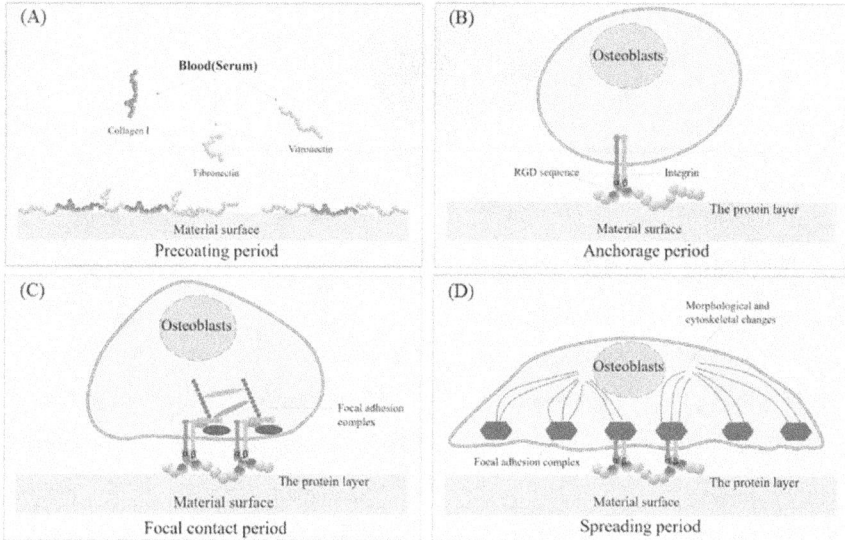

**FIGURE 2.4** The adherence of osteoblastic cells. (A) The pre-coating phase: During which the ECM proteins in the blood are deposited on the substrate. (B) Anchoring period: Integrins recognize RGD sequences and facilitate cell-material anchorage. (C) Focal adhesions phase: Signals trigger the recruitment of a group of cytoskeletal proteins, which leads to focal adhesion formation and maturation. (D) Spreading phase: Cytoskeletal network and actin microfilaments organize and modify cell shape.

### 2.4.6 OSTEOCONDUCTION

One of the most important criteria for biomaterials used for bone regeneration is osteoconduction, the capacity of materials to promote the growth of new bone on their surfaces. In bone defects, osteoconductive materials enable osteoprogenitor cells to migrate, proliferate, differentiate, and deposit extracellular matrix (ECM), which are vital processes in the formation of new bone. Appropriate chemical composition, surface properties, architectural geometry, and other elements are crucial components of osteoconductivity of biomaterials. Calcium phosphates (CaP) based ceramics like tricalcium phosphate (TCP), hydroxyapatite (HA), biphasic calcium phosphate, and bioglass are widely used as osteoinductive biomaterial due to their resemblance to the natural bone material. Through various strategies such as coating and composite, osteoconductivity may be incorporated into non-biological materials such as ceramics, metal, and synthetic polymers. For instance, despite the fact that titanium is often thought to be non-osteoconductive, bone conduction was developed after the creation of a titania layer on its surface with the help of the proper surface treatment [47–49].

### 2.4.7 OSTEOINDUCTION

Biomaterials should have osteoinductive properties, which means that it has the potential to stimulate the differentiation of mesenchymal stem cells into bone-forming

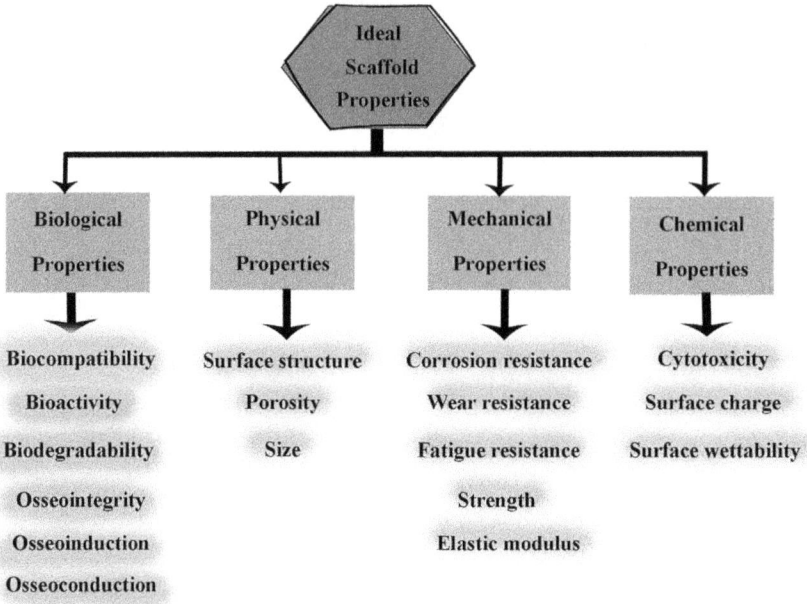

**FIGURE 2.5** Important characteristics of ideal biomaterial.

cells (osteoblasts). Osteoinductive biomaterials have been shown to influence ectopic bone formation in a number of ways, including the following: (i) At the tissue level, they are likely to actively facilitate oxygen, nutrition, and waste transfer; vascularization between the tissue and material; (ii) at cellular level, the development of a biological carbonated apatite layer may initiate stem cell/ osteoprogenitor cell differentiation towards the osteogenic lineage; (iii) on a molecular level, osteoinductive materials may be capable of concentrating osteogenic protein, including bone morphogenetic proteins BMP-2 and BMP-7, due to their strong affinity to these naturally occurring osteoinductive proteins. Enhanced cellular activity on the surface of biomaterials may be facilitated by the enrichment of local growth factors. The most commonly used osteoinductive materials are those made from calcium phosphates, such as tricalcium phosphate (TCP), hydroxyapatite (HA), coralline hydroxyapatite, and biphasic calcium phosphate [48,50].

## 2.4.8 BIOCOMPATIBILITY

One of the criteria for the therapeutic use of biomaterials in biomedical science is biocompatibility. When it comes to biomaterials, the substances used must not have any harmful effects on the surrounding organs, tissues, or biological systems. The responses generated must be ideal for the particular situation and beneficial to the host. Nowadays, biocompatibility involves bioinertia as well as biofunctionality and biostability. It is a fundamental idea that is influenced by a variety of variables, such as the surface chemistry, charges, wettability stiffness, crystallinity, texture, breakdown products, and interactions between the device and the biological environment of the target tissues such as blood contact, inflammatory processes, protein adsorption. Different medical applications require different levels of biocompatibility. For instance, a material used satisfactorily in orthopedic surgery may be inappropriate for cardiovascular applications because of its thrombogenic properties (a property of the material to induce blood clotting) [24]. Furthermore, three crucial qualities of biocompatibility that a potential biomaterial must attain in a wide range of environments are as follows: (1) To elicit the proper biological reaction, they should be noncarcinogenic, nontoxic, non-thrombogenic, nonantigenic, anti-inflammatory, and nonmutagenic; and (2) it is crucial that the device seamlessly integrates with the surrounding tissues from a biomechanical viewpoint; (3) a bioadhesive contact must be developed between the material and the living tissues since primary tissue types need a matrix for adhesion to preserve their distinctive phenotype.

Thus, biocompatibility testing is regarded as a primary requirement for the development and approval of orthopedic materials for clinical use by regulatory agencies. A biomaterial must meet the biocompatibility criteria set by the International Standards Organization (ISO 10993). There are major tests used in the biocompatibility assessment of orthopedic biomaterials [24].

The following are the major variables impacting biocompatibility [50–52]:

1. **Environmental interconnections:** A toxicological or allergic reaction, mutagenic reactions, inflammatory processes, degree of biodegradation, the impact of cytotoxins, and chemical interaction with blood play major role.
2. **Biocompatibility of structure:** This refers to the implant's ability to mechanically adapt to the host tissue as effectively as possible.
3. **Surface biocompatibility:** It refers to the chemical, biological, and morphological compatibility of the implanted surface.
4. **Time length of implant application:** This refers to how long an implant remains in the body. Long-term and short-term implants are the two types.
5. **Implant material:** This comprises the level of aggression between the host tissue and synthetic material.

## 2.4.9 BIOACTIVITY

Bioactivityis the term used to describe any interaction or impact that substances have on cells with the intention of causing or activating certain reactions and behaviors. One of the best-known mechanisms for increasing bioactivity in bone repair and fixation applications is the mineralization and binding between the bone tissue and the implant. Bioactivity is often stronger when its chemical makeup is similar to that of the host tissue. This is because they may facilitate cellular identification and trigger a particular cellular response to encourage tissue development. In order to achieve this goal, the biomaterial's surface may be altered by the addition of extracellular matrix macromolecules including collagen, fibronectin, and laminin. This creates a biomimetic environment that is comparable to the genuine tissue and is capable of influencing cellular behavior and reaction. Bioactive materials, such as calcium phosphate ceramics, have uses in orthopedic implants to improve bone attachment, antimicrobials to reduce infection, and antithrombotics to mitigate thrombus [53,54]. Bioactive materials often employ nature-inspired materials or imitate nature-inspired materials. Bioactive coatings may be created from chemicals that stop blood clotting or start the thrombus's enzymatic breakdown. Cardiovascular implants have been coated with heparin to greatly minimize thrombus and platelet absorption [52].

### 2.4.10 BIODEGRADABLE

The term *biomaterial* may also refer to materials that break down naturally over time. They serve a particular purpose within the body and then gradually deteriorate over time. They have been referred to by a variety of words, such as absorbable, resorbable, and degradable. Different factors, such as the kind of material utilized, its composition, and the production method, have an impact on degradation. The following properties are necessary for biodegradable materials to be successful [52].

### 2.4.11 BIOINERT

"Bioinert" materials are defined as biocompatible materials that mostly do not interfere with biological functions but lack any major foreign body response. Almost 50 years after the first implants made of titanium were developed, it is surprising that we are still implanting "bioinert" materials. Bioinert biomaterials are intended to be corrosion-resistant and to stop the production of particles and reactive ions. Biomaterials should not cause an unacceptable amount of damage to the body, including local and systemic adverse effects, since they were designed to stay in the body for an extended length of time. Due to its advantageous combination of biocompatibility, corrosion resistance, strength, and elastic modulus, as well as its comparably low weight and density to ordinary steel and Co-Cr alloys, titanium is often the material of choice among bioinert materials [44,55,56].

### 2.4.12 BIOFUNCTIONALITY

Biofunctional biomaterials will have a favorable effect on the host and, via their biophysical, biochemical, and/or biological payload, will activate the body's natural healing mechanisms, encouraging functional repair and regeneration. As a result, the cutting edge of technical and scientific research and innovation is focused on biofunctional biomaterials. Such compounds were created to facilitate certain biological reactions or to encourage bone tissue adhesion. The majority of biofunctional materials are also biodegradable or disintegrate over time in a specified manner. Additionally, orthopedic biomaterial surfaces have recently been functionalized to replicate certain extracellular matrix features (such as chemistry and topography). Biofunctional materials may promote tissue attachment or trigger a desired biological response by interacting with a biological environment. HA and bioglass are well-known examples of biofunctional materials. HA has the power to promote mineralization and create a strong bond with bone. Functionalized bioglass has shown a greater rate of bone production than CaPs, which has sparked a lot of research interest in orthopedic materials. Through mechanical support for injured tissue and stimulation of tissue development rates, biofunctional biomaterials are having a significant influence on musculoskeletal illnesses [57,58].

## 2.4.13 CYTOTOXICITY

The biomaterial should have low cytotoxicity. The term *cytotoxicity* refers to a substance's hazardous effects on essential organs and tissues. Unless it is intended to be harmful, a biomaterial shouldn't naturally be toxic. During the preliminary screening of the item, *in vitro* cytotoxicity testing is evaluated. Studies conducted *in vitro* have demonstrated that when the setting reaction in restorative materials (like resin-modified glass ionomer cements and resin composites) isn't complete, the cytotoxic effects are brought on by the release of components from the incompletely set material that changes the metabolic processes of the cell [59]. The toxicity may either be local or systemic, depending on the circumstances. Local toxicity occurs when a biomaterial causes inflammation or necrosis at the site of application. Systemic toxicity appears when injury develops far from the point where the substance was applied. In vitro, cytotoxicity testing is the initial stage in determining if dental materials are biocompatible. Metals like aluminum and iron, which are included in composite resins and RM-GIC in varying concentrations, have been demonstrated to accelerate the cytotoxic cascade [59,60].

## 2.4.14 SURFACE PROPERTIES

A key element in the success of the implant is the biomaterial's surface integrity. Tissue integration with biomaterial-based implants occurs primarily at the point of contact between the implant and the tissue (the scaffold). Surfaces make reactants easily accessible; hence, when surface areas are significant, reaction turnover is larger [57]. To build an ideal scaffold, it is crucial to comprehend which scaffold designs result in the best cell response for a certain application. Taking these properties into account at both the macro- and nanoscales is essential for creating better biomaterials. Surface wettability, dimensionality, composition, charge, porosity, stiffness, and degree of cross-linking are fundamental elements that must be taken into account when designing a substitute for medical applications (R34). Variations in cell adhesion, orientation, and motility as well as the modulation of intracellular signaling pathways that control transcriptional activity and gene expression are all facilitated by different nanostructures of the same material [12,61].

### 2.4.14.1 Surface Chemistry

Cell adhesion and function are greatly influenced by the chemical makeup of biomaterial surfaces. Surface features have a substantial impact on cellular response, which in turn affects how quickly and effectively tissues regenerate.

The directed adsorption of molecules from the surrounding fluid, which results in a conditioned contact and a cell response, is one of the first processes that occur at the surface. Cell proliferation, adhesion, morphogenesis, and differentiation are influenced by the interaction between surface wettability, surface charge, and chemistry [62].

Self-assembled monolayers (SAMs) containing a range of functionalizing groups, including methyl (CH3), COOH, NH2, and OH were used by Yu et al. to illustrate the effect of surface chemistry on human dental pulp stem cells (hDPSCs). When compared to other types of self-assembled monolayers, hDPSCs adhered, proliferated, and showed an osteocyte-like shape on the surface that was changed with $NH_2$ groups. Application of surface changes including $NH_2$ functional groups may improve the osteoconductivity/osteoinductivity, osseointegration, and biocompatibility of dental implants [63].

In a separate study, Patricia Rico et al. determined that the structure and distribution of fibronectin (FN) are directly dependent on substrates with regulated OH group surface density. Cells become increasingly spherical as substrate hydrophilicity increases, suggesting that the presence of OH groups in the samples has a significant impact on their shape [64,65].

The biocompatibility and immunogenicity of a material are greatly affected by its surface chemistry. The current study, taken as a whole, points to the possibility that modifying the surface chemistry of therapeutic materials is a potential strategy for modifying their particular properties.

### 2.4.14.2   Surface Topography

Cell proliferation, migration, adhesion, and differentiation during tissue regeneration are all significantly influenced by the surface topographies of medical implants [65]. This includes factors such as surface roughness, surface pore size, surface groove size, orientation, and distribution. It is widely recognized to have a major influence on the material's bulk characteristics that range from microscale to nanoscale may significantly affect osteoblastic cell adhesion and may be assessed using macroscopic approaches [63,66].

Recent studies have categorized cellular responses to surfaces based on the size of surface patterning. Surface nano-patterning can boost surface energy and enhance protein adsorption, spreading, proliferation, cell migration, gene expression, differentiation, and ultimately tissue-implant integration, whereas micro-patterning mainly affects cell position, growth, morphology, and cytoskeletal reorganization. Roughness and pattern are the two most noticeable characteristics of a surface's topography [67].

### *2.4.14.2.1   Surface Roughness*

The most important aspect of surface topography is surface roughness, which is determined by measuring surface protrusions or depressions and discloses the structure of a material's surface. Osteoblasts prefer rough surfaces, periodontal fibroblasts prefer smooth surfaces, and epithelial cells prefer smooth surfaces, among other cell types with varying preferences for surface texture [67–69]. On rough surfaces, there seems to be an increase in the ability of cells to adhere.

Osteoblastic cell attachment is really hindered when the surface roughness is more than 2.19 m. According to the findings of several studies, nanoscale roughness may significantly improve cell adhesion.

In contrast to osteoblast shape, adhesion, total protein content, or ALP activity, Carine Wirth et al. discovered that nickel-titanium (NiTi) surface roughness had no effect on cell proliferation. Cell growth is stimulated by roughness. As a result, roughness and chemistry impacted cell proliferation, while surface chemistry affected cell adhesion and ALP activity [46].

Cells are capable of detecting micro- and nano-roughness, which affects their ability to adhere, proliferate, and grow. Similar to this, protein adsorption may be impacted by surface shape. Protein adsorption rises when surface roughness is increased because a rougher surface has more surface area. Additionally, it has been discovered that the material's grain size and crystallinity have an effect on how biomaterials adsorb [42].

### 2.4.14.2.2   Surface Pattern

Surface patterns are classified as either anisotropic or isotropic depending on how the cells react to them. A direction-dependent and regularly oriented surface pattern, such as ridges and grooves, is known as an anisotropic surface. An isotropic pattern is defined as a surface without any directional orientation. In addition to protrusions, pits, pillars, and round, star, triangular, and square formations, this surface exhibits micro- and nanoscale topographic characteristics. The grooves often operate as a guiding mechanism for cell migration, forcing the cells to elongate and align themselves in the direction of the grooves, a process known as contact guidance [70].

The impact of other isotropic patterns, such as pillars and pits, has also been explored. Cell proliferation is assumed to increase with declining pit diameter, whereas cell attachment is expected to be greater on microscale pillars than on macroscale pits.

In order to promote the production of bone proteins in the absence of an osteogenic inducer, Dalby et al. created hMSCs on the surface of poly(methyl methacrylate) scaffolds with varied disordered nanoscales. They hypothesized that controlled-disordered nanoscales (100 nm depth, 120 nm diameter, and 300 nm center-center spacing) may encourage higher expression of bone ECM components than random or highly ordered nanopits. In addition to this, deeper nanopits contributed to the cell's reactivity as well as hMSC osteogenesis [71].

### 2.4.14.3   Surface Charge

The influence of surface charge on the incidence of cell attachment is now receiving a lot of investigation. The behavior of cells might be impacted by surface charges. More osteoblastic cells are attached to the positively charged surface as compared to negatively and neutrally charged surfaces. It is necessary to conduct more experiments to have a better understanding of material surface charge. Biocompatibility, cell differentiation, and cell affinity may all be improved by using both positive and negative ions on implanted surfaces, as has been proven in a number of studies [62].

### 2.4.14.4   Surface Energy

It has often been discovered that surface energy, which is closely connected to wettability, has a substantial association with biological interaction. When an atom is taken out of a material's bulk and put at its surface, surface free energy is the increase in energy that results. Less sticky surfaces are those with a low surface free energy (SFE) compared to those with a high surface free energy. SFE was shown by Hallab et al. to be more important than surface irregularity for cellular adhesion strength and proliferation, and surface energy components of the different investigated materials were shown to be positively linked with cellular adhesion strength [72].

Bonding at the surface, both primary (metallic, ionic, and covalent) and secondary (van der Waals), may affect the surface free energy (SFE) of the system. When the SFE is non-polar, this indicates that the dangling bonds are van der Waals bonds. If the bonds are mostly ionic or covalent, the SFE will benefit from Lewis acid and base contributions [73].

SFE is divided into two categories: High-energy surfaces (materials that are ionically, covalently, or metallically bonded) like metals and oxides with surface energies ranging from 500 to 5,000 mN m$^{-1}$ and low-energy surfaces (materials bonded by van der Waals bonds) like molecular crystals and plastics with surface energies ranging from 5 to 50 mN m$^{-1}$ [246]. Tissue formation at interfaces is affected by SFE, which in turn affects protein adsorption, cell adhesion, water interactions, and other surface phenomena. The components of SFE for surfaces with a smooth finish may be assessed using contact angle measurements. It has been suggested that a contact angle of between 60° and 70° is ideal for regulating cell behavior and proliferation [70,74].

### 2.4.14.5   Surface Wettability (Hydrophilicity/Hydrophobicity)

Surface wettability is widely acknowledged as a key factor in determining cell responsiveness. The contact angle of a droplet may be measured as it moves over a surface in order to get an accurate reading of this property. Cell adhesion may be impacted as a consequence, and it may have an impact on protein adsorption. For instance, a surface's hydrophilicity increases in direct proportion to the angle at which it makes contact with water. According to prior studies, the more hydrophilic the surface of material films, the stronger the cell adhesion on the surface [42]. For instance, it was discovered that osteoblast adhesion decreased when the surface contact angle rose from 0° to 106°. According to research, fibroblast adhesion is highest when contact angles are between 60° and 80° [75].

Depending on a surface's wettability, different kinds of proteins get adsorbed onto it. On hydrophilic surfaces, for instance, fibronectin protein predominated, while albumin was adsorbed onto hydrophobic surfaces [239]. On moderately wet substrate surfaces (with a contact angle of 70°–80°), cell attachment and spreading often occur. Positive charge and hydrophilicity promote cell attachment, whereas negative charge and hydrophobicity inhibit attachment, according to studies conducted by researchers at the University of California, Berkeley, and the Scripps Research Institute [53].

## 2.4.15   High Corrosion Resistance

Materials for implants exhibit an excellent resistance to corrosion. Eventually, implants deteriorate in human bodily fluids because they are subject to electro-chemical assault from the electrolytically hostile environment [76]. The chlorine, salt, plasma, proteins, and mucin found in blood and other bodily fluids may be quite corrosive. As a result of implanted materials corroding, it has been shown that a large number of poisonous and harmful metal ions, including Fe, Cr, Ni, Co, and Ti, are released into human fluids. These fundamental trace elements in metallic implants would release harmless ions at first. However, as implants deteriorate, these trace elements disperse strongly into the circulation, where they may reach cells and tissues and potentially destroy the implant, necessitating a second operation to repair the damage. When a metal's oxide layer is destroyed, corrosion occurs and a metal ion is released. During the subsequent step of the regeneration process, the surface layer is given a coating treatment. The surface oxide [77] layer's renewal or repassivation period varies for different relevant materials, and this has a significant impact on the degree of corrosion and the release of certain metal ions. When present in the tissues surrounding implants, metallic ions have been shown to have genotoxic, carcinogenic, hypersensitive, allergic, local tissue toxic, and inflammatory effects [8,78].

## 2.4.16   High Wear Resistance

While gliding across body soft tissues, the material should be wear-resistant and have a low coefficient of friction. A reduction in wear resistance or an increase in friction coefficient may lead to the implant loosening, which may eventually result in a significant inflammatory response and bone damage [3,11]. Along with accelerating corrosion and releasing unbiocompatible metal ions, an implant's sensitivity to wear also speeds up deterioration [8,36].

## 2.4.17   Fatigue Resistance

The microstructure of the biomaterial has a significant impact on the features of fatigue. It is brought on by repetitive cyclic loads that, over time, modify these microstructures, leading to fractures and deformations. The implant may fracture, the stress shield may fail, and the implant may become loose. Strong mechanical stresses cause this to happen often with hip prostheses [10]. The negative host tissue reaction to wear debris generated by the fatigue process, however, is one of the main reasons to be worried about the fatigue of biomaterials. A response that is inflammatory and immunological is typically brought on by wear debris. This creates interfacial problems between the implant and the host tissue by causing blood coagulation, leukocytes, macrophages, and in severe cases, big cells, to move in the foreign wear particles. Several biological reactions have already begun at this phase. One of them is a change in the environment to one that is very acidic (pH 3). The processing and heat treatment used in the production process greatly influence the initial microstructure of metallic biomaterials [79].

### 2.4.18 STRENGTH

An artificial organ's chance of breaking is affected by the strength of the materials used to create it. Inadequate strength can end up in implant fracture. Developing a soft fibrous tissue at the interface may increase relative motion between the implant and the bone under stress when the bone-implant interface begins to deteriorate. The patient feels discomfort as a result of this circumstance, and after a while the suffering becomes intolerable and the implant has to be replaced via a revision process [8].

### 2.4.19 POROSITY

Porosity can be defined as the degree of free space in a solid. *In vitro* and *in vivo* bone formation is significantly influenced by the porosity and the size of pores of biomaterial scaffolds. The role of these morphological features on osteogenesis both *in vitro* and *in vivo*, as well as their links to the mechanical properties of the scaffolds, is examined. Decreased porosity promotes osteogenesis in vitro by limiting cell division and inducing cell aggregation. On the contrary, greater bone growth occurs *in vivo* where higher porosity and pore size are present [80]. The scaffold structure must be porous (50–90% porosity), with a minimum pore diameter of 100 nm in order to be designed as a functioning scaffold. For the transport of vital nutrients, extracellular fluid into and out of the cellular matrix, and oxygen, there must be proper interconnection between the pores. For optimal tissue growth, factors such as pore diameter, volume, size, distribution, shape, and pore wall roughness are also taken into account. According to their size, pores can be divided into micropores (0.1–2 nm), mesopores (2–50 nm), and macropores (>50 nm). Hepatocyte and fibroblast proliferation, in particular, requires a pore size of 20 microns, while the dimension is between 20 and 150 microns for soft tissue healing. Researchers suggested that using pore sizes between 200 and 400 microns for bone tissue engineering will be more convenient [77,81].

## 2.5 CONCLUSION

The discussion outlined in the book chapter represents how one must consider certain fundamental properties for choosing the implant that will be used for a specific application. The primary aspects to consider when selecting a suitable scaffold for implantation are biocompatibility, biodegradability, bioactivity, high immunogenicity, high hydrophobicity, surface roughness, cytotoxicity, physical compatibility, and mechanical compatibility. Increased wear resistance minimizes the risk of implant loosening and revision surgery. A biomaterial shouldn't exhibit mechanical incompatibility and should have an elastic modulus comparable to the bone to prevent the stress-shielding effect. A potential strategy to control cellular behavior in terms of proliferation, adhesion, and differentiation involves the surface modification of scaffolds by bioactive substances to create substrates with desired properties. Therefore, more in-depth knowledge of the long-term in vivo fate of surface-engineered biomaterials may advance this field and expedite the process of translating it for routine medical applications.

## REFERENCES

1. Ratner, Buddy D., and Guigen Zhang. "A history of biomaterials." In *Biomaterials science*, pp. 21–34. Academic Press, 2020.
2. Migonney, Véronique. "History of biomaterials." *Biomaterials* (2014): 1–10.
3. Yang, Ke, Changchun Zhou, Hongsong Fan, Yujiang Fan, Qing Jiang, Ping Song, Hongyuan Fan, Yu Chen, and Xingdong Zhang. "Bio-functional design, application and trends in metallic biomaterials." *International Journal of Molecular Sciences* 19, no. 1 (2017): 24.
4. Swain, Subhasmita, R. D. K. Misra, C. K. You, and Tapash R. Rautray. "TiO$_2$ nanotubes synthesised on Ti-6Al-4V ELI exhibits enhanced osteogenic activity: A potential next-generation material to be used as medical implants." *Materials Technology* 36, no. 7 (2021): 393–399.
5. Zhang, Xingdong, and David Williams, eds. *Definitions of biomaterials for the twenty-first century*. Elsevier, 2019.
6. Gobbi, Silvio Jose, Vagner Joao Gobbi, and Ynglid Rocha. "Requirements for selection/development of a biomaterial." *Biomedical Journal of Scientific & Technical Research* 14, no. 3 (2019): 1–6.
7. Binyamin, Gary, Bilal M. Shafi, and Carlos M. Mery. "Biomaterials: a primer for surgeons." In *Seminars in pediatric surgery*, vol. 15, no. 4, pp. 276–283. WB Saunders, 2006.
8. Los, Marek J., Andrzej Hudecki, and Emilia Wiechec, eds. *Stem cells and biomaterials for regenerative medicine*. Academic Press, 2018.
9. Basu, Bikramjit. "Biomaterials science and implants." (2020).
10. Yliperttula, Marjo, Bong Geun Chung, Akshay Navaladi, Amir Manbachi, and Arto Urtti. "High-throughput screening of cell responses to biomaterials." *European Journal of Pharmaceutical Sciences* 35, no. 3 (2008): 151–160.
11. Rinmayee, Praharaj, Snigdha Mishra, and Tapash R Rautray. "The structural and bioactive behaviour of strontium-doped titanium dioxide nanorods." *Journal of the Korean Ceramic Society* 57 (2020): 271–280.
12. Jones, Julian R. "Observing cell response to biomaterials." *Materials Today* 9, no. 12 (2006): 34–43.
13. Rajesh, P., Sankalp Verma, Vivek Verma, K. Balani, A. Agarwal, and R. Narayan. "Host response of implanted biomaterials." *Biosurfaces: A Materials Science and Engineering Perspective* 2015 (2015): 106–125.
14. Bose, Susmita, and Amit Bandyopadhyay. "Introduction to biomaterials." In *Characterization of biomaterials*, pp. 1–9. Academic Press, 2013.
15. Biswal, Trinath, Sushant Kumar, Bad Jena, and Debabrata Pradhan. "Sustainable biomaterials and their applications: A short review." *Materials Today: Proceedings* 30 (2020): 274–282.
16. Ødegaard, Kristin S., Jan Torgersen, and Christer W. Elverum. "Structural and biomedical properties of common additively manufactured biomaterials: A concise review." *Metals* 10, no. 12 (2020): 1677.
17. Swain, Subhasmita, Tapash Ranjan Rautray, and Ramaswamy Narayanan. "Sr, Mg, and Co substituted hydroxyapatite coating on TiO$_2$ nanotubes formed by electrochemical methods." *Advanced Science Letters* 22, no. 2 (2016): 482–487.
18. Bijayinee, Mohapatra, and Tapash R. Rautray. Strontium-substituted biphasic calcium phosphate scaffold for orthopedic applications. *Journal of the Korean Ceramic Society* 57 (2020): 392–400.
19. Brovold, Matthew, Joana I. Almeida, Iris Pla-Palacín, Pilar Sainz-Arnal, Natalia Sánchez-Romero, Jesus J. Rivas, Helen Almeida, et al. "Naturally-derived biomaterials for tissue

engineering applications." *Novel Biomaterials for Regenerative Medicine* (2018): 421–449.

20. Plenk Jr, Hanns. "The role of materials biocompatibility for functional electrical stimulation applications." *Artificial Organs* 35, no. 3 (2011): 237–241.

21. Swain, Subhasmita, Sapna Mishra, Abhishek Patra, Rinmayee Praharaj, and Tapash Rautray. "Dual action of polarised zinc hydroxyapatite-guar gum composite as a next generation bone filler material." *Materials Today: Proceedings* 62 (2022): 6125–6130.

22. Priyadarshini, Itishree, Subhasmita Swain, Janardhan Reddy Koduru, and Tapash Ranjan Rautray. "Electrically polarized Withaferin A and Alginate-incorporated biphasic Calcium Phosphate microspheres exhibit Osteogenicity and antibacterial activity in vitro." *Molecules* 28, no. 1 (2022): 86.

23. Chong, Jiann, Eric Tzyy, Jun Wei Ng, and Ping-Chin Lee. "Classification and medical applications of biomaterials–A mini review." *Bio Integration* (2022).

24. Huzum, Bogdan, Bogdan Puha, Riana Maria Necoara, Stefan Gheorghevici, Gabriela Puha, Alexandru Filip, Paul Dan Sirbu, and Ovidiu Alexa. "Biocompatibility assessment of biomaterials used in orthopedic devices: An overview." *Experimental and Therapeutic Medicine* 22, no. 5 (2021): 1–9.

25. Ajmal, Sidra, Farzan Athar Hashmi, and Iffat Imran. "Recent progress in development and applications of biomaterials." *Materials Today: Proceedings* (2022).

26. Rautray, Tapash R., R. Narayanan, and Kyo-Han Kim. "Ion implantation of titanium based biomaterials." *Progress in Materials Science* 56, no. 8 (2011): 1137–1177.

27. Abramson S., et al. Classes of materials used in medicine. In: Ratner B.D., et al., eds. *Biomaterial science: An introduction to materials in medicine*, pp. 67–233. London: Elsevier Academic Press, 2004.

28. Vedadghavami, Armin, Farnaz Minooei, Mohammad Hossein Mohammadi, Sultan Khetani, Ahmad Rezaei Kolahchi, Shohreh Mashayekhan, and Amir Sanati-Nezhad. "Manufacturing of hydrogel biomaterials with controlled mechanical properties for tissue engineering applications." *Actabiomaterialia* 62 (2017): 42–63.

29. Satpathy, Anurag, Rinkee Mohanty, and Tapash R. Rautray. "Bio-mimicked g uided tissue regeneration/guided bone regeneration membranes with hierarchical structured surfaces replicated from teak leaf exhibits enhanced bioactivity." *Journal of Biomedical Materials Research Part B: Applied Biomaterials* 110, no. 1 (2022): 144–156.

30. Rautray, Tapash R., and Kyo Han Kim. "Synthesis of Mg2+ incorporated hydroxy-apatite by ion implantation." In *Key engineering materials*, vol. 529, pp. 114–118. Trans Tech Publications Ltd, 2013.

31. Park, Joon B., and Joseph D. Bronzino, eds. "Biomaterials: principles and applications." (2002).

32. Bhat, Sujata V. *Biomaterials*. Alpha Science Int'l Ltd., 2005.

33. Swain, Subhasmita, Rabindra Nath Padhy, and Tapash Ranjan Rautray. "Polarized piezoelectric bioceramic composites exhibit antibacterial activity." *Materials Chemistry and Physics* 239 (2020): 122002.

34. Ralls, Alessandro, Pankaj Kumar, Mano Misra, and Pradeep L. Menezes. "Material design and surface engineering for bio-implants." *JOM* 72 (2020): 684–696.

35. Alves, Natália M., Iva Pashkuleva, Rui L. Reis, and João F. Mano. "Controlling cell behavior through the design of polymer surfaces." *Small* 6, no. 20 (2010): 2208–2220.

36. Patel, Nitesh R., and Piyush P. Gohil. "A review on biomaterials: scope, applications & human anatomy significance." *International Journal of Emerging Technology and Advanced Engineering* 2, no. 4 (2012): 91–101.

37. Bhaskar, Birru, Naresh Kasoju, P. Sreenivasa Rao, R. Raju Baadhe, and V. Nagarjuna. *Biomaterials in tissue engineering and regenerative medicine.* Springer Singapore, 2021.
38. Latour, Robert A. "Biomaterials: protein-surface interactions." *Encyclopedia of Biomaterials and Biomedical Engineering* 1 (2005): 270–284.
39. Zeng, Haitong, Krishnan K. Chittur, and William R. Lacefield. "Analysis of bovine serum albumin adsorption on calcium phosphate and titanium surfaces." *Biomaterials* 20, no. 4 (1999): 377–384.
40. Barberi, Jacopo, and Silvia Spriano. "Titanium and protein adsorption: An overview of mechanisms and effects of surface features." *Materials* 14, no. 7 (2021): 1590.
41. Rautray, Tapash R., R. Narayanan, Tae-Yub Kwon, and Kyo-Han Kim. "Surface modification of titanium and titanium alloys by ion implantation." *Journal of Biomedical Materials Research Part B: Applied Biomaterials* 93, no. 2 (2010): 581–591.
42. Guney, Aysun, Filiz Kara, Ozge Ozgen, Eda Ayse Aksoy, Vasif Hasirci, and Nesrin Hasirci. "Surface modification of polymeric biomaterials." *Biomaterials Surface Science* (2013): 89–158.
43. Velasco-Ortega, E., C. A. Alfonso-Rodríguez, L. Monsalve-Guil, A. España-López, A. Jiménez-Guerra, I. Garzón, M. Alaminos, and F. J. Gil. "Relevant aspects in the surface properties in titanium dental implants for the cellular viability." *Materials Science and Engineering: C* 64 (2016): 1–10.
44. Hench, Larry L. "Bioceramics: from concept to clinic." *Journal of the American Ceramic Society* 74, no. 7 (1991): 1487–1510.
45. Wirth, Carine, Brigitte Grosgogeat, Christelle Lagneau, Nicole Jaffrezic-Renault, and Laurence Ponsonnet. "Biomaterial surface properties modulate in vitro rat calvaria osteoblasts response: Roughness and or chemistry?." *Materials Science and Engineering: C* 28, no. 5-6 (2008): 990–1001.
46. Yu, Xiaohua, Xiaoyan Tang, Shalini V. Gohil, and Cato T. Laurencin. "Biomaterials for bone regenerative engineering." *Advanced Healthcare Materials* 4, no. 9 (2015): 1268–1285.
47. LeGeros, Racquel Zapanta. "Properties of osteoconductive biomaterials: calcium phosphates." *Clinical Orthopaedics and Related Research (1976–2007)* 395 (2002): 81–98.
48. Kazimierczak, Paulina, and Agata Przekora. "Osteoconductive and osteoinductive surface modifications of biomaterials for bone regeneration: A concise review." *Coatings* 10, no. 10 (2020): 971.
49. Rautray, T. R., V. Vijayan, and S. Panigrahi. "Synthesis of hydroxyapatite at low temperature." *Indian Journal of Physics* 81 (2007): 95–98.
50. Urist, Marshall R., Arthur Lietze, and Edgar Dawson. "β-tricalcium phosphate delivery system for bone morphogenetic protein." *Clinical Orthopaedics and Related Research®* 187 (1984): 277–280.
51. Helmus, Michael N., Donald F. Gibbons, and David Cebon. "Biocompatibility: meeting a key functional requirement of next-generation medical devices." *Toxicologic Pathology* 36, no. 1 (2008): 70–80.
52. Hubbell, Jeffrey A. "Bioactive biomaterials." *Current Opinion in Biotechnology* 10, no. 2 (1999): 123–129.
53. Teoh, S. H. "Fatigue of biomaterials: a review." *International Journal of Fatigue* 22, no. 10 (2000): 825–837.
54. Swain, Subhasmita, Joo L. Ong, Ramaswamy Narayanan, and Tapash R. Rautray. "Ti-9Mn β-type alloy exhibits better osteogenicity than Ti-15Mn alloy in vitro." *Journal of Biomedical Materials Research Part B: Applied Biomaterials* 109, no. 12 (2021): 2154–2161.

55. Bergmann, Carlos, and Aisha Stumpf. *Dental ceramics: microstructure, properties and degradation*. Springer Science & Business Media, 2013.
56. Nair, Lakshmi S., and Cato T. Laurencin. "Biodegradable polymers as biomaterials." *Progress in Polymer Science* 32, no. 8-9 (2007): 762–798.
57. Hench, Larry L., and Julia M. Polak. "Third-generation biomedical materials." *Science* 295, no. 5557 (2002): 1014–1017.
58. Rehman, Mubashar, Asadullah Madni, and Thomas J. Webster. "The era of biofunctional biomaterials in orthopedics: what does the future hold?." *Expert Review of Medical Devices* 15, no. 3 (2018): 193–204.
59. Katti, Dhirendra S., Rajesh Vasita, and Kirubanandan Shanmugam. "Improved biomaterials for tissue engineering applications: surface modification of polymers." *Current Topics in Medicinal Chemistry* 8, no. 4 (2008): 341–353.
60. Stanislawski, L., X. Daniau, A. Lauti, and M. Goldberg. "Factors responsible for pulp cell cytotoxicity induced by resin-modified glass ionomer cements." *Journal of Biomedical Materials Research* 48, no. 3 (1999): 277–288.
61. Boyan, Barbara D., Thomas W. Hummert, David D. Dean, and Zvi Schwartz. "Role of material surfaces in regulating bone and cartilage cell response." *Biomaterials* 17, no. 2 (1996): 137–146.
62. Yu, Ting-Ting, Fu-Zhai Cui, Qing-Yuan Meng, Juan Wang, De-Cheng Wu, Jin Zhang, Xiao-Xing Kou, et al. "Influence of surface chemistry on adhesion and osteo/odontogenic differentiation of dental pulp stem cells." *ACS Biomaterials Science & Engineering* 3, no. 6 (2017): 1119–1128.
63. Liu, Fang, Jiawei Xu, Linliang Wu, Tiantian Zheng, Qi Han, Yunyun Liang, Liling Zhang, Guicai Li, and Yumin Yang. "The influence of the surface topographical cues of biomaterials on nerve cells in peripheral nerve regeneration: a review." *Stem Cells International* 2021, (2021): 1–13.
64. Salmerón-Sánchez, Manuel, Patricia Rico, David Moratal, Ted T. Lee, Jean E. Schwarzbauer, and Andrés J. García. "Role of material-driven fibronectin fibrillogenesis in cell differentiation." *Biomaterials* 32, no. 8 (2011): 2099–2105.
65. Assender, Hazel, Valery Bliznyuk, and Kyriakos Porfyrakis. "How surface topography relates to materials' properties." *Science* 297, no. 5583 (2002): 973–976.
66. Lee, Ki-Won, Cheol-Min Bae, Jae-Young Jung, Gi-Bong Sim, Tapash Ranjan Rautray, Hyo-Jin Lee, Tae-Yub Kwon, and Kyo-Han Kim. "Surface characteristics and biological studies of hydroxyapatite coating by a new method." *Journal of Biomedical Materials Research Part B: Applied Biomaterials* 98, no. 2 (2011): 395–407.
67. Chen, Shoucheng, Yuanlong Guo, Runheng Liu, Shiyu Wu, Jinghan Fang, Baoxin Huang, Zhipeng Li, Zhuofan Chen, and Zetao Chen. "Tuning surface properties of bone biomaterials to manipulate osteoblastic cell adhesion and the signaling pathways for the enhancement of early osseointegration." *Colloids and Surfaces B: Biointerfaces* 164 (2018): 58–69.
68. Swain, Subhasmita, Chris Bowen, and Tapash Rautray. "Dual response of osteoblast activity and antibacterial properties of polarized strontium substituted hydroxyapatite—Barium strontium titanate composites with controlled strontium substitution." *Journal of Biomedical Materials Research Part A* 109, no. 10 (2021): 2027–2035.
69. Makadia, Hirenkumar K., and Steven J. Siegel. "Poly lactic-co-glycolic acid (PLGA) as biodegradable controlled drug delivery carrier." *Polymers* 3, no. 3 (2011): 1377–1397.
70. Dalby, Matthew J., Nikolaj Gadegaard, Rahul Tare, Abhay Andar, Mathis O. Riehle, Pawel Herzyk, Chris DW Wilkinson, and Richard OC Oreffo. "The control of human

mesenchymal cell differentiation using nanoscale symmetry and disorder." *Nature Materials* 6, no. 12 (2007): 997–1003.

71. Ponsonnet, L., K. Reybier, N. Jaffrezic, V. Comte, C. Lagneau, M. Lissac, and C. Martelet. "Relationship between surface properties (roughness, wettability) of titanium and titanium alloys and cell behaviour." *Materials Science and Engineering: C* 23, no. 4 (2003): 551–560.

72. Hallab, N., K. Bundy, K. O'Connor, R.L. Moses, and J.J. Jacobs. "Evaluation of metallic and polymeric biomaterial surface energy and surface roughness characteristics for directed cell adhesion." *Tissue Engineering* 71(2001): 55–71.

73. Wei, Jianhua, Toshio Igarashi, Naoto Okumori, Takayasu Igarashi, Takashi Maetani, Baolin Liu, and Masao Yoshinari. "Influence of surface wettability on competitive protein adsorption and initial attachment of osteoblasts." *Biomedical Materials* 4, no. 4 (2009): 045002.

74. Gentleman, Molly M., and Eileen Gentleman. "The role of surface free energy in osteoblast–biomaterial interactions." *International Materials Reviews* 59, no. 8 (2014): 417–429.

75. Wei, Jianhua, Masao Yoshinari, Shinji Takemoto, Masayuki Hattori, Eiji Kawada, Baolin Liu, and Yutaka Oda. "Adhesion of mouse fibroblasts on hexamethyl disiloxane surfaces with wide range of wettability." *Journal of Biomedical Materials Research Part B: Applied Biomaterials: An Official Journal of The Society for Biomaterials, The Japanese Society for Biomaterials, and The Australian Society for Biomaterials and the Korean Society for Biomaterials* 81, no. 1 (2007): 66–75.

76. Singh, Raghuvir, and Narendra B. Dahotre. "Corrosion degradation and prevention by surface modification of biometallic materials." *Journal of Materials Science: Materials in Medicine* 18, no. 5 (2007): 725–751.

77. Dolcimascolo, Anna, Giovanna Calabrese, Sabrina Conoci, and Rosalba Parenti. "Innovative biomaterials for tissue engineering." In *Biomaterial-supported tissue reconstruction or regeneration*. IntechOpen, 2019.

78. Growth mechanism of aligned porous oxide layers on titanium by anodization in electrolyte containing Cl, RinmayeePraharaj, Snigdha Mishra, Tapash R Rautray, Materials Today: Proceedings, 62, 6216-6220

79. Ali, Sadaqat, Ahmad Majdi Abdul Rani, Zeeshan Baig, Syed Waqar Ahmed, Ghulam Hussain, Krishnan Subramaniam, Sri Hastuty, and Tadamilla VVLN Rao. "Biocompatibility and corrosion resistance of metallic biomaterials." *Corrosion Reviews* 38, no. 5 (2020): 381–402.

80. Karageorgiou, Vassilis, and David Kaplan. "Porosity of 3D biomaterial scaffolds and osteogenesis." *Biomaterials* 26, no. 27 (2005): 5474–5491.

81. Swain, Subhasmita, and Tapash Ranjan Rautray. "Effect of surface roughness on titanium medical implants." *Nanostructured Materials and Their Applications* (2021): 55–80.

# 3 Biomaterial Surfaces and Their Properties

*Sally Sabra*
Department of Biotechnology, Institute of Graduate Studies and Research, Alexandria University, Alexandria, Egypt

*Mona M. Agwa*
Department of Chemistry of Natural and Microbial Products, Pharmaceutical and Drug Industries Research Institute, National Research Centre, Giza, Egypt

## 3.1 AN OVERVIEW

Biomaterials can be defined as natural or synthetic materials that can be involved in the biomedical sector either for diagnostic or therapeutic purposes in direct contact with cells and they may remain within the living body for long period of time, such as in case of implants [1]. Biomaterials can be easily distinguished from any other material in that they are able to be in direct contact with tissues of the human body or even perform their function without causing any damage [2]. In fact, utilizing biomaterials for biomedical applications is not new. In ancient times, rubber-soaked linen was used as a wound dressing [3]. Romans also have described the usage of devices similar to the urologic catheters that are used these days [4]. In recent history, biomaterials were included for medical applications for the first time when sutures were fabricated from nylon threads in 1941 and cellulose hydrate for hemodialysis in 1943 [4]. In 1952, the first vessel prosthesis was implanted in a human, and then a complete replacement of a mitral valve with an artificial one was successfully conducted in 1960. The real start of "biomaterials for biomedical applications" was during the 1960s. Extensive research was done in order to enhance their functional properties and improve their biocompatibility to fulfill the medical requirements. Nowadays, biomaterials are widely involved in diverse biomedical implants and devices [3].

In general, biomaterials can be man-made or naturally available and can be used to repair or totally replace various human tissues, which means that they are in a very close contact with body fluids to result in a final improvement in the activities of human vital organs [5]. Through the steps of designing a biomaterial, it should be checked for its biocompatibility, physical, and chemical stabilities, whether they are active, inert, or biodegradable [6]. There are many classes of materials that can be included as biomaterials, including metals (the oldest biomaterial), alloys, inorganic classes, ceramics, and composites [4].

According to the origin of the biomaterials, they can be classified into two main classes: natural and synthetic biomaterials. Natural biomaterials include some naturally available proteins, such as gelatin, collagen, and albumin, or polysaccharides such as

DOI: 10.1201/9781003429920-4

alginate, chitosan, heparin, and hyaluronic acid [7,8]. Owing to their demand in massive quantities that could not be afforded by nature, synthetic biodegradable polymers are greatly considered nowadays due to their facile manipulation and large-scale production [9].

Polymers utilized in the fabrication of biomaterials can also be classified into biostable and biodegradable polymers. Biostable polymers can be included in case of permanent implant, while biodegradable polymers can be utilized in case of a temporary implant. Biostable polymers should be biologically and chemically inert. However, most biostable polymers undergo degradation by time due to their presence in the physiological conditions of the human body [4]. Examples of chemical degradation of biostable polymers include hydrolysis of polyamides over a long time [10] and polyethylene terephthalate [10,11]. These chemical reactions that occur over long periods of time usually cause material loss [11].

Biomaterials fabricated from biodegradable polymers contain chemical bonds that can be cleaved under the physiological conditions in the human body. These bonds are usually cleaved due to the presence of high water content in our body, which mediates breaking down of the polymer chain into oligomers and then monomers. In some cases, polymers can be selectively cleaved via an enzymatic reaction inside the body, regardless the location of the implant [3].

Polymers that can be enzymatically degraded are mostly polypeptides that can be degraded by proteases or polysaccharides (amylose or dextran), which can be degraded by amylases. Another important class is biopolyesters, which can be degraded with esterases. Poly(3-Hydroxybutyrate) P(3HB) is an example of a biopolyesters polymer. It is very well known to be highly compatible and biodegradable, besides being very pure without any impurities [12]. Numerous structures have been synthesized from P(3HB) for certain medical applications, including patches for pericard substitution [13], or for sealing vestibular septum defects [14]. Moreover, P(3HB)-based sleeves were developed for tissue healing applications [15]. Another important benefit for P(3HB) is its piezoelectric properties, making it an ideal material for neuronal [16] and bone regeneration [17].

There are essential basic requirements that should be found in the material to be selected as a biomaterial. The utilized materials should be highly biocompatible to prevent immune rejection, possess sufficient mechanical strength, have high physical and chemical stabilities, besides being resistant to rust, and have osseointegration properties in the case of bone implants [18,19] Biomaterials can be applied in diverse fields such as medicine, food processing units, pharmaceutical companies, and others. In the medical sector, biomaterials are extensively involved in dental fixations, implants, and tissue scaffolds, whereas in the pharmaceutical sector, biomaterials can be utilized in the synthesis of tablets, capsules, or smart formulations for drug delivery [20]. For each purpose of the above-mentioned purposes, the selected biomaterial should have certain criteria before being selected. For example, biomaterials included in dental and orthopedic purposes should have enough mechanical vigor, besides having a prolonged shelf-life time. Implants for artificial joints should be resistant to abrasive wear, while implants for replacing a heart valve should have a high mechanical load capacity. For bone replacement, implants are preferred to have high compression strength and elasticity [4] Conversely, biomaterials involved in skin or visceral organ treatment must be flexible, with a high degradation rate [21].

In the United States alone, about 13 million prosthetics/medical devices are implanted every year in different parts or organs of the human body, including cardiac valves; blood vessels stents; implants in elbows, knees, shoulders, ears, eyes; and some orthodontics structures [22,23]. The shape and the composition of the biomaterial implant could be adjusted according to the assigned function. It might be composed of metals, polymers, ceramics, or composites. If the aim of the implant is to stimulate tissue regeneration, biodegradable polymers can be utilized [24]. For cellular adhesion and proliferation purposes, microporous scaffolds can be used. Chemically modified or drug-coated implant surfaces could be useful for local drug delivery [4].

## 3.2   CRITERIA FOR SELECTION OF A SUITABLE BIOMATERIAL

### 3.2.1   BIOCOMPATIBILITY

Biocompatibility can be defined as the potential of a certain material to be used in close contact with living tissue without causing any harmful side effects on them [25,26]. These materials are usually utilized as implants, so they are expected to be non-toxic, without any inflammatory, allergic, or carcinogenic reactions [27]. In addition, they should not release any toxic materials, such as metal ions, in order to not cause rejection of the implant, especially in the case of prolonged usage [28]. The degree of materials biocompatibility is mainly dependent on the reaction of the human body towards the implant. The least interaction between the implant and the human body, the more the level of biocompatibility [2]. However, upon exposure of some implants to human tissues and fluids, some unfavorable interactions might occur, such as blood clotting and adhesion of blood platelets to the outer surface of the biomaterial [29]. In addition, any material assigned for biomedical applications should resist corrosion, as increased corrosion means higher release of toxic ions, and hence increased risk of adverse side effects [30].

Most of the total mass of the human body consists of oxygen, carbon, hydrogen, and nitrogen (96%) [28,31], which are the building units of proteins, lipids, and nucleic acids [30]. The remaining 4% of the body mass is in the form of blood, bones, and minerals such as calcium, phosphorus, magnesium, and potassium. Consequently, any implant fabricated from any of these elements is supposed to be compatible with implantation in the human body [30]. Materials rather than the above-mentioned ones can be utilized, but they should be adjusted in terms of their concentration and composition to be biocompatible [30].

A common example for this category of materials is titanium, which is not normally found in the human body and has no known function in the body [32]. In addition, it is not toxic, even at high doses (up to 0.8 mg daily ingestion), with most of the titanium excreted without digestion or absorption and it has also excellent corrosion resistance [28,33,34]. In most cases, titanium-based implants are not rejected by the human body and can integrate well with human bones, but on the other hand, some *in vitro* studies showed that this element can inhibit the differentiation of mesenchymal stem cells into osteoblasts [35]. The first generation of titanium alloys (Ti-6Al-4V) was stabilized by aluminum or vanadium, but they showed allergic interactions with the human body [36]. The second generation ($\beta$ alloys) was stabilized with $\beta$ phase stabilizers such as molybdenum, zirconium, or tantalum, which were proven to be relatively safer in comparison to the first generation [37].

Another privilege for titanium alloys is that they show good mechanical properties in terms of low density and low Young's modulus, which make these materials behave in a similar manner to that of the bone [38]. Titanium-based alloys are widely utilized to replace hard tissue in dental implants, joints, and artificial bones [38]. Moreover, titanium-based alloys are considered a good candidate for biomaterial implantation due to their potential to form an extremely adherent and thin oxide layer, which is the major driving force beyond titanium high corrosion resistance [38]. Titanium in its pure form is limited to dental implants only, as its mechanical properties are minimal [38]. As a result, if higher mechanical properties are required in the case of knee implants, hip implants, plates, or screws, Ti-6Al-4V alloy should be used [39,40].

As biocompatibility is the most important parameter on which a new biomaterial can be used in medical applications or not, precautions should be taken before selection [41]. There are some common *in vivo* tests that can be done to ensure biocompatibility, including irritation, toxicities (acute toxicity; less than 24 h of exposure, subacute toxicity; 14–28 days of exposure, chronic toxicity; 24 h, 10% of animal lifespan exposure), genotoxicity, immune system response, hemocompatibility, biodegradation, and reproductive toxicity [42].

### 3.2.2 Hemocompatibility

In many circumstances, the biomaterials come in direct contact with the blood components in the case of vascular devices, permanent implants, or extracorporeal devices like blood bags [43,44]. If blood contact is required, then the surface of the developed biomaterial should not stimulate thrombogensis. In addition, the surface energy and load should be considered because they directly affect the fluid-biomaterial interaction. Moreover, materials with high charge density are preferred because they have the potential to minimize protein interaction with the material, and hence can resist thrombosis. There are many techniques by which the surface load can be increased, including introduction of negatively charged groups on the surface of biomaterials or coating of biomaterials with hydrogels, which can significantly increase the hydrophilicity of the surface, leading to a decrease in protein adsorption [45,46]. In particular, developing biomaterials with a high surface load is of great importance, especially in applications requiring long exposure to blood, such as heart valves, venous catheters, and permanent prostheses implants. Conversely, biomaterials with low surface load can be accepted in other temporary applications, such as dental or orthopedic implants [46].

### 3.2.3 Mechanical Properties

Mechanical properties are very important features that help physicians to select the suitable biomaterial for certain application. There is no standard criteria for good mechanical properties; however, it is usually a relative value according to the assigned function that will be conducted [36], but in general, any biomaterial implant should have adequate mechanical strength to withstand all loads and forces that will focus on it, according to the function that will be exerted; otherwise, it will be easily susceptible to fracture [30].

Mechanical properties should be carefully considered in the case of bone applications as biomaterials with low resistance or great difference in their mechanical properties in relation to bones will be fractured due to mechanical incompatibility [29]. For successful bone implants, the replaced biomaterial must possess Young's modulus equivalent to that of the replaced bone. A bone modulus differs from 4 up to 30 Gpa according to its type and direction of measurement [47]. If the implanted material has greater rigidity than that of the bone, it will cause bone resorption in the area around the implant, and hence implant loosening and further subsequent complications [48]. Consequently, an ideal implant should have adequate strength and suitable Young's modulus close to that of the replaced bone in order to prevent its loosening or necessity for another corrective surgery [29].

### 3.2.4 CORROSION RESISTANCE

Biomaterials are physiologically exposed to a certain level of humidity when coming in contact with body fluids [30]. If the implanted material has low corrosion resistance, this will cause some metal ions to be released in the body fluids, which might cause some allergic or toxic reactions [29,49]. These released metal ions will not be toxic upon its instant release, but when they start to corrode, they will cause adverse side effects to the human body [30]. Corrosion will start to occur when the oxide outer layer on the surface of the metal is broken and it begins to be passivated in a process known as regeneration. The rate of biomaterial corrosion and release of metal ions greatly depends on the regeneration time [28,50]. Prolonged existence of metal ions in the tissues surrounding the implants was found to cause hypersensitivity, inflammation, tissue toxicity, or even carcinogenicity. Consequently, biomaterials with high corrosion resistance are a vital requirement for lifetime implants in the human body [29].

### 3.2.5 OSSEOINTEGRATION

Osseointegration is known as the anchoring of an implant with migration and formation of bone tissue at the bone-implant interface without fibrous tissue growth [51,52]. An osseointegrated implant will be able to withstand normal physiological loads without further deformation, loosening, pain, or rejection [40]. As a result, biomaterials involved in bone tissue regeneration should have suitable surface chemistry, topography, and roughness to adhere well with the adjacent bones [53] without formation of fibrous tissue between the bone and the implant, which might cause implant loosening [54].

## 3.3 DIFFERENT CLASSES OF BIOMATERIALS

### 3.3.1 METALS-BASED BIOMATERIALS

Implants fabricated from metals are the main biomaterials utilized in replacing joints. They can be fabricated from titanium or titanium alloys, CoCr alloys, or stainless steel. Metallic implants were reported to have several advantages, including good thermal conductivity, hardness, high strength, corrosion resistance, improved fracture toughness, and biocompatibility [37]. On the other hand, these metallic implants might cause stress shielding and they may be corrosive upon prolonged use, causing complicated side effects. In addition, they have high stiffness and density compared to tissues [55].

### 3.3.2 STAINLESS STEEL

Stainless steel (316 and 316L) was the first material to be utilized in manufacturing artificial bone, as it is facile to be casted to different shapes and sizes. However, 316L is better than 316 stainless steel in resisting corrosion because it contains lower carbon content. Both types of stainless steel can be used to fabricate temporary devices such as screws, hip nails, or fracture plates due to their favorable mechanical and corrosion properties [56]. Furthermore, the performance of stainless-steel alloys or implants can be improved by adding other alloying elements such as nickel or molybdenum, which results in improving their corrosion resistance, especially in aggressive environments inside the human body [53]. It is thought that stainless steel-based biomaterials have good biocompatibility properties; however, they have low fatigue resistance due to their low proportional limits, which might cause formation of fatigue cracks and loosing of the material at the end [57].

### 3.3.3 COBALT-CHROME

Cobalt-based alloys are well known to be one of the safest biomaterials, especially for orthopedic prostheses, owing to their extraordinary potential to resist corrosion under high chloride environment. This effect might be related to presence of high percentages of basic elements in their composition and the formation of a passive layer from chromium oxide $Cr_2O_3$ [58]. In addition, cobalt-based alloys are characterized by having metals with a high melting point in their composition, such as chromium, zirconium, tantalum, tungsten, and titanium [58]. A typical microstructure of a cobalt-based alloy includes a cobalt-rich solid-solution matrix, including carbides (i.e., $Cr_7C_3$ and $M_{23}C_6$) either inside the grains themselves or at the boundaries, where other metals and cobalt might be located within the same carbide particle [58,59]. The first version utilized from cobalt-based alloys in hip implants was fabricated by investment casting technique and it was found to include a high carbon content (~0.2%) [60]. According to the casting and the manufacturing technique, it is possible to develop about three different microstructures, which can strongly affect the properties of the implant [58]. Low-carbon versions of cobalt-based alloys were then fabricated and they exhibited improved corrosion resistance and better mechanical properties when compared to the original casted alloys [61].

### 3.3.4 TITANIUM-BASED ALLOYS

Titanium is one of the most widely utilized biomaterials for the fabrication of biomedical implants. This element or metal has several privileges, such as being light in weight in comparison to other metals and having good mechanical and chemical properties. On the other hand, inclusion of implants fabricated from titanium alone is not preferred for bone plates or screws due to its poor shear strength. As a result, titanium alloys can be used in which titanium can be combined with other metals, taking the advantage of its superior biocompatibility and complete inertness in the physiological conditions of the human body [57]. In addition, titanium-based alloys are characterized by their high strength, low density, moderate elastic modulus of about 110 GPa, rough surface, and high tendency to integrate into bone, which are very suitable features for fabricating an efficient implant with sufficient osseointegration [29,62].

## 3.3.5 Ceramics

Ceramics are carbides, sulfides, nitrides, or oxides of metals or metalloids [63]. Bioceramics can be natural or synthetic and they are utilized as biomaterials because they can make bonds with bone, and hence they might be good alternative for metallic implants with matching physiochemical properties with the human body [64]. Bioceramic porcelain was utilized in the 18th and 19th centuries in dental applications [65], then their applications have massively increased in the medical sector in the 20th century [66]. Moreover, these biomaterials were shown to possess good bio-compatibility, high melting point, poor degradability, poor plasticity, and non-corrosive when compared to metals [67,68]. Synthetic ceramics such as titania, zirconia, alumina, Ca-P-based porous materials, and bioactive glass are widely utilized in many medical applications, including orthopedics, coatings, implants, and medical sensors [69–71]. Moreover, bioceramics are now utilized in the repair and regeneration of soft tissue [72]. Conversely, one of the drawbacks of ceramics is their brittleness because their mechanical properties are greatly dependent on the density. As a result, precautions must be taken during manufacturing, as any void areas found in the implant might affect their longevity [73].

### 3.3.5.1   Alumina ($Al_2O_3$)

Alumina with high purity and density was the first ceramic material clinically utilized, owing to its improved biocompatibility and corrosion resistance, besides high strength and wear resistance, which might be related to its surface energy. High pure alumina can be found in native corundum and bauxite. Interestingly, when alumina-based ceramic material was implanted in the eye sockets of albino rats, there weren't any signs of rejection after eight weeks, and, surprisingly, there was a notable vascular invasion, fibroblast proliferation, and tissue growth in the pores of the implant [74].

According to the American Society for Testing and Material (ASTM), it is recommended for alumina included in any medical implant application to contain pure alumina (99.5%), with less than 0.1% of $SiO_2$ and other alkali oxides such as $Na_2O$. Another criterion that should be taken into consideration is the strength of the implant. For polycrystalline alumina, its strength depends greatly on the porosity and the grain size. Alumina with lower grain size and lower porosity is supposed to have a higher strength. As a result, the ASTM standards (F603-78) recommend a flexural strength of more than 400 MPa, with an elastic modulus of about 380 GPa. Aluminum oxide-based materials have been extensively used in dental and orthopedic implants, besides their involvement in the region of load-bearing hip prostheses [75].

### 3.3.5.2   Zirconia ($ZrO_2$)

Zirconia is one of the most promising biomaterials, owing to its good mechanical strength and fracture toughness [76]. Zirconia in its pure form can be prepared by chemical conversion of zircon (ZrSiO4), which is abundant as a mineral deposit. Moreover, it has a high melting temperature (Tm> 2500°C), besides its high chemical stability [77]. Zirconia is now clinically used in total hip replacement and in the fabrication of THR ball heads [76]. However, the physical properties of alumina are preferred over zirconia in the case of biomedical implantation applications [73].

### 3.3.5.3   Calcium Phosphate Ceramics

Calcium-phosphate-based ceramics can be easily crystallized into salts such as hydroxyapatite or β-whitlockite according to some parameters that should be adjusted, including Ca:P ratio, impurities, presence of water, and temperature. These materials are mainly used in artificial bone applications in the form of either solid or porous coatings or implants. A characteristic feature for calcium-phosphate-based ceramics is their high porosity, which could afford an ideal pore size very similar to that of spongy bone [78]. Moreover, calcium hydroxyapatite as a biomaterial is characterized by superior biocompatibility, as it can form a chemical bond with hard tissues like bones in the form of lamellar cancellous networks within one or two months post-implantation. However, calcium-phosphate-based ceramics are reported to be brittle and have very poor tensile properties, which make it difficult to load these materials in a bending region without compression [79].

### 3.3.5.4   Bioglass Ceramics

Bioglass ceramics are usually composed of certain amounts of $SiO_2$, $P_2O_5$, and CaO, which can efficiently bind to bone upon implantation [80]. Bioactive glasses were discovered by Larry Hench *et al.* in the late 1960s when there was a need to replace inert metals and plastic-based implants with more tolerable materials that can form a binding interaction with the living tissues upon implantation [81]. It was reported that the chemical interaction between bioglass ceramics and bone arises from the slight solubility of these materials and formation of some sort of solid-state interaction between the formed apatite crystals and the bone [82]. Chemical composition is an important factor that regulates the behavior of the bioglass and it strongly determines its structure, degradation rate, and biocompatibility [81]. However, these materials exhibited some limitations, including increased pH due to their high sodium content and their poor sintering potential, which makes it difficult to develop porous three-dimensional (3D) scaffolds from these materials [81]. Furthermore, the properties of the fabricated bioglass can be improved by doping small amounts of some oxides of therapeutic benefits into the composition of the bioglass that might aid in cellular differentiation and proliferation or they may have antibacterial effects [83].

### 3.3.6   Polymers

Polymers can be defined as substances composed of multiple repetitions of one or more type of monomers that are linked together in such a way to afford enhanced properties for implantation applications. Polymers are considered good candidates for implantation as they can be easily fabricated into multiple forms, including viscous liquids, films, fibers, rods, or textiles. In addition, most polymers greatly mimic the components of tissue components, besides being able to form a bond with natural polymers existing in tissues in some circumstances. On the other hand, these polymers are not strong enough to fulfill the required mechanical properties for orthopedic implantation and they also might absorb fluids and swell, which will result in leaching of some secondary components. Another important problem is the sterilization process that can be performed via autoclaving, irradiation, or with ethylene oxide and it may influence the properties of the polymer. Numerous synthetic polymers have been utilized for implantation purposes, such as silicon, acrylic, nylon, polyurethane, and polypropylene [84,85].

### 3.3.7   COMPOSITES

Composites can be defined as the integration of two or more materials that differ in their composition, morphology, and physical properties (i.e., metal and polymer, ceramics, and polymers). They have been widely utilized in many sectors due to their extraordinary mechanical properties arising from the integration of more than one class of biomaterials [86]. In the biomedical field, the included composite must be able to mimic the structure of the tissues, restore the functionality of the damaged tissues until complete healing, or repair and possess adequate mechanical properties for the assigned function. Biocompatible polymers have been widely applied as a matrix with ceramic materials as a filler in tissue regeneration approaches [87]. In general, composites can be classified into biostable composites, partially biodegradable composites, and totally degradable composites. Carbon fiber (CF)/epoxy and glass fiber/epoxy are well-known biostable composites that were utilized in internal bone fixation approaches, but several concerns have been raised regarding toxicities of their monomers [88]. CF/polyether ether ketone (PEEK) is another example of a biostable and inert composite that has mechanical properties similar to metallic devices, with many additional privileges such as high thermal stability, high toughness, improved fatigue, and creep resistance; outstanding self-lubricating potential; excellent abrasion resistance; besides being non-toxic with low elastic modulus (2–6 GPa), which can minimize the stress shielding effect. Another advantage is X-ray radiolucency, making it transparent during magnetic resonance imaging and computerized tomography scans [89].

Recently, biodegradable composites have gained attention by taking advantage of avoiding another second surgery to remove the implant after its function. Polymers included in this type of composite are usually FDA-approved biodegradable polymers such as polylactic acid (PLA), poly(lactic-*co*-glycolic acid (PLGA), and polycaprolactone (PCL). Natural biodegradable polymers such as chitosan, collagen, or PHB can also be included. However, these polymers cannot be utilized alone to produce implants because their mechanical properties are not adequate and, hence, they should be reinforced by other suitable materials such as CF or bioceramic materials, resulting in the formation of partially biodegradable or totally biodegradable composites [88].

## 3.4   POSSIBLE APPROACHES FOR SURFACE MODIFICATION FOR BIOMEDICAL APPLICATIONS

There are several approaches by which the surface of the biomaterials can be modified [46,90,91]. The modifications that can be conducted on the surface of the biomaterial could be physical or chemical. Chemical modifications include oxidation, hydrolysis, or quaternization, which can change the chemistry of the biomaterial surface. Another approach is grafting of a water-soluble polymer on the surface of the biomaterial, which might reduce protein adsorption and cell adhesion on its surface. Non-ionic polyethylene oxide (PEO), PEO co-polymers with variable molecular weights, or poly(vinyl pyrrolidone) are commonly utilized for this purpose [92]. Biomaterials or composites coated with these polymers have been widely utilized in the medical sector, owing to their biocompatibility, commercial availability, and protein repulsion properties [45,60,93].

Another approach by which the interaction between the body or blood and the biomaterial can be improved is minimizing any possible thrombogenic properties for the biomaterial. Heparinization of the surface of the biomaterial can be performed via adsorption of heparin or heparin-like polysaccharides, which have an important role in preventing blood coagulation. In addition, several recent studies reported that coating of biomaterials with ionomers containing sulfonic-acid residues might be beneficial regarding their anticoagulant properties when they are in contact with blood [94].

Some biomedical implants are fabricated from polymers. These polymers, besides participating in the implant structure, can also serve as a drug carrier that can afford protection of the drug from the surrounding environment and guarantee a controlled drug release into the required organ [4]. This type of implant is usually fabricated by mixing the polymer solution with the active agent or the drug, and then they are sprayed onto the surface of the implant or could be applied by dip coating. Incorporated drugs are then released by a diffusion mechanism or via polymer degradation if the polymer is biodegradable [4]. In other cases, the drug or the active molecule can be coupled onto the implant surface via chemical conjugation in order to endure local release of active molecules so as to inhibit inflammation [95], cellular proliferation [96], or thromboses [97]. Moreover, some implants can be coated with specific peptides or other signaling molecules such as cytokines or growth factors for better cellular recognition or adhesion [98].

A recent approach in improving the multi-functionality of the fabricated biomaterials is the design of a bio-specific polymer surface. This can be achieved by outer surface grafting or inner incorporation of membrane receptors or signaling peptides to develop hybrid composite biomaterials. This approach is usually utilized to afford site-specific recognition or selective adsorption at certain cell types, while excluding their adhesion to other undesired cell types [60,91].

## 3.5 CONCLUSION

For successful selection of a biomaterial to be utilized as an implant, we must adhere to some leading properties according to the function that the implant will perform in order to achieve longevity with better life quality. A suitable biomaterial should be efficient (doing well) and safe (doing no harm). Biocompatibility is the main feature that should be considered, as incompatible biomaterials may cause adverse side effects causing rejection of the body for the implant. In addition, an ideal implant must have suitable mechanical properties with an equivalent Young's modulus to the bone to be replaced. Another important criterion is the corrosion resistance, in which biomaterials with high corrosion resistance are preferred because they avoid the release of metal ions from the implant, which might cause toxicity with time. In addition, the utilized material should be osseointegrated in order to stimulate bone tissue formation. There are different classes of biomaterials that can be included in implants, including metals-based biomaterials, stainless steel, cobalt-chrome, titanium-based materials, ceramics, polymers, and composites. It can be concluded that there is no preferred class of biomaterials; however, the selection is mainly based on several parameters such as the required function, the site of treatment, and the duration of the treatment if it is temporary or it will last the life span.

## REFERENCES

1. Buddy D. Ratner, Allan S. Hoffman, Frederick J. Schoen, Jack E. Lemons, Biomaterials Science, 2nd edition, 2004. https://shop.elsevier.com/books/biomaterials-science/ratner/ 978-0-08-047036-8 (accessed January 13, 2024).

2. D.F. Williams, On the mechanisms of biocompatibility, Biomaterials. 29 (2008) 2941–2953. 10.1016/j.biomaterials.2008.04.023.

3. A. Lendlein, Polymers as implant materials, Chemie Unserer Zeit. 33 (1999) 279–295. 10.1002/ciuz.19990330505.

4. K. Sternberg, Current requirements for polymeric biomaterials in otolaryngology, GMS Curr. Top. Otorhinolaryngol. Head Neck Surg. 8 (2009) Doc11. 10.3205/cto000063.

5. J.R. Davis, Materials for Medical Devices, ASM International, 2012.

6. M. Navarro, A. Michiardi, O. Castaño, J. Planell, Biomaterials in orthopaedics, J. R. Soc. Interface. 5 (2008) 1137–1158. 10.1098/rsif.2008.0151.

7. F. Croisier, C. Jérôme, Chitosan-based biomaterials for tissue engineering, Eur. Polym. J. 49 (2013) 780–792. 10.1016/j.eurpolymj.2012.12.009.

8. K.Y. Lee, D.J. Mooney, Alginate: Properties and biomedical applications, Prog. Polym. Sci. 37 (2012) 106–126. 10.1016/j.progpolymsci.2011.06.003.

9. H. Tian, Z. Tang, X. Zhuang, X. Chen, X. Jing, Biodegradable synthetic polymers: Preparation, functionalization and biomedical application, Prog. Polym. Sci. 37 (2012) 237–280. 10.1016/j.progpolymsci.2011.06.004.

10. S. Heumann, A. Eberl, H. Pobeheim, S. Liebminger, G. Fischer-Colbrie, E. Almansa, A. Cavaco-Paulo, G.M. Gübitz, New model substrates for enzymes hydrolysing polyethyleneterephthalate and polyamide fibres, J. Biochem. Biophys. Methods. 69 (2006) 89–99. 10.1016/j.jbbm.2006.02.005.

11. R.N. King, D.J. Lyman, Polymers in contact with the body, Environ. Health Perspect. 11 (1975) 71–74. 10.1289/ehp.751171.

12. M. Yasin, B.J. Tighe, Strategies for the design of biodegradable polymer systems: Manipulation of polyhydroxybutyrate-based materials, Plast. Rubber Compos. Process. Appl. 19 (1993) 15–27.

13. T. Malm, S. Bowald, A. Bylock, C. Busch, Prevention of postoperative pericardial adhesions by closure of the pericardium with absorbable polymer patches. An experimental study, J. Thorac. Cardiovasc. Surg. 104 (1992) 600–607.

14. T. Malm, S. Bowald, S. Karacagil, A. Bylock, C. Busch, A new biodegradable patch for closure of atrial septal defect. An experimental study, Scand. J. Thorac. Cardiovasc. Surg. 26 (1992) 9–14. 10.3109/14017439209099047.

15. V. Fasiku, S.J. Owonubi, E. Mukwevho, B. Aderibigber, E.R. Sadiku, O. Agboola, Y. Lemmer, W.K. Kupolati, K. Selatile, G. Makgatho, Polyhydroxyesters as scaffolds for tissue engineering, 2018. https://api.semanticscholar.org/CorpusID:116606761.

16. A. Hazari, G. Johansson-Rudén, K. Junemo-Bostrom, C. Ljungberg, G. Terenghi, C. Green, M. Wiberg, A new resorbable wrap-around implant as an alternative nerve repair technique, J. Hand Surg. Br. 24 (1999) 291–295. 10.1054/jhsb.1998.0001.

17. P.A Holmes, Applications of PHB – A microbially produced biodegradable thermoplastic, Phys. Technol. 16 (1985) 32. 10.1088/0305-4624/16/1/305.

18. A.G. Kadam, S.A. Pawar, S.A. Abhang, A review on finite element analysis of different biomaterials used in orthopedic implantation, Int. Res. J. Eng. Technol. 4 (2017) 2192–2195. https://www.irjet.net/archives/V4/i4/IRJET-V4I4559.pdf.

19. J. Lévesque, H. Hermawan, D. Dubé, D. Mantovani, Design of a pseudo-physiological test bench specific to the development of biodegradable metallic biomaterials., Acta Biomater. 4 (2008) 284–295. 10.1016/j.actbio.2007.09.012.

20. K. Tappa, U. Jammalamadaka, Novel biomaterials used in medical 3D printing techniques., J. Funct. Biomater. 9 (2018). 10.3390/jfb9010017.

21. U. Kamachi Mudali, T.M. Sridhar, R.A.J. Baldev, Corrosion of bio implants, Sadhana - Acad. Proc. Eng. Sci. 28 (2003) 601–637. 10.1007/BF02706450.

22. S. Bose, D. Ke, H. Sahasrabudhe, A. Bandyopadhyay, Additive manufacturing of biomaterials, Prog. Mater. Sci. 93 (2018) 45–111. 10.1016/j.pmatsci.2017.08.003.

23. X. Wang, Overview on biocompatibilities of implantable biomaterials, in: R. Pignatello (Ed.), Advances in Biomaterials Science and Biomedical Applications, IntechOpen, Rijeka, 2013: p. Ch. 5.

24. R. Langer, Biomaterials in drug delivery and tissue engineering: One laboratory's experience, Acc. Chem. Res. 33 (2000) 94–101. 10.1021/ar9800993.

25. Y.S. Hedberg, B. Qian, Z. Shen, S. Virtanen, I.O. Wallinder, In vitro biocompatibility of CoCrMo dental alloys fabricated by selective laser melting, Dent. Mater. 30 (2014) 525–534. 10.1016/j.dental.2014.02.008.

26. D.R. Plummer, R.A. Berger, W.G. Paprosky, S.M. Sporer, J.J. Jacobs, C.J. Della Valle, Diagnosis and management of adverse local tissue reactions secondary to corrosion at the head-neck junction in patients with metal on polyethylene bearings, J. Arthroplasty. 31 (2016) 264–268. 10.1016/j.arth.2015.07.039.

27. W.P. Freire, M.V.L. Fook, E.F. Barbosa, C. dos S. Araújo, R.C. Barbosa, Í.M.F. Pinheiro, Biocompatibility of dental restorative materials, Mater. Sci. Forum. 805 (2015) 19–25. 10.4028/www.scientific.net/MSF.805.19.

28. Q. Chen, G.A. Thouas, Metallic implant biomaterials, Mater. Sci. Eng. R Reports. 87 (2015) 1–57. 10.1016/j.mser.2014.10.001.

29. M. Geetha, A.K. Singh, R. Asokamani, A.K. Gogia, Ti based biomaterials, the ultimate choice for orthopaedic implants – A review, Prog. Mater. Sci. 54 (2009) 397–425. 10.1016/j.pmatsci.2008.06.004.

30. R.I.M. Asri, W.S.W. Harun, M. Samykano, N.A.C. Lah, S.A.C. Ghani, F. Tarlochan, M.R. Raza, Corrosion and surface modification on biocompatible metals: A review., Mater. Sci. Eng. C. Mater. Biol. Appl. 77 (2017) 1261–1274. 10.1016/j.msec.2017.04.102.

31. A.H. Bryan, Trace elements in human and animal nutrition, Am. J. Public Health. 47 (1957) 496– 496. 10.2105/ajph.47.4_pt_1.496.

32. I. Pais, M. Fehér, E. Farkas, Z. Szabó, I. Cornides, Titanium as a new trace element, Commun. Soil Sci. Plant Anal. 8 (1977) 407–410. 10.1080/00103627709366732.

33. S. Yaghoubi, C.W. Schwietert, J.P. McCue, Biological roles of titanium, Biol. Trace Elem. Res. 78 (2000) 205–217. 10.1385/BTER:78:1-3:205.

34. J.E. Lemons, K.M. Niemann, A.B. Weiss, Biocompatibility studies on surgical-grade titanium-, cobalt-, and iron-base alloys, J. Biomed. Mater. Res. 10 (1976) 549–553. 10.1002/jbm.820100411.

35. M.L. Wang, R. Tuli, P.A. Manner, P.F. Sharkey, D.J. Hall, R.S. Tuan, Direct and indirect induction of apoptosis in human mesenchymal stem cells in response to titanium particles., J. Orthop. Res. Off. Publ. Orthop. Res. Soc. 21 (2003) 697–707. 10.1016/S0736-0266(02)00241-3.

36. M. Niinomi, Metallic biomaterials, J. Artif. Organs. 11 (2008) 105–110. 10.1007/s1 0047-008-0422-7.

37. M. Niinomi, Recent metallic materials for biomedical applications, Metall. Mater. Trans. A. 33 (2002) 477–486. 10.1007/s11661-002-0109-2.

38. V.S. de Viteri, E. Fuentes, Titanium and Titanium Alloys as Biomaterials, in: J. Gegner (Ed.), Tribology, IntechOpen, Rijeka, 2013. 10.5772/55860.

39. B. Stadlinger, S.J. Ferguson, U. Eckelt, R. Mai, A.T. Lode, R. Loukota, F. Schlottig, Biomechanical evaluation of a titanium implant surface conditioned by a hydroxide ion solution., Br. J. Oral Maxillofac. Surg. 50 (2012) 74–79. 10.1016/j.bjoms.2010.11.013.

40. K. Subramani, R. Mathew, P. Pachauri, Titanium surface modification techniques for dental implants—From microscale to nanoscale, in: K. Subramani and W. Ahmed (Eds.), Emerging Nanotechnologies in Dentistry, Elsevier, 2018: pp. 99–124.

41. J. Newman, Fundamental considerations for biomaterial selection, IEEE Potentials. 26 (2007) 12–15. 10.1109/MP.2007.343034.

42. J.M. Anderson, Fundamental biological requirements of a biomaterial, in: An Introd. to Biomater., CRC Press, 2006: pp. 3–15.

43. S. Dumitriu, Polymeric biomaterials, 2nd ed., r, Marcel Dekker New York, New York SE - XIV, 1168 str. : ilustr. ; 29 cm, 1994. LK - https://worldcat.org/title/443119903.
44. J.M. Courtney, N.M. Lamba, S. Sundaram, C.D. Forbes, Biomaterials for blood-contacting applications, Biomaterials. 15 (1994) 737–744. 10.1016/0142-9612(94)90026-4.
45. N. Angelova, D. Hunkeler, Rationalizing the design of polymeric biomaterials, Trends Biotechnol. 17 (1999) 409–421. 10.1016/s0167-7799(99)01356-6.
46. D.L. Wise, Biomaterials and bioengineering handbook, Sens. Rev. 21 (2001) 323–324. 10.1108/sr.2001.21.4.323.2.
47. L.E. Murr, Open-cellular metal implant design and fabrication for biomechanical compatibility with bone using electron beam melting., J. Mech. Behav. Biomed. Mater. 76 (2017) 164–177. 10.1016/j.jmbbm.2017.02.019.
48. A.A. Al-Tamimi, C. Peach, P.R. Fernandes, A. Cseke, P.J.D.S. Bartolo, Topology optimization to reduce the stress shielding effect for orthopedic applications, Procedia CIRP. 65 (2017) 202–206. 10.1016/j.procir.2017.04.032.
49. N.J. Hallab, S. Anderson, T. Stafford, T. Glant, J.J. Jacobs, Lymphocyte responses in patients with total hip arthroplasty, J. Orthop. Res. Off. Publ. Orthop. Res. Soc. 23 (2005) 384–391. 10.1016/j.orthres.2004.09.001.
50. T. Hanawa, In vivo metallic biomaterials and surface modification, Mater. Sci. Eng. A. 267 (1999) 260–266. 10.1016/s0921-5093(99)00101-x.
51. T. Albrektsson, P.I. Brånemark, H.A. Hansson, J. Lindström, Osseointegrated titanium implants. Requirements for ensuring a long-lasting, direct bone-to-implant anchorage in man, Acta Orthop. Scand. 52 (1981) 155–170. 10.3109/17453678108991776.
52. L. Carlsson, T. Röstlund, B. Albrektsson, T. Albrektsson, P.I. Brånemark, Osseointegration of titanium implants, Acta Orthop. Scand. 57 (1986) 285–289. 10.3109/17453678608994393.
53. M. Viceconti, R. Muccini, M. Bernakiewicz, M. Baleani, L. Cristofolini, Large-sliding contact elements accurately predict levels of bone-implant micromotion relevant to osseointegration, J. Biomech. 33 (2000) 1611–1618. 10.1016/s0021-92 90(00)00140-8.
54. M.B. Nasab, M.R. Hassan, B. Bin Sahari, Metallic biomaterials of knee and hip–a review, Trends Biomater. Artif. Organs. 24 (2010) 69+. https://link.gale.com/apps/doc/A308129449/HRCA?u=googlescholar&sid=googleScholar&xid=3b4d2143.
55. R.M. Pilliar, Metallic biomaterials BT - Biomedical materials, in: R. Narayan (Ed.), Springer US, Boston, MA, 2009: pp. 41–81. 10.1007/978-0-387-84872-3_2.
56. D. Khang, J. Lu, C. Yao, K.M. Haberstroh, T.J. Webster, The role of nanometer and sub-micron surface features on vascular and bone cell adhesion on titanium, Biomaterials. 29 (2008) 970–983. 10.1016/j.biomaterials.2007.11.009.
57. N.M. Alves, I. Pashkuleva, R.L. Reis, J.F. Mano, Controlling cell behavior through the design of polymer surfaces, Small. 6 (2010) 2208–2220. 10.1002/smll.201000233.
58. H. Yildiz, F.-K. Chang, S. Goodman, Composite hip prosthesis design. II. Simulation, J. Biomed. Mater. Res. 39 (1998) 102–119. 10.1002/(SICI)1097-4636(199801)39:1 <102::AID-JBM13>3.0.CO;2-H.
59. J.J. Ramsden, D.M. Allen, D.J. Stephenson, J.R. Alcock, G.N. Peggs, G. Fuller, G. Goch, The design and manufacture of biomedical surfaces, CIRP Ann. 56 (2007) 687–711. 10.1016/j.cirp.2007.10.001.
60. D.L. Elbert, J.A. Hubbell, Surface treatments of polymers for biocompatibility, Annu. Rev. Mater. Sci. 26 (1996) 365–394. 10.1146/annurev.ms.26.080196.002053.
61. K.R. St John, L.D. Zardiackas, R.A. Poggie, Wear evaluation of cobalt-chromium alloy for use in a metal-on-metal hip prosthesis., J. Biomed. Mater. Res. B. Appl. Biomater. 68 (2004) 1–14. 10.1002/jbm.b.10053.
62. B.-H. Lee, C. Lee, D.-G. Kim, K. Choi, K.H. Lee, Y. Do Kim, Effect of surface structure on biomechanical properties and osseointegration, Mater. Sci. Eng. C. 28 (2008) 1448–1461. 10.1016/j.msec.2008.03.015.

63. M.M. Subedi, Ceramics and its dimensions ceramics, Himal. Physics. 4 (2013) 80–82.
64. S. Pina, R. Rebelo, V.M. Correlo, J.M. Oliveira, R.L. Reis, Bioceramics for osteochondral tissue engineering and regeneration, Adv. Exp. Med. Biol. 1058 (2018) 53–75. 10.1007/978-3-319-76711-6_3.
65. J. Chevalier, L. Gremillard, Ceramics for medical applications: A picture for the next 20 years, J. Eur. Ceram. Soc. 29 (2009) 1245–1255. 10.1016/j.jeurceramsoc.2008.08.025.
66. S. Punj, J. Singh, K. Singh, Ceramic biomaterials: Properties, state of the art and future prospectives, Ceram. Int. 47(2021) 28059–28074. doi: 10.1016/j.ceramint.2021.06.238.
67. M. Kaur, K. Singh, Review on titanium and titanium based alloys as biomaterials for orthopaedic applications, Mater. Sci. Eng. C. Mater. Biol. Appl. 102 (2019) 844–862. 10.1016/j.msec.2019.04.064.
68. H.E. Jazayeri, M. Rodriguez-Romero, M. Razavi, M. Tahriri, K. Ganjawalla, M. Rasoulianboroujeni, M.H. Malekoshoaraie, K. Khoshroo, L. Tayebi, The cross-disciplinary emergence of 3D printed bioceramic scaffolds in orthopedic bioengineering, Ceram. Int. 44 (2018) 1–9. 10.1016/j.ceramint.2017.09.095.
69. K. Shanmugam, R. Sahadevan, Bioceramics—An introductory overview, in: S. Thomas, P. Balakrishnan, M.S. Sreekala (Eds.), Fundam. Biomater. Ceram., Elsevier, 2018: pp. 1–46. 10.1016/B978-0-08-102203-0.00001-9.
70. A. Balasubramanian, S. Gurumurthy, B. Balasubramanian, Biomedical applications of ceramic nanomaterials: A review, Int. J. Pharm. Sci. Res. 8(2017) 4950–4959, doi: 10.13040/IJPSR.0975-8232.8(12).4950-59.
71. S.S. Danewalia, K. Singh, Bioactive glasses and glass-ceramics for hyperthermia treatment of cancer: state-of-art, challenges, and future perspectives, Mater. Today. Bio. 10 (2021) 100100. 10.1016/j.mtbio.2021.100100.
72. S. Kargozar, R.K. Singh, H.-W. Kim, F. Baino, "Hard" ceramics for "Soft" tissue engineering: Paradox or opportunity?, Acta Biomater. 115 (2020) 1–28. 10.1016/j.actbio.2020.08.014.
73. A. Aherwar, A. K. Singh, A. Patnaik, Current and future biocompatibility aspects of biomaterials for hip prosthesis, AIMS Bioeng. 3 (2015) 23–43. 10.3934/bioeng.2016.1.23.
74. A. Noiri, F. Hoshi, H. Murakami, K. Sato, S.I. Kawai, K. Kawai, Biocompatibility of a mobile alumina-ceramic orbital implant, Folia Jpn. Ophthalmol. Clin. 53 (2002) 476–480.
75. A.A. Alamdari, M. Hashemkhani, S. Hendessi, P.T. Guner, H.Y. Acar, I.H. Kavakli, U. Unal, A. Motallebzadeh, In vitro antibacterial and cytotoxicity assessment of magnetron sputtered Ti1.5ZrTa0.5Nb0.5W0.5 refractory high-entropy alloy doped with Ag nanoparticles, Vacuum. 203 (2022) 111286. 10.1016/j.vacuum.2022.111286.
76. K. Yamada, S. Nakamura, T. Tsuchiya, K. Yamashita, Electrical properties of polarized partially stabilized Zirconia for biomaterials, Key Eng. Mater. 216 (2001) 149–152. 10.4028/www.scientific.net/KEM.216.149.
77. W.J. Ma, A.J. Ruys, R.S. Mason, P.J. Martin, A. Bendavid, Z. Liu, M. Ionescu, H. Zreiqat, DLC coatings: Effects of physical and chemical properties on biological response, Biomaterials. 28 (2007) 1620–1628. 10.1016/j.biomaterials.2006.12.010.
78. J. Ruan, G.M. Helen, Biocompatibility evaluation in vitro. Part I: Morphology expression and proliferation of human and rat osteoblasts on the biomaterials, J. Cent. South Univ. Technol. 8 (2001) 1–8. 10.1007/s11771-001-0015-6.
79. A. Piattelli, P. Trisi, A light and laser scanning microscopy study of bone/hydroxyapatite-coated titanium implants interface: histochemical evidence of unmineralized material in humans, J. Biomed. Mater. Res. 28 (1994) 529–536. 10.1002/jbm.820280502.
80. J. Wilson, G.H. Pigott, F.J. Schoen, L.L. Hench, Toxicology and biocompatibility of bioglasses, J. Biomed. Mater. Res. 15 (1981) 805–817. 10.1002/jbm.820150605.
81. H.R. Fernandes, A. Gaddam, A. Rebelo, D. Brazete, G.E. Stan, J.M.F. Ferreira, Bioactive glasses and glass-ceramics for healthcare applications in bone regeneration and tissue engineering, Materials (Basel). 11 (2018). 10.3390/ma11122530.

82. C.V.M. Rodrigues, P. Serricella, A.B.R. Linhares, R.M. Guerdes, R. Borojevic, M.A. Rossi, M.E.L. Duarte, M. Farina, Characterization of a bovine collagen-hydroxyapatite composite scaffold for bone tissue engineering., Biomaterials. 24 (2003) 4987–4997. 10.1 016/s0142-9612(03)00410-1.

83. A. G. Clare, "The Unique Nature of Glass," in Bio-Glasses, Wiley, 2012: pp. 1–12.

84. E.T. Thostenson, Z. Ren, T.-W. Chou, Advances in the science and technology of carbon nanotubes and their composites: A review, Compos. Sci. Technol. 61 (2001) 1899–1912. 10.1016/S0266-3538(01)00094-X.

85. J. A. Davidson, F. S. Georgette, and S. of Manufacturing Engineers, State of the Art Materials for Orthopedic Prosthetic Devices. Society of Manufacturing Engineers, Southfield, Michigan, United States, 1987.

86. L.-J. Zhang, X.-S. Feng, H.-G. Liu, D.-J. Qian, L. Zhang, X.-L. Yu, F.-Z. Cui, Hydroxyapatite/collagen composite materials formation in simulated body fluid environment, Mater. Lett. 58 (2004) 719–722. 10.1016/j.matlet.2003.07.009.

87. R. Fazel-Rezai, Biomedical Engineering - From Theory to Applications, InTech, Rijeka, 2011. 10.5772/2629.

88. M. Wang, Q. Zhao, Biomedical Composites, in: R. Narayan (Ed.), Encycl. Biomed. Eng., Elsevier, Oxford, 2019: pp. 34–52. 10.1016/B978-0-12-801238-3.99868-4.

89. H. Ma, A. Suonan, J. Zhou, Q. Yuan, L. Liu, X. Zhao, X. Lou, C. Yang, D. Li, Y. Zhang, PEEK (Polyether-ether-ketone) and its composite materials in orthopedic implantation, Arab. J. Chem. 14 (2021) 102977. 10.1016/j.arabjc.2020.102977.

90. N.P. Desai, J.A. Hubbell, Solution technique to incorporate polyethylene oxide and other water-soluble polymers into surfaces of polymeric biomaterials, Biomaterials. 12 (1991) 144–153. 10.1016/0142-9612(91)90193-E.

91. B.D. Ratner, D.G. Castner, eds., Surface Modification of Polymeric Biomaterials, Springer US, Boston, MA, 1997. 10.1007/978-1-4899-1953-3.

92. C. Nojiri, T. Okano, H.A. Jacobs, K.D. Park, S.F. Mohammad, D.B. Olsen, S.W. Kim, Blood compatibility of PEO grafted polyurethane and HEMA/styrene block copolymer surfaces, J. Biomed. Mater. Res. 24 (1990) 1151–1171. 10.1002/jbm.820240903.

93. D. Nayak, S. Ghosh, A. Venimadhav, Structural and electrochemical kinetics of Na–Fe–Mn–O thin-film cathode: A synergistic effect of deposition conditions, Ionics (Kiel). 27 (2021) 2421–2430. 10.1007/s11581-021-04043-8.

94. C. Douzon, F.M. Kanmangne, H. Serne, D. Labarre, M. Jozefowicz, Heparin-like activity of insoluble sulphonated polystyrene resins. Part III: Binding of dicarboxylic amino acids, Biomaterials. 8 (1987) 190–194. 10.1016/0142-9612(87)90062-7.

95. D.P. Paulson, W. Abuzeid, H. Jiang, T. Oe, B.W. O'Malley, D. Li, A novel controlled local drug delivery system for inner ear disease, Laryngoscope. 118 (2008) 706–711. 10.1097/MLG.0b013e31815f8e41.

96. M. Oberhoff, W. Kunert, C. Herdeg, A. Küttner, A. Kranzhöfer, B. Horch, A. Baumbach, K.R. Karsch, Inhibition of smooth muscle cell proliferation after local drug delivery of the antimitotic drug paclitaxel using a porous balloon catheter, Basic Res. Cardiol. 96 (2001) 275–282. 10.1007/s003950170058.

97. A.B. Seabra, R. da Silva, G.F.P. de Souza, M.G. de Oliveira, Antithrombogenic polynitrosated polyester/poly(methyl methacrylate) blend for the coating of blood-contacting surfaces, Artif. Organs. 32 (2008) 262–267. 10.1111/j.1525-1594. 2008.00540.x.

98. C.A. Simmons, E. Alsberg, S. Hsiong, W.J. Kim, D.J. Mooney, Dual growth factor delivery and controlled scaffold degradation enhance in vivo bone formation by transplanted bone marrow stromal cells, Bone. 35 (2004) 562–569. 10.1016/j.bone.2 004.02.027.

# 4 Smart Biomaterials and Their Applications

*Sangita Mangaraj, Subhasmita Swain, and Tapash R. Rautray*

Biomaterials and Tissue Regeneration Lab, CETMS, Institute of Technical Education and Research, Siksha 'O'Anusandhan (Deemed to be University), Bhubaneswar, Odisha, India

## 4.1 INTRODUCTION

The field of material science is an emerging and visionary research field that plays a vital role in the development of human society due to its various applications [1]. Material has to undergo various changes for the fulfillment of outer space, which further ended up on newly developed techniques, and easy processes, rather than the old conventional method. In the past, humans used materials for their betterment and, as a result, their regular lifestyle increased. Also, the progress of human beings and the ways of life separated based on the emergence of novel materials such as the Stone Age. The most progressive age was the Bronze Age because of the initiation of metallurgy. Numerous innovative, excellent, and relatively inexpensive materials have been developed through the advancement of materials science and are now used in a variety of engineering fields [2]. The materials evolved for multipurpose applications over the past ten years necessitated the enhancement of various classifications and characteristics. Hence, the age of smart biomaterials started. A smart biomaterial is currently being considered the next advanced version of evolution. The new paradigm of advanced material or smart material will have a significant influence on human society [3]. For instance, some of them are capable of adapting their features to the environment, whereas others possess sensory abilities; some can rectify themselves instantly and some can degrade on their own. The exceptional qualities of smart materials will have a significant effect on every feature of society. The progress and creation of novel biomaterials are the focus of billions of research worldwide. These biomaterials might exhibit exceptional qualities in a variety of settings, from clinical treatments to difficult diagnostics, from space to industry, the demand for vaccines, protective apparel, diagnostic equipment, and emergency medical care has particularly in the wake of the COVID-19 outbreak, which has caused harm to people all over the world. A massive change is necessary to satisfy these increasing demands by shifting away from conventional biomaterials to intelligent ones [4].

DOI: 10.1201/9781003429920-5

Numerous studies have been conducted over the past 200 years to create the latest kinds of useful materials that have become categorized into multiple families and groups: [a] metals, [b] ceramics [c], polymers, and [d] smart biomaterials [5].

## 4.2 WHAT IS SMART BIOMATERIALS

Smart biomaterials are a type of part that is crucial to degressive material that can recognize specific external signals and activate themselves to carry out a predetermined task. The terms adaptive, intelligent, and smart are often used to describe these materials, which include sensors and actuators [6,7].

It is capable of recognizing a stimulus, reacting to it in a definite way within a brief window of time, and returning to its initial condition as quickly as the stimulus is taken away [8].

Smartly customized qualities and operations that can promote the regeneration and repair of tissues by specific stimuli can be internal and external, such as pH, temperature, ionic strength, and magnetism, are referred to as "smart biomaterials" and "smart constructs." These biomaterials and structures have informative or prompting effects on cells and tissues. This is accomplished by designing the material's adaptability to both internal and external stimuli [9]. Therefore, for performance requirements, "smart" materials are typically installed in systems whose essential qualities could be advantageously transformed [1]. In 1990, Takagi et al. defined it as intelligent materials because of the way it reacts to surrounding atmospheric changes at most ideal conditions and reconstructs itself based on the environmental condition [10]. Additionally, they possess several characteristics that differentiate them from common biomaterials, as illustrated in Figure 4.1.

**Transiency** means it can respond to several environments and have distinct characteristics based on each circumstance. **Urgency** is nothing, but they react to the environmental effect [stimuli] without losing time. Intelligent is also called **self-actuation**. It means this potential is contained within the matter. Self-actuation is the term for some materials that have unique intrinsic characteristics that are not brought about by or triggered by external forces. **Selectivity**, which means the

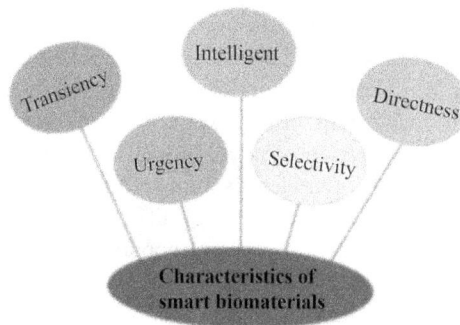

**FIGURE 4.1**  Characteristics of smart biomaterials.

reaction is distinct and easy to predict. **Directness** means action and the reactions are accumulated at one location [6,7,11–13].

Smart biomaterials are categorized into two types, and this classification is based on their properties, i.e., **active** and **passive** biomaterials [14,15].

According to Fairweather et al., in 1998, smart materials are recognized as active materials because of the ability to change their dimensional or physical features in response to the presence of magnetic, electric, or thermal fields, and hence acquiring an innate ability to transmit and receive energy. Magnetostrictive and piezoelectric materials are examples of such materials and those materials that are not active are referred to as passive smart materials. The term *passive* refers to smart materials that can transmit certain energy forms, for example, fiber optics. Despite being intelligent, they are lacking in the innate ability to sequence energy [3,8,16].

There is a significant amount of curiosity about smart structures as a result of current developments in smart materials for dispersed sensors and actuators. Typically, smart materials are used as actuators and sensors corresponding to "stimulus" and "response." Smart materials have smart structures. So, the working mechanism of SBM (smart biomaterials) is that when the SBM is integrated into traditional structures, it starts to sense, then actuate, and further processing happens. That means the structure of smart biomaterials senses the outer stimuli without wasting any time under different kind of factors, i.e., pressure, magnetic field, temperature, stress, etc. and then it undergoes verification, followed by a proper course of action that leads to a solution. It stores all the data regarding the process for future applications [17–19].

The sensory structures have sense and no actuator, and likewise, the adaptive structure has an actuator but no sense and is an active structure; characteristics of both structures are available. An advanced or smart structure is called an active structure because it is the combination of both the sensory and adaptive structure, which means it has both a sensor and actuator. So, advanced smart biomaterial is installed logic and electronics along with sensor and actuator [1].

## 4.3 HISTORICAL OVERVIEW OF SMART OR INTELLIGENT MATERIALS

The history of intelligent biomaterials began in 1932 when the rubbery elasticity of the AuCd alloy at room temperature was discovered. Chang and Read pioneered the alloy shape-memory effect in 1951. William J. Buehler discovered the NiTi alloy in 1963, which was a major advance in the discipline of smart biomaterials. Several research studies have documented the use of NiTi in medical applications because of the invention of the NiTi alloy, mainly in the early 1970s as NiTi was used to create the first practical device in the early 1990s, which made cardiovascular biology and radiology far more advanced. From 1990 onwards, studies of multi-stimuli materials predicted the emergence of the next phase of intelligent biomaterials. In the 1990s, S. Hayashi reportedly developed a thermoplastic polyurethane shape-memory polymer with multiple stimuli, i.e., thermal and moisture responsive. The production of an expensive form of memory alloys and liquid crystalline elastomers was made possible by two-way shape-memory semi-crystalline network polymers. Furthermore, it was claimed that 3D printing might be used to make multi-shape-memory polymeric composites, enlarging the possibilities for intelligent biomaterials for next-generation biomedical engineering purposes. The area of intelligent biomaterial systems has generally advanced significantly in the past few years; some of those are now in use in healthcare situations, while others are still being developed. To trigger certain biological reactions, biomaterials evolved from inert constituents to bioactive ones in the 1980s and 1990s. A historical overview and the evolution of smart biomaterials are illustrated in Figure 4.2 [20–22].

Smart materials differ from common or traditional materials due to their special, controlled, useful qualities. The majority of common materials have set features, so introducing any kind of additional unit may affect their qualities. These characteristics, however, are advantageous in smart materials. These materials are capable of reacting to certain stimuli and subsequently displaying new characteristics. The easy and rapid action of smart materials is another wonderful feature, in contrast to the complicated, stretched, and the difficult reaction of traditional or common materials [23].

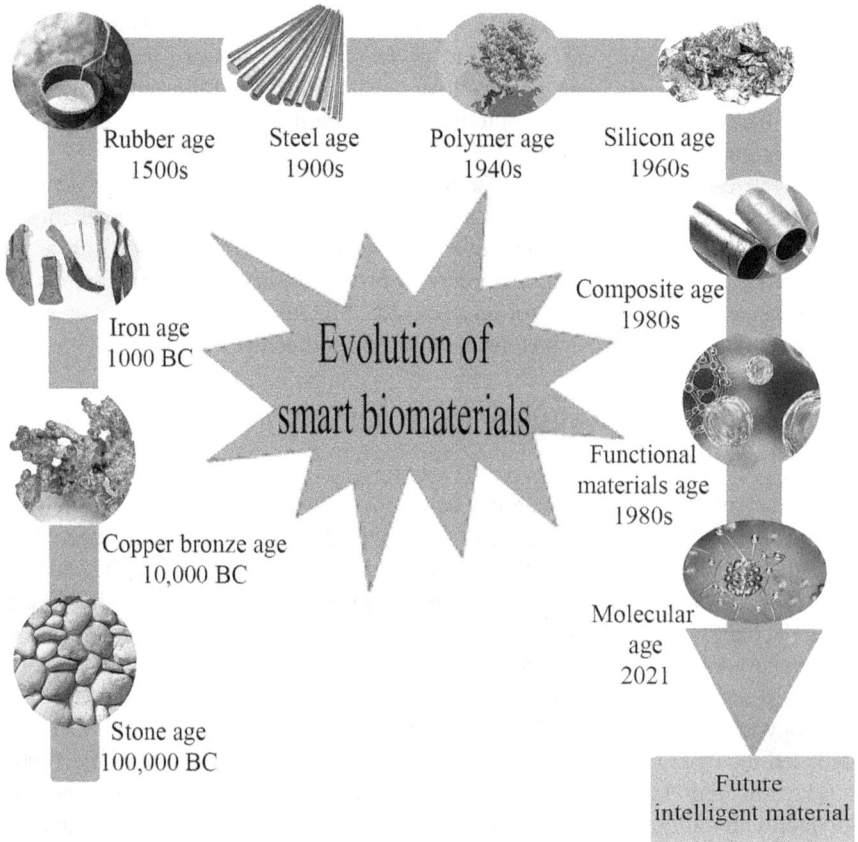

Rubber age
1500s

Steel age
1900s

Polymer age
1940s

Silicon age
1960s

Iron age
1000 BC

Evolution of
smart biomaterials

Composite age
1980s

Functional
materials age
1980s

Copper bronze age
10,000 BC

Molecular
age
2021

Stone age
100,000 BC

Future
intelligent material

**FIGURE 4.2**   History of smart biomaterials.

## 4.4  POSITIVE AND NEGATIVE ASPECTS OF SBM

| Positive Aspects | Negative Aspects |
|---|---|
| Densely packed energy | High cost |
| Improved toughness and endurance | Extremely sensible needs good storage |
| Fantastic bandwidth | Not easily accessible |
| Lowers the cost of manufacture | Unknown long-term consequences |
| Control over the dimensions and shape | Certain smart materials are experiencing corrosion and usefulness loss, much like other conventional materials |
| If there is damage, it will fix itself | |
| Huge amount of volume varies with the temperature | |
| Extreme reactivity to stimuli | |
| Rapid reaction under any circumstances | |
| They adapt to their surroundings on their own | |

## 4.5  REQUIREMENT OF SMART BIOMATERIAL

To get the best clinical outcomes, there is a need to move away from conventional techniques that substitute tissue with permanent implants, to more biologically adaptive approaches that concentrate on reviving, initiating, and reorganizing the tissue's structure and function. This can satisfy long-term repair needs and promote better overall health and well-being [24].

Despite the promise of smart biomaterials and their potential to revolutionize clinical practice, there have been limited success stories in their application. However, recent advancements in the field have enabled us to better understand the physical and chemical properties of these materials and have opened new possibilities for their use in medicine.

Smart biomaterials are revolutionizing the way we interact with biological systems. These materials have been developed to interact with cells and tissues uniquely, enabling them to respond to stimuli and induce changes within the body. Their characteristics, such as instructive/inductive or triggering/stimulating effects, can be used to control cell behavior and tissue responses, making them valuable tools for medical research and therapy [25–27].

Smart biomaterials have many impressive features such as their response to internal or external triggers such as magnetism, ionic strength, pH, and temperature. They possess the ability to influence and induce cell and tissue behavior. Such a feature has been essential for advancing various medical treatments and therapies [22].

Only one of the essential criteria for a smart biomaterial is biocompatibility. Rather than making tissue engineering structures *ex vivo* more complicated, several researchers are working on creating biomaterials that actively support the body's natural ability to heal and self-repair. To encourage cellular intrusion, adhesion, and division, interactive smart biomaterials should be able to imitate the extracellular matrix's (ECM) function [28–30].

## 4.6 DIFFERENT LEVELS OF SMART BIOMATERIALS

Smart biomaterials can be classified based on how much contact they have with their surroundings as well as the resulting biological reactions. This categorization aids in defining a biomaterial's level of intelligence and acknowledges the development of the idea of "intelligence" so that the potential of biomaterials to develop multiple types of functional qualities may therefore be identified by defining a level or extent of intelligence. Hence, creating a level or extent of intelligence can assist in clearing any confusion, particularly for innovative biomaterials capable of responding to various stimuli. Different levels of smart biomaterial [Figure 4.3] are described below [31].

**First Level [Inert]:** inert represents the initial level of intelligence. It is described as the capacity of a biomaterial for simply being bioinert or bio-compatible, or "to do no damage," without conferring any additional biological benefits. Even at the most basic level, the ability to utilize a substance inside the body suggests some amount of intelligence. The host accepts the biomaterial despite being harmful or irritating since inert biomaterials lack treatments or biological relationships [32].

**Second Level [Activeness]:** activeness is the second level of intelligence. It is described as a biomaterial's capacity to deliver a bioactive solution, i.e., open loop. A purposeful one-way interaction with biological mechanisms or the environment is

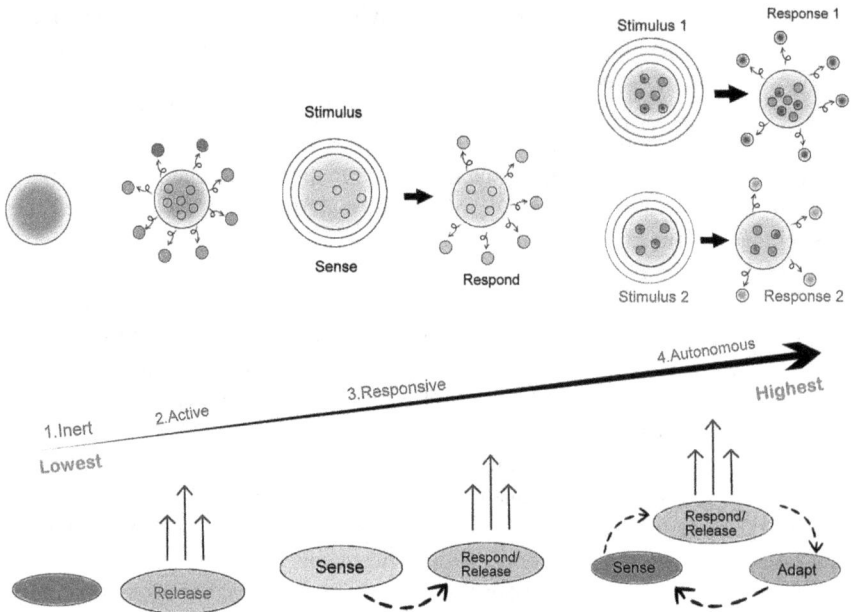

**FIGURE 4.3**  Schematic diagram of four phases for a biomaterial to be smart.

what active biomaterials are meant to provide. This degree of intelligence might imitate the conventional initiation of a biological function [33].

**Third Level [Responsiveness]:** the third level of intelligence is responsiveness. It is described as a biomaterial's capacity to detect a trigger/stimulus and respond by generating certain therapeutic components. A closed-loop feedback control system is analogous to this [34–36].

**Fourth Level [Autonomous]:** the fourth and most advanced level of intelligence is autonomous. These future biomaterials are thought to be self-sufficient. Such biomaterials not only provide medical therapies after being stimulated by the right stimuli but also communicate with each other in a complicated manner by detecting, reacting, and evolving to certain signals. These innovations should ideally be able to detect a specific illness in its early phases, convey its occurrence far from the body, and manage it at various phases before any adverse effects occur. Autonomous biomaterials may perform a variety of tasks without human assistance by becoming living, customizable, and adaptable independent organisms [36–38].

Overall, the research field has identified reactive, adaptive, and autonomous levels as the present generation of "smart biomaterials," i.e., biomaterials that interact with biological processes and microenvironments in a modified manner. The current description of smart biomaterials includes autonomous materials that can sense, respond to, and adapt to changes in the biological environment, as well as inert materials that can be employed within the body without causing any harm [39].

## 4.7   DESIGNING OF SMART BIOMATERIALS

It is crucial to design biofunctional materials for various biological applications. Combining synthetic and natural components to create hybrid smart biomaterials presents new possibilities for revolutionizing medicine and healthcare. By better understanding the properties and mechanisms of these materials, scientists can develop more advanced biomedical solutions that could benefit countless lives around the world [40].

So, advanced processing methods are revolutionizing the way biomaterials are manufactured through the use of sophisticated fabrication techniques [such as additive manufacturing]; scientists and engineers can create biomaterials with a lower degree of complexity and greater smartness. These materials can not only improve their interactions with the environment but also gain new capabilities such as self-repair, sensing, and responding to stimuli. With this technology, biomaterials can now be tailored to specific needs and applications and create more efficient products. This opens up a world of possibilities in terms of what can be achieved with biomaterials, from medical implants to consumer products [32,41,42].

For example, biocompatible metals such as titanium oxides can be transformed through 3D printing technology and shaped to create porous structures with developed functions for osseointegration. The degree of "smartness" associated with these metals is relatively inert, meaning that they are biocompatible yet not overly reactive. With the help of 3D printing, these materials can be used to create scaffolds for medical use efficiently and cost effectively [41].

Contemporary fabrication methods have gained massive attention in recent years, with 3D and 4D bioprinting and electrospinning becoming increasingly popular. These methods allow for the creation of intricate products with a variety of characteristics and properties that were previously impossible to obtain [42]. With the advancements in these technologies, it is now possible to fabricate everything from medical implants to complex 3D structures with unprecedented precision. These techniques give specialized shapes and sizes with constrained micro-structures, i.e., the structure is complex in shape, unique nano- or micro-topography, and a high level of orientation/alignment. Controlling certain chemical and physical factors throughout formulation, fabrication, processing, and production is necessary for the creation of advanced functional biomaterials [43–46]. The 3D structure of smart material is primarily influenced by the distinctive properties of the monomers, polymers, oligomers, and production techniques. For the creation of 3D network biomaterials, particularly hydrogels with various architectures and characteristics, a broad range of techniques are available, including radiation-induced cross-linking, physical-chemical cross-linking, etc. [47].

All these methods aim to replicate the shapes and characteristics required to mimic biological tissues in terms of their nano- or micro-structures, mechanical properties, chemistry, charge, etc. This technology is paving the way for groundbreaking treatments for a variety of diseases and medical conditions [32,44].

## 4.8   SMART BIOMATERIAL CLASSIFICATION FOLLOWING SPECIFIC STIMULI

According to the behavior of specific stimuli, smart materials are divided into several categories. The controlled application of external stimuli, such as pH, temperature, stress, electric field, magnetic field, etc., can drastically change the properties of smart materials. Many types of smart substances are accessible based on their various qualities concerning specific stimuli, as described below [32].

### 4.8.1   Piezoelectric Materials

These are becoming increasingly popular due to their ability to convert mechanical and electrical signals. Through direct piezoelectricity, these materials can transform the mechanical signal into an electrical charge, while converse piezoelectricity allows them to turn an electric charge into a mechanical signal. This versatility makes them ideal for various applications, from powering medical implants to making musical instruments [48–50].

### 4.8.2   Thermoresponsive Materials

These are an important class of intelligent biomaterials, i.e., either shape-memory alloys or shape-memory polymers that can sense and respond to heat. These materials show a phase change at a particular temperature, exhibiting unique properties. It can deform them and restore them to their original shape by heat and produce an actuating force as a result. The applications of these materials range from smart drug delivery systems to self-healing structures and many more [8,51].

### 4.8.3 Magnetic Responsive Biomaterials

An intriguing class of stimuli-responsive biomaterials is magnetically responsive composite materials, which are an applied magnetic field from outside that may produce precise and regulated stimuli with great tissue penetration and spatial precision. These are also described as stimulants. Due to their special intrinsic physical and chemical characteristics, magnetic NPs (MNPs) are a developing platform receiving significant attention for their uses in the biomedical field, which includes magnetic hyperthermia, drug delivery, magnetic resonance imaging [MRI], etc. [23].

### 4.8.4 pH-Sensitive Biomaterial

pH is one of the essential physiological factors that determines or suggests a certain situation in the microenvironment of cells and tissues. These are the substances whose color changes in response to acidity changes. This is relevant to coatings that can alter the color to show signs of corrosion in the metal behind them [8]. One significant class of smart materials that may alter color in response to a particular pH fluctuation is pH-sensitive polymers. They might be acidic or basic and respond according to that. They are utilized nowadays in a variety of contexts, including the delivery of drugs in medicine, surface improvement, and filtering [2].

### 4.8.5 Enzyme-Responsive Biomaterial

An emerging category of smart materials called enzyme-responsive materials undergoes macroscopic modifications as a result of selective enzymatic activity. The excellent catalytic effectiveness and intrinsic biocompatibility of enzyme-responsive systems is the primary factor that makes them a selection of material for biomedical use.

To create enzyme-responsive materials, a variety of enzymes, including kinases, protease, phosphatase, and endonuclease can be examined.

Enzyme-responsive biomaterial can be used to produce biosensors with extremely high sensitivity [52].

## 4.9 CURRENT DISCOVERIES IN SMART BIOMATERIALS

Increasingly, people are paying attention to "smart" biomaterials, specifically those whose characteristics may be controlled in response to external stimuli or environmental changes (such as "smart" hydrogels, "smart" nanomaterials, "smart" bio conjugates," and "shape-memory" materials).

### 4.9.1 Smart Hydrogels

Scientists have worked hard over many years to design hydrogels by modifying their physical and chemical features, thus characterizing them as "smart" hydrogels.

Hydrogels are a revolutionary class of biomaterials as they were the first biomaterials to be rationally designed for human use, due to their unique properties [53,54].

Hydrogels are incredibly versatile and can be used to create implants, medical devices, tissue engineering scaffolds, drug delivery systems, and much more. Thanks to their ability to mimic the natural environment of cells in the body, hydrogels can be used as an ideal platform for biomedical research and development. Also, in response to outer side stimuli, hydrogels can significantly change their performance of swelling, mechanical strength, sol-gel transition, porosity, and network topology [55].

What's more intriguing is that certain smart hydrogels can react to specific molecules, including antigens, and enzymes without the need for an artificial stimulus. They have a high percentage of water, which contributes to their good biocompatibility. The water amount in the equilibrium swollen state is a result of the interplay between hydration (thermodynamic force) and the 3D network's retractive force. In reaction to a minor external trigger, smart hydrogels experience sudden alterations to their chemical and physical characteristics and macroscopic modifications as well. Therefore, this is a favorable material as it can respond to different triggers and efficiently control the volume transition. Its scalability, reproduction capability, and predictability make it particularly appealing. On top of that, its ability to return to its original shape once the trigger has been removed is highly beneficial.

By incorporating smart hydrogels into drug delivery systems, many dosing intervals may be reduced to a single one. By keeping the medications from building up in unsuitable locations in the body, this technique helps maintain a constant therapeutic concentration and minimizes the possibility of adverse effects [56].

Smart hydrogels are an ideal choice for drug-based, long-term release system due to their easy preparation process. Smart hydrogels are increasingly being recognized for their potential applications in bioscience. They can be used for biomedical purposes, such as smart valves, and tissue engineering to regulate cell adhesion and create injectable systems, controlled drug delivery systems for the loading and releasing of bioactive molecules, biosensors like glucose sensors, and actuators are emerging to have great potential in the biotechnology arena [57–59].

Hybrid hydrogels made from synthetic DNA and proteins have promising future applications in biomaterials. According to the research by Rita Lima et al., thermo-responsive hydrogels containing GO and rGO were developed and injected *in situ* as chitosan-agarose mixes for application in cancer therapy [60].

### 4.9.2 Application of Smart Hydrogels

The application of hydrogels includes a wide variety of purposes, but the most common ones are contact lenses, drug delivery systems, medicinal wound dressing, and advanced tissue engineering. Wichterle and Lim were the first to describe a poly-2-hydroxy ethyl methacrylate (PHEMA) hydrogel in a 1960 publication, making PHEMA the first synthetic biocompatible material suitable for use in contact lenses. Most lenses are for correcting vision. They are often categorized as either "hard" or "soft," depending on their flexibility. In hard lenses, they may last longer, but they aren't very much liked by users and can have a longer adjustment period. While hydrogels form the basis of soft contact lenses, hydrophobic materials like poly(methyl methacrylate) (PMMA) and poly(hexa-fluoroisopropyl methacrylate) (HFIM) provide the backbone of hard lenses [61,62].

In addition to that, hydrogels have several applications in wound care, which help in eradicating dead tissue or debris moist dressings, and paste components. However, they may be used on dry wounds since they form gels without the addition of additional wound fluids because they are 70% water [63].

The unique features of hydrogels have made them outstanding for the delivery of drugs. The high porosity nature of the hydrogel can be controlled by water affinity and the density of cross-linking. Because of this porous structure, they are loaded with drugs that are subsequently released. Hydrogels' potential for sustained release results in maintaining a high local concentration of an active pharmaceutical component for a long period is one of the many ways in which they benefit drug delivery applications [64].

Hydrogels have recently been used in the field of tissue engineering, where they may serve as either space-filling agents, delivery agents for bioactive compounds, or three-dimensional structures that organize cells and provide stimuli to enable the production of essential tissue [65].

### 4.9.3 Smart Nanomaterials

Ever since nanotechnology was first introduced, a wide range of nanomaterials has been created and implemented in various applications [Figure 4.4]. Initially, passive structures such as nanoparticles were the main focus. However, the second generation saw active nanostructures being developed, which opened the door to numerous possibilities [66].

In recent times, innovative, stimulus-responsive nanomaterials have experienced a rapid rise in utilization across several facets of medicine, including gene therapy, treatment for cancer, tissue regeneration, biosensors, antibiotics, and biomedical imaging. Anisotropic (direction-independent) and isotropic (direction-dependent) nanomaterials come in a variety of forms, and they can alter morphology in response to external influences [67].

Surface chemistry, biodegradability, drug release, and stimuli-sensitivity like temperature, light, and others make smart nanomaterials useful in biomedical applications. Nevertheless, there are difficulties to release drugs in chemotherapy due to issues including nanocarrier size/shape and biocompatibility with certain cell membranes. The applications of smart nanomaterials are listed below [59,68].

### 4.9.4 Smart Bioconjugations

Biomolecules may be enhanced by synthetic polymer bioconjugations. Several researchers have bonded smart polymers to proteins to provide nano-scale switches for affinity separations, DNA motors, diagnostic devices, biosensors, cell culture processes, drug delivery, and enzyme bioprocesses, including tissue engineering. In

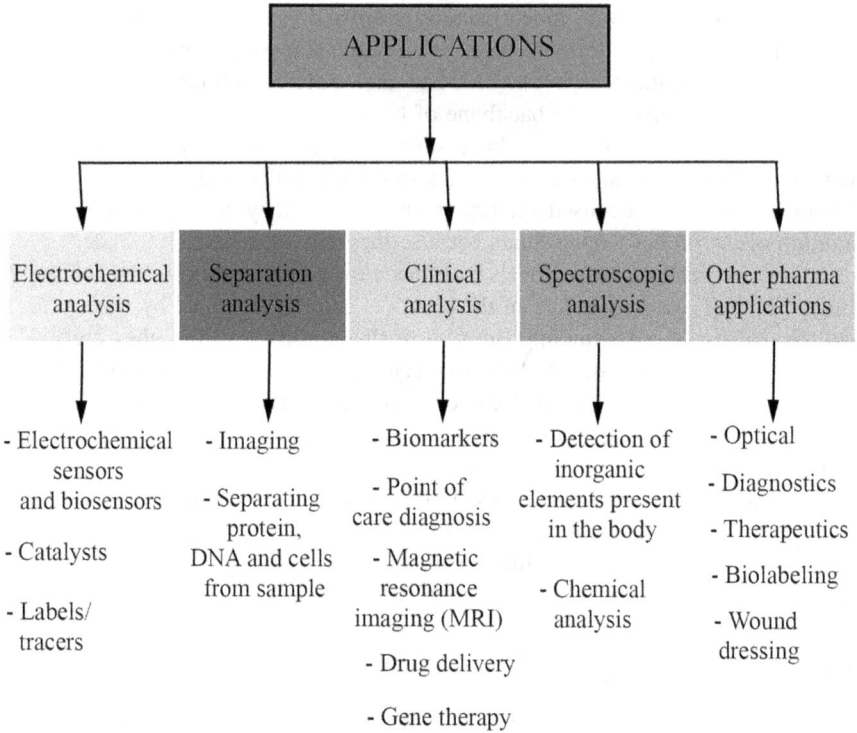

**FIGURE 4.4**  Applications of smart nanomaterials.

addition to proteins, smart polymers might also be used to associate polysaccharides, peptides, lipids, DNA, etc. [69].

Smart conjugation is an advanced method for improving the use of normally inert polymers by endowing them with novel characteristics and functionalities. When coupled to biomolecules like enzymes, plasmid vectors, stimuli-responsive polymers, and liposomes, they may provide reversible solubilization/precipitation. This novel class of "smart" biomaterials exhibits stimuli-responsive activities and characteristics [70,71].

Diagnostic technologies, molecular switching, drug/gene delivery, bioseparations, and immunoassays are just some of the many medical fields that have discovered uses for smart conjugation. Protein-reactive initiators are utilized in the construction of well-defined polymers because they speed up the process and make it possible to create polymers that can couple to proteins without undergoing any further transformation [72,73].

### 4.9.5  SHAPE-MEMORY BIOMATERIALS

The term *shape-memory materials* (SMMs) refers to a class of smart materials that may alter their physical form in response to mechanical or environmental stimuli. In recent years much interest has been shown in shape-memory materials, i.e., shape-memory alloys, shape-memory ceramics, shape-memory polymers, super elastic alloys, etc. As the name suggests, these smart materials are grabbing attention because of their ability to remember a given shape and then return to that shape in response to external inputs [74,75].

Shape-memory materials, in contrast to most smart biomaterials, can regain their previous shape after being quasi-plastically distorted, while most smart biomaterials can only alter their physical or chemical characteristics. Hence, biomaterials belonging to this class that respond to external stimuli might be well suited for incorporation into a larger smart device in which the material itself serves as the sensing and reacting device [76–78].

Shape-memory polymers (SMPs) are different from shape-memory alloys (SMAs) in that they are lighter, easier to make, more flexible, and biodegradable. SMPs facilitate the ability of the scaffolds to connect with the bone tissue around them. One of the most common shape-memory materials is equiatomicNiTi, which is also called nitinol. It was first discovered by Buehler et al. [79]. As a result of their unique properties, shape-memory materials have recently gained significant attention in the biomedical sector, particularly in pharmacy and biotechnology, on cell culture stages, adhesive gadgets, micro-electromechanical systems, and self-healing.

To be considered a "smart material," SMAs must include a minimum of two essential stages that may be switched between changes in stress or temperature. The two phases are austenite and martensite. They may sustain their fundamental structure after being distorted, thanks to their ability to recall it in the austenite phase. When heated above a specific temperature, a special family of metal alloys known as shape-memory alloys can recuperate from apparently irreversible deformation over time. The name "shape-memory modification" refers to the modification of the atomic structure and microstructure of a solid [21].

Shape-memory polymers are a category of memory polymers that may be regulated to take a specific form in reaction to certain inputs (SMPs). Since SMPs depend primarily on molecular structures and not on chemical composition, the form may be adjusted by altering the monomer-to-co-monomer ratio. To provide functional tissues with a less surgical delivery method, Montgomery et al. created injectable shape-memory biomaterials [80].

The advantages of shape-memory alloys are [i] superior purity and [ii] consistency, [iii] low-priced, [iv] single-batch melting, [v] microstructure with a low density, and [v] a large void content. One product of SMA fabrication is the bioengineered robotic hand. It's a small, light, and potent piece of electronics that can be used as an artificial muscle where aluminum and shape-memory alloys (SMAs) are its compositions [81].

### 4.9.6 Carbon Nanotube as Smart Biomaterials

The unique atomic structure of carbon nanotubes (CNTs), which consists of graphene sheets coiled into cylindrical tubes, gives them extraordinary thermal, mechanical, electrical, and biological characteristics. CNTs have been found in a variety of technological contexts because of their exceptional features. Using CNTs to create drug delivery systems, high-performance composites, and hybrid biosensors for implants is now the primary emphasis of CNT research [82].

Carbon nanotubes have the potential to be used as anti-tumor agents, and when combined with traditional medications, they can considerably improve the chemotherapeutic efficacy of conventional treatments via the use of improved drug delivery. There have been reports indicating that paclitaxel-loaded PEG-CNTs have therapeutic potential for the treatment of cancer. CNT-based biosensors have also been utilized to measure the concentration of compounds that could help treat cancer. In this sense, H. Zhou et al. [2018] created an analytical apparatus for detecting methotrexate (MTX) in whole blood as a model system for sensing other anticancer drugs. In their research, Zanganeh et al. [2016] suggested an electrical biosensor for the detection of cancer cells that makes use of vertically aligned, amine-functionalized CNTs bonded to folic acid (FA-VACNTs) molecules [83,84].

Because of their exceptional mechanical skills, most CNTs are included in bulk composites as structural reinforcement. CNTs with structural composite are included in aerospace engineering and automobile engineering; because of their hydrophobic properties, these are used as a coating agent for marine turbines to increase their resistance to bio-fouling. The nanometric size of CNT and chemical compatibility with biomolecules became an advantage in various applications, i.e., as drug delivery, as an implant, as a substrate, and as a biosensor [85].

For biological application, CNTs must be water soluble since this facilitates their adsorption, diffusion, and subsequent removal and this water solubility may be enhanced by functionalization with the correct chemical structure [86].

CNTs have the potential for use in biosensor production due to their unique features. They have a high level of biological function, rapidly transport electrons to a broad spectrum of electroactive molecules, and are readily functionailazible, all of which improve their ability to dissolve and biocompatibility. To achieve this objective, CNTs may serve as nano-scale electrodes, electronic components, or platforms onto which biomolecules can be anchored by interacting with other molecules [87].

The steps involved in making a biosensor out of carbon nanotubes (CNTs) [Figure 4.5] are as follows: attaching functional molecules in the CNT wall, improving the interaction between the target and nanotube, making the biosensor, and using it for further application [88].

The biosensor made use of single-wall carbon nanotubes (SWCNTs) and multiwall carbon nanotubes (MWCNTs) doped with nitrogen as active sensing components (N-MWCNTs). Both kinds of CNTs were synthesized by the team using CVD synthesis [86,87]. Since CNTs were first mentioned as potential drug delivery vehicles in 2004, several methods have been developed to further develop their usage in this capacity. Functionalization, encapsulation, coating with polymers, liposomes, and fabrication of structures, including irregular meshes, bucky papers, membranes, and hydrogels, are some of the most recently recognized methods for utilizing CNTs in this sector. Because of their high aspect ratio, remarkable strength, and stiffness, CNTs are a great choice for enhancing the mechanical characteristics of composites. And HA has low mechanical properties, as a result of which HA becomes the most common component in CNT-reinforced biocomposites used in the healthcare and dentistry

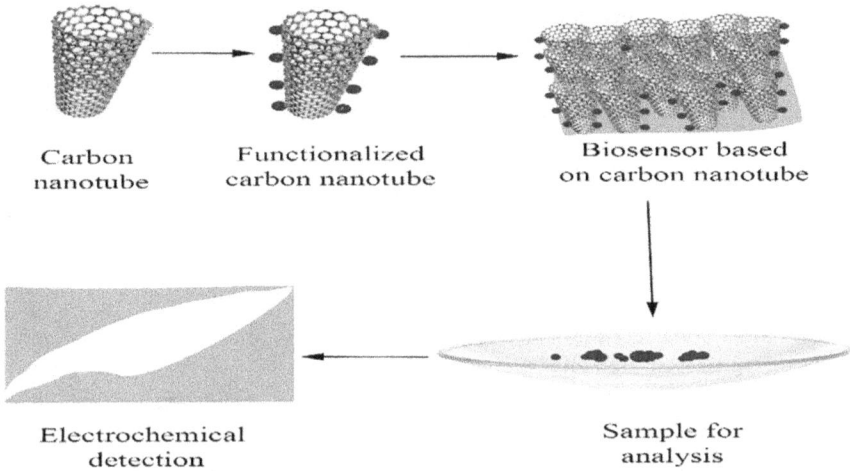

**FIGURE 4.5** Making of biosensor out of carbon nanotubes.

industries [88,89]. Biosensors based on CNTs have been investigated for their ability to detect a wide variety of chemicals, including drugs, glucose, urea, and neurotransmitters, as well as illnesses including dengue, cancer, and avian influenza. The potential of CNTs as drug delivery systems is enhanced by their amazing features, like their drug transporter capacity and how well they penetrate through the cellular membrane. More than 60% of the research into CNTs as drug delivery systems [Figure 4.6] has focused on cancer therapy. CNTs are increasingly being used as composites in the manufacturing of implants. CNTs are increasingly being used in composites for the manufacturing of devices. CNTs are a promising choice for

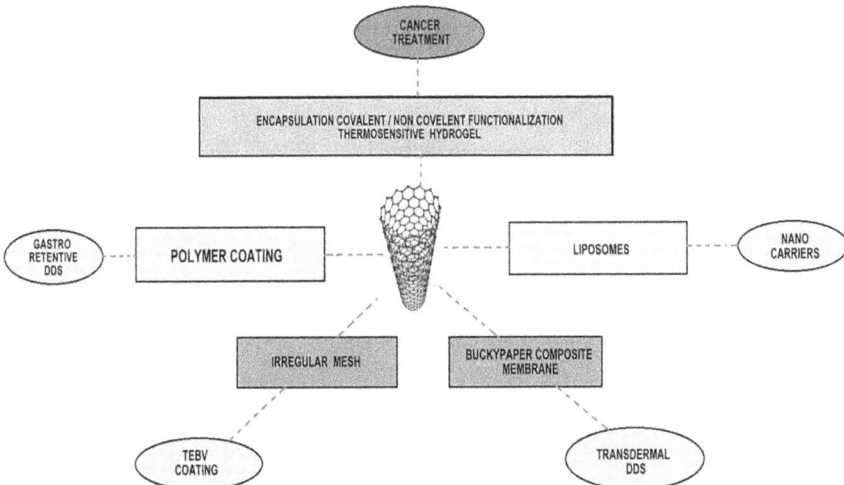

**FIGURE 4.6** Application of CNT as a drug delivery carrier.

enhancing the mechanical characteristics of composites used in the fabrication of dental and orthopedic implants due to their high dimension ratio, porous structure, and outstanding strength and toughness. CNTs have been employed as a filler matrix in polymer and HA composites, as well as a coating agent for titanium alloys [90].

### 4.9.7 DIAMOND-LIKE CARBON AS A SMART BIOMATERIAL

DLC is a kind of amorphous $sp^2$ and $sp^3$ carbon that has hybridized carbon atoms. This material is well known for mechanical and industrial uses, but it is gaining popularity for its novel application in the development and construction of smart and multifunctional textile-based systems by properly functionalizing and modifying the surface of the cellulosic substrates [91,92].

Diamond-like carbon [DLC] is a potential material for applications in medicine because of its outstanding mechanical characteristics, hemocompatibility, corrosion resistance, and biocompatibility. DLC films with a wide range of atomic bond topologies and chemical compositions have been developed in medical and dentistry fields. DLC coating allowed cells to grow without toxicity or inflammatory action. Wear, corrosion, and the accumulation of debris were all reduced by the use of DLC coatings in orthopedic devices. The adhesion of platelets and activation were both decreased by a DLC coating, which contributed to the overall decrease in thrombogenicity. A DLC coating has several characteristics that are beneficial for constructing textiles, i.e., antibacterial, resistant to flame, and superhydrophobic properties [93].

DLC films have great potential as a biomaterial for use in orthopedics because of their toughness, inertness, low friction coefficient, wear and corrosion resistance, and biocompatibility. The DLC coating has a built-in system for cleaning itself, in which a drop of water may roll over to the improved superhydrophobic surface, taking any debris with it. To provide the necessary outstanding performance, the characteristic of flame-retardant properties to cotton fibers while still retaining its natural qualities, DLC thin films with plasma deposition are now suggested as an effective, sustainable, and economical approach. In light of its outstanding qualities of biological compatibility, chemical inertness, mechanical reliability, wear safety, and corrosion resistance, it is clear that this material is cut beyond the rest. Over the last decade, DLC coatings' protective properties throughout a wide range of cardiovascular applications have been actively explored [94,95]. Cells such as fibroblasts, macrophages, and osteoblasts may proliferate healthily on the DLC film surfaces without experiencing inflammation or toxicity. There are two main considerations. First, depending on the application, a DLC coating may display an assortment of atomic bond structures and material characteristics. Therefore, the contradicting findings should be interpreted with caution. DLC coating adherence to biomaterials is a further concern that has to be solved. The application of DLC in various fields is illustrated in Figure 4.7 [96].

## 4.10 APPLICATION OF SMART BIOMATERIAL IN VARIOUS FIELDS

Smart biomaterials are driving the development of "smarter" biomedical equipment. New biomaterials with multi-stimuli-responsive qualities will lead to more

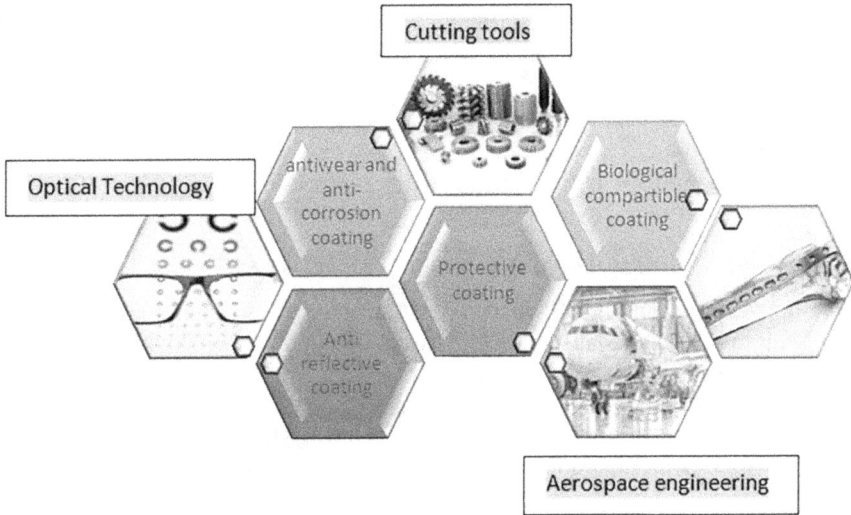

**FIGURE 4.7**   Application of DLC in various fields.

observation and dispersion of treatment and its attendant benefits, as well as more cost-effective healthcare processes.

In addition to that, developments in smarter biomaterials may allow for more accuracy in clinical treatment; these, together with developments in miniaturization and progress toward early diagnosis, are driving the creation of therapy methods that involve just a limited amount of tissue disruption. In addition, improved availability of tissue or organ replacements that have been designed may significantly reduce the need for organ replacement and make it easier to apply current medicines that can assist patients who have irreversible organ failure [97,98].

### 4.10.1   Smart Biomaterial for Tissue Engineering

Smart biomaterials have a significant impact on the tissue engineering field. In the overall framework of cell biology, the latest developments in tissue engineering have been focused on the production of smart materials that respond to external stimuli [99]. Incorporating particular functional groups into biomaterials to provide control over their chemical, physical, and biological characteristics is one strategy for the construction of smart biomaterials for tissue engineering. However, this promotes various functions such as the adhesion of cells, along with growth and migration. In the field of tissue engineering, hydrogels [3D polymeric networks] are hydrophilic in nature, expanded in an aqueous environment, and are utilized as scaffolds [100,101].

These materials may be controlled to resemble the extracellular matrix [ECM] of natural tissue. In addition to that, chemically reactive compounds like dopamine-based groups and also cell-driving groups like collagen, fibronectin, and proteoglycans, have been included in smart biomaterials for gluing the tissue of cells, bonding between them, and concurrently pushing the tissue to repair itself. Wang et al. have developed a tissue adhesive made from chondroitin, which guides more effective tissue restoration. Moreover, the use of 3D printing may facilitate the easy and repeatable manufacturing of biomimetic scaffolds that will enhance tissue integration and production with the host after implantation, which might be useful in the production of smart materials for tissue engineering applications [102–104].

### 4.10.2 SMART BIOMATERIALS FOR DRUG DELIVERY AND MEDICAL DEVICES

Drug delivery systems [DDS] that make utilization of "smart biomaterials" may have their actions modulated by endogenous stimuli like temperature, or physiological factors that fluctuate with illness types and route of dosage. There is an increasing demand for DDS made using smart materials due to the rise of selective medicine and oligonucleotide-based therapies [100,105].

There is an increasing demand for DDS made using smart materials due to the rise of selective medicine, and oligonucleotide-based therapies. By increasing the specificity of the release of drugs from the drug delivery system in complicated diseases, multi-stimulus responsive biomaterials play a vital role in the application for facilitating customized therapy. Nano-DDS are commonly designed using pH-responsive biomaterials like polymers or lipids with functional groups that are ionizable like zwitterions or tertiary amines to assist in the efficient delivery of nucleic acids within the cells [106–108].

One of the biggest problems in drug distribution is getting patients to take their medication as prescribed. One approach might be using DDS with a modular architecture to provide continuous delivery of medicine at therapeutic levels. Designing DDS that need the performance of complicated tasks has benefited greatly from the use of smart biomaterials due to the benefits provided through their morphological and mechanical designs, along with their material qualities [109–111].

Recent developments in the design of biomaterials, the process of functionalization, and 3D printing, have opened the door for smart biomaterials to play a significant role in the development and production of medical devices. Stimuli-responsive materials are integrated into the device's design to create revolutionary applications in biomaterial science. It can be used as a surface coating on smart medical devices or included as a part of the design of the device [112].

### 4.10.3 SMART BIOMATERIALS IN IMMUNE ENGINEERING

The immune system is very important for health and disease management in humans. However, it may also respond negatively to biomaterials, recognizing their initial origin and frequently resulting in adverse immune responses that might hinder the success of any transplant, drug delivery system, or medical device [113,114]. The inability to fully comprehend the complex interactions between immune system components and biomaterials is a significant obstacle. However, with the recent development of intelligent biomaterials, immune systems may be controlled. In recent decades, there has been tremendous growth in the production

translational science and clinical point of view." *Advanced Drug Delivery Reviews* 65, no. 4 (2013): 581–603.

23. Thangudu, Suresh. "Next generation nanomaterials: Smart nanomaterials, significance, and biomedical applications." *Applications of Nanomaterials in Human Health* (2020): 287–312.

24. Hutmacher, Dietmar W. "Regenerative medicine will impact, but not replace, the medical device industry." *Expert Review of Medical Devices* 3, no. 4 (2006): 409–412.

25. Anderson, Daniel G., Jason A. Burdick, and Robert Langer. "Smart biomaterials." *Science* 305, no. 5692 (2004): 1923–1924.

26. Boyan, Barbara D., and Zvi Schwartz. "Are calcium phosphate ceramics' smart'-biomaterials?." *Nature Reviews Rheumatology* 7, no. 1 (2011): 8–9.

27. Yuan, Huipin, Hugo Fernandes, Pamela Habibovic, Jan De Boer, Ana MC Barradas, Ad De Ruiter, William R. Walsh, Clemens A. Van Blitterswijk, and Joost D. De Bruijn. "'Smart'biomaterials and osteoinductivity." *Nature Reviews Rheumatology* 7, no. 4 (2011): 1–2.

28. Rosso, Francesco, Antonio Giordano, Manlio Barbarisi, and Alfonso Barbarisi. "From cell–ECM interactions to tissue engineering." *Journal of Cellular Physiology* 199, no. 2 (2004): 174–180.

29. Swain, Subhasmita, R. D. K. Misra, C. K. You, and Tapash R. Rautray. "$TiO_2$ nanotubes synthesised on Ti-6Al-4V ELI exhibits enhanced osteogenic activity: A potential next-generation material to be used as medical implants." *Materials Technology* 36, no. 7 (2021): 393–399.

30. Rautray, T. R., and K-H. Kim. "Nanoelectrochemical coatings on titanium for bioimplant applications." *Materials Technology* 25, no. 3–4 (2010): 143–148.

31. Kirk, Donald E. *Optimal control theory: an introduction.* Courier Corporation, 2004.

32. Montoya, Carolina, Yu Du, Anthony L. Gianforcaro, Santiago Orrego, Maobin Yang, and Peter I. Lelkes. "On the road to smart biomaterials for bone research: Definitions, concepts, advances, and outlook." *Bone Research* 9, no. 1 (2021): 12.

33. Santin, Matteo, and Gary J. Phillips, eds. *Biomimetic, bioresponsive, and bioactive materials: An introduction to integrating materials with tissues.* John Wiley & Sons, 2012.

34. Morris, Eliza, Michael Chavez, and Cheemeng Tan. "Dynamic biomaterials: toward engineering autonomous feedback." *Current Opinion in Biotechnology* 39 (2016): 97–104.

35. Badeau, Barry A., and Cole A. DeForest. "Programming stimuli-responsive behavior into biomaterials." *Annual Review of Biomedical Engineering* 21 (2019): 241–265.

36. Ahmed, W., Z. Zhai, and C. Gao. "Adaptive antibacterial biomaterial surfaces and their applications." *Materials Today Bio* 2 (2019): 100017.

37. Tibbitt, Mark W., and Robert Langer. "Living biomaterials." *Accounts of Chemical Research* 50, no. 3 (2017): 508–513.

38. Wang, Yingjun. "Bioadaptability: an innovative concept for biomaterials." *Journal of Materials Science & Technology* 32, no. 9 (2016): 801–809.

39. Tibbitt, Mark W., Christopher B. Rodell, Jason A. Burdick, and Kristi S. Anseth. "Progress in material design for biomedical applications." *Proceedings of the National Academy of Sciences* 112, no. 47 (2015): 14444–14451.

40. Khan, Ferdous, and Masaru Tanaka. "Designing smart biomaterials for tissue engineering." *International Journal of Molecular Sciences* 19, no. 1 (2017): 17.

41. Wang, Han, Kexin Su, Leizheng Su, Panpan Liang, Ping Ji, and Chao Wang. "The effect of 3D-printed Ti6Al4V scaffolds with various macropore structures on

osteointegration and osteogenesis: A biomechanical evaluation." *Journal of the Mechanical Behavior of Biomedical Materials* 88 (2018): 488–496.

42. Qasim, Muhammad, Dong Sik Chae, and Nae Yoon Lee. "Advancements and frontiers in nano-based 3D and 4D scaffolds for bone and cartilage tissue engineering." *International Journal of Nanomedicine* 14 (2019): 4333–4351.

43. Mestre, Rafael, Tania Patiño, Xavier Barceló, Shivesh Anand, Ariadna Pérez-Jiménez, and Samuel Sánchez. "Force modulation and adaptability of 3D-bioprinted biological actuators based on skeletal muscle tissue." *Advanced Materials Technologies* 4, no. 2 (2019): 1800631.

44. Metwally, Sara, Sara Ferraris, Silvia Spriano, Zuzanna J. Krysiak, Łukasz Kaniuk, Mateusz M. Marzec, Sung Kyun Kim, et al. "Surface potential and roughness controlled cell adhesion and collagen formation in electrospun PCL fibers for bone regeneration." *Materials & Design* 194 (2020): 108915.

45. Lee, Dong Joon, Jane Kwon, Yong-Il Kim, Xiaoyu Wang, Te-Ju Wu, Yan-Ting Lee, Steven Kim, Patricia Miguez, and Ching-Chang Ko. "Effect of pore size in bone regeneration using polydopamine-laced hydroxyapatite collagen calcium silicate scaffolds fabricated by 3D mould printing technology." *Orthodontics & Craniofacial Research* 22 (2019): 127–133.

46. Pei, Xuan, Liang Ma, Boqing Zhang, Jianxun Sun, Yong Sun, Yujiang Fan, Zhongru Gou, Changchun Zhou, and Xingdong Zhang. "Creating hierarchical porosity hydroxyapatite scaffolds with osteoinduction by three-dimensional printing and microwave sintering." *Biofabrication* 9, no. 4 (2017): 045008.

47. Yoshida, Ryo, Katsumi Uchida, Yuzo Kaneko, Kiyotaka Sakai, Akihiko Kikuchi, Yasuhisa Sakurai, and Teruo Okano. "Comb-type grafted hydrogels with rapid deswelling response to temperature changes." *Nature* 374, no. 6519 (1995): 240–242.

48. Minary-Jolandan, Majid, and Min-Feng Yu. "Nanoscale characterization of isolated individual type I collagen fibrils: polarization and piezoelectricity." *Nanotechnology* 20, no. 8 (2009): 085706.

49. Swain, Subhasmita, Rabindra Nath Padhy, and Tapash Ranjan Rautray. "Polarized piezoelectric bioceramic composites exhibit antibacterial activity." *Materials Chemistry and Physics* 239 (2020): 122002.

50. Swain, Subhasmita, and Tapash Ranjan Rautray. "Assessment of Polarized Piezoelectric SrBi4Ti4O15 Nanoparticles as an alternative antibacterial agent." *bioRxiv* (2021): 2021-01.

51. Li, Jianyu, and David J. Mooney. "Designing hydrogels for controlled drug delivery." *Nature Reviews Materials* 1, no. 12 (2016): 1–17.

52. Zhang, Jieyu, Xian Jiang, Xiang Wen, Qian Xu, Hao Zeng, Yuxing Zhao, Min Liu, Zuyong Wang, Xuefeng Hu, and Yunbing Wang. "Bio-responsive smart polymers and biomedical applications." *Journal of Physics: Materials* 2, no. 3 (2019): 032004.

53. Erol, Ozan, Aishwarya Pantula, Wangqu Liu, and David H. Gracias. "Transformer hydrogels: a review." *Advanced Materials Technologies* 4, no. 4 (2019): 1900043.

54. Mantha, Somasundar, Sangeeth Pillai, Parisa Khayambashi, Akshaya Upadhyay, Yuli Zhang, Owen Tao, Hieu M. Pham, and Simon D. Tran. "Smart hydrogels in tissue engineering and regenerative medicine." *Materials* 12, no. 20 (2019): 3323.

55. Mahinroosta, Mostafa, Zohreh Jomeh Farsangi, Ali Allahverdi, and Zahra Shakoori. "Hydrogels as intelligent materials: A brief review of synthesis, properties and applications." *Materials Today Chemistry* 8 (2018): 42–55.

56. Ooi, H. W., S. Hafeez, C. A. Van Blitterswijk, L. Moroni, and M. B. Baker. "Hydrogels that listen to cells: a review of cell-responsive strategies in biomaterial design for tissue regeneration." *Materials Horizons* 4, no. 6 (2017): 1020–1040.

57. Swain, Subhasmita, Tae Yub Kwon, and Tapash R. Rautray. "Fabrication of silver doped nano hydroxyapatite-carrageenan hydrogels for articular cartilage applications." *bioRxiv* (2021): 2020-12.
58. Rautray, Tapash R., Bijayinee Mohapatra, and Kyo-Han Kim. "Fabrication of strontium–hydroxyapatite scaffolds for biomedical applications." *Advanced Science Letters* 20, no. 3-4 (2014): 879–881.
59. Amukarimi, Shukufe, Seeram Ramakrishna, and Masoud Mozafari. "Smart biomaterials—a proposed definition and overview of the field." *Current Opinion in Biomedical Engineering* 19 (2021): 100311.
60. Lima-Sousa, Rita, Duarte de Melo-Diogo, Cátia G. Alves, Cátia SD Cabral, Sónia P. Miguel, António G. Mendonça, and Ilídio J. Correia. "Injectable in situ forming thermo-responsive graphene based hydrogels for cancer chemo-photothermal therapy and NIR light-enhanced antibacterial applications." *Materials Science and Engineering: C* 117 (2020): 111294.
61. Wheeler, J. C., J. A. Woods, M. J. Cox, R. W. Cantrell, F. H. Watkins, and R. F. Edlich. "Evolution of hydrogel polymers as contact lenses, surface coatings, dressings, and drug delivery systems." *Journal of Long-Term Effects of Medical Implants* 6, no. 3-4 (1996): 207–217.
62. Wichterle, Otto, and Drahoslav Lim. "Hydrophilic gels for biological use." *Nature* 185, no. 4706 (1960): 117–118.
63. Jones, Vanessa, Joseph E. Grey, and Keith G. Harding. "Wound dressings." *BMJ* 332, no. 7544 (2006): 777–780.
64. Hoare, Todd R., and Daniel S. Kohane. "Hydrogels in drug delivery: Progress and challenges." *Polymer* 49, no. 8 (2008): 1993–2007.
65. Caló, Enrica, and Vitaliy V. Khutoryanskiy. "Biomedical applications of hydrogels: A review of patents and commercial products." *European Polymer Journal* 65 (2015): 252–267.
66. Kargozar, Saeid, Francesco Baino, Sepideh Hamzehlou, Michael R. Hamblin, and Masoud Mozafari. "Nanotechnology for angiogenesis: opportunities and challenges." *Chemical Society Reviews* 49, no. 14 (2020): 5008–5057.
67. Lu, Chunliang, and Marek W. Urban. "Stimuli-responsive polymer nano-science: Shape anisotropy, responsiveness, applications." *Progress in Polymer Science* 78 (2018): 24–46.
68. Pham, Son H., Yonghyun Choi, and Jonghoon Choi. "Stimuli-responsive nanomaterials for application in antitumor therapy and drug delivery." *Pharmaceutics* 12, no. 7 (2020): 630.
69. Ebara, Mitsuhiro, Yohei Kotsuchibashi, Ravin Narain, Naokazu Idota, Young-Jin Kim, John M. Hoffman, Koichiro Uto, and Takao Aoyagi. *Smart biomaterials.* Springer, 2014.
70. Abd-El-Aziz, Alaa S., Markus Antonietti, Christopher Barner-Kowollik, Wolfgang H. Binder, Alexander Böker, Cyrille Boyer, Michael R. Buchmeiser, et al. "The next 100 years of polymer science." *Macromolecular Chemistry and Physics* 221, no. 16 (2020): 2000216.
71. Olson, Rebecca A., Angie B. Korpusik, and Brent S. Sumerlin. "Enlightening advances in polymer bioconjugate chemistry: light-based techniques for grafting to and from biomacromolecules." *Chemical Science* 11, no. 20 (2020): 5142–5156.
72. Burridge, Kevin M., Richard C. Page, and Dominik Konkolewicz. "Bioconjugates–From a specialized past to a diverse future." *Polymer* 211 (2020): 123062.
73. Swain, Subhasmita, and Tapash Ranjan Rautray. "Effect of surface roughness on titanium medical implants." *Nanostructured Materials and Their Applications* 1 (2021): 55–80.

74. Benafan, Othmane, Glen S. Bigelow, and Avery W. Young. "Shape memory materials database tool—a compendium of functional data for shape memory materials." *Advanced Engineering Materials* 22, no. 7 (2020): 1901370.

75. Lendlein, Andreas, and Oliver EC Gould. "Reprogrammable recovery and actuation behaviour of shape-memory polymers." *Nature Reviews Materials* 4, no. 2 (2019): 116–133.

76. Fan, Jianfeng, Mengwen Yan, Jiarong Huang, Liming Cao, and Yukun Chen. "Fabrication of smart shape memory fluorosilicon thermoplastic vulcanizates: the effect of interfacial compatibility and tiny crystals." *Industrial & Engineering Chemistry Research* 58, no. 33 (2019): 15199–15208.

77. Lendlein, Andreas, Maria Balk, Natalia A. Tarazona, and Oliver EC Gould. "Bioperspectives for shape-memory polymers as shape programmable, active materials." *Biomacromolecules* 20, no. 10 (2019): 3627–3640.

78. Sabahi, Nasim, Wenliang Chen, Chun-Hui Wang, Jamie J. Kruzic, and Xiaopeng Li. "A review on additive manufacturing of shape-memory materials for biomedical applications." *JOM* 72 (2020): 1229–1253.

79. Bueh, I. W. J. "er, JW Giifi'ich, and RC Wiley." *Journal of Applied Physics* 34 (1963): 1475.

80. Delaey, Jasper, Peter Dubruel, and Sandra Van Vlierberghe. "Shape-memory polymers for biomedical applications." *Advanced Functional Materials* 30, no. 44 (2020): 1909047.

81. Dye, David. "Towards practical actuators." *Nature Materials* 14, no. 8 (2015): 760–761.

82. de Menezes, Beatriz Rossi Canuto, Karla Faquine Rodrigues, Beatriz Carvalho da Silva Fonseca, Renata Guimarães Ribas, Thais Larissa do Amaral Montanheiro, and Gilmar Patrocínio Thim. "Recent advances in the use of carbon nanotubes as smart biomaterials." *Journal of Materials Chemistry B* 7, no. 9 (2019): 1343–1360.

83. Zanganeh, Somayeh, Fatemeh Khodadadei, S. Rafizadeh Tafti, and Mohammad Abdolahad. "Folic acid functionalized vertically aligned carbon nanotube (FA-VACNT) electrodes for cancer sensing applications." *Journal of Materials Science & Technology* 32, no. 7 (2016): 617–625.

84. Zhou, Haifeng, Guoxia Ran, Jean-Francois Masson, Chan Wang, Yuan Zhao, and Qijun Song. "Novel tungsten phosphide embedded nitrogen-doped carbon nanotubes: A portable and renewable monitoring platform for anticancer drug in whole blood." *Biosensors and Bioelectronics* 105 (2018): 226–235.

85. Liu, Zhuang, Joshua T. Robinson, Scott M. Tabakman, Kai Yang, and Hongjie Dai. "Carbon materials for drug delivery & cancer therapy." *Materials Today* 14, no. 7–8 (2011): 316–323.

86. Fu, Yangxi, Victor Romay, Ye Liu, Bergoi Ibarlucea, Larysa Baraban, Vyacheslav Khavrus, Steffen Oswald, et al. "Chemiresistive biosensors based on carbon nanotubes for label-free detection of DNA sequences derived from avian influenza virus H5N1." *Sensors and Actuators B: Chemical* 249 (2017): 691–699.

87. Peng, Huisheng, Qingwen Li, and Tao Chen, eds. *Industrial applications of carbon nanotubes.* William Andrew, 2016.

88. Aslani, Mahshid, M. A. S. O. U. M. E. H. Meskinfam, and H. Aghabozorg. "In situ biomimetic synthesis of gelatin-carbon nanotube-hydroxyapatite biocomposites as bone filler." *Oriental Journal of Chemistry* 33 (2017): 235–241.

89. Afroze, Jannatul D., Md Jaynul Abden, and Md A. Islam. "An efficient method to prepare magnetic hydroxyapatite–functionalized multi-walled carbon nanotubes nano-composite for bone defects." *Materials Science and Engineering: C* 86 (2018): 95–102.

90. Swain, Subhasmita, Chris Bowen, and Tapash Rautray. "Dual response of osteoblast activity and antibacterial properties of polarized strontium substituted hydroxyapatite—

Barium strontium titanate composites with controlled strontium substitution." *Journal of Biomedical Materials Research Part A* 109, no. 10 (2021): 2027–2035.

91. Caschera, Daniela, Roberta Grazia Toro, Barbara Cortese, Fulvio Federici, Domenico Lombardo, and Pietro Calandra. "Diamond-like carbon: a versatile material for developing innovative smart textiles applications. A short review." *Atti della Accademia Peloritana dei Pericolanti - Classe di Scienze Fisiche, Matematiche e Naturali* 97, no. S2 (2019): 27.

92. Praharaj, Rinmayee, Snigdha Mishra, R. D. K. Misra, and Tapash R. Rautray. "Biocompatibility and adhesion response of magnesium-hydroxyapatite/strontium-titania (Mg-HAp/Sr-TiO2) bilayer coating on titanium." *Materials Technology* 37, no. 4 (2022): 230–239.

93. Saikko, Vesa, Tiina Ahlroos, Olof Calonius, and Jaakko Keränen. "Wear simulation of total hip prostheses with polyethylene against CoCr, alumina and diamond-like carbon." *Biomaterials* 22, no. 12 (2001): 1507–1514.

94. Caschera, D., R. G. Toro, F. Federici, C. Riccucci, G. M. Ingo, G. Gigli, and B. Cortese. "Flame retardant properties of plasma pre-treated/diamond-like carbon (DLC) coated cotton fabrics." *Cellulose* 22, no. 4 (2015): 2797–2809.

95. Rautray, T. R., V. Vijayan, and S. Panigrahi. "Synthesis of hydroxyapatite at low temperature." *Indian Journal of Physics* 81 (2007): 95–98.

96. Roy, Ritwik Kumar, and Kwang-Ryeol Lee. "Biomedical applications of diamond-like carbon coatings: A review." *Journal of Biomedical Materials Research Part B: Applied Biomaterials: An Official Journal of The Society for Biomaterials, The Japanese Society for Biomaterials, and The Australian Society for Biomaterials and the Korean Society for Biomaterials* 83, no. 1 (2007): 72–84.

97. Salim, Samar A., Taher A. Salaheldin, Mohamed M. Elmazar, A. F. Abdel-Aziz, and Elbadawy A. Kamoun. "Smart biomaterials for enhancing cancer therapy by overcoming tumor hypoxia: a review." *RSC Advances* 12, no. 52 (2022): 33835–33851.

98. Swain, Subhasmita, Rabindra Nath Padhy, and Tapash Ranjan Rautray. "Electrically stimulated hydroxyapatite–barium titanate composites demonstrate immunocompatibility in vitro." *Journal of the Korean Ceramic Society* 57, no. 5 (2020): 495–502.

99. Furth, Mark E., Anthony Atala, and Mark E. Van Dyke. "Smart biomaterials design for tissue engineering and regenerative medicine." *Biomaterials* 28, no. 34 (2007): 5068–5073.

100. Kowalski, Piotr S., Chandrabali Bhattacharya, Samson Afewerki, and Robert Langer. "Smart biomaterials: recent advances and future directions." *ACS Biomaterials Science & Engineering* 4, no. 11 (2018): 3809–3817.

101. Swain, Subhasmita, Joo L. Ong, Ramaswamy Narayanan, and Tapash R. Rautray. "Ti-9Mn β-type alloy exhibits better osteogenicity than Ti-15Mn alloy in vitro." *Journal of Biomedical Materials Research Part B: Applied Biomaterials* 109, no. 12 (2021): 2154–2161.

102. Place, Elsie S., Nicholas D. Evans, and Molly M. Stevens. "Complexity in biomaterials for tissue engineering." *Nature Materials* 8, no. 6 (2009): 457–470.

103. Bhagat, Vrushali, and Matthew L. Becker. "Degradable adhesives for surgery and tissue engineering." *Biomacromolecules* 18, no. 10 (2017): 3009–3039.

104. Rautray, Tapash R., and Kyo Han Kim. "Synthesis of Mg2+ incorporated hydroxyapatite by ion implantation." In *Key engineering materials*, vol. 529, pp. 114–118. Trans Tech Publications Ltd, 2013.

105. Swain, Subhasmita, and Tapash R. Rautray. "Estimation of trace elements, antioxidants, and antibacterial agents of regularly consumed Indian medicinal plants." *Biological Trace Element Research* 199, no. 3 (2021): 1185–1193.

106. Torchilin, Vladimir P. "Multifunctional, stimuli-sensitive nanoparticulate systems for drug delivery." *Nature Reviews Drug Discovery* 13, no. 11 (2014): 813–827.

107. Praharaj, Rinmayee, Snigdha Mishra, and Tapash R. Rautray. "The structural and bioactive behaviour of strontium-doped titanium dioxide nanorods." *Journal of the Korean Ceramic Society* 57 (2020): 271–280.

108. Mishra, Saswati, and Tapash R. Rautray. "Silver-incorporated hydroxyapatite–albumin microspheres with bactericidal effects." *Journal of the Korean Ceramic Society* 57 (2020): 175–183.

109. Alvarez, Mario M., Joanna Aizenberg, Mostafa Analoui, Anne M. Andrews, Gili Bisker, Edward S. Boyden, Roger D. Kamm, et al. "Emerging trends in micro-and nanoscale technologies in medicine: From basic discoveries to translation." *ACS Nano* 11, no. 6 (2017): 5195–5214.

110. Mishra, Saswati, and Tapash R. Rautray. "Fabrication of Xanthan gum-assisted hydroxyapatite microspheres for bone regeneration." *Materials Technology* 35, no. 6 (2020): 364–371.

111. Mohapatra, Bijayinee, and Tapash R. Rautray. "Facile fabrication of Luffa cylindrica-assisted 3D hydroxyapatite scaffolds." *Bioinspired, Biomimetic and Nanobiomaterials* 10, no. 2 (2021): 37–44.

112. Swain, Subhasmita, Shubha Kumari, Priyabrata Swain, and Tapash Rautray. "Polarised strontium hydroxyapatite–xanthan gum composite exhibits osteogenicity in vitro." *Materials Today: Proceedings* 62 (2022): 6143–6147.

113. Anderson, James M., Analiz Rodriguez, and David T. Chang. "Foreign body reaction to biomaterials." In *Seminars in immunology*, vol. 20, no. 2, pp. 86–100. Academic Press, 2008.

114. Praharaj, Rinmayee, Snigdha Mishra, and Tapash R. Rautray. "Growth mechanism of aligned porous oxide layers on titanium by anodization in electrolyte containing Cl." *Materials Today: Proceedings* 62 (2022): 6216–6220.

115. Wick, Georg, Cecilia Grundtman, Christina Mayerl, Thomas-Florian Wimpissinger, Johann Feichtinger, Bettina Zelger, Roswitha Sgonc, and Dolores Wolfram. "The immunology of fibrosis." *Annual Review of Immunology* 31 (2013): 107–135.

116. Lee, Ki-Won, Cheol-Min Bae, Jae-Young Jung, Gi-Bong Sim, Tapash Ranjan Rautray, Hyo-Jin Lee, Tae-Yub Kwon, and Kyo-Han Kim. "Surface characteristics and biological studies of hydroxyapatite coating by a new method." *Journal of Biomedical Materials Research Part B: Applied Biomaterials* 98, no. 2 (2011): 395–407.

117. Rautray, Tapash R., Subhasmita Swain, and Kyo-Han Kim. "Formation of anodic TiO2 nanotubes under magnetic field." *Advanced Science Letters* 20, no. 3-4 (2014): 801–803.

118. Swain, Subhasmita, Tapash Ranjan Rautray, and Ramaswamy Narayanan. "Sr, Mg, and Co substituted hydroxyapatite coating on TiO2 nanotubes formed by electrochemical methods." *Advanced Science Letters* 22, no. 2 (2016): 482–487.

119. Swain, S., and T. R. Rautray. "Silver doped hydroxyapatite coatings by sacrificial anode deposition under magnetic field." *Journal of Materials Science: Materials in Medicine* 28 (2017): 1–5.

120. Rautray, Tapash R., and Kyo Han Kim. "Synthesis of silver incorporated hydroxyapatite under magnetic field." In *Key engineering materials*, vol. 493, pp. 181–185. Trans Tech Publications Ltd, 2012.

121. Mohapatra, Bijayinee, and Tapash R. Rautray. "Strontium-substituted biphasic calcium phosphate scaffold for orthopedic applications." *Journal of the Korean Ceramic Society* 57 (2020): 392–400.

# 5 Smart Materials for Bioimplant

*Hani Nasser Abdelhamid*

Advanced Multifunctional Materials Laboratory, Department of Chemistry, Faculty of Science, Assiut University, Assiut, Egypt

Nanotechnology Research Centre (NTRC), The British University in Egypt, Cairo, Egypt

## 5.1 INTRODUCTION

Biomaterials have been employed to augment body functions and/or replace damaged tissues [1–5]. Specific interactions with bone cells are driven by a variety of biomaterial qualities and traits, including porosity, roughness, chemistry, surface charge, and mechanical aspects; studying materials with features that can replicate a more realistic and ideal environment for encouraging cell activities and improving bone modeling and remodeling both in vitro and in vivo; to achieve synergistic effects and better performance; multifunctional active biomaterials attempt to combine the effects of numerous active additives that are included inside the same biomaterial. For instance, mixed polymers can reduce the impact of acid-producing bacteria and promote mineral formation at the bonded dentin/restoration interface. A composite made of a matrix of two different materials can increase the strength and resistance to fatigue of dental restorations. Biomaterials advanced several applications including antibacterial agents, regeneration of tissue, drug delivery, antioxidant, and anticancer therapeutics [1,2]. Biomedical applications such as bone repair and regeneration require biomaterials containing certain elements, such as calcium (Ca), strontium (Sr), zinc (Zn), and silica ($SiO_2$). These elements promote and enhance the process of regeneration.

The term *smart biomaterials* was first coined in 2004, describing materials "that respond to specific cellular signals" [6]. The term *intelligent* or *smart* biomaterials refers to a new category of biomaterials that have been produced. These materials are extremely responsive to subtle changes in the settings in which they are placed [6–8]. Smart materials enhanced osseointegration (the process of fusion between the living bone and synthetic implant). They exhibit good properties, such as biodegradability [9,10]. They can indirectly be activating the innate immune system to generate regenerative cues via macrophage polarization from an inflammatory (M1) phenotype to a reparative/regenerative (M2) phenotype. They were applied for several applications, including tissue engineering, drug/gene delivery, and antimicrobial agents [11,12]. Smart biomaterials were performed into different forms such as hydrogels (water-swollen polymeric three-dimensional (3D) materials) [13], aerogels (IUPAC [International Union of Pure and Applied Chemistry] defined it as "aerogel are non-fluid networks composed of interconnected colloidal particles as dispersed phase in a gas (typically air)") [14,15], and thin films [16].

## 5.2 SMART MATERIALS

The phrases "smart biomaterials" and "intelligent biomaterials" should be distinguished from one another to bring clarity to the use of the term *smart biomaterials*. These two phrases are frequently used interchangeably. The terms *smart* and *intelligent* are, however, not synonymous. The capacity to apply previously learned information is what is meant by "smart," while the ability to acquire new knowledge is what is meant by "intelligent" [17]. Self-calibration, self-diagnosis, and self-validation are all examples of intelligent functions that smart biomaterials cannot do. Since the application of artificial intelligence (AI) in medicine is still in its infancy, each idea must be defined precisely so that it can be translated into the biomaterials sector with no confusion. Therefore, it is necessary to define the degree to which a biomaterial may be considered intelligent [18].

There are four levels of smartness for biomaterials (Figure 5.1). Inert, active, responsive, and autonomous are the four increasing degrees of materials' smartness that may be achieved via the use of smart biomaterials. Inert biomaterials provide "merely" biocompatibility and are harmless, meaning that they do not cause any hazardous reactions in or to the body. Therapeutics can be released in a unidirectional and uncontrolled manner from active biomaterials. Responsive

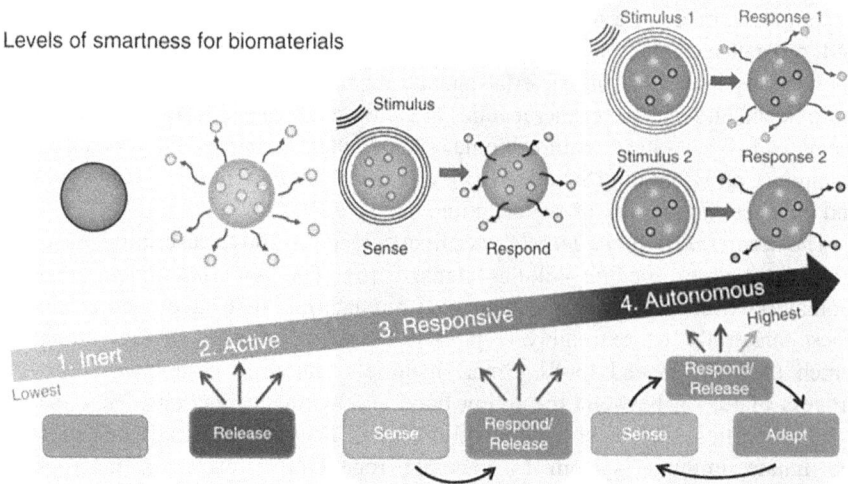

**FIGURE 5.1** The four levels for smart materials. Figure reprinted from the Open Access Ref. [7].

biomaterials can detect certain signals that are produced either by the surrounding environment or by biological processes and subsequently release medicines. To achieve a high level of materials performance, sophisticated and/or alternative kinds of therapies, autonomous biomaterials can recognize a signal, release a specific payload, and adjust their characteristics to changing situations.

Material inertness is the initial and most basic level of intelligence. A biomaterial can only be biocompatible or bioinert; that is, "to not harm," but not to exert any further biological advantages (Figure 5.1). The second level on the ladder of material's smart is activity. It is described as the capability of a biomaterial to release a bioactive therapeutic in a direction that only goes in one direction (i.e., an open loop) (Figure 5.1). The third level of intelligence is responsiveness. The capacity of a biomaterial to detect a stimulus and react or respond to that stimulus by releasing certain therapeutic molecules is the definition of this property [19]. In common parlance, these types of biomaterials are often referred to as stimulus-responsive [1] or bioresponsive [20]. The fourth level of intelligence is called autonomy, and it is currently the level with the highest level of smarts. These futuristic sorts of biomaterials are believed to be "self-sufficient" in that they can independently adapt their characteristics and treatments in response to changes in the surroundings in which they are located and the biological processes that are taking place (Figure 5.1) [21]; in many cases, regarded as "living biomaterials" are autonomous biomaterials (not to be confused with biomaterials that encase living cells or organisms) [22,23].

Smart materials solve the challenges of conventional materials. Because of the signature "burst effect," active biomaterials are prone to uncontrolled leaching or release of the bioactive component, which can reduce the therapy's efficacy and shorten its duration. This is one of the active biomaterials' most significant drawbacks. In the field of dentistry, for instance, fluoride-releasing composites are among the most used antimicrobial treatments [24]. The majority of these systems let fluoride out into the environment via diffusion. Consequently, the antibacterial characteristics are depleted rather fast, in less than two years, which limits their performance over the long run and ultimately leads to a loss of the material's mechanical integrity.

## 5.3  SURFACE MODIFICATION AND COATING

Smart materials and their composites can be prepared via several methods (Figure 5.2). They can be extracted or separated from natural sources. They can also be synthesis via top-down and bottom-up procedures (Figure 5.2). The materials are usually proceed into different forms via several methods, such as 3D printing and electrospinning (Figure 5.2). Polymer-based biomaterials can be prepared via several methods including polymerization, cross-linking copolymerization, polymer-polymer interactions, interpenetrating networks (IPNs), and physically cross-linked reactions. Biopolymer-based hydrogels with more than one biopolymer or hybrid with inorganic materials can be synthesized by one of three methods: blending, in-situ precipitation, or grafting. Blending is the most prevalent synthesis procedure. In the grafting approach, the materials are chemically linked, whereas the blending method relies on

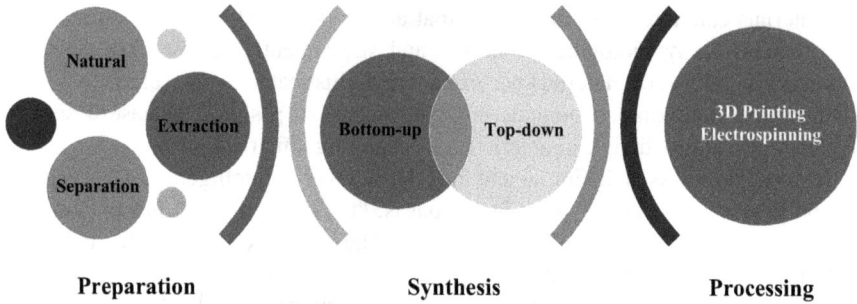

**FIGURE 5.2** Methods for biomaterial preparation, synthesis, and processing.

the physical inclusion of materials in the polymer matrices. Please note that the blending method is also known as the mixing method. Although the in-situ precipitation approach is more cost-effective, it does not perform as well with materials such as biopolymers because some chemical reagents can destroy the produced polymer network during the reaction. However, having control of the synthesis procedure is important to develop effective and smart biomaterials.

The materials after synthesis or preparation may require further modification. Surface modification can be achieved via covalent and non-covalent bonds (Figure 5.3). The modification aims usually to create active sites as antibacterial agents and improve the surface properties such as wettability, porosity, topography, and electrical charge (Figure 5.3). A chemical reaction may be performed on the surface of the materials to induce covalent modifications. For example, the surface modification of peptide-based materials can be achieved via synthetic approaches, e.g., bio-orthogonal chemistry. Current methods have evolved to enable the fabrication and functionalization of biomaterials to facilitate the capture or encapsulation of biologically active components. Cross-linked polymer hydrogels and encapsulation in nanoparticles are the two methods that are utilized the most frequently when it comes to encapsulating drugs and other parametrical agents such as growth factors inside a carrier. Secondary structures, e.g., resilin and elastin, are responsible for the material's resilience, whereas covalent bonds are responsible for the material's stability.

Contemporary fabrication methods, such as three-dimensional (3D) and four-dimensional (4D) bioprinting, or electrospinning are used to obtain products of

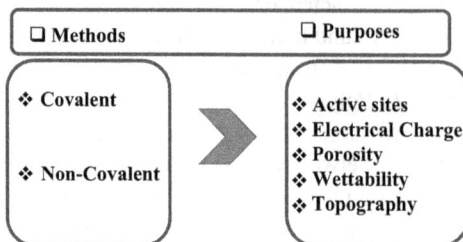

**FIGURE 5.3** Methods and purposes used for surface modification.

customized shapes and sizes with controlled microstructure (complex shapes), distinct nano- or micro-topography, and a high degree of orientation or alignment [25]. These techniques seek to imitate the forms and attributes that are necessary to copy biological tissues in terms of their nano- or microstructures, mechanical properties, chemistry, and charge [26].

The surface coating strategies may also positively influence the other aspects of the smart biomaterials, such as material-cell interactions, elastic modulus, and antibacterial performance. Generally, the surface undergoes changes to improve the antibacterial activity of the materials, increase biocompatibility, induce the growth of apatite, create selective interaction with specific biomolecules, decrease the rate of degradation, and prevent or inhibit the corrosion (Figure 5.4). There are several methods for surface coating. Tungsten was sputtered onto a shape memory alloy made of TiNi using DC magnetron sputtering [27]. On the surface of pure magnesium, Gan *et al.* constructed bioactive Ca-P coatings using the micro-arc oxidation (MAO) technique to achieve a self-sealing structure for the coatings [28]. The layer-by-layer (LBL) electrospun method can be used for surface modification of membranes offering nanofibrous scaffolds [29]. Other methods were reported for surface modification, including plasma processing [30] and argon plasma-induced graft polymerization [31].

Modifications of the biomaterial surface in terms of chemistry, wettability, topography (porosity, roughness, wettability), stiffness, electrical charge, porosity, and leaching of ions, among many others, are common strategies that can be used to augment the activity (or increase the degree of smartness) at the host/biomaterial

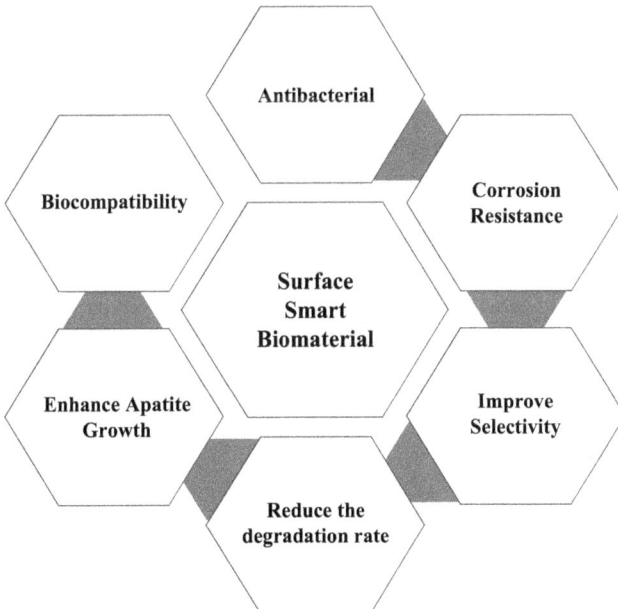

**FIGURE 5.4** Purposes for surface modification of smart biomaterials.

interface [31]. The main target of the surface change is to generate certain chemical and physical conditions that give more favorable biological and environmental reactions. Surface modification is essential for antibacterial biomaterials. Smart antibacterial and anti-biofilm surfaces should be sensitive to the environment in which bacteria are infecting. Being switchable between multiple antibacterial capabilities is important in terms of shape and function that can be achieved via bio-inspiration [32].

The surface modification is usually characterized using surface analysis techniques for surfaces such as X-ray photoelectron spectroscopy (XPS), zeta potential, scanning electron microscopy (SEM), and atomic force microscopy (AFM). The changes in the composition of the surface are usually tracked using XPS analysis. The adsorption process of a biomolecules can be determined using quartz crystal microbalance (QCM). The surface imaging can be recorded using SEM and AFM. The surface charge can be evaluated using zeta potential techniques.

## 5.4  ACTIVE BIOMATERIALS

The functions of bioresponsive biomaterials can be activated by a variety of stimuli coming from either inside the body (also known as "in-body" stimuli) or outside the body (also known as "out-body" stimuli, Figure 5.5). In-body sources, also known as internal sources, are described as signals or stimuli that are discovered within the body in the microenvironments that are close to the biomaterial (Figure 5.5). The terms *out-body* or *external sources* refer to signals or stimuli that are located outside the body and are not in direct touch with the biomaterial (Figure 5.5). Physical, chemical, and biological (or physiological) signals are the three primary classifications that are used to classify both internal and exterior signals. Mechanical stress, surface topography, and surface charge are examples of internal physical stimuli, whereas light, temperature, electrical and magnetic fields are examples of external physical stimuli [33].

The pH levels of the surrounding environment, various ionic variables, and particular compounds like glycoproteins and glucose are examples of chemical stimuli. Enzymes, bioconjugates, antigens, reactive oxygen species (ROS), and

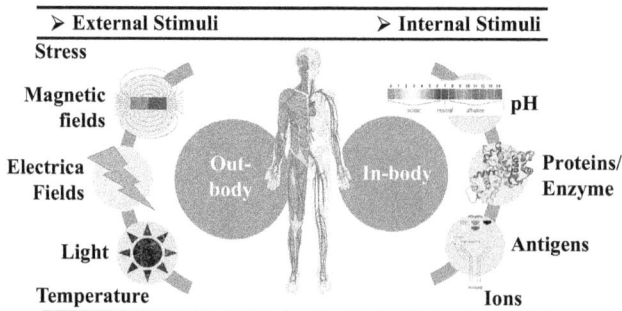

**FIGURE 5.5**  Different types of internal and external stimuli used for biomaterials.

other biochemical agents (such as viruses or bacteria) are examples of biological or physiological stimuli. Biological stimuli can also be referred to as physiological stimuli.

The release of therapies can be triggered by a variety of stimuli. The changes in the characteristics of biomaterials can be induced via several pathways to improve the interaction of biomaterials with cells and biological processes to achieve better therapeutic results. The materials can be classified as follows:

1. **Piezoelectric materials:** there are two types: direct and reverse piezo-electric materials. Direct piezoelectricity refers to the process in which mechanical stress is converted into electrical charges, whereas reverse piezoelectricity refers to the process in which electrical charges are converted into mechanical signals. Both piezoelectric polymers and piezoelectric ceramics might be considered two of the most important examples of piezoelectric biomaterials. Polyvinylidene fluoride (PVDF) is a piezoelectric polymer that is widely used because of its flexibility and biocompatibility. It is one of the most common materials used in bone regeneration engineering [34]. Ceramics such as potassium sodium niobite (KNN) are examples of ceramic-based piezoelectric materials. Piezoelectric films of PVDF may be activated to generate calcium phosphate minerals when subjected to external cyclic loading [35]. When osteoblastic cells (MC3T3-E1) are subjected to an electric field, a protein called TGF may be produced through the calcium/calmodulin pathway. This protein is necessary for cell proliferation, differentiation, and the formation of extracellular matrix (ECM). This ultimately results in temporary inflammation and tissue healing [35].

2. **Magnetic stimulation:** it can be achieved via static magnets or dynamic electromagnetic fields. These biomaterials can enhance the healing of bone fractures and promote bone formation.

3. **Shape memory biomaterials:** biomaterials with shape memory (SMMs), are materials that, when subjected to a certain stimulus, can regain their previous shape after undergoing a large and obvious plastic deformation. Now, the family of SMMs consists of shape-memory hybrids, shape-memory alloys (SMAs), and shape-memory polymers (SMPs); for example, nickel-titanium alloy containing deformable martensite and the "memory" austenite configurations.

4. **pH-/thermo-responsive biomaterials:** they are sensitive to pH and temperature. Physiological stimuli, such as exposure to body temperature at 37°C or pH at 7.4, can cause thermo-responsive hydrogels to undergo a sol-to-gel change. This transformation is induced by the correct physiological stimuli. Recent research has shown that certain biomaterials can adjust their behavior in response to changes in pH that occur inside an inflammatory environment. When it comes to pH-sensitive polymers, the presence of ionizable weak acidic or basic moieties that are connected to a hydrophobic backbone is the most important component of the system [36]. During the process of ionization, the coiled chains respond to the

electrostatic repulsions caused by the newly created charges via substantially extending their length. Temperature-responding polymers have a fine hydrophobic-hydrophilic balance in their structures, and when the temperature is changed slightly around the critical T, the chains of the polymer respond by collapsing or expanding in response to the new adjustments of the hydrophobic and hydrophilic interactions that occur between the polymeric chains and the aqueous media.

5. **Enzyme-responsive biomaterials:** enzymes are essential components in a wide variety of bodily functions, including development, blood coagulation, wound healing, respiration, and digestion. An enzyme activity or expression imbalance can put a person at risk for several severe illnesses, including cancer, cardiovascular problems, inflammation, and degenerative arthritis [37]. Enzyme-responsive materials (ERMs) are materials that may be changed in response to the selective catalytic activity of certain enzymes [38]. Materials that are sensitive to enzymes go through reversible macroscopic transformations that are prompted by the selective catalysis of enzymes [38]. Because of this, every ERM is made up of two different parts: (i) an enzyme-sensitive component, which is a substrate or substrate mimic based on a biomolecule (such as a peptide, a lipid, or a polynucleotide); and (ii) a component that directs and controls changes in non-covalent interactions that cause macroscopic transitions. Further information regarding these materials can be found in review [38].

## 5.5   SMART BIOMATERIAL TYPES

Based on the types of materials, smart biomaterials can be classified into several classes. The following sections discuss some of these types, showing examples for each category.

### 5.5.1   NATURAL-DERIVED BIOMATERIALS

There are several natural-derived biomaterials, including proteins [39], cellulose [40], chitosan [41], and hydroxyapatite [42]. They can be categorized into organic and inorganic-based materials. Natural-derived polymers are widely used including structural proteins (e.g., collagen, laminin, elastin, and fibronectin), and biopolymers (e.g., cellulose, hyaluronic acid, chitosan, and alginate). Most of these biomaterials exhibit high biocompatibility, biodegradability, and ease to proceed into different forms and custom shapes. Readers can find more details of these materials in reviews such as references [43,44]. Polysaccharides, in contrast to synthetic polymers, have several drawbacks, such as a low level of resistance to mechanical stress, which restricts their use in biological applications. As a result of this, polysaccharides are frequently functionalized with other types of polymers or inorganic nanoparticles to get around the constraints of the material.

Cellulose is widely used as a biomaterial [40]. Bacterial cellulose was reviewed as a smart biomaterial [45]. Cellulose is widely acknowledged as the natural biopolymer that may be found in the greatest quantity over the entire world. With the main productivity of plant biomass in the range of 100–125 Gt per year, plant-derived cellulose has attracted great interest as a feedstock for biobased production. Cellulose was used for several applications, such as environmental-based technologies, biomedicine, and energy [46,47].

Because of their ability to stimulate osteoblast proliferation and the production of osteogenic-related proteins and genes, hydroxyapatite materials have a high level of biocompatibility. The influence of the forms of hydroxyapatite (HA) microparticles on the behaviors of cells was investigated by Yang *et al.* [42]. The spherical form of hydroxyapatite performed much better than the rod-shaped kind. Liu *et al.* used so-gel based on water to deposit a thin layer of hydroxyapatite coating onto sandblasted 316 L stainless steel substrates [48]. The material after modification displayed a high level of bioactivity. A coating of hydroxyapatite (HA) was placed on porous titanium before subjecting it to an alkali heat treatment [49]. The authors observed that osteoblasts were able to develop across the different titanium fibers on the modified surface. HA-modified materials had high biological compatibility, enabling the combination of bone tissue with porous titanium. Li *et al.* evaluated the effect of osteoblast activity on porous titanium with a hydroxyapatite (HA)/collagen (COL) coating. Their findings revealed that the addition of collagen similar to that found in humans increased the biocompatibility of titanium implants [49].

He and his colleagues created a unique polyethylene film by coupling collagen to the surface of the film after it had been aminolyzed [50]. This provided the film with a different performance in terms of cell growth and proliferation. Liu *et al.* grafted tobramycin, an aminoglycoside antibiotic, onto the collagen film via a cross-linked procedure using 1-ethyl-3-(3-dimethyl aminopropyl) carbodiimide and N-hydroxysuccinimide [51]. This was done to inhibit the growth of bacteria for applications such as corneal repair. Their subsequent cell investigations revealed that the treated film demonstrated enhanced antibacterial activity in vitro, as well as good adhesion and proliferation capabilities for human corneal epithelial cells. Within the first three months, there is no indication of a corneal rejection response, neovascularization, or keratoconus. The preparation of this film, which can be done in a big quantity and at a reasonable cost, should allow for its use in corneal restoration at some point [51].

## 5.5.2 Metallic-Based Biomaterials

Smart metallic biomaterials should have a low elastic modulus, favorable corrosion resistance, and high biocompatibility. Most of the reported materials exhibit the potential for a variety of clinical applications. Stainless steels (iron alloy) and alloys of titanium and magnesium are examples of the types of implant structural materials. These bioinert materials are generally enclosed by fibrous tissue once they have been implanted during *in vivo* investigation. This tissue serves to isolate the bioinert materials from their surroundings and inhibits the bioinert materials from adhering to bodily tissues. Additionally, the presence of such materials may cause clotting of the blood as well as an infection caused by bacteria. On the surfaces of these implants, several functional coating layers are applied to overcome the difficulties described above.

Gold surface was modified via self-assembling thiols that ended with methyl, carboxyl, hydroxyl, and amino group [52]. The materials were investigated for base fibroblast growth factor (bFGF) adsorption and cell responsiveness. The findings of the XPS experiment indicate that the chemical components found on the surface of the self-assembled structure are comparable to those found in the theoretical calculations. The findings of the tests conducted using an AFM indicate that the assembly of molecules on a surface does not generate a substantial difference between the surface's roughness before and after the assembly of molecules. The hydrophilicity of the end groups of the self-assembled monolayer contributes to an increase in the water surface contact angle. According to the surface zeta potential, the COOH group has the greatest electronegativity compared to other surfaces. The substrates that were produced are modified so that they can absorb bFGF. The findings of the tests performed with a QCM indicate that the levels of bFGF adsorbing onto the various self-assembled surfaces are as follows: carboxylic (–COOH)> hydroxyl (–OH)> methyl (–CH$_3$)> amino (–NH$_2$). Experiments conducted using cell cultures have shown that attaching endothelial cells to various surfaces results in distinct changes to their morphologies. Endothelial cells proliferate at the fastest rate on the self-assembled surface with –COOH, whereas they proliferate at the slowest rate on the self-assembled surface with –CH$_3$ [52].

A bioactive porous structure coating constituted of sodium titanate (Na$_2$Ti$_5$O$_{11}$) and titanium oxide (rutile type of TiO$_2$) with redundant Ca ion was generated on porous titanium surfaces with the application of alkali heat treatment [53]. The authors reported that the hydroxyl functionalized surface with redundant Ca ions allowed bonelike apatite to nucleate and grow smoothly. This was because there were redundant Ca ions on the surface.

Cu-rich precipitations were coated on stainless steel to increase the material's capacity to inhibit the growth of bacteria [54]. Stainless steel was treated with a water-soluble polymethacrylate that contained an oxidized 3,4-dihydroxyphenylalanine group [55]. Magnesium (Mg) based alloys are comparable to natural bones in terms of their density and mechanical characteristics. They are suitable candidates for use as biodegradable materials. These alloys have greater strength than polymers and better toughness than ceramic. However, magnesium degrades too quickly *in vivo*, even though it is biodegradable and non-toxic. This can lead to the accumulation of potentially dangerous chemicals and can also cause magnesium's effectiveness to be diminished. Varieties of materials are coated on the surfaces of magnesium-based metals to slow down the rate at which they deteriorate. On top of pure magnesium, Lu *et al.* created a layer consisting of a combination of CaSiO$_3$, MgSiO$_3$, and Mg(OH)$_2$ via hydrothermal treatment [56]. The electrochemical test of Mg alloy showed that the corrosion current density after coating was decreased by about two orders of magnitude compared with that of the bare magnesium. The electrochemical impedance spectroscopy (EIS) measurement also showed that the corrosion-resistant performance of the coated magnesium was significantly enhanced. The coating resulted in an improvement in the material's resistance to corrosion. Additionally, magnesium was also coated with albumin in the presence of various simulated body fluids (SBF, NaCl, PBS) [57]. The authors observed that albumin is a corrosion inhibitor [57].

Metallic nickel-titanium (NiTi), also called Nitinol, biomaterials are used in a wide variety of therapeutic applications, ranging from stents for the heart to implants for the orthopedic system (Figure 5.6) [58,59]. The use of NiTi smart biomaterials has been getting more and more attention due to the remarkable properties that these materials possess, such as a low elastic modulus, the ability to remember previous shapes, and an acceptable level of biocompatibility [58]. It is important to point out that the apatite that is generated over the coated NiTi is often a HAp that is low in Ca because the $Na^+$, $K^+$, and $Mg^{2+}$ ions that are already present in the simulated body fluid (SBF) can substitute $Ca^{2+}$ in the HAp crystal lattice.

The bare metallic biomaterials provide a suitable platform for various eukaryotic cells to attach to its surface, stay alive, grow, and proliferate; however, a better biological performance, in which a higher number of the living cells with more

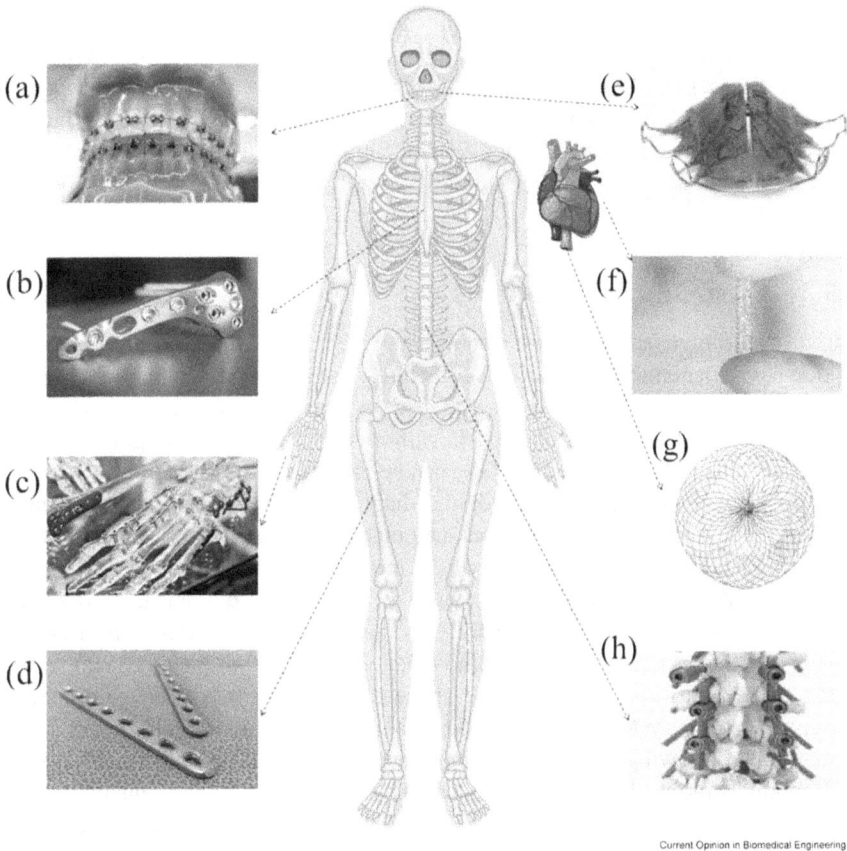

Current Opinion in Biomedical Engineering

**FIGURE 5.6** Biomedical applications of NiTi smart biomaterials: (a) orthodontic wire and brackets, (b) thorax implant, (c) orthopedic superelastic staples and screws, (d) shape memory bone plate, (e) dental palate expander, (f) cardiovascular stent, (g) atrial septal occlusion device, and (h) spacer in the backbone. Figure reprinted with permission from Ref. [58].

elongated morphology and higher proliferation can be obtained when its surface modifies with biocompatible coatings, in particular CaP family materials. The more biocompatible chemical nature of these materials decreased Ni ion release, and enhanced hydrophilicity of the surface with CaP deposition are three potential mechanisms governing such an enhancement. The importance of controlling operating factors and exploiting novel deposition strategies, such production of composite or layered coatings, in further improvement of these properties is stressed.

Metallic biomaterials face several challenges. Some reports stated that NiTi lacks an appropriate *in vitro* bioactivity with no apatite formation on the material's surface during the SBF test (this test depends on soaking the material into a solution containing metals simulated biological environment) [60]. The leaching of Ni ions from the surface of NiTi, the need for decreased elastic modulus, and the desire for improved biological properties, including better material-cell interactions, bio-mineralization, and antibacterial activity, have provided the driving force for a wide variety of surface-modification techniques to address these problems before using NiTi for *in vivo* studies [58]. Dry coating and wet coating procedures have both been used to deposit layers of biocompatible and bioactive materials on the top of NiTi smart biomaterials. The materials used for coating depend on the application that is being targeted. Research and development operations have shown promise, but a significant amount of progress must be made before coated NiTi may be used in clinical settings [58].

### 5.5.3 CARBON-BASED BIOMATERIALS

Carbon-based materials were reported as smart biomaterials [15,61]. Because of its unique physicochemical and biological features, carbon materials are proving to be incredibly fascinating biomaterials for use in the medical field. Carbon-based materials (e.g., carbon dots [C-dots], carbon nanotube [CNTs], graphene [G], graphene oxide [GO], and reduced GO) can be integrated into an increasing number of medical devices that are used for the diagnosis, prevention, and treatment of illnesses. The applications of carbon nanomaterials as biosensors, drug delivery agents, and fillers for composites used in implant production represent the main focus of research investigations. Because of their high biological activity and biocompatibility, as well as their ability to rapidly transfer electrons to a diverse variety of electroactive species, carbon nanomaterials can interact with a vast number of different proteins. Because of these features, it is possible to produce a variety of hybrid systems that have the potential to function as biosensors in the human body, allowing for the detection of viruses, cancer, and other substances.

Reported studies have been conducted on carbon-based biosensors for the detection of illnesses including cancer. These biosensors have also been researched for the detection of drugs, neurotransmitters, glucose, and urea, among other biomarkers that are found in the human body. Other noteworthy qualities of carbon-based nanomaterials, such as their capability to carry drugs and their ability to pass through the membranes of cells, strengthen their potential for application as drug delivery systems. There is an increase in the use of carbon nanomaterials in composites for the manufacturing of implants. Because of their tunable aspect ratio, various structures, outstanding strength and stiffness, and low density, carbon nanomaterials are a good alternative for improving the mechanical characteristics of composites that are used in the fabrication of orthopedic and dental implants.

## 5.6 SMART METAL OXIDE/CHALCOGENIDES-BASED BIOMATERIALS

Metal oxides/chalcogenides of transition metals, e.g., ZnO [62], and $MoS_2$ [63], are used as biomaterials. Among these metal oxides, magnetic nanoparticles are widely used for the synthesis of smart biomaterials [64]. Because of their capacity to be manipulated by an external magnetic field (EMF), magnetically sensitive hydrogels were the subject of a significant amount of published research. Magnetic-based hydrogels can be prepared by combining inorganic (e.g., iron oxide) and organic (e.g., polysaccharides) ingredients during the production process. It is an aim in the synthesis of smart polysaccharide hydrogels to tailor the surface of iron oxide nanoparticles (e.g., $Fe_3O_4$, $Fe_2O_3$, $CoFe_2O_4$). This is due to the introduction of intrinsic functions, such as magnetic behavior, and the enhancement of mechanical properties.

Metal oxide-based biomaterials, e.g., magnetic biomaterials are magnetic response-sensitive materials. Thus, they were applied for biodevices for targeted drug administration, tissue engineering, and hyperthermia therapy. These materials can be used for hyperthermia therapy, tissue engineering, and targeted drug delivery. Furthermore, the properties of magnetic-based biomaterials can be altered by an electromagnetic field. Thus, they may be effective as theragnostic agents, making it possible for both functions, diagnosis and treatment, at the same time.

It is important to consider the magnetic properties such as size and shape during the preparation of the materials. Magnetic particles must be tiny enough to quickly penetrate the lesion region and retain their superparamagnetic behavior. In light of this, the procedure for the synthesis of magnetic materials has to be rethought to regulate the size and morphology of magnetic-based biomaterials, as well as to prevent the aggregation of these particles and their movement through hydrogels. Magnetic-based biomaterials showed a significant advancement in the use of nanotechnology with high mechanical resistance and stability in physiological fluids. Several methods can be used to synthesize magnetic-based biomaterials with desirable properties [65].

## 5.7 SMART ORGANIC-BASED BIOMATERIALS

Polymers are reported under different names such as "smart," "stimuli-responsive," or "environmentally sensitive" [66]. Kim and Matsunaga reviewed thermo-responsive polymers and the potential applications of these polymers as smart biomaterials [67]. The thermo-responsive polymers are the most common type of smart polymer. They provided defining characteristics and properties such as a phase or volume transition that can be reversibly changeable in reaction to a shift in temperature. Temperature, pH, ion concentration, electric and magnetic fields, as well as light, are all examples of stimuli that have the potential to influence the characteristics of smart polymers that can be changed. These sensations have the potential to operate as triggers, which can be activated from the outside or remotely. Conductive polymers show high electroactive properties enabling several applications such as tissue engineering [68]. With the use of these trigger systems, it is possible to build smart polymers for biomimetic structures that have smart capabilities. Polymers can be prepared as injectable hydrogels [69].

Polymeric smart thermo-responsive materials have attracted a lot of attention among these materials due to the relative ease with which they can be handled, prepared, and modified to fulfill the requirement for desired applications. They provide different types of thermo-responsive polymers with various combinations of smart monomers and offer a new class of bio-relevant applications. These thermo-responsive polymers can take advantage of the benefits offered by thermo-responsive polymers. They have properties, such as high mechanical strength, biocompatibility, and biodegradability that can be adjusted and controlled. Polymeric smart materials can be readily transformed in a stable manner using straightforward chemical reactions and physical interactions. Combinations of smart monomers, functional monomers, and moieties can also be used to make unrestricted alterations to the intelligent properties of the material.

Organic polymers are good materials for in-body stimuli. When pH-sensitive polymeric chains are cross-linked to generate hydrogels, the behavior of the resulting hydrogels is affected not only by the characteristics of the ionizable groups, the polymer composition, and the hydrophobicity of the polymer backbone, but also by the density of the cross-linking. This affects the solute permeability in numerous applications in terms of the release of bioactive chemicals. The large molecular-weight solutes provided greater cross-linking density, causing the lower permeability.

Wang et al. used calcium phosphate dehydrate (DCPD) and polymers such as polycaprolactone (PCL) as coating materials for a magnesium-zinc alloy [70]. The coating was successful in protecting the magnesium-zinc alloy. When compared to the corrosion potential ($E_{corr}$) of the DCPD-coated alloy, the corrosion potential ($E_{corr}$) of the DCPD-PCL coated alloy was found to be 0.14 V higher than that of the DCPD-coated alloy. The corrosion current ($i_{corr}$) was approximately 1/3 of what it was for the DCPD-coated alloy. The DCPD-PCL-coated alloy had a real impedance ($Z_{re}$) that was roughly four times as big as that of the DCPD-coated alloy. The results of the immersion test suggested that the DCPD-PCL composite coating maintained its integrity in the SBF after 260 hours of immersion. When compared to the DCPD-coated alloy, the lower amount of hydrogen that was produced was indicative of a decreased rate of deterioration. The deterioration rate quickly rose after 150 h for the DCPD-coated alloy; however, it did not do so until 225 h for the DCPD-PCL coated case. This indicated that the composite coating could delay corrosion for a longer period. As a result, the composite coating can provide a higher level of protection for the magnesium alloy [70].

Dopamine polymerization was given as a great coating material that may be suited to a wide variety of materials [71]. The surface of PCL/gelatin nanofibrous membranes was functionalized with a polydopamine coating [29]. This served to increase both the adhesion force and the cell affinity of the membranes leading to high adipose stem cells' adhesion and osteogenic differentiation [29]. Xu *et al.* reported surface modification of biodegradable Fe plates via heparin grafting followed by polydopamine active layers [72]. As a consequence of the immobilization of heparin, the materials exhibited better anti-thrombus capabilities and the capacity to suppress the proliferation of vascular smooth muscle cells (VSMC) [72].

Taking advantage of many of the benefits offered by thermo-responsive polymers. They show the properties of "on–off" reversible switching and their "on-demand" controllable and repeatable properties in response to changes in temperature. Thus, biomaterials and biomedical applications with biocompatible smart fibers, surfaces, and hydrogels were discussed in this paper, with some examples provided. In addition to the aforementioned instances, ongoing research and development are resulting in the creation of smart thermo-responsive polymers.

Organic-based smart biomaterials exhibit several advantages. Because of this capability, it is possible to perform one-of-a-kind manipulations of smart polymers via an "on-demand" remote control as well as an "on–off" switchable control that is temperature dependent. Biodegradable polymers can solve some of the challenges of conventional polymers [73]. Natural-derived polymers should be further investigated. Synthetic polymers should be avoided to save the environment.

## 5.8  CHALLENGES OF SMART BIOMATERIALS

Smart biomaterials have advanced several applications. However, there are several challenges that should be solved. Smart biomaterials containing metal such as Ni suffer from ion leaching and poor biomineralization. These two challenges need to be addressed by applying surface-modification strategies. There are several strategies for ongoing research and development in smart biomaterial surface and their applications. They can include the use of cutting-edge surface engineering techniques, such as composite and multilayer coatings, which make use of biocompatible particles and components. The optimization of the operating factors should be involved in the employed coating technology. The utilization of one-pot procedure may offer advantages such as low-cost procedures and advantageous pre-treatments and/or the utilization of low-cost surface-modification processes.

Materials' biodegradability and inertness are important aspects of the design of smart materials. At one time, it was believed that inert materials, such as steels, carbon compounds, silicones, and poly(methyl methacrylate), were biocompatible materials. Despite this, the vast majority of them are constrained in some way by the non-degradation features they possess.

## 5.9  CONCLUSIONS

Active biomaterials have been used in tissue engineering, offering enhanced cell adhesion, proliferation, and differentiation. They can also indirectly activate the innate immune system. The surface features of biomaterials, e.g., chemical composition, coating, physical/chemical interactions on the surfaces, and size (micro/nano) of the particles in the surface play a critical role in the material's performance. They are the main factor in the enhancement or induction of tissue regeneration. The promise of smart polymers may be brought to fruition for the benefit of humankind if new, smart polymers and new techniques of molecular synthesis, analysis, and engineering are developed. This will allow for the creation of smart polymers that are unrestricted and will allow the promise of smart polymers to be achieved.

## REFERENCES

1. Ahmed W, Zhai Z, Gao C. Adaptive antibacterial biomaterial surfaces and their applications. Mater Today Bio 2019;2:100017. 10.1016/j.mtbio.2019.100017.
2. Khan HM, Liao X, Sheikh BA, Wang Y, Su Z, Guo C, et al. Smart biomaterials and their potential applications in tissue engineering. J Mater Chem B 2022;10:6859–6895. 10.1039/D2TB01106A.
3. Williams DF. On the nature of biomaterials. Biomaterials 2009;30:5897–5909. 10.1016/j.biomaterials.2009.07.027.
4. Campoccia D, Montanaro L, Arciola CR. A review of the biomaterials technologies for infection-resistant surfaces. Biomaterials 2013;34:8533–8554. 10.1016/j.biomaterials.2013.07.089.
5. Intravaia JT, Graham T, Kim HS, Nanda HS, Kumbar SG, Nukavarapu SP. Smart orthopedic biomaterials and implants. Curr Opin Biomed Eng 2023;25:100439. 10.1016/j.cobme.2022.100439.
6. Anderson DG, Burdick JA, Langer R. Smart biomaterials. Science (80-) 2004; 305:1923–1924. 10.1126/science.1099987.
7. Montoya C, Du Y, Gianforcaro AL, Orrego S, Yang M, Lelkes PI. On the road to smart biomaterials for bone research: definitions, concepts, advances, and outlook. Bone Res 2021;9:12. 10.1038/s41413-020-00131-z.
8. Silva-López MS, Alcántara-Quintana LE. The era of biomaterials: smart implants? ACS Appl Bio Mater 2023. 10.1021/acsabm.3c00284.
9. Ju X-J, Xie R, Yang L, Chu L-Y. Biodegradable 'intelligent' materials in response to physical stimuli for biomedical applications. Expert Opin Ther Pat 2009;19:493–507. 10.1517/13543770902771282.
10. Delaviz Y, Finer Y, Santerre JP. Biodegradation of resin composites and adhesives by oral bacteria and saliva: A rationale for new material designs that consider the clinical environment and treatment challenges. Dent Mater 2014;30:16–32. 10.1016/j.dental.2013.08.201.
11. Kowalski PS, Bhattacharya C, Afewerki S, Langer R. Smart biomaterials: recent advances and future directions. ACS Biomater Sci Eng 2018;4:3809–3817. 10.1021/acsbiomaterials.8b00889.
12. Liu K, Tian Y, Jiang L. Bio-inspired superoleophobic and smart materials: Design, fabrication, and application. Prog Mater Sci 2013;58:503–564. 10.1016/j.pmatsci.2012.11.001.
13. Kopeček J, Yang J. Hydrogels as smart biomaterials. Polym Int 2007;56:1078–1098. 10.1002/pi.2253.
14. Karamikamkar S, Yalcintas EP, Haghniaz R, de Barros NR, Mecwan M, Nasiri R, et al. Aerogel-based biomaterials for biomedical applications: From fabrication methods to disease-targeting applications. Adv Sci 2023. 10.1002/advs.202204681.
15. Dhua S, M J PD, Mishra P. A comprehensive review on multifunctional smart carbon dots (C dots) based aerogel. Food Chem Adv 2023;3:100341. 10.1016/j.focha.2023.100341.

16. Balint R, Cassidy NJ, Cartmell SH. Conductive polymers: Towards a smart biomaterial for tissue engineering. Acta Biomater 2014;10:2341–2353. 10.1016/j.actbio.2014.02.015.

17. Yurish S. Sensors: Smart vs. Intelligent. Sensors & Transducers 2010;114:I–VI.

18. Smart biomaterials, smarter medicine? EBioMedicine 2017;16:1–2. 10.1016/j.ebiom.2017.02.001.

19. Morris E, Chavez M, Tan C. Dynamic biomaterials: toward engineering autonomous feedback. Curr Opin Biotechnol 2016;39:97–104. 10.1016/j.copbio.2016.02.032.

20. Tibbitt MW, Rodell CB, Burdick JA, Anseth KS. Progress in material design for biomedical applications. Proc Natl Acad Sci 2015;112:14444–14451. 10.1073/pnas.1516247112.

21. Tibbitt MW, Langer R. Living biomaterials. Acc Chem Res 2017;50:508–513. 10.1021/acs.accounts.6b00499.

22. Wang Y. Bioadaptability: An innovative concept for biomaterials. J Mater Sci Technol 2016;32:801–809. 10.1016/j.jmst.2016.08.002.

23. Sankaran S, Zhao S, Muth C, Paez J, del Campo A. Toward light-regulated living biomaterials. Adv Sci 2018;5:1800383. 10.1002/advs.201800383.

24. Wiegand A, Buchalla W, Attin T. Review on fluoride-releasing restorative materials—Fluoride release and uptake characteristics, antibacterial activity and influence on caries formation. Dent Mater 2007;23:343–362. 10.1016/j.dental.2006.01.022.

25. de León EH-P, Valle-Pérez AU, Khan ZN, Hauser CAE. Intelligent and smart biomaterials for sustainable 3D printing applications. Curr Opin Biomed Eng 2023;26:100450. 10.1016/j.cobme.2023.100450.

26. Morley CD, Ellison ST, Bhattacharjee T, O'Bryan CS, Zhang Y, Smith KF, et al. Quantitative characterization of 3D bioprinted structural elements under cell generated forces. Nat Commun 2019;10:3029. 10.1038/s41467-019-10919-1.

27. Li H, Zheng Y, Pei YT, De Hosson JTM. TiNi shape memory alloy coated with tungsten: a novel approach for biomedical applications. J Mater Sci Mater Med 2014;25:1249–1255. 10.1007/s10856-014-5158-8.

28. Gan J, Tan L, Yang K, Hu Z, Zhang Q, Fan X, et al. Bioactive Ca–P coating with self-sealing structure on pure magnesium. J Mater Sci Mater Med 2013;24:889–901. 10.1007/s10856-013-4850-4.

29. Ge L, Li Q, Huang Y, Yang S, Ouyang J, Bu S, et al. Polydopamine-coated paper-stack nanofibrous membranes enhancing adipose stem cells' adhesion and osteogenic differentiation. J Mater Chem B 2014;2:6917–6923. 10.1039/C4TB00570H.

30. Karthik C, Rajalakshmi S, Thomas S, Thomas V. Intelligent polymeric biomaterials surface driven by plasma processing. Curr Opin Biomed Eng 2023;26:100440. 10.1016/j.cobme.2022.100440.

31. Affrossman S, Scott A, O'Neill A, Stamm M. Topography and surface composition of thin films of blends of polystyrene with brominated polystyrenes: effects of varying the degree of bromination and annealing. Macromolecules 1998;31:6280–6288. 10.1021/ma971676q.

32. Li X, Wu B, Chen H, Nan K, Jin Y, Sun L, et al. Recent developments in smart antibacterial surfaces to inhibit biofilm formation and bacterial infections. J Mater Chem B 2018;6:4274–4292. 10.1039/C8TB01245H.

33. Lee HP, Gaharwar AK. Light-responsive inorganic biomaterials for biomedical applications. Adv Sci 2020;7:2000863. 10.1002/advs.202000863.

34. Fukada E. History and recent progress in piezoelectric polymers. IEEE Trans Ultrason Ferroelectr Freq Control 2000;47:1277–1290. 10.1109/58.883516.

35. Jacob J, More N, Kalia K, Kapusetti G. Piezoelectric smart biomaterials for bone and cartilage tissue engineering. Inflamm Regen 2018;38:2. 10.1186/s41232-018-0059-8.

36. Jiang S, Chen Y, Duan G, Mei C, Greiner A, Agarwal S. Electrospun nanofiber reinforced composites: a review. Polym Chem 2018;9:2685–2720. 10.1039/C8PY00378E.
37. Hu Q, Katti PS, Gu Z. Enzyme-responsive nanomaterials for controlled drug delivery. Nanoscale 2014;6:12273–12286. 10.1039/C4NR04249B.
38. Ulijn R V. Enzyme-responsive materials: a new class of smart biomaterials. J Mater Chem 2006;16:2217. 10.1039/b601776m.
39. Kopeček J. Hydrogel biomaterials: A smart future? Biomaterials 2007;28:5185–5192. 10.1016/j.biomaterials.2007.07.044.
40. Bonetti L, De Nardo L, Farè S. Thermo-responsive Methylcellulose Hydrogels: from design to applications as smart biomaterials. Tissue Eng Part B Rev 2021;27:486–513. 10.1089/ten.teb.2020.0202.
41. Budiarso IJ, Rini NDW, Tsalsabila A, Birowosuto MD, Wibowo A. Chitosan-based smart biomaterials for biomedical applications: progress and perspectives. ACS Biomater Sci Eng 2023;9:3084–3115. 10.1021/acsbiomaterials.3c00216.
42. Yang H, Zeng H, Hao L, Zhao N, Du C, Liao H, et al. Effects of hydroxyapatite microparticle morphology on bone mesenchymal stem cell behavior. J Mater Chem B 2014;2:4703–4710. 10.1039/C4TB00424H.
43. Yang Q, Peng J, Xiao H, Xu X, Qian Z. Polysaccharide hydrogels: Functionalization, construction and served as scaffold for tissue engineering. Carbohydr Polym 2022;278:118952. 10.1016/j.carbpol.2021.118952.
44. Yang Y, Xu L, Wang J, Meng Q, Zhong S, Gao Y, et al. Recent advances in polysaccharide-based self-healing hydrogels for biomedical applications. Carbohydr Polym 2022;283:119161. 10.1016/j.carbpol.2022.119161.
45. Gregory DA, Tripathi L, Fricker ATR, Asare E, Orlando I, Raghavendran V, et al. Bacterial cellulose: A smart biomaterial with diverse applications. Mater Sci Eng R Reports 2021;145:100623. 10.1016/j.mser.2021.100623.
46. Abdelhamid HN, Mathew AP. Cellulose-based nanomaterials advance biomedicine: a review. Int J Mol Sci 2022;23:5405. 10.3390/ijms23105405.
47. Abdelhamid HN., Mathew AP. A review on cellulose-based materials for bio-medicine. Preprints 2022:2022010035. 10.20944/preprints202201.0035.v1.
48. Liu D-M, Yang Q, Troczynski T. Sol–gel hydroxyapatite coatings on stainless steel substrates. Biomaterials 2002;23:691–698. 10.1016/S0142-9612(01)00157-0.
49. Li T, Kou HC, Li RL, Xu GS, Lu TL, Li JS. Effect of osteoblast activity of porous titanium with HA-coating by adding human-like collagen. Mater Res Innov 2014;18:S4–952-S4-957. 10.1179/1432891714Z.000000000813.
50. He X, Zhai Z, Wang Y, Wu G, Zheng Z, Wang Q, et al. New method for coupling collagen on biodegradable polyurethane for biomedical application. J Appl Polym Sci 2012;126:E354–E361. 10.1002/app.36742.
51. Liu Y, Ren L, Long K, Wang L, Wang Y. Preparation and characterization of a novel tobramycin-containing antibacterial collagen film for corneal tissue engineering. Acta Biomater 2014;10:289–299. 10.1016/j.actbio.2013.08.033.
52. Wu G, Wang J, Chen X, Wang Y. Impact of self-assembled monolayer films with specific chemical group on bFGF adsorption and endothelial cell growth on gold surface. J Biomed Mater Res Part A 2014;102:3439–3445. 10.1002/jbm.a.35007.
53. Li J, Wang X, Hu R, Kou H. Structure, composition and morphology of bioactive titanate layer on porous titanium surfaces. Appl Surf Sci 2014;308:1–9. 10.1016/j.apsusc.2014.03.068.
54. Nan L, Cheng J, Yang K. Antibacterial behavior of a Cu-bearing type 200 stainless steel. J Mater Sci Technol 2012;28:1067–1070. 10.1016/S1005-0302(12)60174-1.

55. Faure E, Lecomte P, Lenoir S, Vreuls C, Van De Weerdt C, Archambeau C, et al. Sustainable and bio-inspired chemistry for robust antibacterial activity of stainless steel. J Mater Chem 2011;21:7901. 10.1039/c1jm11380a.

56. Lu Y, Tan L, Xiang H, Zhang B, Yang K. Study on corrosion resistance of pure magnesium with CaSiO 3 contained coating in NaCl solution. Acta Metall Sin (English Lett 2012;25:287–294.

57. Wan P, Lin X, Tan L, Li L, Li W, Yang K. Influence of albumin and inorganic ions on electrochemical corrosion behavior of plasma electrolytic oxidation coated magnesium for surgical implants. Appl Surf Sci 2013;282:186–194. 10.1016/j.apsusc.2013.05.100.

58. Safavi MS, Bordbar-Khiabani A, Walsh FC, Mozafari M, Khalil-Allafi J. Surface modified NiTi smart biomaterials: Surface engineering and biological compatibility. Curr Opin Biomed Eng 2023;25:100429. 10.1016/j.cobme.2022.100429.

59. Shabalovskaya S, Anderegg J, Van Humbeeck J. Critical overview of Nitinol surfaces and their modifications for medical applications. Acta Biomater 2008;4:447–467. 10.1016/j.actbio.2008.01.013.

60. Wang H, Sun T, Chang L, Liu F, Liu B, Zhao C, et al. Preparation of Ca doping ZrO2 coating on NiTi shape memory alloy by cathodic plasma electrolytic deposition and its structure, in-vitro bioactivity and biocompatibility analysis. Surf Coatings Technol 2017;325:136–144. 10.1016/j.surfcoat.2017.06.067.

61. Menezes BRC de, Rodrigues KF, Fonseca BC da S, Ribas RG, Montanheiro TL do A, Thim GP. Recent advances in the use of carbon nanotubes as smart biomaterials. J Mater Chem B 2019;7:1343–1360. 10.1039/C8TB02419G.

62. Wang Y, Zhang P, Zhao Y, Dai R, Huang M, Liu W, et al. Shape memory composites composed of polyurethane/ZnO nanoparticles as potential smart biomaterials. Polym Compos 2020;41:2094–2107. 10.1002/pc.25523.

63. Cetinel S, Shen W-Z, Aminpour M, Bhomkar P, Wang F, Borujeny ER, et al. Biomining of MoS2 with Peptide-based smart biomaterials. Sci Rep 2018;8:3374. 10.1038/s41598-018-21692-4.

64. Fragal EH, Fragal VH, Silva EP, Paulino AT, da Silva Filho EC, Mauricio MR, et al. Magnetic-responsive polysaccharide hydrogels as smart biomaterials: Synthesis, properties, and biomedical applications. Carbohydr Polym 2022;292:119665. 10.1016/j.carbpol.2022.119665.

65. Liao J, Huang H. Review on magnetic natural polymer constructed hydrogels as vehicles for drug delivery. Biomacromolecules 2020;21:2574–2594. 10.1021/acs.biomac.0c00566.

66. Hoffman AS, Stayton PS. Applications of "Smart Polymers" as Biomaterials. Biomaterials Science, Elsevier; 2020, pp. 191–203. 10.1016/B978-0-12-816137-1.00016-7.

67. Kim Y-J, Matsunaga YT. Thermo-responsive polymers and their application as smart biomaterials. J Mater Chem B 2017;5:4307–4321. 10.1039/C7TB00157F.

68. Palza H, Zapata P, Angulo-Pineda C. Electroactive smart polymers for biomedical applications. Materials (Basel) 2019;12:277. 10.3390/ma12020277.

69. Lim HL, Hwang Y, Kar M, Varghese S. Smart hydrogels as functional biomimetic systems. Biomater Sci 2014;2:603–618. 10.1039/C3BM60288E.

70. Wang H, Zhao C, Chen Y, Li J, Zhang X. Electrochemical property and in vitro degradation of DCPD–PCL composite coating on the biodegradable Mg–Zn alloy. Mater Lett 2012;68:435–438. 10.1016/j.matlet.2011.11.029.

71. Lee H, Dellatore SM, Miller WM, Messersmith PB. Mussel-inspired surface chemistry for multifunctional coatings. Science (80-) 2007;318:426–430. 10.1126/science.1147241.

72. Xu X, Li M, Liu Q, Jia Z, Shi Y, Cheng Y, et al. Facile immobilization of heparin on bioabsorbable iron via mussel adhesive protein (MAPs). Prog Nat Sci Mater Int 2014;24:458–465. 10.1016/j.pnsc.2014.09.001.

73. Furth ME, Atala A, Van Dyke ME. Smart biomaterials design for tissue engineering and regenerative medicine. Biomaterials 2007;28:5068–5073. 10.1016/j.biomaterials.2007.07.042.

# 6 Classification of Biomaterials and Surface Strategies

*Gisela Strohle, Chloe Tan, and Huiyan Li*
School of Engineering, University of Guelph, Guelph,
Ontario, Canada

*Amar K. Mohanty and Manjusri Misra*
School of Engineering, University of Guelph, Guelph,
Ontario, Canada

Bioproducts Discovery and Development Centre,
Department of Plant Agriculture, University of Guelph,
Guelph, Ontario, Canada

## 6.1 INTRODUCTION

Over the years, much research has focused on the development of biomaterials to suffice the demand of the biomedical, biotechnological, and environmental fields for various applications. For example, over the last few decades, the biomedical industry has been required to reduce the production of metal-on-metal implants (i.e., CoCrMo-on-CoCrMo). Metal-on-metal joint replacements can cause excessive localized cobalt-alloy wear [1]. Localized wear may result in the release of fine debris (cobalt and chromium ions) into the surrounding tissue, which can trigger an adverse immune response [1–3]. Therefore, cast high-carbon CoCrMo alloys were developed to replace metal-on-metal joint replacement pieces [1]. What makes cast high-carbon CoCrMo alloys popular is that they possess both high hardness and wear resistance [1,2]. The alloy's high hardness and wear resistance are due to the presence of hard carbide inclusions distributed throughout the alloy's microstructure and its high work-hardening coefficient [1,2].

Additionally, there is an increasing demand for eco-friendly materials and systems aiming to reduce environmental pollution [4,5]. The use of plastics has increased significantly in the last few decades in many industry sectors [6]. Most of the plastics used in health care, transportation, and construction are not biodegradable [6]. This means that they can accumulate in landfills and endanger ecosystems [4,5]. Approximately 580,000 plastic items are found in the ocean per square kilometer [7]. This alarming figure has prompted an urgent demand for strategies to reduce plastic pollution [4,5]. A recent review by the United

DOI: 10.1201/9781003429920-7

Nations revealed that over 600 marine species are affected by the effects of plastic waste [7]. Therefore, research into the development of biomaterials should not only satisfy the requirements of a specific application but also address environmental pollution.

Even though biomaterials can offer many advantages for a specific application, it is difficult to satisfy all required properties (bulk and surface) with the use of a single biomaterial [8]. Therefore, the modification of biomaterial surfaces has grown significantly over the years. Nowadays, surface functionalization is a fundamental component of the design and development of various engineered systems and devices for biomedical, environmental, and biotechnological fields. Several surface modification techniques can be applied to any group of biomaterials (metals, ceramics, and biodegradable and non-biodegradable polymers), depending on their size, architecture, complexity, and structure [8]. The choice of biomaterial and surface modification is greatly influenced by the intended application. Many surface functionalization techniques have been optimized and created over the years, and they can be categorized into three major groups: biological, physical, and chemical [9]. These groups represent either the applied technique or the modifier [9]. Physical surface modification can be implemented by machine-assisted methods and include grinding and polishing [10]. In chemical surface functionalization, the biomaterial surface charge and structure can be tailored to a specific application, by attaching functional groups to its surface through adsorption or covalent bonding, by treating the surface with an alkaline/acid solution, or by applying sol-gel techniques [11]. Finally, biological/biochemical surface modification can be applied to immobilize biological entities onto the biomaterial's surface [12]. Figure 6.1 illustrates the three categories of surface functionalization for a biomaterial and the influence they have on the surface properties [8].

In this chapter, metallic, ceramic, non-biodegradable, and biodegradable polymer biomaterials are introduced. Each of these groups is further divided into natural, synthetic, and hybrid biomaterials. Then, we focus on the most commonly used techniques for the functionalization of biomaterials for each classification using organic and inorganic, synthetic and natural complexes, and biomolecules. Finally, we provide an overview of the structural and functional modification strategies for each classification of biomaterials and discussed their toxicity, performance, and applications. While most examples of biomaterials and respective surface functionalization used in this chapter have biomedical applications, they can also be extended to other fields.

## 6.2  CLASSIFICATION OF BIOMATERIALS

### 6.2.1  METALS

Metals are used as biomaterials due to their properties that include resistance to fractures, electrical conductivity, formability, and high strength [1,13]. They are favored in applications like cardiovascular surgery, orthopedics, and dentistry because they are easy to fabricate [1]. However, the effectiveness of metal biomaterials depends on their

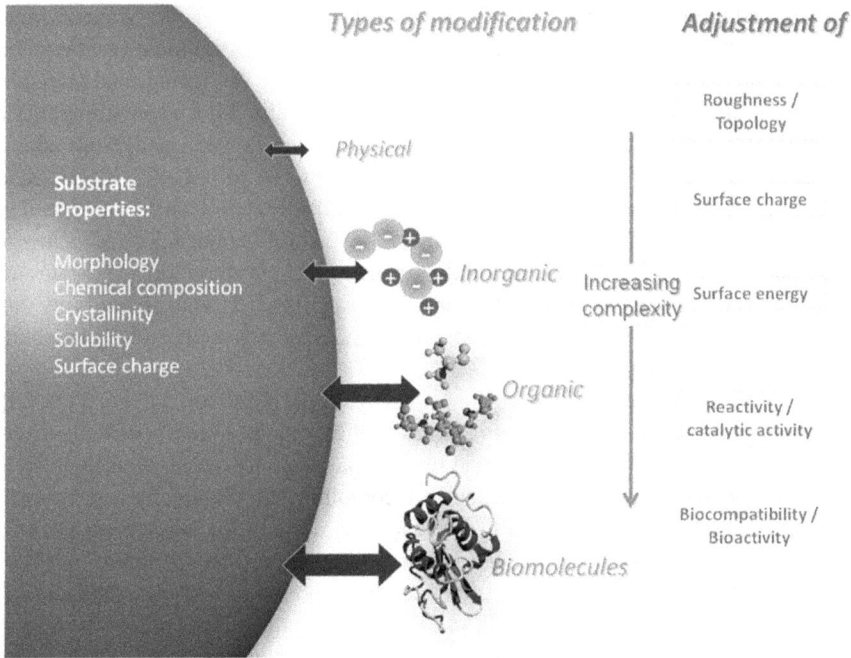

**FIGURE 6.1** Surface modification of biomaterials and its effect on the surface properties. Reproduced from Ref. [8] with permission.

manufacturing process [1]. Therefore, it is imperative to understand the impact that the manufacturing procedure has on metal properties, to prevent life-threatening mechanical failures [1]. Cobalt-based alloys, titanium and its alloys, and stainless steel are the most common metallic biomaterials [1,13,14].

Metals are held together by metallic bonding and have a high coordination number, N, of 12 or 8. The interatomic bonds are non-directional, and electrons move more freely within the metal crystal lattices in comparison to covalently or ionically bonded materials [1]. This aspect of metals is what causes them to have high electrical and thermal conductivity [1,13,13–15]. A metal's ability to plastically deform is also beneficial as it can blunt sharp discontinuities and decrease stress concentrations locally, which is why metals tend to possess high fracture toughness [1,13,13–15].

The relationship between stress and strain for metals is linear at small deformations, except when the temperature is close to the melting temperature [1]. Factors like atomic size, valence, and chemical affinity between elements under specific external conditions (specified temperature or pressure) determine the equilibrium arrangement of atoms in solid metals [1]. Superelasticity is another characteristic of some metallic biomaterials, namely, Ni-Ti alloys (also known as Nitinol) [1]. This characteristic is a result of the extremely low apparent elastic modulus that N-Ti alloys have, which allows it to completely recover from a strain, once the stress has been removed, thereby returning the material to its original structure [16].

Metals also possess high ductility if properly processed [1]. To increase yield strength, methods such as strain hardening, alloying for substitutional or interstitial solid solution strengthening, precipitation or second phase hardening, and strengthening by grain refinement can be used [1,17]. Additionally, high yield strength is a favorable trait for certain biomedical applications, to prevent dangerous shape changes or fractures in the metal [1,15]. The consequence of having a higher yield strength is that the material is less ductile [1,18]. A decrease in ductility may lead to easier fatigue crack initiation so one must be careful when selecting material processing procedures to gain optimal properties [1]. Furthermore, the rates of corrosion and the negative impacts that they can have on biocompatibility are important factors for the selection of metallic biomaterials for applications in medical surgery and dentistry [1].

Metallic biomaterials fall under either the hybrid or synthetic category [19,20]. Examples of synthetic metallic biomaterials include titanium and cobalt alloys, in addition to stainless steel [1,13]. Hybrid biomaterials that use metals tend to do so to improve the performance of an implant, by combining favorable characteristics of different biomaterial categories [20]. For example, a study for making craniofacial implants tested the addition of bioceramic parts with three different powders embedded into a metallic scaffold, to better osseointegration and vascularization at the defect site [20]. Figure 6.2 shows an example of a cranial implant for the human skull designed with calcium phosphate (CaP) tiles connected with a titanium frame [21].

## 6.2.2 Ceramics

Unlike metals, in ceramic biomaterials, the ionic or covalently bound atoms form a much tighter arrangement [22]. Bioceramics are made of metal and non-metal

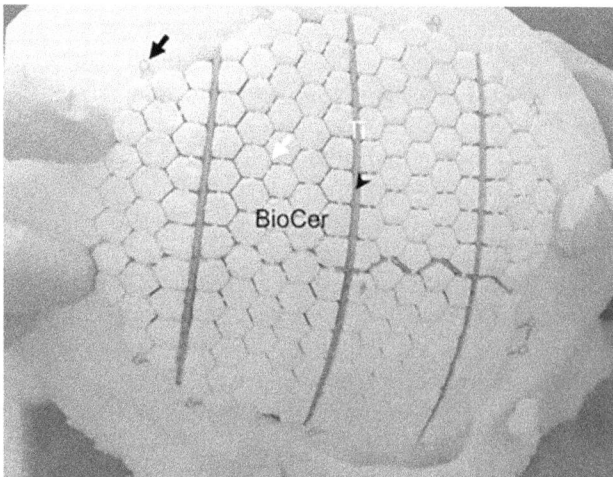

**FIGURE 6.2** Cranial implant on a human skull made of calcium phosphate tiles and a titanium frame. Reproduced from Ref. [21] with permission.

compounds and may be partly crystalline or crystalline [23,24]. They tend to be brittle, hard, and poor conductors of electricity and heat; have high melting points; high compression strengths; and low chemical reactivities (biocompatible) [24–26]. These properties are greatly attributed to the nature of their chemical bonds. Ceramic biomaterials which have atoms held together by ionic bonds will not plastically deform (or have minimal plastic deformation) because the bonds only permit minimal slip to take place [22,27,28]. This property, attributed to the chemical bond type, also makes ceramics susceptible to cracks, initiating fracture with ease where there is stress concentration. The various methods involved in the process of forming ceramics (e.g., liquid casting, vitrification) lead to variations in their microstructures, resulting in diverse mechanical properties [23]. For example, silicon dioxide ($SiO_2$) can be used in different processes to produce either crystalline or glass-ceramic structures [26]. Due to various factors, such as difficulty gripping without causing failure and failure with about 0.1% strain [29], the ceramics' mechanical properties are not assessed using standard testing methods. A bending test (three- or four-point bending) is performed instead to determine the bioceramics' flexural strength [30].

Ceramic biomaterials used in biomedical applications can be grouped into bioinert, bioactive, and bioresorbable ceramics based on the interactions of these biomaterials with the host physiological environment [24,31]. Relatively bioinert ceramics exhibit very little biological response when interacting with tissues [24]. They are known to withstand corrosion in physiological conditions [24], conserve mechanical and physical properties when implanted, and used for structural support (e.g., dental implants, femoral cups, and heads). Examples of synthetic bioinert ceramics include alumina ($Al_2O_3$) [24,32], zirconia ($ZrO_2$) [31,33], and titania nitride (TiN) [24,31]. Bioresorbable (biodegradable) ceramics are also widely used in various biomedical applications because of their excellent biocompatibility (e.g., bone replacement, drug delivery) [24]. They can be selected as substitutes for biodegradable polymeric and metallic materials, as these materials might cause inflammation by releasing molecules that lower the pH [34–37]. The degradation of bioresorbable ceramics is highly dependent on the porosity of the material being used [31]. Examples of bioresorbable ceramics include CaP and hydroxyapatite (HAp or HA) [24]. Bioactive ceramic materials (glass-ceramics/bioglasses) have some properties of bioresorbable and bioinert ceramic biomaterials and are designed to form a bond with hossues [24]. These biomaterials are very brittle since glass does not form a crystalline structure. However, a very fine grain structure (0.1–1 m) can be obtained when crystallizing glass-ceramic in a controlled environment by precipitating intentionally placed metals to nucleate the glass. A grain structure of less than 1 m helps in cell attachment and improves bioactivity [38,39].

Ceramic biomaterials can be natural or synthetic, but hybrid systems can also be created [24,31]. Various agro-food waste such as the ashes from sugarcane leaves and rice rusks, as well as corn cobs, are being used to make natural silica nanoparticles and glass-ceramics [40–42]. Food waste such as eggshells and banana peels contain calcium and potassium that can also be utilized to produce various ceramic biomaterials, such as biphasic CaP, mesoporous bioglasses, and Has [41].

Additionally, hybrid ceramic biomaterials have been widely used in dentistry. They combine the mechanical and optical properties of resins/polymers/nanoparticles and ceramics to possess lower hardness and catastrophic fracture behavior [43,44].

### 6.2.3 POLYMERS

Polymers have a long chain of repeating units composed of several smaller molecules [45]. Due to their length and structure, they can have various properties. Polymer properties vary based on their molecular weight, degree of crystallinity, side chains, chemical composition, and the length of their backbone [46]. The atoms in the polymer long-chain backbone are held together by covalent bonds [47]. In a 3D network polymer, secondary bonding forces such as hydrogen bonds and van der Waals also exist among the chains and give the material extra strength [48]. The side chains in a polymer are flexible and can influence the ordering of the long chains [49]. As the length of the chain increases, it can cause polymers to transition from gas, to liquid, and to solid forms. The properties of a polymer are closely associated with its structure. Compared to ceramics and metals, polymers generally have better deformability after yield, greater strains before failure, and lower elastic modulus [50].

Human activities have polluted the environment [51,52]. Due to the lack of decomposition capabilities of some plastic materials, many of them end up in landfills and oceans [51,52]. Recent research has directed attention to the development of environmentally friendly materials and sustainable systems to address environmental pollution [4,5]. The degradation of polymers carried out anaerobically and aerobically by microorganisms in aquatic or terrestrial ecosystems is known as biodegradation [53]. A polymer that undergoes this process experiences changes to its chemical structure, in response to environmental conditions [53,54]. This process can result in the loss of certain properties, depending on the application and test methods used [54]. Therefore, a polymer can be biodegradable and non-biodegradable, and within these categories, they are either biobased or fossil-based [55,56].

### 6.2.4 BIODEGRADABLE POLYMERS

Plastic materials are known to greatly contribute to environmental pollution [51–53]. They are organic compounds of long-chain polymers that have a high molecular mass [54]. A polymer's complete biodegradation leads to the formation of water and carbon dioxide [56]. However, a polymer's biodegradation efficiency can be influenced by biotic and abiotic factors such as the microorganism type, presence/absence of oxygen, temperature, pH, media, physical stress, UV radiation, and the polymer's characteristics [53,56]. Polymer characteristics that affect biodegradation include polymer crystallinity, cross-linking, flexibility, molecular weight, functional groups, and the presence/absence of additives [56,57]. For example, the biodegradation of polylactic acid (PLA), which is a biobased biodegradable polymer, increases as its molecular weight decreases [57]. The use of biodegradable polymers is expected to increase significantly as a substitute for non-degradable polymers such as polypropylene (PP) and polyethylene (LDPE,

HDPE) [54]. While non-degradable polymeric materials can be broken down into smaller pieces due to the action of abiotic factors, this process, known as disintegration, differs from biodegradation [58,59]. During the biodegradation of a material, both processes occur: first, the material disintegrates into smaller pieces and then biodegradation begins [59]. There are four steps involved in the polymer biodegradation process: biodeterioration, depolymerization, bio-assimilation, and mineralization [60]. First, microorganisms create biofilms and degrade the polymer surface (biodeterioration). Then, they deliver extracellular enzymes to depolymerize long polymer chains (depolymerization) [60]. Third, the microbial cells take in the products of hydrolysis and oxidation and consume them as metabolites (bio-assimilation). Finally, these molecules are transformed into carbon dioxide, methane, nitrogen, and water (mineralization) [60]. For example, the biodegradation of PLA occurs by hydrolysis of the polymer ester linkage, which causes the average molecular weight to be reduced; then the microorganisms in soil digest the lactic acid oligomers and convert them into carbon dioxide and water [61].

Biobased biodegradable polymers include PLA, bio-poly (butylene succinate) (Bio-PBS), polyhydroxyalkanoates (PHA), and starch [55]. Not all biodegradable polymers are biobased. Some biodegradable polymers are fossil-based such as poly($\varepsilon$-caprolactone) (PCL), poly (butylene succinate) (PBS), and polybutylene adipate terephthalate (PBAT) [55]. Although the term biodegradable refers to a material's ability to biodegrade, there are no specific time limits for it to occur [54,56]. For a biodegradable polymer to be compostable, most of its material needs to be biodegraded ($\geq$ 90%) within 180 days [61]. In a composting facility, not all biodegradable polymeric materials will be compostable because of the time required for biodegradation [56].

Mechanical and thermal properties of biodegradable polymeric materials greatly depend on the molecular weight, degree of crystallinity, functional groups, and presence/absence of additives [62]. PLA is a biodegradable polymer of linear polyester, which is semicrystalline and can be made from the fermentation of lactic acid from sources such as potato, corn, and whey [62]. However, due to its relatively brittle nature, it needs to be plasticized to produce food packaging films [62,63]. Various plasticizers can be used to improve the processability and flexibility of PLA. Some of these include poly(ethylene glycol) (PEG), citrate esters, glucose monoesters, oligomeric lactic acid, and glycerol [63]. Adding a plasticizer can reduce the melting and transition temperature of PLA [64,65]. It can also reduce the PLA's tensile strength and modulus, but increase elongation at break [64,65]. Additionally, the incorporation of PEG into PLA has been shown to slow its biodegradation rate [62].

Biodegradable polymeric materials can be synthetic, natural, or hybrid. Examples of synthetic biodegradable polymers include polyesters (PLA, PCL), polyphosphazene (PPHOS), and polyanhydrides [66]. The initial step in the biodegradation of synthetic carbon-chain and hetero-chain polymers happens via peroxidation and hydrolysis, respectively [67,68]. Natural biodegradable polymers include polysaccharides such as cellulose and starch, and polyglutamic acid [66]. In biodegradable hybrid polymers, fillers can be combined with biobased polymers to reinforce the polymer matrix [69]. The addition of a plasticizer to PLA, as explained above, is an example of a biodegradable hybrid polymer [62].

## 6.2.5 Non-Biodegradable Polymers

Non-biodegradable polymers are not influenced by the environment and do not undergo modification in their configuration during the processing period [70]. Non-biodegradable fossil-based plastics such as polyethylene (PE), PP, and polyvinyl chloride (PVC), are used for plastics, due to their low cost, ease of production, and diverse properties [71,72]. These properties include being impervious to water and resisting biodegradation [71]. However, due to their high greenhouse gas emission and the limitation of fossil fuel resources, there is a growing market for biobased plastics [71]. Non-biodegradable bioplastics are desirable because they have a reduced carbon footprint during production and are resistant to microbial degradation [71,73]. Further, they can be derived from renewable resources and used for conventional applications requiring durability [71]. Currently, some of the commonly used plastics have analogous biobased alternatives. An example of this is bio-PE, which is non-biodegradable and possesses the same properties as its fossil-based parallel [73,74]. Non-biodegradable polymers are still prevalent in biomedical applications such as tendon or ligament repairs, implants, bio-scaffolds, and nanomedicine, just to name a few [70,75,75–77].

As with biodegradable polymers, non-biodegradable polymeric materials' mechanical and thermal properties vary with the polymer crystallinity degree, the presence of additives, molecular weight, and the type of functional groups present in the polymer chain. For example, a study used a hybrid-braid structure consisting of polyethylene terephthalate (PET) and PLA multifilament yarns to better suit the mechanical demands of specific tendons [75]. Using both non-degradable and degradable plastics allows for the device to have the strength and durability present in non-biodegradable plastics, while also possessing the biocompatibility and degradability seen in biodegradable plastics, to promote long-term tissue ingrowth and native integration [75]. Another study used ethylene-vinyl acetate (EVA), a fossil-based non-biodegradable polymer, in ocular delivery implants [70]. EVA, a versatile and Food and Drug Administration (FDA) approved transparent thermoplastic, is among the most commonly employed non-biodegradable polymers for creating drug delivery devices for implantation [70]. EVA contains between 1 to 40% of vinyl acetate (VA) monomer [70]. The VA amount present in EVA impacts its properties [70,78]. For instance, increasing the amount of VA causes higher attachment, impact endurance, polarity, elasticity, and compatibility with other polymers [70,79,80]. It should be noted that an increase in VA also causes a decrease in stiffness, melting point, and crystallinity [70,79,80]. EVA in implants is employed as a limiting film that restricts the permeability of a pharmaceutical [70].

Non-biodegradable polymers can be of synthetic, natural, or hybrid origin. The hybrid structure of non-biodegradable (PET) and biodegradable (PLA) polymers cited above is an example of a hybrid non-biodegradable polymer. An example of a synthetic non-biodegradable polymer is poly(dimethylsiloxane) (PDMS) [76]. PDMS is a non-biodegradable polymer that is low in cost, inert, biocompatible, non-toxic, flexible, gas-permeable, visually transparent, and easily manufactured [76]. Due to its methyl groups, PDMS has a hydrophobic surface [76]. This makes PDMS challenging for cell adhesion and cell growth [76].

## 6.3  SURFACE MODIFICATIONS OF BIOMATERIALS

The main idea behind the concept of surface modification is to retain the biomaterial's key properties while altering its outermost surface, to influence its response to the environment, to which the biomaterial is being introduced. Biomaterials can have their surface modified by using physical, chemical, and biological methods [81,82]. The process of surface modification involves physiochemically changing the compounds, atoms, charges, or molecules on the biomaterial surface, coating the surface with a different biomaterial, altering the surface's texture/design, or immobilizing a biological entity/biomolecule. Commonly used strategies for the surface modification of each biomaterial classification, and the purpose of these modifications, will be discussed in the upcoming sections.

### 6.3.1  METALS

Metals undergo surface modifications to alter or suppress unwanted characteristics, to suit their application; this is necessary since bio-functions cannot be added during the metal manufacturing process [83–85]. For example, surface modifications to body implant biomaterials that prevent corrosion are vital, as corroded implants release excessive and dangerous ions such as Fe, Ti, and Co, which can cause clinical failure and allergic reactions [83,84]. One of the best ways to prevent this is using coating depositions onto the metal surface [83]. For instance, HA is a ceramic biomaterial that is often used for coating metallic biomaterials because it has a similar chemical composition and biocompatibility to bone tissue [83,85]. HA on metallic biomaterials provides a barrier against the dissolution of metal ions from the substrate and also promotes mechanical properties like improved load-bearing ability and substrate-coating adhesion [85,86]. Substrate-coating adhesion is used to attach the coat to the substrate [85]. The coating molecules interact with the substrate via hydrogen bonds, electrostatic forces, and/or van der Waals interactions [85,87]. Figure 6.3 shows examples of titanium surface modification strategies for use in dental implants to improve osseointegration and reduce negative clinical outcomes such as peri-implatitis [88].

Other coating strategies for the surface modification of metals include sol-gel preparation, dip-coating, and electrochemical deposition [85]. Sol-gel is a popular strategy because it is cost-effective, simple, and can produce thicker coatings with complex and irregular shapes [85]. HAp can also be used in sol-gel preparations by combining calcium and phosphorus precursors [89]. This process involves hydrolysis and condensation chemical reactions [90]. In sol-gel preparation, the liquid colloidal solution, sol, is transformed into a solid three-dimensional matrix, the gel [90]. Once the sol is prepared, the metal is submerged in the sol, and then withdrawn at uniform speeds [91]. To dry the HAp sol coating, the coated metal is sintered during the final heat treatment process [89]. This generates the crystalline phase of the HAp coating, densifies the gel film, and removes porosity that occurs during this process [89]. Mechanical interlocking occurs once the film penetrates the roughness on the substrate's surface and dries [85]. With that being said, applying HAp in a sol onto a titanium surface is not a long-term solution in the case of

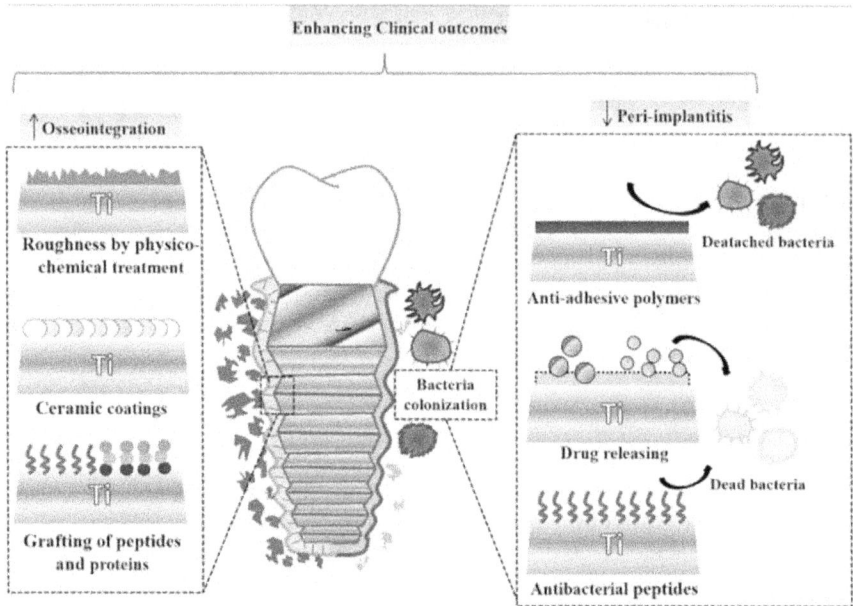

**FIGURE 6.3** Surface functionalization strategies of titanium dental implant to improve osseointegration and reduce negative clinical outcomes such as peri-implantitis. Reproduced from Ref. [88] with permission.

implants, because the HAp coating does not chemically bond to the titanium surface [92]. Only organo-phosphates, organo-silanes, and photo-responsive chemicals are anticipated to bind to titanium [93].

The dip-coating surface modification strategy has three steps: dipping, withdrawing, and drying [85]. The purpose of using dip-coating for the surface modification of metals is to produce uniform depositions by using low processing temperatures [85]. In dip-coating, the substrate is dipped and withdrawn at a fixed speed [85]. The substrate is then left to dry and often heated to increase the adhesion strength between the metal and the coating [85]. Finally, electrochemical deposition has the advantages of being performed at relatively low temperatures and forming a wide range of coating thicknesses [85]. Electrodeposition is conducted by anodic or cathodic systems, with cathodic systems being more effective in modern medical applications [85].

Chemical surface modification of metals is done to alter the surface roughness and characteristics like wettability, surface structure, or surface energy [11]. Some examples of chemical surface treatment methods used for metallic biomaterials are acid-etching and alkali treatment [93]. In biomedical applications, acid-etching is a surface modification technique that aims to create a rougher surface on the implant to promote bio-integration [94]. Alkali treatment, on the other hand, is used to increase the bioreactivity of implant metals such as titanium and its alloys [84].

Biological surface modification of metals uses techniques to immobilize biomolecules such as absorption, covalent bonding, and layer-by-layer (LBL) [95]. In adsorption-based biomolecule immobilization, the biomolecule is immobilized onto the metal surface through van der Waals or electrostatic forces [95]. Since these interactions are rather weak, the biomolecules absorbed by the surface can be quickly detached [95]. The LBL strategy of surface modification uses a self-assembly technique to immobilize organic bioactive molecules onto the metal surface and involves the alternate adsorption of oppositely charged materials [95].

For covalent attachment of biomolecules, salinization is the popular strategy [95]. This is seen with the implementation of biomolecules onto titanium surfaces, to promote initial biological response [95]. When considering titanium-based implants, four main classes of biomolecules can be immobilized: peptides, bone morphogenic proteins (BMPs), non-BMPs growth factors, and extracellular matrix components (ECM) [95]. ECM components such as type 1 collagen and hyaluronic acid (hyaluronan) can be bound onto the implant substrate via covalent bonding [95–97]. Here, the implant's surface is functionalized with carboxyl, hydroxyl, and amine groups so it can react with these biomolecules [95]. This allows for the stabilization of these molecules onto the implant's surface [95]. The silane coating on the implants serves as a linkage between the implant surface and the biomolecule [95]. Silanes like bistriethoxysilylethane (BTSE) are used to bind biomolecules to a metallic surface because they provide functional moieties that allow for further attachment [98]. Here, hyaluronan can be bounded to the titanium surface by cyanoborohydride-promoted covalent linking of periodate-treated hyaluronan to surface amino groups [96,99]. Plasma deposition is used here for particle deposition [96,99]. Hyaluronan itself consists of a disaccharide unit N-acetyl-d-glucosamine-d-glucuronate that repeats and is linked by $\beta$1-4 and $\beta$1-3 bonds [96,99]. From the hyaluronic acid's structure, the carbon single bonds present chemical cues to its external environment that differ from an uncoated titanium implant [96]. The addition of a hyaluronic acid coating via covalent bonding has been shown to improve bone regeneration compared to uncoated titanium *in vivo* in animals [99].

## 6.3.2 CERAMICS

Similar to metals, the purpose of ceramic surface functionalization varies according to its applications. For example, a ceramic's surface can be tailored by attaching organic molecules onto the surface to control the interactions between biological moieties and material, improve the stability of nanoparticles, or to enhance the material's biocompatibility [100–102]. Physical surface modification strategies of ceramics involve mechanical roughening/polishing, laser treatment, or grinding [10,82,103].

The most common ceramic surface modification strategies are either chemical or biological. Like metals, chemical functionalization of ceramics involves sol-gel preparation, treatment with an acid or alkaline solution, and attachment of functional groups [8]. Other techniques such as ion bombardment and slurry-based methods can also be used [8]. Commonly used chemical modification strategies enhance the surface characteristics of ceramics with silanes or PEG [102,104]. However, functionalization with organophosphorus compounds (phosphonates and phosphates) is gaining widespread interest [8]. One to multiple attachments on a ceramic surface can be formed with the use of organophosphorus compounds [105]. They are an ideal alternative to organo-silanes due to their stable hydrolysis in a wide pH range and physiological conditions [106]. Currently, organophosphorus compounds are being utilized on various surfaces such as silica, CaP, titania, and iron oxide [106–109].

Silanization is the method of choice for introducing functional groups to surface of ceramics [100,110]. Its simplicity (no harsh conditions or expensive equipment are needed) and versatility (it can be applied to different ceramic architectures, e.g., nanoporous membranes and colloidal particles) make this functionalization strategy the most commonly used [8,100]. The process of surface silanization involves the activation of the silane molecules by hydrolysis [8]. Then, a condensation reaction between the silanol groups of the silane and the hydroxyl groups of the surface leads to the formation of a covalent bond with the surface [8]. Figure 6.4 illustrates the process of silanization on oxide surfaces [111]. An example of silanization for tailoring the surface of a bioceramic is the functionalization of mesoporous silica with amine and sulfonic functional groups to control an antibiotic (cephalexin) release [112]. The antibiotic release profile can be greatly influenced by the functional group attached to the silica surface. The presence of amino groups on the silica surface resulted in the slow release of cephalexin (after 30 h) while functionalization with sulfonic groups caused fast drug delivery (1–20 h) [112].

Biological surface modification of ceramics involves covalent, affinity, or adsorption-based biomolecule immobilization [8]. Adsorption, just like with metals,

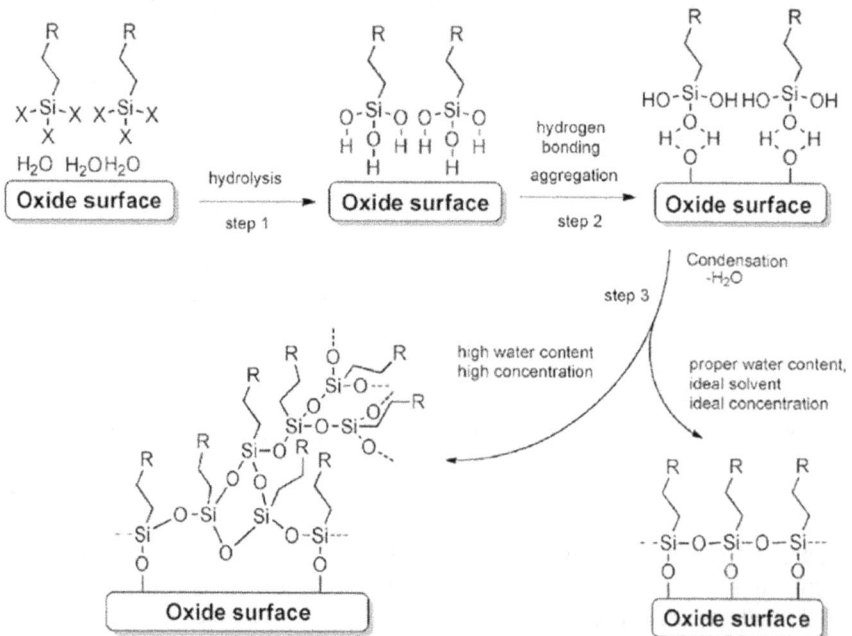

FIGURE 6.4 The three steps involved in the process of salinization of oxide surfaces. Proper water content leads to ideal condensation between the silanol groups of the silane and the hydroxyl groups of the surface. Reproduced from Ref. [111] with permission.

relies on electrostatic and/or hydrophobic interactions, hydrogen bonds, and Van der Walls forces [8]. The goal of using an adsorption biomolecule immobilization strategy on ceramics is to avoid destructive or loss of biological activity [8]. However, external conditions can cause biomolecule desorption/leach from the surface due to weak chemical interactions [113]. One example of this strategy is the protein adsorption onto CaP surface [114,115]. The surface of a CaP implant can be modified with carboxylate and amino groups to immobilize certain biomolecules onto it [114,116,117]. The goal here is to increase the electrostatic interactions between CaP and proteins. On the ceramic's surface, the charged functional groups of CaP can bind to the charged functional groups on the protein molecules such as carboxyl, carbonyl, amino, and aromatic groups [114,118]. In addition to electrostatic interactions, other physicochemical factors include surface coarseness, penetrability, topography, pore size, and particle size govern the interplay between proteins and the CaP surface [114]. Knowing this, one can increase the number of binding sites for protein adsorption by decreasing the size of CaP particles [119] as long as steric hindrance does not play a role in hindering binding sites.

The affinity between aspartic-acid-treated HAp and lysosome is an example of an affinity biological surface modification strategy. When HAp is precipitated in the presence of aspartic acid, aspartic acid binds to the calcium ions in HAp via carboxylate groups [114,120,121]. The other unbounded carboxylate group is exposed at the aspartic-acid-treated HAp(Asp-HAp) surface. This, in addition to the increased phosphate ions from HAp, contributes to the total negative charge of Asp-Hap [114,120,121]. The negatively charged surface of Asp-HAp has a high binding affinity towards the ammonium ion on lysozyme, which is positively charged [8,120]. Functionalization of HAp with Asp can also be applied in bone tissue engineering because some of the most relevant growth factors in biological bone formation such as BMP, have a positive charge [8].

Covalent immobilization uses cross-linker molecules to couple reactive functional groups on the surface of the ceramic biomaterial to biological entities [122]. The most commonly used cross-linkers on ceramic surfaces for biomolecule immobilization are glutaraldehyde and carbodiimide [8,123,124]. This strategy is widely used to immobilize enzymes, proteins, and oligonucleotides for assay technologies, biosensors, and imaging [8]. Because the bond between biomolecules and the ceramic surface is strong and, in most cases, irreversible, this strategy cannot be used for the controlled release of drugs [8]. An example of covalent immobilization on iron oxide surfaces is the immobilization of human serum albumin on amino-functionalized magnetic nanoparticles with glutaraldehyde [123].

### 6.3.3 Biodegradable and Non-Biodegradable Polymers

Polymers require functionalization to improve biocompatibility or to broaden their range of applications [125,126]. Surface wettability is an important factor in determining the type of polymer surface functionalization strategy to use [125]. This is especially true when attempting to join two materials of different surface wettabilities [125]. For instance, to join a hydrophilic metal such as titanium, to a biodegradable polymer such as PLA, that is hydrophobic, the bulk material needs to have its surface modified, to gain similar hydrophilic properties [125]. Typically, proteins bind better on water repealing surfaces whereas cells are usually a fixed and multiply on hydrophylic surfaces [127]. Figure 6.5 shows an example of PDMS surface functionalization, a non-biodegradable polymer with a hydrophobic surface, with hydrophilic poly-D-lysine to improve neuronal cell adhesion [128].

Physicochemical methods of polymer surface modification involve polishing, noncovalent and covalent coating, vapor deposition, etching, and grit blasting [82,130,129]. These techniques can be further grouped into gas phase deposition, liquid/bulk phase deposition, and a combination of those two [12]. Most of these techniques can modify the polymer surface by introducing new elements or molecules through physicochemical interactions (hydrogen bonding, hydrophobic, and/or van der Waals interactions) [12]. The gas phase refers to the use of gases that contain free radicals, ions, molecules, and electrons to modify the biodegradable polymer surface. In addition, electromagnetic radiation, such as gamma rays, ultraviolet rays, and visible light, can also be used [12]. A widely used technique for the surface modification of biodegradable polymers is chemical vapor deposition (CVD) [12]. This process involves vaporizing the biodegradable polymer mono-mers and depositing a film on its surface [130]. A variety of materials with high purity can be deposited at a high rate with this technique, but the precursors are volatile around room temperature, are highly toxic or corrosive, and have hazardous by-products [12,131].

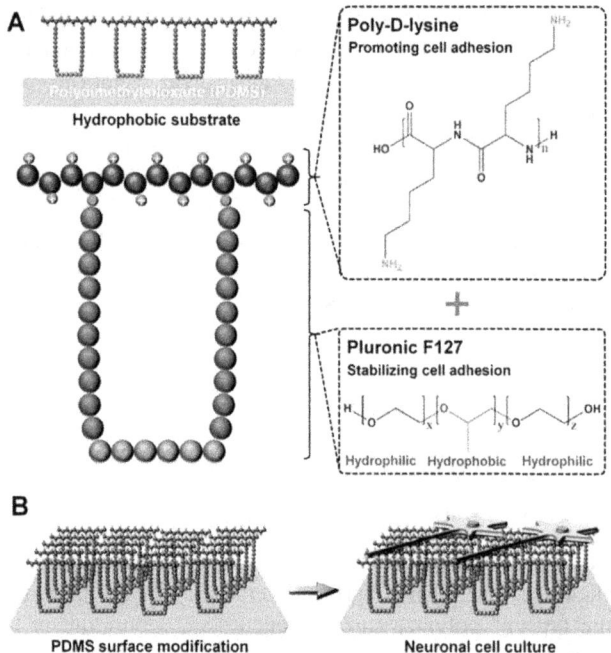

**FIGURE 6.5** Hydorphobic PDMS surface was modified with hydrophilic poly-D-lysine to enhance neuronal cell adhesion. Reproduced from Ref. [128] with permission.

Liquid or bulk phase physicochemical techniques take advantage of adsorption and desorption processes [12]. Adsorption uses diffusion, attachment, and rearrangement mechanisms for polymer surface modification [12,132]. This process helps to improve the surface's flocculation and particle dispersion properties [133]. Coating graphene oxide (GO) onto aminolyzed poly(l-lactide) (PLLA) nanofibrous scaffolds is a good example of this process. GO is coated on the surface of PLLA because graphene substrates are electroconductive when applied with an electric field, and can modulate cell interaction, mobility, and adhesion [134–136]. The amino functional groups on the PLLA surface allow for the stable coating of GO onto the polymer nanofibrous scaffold [134,137]. Amino groups become positively charged in acidic conditions, so the negatively charged GO can be easily coated through electrostatic interactions [134]. With the addition of a GO coat, PLLA demonstrated increased hydrophilicity (reduced water contact angle) due to the presence of hydrophilic hydroxyl, ether, and carboxylic groups on the surface [134].

The process of desorption involves the release of molecules, ions, or atoms from the polymer's surface [12,138]. Polymer desorption can be driven by adding a second element with a strong affinity for the molecule at the polymer surface, which can then displace it [12]. In addition, desorption can occur by altering the polymer's surface solubility and by decreasing the forces used for adsorption [12,132]. In some surface modification processes, avoidance of desorption is preferable. Chemical grafting of non-biodegradable polymers surface is a strategy that promotes long-term chemical stability and avoids desorption [129]. Chemical grafting falls into the next category of physicochemical surface modification.

A combination of the gas and liquid/phases can be accomplished by grafting and patterning [12]. Grafting is a technique used to link monomers by covalent bonding, and patterning (microlithography) to develop patterned 3D polymer surfaces. Grafting polymer brushes onto a polymer substrate has an impact on surface properties like wettability, friction adhesion, and biocompatibility [129]. Chemical grafting strategies of non-biodegradable polymer surface modification include "grafting-to," "grafting-from," and "grafting-through" [129]. "Grafting-to" is a rapid and simple technique; the synthetic steps are not complex [129]. In the grafting-from approach, also referred to as surface-initiated polymerization, polymer chains propagate from surface-attached initiators [129]. Lastly, the "grafting-through" strategy of polymer surface modification is one of the most effective surface modification procedures [129]. This is because the amount of grafted polymer does not depend on the reactions being performed, making the process robust against process variations [139]. In this procedure, polymer chains grow from surface-linked double bonds [129,139]. An example of a combination of gas/liquid surface modification strategy for polymers is the polydopamine coating on PDMS [76]. PDMS first undergoes an oxygen plasma treatment before introducing polar functional groups like silanol to its surface [76,140]. The plasma treatment of PDMS results in surface roughness and increases the bond strength between the polydopamine coating layer and the PDMS scaffold surfaces [76,125,141]. The polydopamine-functionalized PDMS surface has an increased number of free amine groups from the polydopamine [76,142], which can be used to promote cell adhesion and proliferation [76,143].

Biological or biochemical surface modification of polymers can immobilize biomolecules through physical adsorption, entrapment, and covalent attachment [12,81]. Once the molecules are adsorbed, they can then be further linked to one another. Hydrogels and microcapsules are commonly used strategies for physical entrapment systems. For covalent attachment of biomolecules, conjugates are used on solid surfaces [12,81]. Biomolecule immobilization onto the polymer surface can be permanent, of short duration, or of long duration [12,81]. For example, when a biodegradable polymer is used as a support material to a biomolecule, the immobilized biomolecule will be released once the polymer matrix starts to break down and degrade. Cell adhesion peptides have been immobilized onto poly(lactic-co-glycolic acid) for use as scaffolds in tissue engineering [144,145]. Here, depending on the degradation rate of the biodegradable polymer, cells may have time to bind and restore the tissue [145].

## 6.4 PERFORMANCE AND TOXICOLOGICAL ASSESSMENT OF SURFACE STRATEGIES

Surface modification strategies of biomaterials do not come without drawbacks. Due to the increasing number of people with joint diseases and the aging population, the demand for devices and biomaterials that can restore function and improve quality of life is increasing. When a device or biomaterial is implanted into the body, the implanted biomaterial will start slowly breaking down into particles and debris in response to the biological environment. This might lead to complications as these particles can accumulate in the tissues surrounding the implant site. For example, even with the latest surface modification techniques, it is not possible to create a surface coating that is suitable for long-term metal implants [93]. Corrosion still plays a significant role in the failure of metal implants, and it can result in toxicity and tissue injury.

In the case of ceramic biomaterials, the performance of a surface modification strategy is highly dependent on the bioceramic of choice and its purpose. If its intended use is for drug delivery, CaP nanoparticles perform better than silica and iron oxide because of their biodegradability and low toxicity [146]. Additionally, the linker (coupling agent) molecular structure used to modify the bioceramic surface has a great impact in determining its toxicity [114]. As discussed in the above sections, when a bioceramic surface is intended to interact with the human body, the protein's functional groups (thiols, amines, hydroxyls, or carboxylate) attach to the bioceramic surface through linkers [147]. For example, bioceramics can form strong covalent bonds with proteins through silanes, catechol, and mercaptans linkers [8]. Silane linkers, such as aminopropyl-triethoxysilane (APS), are widely used in biomedical applications as silanization uses relatively simple chemical reactions. However, there are still questions about the stability of silane linkers overtime under biological conditions [110,148]. There is a concern about possible toxicity from free silanes and leakage of their by-products [8]. Glutaraldehyde, another coupling agent, has potential toxicity [149]. Compared to adsorption, covalent biomolecule immobilization has the drawback of potentially using toxic substances [8].

The performance and toxicity of polymer surface modification strategies can also depend on the molecule selected for functionalization and its purpose. For example, exposure to PLA, a synthetic biodegradable polymer, can cause a bodily inflammatory response. However, functionalization of its surface with polydopamine reduced *in vivo* toxicity-reduced immunogenicity and tissue inflammation [150]. The catechol group in dopamine can be used to immobilize many different molecules [151], but dopamine is highly toxic and its use should not be overlooked.

*In vitro* assays, such as cell and simulated body fluid tests, are the first step in evaluating the surface modification strategy's toxicity, performance, biocompatibility, and inflammatory response [8]. They need to occur before *in vivo* and clinical tests are performed [152]. Additionally, special attention must be paid to the selection of cells for *in vitro* testing, as cellular responses can be highly heterogeneous based on the selected cell line, cell culture environment, and protocol [153,154]. In simulated body fluid assays, the integration of a modified surface within bone tissue can also be assessed after its immersion in a simulated body fluid [8]. Its bioactivity is determined by the ability of modified biomaterial surfaces to form apatite-like materials [152].

## 6.5 APPLICATIONS OF SURFACE STRATEGIES

Biomaterials can be functionalized to overcome a deficiency while preserving their bulk properties. The ability to tailor the surface of a material to obtain unique interactions and responses allows their use in many applications. Surface functionalization of metallic, ceramic, and polymer biomaterials has been used widely in bioengineering and biomedical applications. Some examples of these applications include biosensors, implants, tissue engineering, lab-on-chips, bioreactors, diagnostic tools, food packaging, and clean water systems.

Medical biosensors are developed by taking advantage of the specificity of some biomolecules against certain compounds. Various surface functionalization techniques that combine biological entities (antibodies, enzymes, cell receptor ligands, aptamers, and other affinity complexes) onto biomaterial surfaces have been suggested to enhance the sensitivity and selectivity of these devices. Platinum has been electrochemically deposited onto alumina functionalized with amino groups via silanization for the non-enzymatic detection of glucose [155]. Functionalized biomaterials can also be used as carriers for drug delivery systems. The design and development of effective drug release systems are typically focused on the control and reproducible characteristics of their release profile. Based on this, biomaterial surface modification strategies have been studied for their effect on the delivery of many drug models. For example, tailoring poly(amidoamine) polymer's surface with PEG to neutralize/replace cationic groups is efficient in decreasing the uptake of this dendrimer and carried drug by the liver, macrophages, and spleen [156].

In orthopedic, orthodontic, and cardiovascular applications, metals such as titanium and cobalt-based alloys need to have their surfaces modified to mitigate implant complications like inflammation and function loss. For instance, having higher amounts of cobalt ions in the body has been linked to a decline in cognitive function, anorexia, and extreme muscle fatigue. In tissue engineering, 3D porous scaffolds made of bioceramics (e.g., CaP, zirconia, and silica) can be functionalized to evoke a determined biological response, to guide cell function in migration and vascularization [157,158].

Biomaterials can also be functionalized to immobilize biomolecules, such as antibodies, aptamers, enzymes, and nucleotides, on lab-on-chip devices. For example, a substrate can be grafted with aminooxy silanes for the binding of oligonucleotides via oxime bond. Similar approaches can be applied for diagnostic tools. Aptamers can be immobilized on the functionalized biomaterial substrate to recognize antigens specific to a certain disease. Figure 6.6 shows an example of an aptamer immobilized on the carboxylic functionalized surface of reduced GO for the specific capture of cancer cells [159]. On enzyme-immobilized membrane bioreactors, functionalized biomaterials can assist in enzyme immobilization [160]. For example, chitosan beads can be functionalized with amino acids for the immobilization of lipase. This way, the enzyme can be stabilized to retain its bioactivity [160].

For clean water system purposes, biomaterial nanoparticles can be functionalized to act as a catalytic agent to degrade pollutants or adsorbents, to remove pollutants in water. For example, amino-functionalized magnetic nanoparticles can be employed to depollute water from heavy metals, as the amino group is highly capable of chelating metals [161]. These functionalized nanoparticle adsorbents can be reused by displacing the metal ions via desorption [161].

There are many possible applications of surface strategies, as the surface of a biomaterial can be tuned to perform distinct functions. Therefore, due to the availability of various biomaterials with distinct surface functionalities, surface functionalization has become a vital component of the design process for new materials and the development of new devices.

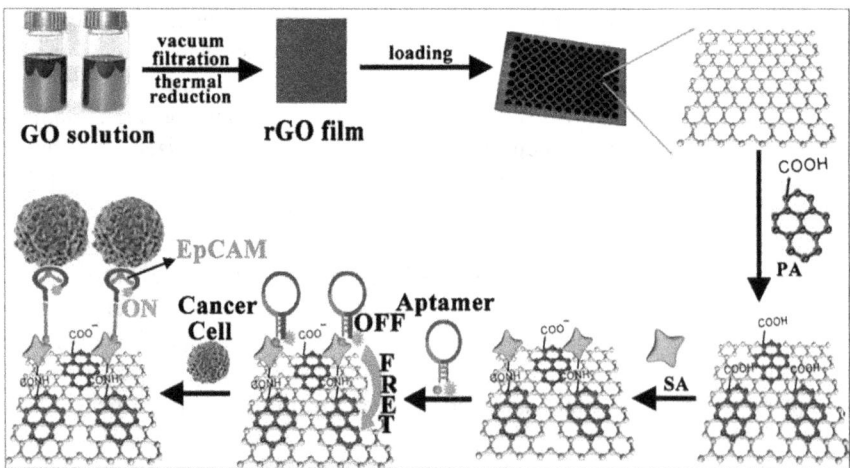

**FIGURE 6.6** Aptamer immobilization on functionalized reduced graphene oxide surface for the specific capture of cancer cells. Reproduced from Ref. [160] with permission.

## 6.6 SUMMARY AND FUTURE PERSPECTIVES

Different classes of biomaterials (metallic, ceramic, biodegradable, and non-biodegradable polymer materials) have been extensively investigated for many applications due to their specific properties (chemical, mechanical, and thermal). Each class can be further divided into natural, synthetic, or hybrid materials. Natural biomaterials generally offer good biocompatibility. Synthetic biomaterials can be modified to offer good biocompatibility. Surface modification of biomaterials broadens their applications in biomedical, biotechnological, and environmental fields. Due to the significant impact that the surface has on a material's performance and properties, a wide range of techniques have been developed and applied for the functionalization of various types of biomaterials. Various surface modification strategies that make use of biomolecules and organic/inorganic modifiers have been applied to biomaterials of different architectures and compositions. Surface functionalization strategies can be biological, physical, and chemical, and can be chosen based on the desired application. Currently, cell-based, biochemical, or microbiological assays are performed for assessing toxicological responses to surface-functionalized biomaterials. Although surface functionalization is a powerful tool to enhance and produce biomaterials for various applications, further research is necessary to improve the selectivity and efficiency of the functionalized surface.

## ACKNOWLEDGMENTS

This study was financially supported by (i) the Ontario Ministry of Agriculture, Food and Rural Affairs (OMAFRA)/University of Guelph – Bioeconomy for Industrial Uses Research Program (Project Nos. 030728); (ii) the Natural Sciences and Engineering Research Council of Canada (NSERC), Canada Research Chair (CRC) program Project No. 460788 and NSERC Discovery Gtants (Project No.401716).

## REFERENCES

1. Pilliar RM. Metallic biomaterials. Biomedical Materials. 2021:1–47.
2. Milošev I. CoCrMo Alloy for biomedical applications. In: Djokić SS, editor. Biomedical applications. Boston, MA: Springer US; 2012. pp. 1–72.
3. Waddell JP. Biological, material, and mechanical considerations of joint replacement. Canadian Journal of Surgery. 1994;37(4):333.
4. Matjašič T, Simčič T, Medvešček N, Bajt O, Dreo T, Mori N. Critical evaluation of biodegradation studies on synthetic plastics through a systematic literature review. The Science of the Total Environment. 2021;752:141959. doi: 10.1016/j.scitotenv.2020.141959.
5. Polman EMN, Gruter G-JM, Parsons JR, Tietema A. Comparison of the aerobic biodegradation of biopolymers and the corresponding bioplastics: a review. The Science of the Total Environment. 2021;753:141953. doi: 10.1016/j.scitotenv.2020.141953.
6. Geyer R, Jambeck JR, Law KL. Production, use, and fate of all plastics ever made. Science Advances. 2017;3(7):e1700782-e. doi: 10.1126/sciadv.1700782.
7. Willis K, Maureaud C, Wilcox C, Hardesty BD. How successful are waste abatement campaigns and government policies at reducing plastic waste into the

marine environment? Marine Policy. 2018;96:243–249. doi: 10.1016/j.marpol.2017. 11.037.

8. Treccani L, Yvonne Klein T, Meder F, Pardun K, Rezwan K. Functionalized ceramics for biomedical, biotechnological and environmental applications. Acta Biomaterialia. 2013;9(7):7115–7150. doi: 10.1016/j.actbio.2013.03.036.

9. Prakash S, Karacor MB, Banerjee S. Surface modification in microsystems and nanosystems. Surface Science Reports. 2009;64(7):233–254. doi: 10.1016/j.surfrep. 2009.05.001.

10. Holthaus MG, Treccani L, Rezwan K. Comparison of micropatterning methods for ceramic surfaces. Journal of the European Ceramic Society. 2011;31(15):2809–2817. doi: 10.1016/j.jeurceramsoc.2011.07.020.

11. Nouri A, Rohani Shirvan A, Li Y, Wen C. Surface modification of additively manufactured metallic biomaterials with active antipathogenic properties. Smart Materials in Manufacturing. 2023;1:100001. doi: 10.1016/j.smmf.2022.100001.

12. Michael FM, Khalid M, Walvekar R, Siddiqui H, Balaji AB. Surface modification techniques of biodegradable and biocompatible polymers. In: Shimpi, Navinchandra Gopal. Biodegradable and biocompatible polymer composites: processing, properties and applications. Woodhead Publishing series in composites science and engineering. Duxford: Woodhead Publishing, an imprint of Elsevier; 2018.

13. Eliaz N. Corrosion of metallic biomaterials: a review. Materials. 2019;12(3):407.

14. Hussein MA, Mohammed AS, Al-Aqeeli N. Wear characteristics of metallic biomaterials: a review. Materials. 2015;8(5):2749–2768.

15. Chen Q, Thouas GA. Metallic implant biomaterials. Materials Science and Engineering: R: Reports. 2015;87:1–57. doi: 10.1016/j.mser.2014.10.001.

16. Acet M, Mañosa L, Planes A. Chapter four - magnetic-field-induced effects in martensitic heusler-based magnetic shape memory alloys. In: Buschow KHJ, editor. Handbook of magnetic materials. Elsevier; 2011. pp. 231–289.

17. Han SZ, Choi E-A, Lim SH, Kim S, Lee J. Alloy design strategies to increase strength and its trade-offs together. Progress in Materials Science. 2021;117:100720. doi: 10.1016/j.pmatsci.2020.100720.

18. Gao YF, Zhang W, Shi PJ, Ren WL, Zhong YB. A mechanistic interpretation of the strength-ductility trade-off and synergy in lamellar microstructures. Materials Today Advances. 2020;8:100103. doi: 10.1016/j.mtadv.2020.100103.

19. Sharma GK, Kukshal V, Shekhawat D, Patnaik A. Fabrication and characterization of metallic biomaterials in medical applications: A review. Advanced Materials and Manufacturing Processes. 2021:95–105.

20. Rahmani R, Kamboj N, Brojan M, Antonov M, Prashanth KG. Hybrid metal-ceramic biomaterials fabricated through powder bed fusion and powder metallurgy for improved impact resistance of craniofacial implants. Materialia. 2022;24:101465. doi: 10.1016/j.mtla.2022.101465.

21. Omar O, Engstrand T, Linder LKB, Åberg J, Shah FA, Palmquist A, et al. In situ bone regeneration of large cranial defects using synthetic ceramic implants with a tailored composition and design. Proceedings of the National Academy of Sciences - PNAS. 2020;117(43):26660–26671. doi: 10.1073/pnas.2007635117.

22. Ogata S, Li J, Hirosaki N, Shibutani Y, Yip S. Ideal shear strain of metals and ceramics. Physical Review B, Condensed Matter and Materials Physics. 2004;70(10):104104.1-.7. doi: 10.1103/PhysRevB.70.104104.

23. Varshney S, Nigam A, Singh A, Samanta SK, Mishra N, Tewari RP. Antibacterial, structural, and mechanical properties of MgO/ZnO nanocomposites and its HA-based bio-ceramics; synthesized via physio-chemical route for biomedical applications. Materials Technology (New York, NY). 2022;37(13):2503–2516. doi: 10.1 080/10667857.2022.2043661.

24. Punj S, Singh J, Singh K. Ceramic biomaterials: properties, state of the art and future prospectives. Ceramics International. 2021;47(20):28059–28074. doi: 10.1016/j.ceramint.2021.06.238.
25. Sáenz A, Rivera-Muñoz E, Brostow W, Castaño VM. Ceramic biomaterials: an introductory overview. Journal of Materials Education. 1999;21(5-6):297–306.
26. Subedi M. Ceramics and its importance. The Himalayan Physics. 2013;4(4):80–82.
27. Wong JY, Bronzino JD, Peterson DR. Biomaterials: principles and practices. Boca Raton, FL: CRC Press; 2013.
28. Festas AJ, Ramos A, Davim JP. Medical devices biomaterials – a review. Proceedings of the Institution of Mechanical Engineers, Part L: Journal of Materials: Design and Applications. 2020;234(1):218–228. doi: 10.1177/1464420719882458.
29. Piddock V, Qualtrough AJE. Dental ceramics—an update. Journal of Dentistry. 1990;18(5):227–235. doi: 10.1016/0300-5712(90)90019-B.
30. Seghi RR, Daher T, Caputo A. Relative flexural strength of dental restorative ceramics. Dental Materials. 1990;6(3):181–184. doi: 10.1016/0109-5641(90)90026-B.
31. Regi Mv, Esbrit P, Salinas AJ. Degradative effects of the biological environment on ceramic biomaterials. In: Sakiyama-Elbert Shelly E, Wagner William R, Zhang G, Yaszemski Michael J, editors. Biomaterials science. 4th ed. Academic Press; 2020. pp. 955–971.
32. Balasubramanian S, Gurumurthy B, Balasubramanian A. Biomedical applications of ceramic nanomaterial: A review. International Journal of Pharmaceutical Sciences and Research. 2017;8(12):4950. doi: 10.13040/IJPSR.0975-8232.8(12).4950-59.
33. Hashim D, Cionca N, Courvoisier DS, Mombelli A. A systematic review of the clinical survival of zirconia implants. Clinical Oral Investigations. 2016;20(7):1403–1417. doi: 10.1007/s00784-016-1853-9.
34. Shi W, Fuad ARM, Li Y, Wang Y, Huang J, Du R, et al. Biodegradable polymeric nanoparticles increase risk of cardiovascular diseases by inducing endothelium dysfunction and inflammation. Journal of Nanobiotechnology. 2023;21(1):65. doi: 10.1186/s12951-023-01808-3.
35. Zindani D, Kumar K, Davim JP. Metallic biomaterials—A review. In: Davim JP, editor. Mechanical behaviour of biomaterials. Woodhead Publishing; 2019. pp. 83–99.
36. Prasad K, Bazaka O, Chua M, Rochford M, Fedrick L, Spoor J, et al. Metallic biomaterials: current challenges and opportunities. Materials. 2017;10(8):884. doi: 10.3390/ma10080884.
37. Jiang W-W, Su S-H, Eberhart RC, Tang L. Phagocyte responses to degradable polymers. Journal of Biomedical Materials Research Part A. 2007;82A(2):492–497. doi: 10.1002/jbm.a.31175.
38. Jones JR, Brauer DS, Hupa L, Greenspan DC. Bioglass and bioactive glasses and their impact on healthcare. International Journal of Applied Glass Science. 2016;7(4):423–434. doi: 10.1111/ijag.12252.
39. Li H, Wu Z, Zhou Y, Chang J. Bioglass for skin regeneration. In: García-Gareta E, editor. Biomaterials for skin repair and regeneration. Woodhead Publishing; 2019. pp. 225–250.
40. Alshatwi AA, Athinarayanan J, Periasamy VS. Biocompatibility assessment of rice husk-derived biogenic silica nanoparticles for biomedical applications. Materials Science & Engineering C. 2015;47:8–16. doi: 10.1016/j.msec.2014.11.005.
41. Sharma G, Kaur M, Punj S, Singh K. Biomass as a sustainable resource for value-added modern materials: a review. Biofuels, Bioproducts and Biorefining. 2020;14(3):673–695. doi: 10.1002/bbb.2079.
42. Ismail H, Mohamad H. Bioactivity and biocompatibility properties of sustainable wollastonite bioceramics from rice husk ash/rice straw ash: A review. Materials. 2021;14(18):5193. doi: 10.3390/ma14185193.

43. Guo S. Reactive hot-pressed hybrid ceramic composites comprising SiC(SCS-6)/Ti composite and ZrB2-ZrC ceramic. Journal of the American Ceramic Society. 2016;99(10):3241–3250. doi: 10.1111/jace.14372.

44. Chen R, Zhang Y, Xie Q, Chen Z, Ma C, Zhang G. Transparent polymer-ceramic hybrid antifouling coating with superior mechanical properties. Advanced Functional Materials. 2021;31(19):2011145-n/a. doi: 10.1002/adfm.202011145.

45. Young RJ, Lovell PA. Introduction to polymers. 3rd ed. Boca Raton: CRC Press; 2011.

46. Brady J, Dürig T, Lee PI, Li J-X. Polymer properties and characterization. In: Qiu Y, Chen Y, Zhang GGZ, Liu L, and Porter W, editors. Developing solid oral dosage forms: pharmaceutical theory and practice. 2nd ed. London: Elsevier Science; 2017.

47. Holt AP, Bocharova V, Cheng S, Kisliuk AM, White BT, Saito T, et al. Controlling interfacial dynamics: Covalent bonding versus physical adsorption in polymer nanocomposites. ACS Nano. 2016;10(7):6843–6852. doi: 10.1021/acsnano.6b02501.

48. Jiang Z, Bhaskaran A, Aitken HM, Shackleford ICG, Connal LA. Using synergistic multiple dynamic bonds to construct polymers with engineered properties. Macromolecular Rapid Communications. 2019;40(10):e1900038-n/a. doi: 10.1002/marc.201900038.

49. Zhao X, Xue G, Qu G, Singhania V, Zhao Y, Butrouna K, et al. Complementary semiconducting polymer blends: influence of side chains of matrix polymers. Macromolecules. 2017;50(16):6202–6209. doi: 10.1021/acs.macromol.7b01354.

50. Roesler J, Harders H, Baeker M. Mechanical behaviour of engineering materials: metals, ceramics, polymers, and composites. Berlin; Heidelberg: Springer Berlin/Heidelberg; 2007.

51. Bahl S, Dolma J, Jyot Singh J, Sehgal S. Biodegradation of plastics: a state of the art review. Elsevier Ltd; 2021. pp. 31–34.

52. Duncan EM, Broderick AC, Fuller WJ, Galloway TS, Godfrey MH, Hamann M, et al. Microplastic ingestion ubiquitous in marine turtles. Global Change Biology. 2019;25(2):744–752. doi: 10.1111/gcb.14519.

53. Meereboer KW, Misra M, Mohanty AK. Review of recent advances in the biodegradability of polyhydroxyalkanoate (PHA) bioplastics and their composites. Green Chemistry: An International Journal and Green Chemistry Resource: GC. 2020;22(17):5519–5558. doi: 10.1039/d0gc01647k.

54. Salazar MdR, Solanilla Duque JF, Saenz-Galindo A, Rodriguez-Herrera R. Biodegradable polymers: concepts and applications. Milton: Taylor & Francis Group; 2023.

55. Mohanty AK, Vivekanandhan S, Pin JM, Misra M. Composites from renewable and sustainable resources: Challenges and innovations. Science. 2018;362(6414):536–542.

56. Kijchavengkul T, Auras R. Compostability of polymers. Polymer International. 2008;57(6):793–804. doi: 10.1002/pi.2420.

57. Castro-Aguirre E, Auras R, Selke S, Rubino M, Marsh T. Insights on the aerobic biodegradation of polymers by analysis of evolved carbon dioxide in simulated composting conditions. Polymer Degradation and Stability. 2017;137:251–271. doi: 10.1016/j.polymdegradstab.2017.01.017.

58. Narayan R. Biobased & biodegradable plastics: rationale, drivers, and technology exemplars. Washington, DC: American Chemical Society; 2012.

59. Mohee R, Unmar GD, Mudhoo A, Khadoo P. Biodegradability of biodegradable/degradable plastic materials under aerobic and anaerobic conditions. Waste Management (Elmsford). 2008;28(9):1624–1629. doi: 10.1016/j.wasman.2007.07.003.

60. Barron A, Sparks TD. Commercial marine-degradable polymers for flexible packaging. iScience. 2020;23(8):101353. doi: 10.1016/j.isci.2020.101353.

61. Rudnik E. Compostable polymer materials. 2nd ed. Marrickville: Elsevier; 2019.

62. Ozkoc G, Kemaloglu S. Morphology, biodegradability, mechanical, and thermal properties of nanocomposite films based on PLA and plasticized PLA. Journal of Applied Polymer Science. 2009;114(4):2481–2487. doi: 10.1002/app.30772.

63. Pillin I, Montrelay N, Grohens Y. Plastification of PLA: Is the interaction parameter a significant factor? Polymer (Guilford). 2006;47(13):4676.

64. Martin O, Averous L. Poly(lactic acid): plasticization and properties of biodegradable multiphase systems. Polymer (Guilford). 2001;42(14):6209–6219. doi: 10.1016/S0032-3861(01)00086-6.

65. Ljungberg N, Wesslén B. The effects of plasticizers on the dynamic mechanical and thermal properties of poly(lactic acid). Journal of Applied Polymer Science. 2002;86(5):1227–1234. doi: 10.1002/app.11077.

66. Bose RJC, Kim M, Chang JH, Paulmurugan R, Moon JJ, Koh W-G, et al. Biodegradable polymers for modern vaccine development. Journal of Industrial and Engineering Chemistry (Seoul, Korea). 2019;77:12–24. doi: 10.1016/j.jiec.201 9.04.044.

67. Zumstein MT, Schintlmeister A, Nelson TF, Baumgartner R, Woebken D, Wagner M, et al. Biodegradation of synthetic polymers in soils: Tracking carbon into CO2 and microbial biomass. Science Advances. 2018;4(7):eaas9024-eaas. doi: 10.1126/sciadv.aas9024.

68. Eubeler JP, Bernhard M, Knepper TP. Environmental biodegradation of synthetic polymers II. Biodegradation of different polymer groups. TrAC, Trends in Analytical Chemistry (Regular ed). 2010;29(1):84–100. doi: 10.1016/j.trac.2009.09.005.

69. Fitriani F, Bilad MR, Aprilia S, Arahman N. Biodegradable hybrid polymer film for packaging: A review. Journal of Natural Fibers. 2023;20(1). doi: 10.1080/15440478. 2022.2159606.

70. García-Estrada P, García-Bon MA, López-Naranjo EJ, Basaldúa-Pérez DN, Santos A, Navarro-Partida J. Polymeric implants for the treatment of intraocular eye diseases: trends in biodegradable and non-biodegradable materials. Pharmaceutics. 2021. doi: 10.3390/pharmaceutics13050701.

71. Andreeßen C, Steinbüchel A. Recent developments in non-biodegradable biopolymers: Precursors, production processes, and future perspectives. Applied Microbiology and Biotechnology. 2019;103(1):143–157. doi: 10.1007/s00253-018-9483-6.

72. Steinbuchel A. Non-biodegradable biopolymers from renewable resources: perspectives and impacts. Current Opinion in Biotechnology. 2005;16(6):607–613. doi: 10.1016/j.copbio.2005.10.011.

73. Rahman MH, Bhoi PR. An overview of non-biodegradable bioplastics. Journal of Cleaner Production. 2021;294:126218. doi: 10.1016/j.jclepro.2021.126218.

74. Nakajima H, Dijkstra P, Loos K. The recent developments in biobased polymers toward general and engineering applications: polymers that are upgraded from biodegradable polymers, analogous to petroleum-derived polymers, and newly developed. Polymers. 2017;9(10):523.

75. Peixoto T, Carneiro S, Fangueiro R, Guedes RM, Paiva MC, Lopes MA. Engineering hybrid textile braids for tendon and ligament repair application. Journal of Applied Polymer Science. 2022;139(17):52013. doi: 10.1002/app.52 013.

76. Razavi M, Thakor AS. An oxygen plasma treated poly(dimethylsiloxane) bioscaffold coated with polydopamine for stem cell therapy. Journal of Materials Science: Materials in Medicine. 2018;29(5):54. doi: 10.1007/s10856-018-6077-x.

77. Talon I, Schneider A, Ball V, Hemmerlé J. Functionalization of PTFE materials using a combination of polydopamine and platelet-rich fibrin. Journal of Surgical Research. 2020;251:254–261. doi: 10.1016/j.jss.2019.11.014.

78. Koutsamanis I, Roblegg E, Spoerk M. Controlled delivery via hot-melt extrusion: A focus on non-biodegradable carriers for non-oral applications. Journal of Drug Delivery Science and Technology. 2023;81. doi: 10.1016/j.jddst.2023.104289.

79. Wang K, Deng Q. The thermal and mechanical properties of Poly(ethylene-co-vinyl acetate) random copolymers (PEVA) and its covalently crosslinked analogues (cPEVA). Polymers. 2019;11(6):1055.

80. Chalykh AE, Stepanenko VY, Shcherbina AA, Balashova EG. Adhesive properties of ethylene and vinyl acetate copolymers. Polymer Science Series D. 2009;2(1):8–15. doi: 10.1134/S199542120901002X.

81. Frey SJ, Hoffman AS, Hubbell JA, Kane RS. Surface-immobilized biomolecules. In: Sakiyama-Elbert Shelly E, Wagner William R, Zhang G, Yaszemski MJ, editors. Biomaterials science - an introduction to materials in medicine. Academic Press; 2020.

82. Ratner BD, Hoffman AS, McArthur SL. Physicochemical surface modification of materials used in medicine. In: Sakiyama-Elbert Shelly E, Wagner William R, Zhang G, Yaszemski Michael J, editors. Biomaterials science. 4th ed. Academic Press; 2020. p. 1.

83. Asri RIM, Harun WSW, Samykano M, Lah NAC, Ghani SAC, Tarlochan F, et al. Corrosion and surface modification on biocompatible metals: A review. Materials Science and Engineering: C. 2017;77:1261–1274. doi: 10.1016/j.msec.2017.04.102.

84. Thanigaivel S, Priya AK, Balakrishnan D, Dutta K, Rajendran S, Soto-Moscoso M. Insight on recent development in metallic biomaterials: Strategies involving synthesis, types and surface modification for advanced therapeutic and biomedical applications. Biochemical Engineering Journal. 2022;187(Complete). doi: 10.1016/j.bej.2022.108522.

85. Harun WSW, Asri RIM, Alias J, Zulkifli FH, Kadirgama K, Ghani SAC, et al. A comprehensive review of hydroxyapatite-based coatings adhesion on metallic biomaterials. Ceramics International. 2018;44(2):1250–1268. doi: 10.1016/j.ceramint.2017.10.162.

86. Asri RIM, Harun WSW, Hassan MA, Ghani SAC, Buyong Z. A review of hydroxyapatite-based coating techniques: Sol–gel and electrochemical depositions on biocompatible metals. Journal of the Mechanical Behavior of Biomedical Materials. 2016;57:95–108. doi: 10.1016/j.jmbbm.2015.11.031.

87. Packham DE. Surface energy, surface topography and adhesion. International Journal of Adhesion and Adhesives. 2003;23(6):437–448. doi: 10.1016/S0143-74 96(03)00068-X.

88. Asensio G, Vizquez-Lasa B, Rojo L. Achievements in the topographic design of commercial Titanium dental implants: Towards anti-Peri-implantitis surfaces. Journal of Clinical Medicine. 2019;8(11):1982. doi: 10.3390/jcm8111982.

89. Jaafar A, Hecker C, Árki P, Joseph Y. Sol-gel derived hydroxyapatite coatings for titanium implants: A review. Bioengineering. 2020;7(4):127.

90. Cristian C-F, Yamile P-P, Manuel Alejandro E-G, Erbin GU-C, Juan Antonio J-M, Alejandro A-O. Surface science engineering through Sol-Gel process. In: Gurrappa I, editor. Applied surface science. Rijeka: IntechOpen; 2019. Ch. 3.

91. Catauro M, Papale F, Bollino F. Characterization and biological properties of TiO2/PCL hybrid layers prepared via sol–gel dip coating for surface modification of titanium implants. Journal of Non-Crystalline Solids. 2015;415:9–15. doi: 10.1016/j.jnoncrysol.2014.12.008.

92. Wang X, Cai S, Liu T, Ren M, Huang K, Zhang R, et al. Fabrication and corrosion resistance of calcium phosphate glass-ceramic coated Mg alloy via a PEG assisted sol–gel method. Ceramics International. 2014;40(2):3389–3398. doi: 10.1016/j.ceramint.2013.09.093.

93. Rasouli R, Barhoum A, Uludag H. A review of nanostructured surfaces and materials for dental implants: surface coating, patterning and functionalization for improved performance. Biomaterials Science. 2018;6(6):1312–1338. doi: 10.1039/C8BM00021B.

94. Sadati Tilebon SM, Emamian SA, Ramezanpour H, Yousefi H, Özcan M, Naghib SM, et al. Intelligent modeling and optimization of titanium surface etching for dental implant application. Scientific Reports. 2022;12(1):7184. doi: 10.1038/s415 98-022-11254-0.

95. Lupi SM, Torchia M, Rizzo S. Biochemical modification of Titanium oral implants: evidence from In Vivo studies. Materials. 2021. doi: 10.3390/ma14112798.

96. Lupi SM, Rodriguez y Baena A, Cassinelli C, Iviglia G, Tallarico M, Morra M, et al. Covalently-linked hyaluronan versus acid etched Titanium dental implants: A crossover RCT in humans. International Journal of Molecular Sciences. 2019. doi: 10.3390/ijms20030763.

97. Scarano A, Lorusso F, Orsini T, Morra M, Iviglia G, Valbonetti L. Biomimetic surfaces coated with covalently immobilized collagen Type I: An X-Ray photoelectron spectroscopy, Atomic Force Microscopy, micro-CT and Histomorphometrical Study in Rabbits. International Journal of Molecular Sciences. 2019. doi: 10.3390/ijms2003 0724.

98. Liu X, Yue Z, Romeo T, Weber J, Scheuermann T, Moulton S, et al. Biofunctionalized anti-corrosive silane coatings for magnesium alloys. Acta Biomaterialia. 2013;9(10):8671–8677. doi: 10.1016/j.actbio.2012.12.025.

99. Morra M, Cassinelli C, Cascardo G, Fini M, Giavaresi G, Giardino R. Covalently-linked hyaluronan promotes bone formation around Ti implants in a rabbit model. Journal of Orthopaedic Research. 2009;27(5):657–663. doi: 10.1002/jor.20797.

100. Neouze MA, Schubert US. Surface modification and functionalization of metal and metal oxide nanoparticles by organic ligands. Monatshefte für Chemie. 2008;139(3):183–195. doi: 10.1007/s00706-007-0775-2.

101. Sperling RA, Parak WJ. Surface modification, functionalization and bioconjugation of colloidal inorganic nanoparticles. Philosophical Transactions of the Royal Society of London Series A: Mathematical, Physical, and Engineering Sciences. 2010;368(1915):1333–1383. doi: 10.1098/rsta.2009.0273.

102. Pei J, Hall H, Spencer ND. The role of plasma proteins in cell adhesion to PEG surface-density-gradient-modified titanium oxide. Biomaterials. 2011;32(34):8968–8978. doi: 10.1016/j.biomaterials.2011.08.034.

103. Hao L, Lawrence J, Chian KS. Osteoblast cell adhesion on a laser modified zirconia based bioceramic. Journal of Materials Science Materials in Medicine. 2005;16(8):719–726. doi: 10.1007/s10856-005-2608-3.

104. Cauda V, Schlossbauer A, Bein T. Bio-degradation study of colloidal mesoporous silica nanoparticles: Effect of surface functionalization with organo-silanes and poly (ethylene glycol). Microporous and Mesoporous Materials. 2010;132(1):60–71. doi: 10.1016/j.micromeso.2009.11.015.

105. Pawsey S, Yach K, Reven L. Self-assembly of carboxyalkylphosphonic acids on metal oxide powders. Langmuir. 2002;18(13):5205–5212. doi: 10.1021/la015749h.

106. Adden N, Gamble LJ, Castner DG, Hoffmann A, Gross G, Menzel H. Phosphonic acid monolayers for binding of bioactive molecules to Titanium surfaces. Langmuir. 2006;22(19):8197–8204. doi: 10.1021/la060754c.

107. Meng H, Liong M, Xia T, Li Z, Ji Z, Zink JI, et al. Engineered design of mesoporous silica nanoparticles to deliver doxorubicin and P-Glycoprotein siRNA to overcome drug resistance in a cancer cell line. ACS Nano. 2010;4(8):4539–4550. doi: 10.1021/nn100690m.

108. Burtea C, Laurent S, Mahieu I, Larbanoix L, Roch A, Port M, et al. In vitro biomedical applications of functionalized iron oxide nanoparticles, including those not related to magnetic properties. Contrast Media and Molecular Imaging. 2011;6(4):236–250. doi: 10.1002/cmmi.423.

109. Panzavolta S, Torricelli P, Bracci B, Fini M, Bigi A. Functionalization of biomimetic calcium phosphate bone cements with alendronate. Journal of Inorganic Biochemistry. 2010;104(10):1099–1106. doi: 10.1016/j.jinorgbio.2010.06.008.

110. Zhu M, Lerum MZ, Chen W. How to prepare reproducible, homogeneous, and hydrolytically stable aminosilane-derived layers on silica. Langmuir. 2012; 28(1):416–423. doi: 10.1021/la203638g.

111. Laurenti M, Stassi S, Canavese G, Cauda V. Surface engineering of nanostructured ZnO surfaces. Advanced Materials Interfaces. 2017;4(2):1600758-n/a. doi: 10.1002/admi.201600758.

112. Basaldella EI, Legnoverde MS. Functionalized silica matrices for controlled delivery of cephalexin. Journal of Sol-Gel Science and Technology. 2010;56(2):191–196. doi: 10.1007/s10971-010-2293-7.

113. Sassolas A, Blum LJ, Leca-Bouvier BD. Immobilization strategies to develop enzymatic biosensors. Biotechnology Advances. 2012;30(3):489–511. doi: 10.1016/j.biotechadv.2011.09.003.

114. Lee W-H, Loo C-Y, Rohanizadeh R. A review of chemical surface modification of bioceramics: Effects on protein adsorption and cellular response. Colloids and Surfaces, B, Biointerfaces. 2014;122:823–834. doi: 10.1016/j.colsurfb.2014.07.029.

115. Kidoaki S, Matsuda T. Mechanistic aspects of protein/material interactions probed by atomic force microscopy. Colloids and Surfaces B: Biointerfaces. 2002;23(2):153–163. doi: 10.1016/S0927-7765(01)00232-6.

116. Lee W-H, Loo C-Y, Zavgorodniy AV, Rohanizadeh R. High protein adsorptive capacity of amino acid-functionalized hydroxyapatite. Journal of Biomedical Materials Research Part A. 2013;101A(3):873–883. doi: 10.1002/jbm.a.34383.

117. Lee W-H, Loo C-Y, Zavgorodniy AV, Ghadiri M, Rohanizadeh R. A novel approach to enhance protein adsorption and cell proliferation on hydroxyapatite: citric acid treatment. RSC Advances. 2013;3(12):4040–4051. doi: 10.1039/C3RA22966A.

118. Wang K, Zhou C, Hong Y, Zhang X. A review of protein adsorption on bioceramics. Interface Focus. 2012;2(3):259–277. doi: 10.1098/rsfs.2012.0012.

119. Lee W-H, Zavgorodniy AV, Loo C-Y, Rohanizadeh R. Synthesis and characterization of hydroxyapatite with different crystallinity: Effects on protein adsorption and release. Journal of Biomedical Materials Research Part A. 2012;100A(6):1539–1549. doi: 10.1002/jbm.a.34093.

120. Lee W-H, Loo C-Y, Van KL, Zavgorodniy AV, Rohanizadeh R. Modulating protein adsorption onto hydroxyapatite particles using different amino acid treatments. Journal of the Royal Society Interface. 2012;9(70):918–927. doi: 10.1098/rsif.2011.0586.

121. Palazzo B, Walsh D, Iafisco M, Foresti E, Bertinetti L, Martra G, et al. Amino acid synergetic effect on structure, morphology and surface properties of biomimetic apatite nanocrystals. Acta Biomaterialia. 2009;5(4):1241–1252. doi: 10.1016/j.actbio.2008.10.024.

122. Talbert JN, Goddard JM. Enzymes on material surfaces. Colloids and Surfaces, B, Biointerfaces. 2012;93:8–19. doi: 10.1016/j.colsurfb.2012.01.003.

123. Can K, Ozmen M, Ersoz M. Immobilization of albumin on aminosilane modified superparamagnetic magnetite nanoparticles and its characterization. Colloids and Surfaces, B, Biointerfaces. 2009;71(1):154–159. doi: 10.1016/j.colsurfb.2009.01.021.

124. Cheng K, Blumen SR, MacPherson MB, Steinbacher JL, Mossman BT, Landry CC. Enhanced uptake of porous silica microparticles by bifunctional surface modification with a targeting antibody and a biocompatible polymer. ACS Applied Materials & Interfaces. 2010;2(9):2489–2495. doi: 10.1021/am100530t.

125. Michael FM, Khalid M, Walvekar R, Siddiqui H, Balaji AB. Surface modification techniques of biodegradable and biocompatible polymers. Biodegradable and Biocompatible Polymer Composites: Processing, Properties and Applications; Shimpi, NG, Ed. 2018:33–54.

126. Nemani SK, Annavarapu RK, Mohammadian B, Raiyan A, Heil J, Haque MA, et al. Surface modification of polymers: methods and applications. Advanced Materials Interfaces. 2018;5(24):1801247. doi: 10.1002/admi.201801247.

127. Arima Y, Iwata H. Effect of wettability and surface functional groups on protein adsorption and cell adhesion using well-defined mixed self-assembled monolayers. Biomaterials. 2007;28(20):3074–3082. doi: 10.1016/j.biomaterials.2007.03.013.

128. Liu W, Han K, Sun M, Wang J. Enhancement and control of neuron adhesion on polydimethylsiloxane for cell microengineering using a functionalized triblock polymer. Lab on a Chip. 2019;19(19):3162–3167. doi: 10.1039/c9lc00736a.

129. Sun W, Liu W, Wu Z, Chen H. Chemical surface modification of polymeric biomaterials for biomedical applications. Macromolecular Rapid Communications. 2020;41(8):e1900430-n/a. doi: 10.1002/marc.201900430.

130. Liu X, Holzwarth JM, Ma PX. Functionalized synthetic biodegradable polymer scaffolds for tissue engineering: functionalized synthetic biodegradable polymer scaffolds. Macromolecular Bioscience. 2012;12(7):911–919. doi: 10.1002/mabi.2 01100466.

131. Wu F, Misra M, Mohanty AK. Challenges and new opportunities on barrier performance of biodegradable polymers for sustainable packaging. Progress in Polymer Science. 2021;117:101395. doi: 10.1016/j.progpolymsci.2021.101395.

132. Källrot N, Linse P. Dynamic study of single-chain adsorption and desorption. Macromolecules. 2007;40(13):4669–4679. doi: 10.1021/ma0702602.

133. Gregory J, Barany S. Adsorption and flocculation by polymers and polymer mixtures. Advances in Colloid and Interface Science. 2011;169(1):1–12. doi: 10.1 016/j.cis.2011.06.004.

134. Zhang K, Zheng H, Liang S, Gao C. Aligned PLLA nanofibrous scaffolds coated with graphene oxide for promoting neural cell growth. Acta Biomaterialia. 2016;37:131–142. doi: 10.1016/j.actbio.2016.04.008.

135. Meng S. Nerve cell differentiation using constant and programmed electrical stimulation through conductive non-functional graphene nanosheets film. Tissue Engineering and Regenerative Medicine. 2014;11(4):274–283. doi: 10.1007/s13 770-014-0011-1.

136. Heo C, Yoo J, Lee S, Jo A, Jung S, Yoo H, et al. The control of neural cell-to-cell interactions through non-contact electrical field stimulation using graphene electrodes. Biomaterials. 2011;32(1):19–27. doi: 10.1016/j.biomaterials.2010.08.095.

137. Amani H, Arzaghi H, Bayandori M, Dezfuli AS, Pazoki-Toroudi H, Shafiee A, et al. Controlling cell behavior through the design of biomaterial surfaces: A focus on surface modification techniques. Advanced Materials Interfaces. 2019;6(13):1900572. doi: 10.1 002/admi.201900572.

138. Douglas JF, Johnson HE, Granick S. A simple kinetic model of polymer adsorption and desorption. Science (American Association for the Advancement of Science). 1993;262(5142):2010–2012. doi: 10.1126/science.262.5142.2010.

139. Henze M, Mädge D, Prucker O, Rühe J. "Grafting Through": mechanistic aspects of radical polymerization reactions with surface-attached monomers. Macromolecules. 2014;47(9):2929–2937. doi: 10.1021/ma402607d.

140. Tan SH, Nguyen N-T, Chua YC, Kang TG. Oxygen plasma treatment for reducing hydrophobicity of a sealed polydimethylsiloxane microchannel. Biomicrofluidics. 2010;4(3). doi: 10.1063/1.3466882.

141. Matějíček J, Vilémová M, Mušálek R, Sachr P, Horník J. The influence of interface characteristics on the adhesion/cohesion of plasma sprayed Tungsten coatings. Coatings. 2013;3(2):108–125.

142. Lee JH, Jung HW, Kang I-K, Lee HB. Cell behaviour on polymer surfaces with different functional groups. Biomaterials. 1994;15(9):705–711. doi: 10.1016/0142-9612(94)90169-4.

143. Chuah YJ, Kuddannaya S, Lee MHA, Zhang Y, Kang Y. The effects of poly (dimethylsiloxane) surface silanization on the mesenchymal stem cell fate. Biomaterials Science. 2015;3(2):383–390.

144. Ryu GH, Yang W-S, Roh H-W, Lee I-S, Kim JK, Lee GH, et al. Plasma surface modification of poly (d,l-lactic-co-glycolic acid) (65/35) film for tissue engineering. Surface & Coatings Technology. 2005;193(1):60–64. doi: 10.1016/j.surfcoat.2004. 07.062.

145. Gentile P, Chiono V, Carmagnola I, Hatton PV. An overview of poly(lactic-co-glycolic) acid (PLGA)-based biomaterials for bone tissue engineering. International Journal of Molecular Sciences. 2014;15(3):3640–3659. doi: 10.3390/ijms15033640.

146. Bose S, Tarafder S. Calcium phosphate ceramic systems in growth factor and drug delivery for bone tissue engineering: A review. Acta Biomaterialia. 2012;8(4):1401–1421. doi: 10.1016/j.actbio.2011.11.017.

147. Yewle JN, Wei Y, Puleo DA, Daunert S, Bachas LG. Oriented immobilization of proteins on hydroxyapatite surface using bifunctional bisphosphonates as linkers. Biomacromolecules. 2012;13(6):1742–1749. doi: 10.1021/bm201865r.

148. Szczepanski V, Vlassiouk I, Smirnov S. Stability of silane modifiers on alumina nanoporous membranes. Journal of Membrane Science. 2006;281(1):587–591. doi: 10.1016/j.memsci.2006.04.027.

149. Gough JE, Scotchford CA, Downes S. Cytotoxicity of glutaraldehyde crosslinked collagen/poly(vinyl alcohol) films is by the mechanism of apoptosis. Journal of Biomedical Materials Research. 2002;61(1):121–130. doi: 10.1002/jbm.10145.

150. Hong S, Kim KY, Wook HJ, Park SY, Lee KD, Lee DY, et al. Attenuation of the in vivo toxicity of biomaterials by polydopamine surface modification. Nanomedicine (London, England). 2011;6(5):793–801. doi: 10.2217/NNM.11.76.

151. Ye Q, Zhou F, Liu W. Bioinspired catecholic chemistry for surface modification. Chemical Society Reviews. 2011;4(7):4244–4258. doi: 10.1039/c1cs15026j.

152. Kokubo T, Takadama H. How useful is SBF in predicting in vivo bone bioactivity? Biomaterials. 2006;27(15):2907–2915. doi: 10.1016/j.biomaterials.2006.01.017.

153. Mano SS, Kanehira K, Sonezaki S, Taniguchi A. Effect of polyethylene glycol modification of TiO2 nanoparticles on cytotoxicity and gene expressions in human cell lines. International Journal of Molecular Sciences. 2012;13(3):3703–3717. doi: 10.3390/ijms13033703.

154. Pearce AI, Richards RG, Milz S, Schneider E, Pearce SG. Animal models for implant biomaterial research in bone: a review. European Cells & Materials. 2007;13:1–10. doi: 10.22203/eCM.v013a01.

155. Yuan JH, Wang K, Xia XH. Highly ordered platinum-nanotubule arrays for amperometric glucose sensing. Advanced Functional Materials. 2005;15(5):803–809. doi: 10.1002/adfm.200400321.

156. Sur S, Rathore A, Dave V, Reddy KR, Chouhan RS, Sadhu V. Recent developments in functionalized polymer nanoparticles for efficient drug delivery system. Nano-Structures & Nano-Objects. 2019;20:100397. doi: 10.1016/j.nanoso.2019.100397.

157. Duan B, Wang M. Customized Ca–P/PHBV nanocomposite scaffolds for bone tissue engineering: design, fabrication, surface modification and sustained release of growth factor. Journal of the Royal Society Interface. 2010;7(Suppl-5):S615–S629. doi: 10.1098/rsif.2010.0127.focus.
158. Kim H-W, Kim H-E, Knowles JC. Hard-tissue-engineered zirconia porous scaffolds with hydroxyapatite sol-gel and slurry coatings. Journal of Biomedical Materials Research. 2004;70B(2):270–277. doi: 10.1002/jbm.b.30032.
159. Qian W, Miao Z, Zhang X-J, Yang X-T, Tang Y-Y, Tang YY, et al. Functionalized reduced graphene oxide with aptamer macroarray for cancer cell capture and fluorescence detection. Mikrochimica Acta (1966). 2020;187(7):407. doi: 10.1007/s00604-020-04402-8.
160. Fang Y, Huang X-J, Chen P-C, Xu Z-K. Polymer materials for enzyme immobilization and their application in bioreactors. BMB Reports. 2011; 44(2):87–95. doi: 10.5483/BMBRep.2011.44.2.87.
161. You J, Wang L, Zhao Y, Bao W. A review of amino-functionalized magnetic nanoparticles for water treatment: Features and prospects. Journal of Cleaner Production. 2021;281:124668. doi: 10.1016/j.jclepro.2020.124668.

# 7 Corrosion, Degradation, and Material Release by Biomaterials

*Sanjay Sharma and Santosh G*
Department of Mechanical Engineering, NMAM Institute of
Technology, Karnataka, India

## 7.1 INTRODUCTION

Various types of materials are accessible and used in different medical devices. These materials are used in combination with medical interference. The commonly used term for such materials is "biomaterials". One of the definitions of biomaterials can be "non-transferable materials which have interface with the biological system to assess, treat, supplement or replace any body tissue, organ or function of the body". The biomaterials field is of enormous importance for human beings as the very continuation and long life of some of the less fortunate human beings, sometimes even at the time of birth, are born with heart disease and also for the aged people who have disabilities and adopt biomedical implants to lead a happy time for the rest of their life span. The aged people need the help of experienced physicians for several treatments, as the parts of the human body that have performed their expected tasks for long years are now becoming worn out. One of the major illnesses is arthritis which is generally faced by the aged and now even young people are also affected by these illnesses and it damages the life of those affected, leading to immobility and unbearable pain. However, the cause of this disease remains unknown even today, in spite of tremendous scientific advancements. Apart from diseased people, young and dynamic people like sports persons often need replacements due to fracture and excessive strain. The concepts of biomaterials and surgeries came as early as 600 B.C., as many Hindu surgeons used to restore the missing body parts, and as it was during that period Saint Sushrutha repaired an injured nose with a patch of living flesh. Moving to the east, back 4000 years ago, gold and iron were used for dental applications and wood for toe replacements by the Egyptians and Romans, but they had less knowledge of corrosion. After World War II, nylon, teflon, silicon, steel, and titanium were some of the materials used as implant material. At the present era, advancement in the knowledge of materials and surgical procedures and various classes of materials have been developed, such as metals, alloys, polymers, ceramics, and composites, which have been widely used to fabricate the bio-implants. Bio-implants are used in dental, orthopedics, plastic and reconstructive surgery,

DOI: 10.1201/9781003429920-8

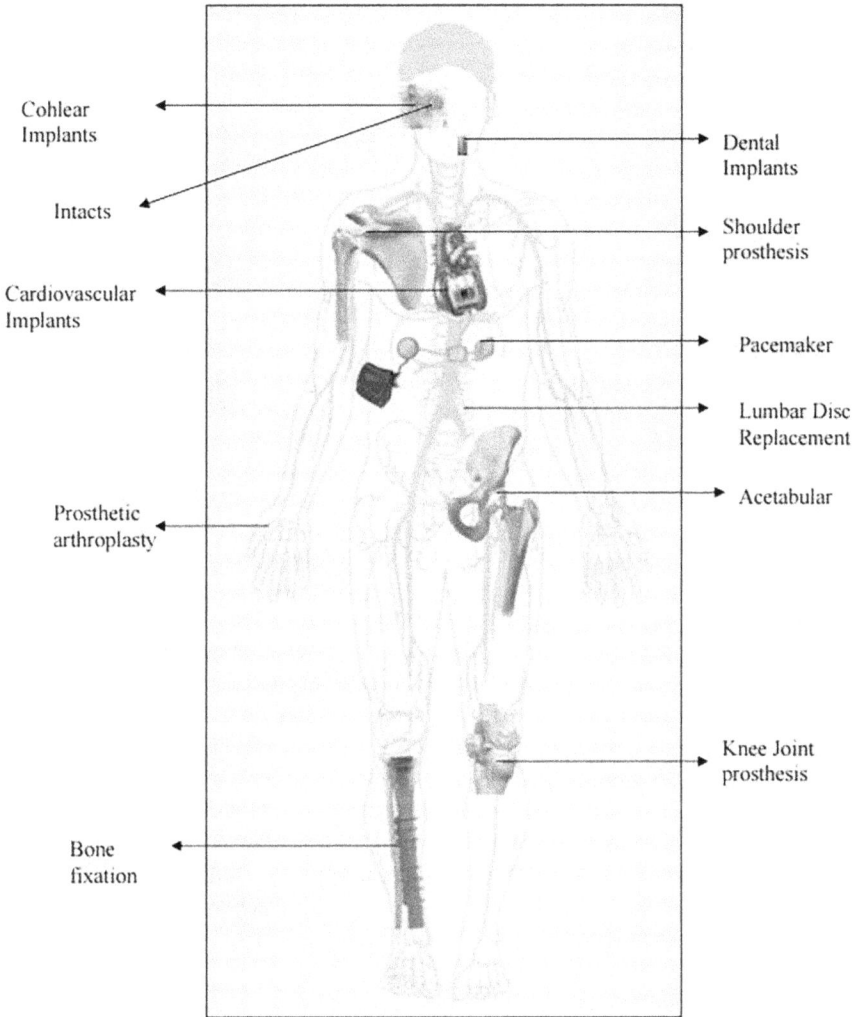

**FIGURE 7.1**  Biomaterials used in different parts of the human biological system [4].

ophthalmology, cardiovascular surgery, neurosurgery, immunology, histopathology, experimental surgery, and veterinary medicine, as shown in Figure 7.1 [1–4].

## 7.2  BIOCOMPATIBILITY

Biocompatibility is an important necessity of a biomaterial. The material implanted should not induce any undesirable consequences, such as allergy, inflammation, or toxicity, either immediately following surgery or in the post-operative period. Apart from fatigue strength and fracture toughness, biomaterials should have sufficient mechanical strength to withstand the forces to which they are subjected so that they do not fracture. More importantly, a bioimplant should have very high wear and

| REQUIREMENT OF IMPLANTS |
|---|

| Compatibility | Mechanical properties | Manufacturing |
|---|---|---|
| 1. Tissue reactions | Elasticity | Fabrication methods |
| 2. Change in properties | Yield Stress | Consistency and conformity to all requirement |
| a. Mechanical | Ductility | Quality of raw materials |
| b. Physical | Toughness | Superior techniques to obtain excellent surface finish or texture |
| c. Chemical | Time dependent deformation | |
| 3. Degradation leads to | Creep | Capability of material to get safe efficient sterilization |
| a. Local changes | Ultimate Strength | Cost of product |
| b. Harmful systematic effects | Fatigue strength | |
| | Hardness | |
| | Wear resistance | |

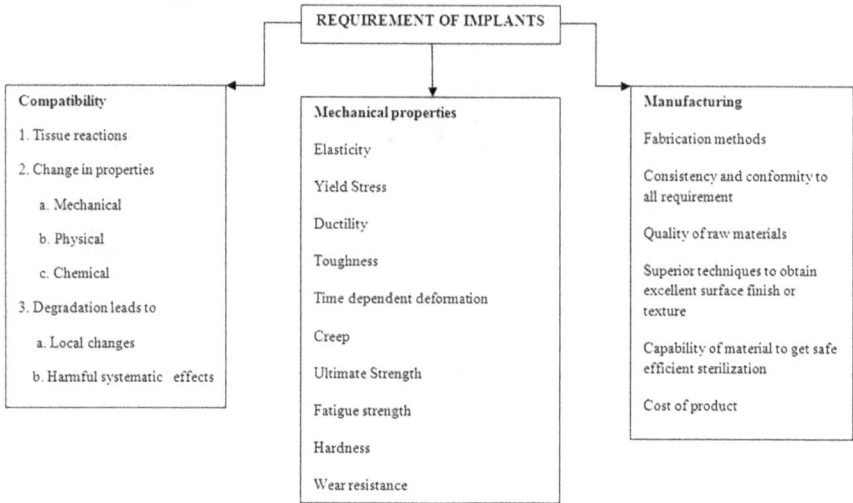

FIGURE 7.2   Biomaterial requirements in medical applications [1].

corrosion resistance in highly corrosive body environments and varying loading conditions. Manufacturing methods also must have superior quality and techniques with excellent surface finish and texture. Biomaterial used for production implants must have the capability to have safe, efficient sterilization and be cost effective. A biomaterial ought to last longer and shouldn't stop functioning until the human has passed away. It is clear from this criterion that elderly patients must serve a minimum of 10 to 20 years, while younger patients must serve a minimum of more than 20 years. Figure 7.2 illustrates the main criteria for biomaterials. A good biomaterial should be non-toxic, non-carcinogenic, and not cause chronic inflammation or interfere with biological processes. Consequently, the effectiveness of a biomaterial or an implant depends greatly on these key parameters [1]:

   i. A biocompatible chemical composition to keep away from unpleasant tissue reactions.
  ii. Outstanding resistance to deterioration (for example, metal corrosion resistance, biological decomposition of polymers).
 iii. Sufficient strength to sustain fatigue loading applied to the joint.
  iv. Allow modulus to reduce bone constriction.
   v. Better resistance to wear decreases the generation of wear debris.
  vi. Health condition of receiver and the proficiency of the surgeon.

## 7.3   NATURE OF ENVIRONMENT IN HUMAN BODY

Biomaterials react with body fluids, various cells, and enzymes that can generate toxicity and allergy and, finally, failure of the implant. The human body contains 50% to 70% water of its total mass and the body environment is aggressive to metals and alloys. Therefore, the relationship between a metallic implant's environment and its

| Dependent on: | Mechanisms of Material Degradation: | Material Properties Adversely Affected: |
|---|---|---|
| 1. Type of material<br>2. Static/dynamic stress<br>3. Projected device life<br>4. Interactions with other device components | 1. Corrosion<br>2. Dissolution<br>3. Chemical modification<br>4. Swelling<br>5. Leaching<br>6. Wear | 1. Strength<br>2. Fracture Toughness<br>3. Stiffness<br>4. Surface roughness<br>5. Wear Resistance<br>6. Chemical stability |

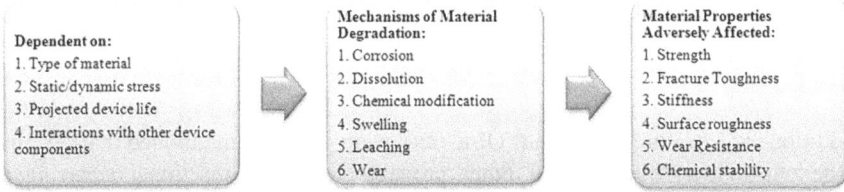

**FIGURE 7.3**   The effect of body environment on metallic bio-implants [1].

host tissue is crucial to consider. Figure 7.3 shows the effect of the body environment on bio-implants [4,5].

Extracellular and intracellular fluids are the two main categories into which the body's water may be broken down. Extracellular fluids (ECFs) include lymph, transcellular fluids (such as joint fluids), interstitial fluid, which surrounds cells, and plasmas, which are found in blood vessels. Cellular water is known as intracellular fluid (ICF). The body has a system called homeostasis that maintains regularity and consistency in the distribution and quantity of physiological fluids and electrolytes. Electrolytes are crucial to how the body functions. Electrolytes carry out a number of functions, including regulating osmolarity, metabolism, and the potential of cell membranes in physiological fluids. In electrolytes, cations comprise the ions hydrogen, sodium, potassium, calcium, and magnesium, whereas anions include the ions hydroxide, bicarbonate, chloride, phosphate, and sulfate. Dissolved salts are the most essential components for implant corrosion in vivo. For practically all metals, chloride ions (and other halides) accelerate corrosion and impede different corrosion prevention measures. Temperature and pH are two more major elements that influence biomaterial corrosion. Under normal conditions, body fluids have a temperature of 370°C. This is the temperature that stands constant throughout the life expectancy of an implant in terms of corrosion.

The corrosion of metallic biomaterials can release undesirable metal ions and corrosion products that are non-bio-compatible to the human body, may reduce the life of the implant, and in turn shorten the life of human beings. The two major cathodic reactions in corrosion in the general human body environment are oxygen evolution reaction (OER) along with hydrogen evolution reaction (HER). The HER in acid solution is:

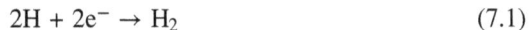

$$2H + 2e^- \rightarrow H_2 \tag{7.1}$$

whereas in alkaline solution it is:

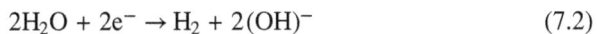

$$2H_2O + 2e^- \rightarrow H_2 + 2(OH)^- \tag{7.2}$$

The OER in acidic solution is:

$$O_2 + 4H^+ + 4e^- \rightarrow 2H_2O \tag{7.3}$$

whereas in the alkaline solution it is:

$$4OH^- \rightarrow O_2 + 4e^- + 2H_2O \qquad (7.4)$$

Additionally, to the HER and OER reduction actions mentioned earlier, the equations below show several other probable reduction reactions occurring at implant surfaces. The reactions include peroxide of hydrogen ($H_2O_2$) and hydroxide radicals ($HO_2^\bullet$ and $OH\bullet$). As a result, it is clear that some intermediate species have a substantial impact on the biological system and enhance oxidative stress in cells and reduction processes [3]. Other probable reduction processes in the body environment include the decline of disulfide bonds and other protein-like molecules, as well as some local redox environment around metallic implants that can undergo a cell redox state [6].

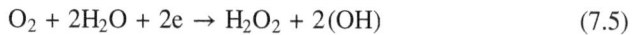

$$O_2 + 2H_2O + 2e \rightarrow H_2O_2 + 2(OH) \qquad (7.5)$$

$$H_2O_2 + 2e^- \rightarrow 2(OH) \qquad (7.6)$$

$$O_2 + H_2O + e^- \rightarrow HO_2^\bullet + (OH) \qquad (7.7)$$

$$HO_2 + H_2O \rightarrow H_2O_2 + OH^\bullet \qquad (7.8)$$

Since enzymes, which are dependent on metabolism and sensitive to pH, are a necessary component of homeostasis, the pH balance of acids and bases is also crucial. In the human body, blood plasma typically has a pH between 7.35 and 7.45. Acidosis is a condition where the blood pH falls below normal, whereas alkalosis is a condition where the blood pH rises over normal. Buffers can endure changes in pH. The pH of the body may be controlled by two separate processes: one chemical and the other physiological. Rapid-acting buffers are chemicals that enter biological fluids and react with any excess acid or alkali to stop abrupt changes in hydrogen ion concentration and pH level. Examples of these chemicals are protein, phosphate, and bicarbonate systems. If chemical buffers are unable to stabilize the pH immediately, physiological buffers, such as the respiratory and urine response systems, operate as an extra line of defense against hazardous pH fluctuations. The emission of bases, acids, or $CO_2$ is controlled by the buffer system. In contrast to the respiratory response system, which buffers in a matter of minutes, the urine response system takes the longest to do so [7].

After fixing an implant in the body, bone marrow decay is frequently accompanied by severe illness, which will impair therapy and create an electro-chemical shift in the body's equilibrium state [8]. Furthermore, the pH of the bodily fluid might decrease from 7.4 to 5.5, and it may take 10 to 15 days to return to normal. Bacterial infection may result in a larger range of pH surrounding the implant surface, ranging from acidic to alkaline. After implantation, the proteins will be adsorbed on the oxide surface, and space filled with biofluid will exist next to the implant surface as time passes (nm thickness). This will result in osteoinduction, which happens because of proliferation of cells, revascularization, and eventual gap closing. Tissue and the implant will progressively develop a solid relationship. The corrosion products that are discharged from the implant because of the pathological alterations indicated above are caused by biological activity of the implant of any form and size, and they can vary throughout the implant's surface, potentially resulting in the growth of electrochemical cells.

In medical terms, there are different detrimental reactions that may take place at the implantation of medical devices:

    i. Protein adsorption.
   ii. Hypersensitivity to foreign bodies and allergic reactions.
  iii. Coagulation and hemolysis.
   iv. Cytotoxicity.
    v. Mutagenicity and carcinogenesis.

The potential-pH or Pourbaix diagram, developed by Marcel Pourbaix [9], is one widely used technique for illustrating the stability of metals in aqueous solutions. The Black created the original Pourbaix diagram for bodily fluids [10]. This figure illustrates the variety and complexity of situations that biomaterials may encounter in vivo. The OER line, which indicates whether it is oxygen-rich solutions or electrolytes close to oxidizing materials, is the top limit of water stability. Since the human body is oxygen-saturated, the stability domains of saliva, intracellular fluid, and interstitial fluid are close to the oxygen evolution line. The HER line represents the bottom limit of water stability. The human body's secretions from ductless glands, bile, the lower gastrointestinal system, and urine all have stability domains that are generally above the hydrogen evolution line. Between these two lines, where aqueous corrosion may happen, is nothing more than the water stability domain. The Pourbaix diagram makes it evident that different pH levels and oxygen concentrations predominate in various areas of the body. As a result, a metal that will function effectively in one area of the body where it may be immune or passive might have unfavorable levels of corrosion in an additional area.

The Pourbaix diagram has various drawbacks:

    i. Since they are equilibrium diagrams, we can only infer what cannot occur from them; they do not represent kinetics.
   ii. According to the Pourbaix diagram, a metal can corrode under certain pH and potential conditions; however, this is not proven in practice.
  iii. They speak of pure water, yet bodily fluids include additional ions that might interfere with equilibrium.
   iv. Prior knowledge of the particular system is necessary since their shape is influenced by the species that are taken into account.
    v. The pH at the metal surface may be considerably different from the pH of the bulk solution.

**TABLE 7.1**

**Composition of Various Simulated Body Fluids [8]**

| Component | PBS (g.L$^{-1}$) | Ringer's (g.L$^{-1}$) | Hank's (g.L$^{-1}$) |
|---|---|---|---|
| NaCl | 8.0 | 8.6 | 8.0 |
| CaCl$_2$ | – | 0.3 | 0.14 |
| KCl | 0.2 | 0.3 | 0.4 |
| MgCl$_2$.6H$_2$O | – | – | 0.1 |
| MgSO$_4$.7H$_2$O | – | – | 0.1 |
| NaHCO$_3$ | – | – | 0.4 |
| NaH$_2$PO$_4$ | 1.2 | – | – |
| Na$_2$HPO$_4$.12H$_2$O | – | – | 0.12 |
| KH$_2$PO$_4$ | 0.2 | – | 0.1 |
| Phenol red | – | – | 0.02 |
| Glucose | – | – | 1.0 |

Before applications of biomedical implants, it must be subjected to both in vitro and in vivo studies. Studies which are performed in simulated body condition that give an overview of the behavior of the material under the given condition is known as in vitro studies and it is evident that it cannot be taken as the ultimate test to suggest a material as an implant. The tests that are performed using animal models to evaluate the actual performance of the materials are known as in vivo studies and these tests are necessary in order to get it approved by the Food and Drug Administration, USA (FDA). For corrosion testing to be in vitro condition, the environment is re-created as inside a body. Different solutions are available for testing, for orthopedic biomaterials Ringer's and Hank's mixtures are used to carry out experiments, and for dental biomaterials artificial saliva is used [11]. Their composition is given in Tables 7.1 and 7.2.

Different standards are used for evaluating corrosion studies of substances under various situations. The most common is ASTM (American Society for Testing and Materials). Various ASTM norms for evaluating the corrosion study procedures of metallic biomaterials are shown in Table 7.3.

## 7.4 CORROSION OF METALLIC IMPLANT MATERIALS

Since the structural integrity of the implant might be compromised during the deterioration process and because the host may be exposed to the degradation products, in situ degradation of metal-alloy implants is not ideal. Wear, electrochemical dissolution, or a cooperative interaction of the two can all lead to degradation. General corrosion, which affects the whole implant's surface evenly, and localized corrosion, which affects either areas of the device that are protected from tissue fluids (crevice corrosion) or seemingly random spots on the surface (pitting corrosion), are examples of electrochemical processes.

**TABLE 7.2**
**Composition in Different Artificial Saliva [9]**

| Components | Xialine 1(g · L$^{-1}$) | Xialine 2 (g · L$^{-1}$) | Saliveze (g · L$^{-1}$) |
|---|---|---|---|
| Xanthan gum | 0.90 | 0.20 | – |
| Sodium carboxymethylcellulose | – | – | 10.0 |
| Potassium chloride | 1.20 | 1.20 | 0.620 |
| Sodium chloride | 0.850 | 0.850 | 0.870 |
| Magnesium chloride | 0.051 | 0.051 | 0.060 |
| Calcium chloride | 0.13 | 0.13 | 0.17 |
| Dipotassium hydrogen orthophosphate | 0.131 | 0.130 | 0.801 |
| Potassium dihydrogen orthophosphate | – | – | 0.300 |
| Sodium fluoride | – | – | 0.004 |
| Sorbitol | – | – | 29.95 |
| Methyl p-hydroxybenzoate | 0.350 | 0.351 | 1.00 |
| Spirit of lemon | – | – | 5 ml |

**TABLE 7.3**
**Standards for Testing Corrosion Resistance of Biomaterials [10]**

| Standard | Specifications |
|---|---|
| ASTM F746 | Standard Test Method for Pitting or Crevice Corrosion of Metallic Surgical Implant Materials |
| ASTM F897 | Standard Test Method for Measuring Fretting Corrosion of Osteosynthesis Plates and Screws |
| ASTM F1089 | Standard Test Method for Corrosion of Surgical Instruments |
| ASTM F1801 | Standard Practice for Corrosion Fatigue Testing of Metallic Implant Materials |
| ASTM F2129 | Standard Test Method for Conducting Cyclic Potentiodynamic Polarization Measurements to Determine the Corrosion Susceptibility of Small Implant Devices |
| ASTM G5 | Standard Reference Test Method for Making Potentiodynamic Anodic Polarization Measurements |
| ASTM G31 | Standard Guide for Laboratory Immersion Corrosion Testing of Metals |
| ASTM G48 | Standard Test Methods for Pitting and Crevice Corrosion Resistance of Stainless Steels and Related Alloys by Use of Ferric Chloride Solution |
| ASTM G61 | Standard Test Method for Conducting Cyclic Potentiodynamic Polarization Measurements for Localized Corrosion Susceptibility of Iron-, Nickel-, or Cobalt-Based Alloys |
| ASTM G71 | Standard Guide for Conducting and Evaluating Galvanic Corrosion Tests in Electrolytes |

Stress corrosion cracking, corrosion fatigue, and fretting corrosion are examples of electrochemical and mechanical processes that may combine, leading to an early structural collapse and a quicker release of metal ions and particles. Particulate deterioration and wear products in the tissue surrounding an implant that could ultimately lead to periprosthetic bone loss are clinical indicators of the relevance of metal implant deterioration. Additionally, several publications have noted elevated local and systematic trace metal concentrations in connection with metal implants [12]. The chance of the implant breaking due to corrosion is limited but modest. Grossly corroded ferrous metals caused by prolonged inflammatory reactions were a frequent occurrence with early implantation [13]. In many situations, mild corrosion might also cause symptoms that necessitate removing the implant.

The signs and symptoms might be anything from a localized softness at the site of the corroded region to sensitive discomfort, reddening, and swelling all over the area surrounding the device [12]. These signs show that the implant is causing the tissue to respond. Three things can happen when metallic implants corrode: electrical currents can change cell behavior; the corrosion process can change the chemical environment (pH, pO2); and metallic ions can change cellular metabolism. The third one is the most severe of the three. Implant corrosion products lead to fibrosis and alterations in the surrounding bone. In metallurgy, the degradation process may be thought of in reverse. When put in a solution, the majority of pure metals frequently transform back into soluble ionic species, oxides, or hydroxides. Fe2O3 and –FeOOH are the main corrosion products, according to an X-ray examination of a 29-year-old low-alloy steel bone plate's corrosion products [14].

## 7.5   TYPES OF CORROSION IN BIOMATERIALS

Metallic biomaterials can experience a variety of kinds of corrosion, mostly general and localized corrosion, including corrosion exhaustion, galvanic corrosion, pitting, crevice corrosion, fretting corrosion, and strain-corrosion splitting.

### 7.5.1   GALVANIC CORROSION

Galvanic corrosion, also called as bi-metallic degradation, happens when there is physical contact with two dissimilar metals in contact with an electrolyte. Electrolyte may be serum or body fluid, particularly in an acidic, i.e., in low pH situation leading to irritation. Galvanic corrosion can be localized or uniform. While making surgical implants contact between two dissimilar metals is unavoidable. For example, galvanic corrosion occurs if bone screw and bone plate are produced of different metals or alloys. Also, in dental implants, the crown is made of noble metal alloys; thus, one that is more noble will become a cathode and one that is less noble will become an anode according to the electrochemical series. Thus, degradation takes place at the anode. On the other hand, galvanic corrosion might not take place when the potential difference between the two dissimilar metals is quite small [11–16].

### 7.5.2   PITTING CORROSION

Pitting is a type of very limited corrosion that develops as a result of the implant's cavities being created and the constrained disintegration of the passive layer. Pitting frequently happens in environments where halides are present. Due to the presence of proteins in tissue and serum, which can accelerate pitting corrosion, chloride is the most aggressive ion for pitting other than halides. Metals have a typical pitting propensity, according to earlier investigations. Stable pits arise when potential is noble to pitting potential. Because stable pitting potential prevents pit propagation, meta-stable pits form at potentials lower than the pitting potential [6]. Compared to Ti- and Co-based alloys, stainless steel is more susceptible to pitting in halide solutions. Pitting typically occurs on the underside of screw heads in bio-implants. Resistance to pitting is measured by Cr and Mo content [13,17].

### 7.5.3 CREVICE CORROSION

Crevice corrosion, which is corrosion related to the inculcate's microstructure, typically takes place in small cracks on the metal's surface where there is little mass transport. Either a metal with a metal crack or a metal with a non-metal crack can experience it. Crevice corrosion is caused by three main factors: a decrease in pH, a loss in oxygen, and an increase in Cl- ions (Table 7.4).

Such change in the surroundings causes the de-passivation of the metal in the crevice. Stainless steel is susceptible out of the blend group considered here to

## TABLE 7.4
## Pitting Potential of Biomaterials Determined by Various Methods [13]

| Material | Electrolyte | Pitting Potential, mV | Experimental Technique |
|---|---|---|---|
| 316 L | Deaerated, pH-7.4, Hank's solution, 37°C | 280 | Potentiodynamic |
| | 0.9% NaCl solution, 40°C | 400 | Do |
| | Do | 130 | ASTM F746 |
| Coated 316 L | Deaerated, pH-7.4, Hank's solution | 449 to 567 | ASTM G61 |
| 316 L | Deaerated Hank's solution | 352 | Do |
| | Deaerated, pH-7.4, Hank's solution, 37°C | 280 | Do |
| | Artificial saliva at 40°C | 400 | ASTM G61 |
| | Hank's solution, at 37°C | 350 to 400 | ASTM G61 |
| Ti-6A1-4V | Hank's solution, at 37°C | >1000 | Do |
| | Deaerated Hank's solution, 37°C | 1900 | Do |
| Co-Cr | Deaerated Hank's solution, 37°C | 650 | Do |
| | 0.9% NaCl at 40°C | 100 to 200 | ASTM F746 |
| Ni-Ti | 0.9% NaCl at 40°C | 400 | ASTM G61 |
| | Artificial saliva at 40°C | ~ 1000 | ASTM G61 |
| | Hank's solution at 37°C | 650 | ASTM G61 |

crevice-persuaded restricted corrosion. Crevice corrosion of Ti in chloride containing solutions takes place at elevated temperatures, but an in vivo environment such as temperature is not seen. For metallic biomaterials, crevice corrosion may occur on mating surfaces; for example, in the interface of head-stem in hip joints along with interface of screw-hole at bone plates [18].

### 7.5.4 CORROSION FATIGUE

The failure of metals owing to fracture, which results from the combined interaction of cyclic stress and electrochemical processes, is known as corrosion fatigue. For metals used in load-bearing surgical implants or applications involving cyclic motion, this is crucial. The collapse might occur with less load and over a shorter time if the metal was exposed to corrosive environments. Crevice initiation and crack development strategies were linked to resistance to corrosion exhaustion. Corrosion pits can cause fatigue to form if they are present. Corrosion fatigue is a significant cause of mechanical failures in orthopedic implants. Some researches claimed that nitrogen implantation and heat treatments might be used to increase the corrosion exhaustion in metallic inculcates [18,19].

### 7.5.5 FRETTING CORROSION

Fretting, also known as tribo-corrosion, is the combined action of corrosive assault and mechanical wear on a material's surface. Fretting corrosion happens when there is a tiny oscillation or relative movement between two different surfaces, such as the screw head of a prosthetic device and the bone plate. Even in the absence of a corrosive environment, fretting can still happen. Although the passive layer shields the biomedical implant, it has been noticed that with time, fretting corrosion causes the passive layer to break down locally. Biomaterials' fretting resistance may be improved by surface modification methods as plasma surface alloying, plasma ion nitriding, and nitrogen ion implantation [20]. The various types of corrosion, together with their locations and implant shapes, are depicted in Table 7.5 and Figure 7.4 [21].

**TABLE 7.5**

**Location, Material, Type of Corrosion [21]**

| Type of Corrosion | Material | Implant Location |
|---|---|---|
| Galvanic | 304SS/316SS, CoCr+Ti6Al4V, 316SS/Ti6Al4V Or CoCrMo | Oral implants Screws and nuts |
| Pitting | 304 Stainless Steel, Cobalt alloy | Orthopedic and dental alloy |
| Crevice | 316 L SS | Bone plates and screws |
| Corrosion fatigue | 316 SS, CoCrNiFe | Bone cement |
| Fretting | Ti6Al4V, CoCrSS | Ball joints |

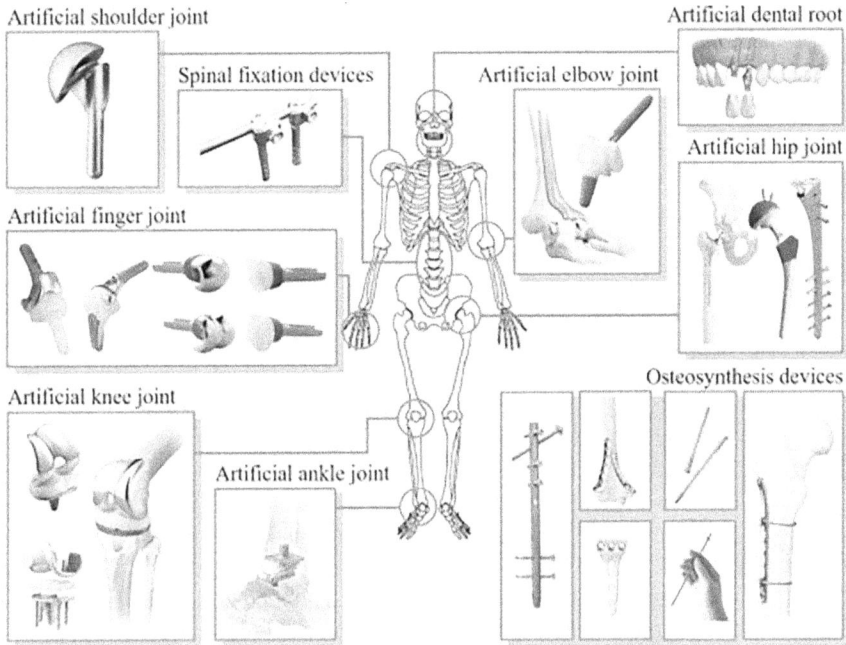

**FIGURE 7.4** Shape of implants and its location [21].

## 7.6 CORROSION OF VARIOUS BIOMATERIALS

Various materials have been used in bio-implants based on their application and required properties. Also, many material engineers or scientists researching on improvement in the stability and other properties. Metals, polymers, ceramics, and composites are the four general groups into which engineering materials may be categorized.

### 7.6.1 METALS

Heavy loads are best supported by metals with high yield points, moduli, and ductility. Metal implants are frequently used to replace hard tissue. Several authors have investigated the use of metals in implants. The two basic functions of metallic implants are to: (i) replace a component of the body, as do prostheses like skull plates and total joint replacements; and (ii) act as fixation devices, which hold damaged bones and other tissues in place while they recover naturally. The most often used metals and alloys are titanium, cobalt-chromium, and stainless steel. Due to their exceptional mechanical characteristics, stainless steel and cobalt-chrome alloys were first preferred for bone replacement applications. The high mechanical strength of such metallic implants, however, resulted in stress shielding and bone resorption since the surrounding bone's elastic modulus did not match the implant's. An emphasis on titanium and its alloys resulted from this problem, information on how corrosion affects mechanical strength, and information on hazardous by-products. Because titanium and titanium alloys have a substantially lower elasticity modulus and a higher resistance to corrosion, they offer a benefit over other metals [21].

## 7.6.2 POLYMERS

Monomer units are repeated in long, high-molecular-weight chains to form polymers. In addition to the chemical makeup, other factors that affect physio-chemical qualities include molecular weight distribution and degree of cross-linking. Since polymers' physical characteristics closely resemble those of soft tissue, they may be used to replace skin, tendons, cartilage, vessel walls, and drug delivery systems, among other things. A joint prosthesis articulating surface is made of ultra-high-molecular-weight polyethylene (UHMWPE), and polycaprolactone has been utilized to make resorbable screws, plates, and sutures for fracture repair. The tremendous diversity of polymers that are accessible for implantation purposes is due in large part to the manufacturing options' wide range. A variety of advantages over metals are also offered by polymeric materials. For example, the extracellular fluid's isotonic saline solution is extremely hostile to metals, yet has little to no effect on polymeric components. Stress shielding and bone resorption are reduced as a result of a smaller difference in elastic modulus between the polymers and the bone. The majority of polymers may also be designed to degrade, which implies that throughout the process of deterioration, the host tissue progressively replaces them, negating the need for subsequent surgery. The usage of these materials may be limited as a result of this technique, which can occasionally result in the loss of mechanical qualities and, in addition, undesirable tissue responses owing to released breakdown products [22].

## 7.6.3 CERAMICS

Body components that are sick or injured can be repaired and rebuilt using bioceramics. Despite the fact that many ceramics are known, only a select handful are appropriate and biocompatible. There is a wide variety of inorganic/non-metallic compositions that are available, including those that are bioinert (e.g., alumina, zirconia), resorbable (e.g., tricalcium phosphate), bioactive (e.g., hydroxyapatite, bioactive glasses, and glass ceramics), and porous for tissue in growth. Ceramics are more frequently employed in circumstances where wear resistance is crucial because they are tough, chemically stable, stiff, and hard. Major uses include repairs for periodontal disease and knee, hip, and tendon replacements. These materials offer a surface that is so biologically compatible with bone-forming cells that the cells directly develop bone by adding new layers to the material's surface. Despite all of its benefits, ceramics have low mechanical attributes and strength [23–26].

## 7.6.4 COMPOSITES

The term *composite* refers to a mixture of two or more materials that are joined so that stress is transferred across the phase boundary. As a result, although having two phases, solid and void-porous materials are typically not called composites since stress is not transmitted to voids. Obviously, composite materials are made to offer a variety of qualities that cannot be combined in a single-phase substance. When contrasting human tissue with the various metals, polymers, and ceramics, there is a significant mismatch between their specific qualities. The creation of composites has thus received a lot of attention in an effort to blend several types of materials with advantageous features while avoiding some of their disadvantages. For instance, in replacements of hips, hydroxyapatite (HAP) or bioglass may be coated on top of a bioinert substance like alumina to facilitate direct bone attachment [24–26].

## 7.6.5  ALLOYS

An implant material must have the requisite functional qualities, particularly mechanical ones, and must also be biocompatible in order to be regarded as appropriate for use in prostheses. Stainless steels, cobalt-chromium alloys, and titanium alloys are the three main categories of metallic implants used today [27–30].

### 7.6.5.1  Stainless Steels

Due to biocompatibility of stainless steel, it has long been regarded as a reliable substance for use in permanent surgical implants. The 316 L variety, which has a highly corrosion-resistant surface due to the creation of a protective layer of chromium oxide, is frequently employed for this purpose. The environment inside the body is complicated, though, and corrosion might result in the release of dangerous substances. Permanent prosthetic devices can be made of tougher, more corrosion-resistant materials, which are frequently preferred over stainless steel. Despite this, stainless steels still have enough chromium to withstand corrosion through the chromium oxide passive layer [28].

### 7.6.5.2  Cobalt-Chromium Alloys

CoCr alloys were created for the aircraft sector many years ago, and the development of a chromium oxide surface layer gives them their inertia. These alloys are a common option for orthopedic implants because of their good mechanical qualities. The most common alloys are CoCrMo or CoNiCrMo; however, they may also contain tungsten or iron (Fe). In prosthetic devices, CoCr alloys are both wrought and cast, and each has unique qualities. They are frequently used as parts in modular prosthetic joints, such as those for the hip or knee, and are particularly ideal for bearing surfaces (typically in opposition to ultra-high-molecular-weight polyethylene) [29,30].

### 7.6.5.3  Titanium and Titanium Alloys

Titanium started to be widely used in a variety of industrial areas towards the end of the 20th century. It appeared to be the ideal material for surgical implants due to its high inertia, which results from the development of a thin titanium oxide layer on its surface, as well as its low weight and outstanding biocompatibility. Among all surgical implant materials, titanium has the best strength-to-weight ratio. In prosthetic devices, both pure titanium and Ti6Al4V are widely used, with the choice of material based on the particular functional needs [26–30].

## 7.7  METHODS OF CORROSION PREVENTION

### 7.7.1  SURFACE TREATMENT

Depending on the sort of synthetic material implanted in the body, different tissues will react differently to it. Depending on how the tissue responds to the implant's surface, the process by which it engages with the implant will vary. Investigating other options is required when metallic implants malfunction. Material surface modification has drawn a lot of interest because it makes it possible to independently optimize both bulk and surface characteristics. Surface alteration methods can be used to change materials that have acceptable bulk attributes like strength, toughness, and density, but lack particular surface properties like corrosion or wear resistance [2].

#### 7.7.1.1  Ion Implantation

Surfaces are made harder and more corrosion- and wear-resistant thanks to ion implantation. Ion implantation was a well-established commercial method by the early to mid-1970s. This method has been often employed to change the electronic properties of semiconductor devices. After achieving success in the semiconductor industry, the ion implantation process has been used to biomaterials to enhance their corrosion and wear resistance. To improve the wear-accelerated corrosion of orthopedic implant materials, Buchanan et al. first proposed using ion implantation as a surface modification technique. Ion implantation is the method of employing high-velocity ions to introduce a negligibly small number of atoms of any element onto a material's surface without altering the material's surface finish or underneath bulk properties, and without taking into account thermodynamic constraints. Titanium and its alloys are only sometimes used in orthopedic applications because to their poor wear characteristics. However, ion implantation has proven to be remarkably effective in enhancing the resistance to wear of titanium surfaces [22].

#### 7.7.1.2  Passivation

When a thin, corrosion-resistant layer of reaction products forms on the surface of a metal or alloy, it gives the material strong corrosion-resistant properties and reduces the rate at which metal ions dissolve. This process is known as passivation. The fundamental objective of passivation is to enhance the passive protective layer by altering its composition, structure, and thickness as well as by eliminating weak areas such as non-metallic contaminants. Kamachi et al. [31,32] have published comprehensive mechanistic and electrochemical characterizations. Many authors have studied how metals respond to alkali and acid treatments. In order to produce hydroxyapatite for orthopedic applications, titanium must first undergo an alkali treatment followed by a heat treatment.

### 7.7.1.3   Electro Polishing and Thermal Methods

The corrosion resistance of metals has been greatly improved by low cost, straightforward processes including electro polishing and heat techniques. For intricate metal implants like mechanical heart valve frames and tiny glaucoma implants, Eliaz and Nissan in particular created an efficient electro polishing procedure. The typical ASTM procedure, ultrasonic bath polishing, pulsed voltage polishing, and other polishing techniques were all compared and assessed in the study. The polished portions were investigated using several methods, including stereomicroscopy, optical microscopy, atomic force microscopy, noncontact surface profilometry, and X-ray diffraction, while current (I) and voltage (V) curves were created for various solutions and bath temperatures. The study discovered that electro polishing in an ultrasonic bath was helpful when a rough, patterned surface was required for osseointegration reasons, whereas pulse polishing efficiently minimized the erosion effects of gas bubbles in solution. Histopathology and preliminary animal investigations both revealed that the polished surfaces only caused a mild body reaction, which is ideal for such applications [33–39].

### 7.7.1.4   Bioceramic Coatings on Implants

The bulk of implants used today for clinical purposes requiring load-bearing qualities are comprised of metal, which might have negative consequences. Applying bioceramic coatings to metal implants might be a solution to this issue. Although there is still work to be done, this method is now used in both dental implants and hip joint prostheses. However, there are now a number of commercially marketed metallic implants with bioceramic coatings, and work is being done to overcome any remaining issues [30,36].

Multiple techniques are available to apply bioceramic coatings on metallic substrates, but the procedure is complicated. These coatings' quality and endurance, which are influenced by elements including purity, particle size, chemical composition, layer thickness, and surface shape, are crucial to their clinical effectiveness. Bioceramic coatings offer an efficient barrier that prevents the kinetics of ion release in the live organism, in addition to limiting the release of metal ions from the implant [26,32,33].

For this reason, hydroxyapatite ($Ca10(PO4)6(OH)2$), also known as HAP, is frequently utilized because it promotes bone ingrowth and reduces motion at the implant-bone interface. Excellent biological characteristics of HAP include its lack of toxicity, inability to cause an inflammatory response, and lack of fibrous or immunological responses [40].

## 7.8   QUALITY CONTROL

The following observations are related to quality control measures [40]:

  i. It is advised that producers adopt suitable metallurgical standards, exercise caution during implant production, and maintain proper testing facilities in order to improve standards and quality control.

ii. Pits, fissures, excessive grain size, inclusions, and porosity in implants can be reduced by design changes. Pitting in implant hardware has been virtually eradicated by the use of vacuum melting and remelting.

iii. By lowering the carbon concentration, the possibility of intergranular corrosion, which can happen when chromium carbide precipitates at the grain boundary in stainless steel with a carbon level over 0.03% has all but disappeared. However, reducing the carbon percentage also causes stainless steel's ultimate tensile strength to decrease.

iv. After welding, a proper heat treatment can restore the proper compositional distribution and stop intergranular assault.

v. It's crucial to avoid implanting several metals in the same area. Matching components from the same batch of the same alloy variety must be delivered during the production process. The use of equipment made of the same material as the implant must also be assured.

## 7.9   RESEARCH AND DEVELOPMENT

i. It's crucial to create alloys with strong wear resistance and a high rate of repassivation in order to avoid fretting.

ii. Galvanic corrosion may not occur but might be improved by joining two metals that are far apart in the galvanic series, such as titanium and chromium. There is a chance of worsening pitting or crevice corrosion, though.

iii. This kind of corrosion can be avoided by choosing an alloy with an open circuit or rest potential that is lower than the critical potential for pitting.

## 7.10   SCOPE FOR FUTURE DEVELOPMENT AND CONCLUSION

Despite the use of new, corrosion-resistant super alloys to replace freely corroding implant materials, biomaterial corrosion remains a substantial clinical problem. Destructive corrosion processes continue to occur in some therapeutic settings. It is thought that the corrosion rate can be decreased, and negative clinical consequences can be reduced by concentrating on factors relating to metallurgical processing, modular connection tolerances, surface-processing techniques, and proper material selection. Further research is needed on the electrochemical and mechanical interactions of passivating metal-oxide interfaces, particularly the pressures and movements required to break passivating oxide layers and the implications of repetitive oxide abrasion on the electrochemical behavior of the interface and eventually the implant. Additionally, the consequences for clinical practice of a rise in metal content in bodily fluids and the involvement of particle corrosion products in unfavorable local tissue responses and distant organs of individuals with metal implants need to be further examined. It takes a lot of investigation to determine the metal's chemical structure, the characteristics of its ligands, and its possible toxicity. Biomaterial corrosion may be fought with promising methods including surface engineering, bioceramics, and functionally graded coatings.

## REFERENCES

1. Sridhar TM, Rajeswari S. Biomaterials corrosion. Corrosion Reviews. 2009 Dec; 27 (Supplement): 287–332.
2. Sharma MP. Corrosion of bio-materials. International Journal of Metallurgical & Materials Science and Engineering. 2020 Jun; 10(1): 21–28.
3. Manam NS, Harun WSW, Shri DNA, Ghani SAC, Kurniawan T, Ismail MH, et al. Study of corrosion in biocompatible metals for implants: A review. Journal of Alloys and Compounds. 2017 Apr; 701: 698–715.
4. Eliaz N. Corrosion of metallic biomaterials: A review. Materials. 2019 Jan 28; 12(3): 407.
5. Manivasagam G, Dhinasekaran D, Rajamanickam A. Biomedical implants: Corrosion and its prevention - A review. Recent Patents on Corrosion Science. 2010 Jun 4; 2(1): 40–54.
6. Suman P, Lakshumu Naidu A, Rao PR. Processing and mechanical behaviour of hair fiber reinforced polymer metal matrix composites. International Journal of Mechanical and Production Engineering Research and Development. 2016; 469: 478.
7. Virtanen S. Corrosion of biomedical implant materials. Corrosion Reviews. 2008; 26(2–3): 147–171.
8. Singh G, Vedrtnam A, Bhati H. Fabrication and characterisation of aluminum-graphite composite material. International Journal of Mechanical and Production Engineering Research and Development. 2013; 7: 313–320.
9. Antunes RA, de Oliveira MCL. Corrosion fatigue of biomedical metallic alloys: Mechanisms and mitigation. Acta Biomaterialia. 2012; 8(3): 937–962.
10. Kamachimudali U, Sridhar T, Raj B. Corrosion of bio implants. Sadhana. 2003; 28(3–4): 601–637.
11. Fontana MG. Corrosion Engineering. Tata McGraw-Hill Education. 2005.
12. Hiromoto S. Corrosion of metallic biomaterials. In Metals for Biomedical Devices (Volume 1, Second Edition). Woodhead Publishing. 2019; 131–152.
13. Manivasagam G, Dhinasekaran D, Rajamanickam A. Biomedical implants corrosion and its prevention - A review. Recent Patents on Corrosion Science. 2010; 2: 40–54.
14. Suresh G. Electro chemical behaviour of lens TM deposited Co-Cr-W alloy for bio-medical applications. International Journal Mechanical Production Engineering Resources. 2018 Dec; 41–52.
15. Virtanen, S, Curty C. Metastable and stable pitting corrosion of titanium in halide solutions. Corrosion. 2004; 60(7): 643–649.
16. Nakagawa M, Matsuya S, Udoyh, K. Corrosion behaviour of pure titanium alloys in fluoride-containing solutions. Dental Materials Journal. 2001; 20: 167–305.
17. Aziz-Kerrzo M, Conroy KG, Fenelon AM, Farrell ST, Breslin CB. Electrochemical studies on the stability and corrosion resistance of titanium-based implant materials. Biomaterials. 2001; 22: 1531–1539.
18. Grosgogeat B, Reclaru L, Lissac M, Dalard F. Ti/Ti–6Al–4V implants and dental alloys by electrochemical techniques and auger spectrometry. Biomaterials. 1999; 20: 933–941.
19. Rabkin DJ, Lang EV, Brophy DP. Nitinol properties affecting uses in interventional radiology. Journal of Vascular and Interventional Radiology. 2000; 11: 343–350.
20. Shabalovskaya SA. Surface, corrosion and biocompatibility aspects of nitinol as an implant material. Bio-medical Materials and Engineering. 2002; 12: 69–109.
21. Nakano T. Physical and mechanical properties of metallic biomaterials [Internet]. In Niinomi, M., Metals for Biomedical Devices (Second Edition). Woodhead Publishing, Pages 97–129, 2019. ISBN 9780081026663, https://doi.org/10.1016/B978-0-08-102666-3.00003-1.

22. Pogrebnjak AD, Bratushka SN, Beresnev VM, Levintant-Zayonts N. Shape memory effect and superelasticity of titanium nickelide alloys implanted with high ion doses. Russian Chemical Reviews. 2013; 82: 1135–1159.

23. Seo J, Kim YC, Hu JW. Pilot study for investigating the cyclic behavior of slit damper systems with recentering shape memory alloy (SMA) bending bars used for seismic restrainers. Applied Sciences. 2015; 5: 187– 208.

24. Superelasticity and Shape Memory Alloys. University of Cambridge. 2004–2023, https://www.doitpoms.ac.uk/tlplib/superelasticity/index.php

25. Wadood A. Brief overview on nitinol as biomaterial. Advances in Materials Science and Engineering. 2016; 2016: 1–9.

26. Duerig T, Pelton A, Stöckel D. An overview of nitinol medical applications. Materials Science & Engineering A: Structural Materials: Properties, Microstructure and Processing. 1999; 273–275: 149–160.

27. ASTM F2005-21, Standard Terminology for Nickel-Titanium Shape Memory Alloys; ASTM F2005-05(2015); ASTM: West Conshohocken, PA, USA, 2015, Volume: 13.01, Developed by Subcommittee: F04.12, Pages: 3, DOI: 10.1520/F2005-21, ICS Code: 77.120.40; 77.120.50.

28. ASTM F2063-12 Standard Specification for Wrought Nickel-Titanium Shape Memory Alloys for Medical Devices and Surgical Implants; ASTM F2063-12; ASTM: West Conshohocken, PA, USA, 2012.

29. Vamsi Krishna B, Bose S, Bandyopadhyay, A. Fabrication of porous NiTi shape memory alloy structures using laser engineered net shaping. Journal of Biomedical Materials Research - Part B Applied Biomaterials. 2009; 89: 481–490, doi:10.1002/jbm.b.31238.

30. Vamsi Krishna B, Bose S, Bandyopadhyay A. Laser processing of net-shape NiTi shape memory alloy. Metallurgical and Materials Transactions A: Physical Metallurgy and Materials Science. 2007; 38: 1096–1103.

31. Hamilton RF, Bimber BA, Andani MT, Elahinia M. Multi-scale shape memory effect recovery in NiTialloys additive manufactured by selective laser melting and laser directed energy deposition. Journal of Materials Processing Technology. 2017; 250: 55–64.

32. Ryhänen J, Niemi E, Serlo W, Niemelä E, Sandvik P, Pernu H, Salo T. Biocompatibility of nickel titanium shape memory metal and its corrosion behavior in human cell cultures. Journal of Biomedical Materials Research. 1997; 35: 451–457.

33. Webster, JG (Ed.). Encyclopedia of Medical Devices and Instrumentation, 2nd ed., Volume 1, John Wiley & Sons. 2006.

34. Fathi M, Mortazavi V. A review on dental amalgam corrosion and its consequences. Journal of Research in Medical Sciences. 2004; 1: 42–51.

35. Stoner GE, Senti SE, Gileadi E. Effect of sodium fluoride and stannous fluoride on the rate of corrosionof dental amalgams. Journal of Dental Research. 1971; 50: 1647–1653.

36. Eley, BM. The future of dental amalgam: A review of the literature. Part 1: Dental amalgam structure andcorrosion. British Dental Journal. 1997; 182: 247–249.

37. Joska L, Bystrainsky L, Novák P. The effect of the alloy powder preparation on the corrosion behaviour of dental amalgams. In Proceedings of the 15th International Corrosion Congress, 22–27 September, Granada, Spain, 2002; pp. 2228–2234.

38. Newton T. Dental fillings. Chem. Br. 2002, 38, 24–27.219. Sarkar, N.K.; Eyer, C.S. The microstructural basis of creep of □1 in dental amalgam. Journal of Oral Rehabilitation. 1987; 14: 27–33.

39. Rosenberg AD. In-Vitro Electrochemical Testing of a Microchip-Based Controlled Drug 39. Delivery Device. Master's Thesis. MIT. 2001.

40. Frankental RP, Siconolfi DJ. Anodic corrosion of gold in concentrated chloride solutions. Journal of the Electrochemical Society. 1969; 3: 465–470.

# Section 2

Surface Synthesis and
Engineering of Biomaterials

# 8 Various Processing Routes for Biocompatible Materials
## From Conventional Process to Additive Manufacturing

*Sibani Mahapatra, Sampara Ila Grace Victoria, and Shampa Aich*
Department of Metallurgical and Materials Engineering, Indian Institute of Technology Kharagpur, West Bengal, India

## 8.1 INTRODUCTION

Biocompatible materials refer to substances that are compatible with living tissues and can be used in medical and biological applications without causing harm or adverse reactions. These materials are precisely crafted to mingle with the body in a way that minimizes any detrimental effects, such as inflammation, immune response, or toxicity [1]. Biocompatible materials can include ceramics, composites, metals, natural materials, and polymers, as mentioned in the literature [1–3]. The terms *biocompatible materials* and *biomaterials* are closely related but have slightly different meanings. Conversely, biomaterials have a broader definition, encompassing any material, natural or synthetic, that is used in a medical or biological application to mingle with living systems. This includes materials used in drug delivery systems, diagnostic tools, implants, medical devices, tissue engineering, and more. Biomaterials can be biocompatible, meaning they are compatible with living tissues, or they may have other functions such as promoting cell growth, enabling controlled drug release, or providing mechanical support. In summary, biocompatible materials are a subset of biomaterials [4,5].

### 8.1.1 EXPLORATION OF PROCESSING METHODS FOR BIOCOMPATIBLE MATERIALS

The processing of biocompatible materials involves various routes and techniques to fabricate the desired forms and structures with specific properties, such as mechanical strength, biodegradability, and compatibility with biological systems.

Here is an introduction to some of the common processing routes for biocompatible materials:

### 8.1.1.1 Primary Processing Methods:

- **Casting and Molding:** Casting and molding techniques involve pouring a liquid or molten biocompatible material into a mold and allowing it to solidify, or subsequently, various cool forms can be obtained, including films, fibers, and scaffolds. This process is commonly used for polymers, hydrogels, and certain ceramics.

- **Solvent Casting and Particulate Leaching (SCPL):** This technique involves mixing a biocompatible polymer with a soluble particulate material and casting it into a desired shape. After solidification, the soluble particulates are leached out using an appropriate solvent, leaving behind a porous structure. By varying the solvent composition and casting parameters, the film's thickness, uniformity, and surface properties can be controlled. This method is often used to create porous scaffolds for tissue engineering applications [6,7].

- **Polymer Synthesis:** Polymers with desired properties can be synthesized through steps such as step-growth polymerization, or ring-opening polymerization. By carefully selecting monomers and controlling the polymerization process, biocompatible polymers can be customized for specific instances, for instance, hydrogels for tissue engineering or drug-eluting coatings for medical devices [8].

- **Sol-Gel Processing:** Sol-gel processing refers to the production of materials by synthesizing them from a sol, which is a stable colloidal suspension containing solid particles dispersed within a liquid medium. It involves the transformation of a liquid "sol" into a solid "gel" by controlled hydrolysis and condensation reactions. The sol undergoes gelation, resulting in a three-dimensional network of interconnected particles. This resulting gel can be further processed to obtain the desired shape, such as coating, dipping, or 3D printing. This route is commonly used for producing biocompatible ceramic materials, such as bioactive glasses or ceramics for bone regeneration [9].

- **Electrospinning:** Electrospinning is a technique that produces ultrafine fibers from a polymer solution or melts by subjecting it to a high-voltage electric field. Electrospinning allows the fabrication of nanofibers with high surface area, which can closely resemble the structure of the natural extracellular matrix and facilitate cell attachment and growth. The fibers obtained because of this process have a high surface area to volume. This property facilitates the use of nanofibers for applications including drug delivery systems, tissue scaffolds, and wound dressings [10].

- **Additive Manufacturing/3D Printing:** Additive manufacturing or 3D printing, entails generating three-dimensional objects layer by layer from a digital model. It has gained remarkable focus in the biomedical field for creating drug delivery systems, scaffolds for tissue engineering, and patient-specific implants. There are several categories of 3D printing technologies like fused filament fabrication (FFF) and stereolithography (SLA), which can be used to process biocompatible materials, including polymers, ceramics, and metals [11].

### 8.1.1.2 Secondary Processing Methods:

- **Surface Modification:** Surface modification mechanisms are used to elevate the biocompatibility and functional properties of materials. Techniques such as plasma treatment, chemical grafting, and biomolecule immobilization can be used to modify the surface chemistry, roughness, and topography of materials, improving their interaction with biological systems.
- **Machining and Surface Finishing:** Machining processes, including cutting, milling, and grinding, are used to shape biocompatible materials, particularly metals, and alloys. Surface finishing techniques, such as polishing, electro polishing, and passivation, are then applied to improve the machined components' surface quality, corrosion resistance, and biocompatibility (Figure 8.1).

These are a few examples of the processing routes employed for biocompatible materials. The selection of a specific processing method depends on the material type, desired properties, intended application, and manufacturing scalability. Researchers and engineers continue to develop new and advanced processing

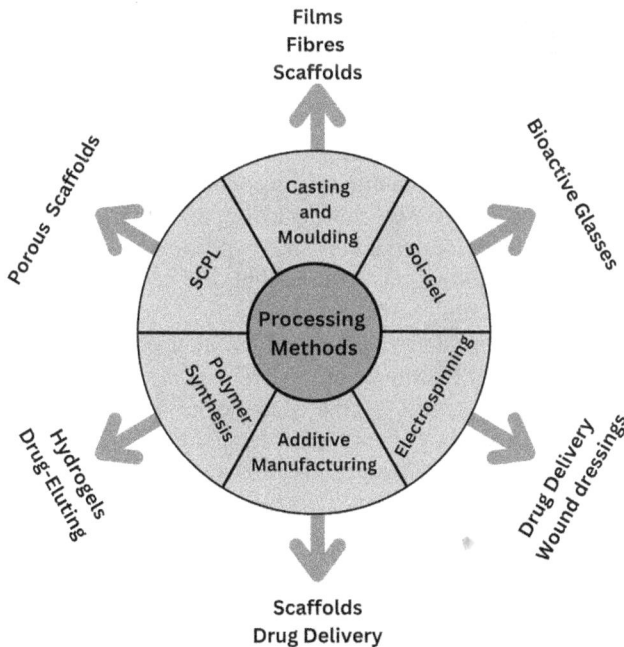

**FIGURE 8.1** Schematic diagram of different processing methods for biocompatible materials.

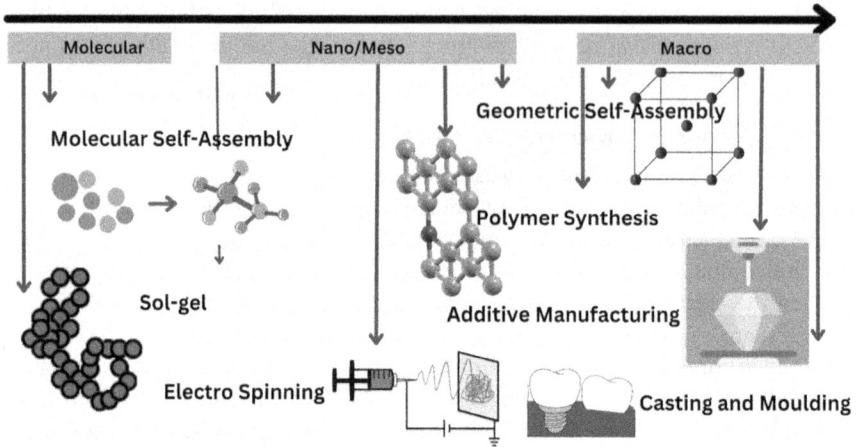

**FIGURE 8.2** The toolbox of different biomaterials processing diverging across numerous size scales from nano to macro.

techniques to expand the possibilities of biocompatible materials in the field of medicine and beyond (Figure 8.2).

## 8.2 HISTORICAL OVERVIEWS

The historical perspective of biocompatible materials processing can be traced back to ancient times when humans began using natural materials for medical purposes. Over the centuries, the understanding of biocompatibility and the development of materials and processing techniques have evolved significantly. In ancient times, natural materials like bone, wood, and metals were used for medical purposes, as evidenced by the practices of ancient Egyptians who used linen and honey to dress wounds, while Mayans used seashells as dental implants [12,13]. Moving on to the 19th century, the use of biocompatible materials started gaining more scientific attention. In the early 1800s, porcelain was introduced for dentures, improving biocompatibility compared to previous materials like ivory or wood [14].

Continuing to the early 20th century, the development of synthetic materials opened new possibilities for biocompatible materials. A significant milestone came in 1909 with the creation of Bakelite, the first synthetic polymer, laying the groundwork for future advancements. As the mid-20th century approached, the concept of biocompatibility gained prominence in medical research and materials science. Scientists began investigating materials like titanium and stainless steel for implants, which demonstrated improved compatibility with the human body compared to earlier options [15]. This progress continued into the late 20th century, driven by advances in material science and processing techniques. Notably, the 1960s saw the successful implementation of the first hip replacement surgery using a metal-on-polyethylene implant, marking a breakthrough [16]. In the 21st century, the demand for biocompatible materials grew alongside advancements in medical technology. To meet specific medical needs, new materials such as bioresorbable polymers,

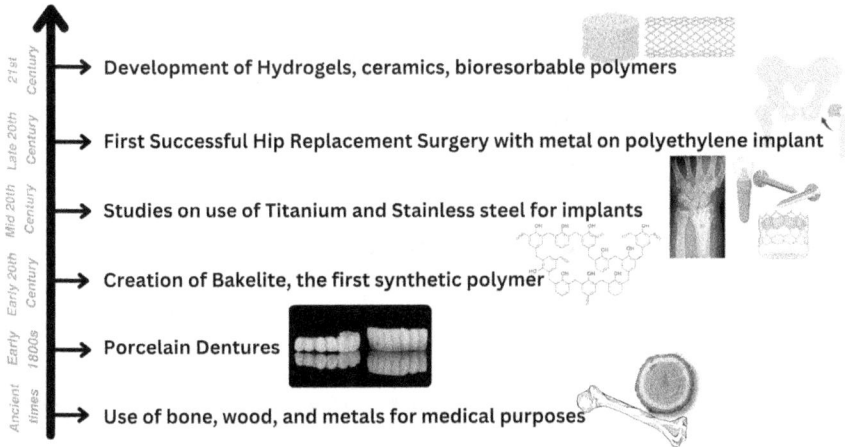

**FIGURE 8.3** History of biocompatible materials from ancient times to the 21st century.

hydrogels, and ceramics were developed [15,17]. Processing techniques like additive manufacturing (3D printing) enabled the fabrication of complex structures [18]. Looking ahead, future perspectives involve enhancing the properties of existing materials and exploring novel biomaterials, such as bioactive glasses and nanocomposites, to further improve biocompatibility [4,19]. Additionally, advancements in surface modifications, nanotechnology, and tissue engineering hold promise for the development of advanced biocompatible materials [12]. These developments will contribute to the continued evolution and application of biocompatible materials in various medical fields (Figure 8.3).

## 8.3 CASTING

Casting is a popular manufacturing method that involves pouring a liquid material into a specially designed mold, which has a hollow cavity in the desired shape. The material then cools down and solidifies in the mold, producing the final product with the desired shape.

There are two critical types of casting: **expendable mold casting** and **non-expendable mold casting**. Expendable mold casting includes methods like lost wax casting, where the mold is made of ceramics or plaster. In this process, a wax prototype is created and covered with a refractory material. Lost wax casting is often used for creating complex geometries, such as net-shaped products used in dental implants. Non-expendable mold casting, on the other hand, typically uses steel molds coated with a refractory material. One example of this method is die casting, where molten material is injected into the mold under high pressure and speed. Die casting offers excellent dimensional precision and can produce thin, intricately shaped structures with a smooth surface finish [20]. Depending on the complexity of the product, single or multiple dies may be used. This method is commonly employed for molding components made of polymers and metals. However, die casting is not suitable for manufacturing biocompatible products like prosthetics. The dimensions and structure of prosthetics often vary from one person to another, making die-casting economically impractical for their production [6]. Therefore, alternative manufacturing methods are preferred for creating bio-compatible products that require customization.

## 8.4   SOL-GEL

The recent attention on biopolymer-silica hybrid materials stems from their assuring characteristics and suitability with living matter. These materials have various applications, including in the realm of biocompatible materials, enzyme and cell immobilization, cement for bone repair and reconstruction, bone substitutes, catalysis, and sensors. The sol-gel process has proven to be an effective technique for synthesizing these hybrid materials, which exhibit unique mechanical properties such as low elastic modulus, superior ductility, and superior mechanical strength.

S. Smitha et al. [21] have studied a silica-methyltrimethoxysilane (MTMS)-gelatin biohybrid that was successfully synthesized using colloidal silica as the forerunner. Surface treatment with methyltrimethoxysilane resulted in hydrophobic $SiO_2$-MTMS-gelatin hybrids. The presence of $CH_3$ and Si-C bonds is accountable for hydrophobicity with no water desorption and organic group oxidation at 530°C, indicating hydrophobicity up to that temperature. Varying MTMS concentrations allowed the tailoring of hydrophobic properties, achieving a maximum contact angle of 95°. The hybrid coatings exhibited high optical transmittance, making them suitable for optical and transparent biocompatible applications. In summary, biopolymer-silica hybrid materials have garnered interest due to their favorable properties and compatibility with living matter. The sol-gel process is an efficient method for synthesizing these materials, resulting in hybrids with unique mechanical properties [9].

With its unique advantages, the sol-gel process offers a range of benefits. Firstly, it utilizes relatively low temperatures, making it a practical and energy-efficient method. Additionally, this process excels at creating very fine powders, resulting in materials with enhanced properties and improved performance. Furthermore, the sol-gel process enables the production of compositions that would be otherwise unattainable through solid-state fusion techniques. However, it is important to acknowledge the existence of certain disadvantages. The cost of the raw materials, specifically the chemicals involved, can be a limiting factor. For instance, obtaining high-purity MgO powder may come at a steep price of $30/kg, and the chemical source magnesium ethoxide can cost as much as $200/kg. Moreover, during the drying phase, the sol-gel process often experiences significant volume shrinkage and the potential for cracking, necessitating careful handling and process optimization.

## 8.5   ELECTROSPINNING

Electrospinning is a versatile technique with diverse applications in the field of biocompatible materials. It involves subjecting a polymer solution or melt to an electric field, resulting in the production of ultrafine fibers. These fibers possess a high surface area-to-volume ratio, making them ideal for various biomedical applications. Electrospinning allows for the creation of different structures and has found notable use in tissue engineering scaffolds, wound dressings, biosensors, and drug delivery systems.

In tissue engineering, electrospun fibers can be utilized to create three-dimensional scaffolds that closely mimic the native extracellular matrix. This provides an optimal environment for cell growth and tissue regeneration. Biocompatible polymers such as polycaprolactone (PCL), poly (lactic-co-glycolic acid) (PLGA), or gelatin are commonly used to construct these scaffolds. By tuning electrospinning parameters such as fiber diameter and alignment, it becomes possible to mimic specific tissue architectures, enabling more effective tissue regeneration [22]. Electro-spun nanofibers also exhibit huge promise as drug carriers because of their high porosity. Drugs can be incorporated within the fibers or deposited on their surfaces, allowing for controlled release by modifying the composition and structure of the fibers [23]. Biocompatible polymers like polyethylene glycol (PEG) and chitosan are frequently employed for drug encapsulation in electro-spun fibers. This approach holds promise for targeted and sustained drug delivery, enhancing therapeutic outcomes [24,25]. Furthermore, electro-spun fibers find application in the development of advanced wound dressings. These dressings possess enhanced properties such as high water absorption, gas permeability, and mechanical strength. Biocompatible polymers like polyvinyl alcohol (PVA), collagen, or elastin can be electro-spun into fibrous mats, providing a favorable environment for wound healing. Additional functional additives, including antimicrobial agents, growth factors, or nanoparticles, can be incorporated into the dressings to promote enhanced healing [25].

In the field of biosensing, electro-spun fibers can be functionalized with biomolecules or nanoparticles to create highly sensitive biosensors. These biosensors are capable of detecting various analytes such as glucose, proteins, or DNA. The substantial surface area of the fibers allows for the efficient immobilization of biomolecules, while the porosity facilitates analyte diffusion and binding. Biocompatible polymers such as polyacrylic acid (PAA) or polyvinylpyrrolidone (PVP) serve as suitable matrix materials for biosensing applications [22]. These examples highlight the diverse and significant applications of electrospinning in the realm of biocompatible materials, demonstrating its potential for advancing biomedical research and improving patient care.

Electrospinning offers various advantages as an indispensable tool in bioengineering. Its simplicity, affordability, and exceptional performance make it highly desirable. The technique enables the production of controllable micro/nano-sized fibers that intimately resemble the native extracellular matrix (ECM) structure.

Although electrospinning offers numerous advantages, it is not without limitations. Firstly, the use of toxic solvents for polymer dissolution poses risks to human health and the environment. Additionally, the dense and interconnected network of nanofibrous structures created by electrospinning hinders cell infiltration, making it difficult to achieve the homogeneous cell distribution necessary for tissue engineering. Uneven cell deposition, restricted cell migration, and lack of cell-binding sites on nanofiber surfaces further impede cell distribution. Moreover, the technique's limited material compatibility hampers the development of innovative approaches tailored to tissue engineering. Lastly, producing scaffolds with cell-containing biomaterials is challenging due to potential cell damage caused by high voltage parameters, making single-step cellular fiber production less feasible.

## 8.6   SELF-ASSEMBLY

Self-assembly has surfaced as a powerful concept in the construction of functional biomaterials, offering new possibilities for the treatment of injuries and diseases. By leveraging spontaneous molecular associations driven by noncovalent interactions, self-assembly enables the formation of complex and adaptable materials with significant biological effects. This section explores the utilization of self-assembly to create innovative systems for regenerative medicine, and biology and highlights the advantages, practical applications, and design principles of self-assembled biomaterials, supported by relevant scientific references.

In comparison to traditional covalent materials, self-assembled biomaterials offer distinct advantages. Their modular nature allows for easy tuning of material composition by adjusting molecular components, leading to enhanced versatility and customization possibilities. The inherent multivalency of self-assembly amplifies biological effects and facilitates higher local concentrations of chemical signals, enabling more efficient cellular responses. Additionally, self-assembled structures can mimic the nanoscale length scales at which cells interact with their environment, promoting better compatibility and interaction with biological systems [26]. Furthermore, self-assembled biomaterials exhibit dynamic and self-healing properties, making them responsive to external conditions and capable of repairing themselves when temporarily disrupted. They are also biodegradable, cytocompatible, and monodisperse, rendering them well suited for various biomedical applications.

### 8.6.1   Types of Self-Assembly Techniques

- **Peptide-Based Self-Assembled Biomaterials:** Peptides, including amyloid proteins, α-helical peptides, and collagen-mimetic peptides, have been extensively utilized in the design of self-assembled biomaterials. Amyloid proteins, which are typically associated with degenerative diseases, can be repurposed to form hydrogels that facilitate cell adhesion by introducing additional cell-adhesive sequences. α-helical peptides can assemble into coiled-coil structures through specific design principles, enabling the creation of hydrogels compatible with cell culture. Collagen-mimetic peptides, inspired by the triple helical structure of collagen, self-assemble into triple helices and form hydrogels that mimic the properties of native collagen [27].
- **Short Aromatic Peptides and Protein-Based Self-Assemble Biomaterials:** Short aromatic peptides, composed of 2–5 amino acids, spontaneously self-assemble into various structures, including nanotubes, vesicles, nanofibers, and hydrogels, driven by hydrophobic interactions, intermolecular hydrogen

bonding, and π-π stacking. These simple and scalable peptides have been employed as biomaterials for cell culture, offering compatibility with cell growth and proliferation. Protein-based biomaterials, such as collagen, elastin, and spider silk, naturally self-assemble and provide greater complexity and functionality than short peptides. Coiled-coil interactions and amphiphilic block co-polypeptides have enabled the creation of protein-based hydrogels with tuneable properties [28].

- **DNA-Based Self-Assemble Biomaterials:** DNA-based biomaterials harness the programmability and specificity of DNA base-pairing interactions. DNA hydrogels can be constructed by designing branched DNA structures that cross-link via enzymatic reactions. These hydrogels exhibit adjustable swelling and mechanical properties, allowing for the controlled release of drugs and efficient protein production. Other DNA-based materials, including i-motif cross-linked gels and Y-shaped DNA monomers, have shown stability and potential for hydrogel formation [29].

The self-assembly of nanostructures in biocompatible materials offers significant advantages in terms of cost-effectiveness, minimal waste generation, and the ability to rapidly produce complex nanostructures. This makes self-assembly an attractive option for large-scale manufacturing of biocompatible materials, as it provides a cost-efficient and sustainable approach. However, one limitation of self-assembly is the constraint on material selection. Not all materials possess the necessary self-assembly properties required for this technique. Only materials with specific characteristics and interactions can successfully undergo self-assembly, restricting the spectrum of materials that can be utilized in the process. This necessitates careful consideration and selection of appropriate materials with the desired self-assembly properties.

## 8.7 CAD/CAM TECHNIQUES

CAD/CAM production has been widely used in the dental industry for over 20 years to create various fixed prosthetic restorations such as crowns, inlays, on-lays, fixed partial dentures (FPDs), and veneers. The process involves three main steps. First, the geometry of the site or product is scanned using advanced technologies to convert it into digital data. This data is then manipulated and processed using computer-aided design (CAD) software. Lastly, the compiled data is utilized for manufacturing the desired structures using computer-aided manufacturing (CAM) techniques [30–32]. There are two main manufacturing processes in CAD/CAM production: **subtractive** and **additive** manufacturing (Figure 8.4).

### 8.7.1 ADDITIVE MANUFACTURING/3D PRINTING

Additive manufacturing (AM) is often employed to produce initial 3D structures in a green state. These structures are built using a mixture of powders and binder materials, which may be organic or inorganic. Subsequently, the green structures undergo debinding to remove the organic binder and sintering to enhance their density. The

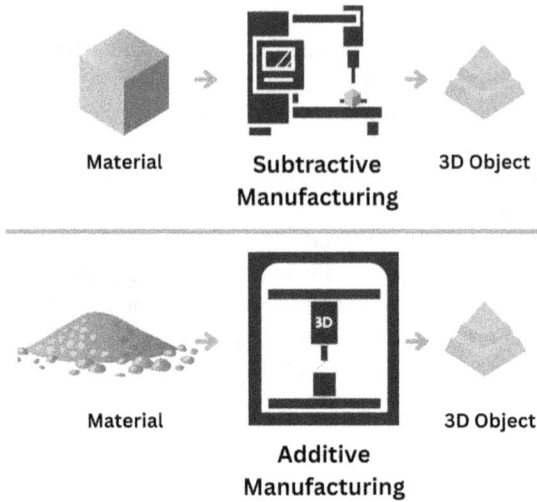

**FIGURE 8.4** Schematic process of subtractive manufacturing and additive manufacturing [33].

debinding process is influenced by the organic components and becomes diffusion limited as the part's thickness increases, resulting in longer debinding times [32]. Careful consideration of the debinding temperature is crucial to avoid crack formation or delamination caused by excessive removal rates of volatile products. Plasticizing agents can be used to mitigate the internal stresses generated during debinding. Additive manufacturing has various applications in dentistry. One of the earliest applications is the creation of medical models for surgical guides, enabling the review of complex or uncommon anatomical features and facilitating surgical planning and practice. These models also play a role in the creation of dental restorations, including veneered materials. Digital orthodontic systems are available that digitally align a patient's teeth, resulting in the generation of a sequence of 3D-printed models utilized for producing aligners like Invisalign®. Additionally, 3D-printing technology enables the production of titanium dental implants with novel porous or rough surfaces, as well as various kinds of refurbishments and constituents made from polymeric or metallic materials, including onlays, inlays, facets, dentures, and crowns [32].

### 8.7.1.1 Different Types of Additive Manufacturing Techniques
In the field of AM, ASTM has defined a set of standards known as ASTM F42, which covers terminology, materials, processes, testing, and more. The ASTM standard F2792 provides a terminological classification of AM processes, here are seven different types of additive manufacturing techniques [34–37]: (Table 8.1).

### 8.7.1.2 Biocompatible Applications of the Additive Manufacturing Method
The field of biomaterials and biomedical devices has been transformed by the arrival of additive manufacturing, commonly known as 3D printing. This innovative technology has opened new opportunities for fabricating intricate structures and

## TABLE 8.1
## Classification of AM Process

| Category | Operating Principle | Examples of Technology | Materials Used |
|---|---|---|---|
| Material Extrusion | The buildup of layers by extruding | Fused Deposition Modeling (FDM) | Thermoplastics, Composites |
| | | Fused Filament Fabrication (FFF) | Polymers, Composites |
| | | Direct Ink Writing (DIW) | Thermosetting polymers |
| Material Jetting | Depositing material droplets | PolyJet | Photopolymers, Waxes, Elastomers |
| | | Multi-Jet (MJ) | Polymers |
| Powder Bed Fusion | Sintering or melting of powder layers using a laser or electron beam | Selective Laser Sintering (SLS) | Polymers, Metals, Ceramics |
| | | Selective Laser Melting (SLM) | Metal Powders, Alloys |
| | | Electron Beam Melting (EBM) | Titanium Alloys, Stainless Steel |
| Vat-Photopolymerization | Curing liquid resin with UV light | Stereolithography (SLA) | Photopolymers |
| | | Digital Light Processing (DLP) | Resins, Polymers |
| Sheet Lamination | Bonding sheets of material | Laminated Object Manufacturing (LOM) | Paper, Plastic Sheets, Metal Foils |
| Directed Energy Deposition | Depositing material using a focused energy source | Laser Engineered Net Shaping (LENS) | Metals, Alloys |
| | | Laser Powder Deposition (LPD) | Titanium, Aluminum, Steel |
| Binder Jetting | Depositing binder to bond powders | Binder Jetting Direct Metal | Sand, Metal Powders, Ceramics |

personalized designs. This chapter explores the diverse applications of additive manufacturing in the areas of orthopedic and dental implants, as well as tissue engineering.

**Orthopedics and Dental Implants:** Additive manufacturing has made significant advancements in the production of orthopedic and dental implants. By utilizing additive manufacturing techniques, implants can be customized to fit the unique anatomy of each patient, resulting in improved functionality, fit, and compatibility. Traditional manufacturing methods, like casting or machining, can only produce standardized implants that do not perfectly match an individual's specific anatomy. Additive manufacturing overcomes these limitations by enabling the creation of patient-specific implants [11,38,39].

Metal-based implants, like titanium, are commonly manufactured using additive manufacturing due to their excellent mechanical properties and biocompatibility [40].

The layer-by-layer fabrication process of additive manufacturing allows producing complex geometries with intricate internal structures. This capability is particularly beneficial for implants that require porous structures to promote bone ingrowth and vascularization [41,42]. By tailoring the porosity and surface characteristics, additive manufacturing enhances the process of osseointegration, thereby improving the long-term stability and performance of orthopedic and dental implants [11]. Furthermore, additive manufacturing has expanded its applications beyond metals [43,44]. Researchers are exploring implementing other materials, such as polymers and ceramics, for implant fabrication. Ceramic implants offer advantages such as exceptional biocompatibility and the ability to mimic natural bone properties [45–47]. Additive manufacturing techniques enable the creation of intricate ceramic structures with controlled porosity, closely resembling the properties of surrounding bone tissue. Similarly, polymer-based implants can be customized to meet specific patient requirements, offering flexibility in design and functionality [39].

**Tissue Engineering:** Additive manufacturing plays a vital role in tissue engineering, which targets to produce functional tissues or organs using bioactive materials and cells [48]. Traditional tissue engineering methods have relied on top-down approaches, where cells and tissue growth factors are assembled into a temporary structure to promote further growth and maturation [49]. While this approach has achieved scientific and clinical success, it has limitations in controlling tissue architecture and nutrient and waste transfer. A promising bottom-up approach called bioprinting has emerged in tissue engineering, lever-aging additive manufacturing principles [50]. Bioprinting allows for the production of intricate three-dimensional structures with micro- and macro-porosity. It enables precise placement of cells, growth factors, and biomaterials, facilitating the development of complex tissue architectures. Bioprinting offers several advantages over traditional tissue engineering, including the ability to create hierarchical structures with precise control over cell placement and tissue organization. Three common bioprinting systems are inkjet bioprinters, micro extrusion bioprinters, and laser-assisted bioprinters. Inkjet bioprinters use air or piezoelectric pressure to deposit droplets containing cells or biomaterials. Micro extrusion bioprinters extrude continuous materials using pneumatic or mechanical systems, enabling the deposition of bio-inks with higher viscosity. Laser-assisted bioprinters employ laser energy to propel cell-containing materials onto the substrate, providing precise control and high-resolution printing.

In tissue engineering, additive manufacturing has been applied to various types of tissues, including soft tissue (e.g., cartilage), hard tissue (e.g., bone), and organ tissue (e.g., liver). The ability to create tissue structures with controlled porosity and tailored mechanical properties has greatly accelerated progress in the field of regenerative medicine.

**Otorhinolaryngology:** There are multiple applications of biomaterials in the field of otorhinolaryngology. These applications include middle ear prostheses, cochlear implants, materials for bone fixation, voice prostheses, and nasal packing following sinus surgery [51]. Ossicular chain reconstruction in the middle ear can be achieved using total or partial ossicular replacement prostheses [52]. Mild to moderate hearing loss can be treated with implantable middle-ear hearing aids [54,53]. Titanium prostheses have gained popularity due to their effectiveness in sound transmission, ease of manipulation, and long-term durability. Among various materials used for ossicular reconstruction, titanium is known to be biocompatible and lightweight [54,55].

A retrospective review comparing titanium and hydroxyapatite prostheses for ossicular chain reconstruction demonstrated favorable functional and anatomical outcomes for both types. However, partial titanium prostheses showed better hearing outcomes [56].

**Drug Delivery:** Drug management for systemic effects and disease treatment involves various techniques based on specifically targeted body locations. The choice of drug delivery method depends on factors such as disease type, desired therapeutic outcome, drug efficacy, and sometimes patient preferences. Additive manufacturing has revolutionized the drug delivery industry by enabling personalized medicine approaches. Material extrusion, particularly through fused deposition modeling (FDM), is commonly used for manufacturing rectal, oral, transdermal, vaginal, and implantable drug products [57]. FDM offers versatility with a wide range of biocompatible materials, making it suitable for producing oral solid tablets [58]. Achieving prompt drug release is crucial for optimal therapeutic effect. 3D-printed tablets with varying infill densities, complex structures, and drug-loading contents have been developed to achieve immediate or extended-release profiles [59]. Materials such as PLA, HP, PVA, HPMC, and Irgacure have been used [60–62], and drugs like caffeine, paracetamol, acetaminophen, and curcumin have been tested.

The findings indicate that all printed samples demonstrated complete drug release within a time frame of fewer than 480 minutes. Moreover, tablets with higher drug-loading content exhibit faster drug release [63] (Tables 8.2 and 8.3).

---

**TABLE 8.2**

**List of Literature on Additive Manufacturing Techniques and Their Biocompatible Applications**

| Type of Additive Manufacturing | Application | Literature |
|---|---|---|
| | Orthodontic appliances | "Fabrication of a resin appliance with alloy components using digital technology without an analog impression" [64] |
| | Provisional crowns | "3D printing restorative materials using a stereolithographic technique: a systematic review" [65] |
| Stereolithography (SLA) | Drug delivery systems | "The Evolution of the 3D-Printed Drug Delivery Systems: A Review" [66] |
| | Bio fabrication of organs | "Biofabrication Strategies for Musculoskeletal Disorders: Evolution towards Clinical Applications" [67] |

*(Continued)*

**TABLE 8.2** *(Continued)*
## List of Literature on Additive Manufacturing Techniques and Their Biocompatible Applications

| Type of Additive Manufacturing | Application | Literature |
| --- | --- | --- |
| | Tissue engineering scaffolds | "3D Printing of Scaffolds for Tissue Regeneration Applications" [68] |
| | Scaffolds | "Development of controlled porosity polymer-ceramic composite scaffolds via fused deposition modelling" [69] |
| | Tablets | "Effect of geometry on drug release from 3D printed tablets" [72] |
| Fused deposition modeling (FDM) | Intrauterine prototypes | "Ethylene vinyl acetate (EVA) as a new drug carrier for 3D printed medical drug delivery devices" [62] |
| | Drug delivery systems | "Polymeric drug delivery systems by additive manufacturing" [70] |
| | Patient-specific anatomical models | "Three-dimensional printing in spine surgery: a review of current applications" [71] |
| | Organ-on-a-chip systems | "Microfluidic Organ-on-a-Chip Technology for Advancement of Drug Development and Toxicology" [72,73] |
| Inkjet bioprinting | Skin tissue engineering | "Bioprinting of Skin Constructs for Wound Healing (He et al., 2018)" |
| | Bioprinting of living cellular constructs | "Bioprinting of 3D tissue models using decellularized extracellular matrix bioink" [74] |
| | Customized implants | "A review of fabrication polymer scaffolds for biomedical applications using additive manufacturing techniques" [75] |
| Selective laser sintering | Bioprinting of vasculature | "Direct 3D bioprinting of prevascularized tissue constructs with complex microarchitecture" [76] |
| | Dental applications | "A Review of 3D Printing in Dentistry: Technologies, Affecting Factors, and Applications" [77] |

## 8.8  CONCLUSION AND ASPECTS OF RESEARCH

The primary aim of this chapter is to critically analyze and evaluate various processing techniques employed for the fabrication of biocompatible materials, ranging from conventional methods to contemporary approaches, along with their respective applications. Within the scope of our in vitro investigation, we have derived the subsequent finding that each process mentioned in the chapter has its

**TABLE 8.3**

**Summarizing the Advantages and Limitations of Bio-Compactible Processing Techniques**

| Technique | Advantages | Disadvantages |
|---|---|---|
| Casting | • Simple and cost-effective.<br>• Can be used for a vast range of biomaterials.<br>• Complex geometries can be made easily on a large scale. | • Cast products are subject to porosity and other defects.<br>• The time of processing is longer.<br>• There is minimal control over material properties. |
| Sol-gel | • The material composition and structure can be controlled precisely.<br>• Extra materials can be added to modify surface properties.<br>• Thin films and coatings can be produced with excellent uniformity. | • Large-scale production is not feasible.<br>• Time of processing is longer.<br>• The equipment used is specialized, and one should have a good amount of experience in using this technique. |
| Electrospinning | • Nano-fibers with a high surface area to volume ratio can be produced quickly.<br>• Fiber diameter and morphology can be controlled.<br>• This method can be used to mimic natural extracellular matrix structures (ECM). | • Only fibrous materials can be used in this method.<br>• Large-scale production is challenging. |
| Self-assembly | • The cost of production is low, and the waste generation is minimum.<br>• Helps produce complex nanostructures quickly. | • Not all materials can be used. Materials with desirable self-assembly properties can only be used.<br>• Complex characterization techniques are required to obtain the desired properties. |
| Additive manufacturing | • Can produce products with highly complex geometry with significant precision.<br>• The amount of waste produced is low. | • Equipment used and the materials can be costly.<br>• Obtained products might have defects in them.<br>• The surface finish and resolution are limited. |

own set of advantages and disadvantages. Considering the broader picture, additive manufacturing, in the near future has a vast scope of development in this field, specifically when low cost of production, low waste, high precision, and bulk production are targeted.

## REFERENCES

1. B.D. (Buddy D.) Ratner, *Biomaterials Science: An Introduction to Materials in Medicine*, Academic Press, 2013.
2. B.D. (Buddy D.) Ratner, *Biomaterials Science: An Introduction to Materials in Medicine*, Elsevier Academic Press, 2004.
3. L. Crawford, M. Wyatt, J. Bryers, B. Ratner, Biocompatibility evolves: Phenomenology to toxicology to regeneration, *Adv Healthc Mater*. 10 (2021). 10.1002/adhm.202002153.
4. T. Xue, S. Attarilar, S. Liu, J. Liu, X. Song, L. Li, B. Zhao, Y. Tang, Surface modification techniques of titanium and its alloys to functionally optimize their biomedical properties: Thematic review, *Front Bioeng Biotechnol*. 8 (2020). 10.3389/fbioe.2020.603072.
5. X. Wang, Overview on biocompatibilies of implantable biomaterials, in: *Advances in Biomaterials Science and Biomedical Applications*, InTech, 2013. 10.5772/53461
6. J.S. Miller, K.R. Stevens, M.T. Yang, B.M. Baker, D.H.T. Nguyen, D.M. Cohen, E. Toro, A.A. Chen, P.A. Galie, X. Yu, R. Chaturvedi, S.N. Bhatia, C.S. Chen, Rapid casting of patterned vascular networks for perfusable engineered three-dimensional tissues, *Nat Mater*. 11 (2012) 768–774. 10.1038/nmat3357.
7. A. Sola, J. Bertacchini, D. D'Avella, L. Anselmi, T. Maraldi, S. Marmiroli, M. Messori, Development of solvent-casting particulate leaching (SCPL) polymer scaffolds as improved three-dimensional supports to mimic the bone marrow niche, *Mater Sci Eng: C*. 96 (2019) 153–165. 10.1016/J.MSEC.2018.10.086.
8. H. Holback, Y. Yeo, K. Park, Hydrogel swelling behavior and its biomedical applications, *Biomed Hydrogels*. (2011) 3–24. 10.1533/9780857091383.1.3.
9. G.J. Owens, R.K. Singh, F. Foroutan, M. Alqaysi, C.M. Han, C. Mahapatra, H.W. Kim, J.C. Knowles, sol-gel based materials for biomedical applications, *Prog Mater Sci*. 77 (2016) 1–79. 10.1016/j.pmatsci.2015.12.001.
10. Q.P. Pham, U. Sharma, A.G. Mikos, Electrospinning of polymeric nanofibers for tissue engineering applications: A review, Https://Home.Liebertpub.Com/Ten. 12 (2006) 1197–1211. 10.1089/TEN.2006.12.1197.
11. S. Bose, D. Ke, H. Sahasrabudhe, A. Bandyopadhyay, Additive manufacturing of biomaterials, *Prog Mater Sci*. 93 (2018) 45–111. 10.1016/J.PMATSCI.2017.08.003.
12. E. Marin, F. Boschetto, G. Pezzotti, Biomaterials and biocompatibility: An historical overview, *J Biomed Mater Res A*. 108 (2020) 1617–1633. 10.1002/JBM.A.36930.
13. M. Vallet-Regí, Evolution of biomaterials, *Front Mater*. 9 (2022). 10.3389/fmats.2022.864016.
14. A. Szczesio-Wlodarczyk, J. Sokolowski, J. Kleczewska, K. Bociong, Ageing of dental composites based on methacrylate resins - A critical review of the causes and method of assessment, *Polymers (Basel)*. 12 (2020). 10.3390/POLYM12040882.
15. A.K. Gaharwar, N.A. Peppas, A. Khademhosseini, Nanocomposite hydrogels for biomedical applications, *Biotechnol Bioeng*. 111 (2014) 441–453. 10.1002/bit.25160/abstract.
16. S.R. Knight, R. Aujla, S.P. Biswas, Total hip arthroplasty – over 100 years of operative history, *Orthop Rev (Pavia)*. 3 (2011) e16–e16. 10.4081/OR.2011.E16.
17. E.S. Place, J.H. George, C.K. Williams, M.M. Stevens, Synthetic polymer scaffolds for tissue engineering, *Chem Soc Rev*. 38 (2009) 1139–1151. 10.1039/b811392k.
18. S.V. Murphy, A. Atala, 3D bioprinting of tissues and organs, *Nat Biotechnol*. 32 (2014) 773–785. 10.1038/nbt.2958.
19. M.N. Rahaman, D.E. Day, B. Sonny Bal, Q. Fu, S.B. Jung, L.F. Bonewald, A.P. Tomsia, Bioactive glass in tissue engineering, *Acta Biomater*. 7 (2011) 2355–2373. 10.1016/j.actbio.2011.03.016.

20. W. Fu, S. Liu, J. Jiao, Z. Xie, X. Huang, Y. Lu, H. Liu, S. Hu, E. Zuo, N. Kou, G. Ma, Wear resistance and biocompatibility of Co-Cr dental alloys fabricated with CAST and SLM techniques, *Materials*. 15 (2022) 3263. 10.3390/MA15093263.
21. S. Smitha, P. Shajesh, P. Mukundan, T.D.R. Nair, K.G.K. Warrier, Synthesis of biocompatible hydrophobic silica-gelatin nano-hybrid by sol-gel process, *Colloids Surf B Biointerfaces*. 55 (2007) 38–43. 10.1016/j.colsurfb.2006.11.008.
22. L.A. Pruitt, A.M. Chakravartula, Mechanics of biomaterials: Fundamental principles for implant design, *Mechanics Biomate: Fundamental Principles Implant Design*. 9780521762212 (2011) 1–681. 10.1017/CBO9780511977923.
23. S. Deng, A. Chen, W. Chen, J. Lai, Y. Pei, J. Wen, C. Yang, J. Luo, J. Zhang, C. Lei, S.N. Varma, C. Liu, Fabrication of biodegradable and biocompatible functional polymers for anti-infection and augmenting wound repair, *Polymers*. 15 (2022) 120. 10.3390/POLYM15010120.
24. E.D. Boland, J.A. Matthews, K.J. Pawlowski, D.G. Simpson, G.E. Wnek, G.L. Bowlin, Electrospinning collagen and elastin: Preliminary vascular tissue engineering, *Frontiers Biosci*. 9 (2004) 1422–1432. 10.2741/1313/PDF.
25. A. Gul, I. Gallus, A. Tegginamath, J. Maryska, F. Yalcinkaya, Electrospun antibacterial nanomaterials for wound dressings applications, *Membranes (Basel)*. 11 (2021) 908. 10.3390/MEMBRANES11120908/S1.
26. N. Stephanopoulos, J.H. Ortony, S.I. Stupp, Self-assembly for the synthesis of functional biomaterials, *Acta Mater*. 61 (2013) 912–930. 10.1016/J.ACTAMAT.2012.10.046.
27. N. Habibi, N. Kamaly, A. Memic, H. Shafiee, Self-assembled peptide-based nanostructures: Smart nanomaterials toward targeted drug delivery, *Nano Today*. 11 (2016) 41–60. 10.1016/J.NANTOD.2016.02.004.
28. S. Lee, T.H.T. Trinh, M. Yoo, J. Shin, H. Lee, J. Kim, E. Hwang, Y.B. Lim, C. Ryou, Self-assembling peptides and their application in the treatment of diseases, *Int J Mol Sci*. 20 (2019) 5850. 10.3390/IJMS20235850.
29. X. Jian, X. Feng, Y. Luo, F. Li, J. Tan, Y. Yin, Y. Liu, Development, preparation, and biomedical applications of DNA-based hydrogels, *Front Bioeng Biotechnol*. 9 (2021). 10.3389/fbioe.2021.661409.
30. S. Chatterjee, J. Xu, T. V. Huynh, K. Abhishek, S. Kumari, A. Behera, Performance of smart alloys in manufacturing processes during subtractive and additive manufacturing: A short review on SMA and metal alloys, *Smart 3D Nanoprinting*. (2022) 267–280. 10.1201/9781003189404-14.
31. J. Abduo, K. Lyons, M. Bennamoun, Trends in computer-aided manufacturing in prosthodontics: A review of the available streams, *Int J Dent*. 2014 (2014). 10.1155/2014/783948.
32. R. Galante, C.G. Figueiredo-Pina, A.P. Serro, ScienceDirect additive manufacturing of ceramics for dental applications: A review, (2019). 10.1016/j.dental.2019.02.026.
33. G. Mittal, S. Paul, 3D printing for hybrid nanocomposites, in: *Smart 3D Nanoprinting*, CRC Press, 2022: pp. 1–22. 10.1201/9781003189404-1.
34. Additive Manufacturing Standards - Standards Products - Standards & Publications - Products & Services, (n.d.). https://www.astm.org/products-services/standards-and-publications/standards/additive-manufacturing-standards.html (accessed July 10, 2023).
35. A. Martínez-García, M. Monzón, R. Paz, Standards for additive manufacturing technologies: Structure and impact, *Addit Manuf*. (2021) 395–408. 10.1016/B978-0-12-818411-0.00013-6.
36. ASTM INTERNATIONAL Helping our world work better The Global Leader in Additive Manufacturing Standards, (n.d.). www.astm.org (accessed July 10, 2023).

37. J. Izdebska-Podsiadły, Classification of 3D printing methods, polymers for 3D printing: Methods, properties, and characteristics. (2022) 23–34. 10.1016/B978-0-12-818311-3.00009-4.
38. R. Zandparsa, Digital imaging and fabrication, *Dent Clin North Am.* 58 (2014) 135–158. 10.1016/J.CDEN.2013.09.012.
39. J. Jockusch, M. Özcan, Additive manufacturing of dental polymers: An overview on processes, materials and applications, *Dent Mater J.* 39 (2020) 345–354. 10.4012/DMJ.2019-123.
40. T.S. Tshephe, S.O. Akinwamide, E. Olevsky, P.A. Olubambi, Additive manufacturing of titanium-based alloys – A review of methods, properties, challenges, and prospects, *Heliyon.* 8 (2022). 10.1016/j.heliyon.2022.e09041.
41. S.A.M. Tofail, E.P. Koumoulos, A. Bandyopadhyay, S. Bose, L. O'Donoghue, C. Charitidis, Additive manufacturing: Scientific and technological challenges, market uptake and opportunities, *Materials Today.* 21 (2018) 22–37. 10.1016/J.MATTOD.2017.07.001.
42. M. Braian, R. Jimbo, A. Wennerberg, Production tolerance of additive manufactured polymeric objects for clinical applications, *Dental Materials.* 32 (2016) 853–861. 10.1016/J.DENTAL.2016.03.020.
43. A.S. Rizkalla, D.W. Jones, Mechanical properties of commercial high strength ceramic core materials, *Dental Materials.* 20 (2004) 207–212. 10.1016/S0109-5641(03)00093-9.
44. C. Gautam, J. Joyner, A. Gautam, J. Rao, R. Vajtai, Zirconia based dental ceramics: Structure, mechanical properties, biocompatibility and applications, *Dalton Trans.* 45 (2016) 19194–19215. 10.1039/C6DT03484E.
45. A. Shenoy, N. Shenoy, Dental ceramics: An update, *J Conserv Dent.* 13 (n.d.). 10.4103/0972-0707.73379.
46. M. Hisbergues, S. Vendeville, P. Vendeville, Zirconia: Established facts and perspectives for a biomaterial in dental implantology, *J Biomed Mater Res B Appl Biomater.* 88B (2009) 519–529. 10.1002/JBM.B.31147.
47. H. Shin, H. Ko, M. Kim, Cytotoxicity and biocompatibility of Zirconia (Y-TZP) posts with various dental cements, *Restor Dent Endod.* 41 (2016) 167–175. 10.5395/RDE.2016.41.3.167.
48. M. Ramasamy, Giri, R. Raja, Subramonian, Karthik, R. Narendrakumar, Implant surgical guides: From the past to the present, *J Pharm Bioallied Sci.* 5 (2013). 10.4103/0975-7406.113306.
49. B.P. Chan, K.W. Leong, Scaffolding in tissue engineering: General approaches and tissue-specific considerations, *Eur Spine J.* 17 (2008) 467–479. 10.1007/S00586-008-0745-3.
50. A. Goyanes, P. Robles Martinez, A. Buanz, A.W. Basit, S. Gaisford, Effect of geometry on drug release from 3D printed tablets, *Int J Pharm.* 494 (2015) 657–663. 10.1016/J.IJPHARM.2015.04.069.
51. J. Spałek, P. Ociepa, P. Deptuła, E. Piktel, T. Daniluk, G. Król, S. Góźdź, R. Bucki, S. Okła, Biocompatible materials in otorhinolaryngology and their antibacterial properties, *Int J Mol Sci.* 23 (2022). 10.3390/ijms23052575.
52. C.H. Stupp, H.F. Stupp, D. Grün, Ossicular chain reconstruction with titanium implants – Initial clinical experiences, *Laryngorhinootologie.* 75 (1996) 335–337. 10.1055/S-2007-997590/BIB.
53. A. Sengupta, S. Basu, M. Janweja, A. Sengupta, A clinical study on audiological evaluation in patients with active squamous variety of chronic otitis media following canal wall down mastoidectomy with ossicular reconstruction, *Indian J Otolaryngol Head Neck Surg.* 71 (2019) 1592–1598. 10.1007/S12070-019-01680-4.

54. P. Wongwiwat, A. Boonma, Y.S. Lee, R.J. Narayan, Bioceramics in ossicular replacement prostheses: A review, *J Long Term Eff Med Implants.* 21 (2011) 169–183. 10.1615/JLONGTERMEFFMEDIMPLANTS.V21.I2.70.
55. A.G.W. Meijer, H.M. Segenhout, F.W.J. Albers, H.J.L. Van De Want, Histopathology of biocompatible hydroxylapatite-polyethylene composite in ossiculoplasty, *ORL.* 64 (2002) 173–179. 10.1159/000058021.
56. M. Ziabką, E. Menaszek, J. Tarasiuk, S. Wroński, Biocompatible nanocomposite implant with silver nanoparticles for otology—In vivo evaluation, *Nanomater.* 8 (2018) 764. 10.3390/NANO8100764.
57. K. Pietrzak, A. Isreb, M.A. Alhnan, A flexible-dose dispenser for immediate and extended release 3D printed tablets, *Eur J Pharm Biopharm.* 96 (2015) 380–387. 10.1016/J.EJPB.2015.07.027.
58. J. Zhang, X. Feng, H. Patil, R. V. Tiwari, M.A. Repka, Coupling 3D printing with hot-melt extrusion to produce controlled-release tablets, *Int J Pharm.* 519 (2017) 186–197. 10.1016/J.IJPHARM.2016.12.049.
59. J. Skowyra, K. Pietrzak, M.A. Alhnan, Fabrication of extended-release patient-tailored prednisolone tablets via fused deposition modelling (FDM) 3D printing, *Eur J Pharm Sci.* 68 (2015) 11–17. 10.1016/J.EJPS.2014.11.009.
60. A. Goyanes, M. Kobayashi, R. Martínez-Pacheco, S. Gaisford, A.W. Basit, Fused-filament 3D printing of drug products: Microstructure analysis and drug release characteristics of PVA-based caplets, *Int J Pharm.* 514 (2016) 290–295. 10.1016/J.IJPHARM.2016.06.021.
61. R.C.R. Beck, P.S. Chaves, A. Goyanes, B. Vukosavljevic, A. Buanz, M. Windbergs, A.W. Basit, S. Gaisford, 3D printed tablets loaded with polymeric nanocapsules: An innovative approach to produce customized drug delivery systems, *Int J Pharm.* 528 (2017) 268–279. 10.1016/J.IJPHARM.2017.05.074.
62. N. Genina, J. Holländer, H. Jukarainen, E. Mäkilä, J. Salonen, N. Sandler, Ethylene vinyl acetate (EVA) as a new drug carrier for 3D printed medical drug delivery devices, *Eur J Pharm Sci.* 90 (2016) 53–63. 10.1016/J.EJPS.2015.11.005.
63. M. Cui, H. Pan, D. Fang, S. Qiao, S. Wang, W. Pan, Fabrication of high drug loading levetiracetam tablets using semi-solid extrusion 3D printing, *J Drug Deliv Sci Technol.* 57 (2020) 101683. 10.1016/J.JDDST.2020.101683.
64. N. Al Mortadi, Q. Jones, D. Eggbeer, J. Lewis, R.J. Williams, Fabrication of a resin appliance with alloy components using digital technology without an analog impression, *Am J Orthod Dentofacial Orthop.* 148 (2015) 862–867. 10.1016/J.AJODO.2015.06.014.
65. A. Della Bona, V. Cantelli, V.T. Britto, K.F. Collares, J.W. Stansbury, 3D printing restorative materials using a stereolithographic technique: A systematic review, *Dental Materials.* 37 (2021) 336–350. 10.1016/J.DENTAL.2020.11.030.
66. I. Bácskay, Z. Ujhelyi, P. Fehér, P. Arany, The evolution of the 3D-printed drug delivery systems: A review, *Pharmaceutics.* 14 (2022) 1312. 10.3390/PHARMACEUTICS14071312.
67. S. Naghieh, G. Lindberg, M. Tamaddon, C. Liu, Biofabrication strategies for musculoskeletal disorders: Evolution towards clinical applications, *Bioengineering.* 8 (2021) 123. 10.3390/BIOENGINEERING8090123.
68. A.V. Do, B. Khorsand, S.M. Geary, A.K. Salem, 3D printing of scaffolds for tissue regeneration applications, *Adv Healthc Mater.* 4 (2015) 1742–1762. 10.1002/ADHM.201500168.
69. S.J. Kalita, S. Bose, H.L. Hosick, A. Bandyopadhyay, Development of controlled porosity polymer-ceramic composite scaffolds via fused deposition modeling, *Materials Sci Eng: C.* 23 (2003) 611–620. 10.1016/S0928-4931(03)00052-3.

70. S. Borandeh, B. van Bochove, A. Teotia, J. Seppälä, Polymeric drug delivery systems by additive manufacturing, *Adv Drug Deliv Rev.* 173 (2021) 349–373. 10.1016/ J.ADDR.2021.03.022.

71. Y. Tong, D.J. Kaplan, J.M. Spivak, J.A. Bendo, Three-dimensional printing in spine surgery: a review of current applications, *Spine J.* 20 (2020) 833–846. 10.1016/ J.SPINEE.2019.11.004.

72. J.D. Caplin, N.G. Granados, M.R. James, R. Montazami, N. Hashemi, Microfluidic organ-on-a-chip technology for advancement of drug development and toxicology, *Adv Healthc Mater.* 4 (2015) 1426–1450. 10.1002/ADHM.201500040.

73. P. He, J. Zhao, J. Zhang, B. Li, Z. Gou, M. Gou, X. Li, Bioprinting of skin constructs for wound healing, *Burns Trauma.* 6 (2018). 10.1186/S41038-017-0104-X.

74. F. Pati, D.W. Cho, Bioprinting of 3D tissue models using decellularized extracellular matrix bioink, *Methods Mol Biol.* 1612 (2017) 381–390. 10.1007/978-1-4939-7021-6_27/FIGURES/2.

75. P. Szymczyk-Ziółkowska, M.B. Łabowska, J. Detyna, I. Michalak, P. Gruber, A review of fabrication polymer scaffolds for biomedical applications using additive manufacturing techniques, *Biocybern Biomed Eng.* 40 (2020) 624–638. 10.1016/ J.BBE.2020.01.015.

76. W. Zhu, X. Qu, J. Zhu, X. Ma, S. Patel, J. Liu, P. Wang, C.S.E. Lai, M. Gou, Y. Xu, K. Zhang, S. Chen, Direct 3D bioprinting of prevascularized tissue constructs with complex microarchitecture, *Biomaterials.* 124 (2017) 106–115. 10.1016/J.BIOMATERIALS. 2017.01.042.

77. Y. Tian, C.X. Chen, X. Xu, J. Wang, X. Hou, K. Li, X. Lu, H.Y. Shi, E.S. Lee, H.B. Jiang, A review of 3D printing in dentistry: Technologies, affecting factors, and applications, *Scanning.* 2021 (2021). 10.1155/2021/9950131.

# 9 Biofilm for Implant Materials

*Mohammad Azadi*
Faculty of Mechanical Engineering, Semnan University, Semnan, Iran

*Mahboobeh Azadi*
Faculty of Materials and Metallurgical Engineering, Semnan University, Semnan, Iran

*Shokouh Dezianian*
Faculty of Mechanical Engineering, Semnan University, Semnan, Iran

*Amir Hossein Beyzavi and Valeh Talebsafa*
Faculty of Materials and Metallurgical Engineering, Semnan University, Semnan, Iran

## 9.1 INTRODUCTION

Implants have been widely used for the support of body in different fields such as cardiovascular surgery, orthopedics, dental, urology, plastic and reconstructive surgery, neurosurgery, and tissue engineering. Besides, biomedical films or coatings are resistant to disinfectants, antibiotics, and the human immune system. Consequently, a surface modification for the implants could play a vital role to improve the anti-infection and biocompatibility properties [1]. For this objective, surface coatings are usually employed to improve corrosion, fatigue, and wear resistance of the implants [2], using different techniques and types. Moreover, the formation of such bio-coatings could be determined by the availability of nutrients, the type of organisms, and finally, the characteristics of substrates [3].

To fabricate these biofilms, various biomaterials could be utilized from polymers to ceramics [3]. In this chapter book, the polymeric materials are focused to improve the surface characteristics of implants. In addition, different chemical and physical techniques may use to fabricate these coatings. Again, through this chapter book, 3D printing is considered for the biofilm fabrication, as an additive manufacturing method.

DOI: 10.1201/9781003429920-11

## 9.2  IMPLANTS

For implants, their materials need necessary requirements such as the biocompatibility, long-term stability, proper strength, sufficient lifetime, and the ability to be sterilized without degradation. Therefore, metal implants, such as titanium alloys or magnesium alloys or stainless steels, have been widely utilized in clinical applications due to their superior wear and fatigue strength [4]. Unfortunately, these metallic materials (especially magnesium alloys) have no proper corrosion behaviors when they are inside human body. Thus, they are needed to be protected by surface coatings, either by passive or active surface finishing or modification [5].

Notably, although stainless steels are corrosion resistant, the implant weight is too high by such materials. Besides, in the human body, the titanium alloys also do not corrode. However, the oxide layer was gradually diffused by the metal ions and this issue caused the accumulation in the tissue [6]. Moreover, the biocompatibility of titanium alloy is not proper due to the existence of nickel in the composition and the potential for an allergic reaction [7].

Nowadays, the use of magnesium alloys in implants is increasing, especially for bone applications due to their biodegradability and similar densities and elastic modulus, compared to the natural bones [8]. However, as mentioned, magnesium alloys degrade very fast in the human physiological environments. This issue leads to a large release of hydrogen and an immediate loss of lifetime and strength [8].

## 9.3  BIOMEDICAL FILMS OR COATINGS

Most medical devices such as stents, orthodontic appliances, pacemakers, and surgical and dental instruments suffer from degradation in the human body. Thus, utilizing biomedical coatings such as hydrophilic, antimicrobial, hydrophobic, and lubrication coatings can protect these structures.

Various methods that can provide unique anticorrosion organic and inorganic coatings on different medical devices [9,10] could be listed as follows:

- Chemical vapor deposition (CVD).
- Physical vapor deposition (PVD).
- Pulsed laser deposition (PLD).
- Electrophoretic deposition (EPD).
- Micro-arc oxidation (MAO).
- Electro-deposition (ED).
- Plasma spray.
- Plasma electrolytic oxidation (PEO).
- Sol-gel processes.

The details of each method are described in the following sentences.

**Chemical vapor deposition (CVD):** In the CVD polymerization, monomers in the vapor phase will be reacted to form a solid layer on the substrate surface. Therefore, the coating and polymerization process happens in a single stage. In this situation, eliminating of dissolve macromolecules is not needed, and insoluble polymers can be also coated on the substrate. Moreover, the CVD process can be applied to any substrates, including inorganic, organic, flexible, rigid, three-dimensional, planar, porous, or dense [11].

**Physical vapor deposition (PVD):** PVD, or in some cases named physical vapor transport (PVT), refers to vast vacuum deposition processes that can be produced coatings of polymers on different substrates. Based on the polymer material type, coatings can be vapored with different methods. As an example, the direct vapor deposition will be used for polymers such as poly-tetrafluoroethylene (PTFE) and polyethylene with weak intermolecular interaction. However, polymer monomer generally evaporates and is deposited on the substrate surface through the polymerization reaction [12].

**Pulsed laser deposition (PLD):** In the PLD method, as a technique of PVD process for forming bio-coatings, a high pulse energy laser beam is focused on the target material which is to be deposited. Such a high energy is leading to fast heating and explosive evaporation. Thus, this non-thermal equilibration mechanism may be useful for creating multi-component homogeneous films with a high deposition rate [13].

**Electrophoretic deposition (EPD):** The EPD method is an effective technique to assemble polymers and other types of materials into 2D, 3D, and intricately shaped implants or other medical devices. It is notable that various bio-coatings such as peptides and proteins can be deposited on various substrates through the EPD process at room temperature, without affecting chemical structures. The EPD process includes two main stages: electrophoresis and deposition. An EPD device involves a power supply, electrodes, and a stable precursor solution/suspension/ (like a solution of charged polymers). At first, an electrical potential will be established between two electrodes, and then, charged polymers diffuse from the precursor solution/suspension (based on the presence of the electric field, called electrophoresis) toward the substrate for depositing. Finally, the gel will change into the coating after a drying procedure [14].

**Micro-arc oxidation (MAO):** The MAO process is an electrochemical surface technique to form oxide protective bio-coatings on some metals like magnesium and titanium. The MAO process is performed in an aqueous electrolyte, which needs intensive bath cooling. This process is based on anodizing reactions, which happen on a metallic surface through micro-arc discharges to form protective layers with a specific phase composition and morphology. Many factors such as the concentration, chemical composition, and temperature of the electrolyte, the anodizing time, the structure, and the type of substrate and electrical parameters affect the properties of deposited films [15].

**Electro-deposition (ED):** The ED process is an attractive, low-cost, and flexible method for fabrication of 2D and 3D, materials such as various coatings and films. The basis of this method is the electro-deposition reaction related to the reduction or deposition of electro-active species on the surface of the cathode. For polymeric coatings, the electrolytic solution contains monomeric and/or polymeric starting materials. Under electrochemical processes, chemical or physical transformation of the dissolved material onto a deposited solid film on the electrode happened [16].

**Plasma spray:** The plasma spray method is a flexible process that can deposit a high-performance coating on various materials. In a plasma spray process, the material will be vaporized or melted through a plasma resource and then sprayed onto the surface to be coated. It is notable that before the deposition process, the surface of substrates is cleaned and roughened for enhancing the adhesive strength of coatings since spray particles adhere to the surface through mechanical agents [17].

**Plasma electrolytic oxidation (PEO):** The PEO method, as a technique of MAO, is a novel surface technology that can manufacture multiple ceramic films on the surface of light metals. This method can deposit dense, thick, and metal oxide bio-coatings on substrates. The coating layer, deposited from the PEO method, is superior to the normal anodic oxidation process. PEO involves electrochemical processes that the electrolysis of an aqueous solution will happen at the surface of electrodes during the electrical discharge process. During the PEO process, surface interactions can be used for cleaning surfaces, and diffusive layers of ceramic, metal, and composite coatings will be formed on the cleaned surface [18].

**Sol-gel processes:** The sol-gel method is a depositing technique that is based on two stages of chemical reactions, hydrolysis, and condensation. It consists of the hydrolysis/condensation of various precursors (usually alkoxides) in water/ inorganic solution to form a colloidal dispersion of particles. Then, the immersion of a substrate into the sol solution with a controlled speed will happen. Finally, a thin film of gel is thus formed on the substrate since the evaporation rate of the solvent is rapid. The deposited film through this method shows anti-fingerprint, hydrophobic, anticorrosion, scratch, and wear resistance [19].

Besides, the biomedical organic coatings contain various types of textile and polymer coatings that could be used [9,10,20–23], such as,

- Polylactic acid (PLA)
- Polyethylene glycol (PEG)
- Polycaprolactone (PCL)
- Hydroxy-ethyl-cellulose (HEC)
- Poly-tetrafluoroethylene (PTFE)
- Acrylonitrile butadiene styrene (ABS)
- Polymethyl methacrylate (PMMA)
- Polyethylene terephthalate glycol (PETG)
- Polyvinylidene fluoride (PVDF)
- Nylon 12
- Polyester-amides
- Polypropylene (PP)
- Polyvinylpyrrolidone (PVP)
- Polydimethylsiloxane (PDMS)
- Polyether ether ketone (PEEK)
- Polyurethanes (PU)
- Thermoplastic polyurethane (TPU)
- Natural bio-polymers

Moreover, as mentioned in the last material type, some natural bio-polymer coatings such as chitosan (CS) and collagen can also be used as bio-coatings [20,21]. Most of these materials are usually cost-effective, non-toxicity, bio-degradable, or biocompatible for the human body and have good performance in cyclic loads [21]. In addition, for bio-polymer coatings, specific fabrication methods like polymer brushes, Langmuir-Blodgett (LB), layer-by-layer (LBL), spin coatings, dip coating, plasma-based coating methods, and hydrogels are utilized [20,22].

## 9.4 CASE STUDY

### 9.4.1 RESEARCH METHOD

In this chapter book, the studied material is AM60 magnesium alloy, as the substrate. Therefore, the commercial ingot of AM60 magnesium alloys was firstly cut into the plates, with the dimensions of $25 \times 8 \times 4$ mm$^3$. Then, various biomaterials were used to fabricate coatings on this substrate. Details of the chemical composition of the studied material, AM60 magnesium alloys as substrates are represented in Table 9.1.

Before coating process, specimens were finalized with the sandpaper up to 1,000 grit. Then, they were washed with acetone to degrease any surface pollution.

Polylactic acid (PLA), acrylonitrile butadiene styrene (ABS), polyethylene terephthalate glycol (PETG), polycaprolactone (PCL), and thermoplastic poly-urethane (TPU) were utilized as coatings for the biomechanics applications. It is notable that all these materials can be used in biomedical applications [24–26].

A process before 3D printing was to use the adhesion, including the PVP glue stick layer. This adhesion type was permanent, washable, acid-free, solvent-free, non-toxic, and dyed. The main chemical composition component of glue was polyvinylpyrrolidone.

Moreover, two different types of PCL filaments were used. The macroscopic view of one of them was white and another had a transparent color. Therefore, they were named PCL-W (in white color) and PCL-T (transparent version).

A device of 3D printer was utilized, which was working according to the technique of fused deposition modeling (FDM). This 3D printer was used in order to fabricate the layers of coatings on the magnesium alloys. Different printing parameters have been utilized to have a proper adhesion between the substrate and coatings. These factors can be seen in Table 9.2 for various filaments with the diameter of 1.75 mm, selected based on the literature [27–31] and the laboratory experiences. The nozzle diameter was 0.2 mm and the 3D printing speed was 5 mm/s.

**TABLE 9.1**
**The Chemical Composition of Utilized AM60 Magnesium Alloy**

| Al | Mn | Zn | Si | Cu | Fe | Mg |
|------|------|------|------|------|------|-------|
| 5.90 | 0.37 | 0.11 | 0.05 | 0.01 | 0.01 | 93.55 |

**TABLE 9.2**

**3D Printing Parameters for Different Filaments**

| Filaments | Infill Pattern | Infill Density | Nozzle Temperature (°C) |
|-----------|----------------|----------------|-------------------------|
| PLA | Linear [0°/90°] | 100% | 180 |
| ABS | Linear [0°/90°] | 100% | 225 |
| PCL-T | Linear [0°/90°] | 100% | 130 |
| PCL-W | Linear [0°/90°] | 100% | 130 |
| TPU | Linear [0°/90°] | 100% | 200 |
| PETG | Linear [0°/90°] | 100% | 210 |

For each coating, two layers were 3D printed on the substrate, with the layer height of 0.2 mm. Therefore, the final coating thickness was considered 0.4 mm.

For the corrosion behavior, electrochemical impedance spectroscopy (EIS) with the Organo flex device was utilized for the samples. The environment was the simulated body fluid (SBF), in which pH was 5.5 at the temperature of $37 \pm 1$ °C. The immersion time was about 1, 3, and 5 h. Notably, this job was similarly done in the literature but for SBF-10X and 24 h of the immersion [23]. In other words, the immersion time for SBF-1X was a study parameter in this chapter book.

In the corrosion text, a saturated calomel electrode was utilized for a reference electrode (SCE), besides a platinum sheet for a counter electrode. The EIS measurements were done in the area of 1.6 $cm^2$ on coatings. The AC voltage was applied as ±10 mV around the open circuit potential. The range of frequency was also 100 kHz to 10 mHz. After the immersion, the field-emission scanning electron microscopy (FESEM) was also done on the sample surface to find the corrosion severity.

### 9.4.2 Results and Discussion

Figure 9.1 shows Nyquist plots for all bio-coatings during various immersion times (including 1, 3, and 5 h) in SBF solution at 37°C. For the magnesium alloy (AM60) without any coating, one semi-circular was found in the Nyquist plot. It was reported that the diameter size of such semi-circular was related to the impedance of specimens [32]. Therefore, for all exposure times, the magnesium alloy showed the lowest impedance compared to other specimens with the lowest diameter size in Nyquist plots.

Figure 9.2 displays Bode plots for different samples. The impedance modulus (Z) was highest for PCL-T bio-coatings at all exposure times. In addition, the lowest Z value among coatings was attributed to PETG coatings. When the exposure time was low, the difference in Z value for most coatings was low. However, by enhancing the exposure time from 1 to 5 h, significant changes in Z values for various bio-coatings were observed. Such an event showed that the exposure time in the corrosive media was an effective factor to change the characteristics of coatings.

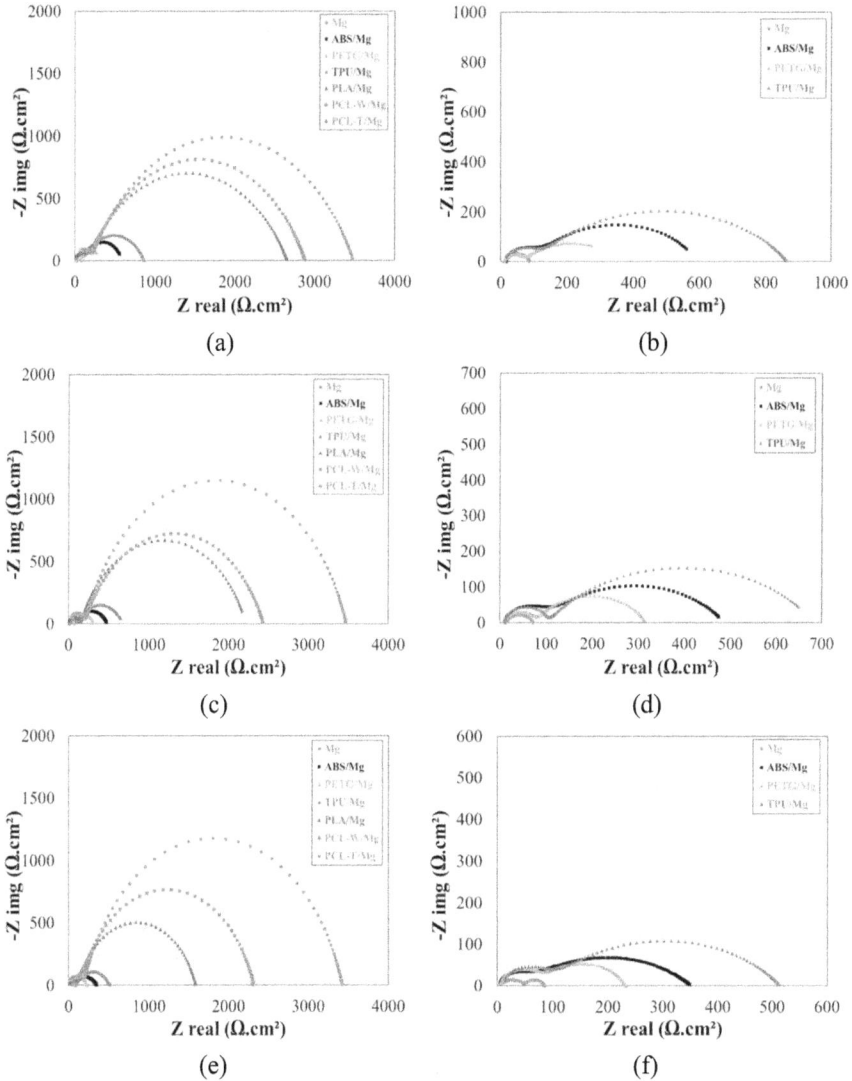

**FIGURE 9.1** Nyquist plots for all bio-coatings at different immersion times in the SBF solution at 37 °C: (a) and (b) after 1 h, (c) and (d) after 3h, and (e) and (f) after 5 h.

Totally, by increasing the time, the shape of both Nyquist and Bode plots did not change significantly for coatings. In this manner, the corrosion mechanism was not changed during the exposure time. A similar behavior was also found in other research [33].

For more details, the Z-view software was utilized to extract the experimental data of EIS measurements and fit proper circuits. Figure 9.3 illustrates such electrical circuits. Based on Nyquist plots, the suggested circuit for the magnesium alloy, at the exposure times of 1 and 3 h, could be contained a capacitor that was in

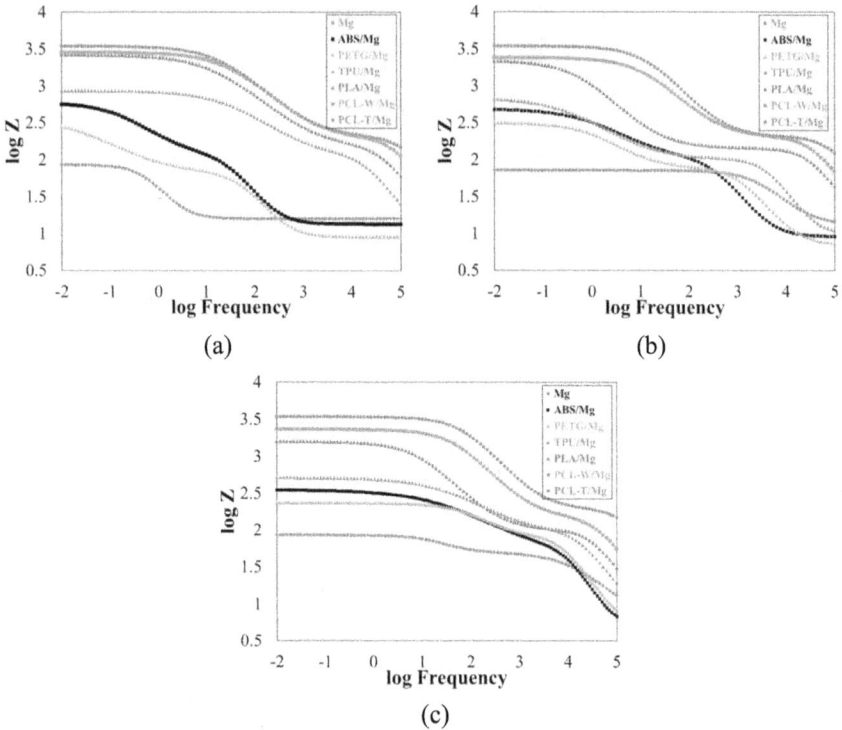

**FIGURE 9.2**  Bode plots for all bio-coatings at various immersion times in the SBF solution at 37 °C: after (a)1 h, (b) 3h, and (c) 5 h.

parallel with the polarization resistance ($R_p$). Both of them were in series with the solution resistance ($R_s$), as shown in Figure 9.3(a). It was found that this capacitor was related to the double-layer property. However, based on the surface roughness of the magnesium alloy, a constant phase element ($CPE_{dl}$) was replaced with a capacitor. A similar report was also found in another research [23].

By printing coatings on the magnesium surface, two semi-circular curves appeared in various Nyquist plots at all exposure times. Thus, in the suggested circuit, two other electrical elements were found, as shown in Figure 9.3(b). A $CPE_c$ and another resistance ($R_c$), both of them, were attributed to coating characteristics.

As reported in Table 9.3, when the exposure time was about 1 h, the highest and the lowest coating resistance ($R_c$) were attributed to PCL-T and PETG bio-coatings, respectively. In other words, both PCL bio-coatings showed a higher corrosion resistance, compared to other types of coatings. Values of $R_c$ for these bio-coatings were about three times higher than $R_c$ for PETG bio-coating. Moreover, the range of the enhancement in the polarization resistance ($R_p$) for all coatings, comparing to the AM60 magnesium alloy (%R), was about 75.8 to 97.9%. The lowest values of $CPE_{dl}$ and $CPE_c$ were attributed to both PCL bio-coatings. It was concluded that when the $CPE_{dl}$ value was low, it could illustrate that higher double-layer thickness and lower surface roughness [34,35]. In this

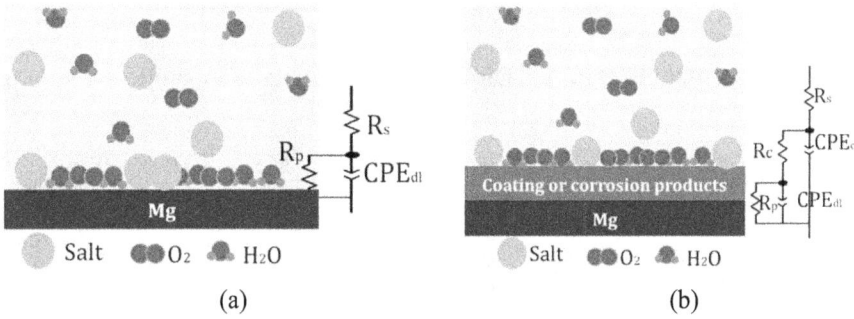

**FIGURE 9.3** (a) The suggested electrical circuit for the magnesium alloy at exposure times of 1 and 3 h, and (b) for the magnesium alloy at the exposure time of 5 h and for all bio-coatings at the immersion time of 1, 3, and 5 h.

**TABLE 9.3**
**The Experimental EIS Data for Different Bio-Coatings after the Immersion of 1 h**

| Specimens | $R_s$ ($\Omega cm^2$) | $CPE_c$ ($\mu F/cm^2$) | $n_c$ | $R_c$ ($\Omega cm^2$) | $CPE_{dl}$ ($mF/cm^2$) | $n_{dl}$ | $R_p$ ($\Omega cm^2$) | %R |
|---|---|---|---|---|---|---|---|---|
| Mg | 16 | – | – | – | 5.00 | 0.92 | 70 | – |
| PCL-W/Mg | 12 | 0.02 | 0.96 | 189 | 0.01 | 0.69 | 2680 | 97.4 |
| PCL-T/Mg | 12 | 0.01 | 0.93 | 196 | 0.01 | 0.69 | 3280 | 97.9 |
| PLA/Mg | 11 | 0.05 | 0.95 | 150 | 0.02 | 0.65 | 2502 | 97.1 |
| TPU/Mg | 14 | 0.08 | 0.99 | 105 | 0.04 | 0.64 | 749 | 90.6 |
| ABS/Mg | 11 | 81.2 | 0.92 | 115 | 1.90 | 0.69 | 466 | 85.0 |
| PETG/Mg | 9 | 90.1 | 0.91 | 61 | 4.10 | 0.79 | 289 | 75.8 |

situation, the highest and lowest values of $CPE_{dl}$ were related to magnesium alloys and both PCL bio-coatings, respectively.

By enhancing the immersion time from 1 to 3 h, values of $R_p$ for all specimens decreased, as shown in Table 9.4. However, the range of R values was about 75.2 to 98.2%, with insignificant change compared to values for 1 h. Such an event demonstrated that the diffusion of corrosive ions through defects of coatings increased by increasing the exposure time. Moreover, values of $R_c$ for all bio-coatings were reduced by increasing the time. However, the change in $R_c$ for PCL-T bio-coating was the lowest. Moreover, values of $CPE_{dl}$ and $CPE_c$ increased when the time increased. An increase in $CPE_{dl}$ could show a reduction in the thickness of the double layer. Similar to data at 1 h, the highest and the lowest $R_c$ were related to PCL-T and PETG bio-coatings, respectively.

Table 9.5 displays that when the immersion time was enhanced to 5 h, the corrosion product was collected on the surface of the magnesium alloy. Thus, two

**TABLE 9.4**

**The Experimental EIS Data for Different Bio-Coatings after the Immersion of 3 h**

| Specimens | $R_s$ ($\Omega cm^2$) | $CPE_c$ ($\mu F/cm^2$) | $n_c$ | $R_c$ ($\Omega cm^2$) | $CPE_{dl}$ ($\mu F/cm^2$) | $n_{dl}$ | $R_p$ ($\Omega cm^2$) | %R |
|---|---|---|---|---|---|---|---|---|
| Mg | 13 | – | – | – | 6.1 | 0.81 | 59 | – |
| PCL-W/Mg | 10 | 0.05 | 0.94 | 164 | 0.02 | 0.73 | 2234 | 97.3 |
| PCL-T/Mg | 10 | 0.02 | 0.92 | 195 | 0.01 | 0.78 | 3275 | 98.2 |
| PLA/Mg | 10 | 0.07 | 0.95 | 134 | 0.21 | 0.73 | 2080 | 97.1 |
| TPU/Mg | 12 | 0.10 | 0.99 | 92 | 0.99 | 0.62 | 584 | 89.8 |
| ABS/Mg | 9 | 10.80 | 0.91 | 101 | 18.20 | 0.62 | 377 | 84.3 |
| PETG/Mg | 7 | 5.50 | 0.89 | 72 | 0.90 | 0.79 | 238 | 75.2 |

**TABLE 9.5**

**The Experimental EIS Data for Different Bio-Coatings after the Immersion of 5 h**

| Specimens | $R_s$ ($\Omega cm^2$) | $CPE_c$ ($\mu F/cm^2$) | $n_c$ | $R_c$ ($\Omega cm^2$) | $CPE_{dl}$ ($\mu F/cm^2$) | $n_{dl}$ | $R_p$ ($\Omega cm^2$) | %R |
|---|---|---|---|---|---|---|---|---|
| Mg | 12 | 55 | 0.81 | 35.9 | 7.2 | 0.71 | 45 | – |
| PCL-W/Mg | 8 | 0.07 | 0.92 | 152 | 5.3 | 0.78 | 2156 | 97.9 |
| PCL-T/Mg | 7 | 0.02 | 0.92 | 192 | 0.02 | 0.80 | 3271 | 98.6 |
| PLA/Mg | 7 | 0.09 | 0.96 | 95.8 | 33.2 | 0.76 | 1487 | 97.0 |
| TPU/Mg | 9 | 0.14 | 0.99 | 87.3 | 88.9 | 0.60 | 416 | 89.3 |
| ABS/Mg | 5 | 0.79 | 0.99 | 50.2 | 20.9 | 0.53 | 296 | 84.9 |
| PETG/Mg | 5 | 0.31 | 0.91 | 79.1 | 55.4 | 0.75 | 147 | 60.7 |

electrical elements would appear for the circuit of the magnesium alloy at this exposure time. A $CPE_c$ and another resistance ($R_c$), both of them, were attributed to corrosion product characteristics that could act as a layer. By increasing the exposure time up to 5 h, values of $R_c$ for most coatings were reduced. Such an observation could illustrate higher water adsorption for bio-coatings by increasing the time [34]. In this manner, higher corrosion degradation would be observed. It was notable that by enhancing the immersion time from 1 to 3 h, $R_c$ for PETG bio-coatings increased insignificantly. Such an event could indicate that the coating resistance increased based on the collection of corrosion products on the surface of coating layers. Then, a similar observation was also found in another article [33]. Consequently, the highest and the lowest $R_c$ were attributed to PCL-T and ABS bio-coatings, respectively. However, the highest and the lowest $R_p$ were related to PCL-T and PETG bio-coatings, respectively. Totally, after 5 h, the range of $R$ values was about 60.7 to 98.6%.

It was notable that changes in values of $R_s$ were not significant for all exposure times. However, by enhancing the immersion time from 1 to 5 h, the solution resistance decreased. Such a behavior was related to changes in present ions and the salt percentage in corrosive media (the SBF solution). A similar observation was also found in other studies [33,34].

Moreover, the value of $n$ showed the inhomogeneity of the surface. When the roughness of the surface was lowest and the sample surface was in an ideal condition with a smooth manner, the value of $n$ was equal to unity. However, by increasing the surface roughness, $n$ would be decreased, and its value approached zero [23,32]. It was found that the corrosion reaction caused the surface degradation that resulted in higher roughness. Therefore, the lower value of $n$ indicated lower corrosion degradation at the surface. For all times, the value of $n_c$ was higher than $n_{dl}$. Thus, such a behavior could show that the corrosion degradation happened under most coatings. In this situation, the corrosive ions diffused from the sample surface of bio-coatings toward the metal surface, and corrosion reactions happened at the interface of coatings and metal surfaces. Therefore, the water adsorption of coatings, present defects such as cracks and pores in coatings, and other characteristics of coatings like the thickness were effective parameters to change the corrosion performance of different coatings. Consequently, the highly resistant bio-coatings could be arranged in the order of PCl, PLA, TPU, ABS, and PETG in the SBF solution at 37°C.

Figure 9.4 displays the Fourier-transform infrared spectroscopy (FTIR) patterns for various bio-coatings, before the corrosion test. In addition, schematics for the monomer of each bio-coating are represented in Figure 9.4. For both PCL and PLA bio-coatings, two bonds of C=O and C–O were found. These bonds acted as the characteristics bond. However, such bonds were also detected in PETG. The range of wavenumber for these bonds was 1,500–1,700 1/cm. The difference for PCL-W and PCL-T were in some peaks in the range of 600–1,000 1/cm, which were attributed to C-H and C-C bonds. Similar peaks in FTIR patterns for various bio-coatings were reported in the other research [36–40]. More details of each bond for all bio-coatings were found in the previous article [23]. It was concluded that polymers with a larger monomer would have resulted in a higher degradation intensity. Moreover, bio-coatings without the C=C bond and cyclic hydrocarbon rings showed higher impedance in the SBF solution.

Figure 9.5 shows the FESEM images of corroded surfaces after 5 h of the exposure in the SBF solution. For the magnesium alloy substrates, corrosion products accumulated on the surface. In addition, many corrosion attacks such as pores and cracks were found on the surface. Such an observation showed severe corrosion attacks for magnesium alloys without any coating.

The morphology of corrosion attacks changed for all coatings compared to the magnesium alloy. The lowest corrosion degradation was related to both PCL coatings. Such a result was consistent with the EIS testing measurements. It was notable that some pits appeared for PCL-W coating, compared with PCL-T. The

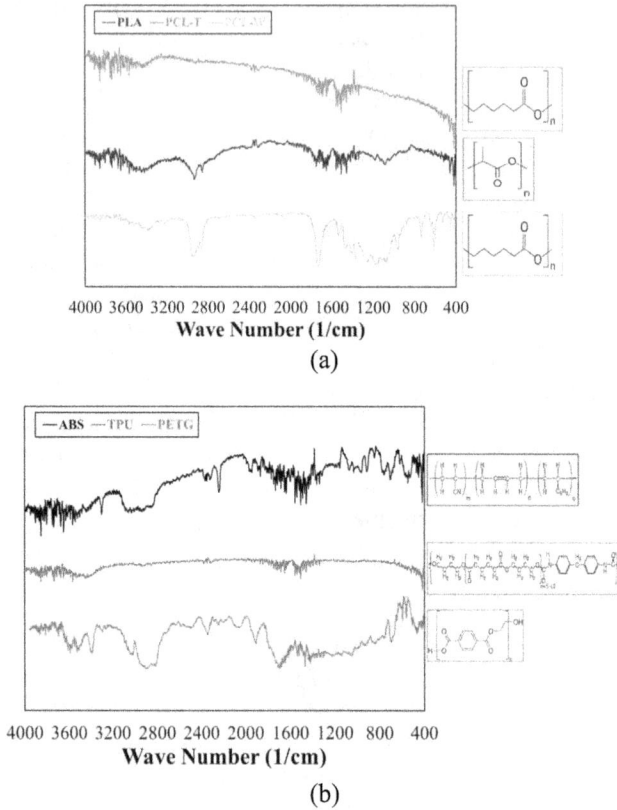

**FIGURE 9.4**  FTIR patterns for various bio-coatings: (a) PLA, PCL-T, and PCL-W, and (b) ABS, TPU, and PETG.

morphology of corrosion products for PLA coating was similar to TPU. However, more cracks were found for TPU coating. The highest attack was also attributed to the PETG coatings. There were some deep cracks between the 3D-printed layers. In this situation, the corrosive ion diffusion was done through these defects and caused the degradation on the surface of coatings.

## 9.5   CONCLUSIONS

This chapter presented a study on the topic of biofilms for implant materials. After a general introduction, information about various implants (mostly metallic ones) and different biomedical films or coatings (mostly polymeric ones) was proposed. Finally, a case study was illustrated on the corrosion behavior of the bio-polymer coatings 3D printed on the AM60 magnesium alloy. Based on the experimental results, the superior characteristics were generally related to PCL coatings for 3D printing on the magnesium alloys.

(a)

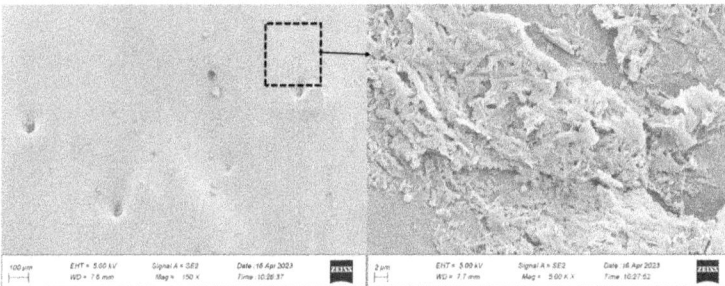

(b)

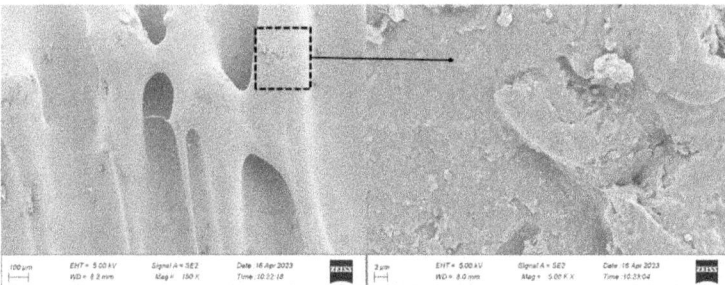

(c)

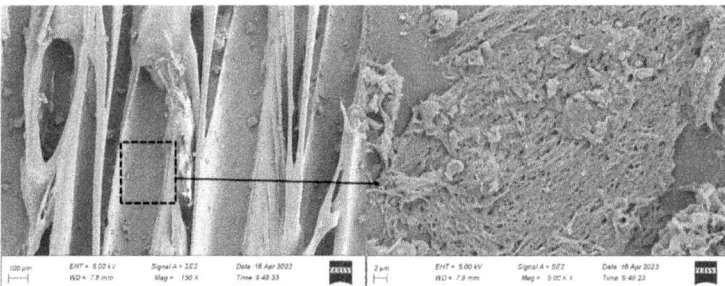

(d)

**FIGURE 9.5** The FESEM images of corroded surfaces from the top view for various specimens including (a) Mg, (b) PCL-W, (c) PCL-T, (d) PLA, (e) TPU, (f) ABS, and (g) PETG.

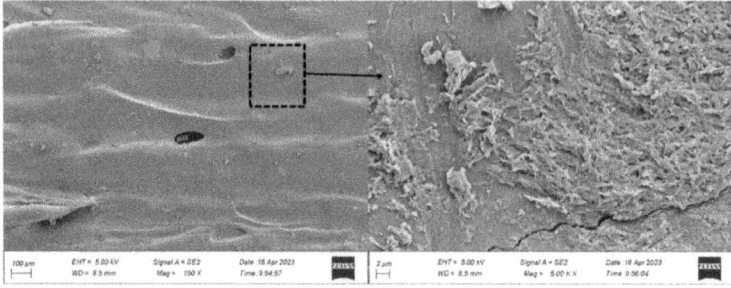

(e)

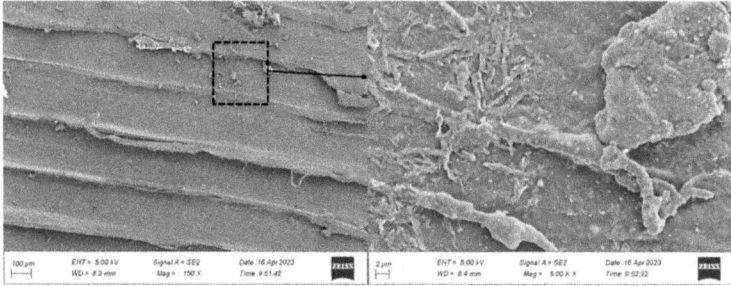

(f)

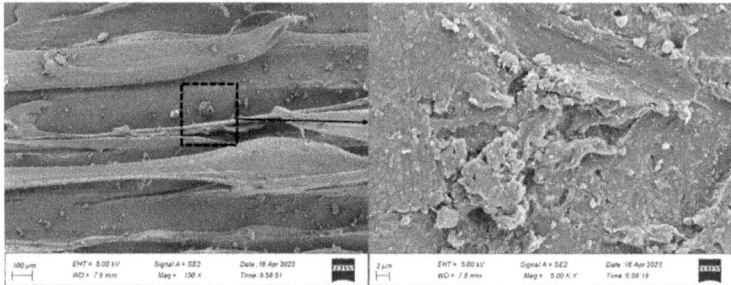

(g)

**FIGURE 9.5** (*continued*)

## REFERENCES

1. P.S.V.V.S. Narayana, P.S.V.V. Srihari, Biofilm resistant surfaces and coatings on implants: A review, *Materials Today: Proceedings*, 18(7) (2019) 4847–4853
2. B. Shokeen, L. Zamani, S. Zadmehr, S. Pouraghaie, R. Ozawa, B. Yilmaz, S. Lilak, S. Sharma, T. Ogawa, A. Moshaverinia, R. Lux, Surface characterization and assessment of biofilm formation on two titanium-based implant coating materials, *Frontier in Dental Medicine*, 2 (2021) 695417
3. V. Nandakumar, S. Chittaranjan, V.M. Kurian, M. Doble, Characteristics of bacterial biofilm associated with implant material in clinical practice, *Polymer Journal*, 45 (2013) 137–152

4. S. Sarfraz, P.H. Mantynen, M. Laurila, S. Rossi, J. Leikola, M. Kaakinen, J. Suojanen, J. Reunanen, Comparison of titanium and PEEK medical plastic implant materials for their bacterial biofilm formation properties, *Polymers*, 14 (2022) 3862

5. D.J. Davidson, D. Spratt, A.D. Liddle, Implant materials and prosthetic joint infection: The battle with the biofilm, *EFORT Open Reviews*, 4(11) (2019) 633–639

6. I. Gotman, Characteristics of metals used in implants, *Journal of Endourology*, 11(6) (1997) 383–389

7. M. Plecko, C. Sievert, D. Andermatt, R. Frigg, P. Kronen, K. Klein, S. Stubinger, K. Nuss, A. Burki, S. Ferguson, U. Stoeckle, B. von Rechenberg, Osseointegration and biocompatibility of different metal implants - A comparative experimental investigation in sheep, *BMC Musculoskeletal Disorders*, 13 (2012) 32

8. C. Shuai, S. Li, S. Peng, P. Feng, Y. Lai, C. Gao, Biodegradable metallic bone implants, *Materials Chemistry Frontiers*, 3(4) (2019) 544–562

9. P. Wyman, Coatings for biomedical applications, *Wood Head Publishing Series in Biomaterials*, 1 (2012) 3–42

10. R. Hauert, C. Valentin Falub, G. Thorwarth, U. Muller, C. Voisard, Coatings for biomedical applications, in *Encyclopedia of Tribology*, (2013) 390–396, Springer.

11. A. Asatekin, M.C. Barr, S.H. Baxamusa, K.K.S. Lau, W. Tenhaeff, J. Xu, K.K. Gleason, Designing polymer surfaces via vapor deposition, *Materials Today*, 13(5) (2010) 26–33

12. H. Usui, Preparation of polymer thin films by physical vapor deposition, in *Functional Polymer Films*, (2011) (Vol. 2), Wiley.

13. R. Delmdahl, A. Wiessner, Pulsed laser deposition for coating applications, *Journal of Physics: Conference Series*, 59 (2007) 28–31

14. X. Cheng, Y. Liu, O. Liu, Y. Lu, Z. Liao, Z. Hadzhieva, L. Chen, S.G.C. Leeuwenburgh, A.R. Boccaccini, F. Yang, Electrophoretic deposition of coatings for local delivery of therapeutic agents, *Progress in Materials Science*, 136 (2023) 101111

15. A. Sobolev, I. Wolicki, A. Kossenko, M. Zinigrad, K. Borodianskiy, Coating formation on Ti-6Al-4V alloy by micro arc oxidation in molten salt, *Materials (Basel)*, 11(9) (2018) 1611

16. F. Beck, Electrodeposition of polymer coatings, *Electrochimica Acta*, 33(7) (1998) 839–850

17. H. Koivuluoto, A review of thermally sprayed polymer coatings, *Journal of Thermal Spray Technology*, 31 (2022) 1750–1764

18. E.I. Meletis, X. Nie, F.L. Wang, J.C. Jiang, Electrolytic plasma processing for cleaning and metal-coating of steel surfaces, *Surface and Coatings Technology*, 150 (2–3) (2002) 246–256

19. J. Lukasz, E. Jolanta, A brief review on selected applications of hybrid materials based on functionalized cage-like silsesquioxanes, *Polymers*, 15(6) (2023) 1452

20. A.J. Nathanael, T.H. Oh, Biopolymer coatings for biomedical applications, *Polymers*, 12 (2020) 3061

21. K.K.A. Mosas, A.R. Chandrasekar, A. Dasan, A. Pakseresht, D. Galusek, Recent advancements in materials and coatings for biomedical implants, *Gels*, 8 (2022) 323

22. J.R. Smith, D.A. Lamprou, Polymer coatings for biomedical applications - A review, *Transactions of the Institute of Metal Finishing*, 92(1) (2014) 9–19

23. A.H. Beyzavi, M. Azadi, M. Azadi, S. Dezianian, V. Talebsafa, Bio-polymer coatings fabricated on AM60 magnesium alloys by fused deposition modeling 3D-printing to investigate electrochemical behavior, *Materials Letters*, 377 (2023) 133935

24. Y. Liu, X. Cao, J. Shi, B. Shen, J. Huang, J. Hu, Z. Chen, Y. Lai, A superhydrophobic TPU/CNTs@SiO$_2$ coating with excellent mechanical durability and chemical stability for sustainable anti-fouling and anti-corrosion, *Chemical Engineering Journal*, 434 (2022) 134605

25. M. Ziąbka, M. Dziadek, E. Menaszek, Biocompatibility of poly(acrylonitrile-butadiene-styrene) nanocomposites modified with silver nanoparticles, *Polymers*, 10(11) (2018) 1257

26. M.H. Hassan, A.M. Omar, E. Daskalakis, Y. Hou, B. Huang, I. Strashnov, B.D.G.P. Bartolo, The potential of polyethylene terephthalate glycol as biomaterial for bone tissue engineering, *Polymers*, 12 (2020) 3045

27. M. Azadi, A. Dadashi, M.S.A. Parast, S. Dezianian, A. Bagheri, A. Kami, M. Kianifar, V. Asghari, A comparative study for high-cycle bending fatigue lifetime and fracture behavior of extruded and additive-manufactured 3D-printed acrylonitrile butadiene styrene polymers, *International Journal of Additive-Manufactured Structures*, 1(1) (2022) 1

28. M. Azadi, A. Dadashi, S. Dezianian, M. Kianifar, S. Torkaman, M. Chiyani, High-cycle bending fatigue properties of additive-manufactured ABS and PLA polymers fabricated by fused deposition modeling 3D-printing, *Forces in Mechanics*, 3 (2021) 100016

29. A. Bagheri, M.S. AgharebParast, A. Kami, M. Azadi, V. Asghari, Fatigue testing on rotary friction-welded joints between solid ABS and 3D-printed PLA and ABS, *European Journal of Mechanics - A/Solids*, 96 (2022) 104713

30. M. Azadi, A. Dadashi, Experimental fatigue dataset for additive-manufactured 3D-printed Polylactic acid biomaterials under fully-reversed rotating-bending bending loadings, *Data in Brief*, 41 (2022) 107846

31. A. Dadashi, M. Azadi, Experimental bending fatigue data of additive-manufactured PLA biomaterial fabricated by different 3D printing parameters, *Progress in Additive Manufacturing*, 8(2) (2022) 255–263

32. H. Mobtaker, M. Azadi, N. Hassani, M. Neek-Amal, M. Rassouli, M.A. Bidi, The inhibition performance of quinoa seed on corrosion behavior of carbon steel in the HCl solution; theoretical and experimental evaluations, *Journal of Molecular Liquids*, 335 (2021) 116183

33. M. Azadi, M.J. Olya, M.E. Bahrololoom, EIS study of epoxy paints in two different corrosive environments with a new filler: Rice husk ash, *Progress in Color, Colorants and Coatings*, 9(1) (2016) 53–60

34. M. Azadi, M. FerdosiHeragh, M.A. Bidi, Electrochemical characterizations of epoxy coatings embedded by modified calcium carbonate particles, *Progress in Color, Colorants and Coatings*, 13(4) (2020) 213–222

35. M.A. Bidi, M. Azadi, M. Rassouli, An enhancement on corrosion resistance of low carbon steel by a novel bio-inhibitor (leech extract) in the $H_2SO_4$ solution, *Surfaces and Interfaces*, 24 (2021) 101159

36. L.Y. Li, L.Y. Cui, R.C. Zeng, S.Q. Li, X.B. Chen, Y. Zheng, M.B. Kannan, Advances in functionalized polymer coatings on biodegradable magnesium alloys–A review, *Acta Biomaterialia*, 79 (2018) 23–36

37. Y. Yao, M. Xiao, W. Liu, A short review on self-healing thermoplastic polyurethanes, *Macromolecular Chemistry and Physics*, 222 (2021) 2100002

38. B.W. Chieng, N.A. Ibrahim, W.M.Z. Wan Yunus, M.Z. Bin Hussein, Effects of graphene nanopletelets on poly(lactic acid)/poly(ethylene glycol) polymer nanocomposites, *Polymers*, 6(1) (2013) 93–104

39. A. Benkaddour, K. Jradi, S. Robert, C. Daneault, Grafting of polycaprolactone on oxidized nanocelluloses by click chemistry, *Journal of Nanomaterials*, 3(1) (2013) 141–157

40. M.I. Mohammed, D. Wilson, E. Gomez-Kervin, J. Wang, Investigation of closed loop manufacturing with acrylonitrile butadiene styrene (ABS) over multiple generations using additive manufacturing, *ACS Sustainable Chemistry & Engineering*, 7(16) (2019) 13955–13969.

# 10 Additive Manufactured Biomaterials Surfaces

*Modupeola Dada and Patricia Popoola*
Chemical, Metallurgical and Materials Engineering, Tshwane
University of Technology, Pretoria, South Africa

## 10.1 INTRODUCTION

Over the past two decades, various methods have been explored for fabricating tissue engineering scaffolds, including solvent casting, thermally induced phase separation, electrospinning, batch foaming, microcellular injection molding, and additive manufacturing. Additive manufacturing offers advantages, such as increased component complexity, bespoke components, and lead-time reduction [1]. However, there is a lack of examples of using multilayered architectural materials as feedstocks for 3D printing. Co-extrusion is a promising technique for creating multilayered structures by combining the properties of immiscible polymers in a stratified architecture. Combining a classical co-extrusion process with layer-multiplying elements (LMEs) allows for the production of a high number of alternating layers of two polymers, yielding individual layer thicknesses down to the micro- and nano-scale [2,3]. This allows for precise control of the size and distribution of individual material layers in a composite filament for FDM 3D printing. Additive manufacturing (AM), rapid prototyping (RP), solid freeform fabrication (SFFF), and layered manufacturing (LM) are all terms for 3D printing. This manufacturing process, which emerged in the 1980s, is characterized by adding feedstock material layer by layer according to predefined tool paths from a computer-aided design (CAD) and reducing material resource consumption. It is particularly suitable for personalized, small-series products that can be costly using conventional processes [4–6]. Additive manufacturing processes have significant impacts on the manufacturing industry, particularly in the medical field, where surgical planning is often used in orthopedic and maxillofacial surgery. 3D-printed patient models and jigs provide surgeons with a physical copy of the patient's affected part, enhancing preoperative planning. AM has emerged as an advanced automated process technology for tissue generation due to its high reproducibility, complexity, and scalability towards biomimetic tissue architectures. Additive manufacturing (AM) has gained attention due to its advantages, such as faster part building and the production of customized implants with minimal waste material [2]. AM involves various techniques, such as fused deposition modeling, stereolithography, selective laser sintering, electron beam melting, and laser-engineered net shaping. Additive manufactured biomedical implants offer benefits such as a perfect fit for each patient, good esthetics, early recovery, and less damage to bones and soft tissues. AM enables the creation of three-dimensional (3D) fiber-based tissue (FBT) and bioink-based tissue (BBT) models that capture various in vivo physiological or pathological responses. The fabrication workflows for FBT and BBT models are divided into extrusion-based bioprinting (EBB), electrohydrodynamic jetting, droplet-based bioprinting (DBB), and laser-based bioprinting (LBB). Electrohydrodynamic-based printing techniques, such as melt electrowriting, offer great potential for printing ordered fiber in a predefined pattern. Recent research studies have introduced the accurate design and fabrication of spatially heterogeneous scaffolds using melt electrowriting [7]. Bone fractures are characterized by the partial or full discontinuity in bone tissues, which are mainly caused by accidents, aging, and stress on the bone, hindering mechanical stability. Healing of bone fractures involves the interaction of various cells, such as osteoclasts, osteogenic cells, osteoblasts, and osteocytes. Primary healing at the fracture site is favored by endochondral ossification (EO), while secondary healing involves intramembranous ossification (IO). Long bones heal by a combination of EO and IO processes, resulting in a woven bone with a disorganized hydroxyapatite (HA) matrix. Internal fixation methods are essential for fast healing, the prevention of improper healing, and healing in improper positions. These methods regulate the functioning of osteoblasts, osteocytes, fibroblasts, and mesenchymal cells [8]. Bone plates have superior compression resistance, stability, and torsional and shear resistance, while intramedullary nails are better for patients with diabetes, vascular disease, and neuropathy with wound complications. Interlocking nailing devices provide superior resistance against nail migration and loss of fixation with enhanced rotational control. Bone screws provide support for the fixation of bone plates and fragments but require high pullout strength and reduced chances of screw loosening. Wires and pins are often added as internal fixation devices for enhanced bone stability. The emergence of bridging callus has led to the development of micromotion devices for fracture fixation devices, which are essential for the healing of fractures [6].

Nonetheless, bone implants can be made of metallic biomaterials like stainless steel, titanium alloys, cobalt-chromium alloys, nickel-titanium (NiTi) alloys, and magnesium-based alloys, which have excellent mechanical properties such as high tensile strength, fatigue strength, better machinability, and biocompatibility. However, metallic bone implants can cause stress shielding due to their higher stiffness, causing reduced bone mass and osteoporosis. Polymer biomaterials have an elastic modulus between 0.5 and 12 GPa, which is close to human bone. They do not interfere with MRI scans and do not require post-treatment removal, as they are bioabsorbable and degrade in the human body with human blood. Various polymers, such as polylactic acid (PLA), polyether ether ketone (PEEK), polycaprolactone (PCL), polyethylene glycol diacrylate (PEGDA), high-density polyethylene (HDPE), polytetrafluoroethylene (PTFE), and polymethylmethacrylate (PMMA), are being studied as implant materials [9]. Hence, surface modification of polymer biomaterials is needed, and it can be achieved through reinforcing additives in the base polymer material or by applying coatings to enhance mechanical properties, osseointegration, bioactivity, and protection against bacterial activity. Surface treatment of AM-fabricated implants may also have a positive impact on the early deposition of organic matrix and bone mineralization. Coatings can enhance corrosion resistance and bone ingrowth; they can also improve wear resistance, surface texture, corrosion resistance, surface hardness, water contact angle, and thermal insulation. Apatite, a synthetic form of hydroxyapatite (HA), is a common biochemical modification of the Ti surface to accelerate the biological fixation of the prosthesis with the environment. HA as a coating material offers osteocompatible and bactericidal properties, accelerating the adsorption of extra-cellular matrix proteins and the proliferation of osteoblasts around the material [10]. The stability over time of a ceramic coating affects the success ratio for clinical applications, which depends on the deposition method. The pulsed laser deposition (PLD) method efficiently ensures the controlled transfer at a stoichiometric level of the composition of the material to be deposited on the selected metal surface, as well as controlling the thickness of the coating and guaranteeing adherence. Controlling material surface roughness is a strategy for directing cell-material interactions and modulating material biocompatibility. However, predicting the specific effect of material surface roughness on cell response remains a challenge. The surface properties of biomaterials are of critical importance for their biological performance. Surface roughness, topography, and chemistry can all influence the adhesion, proliferation, and differentiation of cells, as well as the formation of a protective layer of tissue [11].

A similar revolution is happening in designing and 3D printing patient-specific implants, which can promote bone regeneration by optimizing fluid, cell, and blood vessel integration and circulation to the core of the scaffolds when implanted. The choice of biomaterial depends on the expected mechanical properties for replacement or regeneration, as well as the bioresorption ability of the material. The challenge in bone tissue engineering is to develop a biomaterial-based structure that mimics the multiple physicochemical properties of bone (physiology, morphology, architecture, and mechanical properties). Hence, additive manufacturing techniques are well suited for the 3D reconstruction of complex tissue due to their layer-by-layer building strategy. Integrating biological materials (cells, enzymes, growth factors) in the 3D construct enhances functionality but requires stricter regulation if transferred to the clinic due to the use of biological materials [12]. In vitro, studies on porous Ti6Al4V with bio-inspired surface porosity have shown better cell mineralization and responsiveness. This has led to long-stable osseointegration with reduced stress shielding and improved strength. The high precision of 3D-printed implants allows for the creation of implants of any ideal size and shape. Regulatory authorities like the FDA have approved the Ti6Al4V biomaterial without any observed medical problems during implantation. The 3D printing or DMLS strategy can be used to fabricate implants from different Ti alloys, which are more difficult to manufacture using conventional machining. The mechanical characteristics of titanium alloys or other biomaterials can be altered by 3D printing techniques through phases present in the Ti alloys [13]. High percentages of the $\beta$ phase are found to be beneficial for improving mechanical properties in bone replacement applications. Elasticity parameters of SLS-manufactured Ti alloys and biomaterials have been found to improve significantly, along with other mechanical and clinical benefits. 3D-printed titanium dental implants are suitable for immediate implantation, resulting in decreased rehabilitation time and no need for a second surgery. RAI (Root-Analogue Implant) implants offer better patient acceptance and can be more durable than traditional cylindrical implants. The powder bed AM technique allows for a wide variety of materials or combinations of materials as specimens, requiring only one 3D printing machine. Bone ingrowth can be accelerated by unique porosity, such as differences in pore size, shape, geometry, interconnectedness, and distribution [14]. Porous materials with pores of 300–500 μm can improve osseointegration and shielding effects. Thus, size flexibility is observed in 3D-printed implants, with irregular surfaces with micro- or nano-roughness playing a significant role in cell attachment and viability. Architectured materials are a growing class of materials resulting from the optimized spatial arrangement of multiple constituents at different microstructural scales. These materials are efficient, lightweight, smart, and adaptive, inspiring materials scientists and engineers. Over the past two decades, various methods have been explored for fabricating tissue engineering scaffolds, including solvent casting, thermally induced phase separation, electrospinning, batch foaming, microcellular injection molding, and additive manufacturing [15]. Tissue engineering applications sometimes demand topographic cues at a smaller scale to improve functionality. Combining architectured materials with FDM may improve the functionality of the produced biomaterial. Nonetheless, the development of new architectured composite materials for FDM requires screening in four phases: filament creation from the feedstock, filament physicochemical characterization, filament feeding and melting in the nozzle, molten polymer deposition, road solidification, and formation of the design geometry. The architecture, multi-layered composites obtained can have beneficial and tunable mechanical properties, resulting in enhanced cardiomyocyte morphology and function compared to their single-component counterparts. This facile fabrication scheme could lead to new hybrid materials with hierarchical structures yielding enhanced properties [16].

The electron beam melting (EBM) process is a promising additive manufacturing technology for biometals with higher growth, offering excellent shape control and strength-to-weight ratio. It can be used to fabricate patient-specific parts with porous internal structures for bone scaffolds, facilitating cell growth and maintaining natural functions, ultimately leading to better bone regeneration. Conventionally produced orthopedic implants suffer from certain shortcomings, such as a high elastic modulus, which can cause stress shielding and complications at the tissue/implant interface. Bulk metallic glasses (BMGs) are a promising class of metallic materials with high strength, excellent corrosion and wear resistance, and a relatively low Young's modulus, making them potential candidates for orthopedic applications, particularly load-bearing implants [17]. However, the production of BMGs in larger dimensions and complex geometries has remained challenging due to size restrictions and the need for high cooling rates. The electron beam melting (EBM) and SLM processes offer flexibility in the manufacturing of biomedical implants, allowing customized parts to be produced straightforwardly, allowing for patient-specific implants and complex internal structures with gradient porosities.

Ceramic materials are biocompatible and have a high resistance to corrosion and compression, making them ideal for medical implants. They have low toxicity and can be absorbed back into the body to encourage healthy tissue restoration. Furthermore, polymers are the most common material used in the 3D printing industry due to their diversity, ease of manufacturing, and availability with desired physical and mechanical properties. Polymers are biocompatible, biodegradable, sterilizable, have adequate mechanical and physical properties, and are excellent manufacturable due to their low melting points [18]. Traditional manufacturing methods, such as casting and machining, can be used to some extent to control the surface properties of biomaterials. However, these methods are limited in their ability to create complex and intricate surface features. Additive manufacturing (AM), also known as 3D printing, offers several advantages over traditional manufacturing methods for the fabrication of biomaterials with tailored surface properties. AM allows for the creation of complex and customized implants and devices with precise control over the surface roughness, topography, and chemistry [19,20].

This has led to a growing interest in the use of AM for the development of novel biomaterials with improved surface properties. AM has been used to create a variety of biomaterials with tailored surface properties, including:

- Bioactive materials that promote the adhesion and differentiation of cells.
- Antibacterial materials that resist the growth of bacteria.
- Drug-eluting materials that release drugs at a controlled rate.
- Tissue engineering scaffolds that support the growth of new tissue.

The use of AM for the development of novel biomaterials with improved surface properties is a rapidly developing field with the potential to revolutionize the field of biomaterials.

## 10.2 OPTIMIZATION OF ADDITIVE MANUFACTURING TECHNIQUES TO FABRICATE BIOMATERIAL SURFACES WITH ENHANCED BIOCOMPATIBILITY

By controlling the roughness of the biomaterial's surface, adding bioactive molecules, making micro- and nano-scale features, and using different materials, AM techniques can be used to make biomaterial surfaces with better biocompatibility and more specific functions. The surface roughness of a biomaterial can have a significant impact on its biocompatibility. Rough surfaces have been shown to promote cell adhesion, differentiation, and proliferation [21], and enhance the interactions between the target tissue's cells and the biomaterial, while smooth surfaces can inhibit these processes [22]. According to Khanlou et al. [23], the adaptive neuro-fuzzy system (ANFIS) model can be used to predict surface roughness in titanium biomaterials, which affects biocompatibility. Surface roughening using sandblasted, large-grit, acid-etched (SLA) methods is also very effective. Subsequent processes, such as polishing, sandblasting, and acid etching, can be employed. Surface roughness should be reduced in all conditions, especially in high temperatures. The ANFIS model predicted surface roughness with a 10% error band, suggesting the potential for obtaining proper biological signs on roughened surfaces. Rabel et al. [24] found that texture aspect and surface enlargement significantly impact osteoblast morphology and proliferation, with osteoblast proliferation being a function of cell morphology. Implant surface topography controls cell behavior and proliferation, potentially enabling cell-instructive biomaterials. Hu et al. [25] investigated the interactions between C2C12 myoblasts, human bone marrow stem cells (hMSCs), and silk-tropoelastin biomaterials in a temperature-controlled water vapor annealing controlled beta-sheet crystal formation, resulting in insoluble silk-tropoelastin matrices. Low surface roughness and high stiffness promoted C2C12 cell proliferation, while high roughness favored hMSCs. Hence, a combination of low surface roughness and high stiffness of the substrate appeared to be the most favorable for the proliferation and myogenic differentiation of C2C12 cells. AM techniques can be used to create biomaterial surfaces with a wide range of roughnesses, allowing for the precise control of this important parameter. By optimizing the surface properties of biomaterials using AM techniques, it is possible to create biomaterial surfaces with enhanced biocompatibility and tailored functionality. These surfaces can be used to improve the performance of implants and devices, and they can also be used to promote the regeneration of tissue. Addressing intrinsic material issues is crucial for the success and longevity of metallic devices. According to Bose et al. [26,27], modifying implant surfaces can prevent corrosion, enhance biocompatibility, and improve osseointegration without compromising bulk properties. Incorporating passivating elements (Al, Cr, Ti), and augmenting surface energy, topography, and crystalline structure are the best routes for improved corrosion resistance and implant-host interaction. Metal implants generally have high surface energy and smooth surfaces, creating a poor environment for cell adhesion. Modifying the surface roughness can significantly reduce the fatigue stress of the material. Surface topography can also induce adhesion and favorable differentiation. Surface porosity can increase tissue integration and implant fixation without sacrificing near-net shape manufacturing ability. Control of lattice parameters, specifically pore size, is critical to cell motility and adhesion. AM has exceptional capabilities to modify surface topography and can be implemented into the design of the build. Incorporating elements that self-passivate (Ti, Cr, and AL) is the most common method to prevent corrosion. Titanium-based implants are particularly stable as they form a protective, dense oxide film in situ, which prevents corrosion and enhances biocompatibility. However, metal-on-metal implants are susceptible to wear over time, releasing cobalt and chromium ions, which are known to cause metallosis and osteolysis. One method of AM that has garnered attention for its capability to produce compositionally complex materials and its ability to modify existing surface microstructure is Laser-Engineered Net Shaping (LENSTM). LENS™, which uses a focused laser beam to create a molten pool of substrate material at specific x, y, and z coordinates, allows for a full or gradient composite build. Krishna et al. [28] fabricated porous NiTi alloy samples with 12–36% porosity for implants using laser-engineered net shaping (LENS™). The density increased rapidly with energy input up to 50 J/mm3, and high cooling rates increased reverse transformation temperatures. The porous NiTi alloy samples with 12–36% porosity showed low Young's modulus, high compressive strength, and recoverable strain. The authors stated that the high open pore volume could accelerate healing and improve biological fixation when implanted in vivo, making them a promising biomaterial for hard tissue replacements. Kayane et al. (2022) synthesized Ti-Al-xNb alloys using laser additive manufacturing for biomedical applications. The microstructural evolution and corrosion behavior of the alloys were analyzed using SEM and EDS. The microhardness and corrosion behavior were found to be dependent on laser-processing parameters, with maximum microhardness at 0.061 g/min and 515.8 HVN at 0.041 g/min. Thus, LENS is an innovative fabrication technique that creates porous structures with mechanical properties and enhanced biological fixation, potentially eliminating shortfalls in metal implants. It can create monoblock designs with functional gradients and graded hard and wear-resistant coatings, potentially reducing osteolysis in load-bearing joints [29].

Bioactive molecules, such as proteins and growth factors, can be incorporated into the surface of a biomaterial to enhance its biocompatibility. These molecules can promote cell adhesion, proliferation, and differentiation, and they can also help prevent the formation of scar tissue [30]. Historically, natural products have been crucial in drug discovery, particularly for cancer and infectious diseases. The marine environment offers a unique resource for marine bioactive products, as many marine compounds have chemical characteristics not found in terrestrial products. Aquatic invertebrates, due to their high genetic richness, have been a major source of marine natural products (MNPs) of social value, producing molecules such as enzymes, biopolymers, bioactive compounds, and secondary metabolites. These compounds can find applications in various fields, including pharmaceutics, nutraceuticals, cosmetics, antibiotics, antifouling products, and biomaterials. The most studied marine invertebrates as sources of bioactive compounds include sponges, cnidarians, molluscs, echinoderms, and ascidians. The first compound developed for clinical use is a synthetic analog of the C-nucleoside cytarabine (or Ara-C) isolated from the Caribbean sponge Tethyacrypta, which is still used to treat acute promyelocytic leukemia and non-Hodgkin's lymphoma. Other marine products have been approved as anticancer agents, including eribulinmesylate (Halaven), an analog of halichondrin B from the sponge Halichondriaokadai, plitidepsin (Aplidine), two derivatives of dolastatin 10, and polatuzumab vedotin (POLIVY) from the mollusc Dolabella auricularia [31]. Biomedical advancements have led to the development of innovative biomaterials using bioactive molecules from biological waste from industries like fruit and beverage processing and fish, meat, and poultry. These bioactive molecules are used in scaffolds, dressing materials, drug delivery, tissue engineering, and wound healing. They are environmentally friendly, biodegradable, and biocompatible, with excellent tissue regeneration properties. These cost-effective biomaterials reduce healthcare burdens and offer sustainable waste management solutions [32,33]. AM techniques can be used to incorporate bioactive molecules into the surface of a biomaterial with high precision, ensuring that the molecules are evenly distributed and that they are not released too quickly. AM techniques can be used to incorporate bioactive molecules into the surface of a biomaterial with high precision, ensuring that the molecules are evenly distributed and that they are not released too quickly [12]. This material-oriented technology enables the production of customized scaffolds in a layer-by-layer manner and based on computer models. Bioactive glasses (BGs) have gained significant attention in tissue engineering due to their bioactivity, biocompatibility, and osteogenic, angiogenic, and antibacterial properties. BGs can react with body fluids and form hydroxyapatite, enhancing implant bonding to both hard and soft tissues [16]. Polymer-based BGs scaffolds have been researched due to their bioactivity effects. BG-derived 3D scaffolds are considered ideal porous templates with suitable mechanical properties for use in tissue engineering strategies and regenerative medicine. This review reviews the application of BGs for 3D-printed and bioprinted scaffolds and their usability in tissue engineering. The paper discusses various fabrication methods and in vitro and in vivo studies using BGs alone or in combination with biocompatible polymers as inks for 3D-printed scaffolds [4].

Micro- and nano-scale features can be created on the surface of a biomaterial using AM techniques. The fabrication process involves creating micropatterns on silicon wafers using photolithography and reactive ion etching techniques. The wafers are then dry-etched and cleaned. On silicon microstructures, nanostructured hydroxyapatite (NHAP) is made by changing the surface with acidic moieties and then mineralizing it in supersaturated calcium phosphate solutions. In the process of changing the surface, silanization and succinylation are used. Amine groups are attached to silicon patterns, and surfaces with carboxyl ends are found by measuring the contact angle. By dissolving ingredients in deionized water, you can make supersaturated calcium phosphate solutions, which are then used to soak silicon chips with carboxyl-terminated patterns. The rate of mineralization is monitored every 2 hours, and solutions are refreshed every 24 hours to ensure sufficient ion supplies for mineralization [34]. These features can provide additional binding sites for cells, and they can also influence the way that cells interact with the surface. For example, micro- and nano-scale features have been shown to promote the differentiation of stem cells into specific cell types. Both micrometer- and nanometer-scale features significantly influence cell behaviors, such as morphology, migration, adhesion, proliferation, and differentiation. Materials with multi-scale organization, such as nano-scale silica, hexagons, and honeycombs, are more advantageous in biomedical applications. However, the synthesis of organized structures with controlled features on both micrometer and nanometer length scales has not been reported [35].

The materials used to fabricate a biomaterial can also have a significant impact on its biocompatibility. Some materials, such as titanium and stainless steel, are naturally biocompatible. Metals and their alloys are widely used in biomedical applications due to their excellent corrosion resistance and mechanical properties. Ti6Al4V is extensively studied due to its affordability and lower ductility. Stainless steel, while cheaper, is also biocompatible due to its higher elastic modulus [36]. Biomedical metals and alloys are suitable for use as porous implants in orthopedic applications, but their elastic moduli are larger than those for replacing bones. To prevent stress shielding at the bone-implant interface, the elastic modulus and yield strength of metallic implants must be tuned accordingly. Several methods can enhance the mechanical properties of bone and metal interfaces, such as creating graded metallic porous implants or introducing elements to the alloy structure, such as ß-phase-stabilizing elements like Ta, Nb, Zr, and Mo, while other materials, such as polymers, may require surface modification to improve their biocompatibility. AM techniques can be used to fabricate biomaterial surfaces from a wide range of materials, allowing for precise control of the material properties [37].

Additive manufacturing parameters such as printing speed, temperature, layer thickness, and surface characteristics such as roughness, porosity, and surface chemistry have a significant influence on optimizing biocompatibility. Higher speeds can reduce printing time but may cause rougher surfaces [7]. Temperature affects material properties, such as viscosity, curing, and melting behavior. Layer thickness affects surface roughness and porosity. Thicker layers may have a more noticeable step-like appearance, while thinner layers provide better resolution. Optimizing biocompatibility requires controlling surface chemistry, which indirectly affects physical characteristics. By controlling these parameters, defects can be minimized, surface quality improved, and porosity reduced [38]. Bioactive agents, such as growth factors or antimicrobial agents, can be incorporated during the additive manufacturing process to impart specific functionalities to the biomaterial surfaces by providing specific functionalities and performance in various applications. Common methods include mixing and addition, surface coating, encapsulation and inclusion, and 3D printing with bioinks. Mixing or addition involves mixing the bioactive agent directly into the printing material, while surface coating involves applying a bioactive coating post-printing. Encapsulation or inclusion involves incorporating the bioactive agent into micro/nanostructures or particles, which are then mixed with the printing material. Bioinks, designed for 3D bioprinting, can contain living cells and be dispensed simultaneously, allowing for precise spatial control [39].

Post-processing techniques are essential for optimizing the biocompatibility and functionality of additively manufactured biomaterial surfaces. Post-processing for 3D-printed polymers includes support removal, powder removal, and resin removal, which can consume time. Post-processing can enhance properties like UV resistance, strength, surface quality, heat stability, weather resistance, and esthetics. Post-processing operations can be manual, semi-automated, or automated and can increase manufacturing costs. Common post-processing techniques for 3D-printed polymers are arranged according to printing technology. Post-treatments, such as sandblasting and vibratory finishing, chemical/electro/plasma polishing, high-temperature heat treatments, and coatings, can improve the biological performance of additively manufactured materials. Laser surface treatment can reduce surface roughness, optimizing surface morphology for fatigue behavior and osteoblastic differentiation. Chemical treatments, such as alkali, mixed acids, hydrothermal treatments, and electrochemical oxidation, can produce micro- and nanotextured layers on AM Ti-based structures. These surfaces are characterized by inorganic bioactivity, osteoblast activity, and direct bone bonding. Bone implants should reduce bacterial contamination to prevent infections. Using an SLM process to make a Cu-containing Ti alloy can cause antibacterial activity, but SLM alone can't make a surface with antimicrofouling properties. Electrochemical oxidation in electrolytes with calcium, phosphorous, and strontium, together with silver nanoparticles, can develop a bioactive and antibacterial microtextured surface layer. Finally, additive manufacturing (SLM) and heat treatment in a nitrogen atmosphere can improve the bio-tribological performance of Ti-based implants [40].

Common post-processing techniques include annealing, surface polishing, surface coatings, and sterilization. Annealing enhances mechanical properties, such as strength and toughness, by heating the printed object and cooling it. According to Li et al. [41], stress-relief annealing improves the anodic dissolution behavior of laser-solid-formed Ti-6Al-4V in a 15 wt.%NaCl electrolyte, resulting in improved resistance to dissolution and superior corrosion resistance. This is due to reduced micro-segregation, a stable passive film, a weakened galvanic effect, and selective dissolution of $\alpha$ phase. Longhitano et al. [42] analyzed the microstructure of direct metal laser sintering (DMLS) samples after heat treatments, focusing on the nucleation and growth of the $\beta$ phase. The results show that after anodizing, an oxide film with a barrier and nanoporous layer doped with F ions reduces passive current density, improving corrosion resistance and decreasing ion release into the bloodstream. A decrease in the mechanical strength of diagonally built additive samples after hot isostatic pressing (HIP) treatment, resulting in lower maximum strength but higher ductility. It suggests avoiding horizontal struts during manufacturing and using stress-relief heat treatment and HIP treatment for statically loaded applications and dynamically loaded applications [43].

Bacterial infection is a common issue with biomaterials implanted, causing aseptic loosening, prosthesis failure, and morbidity. Surface properties like roughness and porosity impact bone formation. Hence, there is a need to develop porous and rough additively manufactured materials with high biocompatibility and antimicrobial efficacy. Non-pharmacological techniques are needed to reduce bacterial colonization in complex 3D-printed structures. Surface polishing reduces roughness and enhances biocompatibility by minimizing bacterial and cell adhesion. However, excessive polishing can remove bioactive coatings or alter surface chemistry, so careful control is necessary. Surface coatings, such as biocompatible polymers, bioactive coatings, or antimicrobial coatings, can impart specific functionalities to the biomaterial surface [15]. Hybrid-electrochemical magnetorheological (H-ECMR) polishing was developed by Rajput et al. [44] to improve additively fabricated parts' surface quality. This advanced process uses mechanical abrasion and electrochemical reactions to enhance workpieces without affecting surface topography. According to the authors, H-ECMR finishes reduce finishing time and produce uniform surface quality. A feature-based hybrid H-ECMR finishing process was developed, using paraffin wax to fill holes and pockets. H-ECMR polishing enhanced the bone plates' surface quality, providing mechanical stability during fracture healing.

Sterilization is a critical post-processing step, ensuring compatibility with the printed material and any incorporated bioactive agents. There are different sterilization methods such as gamma irradiation, ethylene oxide gas, and steam autoclaving of additively manufactured biomaterial surfaces [45,46]. Sterilization in biomaterial and medical device manufacturing to prevent complications like infections and rejections. It eliminates or destroys all microbial life forms, including viruses, bacteria, and fungi, with vegetative cells or spores. The statistical definition of sterility is based on the Sterility Assurance Level (SAL), which should be limited to 10–6 for biomedical devices. Sterilization methods should not cause significant changes in material properties, preventing adverse body responses or compromise [84].

It is crucial to consider the potential effects of sterilization on material properties, surface characteristics, and bioactivity [4]. De Maio et al. used graphene nanoplatelets in PLA filaments enabled in 3D-printed devices and sterilized by near-infrared light exposure, killing the SARS-CoV-2 viral particles. This biocompatible material was reported to be ideal for sterilizable personal protective equipment and daily objects for multiple users.

## 10.3 THE MOST SUITABLE BIOMATERIALS FOR ADDITIVE MANUFACTURING OF BIOMATERIAL SURFACES

Human tissue that is damaged or missing is analyzed on a microscopic, physical, and chemical level to figure out how to choose materials that will help restore function or allow for more research and development of biocompatible materials. Additive manufacturing techniques offer flexibility in processing materials, requiring the determination of intrinsic properties before and after 3D printing. The rheology of the material is crucial for successful printing, especially in deposition techniques like fused filaments (FDM) or liquids (LDM). The thermal behavior of the feedstock material is also important, with kinematics, printing, and base temperatures adapted to the material's thermal properties. Design constraints include cell viability, biocompatibility, resorption, and adapted stiffness. Research is ongoing to identify new materials, particularly biomass like zinc, that meet AM processability criteria for reliable 3D printing feedstock materials, as shown in Figure 10.1 [4].

## 10.4 POLYMERS

Polymers are ideal for manufacturing in additive manufacturing (AM) technologies due to their lower melting points and modifiable chemical structures. However, they also need biocompatibility and biodegradability for biomedical applications like implants and natural tissue regeneration. Synthetic polymers are more hydrophobic and mechanically stable, but slower degradation rates may be better for tissue regeneration. The fatigue behavior of 3D-printed polymeric materials is crucial for medical devices, making the choice of materials and combinations challenging. Polymers have been used in surgery since the early 1940s, with nylon sutures and other bulk commodity polymers also adapted for surgery. They are essential components of permanent prosthetic devices like hip implants, artificial lenses, and catheters. New applications in molecular cell biology and developmental biology are driving the development of implantable polymers for controlled drug delivery and gene therapy [47].

Proteins are becoming more popular as possible drugs, and scientists have shown that active proteins can be released slowly from polymer matrices. This has led to the development of protein and peptide therapeutics. Prior to the 1980s, a few protein-based medications were in widespread clinical use. With the advent of molecular biology, a gene encoding a protein could be made synthetically and

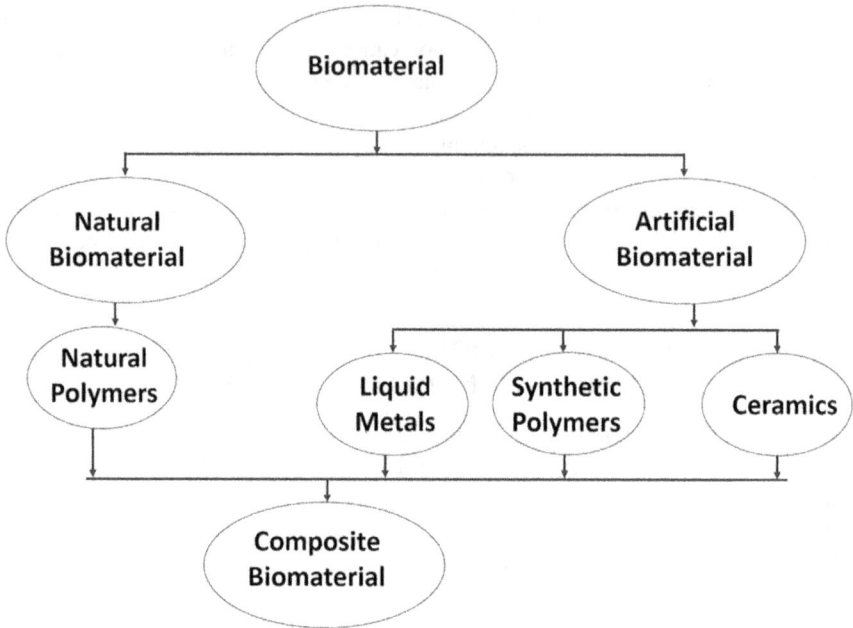

**FIGURE 10.1** Materials used in additive manufacturing for biomaterial surfaces.

introduced into cells to drive production of the desired protein in a relatively controlled fashion [48]. This has led to an increase in the number of protein and peptide therapeutics being found and made. These include antibodies for treating cancer, growth factors for making blood cells multiply, and enzymes for controlling the symptoms of cystic fibrosis. One of the strongest motivations for polymeric delivery systems is that many recently discovered molecules require delivery at a very localized level due to their mechanism of action. The potential for drug-polymer interactions, particularly for protein-based drugs, also affects the choice of polymer. Protein structure can be altered to minimize the occurrence of amino acids susceptible to reaction without altering the bioactivity of the protein, improving stability [49]. Natural polymers such as polynucleotides like DNA and RNA; polysaccharides like cellulose, dextran, amylose, glycosaminoglycans, and chitin; and bioactive proteins like silk, gelatin, collagen, elastin, fibrinogen, keratin, myosin, and actin enhance cell performance but are difficult to engineer due to limited processing abilities, high contamination risk, and batch-to-batch variability. Synthetic polymers, like polylactic acid (PLA), polyglycolic acid (PGA), poly-L-lactide (PLLA), poly ε-caprolactone (PCL), polylactic-glycolic acid (PLGA) copolymers, and polyhydroxy-alkanoates (PHA), are popular scaffold materials due to their chemistry, easy processing, and tailoring ability variability [50,51]. However, these polymers can elicit inflammatory responses and are not bioactive. New polymers are being developed to meet these specific criteria, such as non-toxic proton sponges and cyclodextrins. The molecular recognition component is generally a ligand for an internalizable cell surface receptor.

Polymers as biomaterials are polyethylene (UHMWPE), polyether ether ketone (PEEK), polymethylmethacrylate (PMMA), polyamide (PA), and polyvinyl alcohol (PVA). Molecular structure, crystallinity, and particle size are crucial factors in the final product fabricated by selective laser sintering (SLS), an additive manufacturing method. Low zero viscosity and low surface tension are necessary for suitable coalescence, affecting the quality of the final 3D part. PA6 and PA12, two polyamides with similar structures, have different applicability in SLS processing due to higher hydrogen bonding and higher melt viscosity. Amorphous polymers with a high viscosity above Tg often result in brittle and unstable products. The size and morphology of polymer powders also impact the quality and density of the final part. Amorphous polymer powders have a soft melting point and viscosity that decrease with temperature above the glass transition temperature. They are less easy to flow than semi-crystalline materials, leading to lower consolidation, higher porosity, and less strength. Semi-crystalline polymer powders have an ordered molecular structure with a sharp melting point, allowing for higher consolidation and full density in 3D structures. However, the crystallization rate should be kept relatively slow to avoid part distortion due to freezing shrinkage. SLS can process polymers with one fixed, controlled consolidation temperature or several close temperature peaks with a 5–10°C difference. The mean molecular weight (MW) is another important property of a polymer that ensures its processability with SLS. Studies have reported better laser sinterability of poly-ε-caprolactone (PCL) with a MW of 50.000 g/mol compared to 40.000 g/mol due to shrinkage caused by lower molecular weight and purity differences [52].

## 10.5 CERAMICS

Bioceramics, whether synthetic or natural, are important biomedical materials due to their biocompatibility, low degradability, high melting temperature, non-corrosive properties, and better mechanical properties compared to metal-based biomaterials [53]. They have gained popularity in various applications, including dentistry, orthopedics, calcified tissues, implants, coatings, medical sensors, and soft tissue repair and regeneration. Bioglasses and glass ceramics have also become versatile materials that can be molded according to the user's requirements. The market demand for bioglasses and glass ceramics is growing at approximately 1.5 million per year at a cost of $10 billion [54]. Silica-based bioglasses have been extensively studied, with combinations of borosilicate, borophosphate, and other therapeutic elements providing novel approaches for various applications due to their osteoproductive, osteoconductive, and osteoinductive properties. Bioactive glasses, particularly silica-based materials, play a crucial role in addressing the issues of bone diseases and trauma. Recent advancements in sol-gel processes and supramolecular chemistry have controlled bioglass porosity, leading to new applications in drug delivery systems and regenerative grafts. Combining silica-based glasses with organic components has resulted in improved mechanical properties. Macroscopic sol-gel glass preparation has led to new three-dimensional macroporous scaffolds for tissue engineering techniques and implant placement. Farmani et al. reviewed Li-doped bioactive glasses (Li-BBGs) as substitute materials used in bone regeneration due to their synergistic effect and ease of preparing polymer composites. The authors reported that doping Sr and Li with bioactive glasses can improve bone regeneration, especially in early in vivo osseointegration. Li-releasing bioactive glasses derived from sol-gel processes are suitable for cartilage regeneration, suggesting Li-doped bioactive glasses have great potential in osteochondral regeneration. Li-doped bioactive glasses have also found many applications in soft tissue reconstruction due to their pro-angiogenesis properties. Li-doped bioactive glasses can be considered inorganic angiogenic agents, which can be used in place of expensive and potentially harmful growth factors.

Ceramic-based biomaterials have advantages like density, porosity, high elastic modulus, hardness, and low cost, but they also have demerits like poor sinterability, ductility, and machinability [55]. Sustainable natural bioceramics, such as agro wastes and ashes, can be used as an attractive source of silica for cost-effective biogenic silica nanoparticles, bioglasses, and glass ceramics. Synthetic minerals combined with agro-food waste and ashes can serve as biomaterial substitutes for various biomedical applications [56,57]. The selection of initial constituents and their amounts play a vital role in determining the suitability of biomaterials. Other ceramics, primarily hydroxyapatite (HAp) and tricalcium phosphate (TCP), are used as biomaterials for artificial bone. HAP is a bioceramic that closely resembles the mineral component of natural bone. It has excellent biocompatibility and can enhance bone regeneration. TCP is another bioceramic commonly used in additive manufacturing due to its good mechanical strength and its ability to be gradually reabsorbed in the body. However, their mechanical properties, particularly fracture toughness, limit their use for heavy loads. Nonami et al. fabricated a composite of diopside and HAp to increase fracture toughness, with bending strength and fracture toughness significantly higher than sintered HAp. The composite showed no toxicity in cell culture tests. The authors examined the physical properties and biocompatibility of diopside, prepared by sintering a powder compact with CaO, MgO, and $2SiO_2$. Diopside exhibited 300 MPa bending strength and 3.5 MPa fracture toughness, higher than HAp. Its weight loss in lactic acid and physiological salt solutions was also less than HAp. Diopside formed a uniform junction with newly grown bone, with crystal growth and continuity. The morphological characteristics suggested apatite crystals, potentially useful for artificial bones and dental roots.

Bioceramics, such as polymer/hydroxyapatite composite scaffolds with micro- and nano-bioactive surfaces, have been made with the help of AM technologies. Polycaprolactone-polylactic acid-nanohydroxyapatite (PCL-PLA-nHA) composite porous scaffolds with micro-nanobioactive surfaces were fabricated by Xu et al. [58] using laser sintering and hydrothermal deposition. The composites showed enhanced compressive and tensile strength, modulus, and fracture mode, with a proposed fracture mechanism. The PCL-PLA-nHA scaffold showed higher miner-alization ability, excellent wettability, and better cytocompatibility, indicating its potential in bone tissue engineering applications. Bone scaffolds were fabricated by Cestari et al. [59] using binder jetting 3D printing and a biphasic calcium phosphate (BCP) nanopowder from cuttlebones. The nanopowder was mixed with glass-ceramic powder and pure AP40mod scaffolds. The sintered scaffolds were mainly composed of hydroxyapatite and wollastonite, with the addition of bio-derived powder increasing porosity. Human mesenchymal stem cells were seeded on the scaffolds, and their cellular behavior was similar across different pore geometries and compositions. However, metabolic activity decreased over time, possibly indicating cell differentiation. Overall, all scaffolds promoted fast cell adhesion and proliferation, allowing them to penetrate and colonize the 3D porous structure. High strength and wear resistance are two benefits of additively made bioinert ceramics, which are the oxide ceramics $ZrO_2$ and $Al_2O_3$, but their low bioinductivity makes it difficult to stimulate bone growth. Digital light processing (DLP), stereolithography apparatus (SLA), dielectric wave (DIW), and fluorine deposition (FDM) are other methods commonly used to prepare bioceramics. Wang et al. [60] developed a stable slurry with nano-sized BCP powders and porous BCP bioceramic scaffolds using DLP additive manufacturing technology. The slurry had low viscosity, and optimal photoinitiator concentration was closely associated with energy dosage in DLP. Sintering temperatures (1100, 1200, and 1300°C) had the best mechanical properties, in vitro cytocompatibility, and homogeneous bone-like apatite formation ability. DLP-formed 3D scaffolds with varying pore sizes were successfully fabricated, demonstrating the potential of DLP technology for bone tissue regeneration. Feng et al. [61] fabricated hydroxyapatite (HA) bioceramic scaffolds using digital light processing (DLP)-based additive manufacturing. Key issues, including dispersion, DLP fabrication, sintering, mechanical properties, and biocompatibility, were studied. The optimal dispersant dosage, solid loading, and sintering temperature were found to be 2 wt%, 50 vol%, and 1250°C. The 3D printing of β-TCP green bodies was achieved by Zhou et al. [62] using SLA technology with optimized KH-560 dispersant and solid loading. The resulting slurry showed good fluidity, uniform dispersion, and stability. Sintering schedule optimization was done using DSC-TG analysis, resulting in high-quality bioceramics without cracks or deformation. Increased solid loading reduced porosity and shrinkage, but printing was difficult when higher than 50 wt%. Zhnag et al. fabricated β-TCP bioceramic scaffolds with a gyroid structure using stereolithography 3D printing, ensuring high precision. The authors stated that these scaffolds can be adjusted for cancellous bone sites, promoting cell proliferation and efficient bone repair. PEGDA hydrogel resins with hydroxyapatite, calcium phosphate, and graphite content have been characterized for tissue engineering applications by Kumar et al. [63]. Poly (ethylene glycol) diacrylate/ hydroxyapatite (PEGDA/HAP) composites show high tensile, flexural, and compressive strength improvements, while Poly (ethylene glycol) diacrylate/graphite (PEGDA/Gr) composites show high wear resistance, low friction force, and wettability. These composites are thermally stable up to 187°C and non-toxic, supporting cellular growth and biocompatibility. Verma et al. [64] examined the impact of piezoelectric biocompatible $Na_{0.5}K_{0.5}NbO_3$ (NKN) on hydroxyapatite's dielectric and electrical properties. The composites HA-x NKN were synthesized using a solid-state ceramic method and sintered at 1075°C for two hours. X-ray diffraction and FTIR patterns confirmed the formation of pure HA and NKN phases. The results show ionic conduction as the dominant conduction mechanism, with hydroxyl ions and oxygen vacancies responsible for conduction in the HA-xNKN composite system.

According to Rey et al., fluoridated apatite materials promote osteoblast adhesion and proliferation, forming a bone-like apatite layer upon contact with body fluids. Their low solubility and low resorption rate contribute to their in vivo stability, potentially improving bone-implant bonding. Generally, coatings are an excellent alternative method for controlling the corrosion and degradation of bioceramics for several applications [65,66].

## 10.6  METALS

Metals have been a significant biomaterial group for a long time, and they are widely used in biomedical applications like implants to replace damaged or diseased hard tissues [67]. Key characteristics of metallic implants include adequate mechanical strength, corrosion resistance, biocompatibility, low cost, wide availability, and easy machining. Commonly used metallic biomaterials include gold, silver, stainless steel, CoCr alloys, and Ti alloys. Metals are superior to ceramics and polymeric materials due to their high mechanical strength [60]. However, their mechanical stiffness is higher than that of bone tissues, which is associated with stress shielding. Corrosion is another issue for metal implants in the physiological environment, except for magnesium (Mg). Degradable materials like magnesium and its alloys have been adopted as temporary implants, providing short-term structural support and dissolving after the damage is cured [68,69]. These materials have been applied for clinical applications such as cardiovascular stents, wound-closing devices, and musculoskeletal surgery. Efforts have been made to control the corrosion rate of Mg using various methods, such as purification, alloying, anodizing, and surface coating. Studying the precise controllability according to requirements is still a significant subject for further optimizing the effects of magnesium-based implants [70]. Additionally, the potential roles of other metals in the biomedical area should be considered and widely investigated to enrich metal biomaterials. Acellular and cellular ammonia additive manufacturing techniques for biomaterials are divided into acellular and cellular processes. Acellular techniques include binder jetting, directed energy deposition, material extrusion, material jetting, powder bed fusion, sheet lamination, and vat photopolymerization. Metal AM processes include powder bed fusion and directed energy deposition. Powder bed fusion uses thermal energy to fuse particle powder, allowing for high dimensional accuracy, fabrication without support, and a wide range of powders. Direct energy-based techniques use focused thermal energy to fuse materials, primarily metals. Laser-engineered net shaping (LENS) is a direct laser deposition technique for powder-based metallic biomaterials. According to Harun et al., metal-additive manufacturing processes can process biocompatible metals like Ti, Ti alloys, surgical-grade stainless steel, and cobalt-chromium alloys. Raw spherical powders like Ti6Al4V alloy, CoCrMo alloy, and 316 L stainless steel are used. However, the quality of the metal powder is crucial for efficiency and speed. Due to the monopoly of AM machine manufacturers over metal powders, prices are high. Materials like magnesium and its alloys have great potential for biomaterials due to their similar mechanical properties to natural bones, bio-resorbability, and noninflammability. However, magnesium has a high degradation rate, while iron has a low degradation rate. Zinc has emerged as a promising material for biodegradable implants due to its intermediate degradation rate between magnesium and iron. Recent focus has been on the AM of implants using titanium alloys with non-toxic elements such as niobium, tantalum, zirconium, and tin. Composite materials have also been studied to modify the properties of titanium implants, such as biocompatibility and wear resistance [71]). The development of new materials in AM requires sufficient process parameter optimization to reduce porosity. Experimentally, process simulation can help improve the accuracy of process simulation and speed up parameter optimization. However, not all alloy compositions are suitable for 3D printing, and in situ, alloying of elemental powder blends can be a solution. A process window is needed to minimize unmelted particles and the vaporization of elements with low boiling points, but this can pose challenges due to their multi-material nature [72].

## 10.7 SILK

Cultured silkworms and other worms of the Arachnida class produce silk, a popular textile material. Specialized epithelial cells make these fibrous proteins, which have structural roles in making cocoons, nests, traps, webs, safety lines, and protecting eggs [73]. Silks are composed of $\beta$-sheet structures, with hydrophobic domains forming hydrophobic anti-parallel chains. Silk from silkworms and orb-weaving spiders has great mechanical properties, environmental stability, biocompatibility, controlled proteolytic biodegradability, morphological flexibility, and the ability to immobilize growth factors [74]. Silks, a unique family of structural proteins, are biocompatible, degradable, mechanically superior, and suitable for various biomedical applications. They are amenable to aqueous or organic solvent processing and can be chemically modified. Many polymers that resemble the structure of spider silk have been created [75]. A polymer known as poly(ethylene-alt-alanine), or PEAA, is one of the most promising. It is a copolymer of the amino acids ethylene glycol and alanine. It is a man-made polymer that is not derived from nature. Poly(ethylene-alt-alanine) has special features due to the alternation of ethylene and alanine monomers. It is a biocompatible polymer; poly(ethylene-alt-alanine) is not hazardous to live cells [76]. Hence, it might be a contender for use in medical applications such as implants and medication delivery systems. Additionally, the body may gradually degrade poly(ethylene-alt-alanine), a biodegradable polymer. PEAA possesses many of the same characteristics as spider silk, including strength, toughness, and elasticity, and its structure is remarkably similar to that of spider silk. It can be prepared through various polymerization methods, including ring-opening polymerization and solid-phase synthesis. According to Pal et al. [77], PEA is biocompatible, which means that living tissues tolerate it well and that it has no negative effects or noticeable immunological reactions. For biomaterials designed for implantation or interaction with biological systems, this characteristic is essential. The amino acid residues in PEA give it its hydrophilic characteristics. By enhancing cell adhesion and enabling biological activities at the material interface, this hydrophilicity may improve interactions with water and biomolecules. Depending on the molecular weight, copolymer makeup, and polymerization technique, PEA may have variable mechanical characteristics. It is feasible to get the appropriate mechanical strength and flexibility needed for certain applications by altering these properties. PEA may break down gradually in the presence of certain enzymes or in reaction to the environment [78]. The body may digest or remove the breakdown products, which are typically non-toxic. This quality is especially beneficial for temporary implants or medication delivery systems.

Poly (lactic-co-glycolic acid) (PLGA) is another polymer that resembles the composition of spider silk. Lactic and glycolic acid repeating units make up this biocompatible polymer. Because PLGA is biodegradable, it can be broken down by the body. This makes PLGA a potential substance for use in implants for medicinal purposes. To design a better-controlled drug delivery device, it is crucial to understand the physical, chemical, and biological properties of polylactic acid (PLGA). PLGA can be made in highly crystalline or completely amorphous forms, while PGA is void of methyl side groups and has a highly crystalline structure. PLGA can be processed into various shapes and sizes and is soluble in various solvents. PLGA biodegrades by hydrolysis of its ester linkages, making it more hydrophobic than PGA. The glass transition temperature (Tg), moisture content, and molecular weight are factors that influence the release and degradation rates of incorporated drug molecules [79].

The molecular weight of PLGA, which depends on the type and amount of each monomer component in the copolymer chain, has a direct effect on how crystallized it is. Higher PGA content leads to quicker degradation rates, except for the 50:50 ratio of PLA/PGA, which exhibits the fastest degradation. The degree of crystallinity and melting point of PLGA polymers are directly related to their molecular weight [80]. The development of novel polymers that resemble the composition of spider silk is still ongoing. The possible uses for these novel polymers include textiles, composites, and medical implants, among many others. There are some more polymers that have been created to resemble the composition of spider silk. Poly (glutamic acid) (PGA) is a water-soluble, anionic, biodegradable, and edible biopolymer produced by Bacillus subtilis. It has potential applications in food, pharmaceuticals, healthcare, and water treatment. This review provides information on fermentative production, fermentation conditions, media components, genetic engineering, kinetic studies, and the recovery and purification of PGA. It also discusses current and potential applications and contributes to the development of this commercially and academically interesting biopolymer. One often used anticancer drug is paclitaxel. Clinical trials are now being conducted on paclitaxel-poly (glutamic acid) conjugates, which have shown considerable promise in preclinical studies. Preclinical findings indicate that more paclitaxel is administered selectively to tumor locations when compared to nonconjugated paclitaxel. Similar increases in efficacy and decreases in toxicity are shown when poly (glutamic acid) is conjugated to other families of cancer medications. Making this technique practical requires optimizing poly (glutamic acid) for use in drug delivery applications [81]. A synthetic biodegradable aliphatic polyester known as poly (-caprolactone), or simply polycaprolactone as it is more commonly known, has gained significant attention recently, particularly in the biomedical fields of controlled-release drug delivery systems, absorbable surgical sutures, nerve guides, and three-dimensional (3-D) scaffolds for use in tissue engineering and their resemblance to the composition of spider silk. This polymer has been used to create a variety of polymeric objects, including microspheres, microcapsules, nanoparticles, pellets, implants, and films. It may be spun into filaments to be used in the creation of desired textile constructions. Spinning may be done in a number of ways. The fibers may be made into many shapes and utilized for sutures, implants, and other surgical procedures. Despite the fact that multiple studies have examined various PCL qualities and uses, there hasn't been a thorough investigation of the various ways PCL fibers are made and how they're used in biomedical applications. The current article provides an overview of PCL fiber manufacturing using several techniques, as well as relationships between fiber structure and qualities. There is also discussion of how these fibers may be used in biomedical fields [82]. Poly (vinyl alcohol), or PVA, is a versatile industrial polymer known for its unique qualities and applications. Reversible addition-fragmentation chain transfer (RAFT) polymerization has recently created PVA with a regulated molecular weight and high syndiotacticity. PVA's simple structure has attracted researchers' interest due to its secondary alcohol functionality and solubility in water and organic polar solvents. Its biocompatibility has led to widespread applications in biomedical fields such as orthopedics, embolic materials, and metallic nanoparticles. PVA has also been used as an embedding matrix and mat for metallic nanoparticles, removing arsenic ions from water, and reducing toxicity in $CoFe_2O_4$ nanoparticles. PVA has undergone functionalization and modification by agents for various applications, such as biomedical systems, membrane fuel cells, adsorption and separation, and catalysis in chemical processes. PVA's functionalization is initially defined in conjunction with its uses in biological and biomedical applications, membrane fuel cells, adsorption and separation, and catalysis in chemical processes [83,84]. Poly-L-lysine (PLL) is a biocompatible, biodegradable, and water-solubilizing cationic polymer with good properties. Modified with ferrocene carboxylate, the resulting product showed improved performance as a redox mediator with the glucose oxidase enzyme. It also demonstrated high selectivity and linear range, with a limit of detection of 23 µM and a sensitivity of 6.55 µA/cm2 mM. By attaching N-isopropyl acrylamide (NIPAm) moieties to the amines, poly(L-lysine) (PLL) was largely changed at ambient temperature using the simple free-catalyzedaza-Michael addition procedure. The resultant PLL-g-NIPAm displayed thermosensitivity of the LCST type. The NIPAm content integrated into the macromolecules may be used to adjust the LCST. Importantly, the remaining free lysine residues' capacity to ionize means that depending on the NIPAm concentration, LCST is greatly reliant on pH and ionic strength. PLL-g-NIPAm is a new biodegradable LCST polymer that might be utilized to create pH/temperature-responsive self-assemblies (nanocarriers and/or networks) for possible bioapplications. It can be employed as a "smart" block in block copolymers and/or terpolymers of any macromolecular architecture [85].

These polymers are suitable for a variety of applications because of their varied characteristics. For instance, PGA is a suitable material to utilize in composites since it is robust and durable. Because PCL is biodegradable, it is an excellent material for use in medical implants. PVA is a suitable material to use in textiles since it is water-soluble. PLL is suitable for use in contact lenses since it is biocompatible.

## 10.8    SURFACE MODIFICATION TECHNIQUES CAN BE EMPLOYED TO IMPROVE THE INTEGRATION OF ADDITIVE MANUFACTURED BIOMATERIAL SURFACES

Biomaterials are materials that support the function of or replace tissue or organs in the body. However, these biomaterials often fail due to factors such as high modulus, insufficient wear and corrosion resistance, and a lack of biocompatibility. Corrosion and wear of metallic implants compromise the implant's structural strength and release metallic debris, which can cause chronic inflammation, allergic reactions, and granuloma formation [86]. The biological response to corrosion products is a common reason for revision surgeries, as it can cause inflammation, swelling, and pain. The development of new alloys and surface modification techniques is important to improve corrosion and wear resistance and mechanical properties. There are different surface modification techniques, as shown in Figure 10.2. These techniques try to change the behavior of a metal surface without changing the bulk characteristics of the metal [15].

Such adjustments may be made to the bulk qualities, such as corrosion resistance, and the bulk reactions, such as osseointegration. The possible surface changes may be divided into chemical, physical, thermal, and coating categories [87]. Thermal surface modification techniques are less expensive than chemical techniques and tend to have stronger bonding forces between the substrate and coating. The surface modification would increase a surface layer's compatibility with neighboring tissues as well as its chemical and mechanical durability [88]. Because surface modification primarily affects the top surface of cardiovascular devices, it is regarded as a low-cost approach. There are variances depending on the

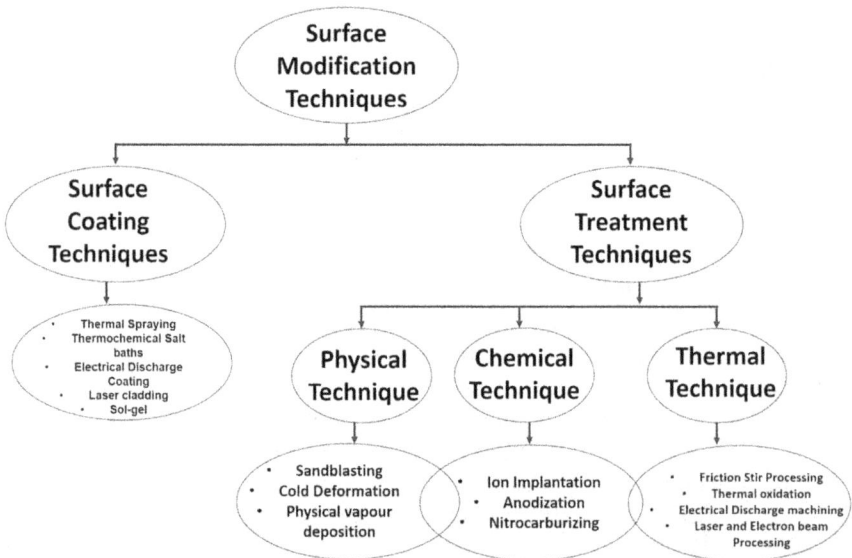

**FIGURE 10.2**    Surface modification techniques for additive manufacturing of biomaterial surfaces.

application and kind of material being dealt with when it comes to surface modification of biomaterials. To prevent germs from entering the gap between the gingival epithelial membrane and the dental implant surface, soft tissue compatibility is required for dental prostheses. Materials used in cardiac stents must be constructed to inhibit cell growth and restenosis. Blood compatibility, or antithrombogenicity, is necessary for cardiovascular devices. The best procedure is determined by the base material, design development, morphology, cost, and end-user application, among other things [89]. Two surface engineering process factors that are commonly discussed are coating thickness and surface engineering process temperature. Wathanyu et al. [90] used cold spraying to put three layers of porous titanium coatings on 316 L stainless steels that had been grit-blasted and ground at different temperatures and pressures of nitrogen gas. The lowest porosity gradient coating had the lowest porosity, while the highest porosity coating had the highest.

Different failure mechanisms were discovered, including adhesion, adhesive, and cohesion failure. High porosity gradient samples had higher bond strength, while low porosity gradient samples had lower bond strength due to cohesion breakdown [91]. Başman et al. [66] investigated the effects of the Ekabor-2 bath on the formation mechanism and properties of boride layers in thermochemical diffusion boriding of AISI 316 L stainless steel. Results showed that thickness increased with temperature and time, with $Fe_2B$ and $FeB$ phases forming. Surface hardness values increased with temperature and time, with borided materials having a formation activation energy of 149.3 kJmol-1. Ekmekci and Efe [92] examined the deposition of Hydroxyapatite (HA) on titanium alloy surfaces using Hydroxyapatite Powder Mixed Electrical Discharge Machining (HA-PMEDM). Nano- and micro-HA powders were mixed into dielectric liquid and tested for various pulse currents and durations. Surface properties were evaluated using surface roughness, SEM, EDS, and XRD. Results showed nanoHA-PMEDM had the highest rate of deposition, with higher layer thicknesses and hardness values. These findings support the use of HA-PMEDM for functional surfaces in medical applications. Wu et al. [93] used laser cladding and plasma electrolytic oxidation to make a porous ceramic coating out of titanium on 316 L stainless steel to make it more bioactive. Characterizations of the coated samples showed anatase and rutile anatase with highly crystalline HA. The LC+PEO composite bio-coating demonstrated superior corrosion resistance and bioactivity, making it a promising solution for enhancing 316 L stainless steel's bioactivity. Garcia-Amaez et al. analyzed the effects of Ca-doped biomaterials in vitro and in vivo using proteomic and histological techniques. The sol-gel route was used to synthesize coatings with 0.5 and 5 wt% $CaCl_2$. Morphology and $Ca^{2+}$ ion release were examined. In vitro and in vivo experiments were conducted using a rabbit model. Ca-doped biomaterials increased the adsorption of coagulation-related proteins, enhancing affinity for proteins involved in inflammation and osteogenic functions. The results showed a good correlation between in vitro and in vivo experiments, making proteomic analysis a useful tool for predicting biomaterials' impact in vivo. Ti-Zr alloys are investigated as an alternative to commercially pure Ti, with a 30–50 mol% Zr proportion offering competitive advantages. Tan et al. [94] evaluated the biological response to Ti-Zr alloys with different compositions and surface characteristics using sandblasting.

Results show similar surface structures for c.p.Ti, Ti-10Zr, and Ti-30Zr, but Ti-50Zr does not form a micro-rough structure. Zareidoost and Yousefpour [95] investigated the impact of crystallographic texture and nanostructuring on the cytocompatibility of an as-cast TZNT-Ag1.5 alloy subjected to cold rolling up to 90% reduction. The study analyzed test specimens using various techniques, including XRD, TEM, HRTEM, SEM, nanoindentation, and OM. Severe cold deformation led to fast grain refinement in the 50–100 nm range, and the fiber texture was formed. The microhardness and reduced Young's modulus increased with cold deformation, while the contact angle showed a more hydrophilic and wettable surface. Osteoblastic cells showed significant differences in cell proliferation and differentiation compared to as-cast alloy test specimens and/or CP-Ti. Gabor et al. [96] focused on creating hybrid layers for surface modification of Ti-6Al-4V alloys for orthopedic and dental applications using vapor deposition. By combining micro-arc oxidation (MAO) and physical vapor deposition (PVD), Ti and ZrTi oxide layers were formed, resulting in high tribological properties, corrosion potential, and polarization resistance.

The hybrid coatings promoted greater osteogenic differentiation of human bone marrow mesenchymal stem cells with a higher activity of alkaline phosphatase. These hybrid oxide layers are promising for coating metallic bone implants. Lau et al. [97] investigated the biofunctionalization of silk biomaterials with recombinantly expressed human perlecan (rDV) domain V to promote endothelial cell interactions and functional endothelium formation using ion implantation. rDV was immobilized on silk using plasma immersion ion implantation (PIII), allowing for strong immobilization without chemical cross-linkers. The results showed rapid endothelial cell adhesion, spreading, and proliferation, indicating the potential of rDV-PIII-silk as a biomimetic vascular graft material. Boopathi et al. [98] investigated the hardness, layer thickness, and corrosion rate of nitro-carburizing coating on AISI 4150 steel using the Quench-Polish-Quench process. It considers coating parameters like temperature, time, and oxidation time. The Taguchi method was used to analyze the effects of these parameters. The Taguchi method examined the impact of coating parameters like nitro-carburizing temperature, time, and oxidizing time on hardness, layer thickness, and corrosion rate. The optimal settings achieved maximum surface hardness (504.71 HV) and layer thickness (13.6 μm), while the minimum corrosion rate (1.27 mm/mile) was achieved with the best parameter settings. Vidal et al. [99,100] used multi-pass upward friction stir processing (UFSP) to make new Mg/HAP/FA composites. Observations of the microstructure and micro-CT reconstruction showed that the HAP and FA particles were spread out well, which made the magnesium grains smaller. In vitro bioactivity behavior showed fluoride-rich apatite layer formation and cytocompatibility, promoting osteoblast-biomaterial interaction. Friction stir processing (FSP) improves the mechanical and tribological properties of metal sheets, but its biocompatibility is still a concern. Ghahramanzadeh et al. [5] found that FSP enhanced tensile properties but decreased elongation. The authors investigated AZ31B magnesium alloy, and they reported that the wear performance was enhanced in ambient air and simulated body fluid (SBF) solutions, with fewer adhesive bonds between friction surfaces. However, the biomaterials were not cytotoxic for human gingival fibroblasts, suggesting their safety for clinical use.

Correa et al. evaluated the impact of temperature on Ti-15Zr-xMo samples'
morphology, crystal structure, chemical composition, roughness, wettability,
Vickers microhardness, and corrosion resistance. Thermal oxidation treatments
were performed at 773 K and 1173 K for 21.6 ks. Oxide layers were preferentially
Ti and Zr, with $TiO_2$ and ZrO influencing phase composition. Vickers microhardness
increased with oxide layer growth, while bulk samples exhibited higher values. The
authors concluded that optimizing thermal oxidation treatment can improve surface
properties, which could lead to more biomedical uses of Ti-15Zr-Mo-based alloys.
Asif et al. looked into the use of eco-friendly and biodegradable surfactant additives in
electro-discharge machining (EDM) for biomedical applications. Results showed an
improvement of 41.7%, 75.3%, 55.3%, 47.4%, and 80.3% in material removal rate
(MRR), tool wear rate (TWR), surface roughness (Ra), Rz, and overcut (OC) with
surfactants. Surface topography shows fewer cracks and holes, and the presence of
oxides and carbides is good for making bioactive surfaces that are harder and more
resistant to corrosion. Polly et al. investigated additively manufactured porous
biomaterials based on triply periodic minimal surfaces (TPMS) with unique
properties like open porosity, large surface area, and surface curvature. The scaffolds
were analyzed by micro-CT and mechanical tests to determine their morphological
characteristics and fatigue resistance. The smallest unit cell design had the highest
quasi-elastic gradient, compressive offset stress, and compression strength. The
fabricated scaffolds showed promising mechanical and morphological properties,
making them suitable for bone substitute applications.

## 10.9  POTENTIAL CHALLENGES AND FUTURE OUTLOOK ASSOCIATED WITH ADDITIVE MANUFACTURED BIOMATERIAL SURFACES

Metallic biomaterials have been considered a successful option for skeletal
implantology, but there are still concerns about biocompatibility. The next
generation of metallic implant materials must address critical challenges such as
metal ion sensitivity, infection prevention, early-stage osseointegration, low
stiffness, and incorporating bio-resorbability. An issue with sophisticated orthopedic repair unmet needs is revision operations. The standard procedure in
arthroplasty surgery is to match the patient to certain implant types and sizes,
which may result in a subpar repair, future failure, revision surgery, pain, and
impairment. A more futuristic perspective contrasts this with the implant, which is
tailored for the patient and has shown to be a superior model for enhanced stability
and decreased failure rates over the last 40 years [14]. Hence, Patient-specific AM
technology is a recent development in the biomedical equipment sector to solve
problems with implants, with metal 3D printing being the main platform for load-
bearing implants. As technology advances, methods and approaches that save time
and money and produce better, more personalized outcomes are becoming more
popular. The application of AM in a circular economy system presents significant
challenges, including addressing the challenges of biocompatibility, addressing the
need for more personalized and cost-effective solutions, and focusing on developing
novel materials with modified bulk properties or chemistry [2]. The three metal ions

most frequently mentioned in clinical investigations are Ni, Co, and Cr. Both SS 316 L and CoCrMo alloys contain these metal ions. There is, however, no straightforward mechanism to exchange these metal ions for alloys or other metals. Due to the necessity for significant experimental knowledge, regulatory channels, and the possibility of failure, innovative alloy design is often not investigated in biological devices for decades. Infection is another significant issue for metallic implants and fracture care instruments, with biofilms forming on the implant surface if not addressed immediately. Revision surgery is difficult, costly, and dangerous to soft tissue and bone. Osseointegrated devices increase the quality of life of amputees who use prosthesis by 17%. An analysis indicated that over 20% of the implants in a cohort of patients with lower extremity amputations had to be removed owing to stability issues, fractures, breakage, and infection. The procedure and maintenance of osseointegration treatment are complex, but the clinical results are compelling [7]. Future research should look into the therapeutic benefits of osseointegration and the cost-effectiveness of traditional complete arch fixed or removable prosthodontic rehabilitation techniques. Infection risks with dental implants include titanium implants, which are not teeth. The clinical assessment of dental implants is crucial in determining if an implant can function in the first two years after implant placement. Failure progression may be rapid for implants that malfunction and cause a complication, disease, or infection. The effect of implant failure on the patient can be substantial, as the implant may have supported many functional prostheses, ranging from a single crown to bridges or full dentures. Technical issues with ceramic alloys, such as titanium and ceramic-artistic combinations, have not been widely discussed in dental writing [11]. The development of metal-earthenware combinations is still in the early stages, but it is reasonable to be cautious until more data about their execution is available. More alloys are also expected to be created and developed to protect patients from infection and the growth of biofilms without creating a local toxicity concern. Thus, the rise in 3D printing for alloy synthesis is attributed to the ease of melting newer alloys on a smaller scale in a metal AM system. Further study is expected to focus on alloy design for metallic implants, as human-use alloys lacking Ni, Co, or Cr can be developed without sacrificing clinical performance [101].

AM is a cutting-edge technology with significant environmental advantages, but few studies have addressed the challenges of adopting AM in achieving circular economy goals, particularly in underdeveloped nations. Continuous research and development are necessary as AM is constantly changing, and massive expenditures are needed to address the lack of a competent workforce and the ongoing processes of hiring and educating the workforce used in AM technologies. By addressing these issues, bottlenecks can be removed, and the benefits of AM's environmental sustainability, including reduced $CO_2$ emissions, environmental protection, less waste, and extended product life, can be realized. The most widely used PBF and DED procedures for processing metallic alloy powders are vulnerable to defect inclusion due to uneven powder flowability and porosity. Only a small number of metal alloys have been certified for processing by additive manufacturing (AM), limiting the adoption of AM techniques for various metal alloy applications. The successful conclusion of a focused gas atomization process research effort may provide a strong position for powder manufacturing companies that are open to integrating new technologies. Adhesion of coatings on polymer surfaces is a significant challenge due to their hydrophobic, non-polar, low surface energy, and chemically inert nature [102]. Covalent bonding is only favorable with cleaned, surface-roughened, hydrophilic, and cross-linked polymer substrates. Moisture absorption by polymer substrates can significantly impact the bonding strength between the coating and polymer substrate. Hydrophilic polymers like sodium polyacrylate and polyacrylic acid can absorb moisture or water vapor, causing swelling, dimensional and mass changes, and hydrolysis, deteriorating mechanical performance. High glass transition temperatures of polymers, such as PLA, PCL, PEG, HDPE, and PMMA, can cause glass transition temperatures to be low, affecting the bonding strength. Surface modification treatments, including physical, chemical, and reactive gas treatments, are preferred to increase surface energy and improve adhesion, wettability, and bonding with coating materials. Physical surface treatments, such as surface roughening, media blasting, and adhesive abrading, can enhance mechanical linkage and surface modification. Chemical surface treatments, such as acid etching and chromatic acid, can create new functional groups capable of bonding with the coating. However, these treatments also generate hazardous chemicals and waste, making reactive gas surface treatments more viable [103].

The rate of immersion and withdrawal during the dip coating process must remain constant to control the flow of coating on the polymer substrate. Rapid withdrawal can result in thick coatings, while rapid withdrawal can result in irregularities. Accurate and durable coating thickness is difficult due to the iterative 3D printing design process, which involves computer tomographic (CT) scans, image reconstruction, design changes, FEA, and printing. The fatigue life of implants due to layer-by-layer deposition of material is inadequate, resulting in plate-like structures with unique surface finishes. The use of 3D printing technology (AM) has the potential to manufacture orthopedic bone screws, but it faces these significant challenges. Quality consistency is crucial for the fabrication of orthopedic bone screws, as it is highly influenced by key parameters such as build orientation, metal powder size, shrinkage, scanning speed, layer height, wall thickness, support material, resolutions, printing speed, proper alignment, printing time, and manual post-processing. Other challenges include the adoption of intelligent manufacturing technologies like the Internet of Things, cloud computing, big data, machine learning (ML), and artificial intelligence (AI) [104]. These technologies can improve processing defects and inconsistent quality of fabricated parts, as well as optimize design, process, and production. Scalability limitations are also a major challenge, as AM techniques require operational skills and knowledge to operate machines. The adaptation of AM technology for mass production, repeatability, and consistency of fabricated orthopedic bone screws is critical.

Narrow range and expensive biomaterials are also challenges in AM technology, as few biocompatible biomaterials have been explored for orthopedic bone screw fabrication. Extrusion-based multi-material AM techniques have the potential to fabricate multi-functional metallic biomaterials, but their development may hinder biomedical adaptation. Lack of real-time manufacturing is another challenge in AM. The communication and connection gap between medical practitioners and design engineers may lead to a lack of real-time manufacturing of orthopedic implants required during surgery. However, the clinical application of real-time manufactured orthopedic implants may be possible with the knowledge and adaptation of AM technology and respective designing software. Magnesium is an essential element found in human bones, blood serum, and blood. It is a biomaterial with good biocompatibility, osteopromotive properties, and natural degradability. However, few research reports have utilized magnesium powder ceramics for orthopedic bone screw fabrication using AM techniques. Hence, Magnesium powder can be a future demand for biodegradable bone screw fabrication using AM technology. Stainless steel and cobalt-chromium should also be explored for various AM techniques [105].

Standardized bone screw geometries are used in fractured human bones, but the traditional geometry may lead to complications and loosening in older patients with less bone mineral density. Customized patient-specific bone screws can provide better stability and rigid internal fixation at the fractured site. Controlling the AM process is challenging due to the miscibility and wetting constraints of various materials and dissimilarities in their properties. Future research should focus on material selection, design, and manufacturing aspects of AM parts, particularly the residual and surface stresses generated during AM of metallic parts. AM processes face more challenges in the manufacturing of biomaterial components due to the varying ink cartridge/spool and printing process parameters. This requires extensive knowledge of 3D printable materials, their chemical compositions, optimal printing parameters, and manufacturing constraints. A limited selection of 3D printable materials and scanty design guidelines on material compatibility and multi-material 3D printability are major limitations in achieving AM biomaterial components. Furthermore, software development requires significant research and development to fully utilize the potential of AM research [18]. Novel techniques for preparation, analysis, and slicing of biomaterial components are still required. Additive manufacturing technology, such as dental layout additive manufacturing (3D printing), has been slower to evolve in dental settings due to its high cost. The broad range of additive materials enables many other dentistry applications, such as producing CoCr substrates for dental prostheses, due to their lower cost and favorable properties. Further research is needed to evaluate the suitability of AM manufacturing on CoCr in dental prostheses for use as a substructure.

## 10.10   CONCLUSION

Additive manufacturing (AM) is a rapidly growing technology that has the potential to revolutionize the field of biomaterials by allowing precise control of surface properties. This control can significantly impact in vivo performance, such as cell adhesion, proliferation, and differentiation. AM can be used for surface modification in various applications, such as implants, drug delivery, and tissue engineering. As AM continues to develop, it will become more versatile and powerful, opening up new possibilities for its use in various biomedical applications. As AM technology continues to develop, it will create biomaterials with more complex and sophisticated surface properties, enabling better interaction with natural tissues and cells. Additionally, AM will enable the development of personalized biomaterials, such as implants and devices designed for individual patients, and new types of biomaterials that are not possible with traditional manufacturing methods, potentially leading to new treatments for various diseases.

## REFERENCES

1. Zaeri, A., Cao, K., Zhang, F., Zgeib, R. and Chang, R.C., 2022. A review of the structural and physical properties that govern cell interactions with structured biomaterials enabled by additive manufacturing. *Bioprinting*, *26*, p. e00201.
2. Sharma, S., Gupta, V. and Mudgal, D., 2022. Current trends, applications, and challenges of coatings on additive manufacturing based biopolymers: A state of art review. *Polymer Composites*, *43*(10), pp. 6749–6781.
3. Sharma, S.K., Saxena, K.K., Dixit, A.K., Singh, R. and Mohammed, K.A., 2022. Role of additive manufacturing and various reinforcements in MMCs related to biomedical applications. *Advances in Materials and Processing Technologies*, pp. 1–18.
4. Germaini, M.M., Belhabib, S., Guessasma, S., Deterre, R., Corre, P. and Weiss, P., 2022. Additive manufacturing of biomaterials for bone tissue engineering–A critical review of the state of the art and new concepts. *Progress in Materials Science*, *130*, p. 100963.
5. Ghahramanzadeh Asl, H., Celik-Uzuner, S., Uzuner, U., Sert, Y. and Küçükömeroğlu, T., 2022. The effect of friction stir process on the mechanical, tribological, and biocompatibility properties of AZ31B magnesium alloy as a biomaterial: A pilot study. *Proceedings of the Institution of Mechanical Engineers, Part H: Journal of Engineering in Medicine*, *236*(12), pp. 1720–1731.
6. González-Estrada, O.A., Comas, A.P. and Ospina, R., 2022. Characterization of hydroxyapatite coatings produced by pulsed-laser deposition on additive manufacturing Ti6Al4V ELI. *Thin Solid Films*, *763*, p. 139592.
7. Pesode, P. and Barve, S., 2022. Additive manufacturing of metallic biomaterials and its biocompatibility. *Materials Today: Proceedings*.
8. Vellayappan, M.V., Duarte, F., Sollogoub, C., Dirrenberger, J., Guinault, A., Frith, J.E., Parkington, H.C., Molotnikov, A. and Cameron, N.R., 2023. Fabrication of architectured biomaterials by multilayer co-extrusion and additive manufacturing. *Advanced Functional Materials*, *33*, p. 2301547.
9. Larsson, L., Marattukalam, J.J., Paschalidou, E.M., Hjörvarsson, B., Ferraz, N. and Persson, C., 2022. Biocompatibility of a Zr-based metallic glass enabled by additive manufacturing. *ACS Applied Bio Materials*, *5*(12), pp. 5741–5753.
10. Abdudeen, A., Abu Qudeiri, J.E., Kareem, A. and Valappil, A.K., 2022. Latest developments and insights of orthopedic implants in biomaterials using additive manufacturing technologies. *Journal of Manufacturing and Materials Processing*, *6*(6), p. 162.
11. Safavi, M.S., Bordbar-Khiabani, A., Khalil-Allafi, J., Mozafari, M. and Visai, L., 2022. Additive manufacturing: an opportunity for the fabrication of near-net-shape NiTi implants. *Journal of Manufacturing and Materials Processing*, *6*(3), p. 65.
12. Mirzaali, M.J., Moosabeiki, V., Rajaai, S.M., Zhou, J. and Zadpoor, A.A., 2022. Additive manufacturing of biomaterials—Design principles and their implementation. *Materials*, *15*(15), p. 5457.
13. Vasudev, H. and Prakash, C., 2023. Surface engineering and performance of biomaterials. *Journal of Electrochemical Science and Engineering*, *13*(1), pp. 1–3.

14. Pesode, P. and Barve, S., 2023. Additive manufacturing of metallic biomaterials: sustainability aspect, opportunity, and challenges. *Journal of Industrial and Production Engineering*, *40*, pp. 1–42.

15. Nouri, A., Shirvan, A.R., Li, Y. and Wen, C., 2023. Surface modification of additively manufactured metallic biomaterials with active antipathogenic properties. *Smart Materials in Manufacturing*, *1*, p. 100001.

16. Simorgh, S., Alasvand, N., Khodadadi, M., Ghobadi, F., Kebria, M.M., Milan, P.B., Kargozar, S., Baino, F., Mobasheri, A. and Mozafari, M., 2022. Additive manufacturing of bioactive glass biomaterials. *Methods*.

17. Kolken, H.M.A., Callens, S.J.P., Leeflang, M.A., Mirzaali, M.J. and Zadpoor, A.A., 2022. Merging strut-based and minimal surface meta-biomaterials: decoupling surface area from mechanical properties. *Additive Manufacturing*, *52*, p. 102684.

18. Selvaraj, S.K., Prasad, S.K., Yasin, S.Y., Subhash, U.S., Verma, P.S., Manikandan, M. and Dev, S.J., 2022. Additive manufacturing of dental material parts via laser melting deposition: A review, technical issues, and future research directions. *Journal of Manufacturing Processes*, *76*, pp. 67–78.

19. Li, J., Cui, X., Lindberg, G.C., Alcala-Orozco, C.R., Hooper, G.J., Lim, K.S. and Woodfield, T.B., 2022. Hybrid fabrication of photo-clickable vascular hydrogels with additive manufactured titanium implants for enhanced osseointegration and vascularized bone formation. *Biofabrication*, *14*(3), p. 034103.

20. Punia, U., Kaushik, A., Garg, R.K., Chhabra, D. and Sharma, A., 2022. 3D printable biomaterials for dental restoration: A systematic review. *Materials Today: Proceedings*, *63*, pp. 566–572.

21. Amani, H., Arzaghi, H., Bayandori, M., Dezfuli, A.S., Pazoki-Toroudi, H., Shafiee, A. and Moradi, L., 2019. Controlling cell behavior through the design of biomaterial surfaces: a focus on surface modification techniques. *Advanced Materials Interfaces*, *6*(13), p. 1900572.

22. Albrektsson, T., Brånemark, P.I., Hansson, H.A. and Lindström, J., 1981. Osseointegrated titanium implants: requirements for ensuring a long-lasting, direct bone-to-implant anchorage in man. *Acta Orthopaedica Scandinavica*, *52*(2), pp. 155–170.

23. Khanlou, H.M., Ang, B.C., Barzani, M.M., Silakhori, M. and Talebian, S., 2015. Prediction and characterization of surface roughness using sandblasting and acid etching process on new non-toxic titanium biomaterial: adaptive-network-based fuzzy inference System. *Neural Computing and Applications*, *26*, pp. 1751–1761.

24. Rabel, K., Kohal, R.J., Steinberg, T., Tomakidi, P., Rolauffs, B., Adolfsson, E., Palmero, P., Fürderer, T. and Altmann, B., 2020. Controlling osteoblast morphology and proliferation via surface micro-topographies of implant biomaterials. *Scientific Reports*, *10*(1), p. 12810.

25. Hu, X., Park, S.H., Gil, E.S., Xia, X.X., Weiss, A.S. and Kaplan, D.L., 2011. The influence of elasticity and surface roughness on myogenic and osteogenic-differentiation of cells on silk-elastin biomaterials. *Biomaterials*, *32*(34), pp. 8979–8989.

26. Bose, S., Robertson, S.F. and Bandyopadhyay, A., 2018. Surface modification of biomaterials and biomedical devices using additive manufacturing. *Actabiomaterialia*, *66*, pp. 6–22.

27. Cappello, E. and Nieri, P., 2021. From life in the sea to the clinic: The marine drugs approved and under clinical trial. *Life*, *11*(12), p. 1390.

28. Krishna, B.V., Bose, S. and Bandyopadhyay, A., 2009. Fabrication of porous NiTi shape memory alloy structures using laser engineered net shaping. *Journal of Biomedical Materials Research Part B: Applied Biomaterials: An Official Journal of the Society for Biomaterials, The Japanese Society for Biomaterials, and the*

*Australian Society for Biomaterials and the Korean Society for Biomaterials*, *89*(2), pp. 481–490.

29. Das, M., Balla, V.K., Kumar, T.S. and Manna, I., 2013. Fabrication of biomedical implants using laser engineered net shaping (LENS™). *Transactions of the Indian Ceramic Society*, *72*(3), pp. 169–174.

30. Hu, Y., Chen, J., Hu, G., Yu, J., Zhu, X., Lin, Y., Chen, S. and Yuan, J., 2015. Statistical research on the bioactivity of new marine natural products discovered during the 28 years from 1985 to 2012. *Marine Drugs*, *13*(1), pp. 202–221.

31. Schöffski, P., Dumez, H., Wolter, P., Stefan, C., Wozniak, A., Jimeno, J. and Van Oosterom, A.T., 2008. Clinical impact of trabectedin (ecteinascidin-743) in advanced/metastatic soft tissue sarcoma. *Expert Opinion on Pharmacotherapy*, *9*(9), pp. 1609–1618.

32. Jana, S., Das, P., Mukherjee, J., Banerjee, D., Ghosh, P.R., Das, P.K., Bhattacharya, R.N. and Nandi, S.K., 2022. Waste-derived biomaterials as building blocks in the biomedical field. *Journal of Materials Chemistry B*, *10*(4), pp. 489–505.

33. Kanyane, L.R., Popoola, A.P.I., Pityana, S. and Tlotleng, M., 2022. Synthesis of Ti-Al-xNb ternary alloys via laser-engineered net shaping for biomedical application: Densification, electrochemical and mechanical properties studies. *Materials*, *15*(2), p. 544.

34. Tan, J. and Saltzman, W.M., 2004. Biomaterials with hierarchically defined micro- and nanoscale structure. *Biomaterials*, *25*(17), pp. 3593–3601.

35. Madou, M.J., 2018. *Fundamentals of microfabrication and nanotechnology, three-volume set*. CRC Press.

36. Gepreel, M.A.H. and Niinomi, M., 2013. Biocompatibility of Ti-alloys for long-term implantation. *Journal of the Mechanical Behavior of Biomedical Materials*, *20*, pp. 407–415.

37. Long, M. and Rack, H.J., 1998. Titanium alloys in total joint replacement—a materials science perspective. *Biomaterials*, *19*(18), pp. 1621–1639.

38. Chua, K., Khan, I., Malhotra, R. and Zhu, D., 2021. Additive manufacturing and 3D printing of metallic biomaterials. *Engineered Regeneration*, *2*, pp. 288–299.

39. Dienel, K.E., van Bochove, B. and Seppälä, J.V., 2019. Additive manufacturing of bioactive poly (trimethylene carbonate)/β-tricalcium phosphate composites for bone regeneration. *Biomacromolecules*, *21*(2), pp. 366–375.

40. Ferraris, S. and Spriano, S., 2021. Porous titanium by additive manufacturing: A focus on surfaces for bone integration. *Metals*, *11*(9), p. 1343.

41. Li, J., Lin, X., Wang, J., Zheng, M., Guo, P., Zhang, Y., Ren, Y., Liu, J. and Huang, W., 2019. Effect of stress-relief annealing on anodic dissolution behaviour of additive manufactured Ti-6Al-4V via laser solid forming. *Corrosion Science*, *153*, pp. 314–326.

42. Longhitano, G.A., Arenas, M.A., Conde, A., Larosa, M.A., Jardini, A.L., de CarvalhoZavaglia, C.A. and Damborenea, J.J., 2018. Heat treatments effects on functionalization and corrosion behavior of Ti-6Al-4V ELI alloy made by additive manufacturing. *Journal of Alloys and Compounds*, *765*, pp. 961–968.

43. Wauthle, R., Vrancken, B., Beynaerts, B., Jorissen, K., Schrooten, J., Kruth, J.P. and Van Humbeeck, J., 2015. Effects of build orientation and heat treatment on the microstructure and mechanical properties of selective laser melted Ti6Al4V lattice structures. *Additive Manufacturing*, *5*, pp. 77–84.

44. Rajput, A.S., Kapil, S. and Das, M., 2023. Surface enhancement of additively manufactured bone plate through hybrid-electrochemical magnetorheological finishing process. *3D Printing and Additive Manufacturing*.

45. Ribeiro, N., Soares, G.C., Santos-Rosales, V., Concheiro, A., Alvarez-Lorenzo, C., García-González, C.A. and Oliveira, A.L., 2020. A new era for sterilization based on

supercritical $CO_2$ technology. *Journal of Biomedical Materials Research Part B: Applied Biomaterials, 108*(2), pp. 399–428.

46. Soares, G.C., Learmonth, D.A., Vallejo, M.C., Davila, S.P., González, P., Sousa, R.A. and Oliveira, A.L., 2019. Supercritical $CO_2$ technology: The next standard sterilization technique?. *Materials Science and Engineering: C, 99*, pp. 520–540.
47. Griffith, L.G., 2000. Polymeric biomaterials. *Actamaterialia, 48*(1), pp. 263–277.
48. He, W. and Benson, R., 2017. Polymeric biomaterials. In *Applied plastics engineering handbook* (pp. 145–164). William Andrew Publishing.
49. Vert, M., 2007. Polymeric biomaterials: Strategies of the past vs. strategies of the future. *Progress in Polymer Science, 32*(8–9), pp. 755–761.
50. Dwivedi, R., Kumar, S., Pandey, R., Mahajan, A., Nandana, D., Katti, D.S. and Mehrotra, D., 2020. Polycaprolactone as biomaterial for bone scaffolds: Review of literature. *Journal of Oral Biology and Craniofacial Research, 10*(1), pp. 381–388.
51. Mondal, D., Griffith, M. and Venkatraman, S.S., 2016. Polycaprolactone-based biomaterials for tissue engineering and drug delivery: Current scenario and challenges. *International Journal of Polymeric Materials and Polymeric Biomaterials, 65*(5), pp. 255–265.
52. Kruth, J.P., Levy, G., Klocke, F. and Childs, T.H.C., 2007. Consolidation phenomena in laser and powder-bed based layered manufacturing. *CIRP Annals, 56*(2), pp. 730–759.
53. Punj, S., Singh, J. and Singh, K., 2021. Ceramic biomaterials: Properties, state of the art and future prospectives. *Ceramics International, 47*(20), pp. 28059–28074.
54. Pina, S., Rebelo, R., Correlo, V.M., Oliveira, J.M. and Reis, R.L., 2018. Bioceramics for osteochondral tissue engineering and regeneration. *Osteochondral Tissue Engineering: Nanotechnology, Scaffolding-Related Developments and Translation, 1058*, pp. 53–75.
55. Chevalier, J. and Gremillard, L., 2009. Ceramics for medical applications: A picture for the next 20 years. *Journal of the European Ceramic Society, 29*(7), pp. 1245–1255.
56. Rieger, W., 2001. Ceramics in orthopedics–30 years of evolution and experience. In *World tribology forum in arthroplasty* (pp. 283–294). Hans Huber Verlag Bern.
57. Romano, G., Almeida, M., Varela Coelho, A., Cutignano, A., Gonçalves, L.G., Hansen, E., Khnykin, D., Mass, T., Ramšak, A., Rocha, M.S. and Silva, T.H., 2022. Biomaterials and bioactive natural products from marine invertebrates: From basic research to innovative applications. *Marine Drugs, 20*(4), p. 219.
58. Xu, Y., Ding, W., Chen, M., Du, H. and Qin, T., 2022. Synergistic fabrication of micro-nano bioactive ceramic-optimized polymer scaffolds for bone tissue engineering by in situ hydrothermal deposition and selective laser sintering. *Journal of Biomaterials Science, Polymer Edition, 33*(16), pp. 2104–2123.
59. Cestari, F., Yang, Y., Wilbig, J., Günster, J., Motta, A. and Sglavo, V.M., 2022. Powder 3D printing of bone scaffolds with uniform and gradient pore sizes using cuttlebone-derived calcium phosphate and glass-ceramic. *Materials, 15*(15), p. 5139.
60. Wang, Y., Chen, S., Liang, H., Liu, Y., Bai, J. and Wang, M., 2022. Digital light processing (DLP) of nano biphasic calcium phosphate bioceramic for making bone tissue engineering scaffolds. *Ceramics International, 48*(19), pp. 27681–27692.
61. Feng, C., Zhang, K., He, R., Ding, G., Xia, M., Jin, X. and Xie, C., 2020. Additive manufacturing of hydroxyapatite bioceramic scaffolds: Dispersion, digital light processing, sintering, mechanical properties, and biocompatibility. *Journal of Advanced Ceramics, 9*, pp. 360–373.
62. Zhou, T., Zhang, L., Yao, Q., Ma, Y., Hou, C., Sun, B., Shao, C., Gao, P. and Chen, H., 2020. SLA 3D printing of high-quality spine shaped β-TCP bioceramics for the hard tissue repair applications. *Ceramics International, 46*(6), pp. 7609–7614.

63. Kumar, M., Ghosh, S., Kumar, V., Sharma, V. and Roy, P., 2022. Tribo-mechanical and biological characterization of PEGDA/bioceramics composites fabricated using stereolithography. *Journal of Manufacturing Processes*, 77, pp. 301–312.

64. Verma, A.S., Kumar, D. and Dubey, A.K., 2019. Dielectric and electrical response of hydroxyapatite–Na0. 5K0. 5NbO3 bioceramic composite. *Ceramics International*, 45(3), pp. 3297–3305.

65. Bakhsheshi-Rad, H.R., Hamzah, E., Kasiri-Asgarani, M., Jabbarzare, S., Iqbal, N. and Kadir, M.A., 2016. Deposition of nanostructured fluorine-doped hydroxyapatite–polycaprolactone duplex coating to enhance the mechanical properties and corrosion resistance of Mg alloy for biomedical applications. *Materials Science and Engineering: C*, 60, pp. 526–537.

66. Başman, G., Arikan, M.M., Arisoy, C. and Şeşen, K., 2023. A kinetic study of thermochemical borided aisi 316 l Stainless steel. *Journal of Scientific Reports-A*, 52, pp. 279–296.

67. Yi, L. and Liu, J., 2017. Liquid metal biomaterials: a newly emerging area to tackle modern biomedical challenges. *International Materials Reviews*, 62(7), pp. 415–440.

68. Gao, W., Wang, Y., Wang, Q., Ma, G. and Liu, J., 2022. Liquid metal biomaterials for biomedical imaging. *Journal of Materials Chemistry B*, 10(6), pp. 829–842.

69. García-Arnáez, I., Romero-Gavilán, F., Cerqueira, A., Elortza, F., Azkargorta, M., Munoz, F., Mata, M., de Llano, J.M., Suay, J., Gurruchaga, M.A.R.I.L.O. and Goñi, I., 2022. Correlation between biological responses in vitro and in vivo to Ca-doped sol-gel coatings assessed using proteomic analysis. *Colloids and Surfaces B: Biointerfaces*, 220, p. 112962.

70. Helsen, J.A. and Jürgen Breme, H., 1998. Metals as biomaterials (p. 522).

71. Alvarez, K. and Nakajima, H., 2009. Metallic scaffolds for bone regeneration. *Materials*, 2(3), pp. 790–832.

72. Touri, M., Kabirian, F., Saadati, M., Ramakrishna, S. and Mozafari, M., 2019. Additive manufacturing of biomaterials– the evolution of rapid prototyping. *Advanced Engineering Materials*, 21(2), p. 1800511.

73. Vepari, C. and Kaplan, D.L., 2007. Silk as a biomaterial. *Progress in Polymer Science*, 32(8–9), pp. 991–1007.

74. Hardy, J.G., Römer, L.M. and Scheibel, T.R., 2008. Polymeric materials based on silk proteins. *Polymer*, 49(20), pp. 4309–4327.

75. Hardy, J.G. and Scheibel, T.R., 2010. Composite materials based on silk proteins. *Progress in Polymer Science*, 35(9), pp. 1093–1115.

76. Porter, D., Vollrath, F. and Shao, Z., 2005. Predicting the mechanical properties of spider silk as a model nanostructured polymer. *The European Physical Journal E*, 16, pp. 199–206.

77. Pal, R.K., Farghaly, A.A., Wang, C., Collinson, M.M., Kundu, S.C. and Yadavalli, V.K., 2016. Conducting polymer-silk biocomposites for flexible and biodegradable electrochemical sensors. *Biosensors and Bioelectronics*, 81, pp. 294–302.

78. Xia, Y. and Lu, Y., 2008. Fabrication and properties of conductive conjugated polymers/silk fibroin composite fibers. *Composites Science and Technology*, 68(6), pp. 1471–1479.

79. Gentile, P., Chiono, V., Carmagnola, I. and Hatton, P.V., 2014. An overview of poly (lactic-co-glycolic) acid (PLGA)-based biomaterials for bone tissue engineering. *International Journal of Molecular Sciences*, 15(3), pp. 3640–3659.

80. Makadia, H.K. and Siegel, S.J., 2011. Poly lactic-co-glycolic acid (PLGA) as biodegradable controlled drug delivery carrier. *Polymers*, 3(3), pp. 1377–1397.

81. Richard, A. and Margaritis, A., 2001. Poly (glutamic acid) for biomedical applications. *Critical Reviews in Biotechnology*, *21*(4), pp. 219–232.

82. Azimi, B., Nourpanah, P., Rabiee, M. and Arbab, S., 2014. Poly (ε-caprolactone) fiber: an overview. *Journal of Engineered Fibers and Fabrics*, *9*(3), p. 155892501400900309.

83. Panda, P.K., Sadeghi, K. and Seo, J., 2022. Recent advances in poly (vinyl alcohol)/ natural polymer based films for food packaging applications: A review. *Food Packaging and Shelf Life*, *33*, p. 100904.

84. Pérez Davila, S., González Rodríguez, L., Chiussi, S., Serra, J. and González, P., 2021. How to sterilize polylactic acid based medical devices? *Polymers*, *13*(13), p. 2115.

85. Stamou, A., Iatrou, H. and Tsitsilianis, C., 2022. NIPAm-based modification of poly (L-lysine): A pH-dependent LCST-type thermo-responsive biodegradable polymer. *Polymers*, *14*(4), p. 802.

86. Arango-Santander, S., 2022. Bioinspired topographic surface modification of biomaterials. *Materials*, *15*(7), p. 2383.

87. Yeo, I.S.L., 2022. Dental implants: enhancing biological response through surface modifications. *Dental Clinics*, *66*(4), pp. 627–642.

88. Chen, M., Wang, X.Q., Zhang, E.L., Wan, Y.Z. and Hu, J., 2022. Antibacterial ability and biocompatibility of fluorinated titanium by plasma-based surface modification. *Rare Metals*, *41*, pp. 689–699.

89. Thanigaivel, S., Priya, A.K., Balakrishnan, D., Dutta, K., Rajendran, S. and Soto-Moscoso, M., 2022. Insight on recent development in metallic biomaterials: Strategies involving synthesis, types and surface modification for advanced therapeutic and biomedical applications. *Biochemical Engineering Journal*, *187*, p. 108522.

90. Wathanyu, K., Tuchinda, K., Daopiset, S., Sirivisoot, S., Kondas, J. and Bauer, C., 2022. Study of the properties of titanium porous coating with different porosity gradients on 316 L stainless steel by a cold spray process. *Journal of Thermal Spray Technology*, *31*(3), pp. 545–558.

91. Douglass, M., Garren, M., Devine, R., Mondal, A. and Handa, H., 2022. Bio-inspired hemocompatible surface modifications for biomedical applications. *Progress in Materials Science*, *130*, p. 100997.

92. Ekmekci, N. and Efe, Y., 2022. The effect of nano and micro hydroxyapatite powder additives on surface integrity in electrical discharge machining of Ti6Al4V alloy. *Surface and Coatings Technology*, *445*, p. 128708.

93. Wu, G.L., Zhang, S., Ye, W.A.N.G., Min, S.U.N., Zhang, Q.L., Kovalenko, V. and Yao, J.H., 2022. Porous ceramic coating formed on 316 L by laser cladding combined plasma electrolytic oxidation for biomedical application. *Transactions of Nonferrous Metals Society of China*, *32*(9), pp. 2993–3004.

94. Tan, T., Zhao, Q., Kuwae, H., Ueno, T., Chen, P., Tsutsumi, Y., Mizuno, J., Hanawa, T. and Wakabayashi, N., 2022. Surface properties and biocompatibility of sandblasted and acid-etched titanium–zirconium binary alloys with various compositions. *Dental Materials Journal*, *41*(2), pp. 266–272.

95. Zareidoost, A. and Yousefpour, M., 2022. Coinciding significance of the crystallographic orientation and nanostructuring on the biocompatibility of TZNT-Ag1. 5 alloy deformed by the cold rolling process. *Journal of Biomedical Materials Research Part B: Applied Biomaterials*, *110*(3), pp. 625–637.

96. Gabor, R., Cvrček, L., Doubková, M., Nehasil, V., Hlinka, J., Unucka, P., Buřil, M., Podepřelová, A., Seidlerová, J. and Bačáková, L., 2022. Hybrid coatings for orthopaedic implants formed by physical vapour deposition and microarc oxidation. *Materials & Design*, *219*, p. 110811.

97. Lau, K., Fu, L., Zhang, A., Akhavan, B., Whitelock, J., Bilek, M.M., Lord, M.S. and Rnjak-Kovacina, J., 2023. Recombinant perlecan domain V covalently immobilized on silk biomaterials via plasma immersion ion implantation supports the formation of functional endothelium. *Journal of Biomedical Materials Research Part A, 111*(6), pp. 825–839.

98. Boopathi, S., 2022. An experimental investigation of Quench Polish Quench (QPQ) coating on AISI 4150 steel. *Engineering Research Express, 4*(4), p. 045009.

99. Vidal, C., Alves, P., Alves, M.M., Carmezim, M.J., Fernandes, M.H., Grenho, L., Inacio, P.L., Ferreira, F.B., Santos, T.G. and Santos, C., 2022. Fabrication of a biodegradable and cytocompatible magnesium/nanohydroxyapatite/fluorapatite composite by upward friction stir processing for biomedical applications. *Journal of the Mechanical Behavior of Biomedical Materials, 129*, p. 105137.

100. Wang, L., Lai, R., Zhang, L., Zeng, M. and Fu, L., 2022. Emerging liquid metal biomaterials: From design to application. *Advanced Materials, 34*(37), p. 2201956.

101. Makhesana, M.A. and Patel, K.M., 2023. Processing of biomaterials by additive manufacturing. In *Advances in additive manufacturing artificial intelligence, nature-inspired, and biomanufacturing* (pp. 273–279). Elsevier.

102. Rehman, M., Yanen, W., Mushtaq, R.T., Ishfaq, K., Zahoor, S., Ahmed, A., Kumar, M.S., Gueyee, T., Rahman, M.M. and Sultana, J., 2022. Additive manufacturing for biomedical applications: a review on classification, energy consumption, and its appreciable role since COVID-19 pandemic. *Progress in Additive Manufacturing, 1*, pp. 1–35.

103. Agarwal, R., Gupta, V. and Singh, J., 2022. Additive manufacturing-based design approaches and challenges for orthopaedic bone screws: a state-of-the-art review. *Journal of the Brazilian Society of Mechanical Sciences and Engineering, 44*(1), p. 37.

104. Nazir, A., Gokcekaya, O., Billah, K.M.M., Ertugrul, O., Jiang, J., Sun, J. and Hussain, S., 2023. Multi-material additive manufacturing: A systematic review of design, properties, applications, challenges, and 3D Printing of materials and cellular metamaterials. *Materials & Design, 226*, p. 111661.

105. Arif, Z.U., Khalid, M.Y., Noroozi, R., Hossain, M., Shi, H.H., Tariq, A., Ramakrishna, S. and Umer, R., 2023. Additive manufacturing of sustainable biomaterials for biomedical applications. *Asian Journal of Pharmaceutical Sciences, 18*, p. 100812.

# 11 Various Additive Manufacturing Techniques for the Fabrication of Biomaterials

*Priyatosh Sahoo*
Institute of Materials Science, Technische Universität Darmstadt, Darmstadt, Germany

*Wrootchit Mishra*
Biological Sciences, Carnegie Mellon University, Pittsburgh, Pennsylvania, USA

## 11.1 INTRODUCTION TO ADDITIVE MANUFACTURING

Additive manufacturing, or AM, is an advanced production approach whereby the desired final product is fabricated in a layer-by-layer format. The parts to be fabricated are first designed in a computer using computer-aided design (CAD) software. The design is then sent to an AM machine, which puts 2D layers on top of each other to create the final 3D model. Using this processing route, parts with very complex geometric designs can be prepared with ease from a variety of substrate materials, such as metals, ceramics, or polymers. The additive manufacturing process is also referred to as 3D printing (3DP) or layered manufacturing (LM) [1,2]. AM is helping shape a wide range of industries such as aerospace, biomedical, automobile, and more [1,3,4]. Additive manufacturing solves the problem of the high cost of customized parts through conventional methods (Figure 11.1).

Conventional manufacturing (CM) processes are designed and developed for very high production output. Their processes are optimized to produce the highest yield within the shortest possible time. The typical processing routes in CM are rather long and tedious due to several intermediate processing steps. Also, there is a very high material loss in the post-processing treatment while cutting and grinding to the final required dimensions. However, it is economical due to the high production volume. In contrast to CM, AM focuses on the production of material and application-specific parts. They are used in certain applications where small

DOI: 10.1201/9781003429920-13

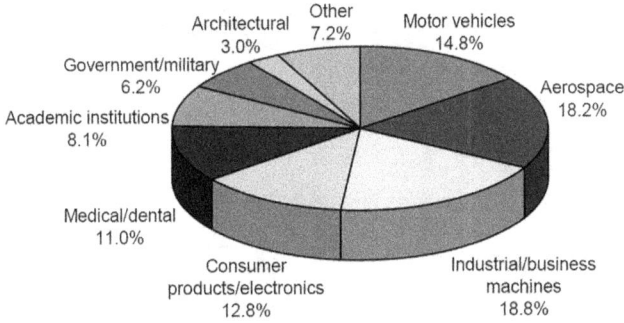

**FIGURE 11.1**   Industrial use of AM in different sectors [5].

quantities of complicated geometries are required rather than high volumes. AM is developed for systems where the biggest concern is application and design rather than the cost and time for production. The production cost through most AM processes is generally independent of the volume of production. This is unlike the CM where the cost of production can be high for small quantities. Recently, many developments have been made in AM with a focus on improving the efficiency of the process and making it more economical to compete with the CM processes.

Biomaterials are the materials group which are used to repair or replace a damaged part in a living being. These materials can be prepared from naturally occurring substances or from synthetic materials. These are made such that they interact with the living organism and become a part of their system and sometimes even perform their work. Many such biomaterials have already been developed which are capable of replacing human organs or tissues and substituting their functions. Over time, research on different material classes including ceramic, composite, and metallic has been made to understand and study their biocompatibility inside the body. Along with excellent bio-compatibility, biomaterials are also expected to have other unique material properties such as mechanical strength or structural stability depending on the specific application. Recently, they are also being developed as a cure for diseases due to their antibiotic and antimicrobial properties [6]. The current aim is to produce biomaterials with multifunctional properties that can be used in a variety of applications (Figure 11.2).

Contrary to all the advantages that additive manufacturing has over conventional manufacturing, there are also some disadvantages and limitations. The cost of production of customized parts by AM may be lower; however, the initial cost to set up the AM machine can sometimes be high. In most cases, special metallic and ceramic powders must be made for the desired application to be used as feedstock materials in AM manufacturing processes that are quite expensive in comparison to cheap polymeric materials. There needs to be different machines for different feedstock materials as it can lead to contamination and degradation of properties. This further increases the cost of production. The AM machines are required to be optimized for each substrate material to obtain and reproduce the desired results.

# Biomaterials

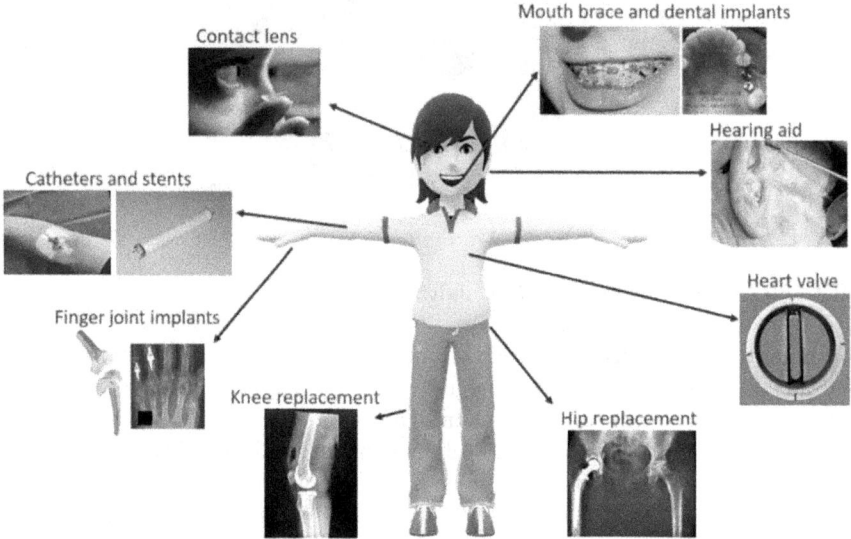

FIGURE 11.2  Different biomaterials used in human body [7].

## 11.2  DIFFERENT ADDITIVE MANUFACTURING TECHNIQUES

### 11.2.1  FUSED DEPOSITION MODELING (FDM)

FDM, one of the earliest and most used additive manufacturing techniques, is lauded for its cost-effectiveness and ability to process a variety of biocompatible thermoplastic materials. From patient-specific prosthetic devices to tissue engineering scaffolds, FDM's applications are broad and varied [8]. It also has the capacity to fabricate biomaterials at ambient conditions, enabling the incorporation of sensitive biomolecules or cells directly into the scaffold during fabrication [9]. However, the dimensional accuracy of FDM is sometimes compromised by the thermal contraction of the extruded material, leading to warping or shrinkage [10] (Figure 11.3).

### 11.2.2  STEREOLITHOGRAPHY (SLA)

Known for its high precision and fine details, SLA is suitable for creating accurate biological models and scaffolds for tissue engineering [12]. SLA technology also enables parts with smooth surface finishes, crucial in biomedical applications where surface roughness can influence cell behavior [13]. A major drawback, however, is the limited range of biocompatible resins currently available, although ongoing research aims to address this [14] (Figure 11.4).

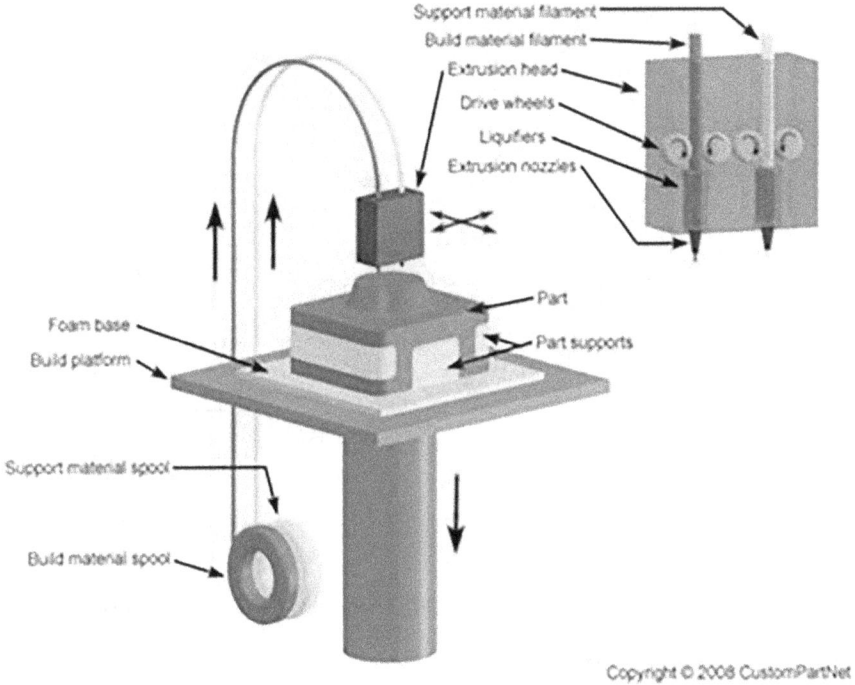

**FIGURE 11.3**   Key components of fused deposition modeling (FDM) apparatus [11].

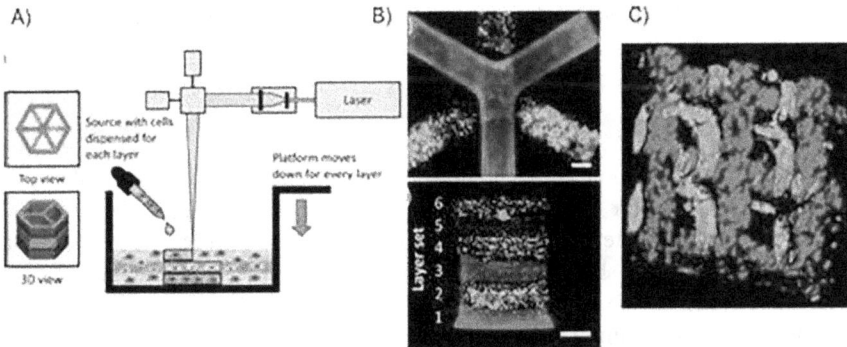

**FIGURE 11.4**   Laser stereolithography using SLA (stereolithography apparatus). (A) Schematic representation of the SLA technique. (B) 3D hydrogels fabricated using SLA, consisting of six layers. (C) Spatially patterned 3D scaffolds produced through laser-assisted stereolithography [15,16].

## 11.2.3   SELECTIVE LASER SINTERING (SLS)/SELECTIVE LASER MELTING (SLM)

Versatile and compatible with a variety of materials including metals, ceramics, and polymers, SLS and SLM provide excellent mechanical properties and enable complex

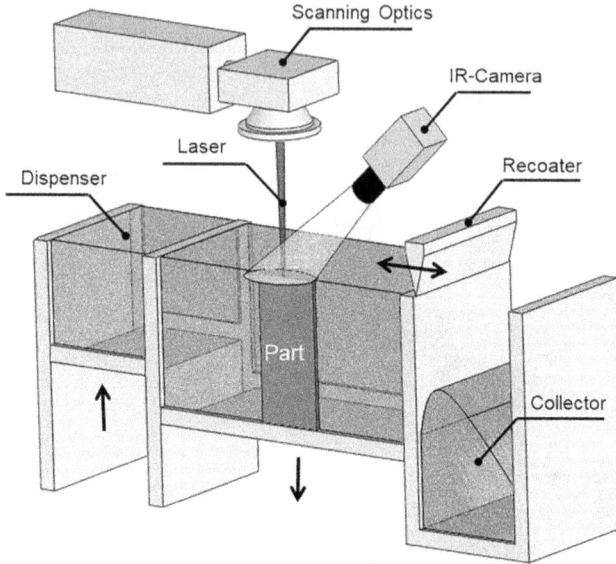

**FIGURE 11.5** Selective laser melting (SLM) technology is an additive manufacturing process that utilizes a high-power laser to selectively melt and fuse metal powders to create complex three-dimensional structures [19].

geometries. This makes them ideal to produce orthopedic implants and dental prostheses [16]. These techniques excel in manufacturing components with intricate internal structures like lattice or porous structures, enhancing tissue integration and vascularization [17]. However, these techniques may lead to high residual stress in the printed parts, necessitating post-processing treatments [18] (Figure 11.5).

## 11.2.4 BIOPRINTING

Bioprinting, a revolutionary technology in regenerative medicine and tissue engineering, allows for the precise spatial placement of living cells, bioactive molecules, and biomaterials to fabricate biomimetic structures [20]. With the potential to address the shortage of organ donors, bioprinting stands at the forefront of regenerative medicine. However, preservation of the bioactivity and functionality of printed cells over time remains a challenge, requiring optimization of bio-ink formulations and printing conditions [21] (Figure 11.6).

## 11.2.5 DIGITAL LIGHT PROCESSING (DLP)

DLP is known for its speed and precision. Its ability to cure an entire layer of resin in a single exposure, results in faster build times compared to SLA, making it advantageous for mass production of small, complex parts [23]. Despite these advantages, materials for DLP are generally more expensive and less diverse compared to those used in SLA [24] (Figure 11.7).

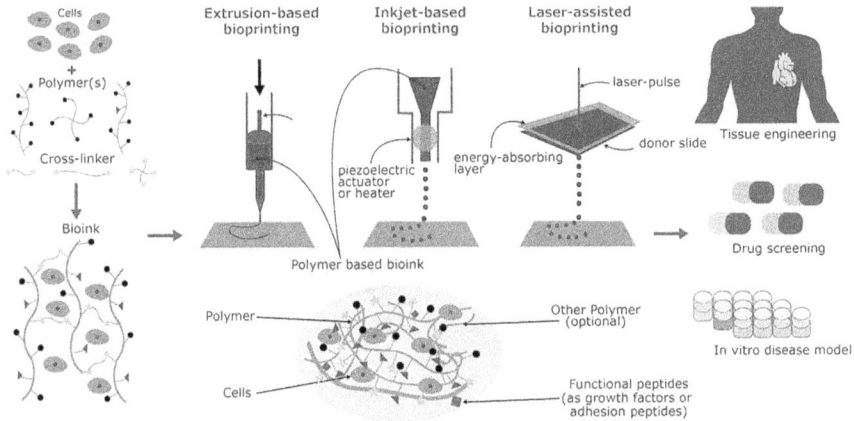

**FIGURE 11.6** Process of 3D bioprinting for tissue and organ fabrication. Showcases the creation of bio-inks. These bio-inks enable the production of functional tissue constructs [22].

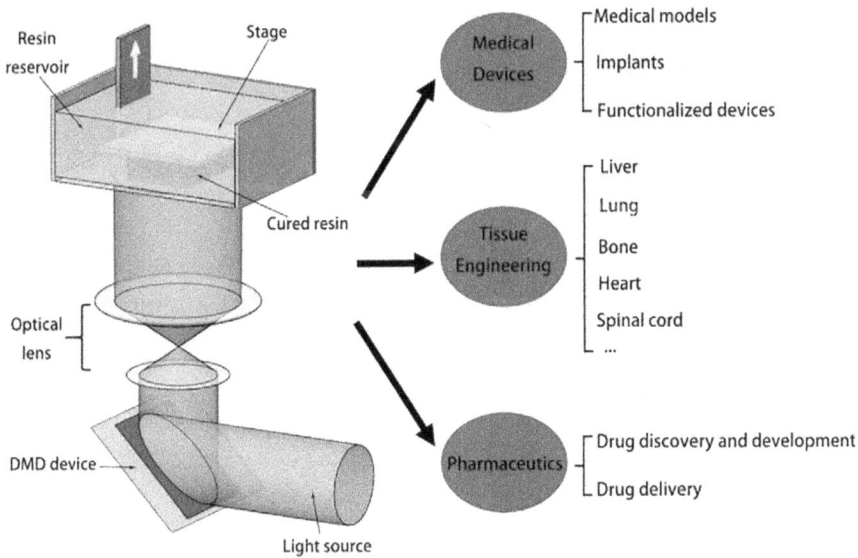

**FIGURE 11.7** DLP-based 3D printing technology, where light from a DMD or LCD panel selectively cures layers of photosensitive resin, creating a 3D structure [25].

## 11.2.6 PolyJet Technology

PolyJet technology offers a unique ability to print multi-material and multi-color parts in a single build, making it extremely versatile [26]. It's particularly useful for producing detailed anatomical models for surgical planning and medical device prototyping. Another advantage of PolyJet is its high resolution and accuracy.

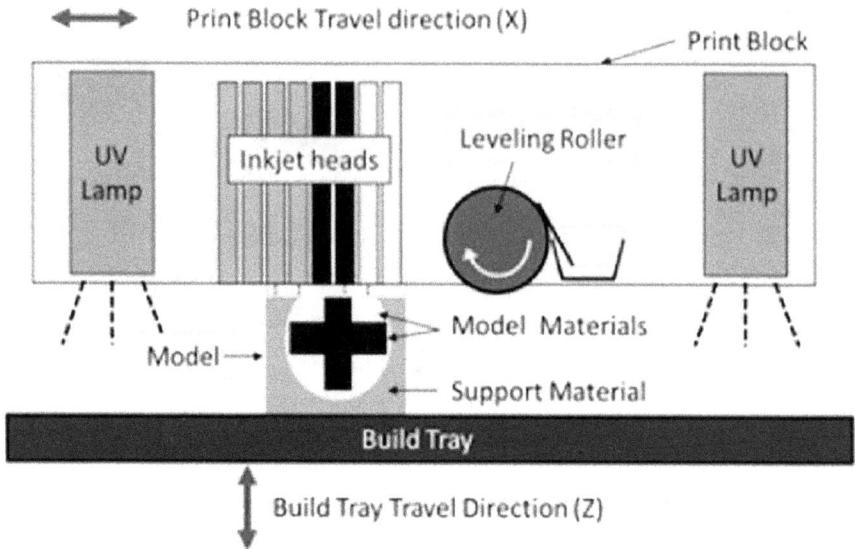

**FIGURE 11.8**  Depicts PolyJet printing process [28].

However, it relies heavily on proprietary materials and the operational cost can be high [27] (Figure 11.8).

### 11.2.7 Electron Beam Melting (EBM)

EBM is a powder bed fusion technique that uses a high-energy electron beam to fuse metal powder layer-by-layer, resulting in components with superior mechanical properties and durability [29]. Its ability to produce components with nearly 100% density makes it especially suitable for the manufacture of load-bearing implants. A limitation, however, is the relatively rough surface finish produced by EBM, which might necessitate post-processing depending on the application [30] (Figure 11.9).

## 11.3 ADDITIVE MANUFACTURING OF IMPORTANT BIOMATERIALS

### 11.3.1 Hydroxyapatite

Hydroxyapatite or HA is a naturally occurring mineral form of the calcium apatite group with the chemical formula $Ca_5(PO_4)_3OH$. The crystal unit cell of HA comprises two entities and is hence commonly denoted as $Ca_{10}(PO_4)_6OH_2$ [32]. The Ca to P ratio in calcium phosphate ceramics is responsible for the different phases. All these different phases show different biocompatibility and are a subject of discussion [33]. HA has a Ca to P ratio of 1.67 and is proven to show excellent biocompatibility. It is found in the human body in bones and teeth. Around 70 wt% of human bone contains a modified form of HA, called bone mineral [34].

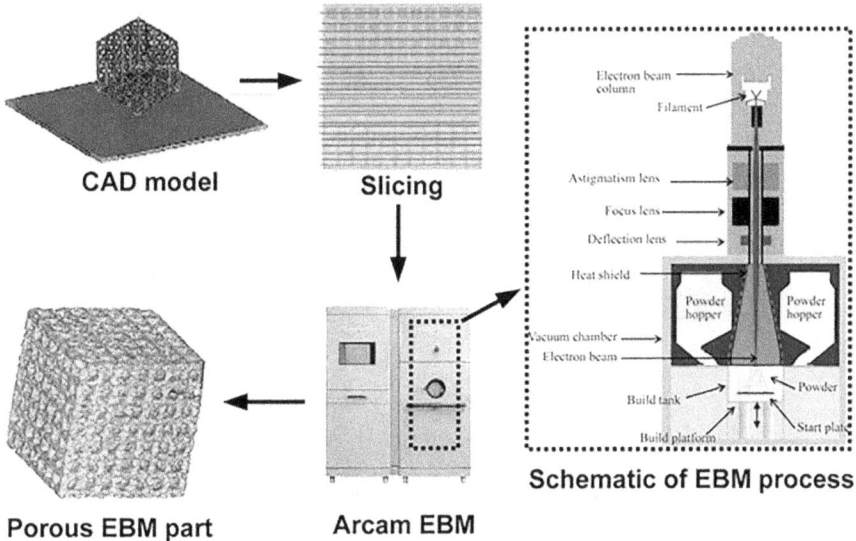

**FIGURE 11.9** Process of additive manufacturing of titanium alloys using electron beam melting (EBM) [31].

Additive manufacturing techniques for bio-ceramics like HA, are rather difficult to develop due to many inherent properties of ceramics. Ceramics in general have a very high melting point which makes them very difficult to melt using conventional heating methods. They also have very complex phase diagrams, and many different undesirable phases tend to form while cooling. The high temperatures involved in the processing of these ceramics result in the formation of pores and cracks. These affect the mechanical properties and structural integrity of these materials. However, these pores also facilitate cell growth and increase the stability of the implants. Therefore, a balance between mechanical and biological properties is important to achieve by regulating the pores.

One of the effective additive manufacturing methods for synthesizing complex HA parts is through vat polymerization. In this technique, photoactive polymer additives are added together with HA powder to prepare a suspension which is then cured layer by layer using a precision laser. After this, the final part is then sintered at high temperatures to remove the polymer additives and for the densification of the part.

Chu et al. [35] developed a method for fabricating HA implants with interconnected channels and beams. They initially prepared a 40-volume % HA suspension to be cast in epoxy molds which were fabricated using vat polymerization. The suspension was then heat treated to remove the binders and achieve full density. Using this technique, they were able to create implants with pore channel sizes in the range of 366–968 μm. However, they argued that this size was due to their machine limitation and even finer channel sizes can also be achieved using a laser with a small spot size to prepare the epoxy molds. They were able to fabricate HA implants with crack-free internal structure and porosity in the range of 26–52%.

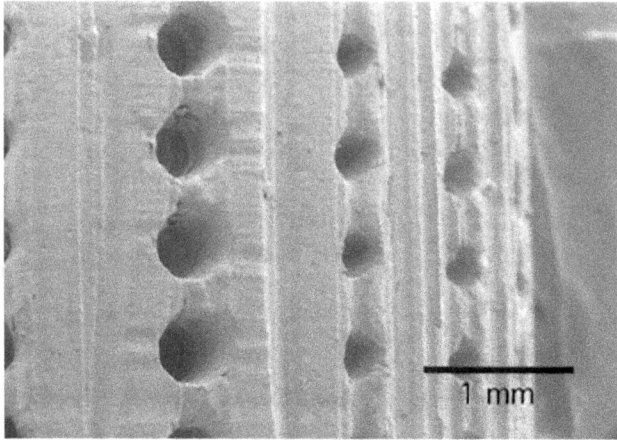

**FIGURE 11.10**   SEM image of HA implant showing the fine channels [35].

Woesz et al. [36] used a similar technique to fabricate HA implants from resin molds prepared by vat polymerization. In their experiments, they test for the biocompatibility of the implants for bone replacement applications in the preosteoblastic cell line MC3T3-E1, which was derived from mouse calvariae. They first prepared HA implants consisting of 450 μm pore size. Then they tested the biocompatibility of the implants by keeping them in a cell culture medium for two weeks. Their studies confirmed that these implants show excellent bio-compatibility. The MC3T3-E1 cells covered all the surfaces and pores of the implants and formed a tissue-like structure (Figure 11.10).

Kim et al. [37] developed a method to create bone scaffolds from HA nanopowders. They initially prepared a scaffold material by using the vat polymeri-zation technique. To create the scaffold, they illuminated a liquid polymer SL5180 resin with an ultraviolet laser. The HA nanopowders were then filled inside the scaffold material, which was later sintered to densify and remove the resin material. They fabricated HA scaffolds with 250 μm internal pore size and 130 μm wide patterned lines. They reported that the shrinkage was isotropic and was about 40% in relation to the original scaffold material during the sintering process. This could be taken into account while fabricating scaffolds of accurate dimensions (Figures 11.11 and 11.12).

## 11.3.2   Ti-6Al-4V or Ti64

Ti-6Al-4V or Ti64 or ASTM Grade 5 is an alpha-beta titanium alloy phase. It is comprised of 90% titanium, 6% aluminum, and 4% vanadium. In Ti64, the alpha phase is stabilized by aluminum, whereas the beta phase is stabilized by vanadium. This alloy shows high specific strength and good corrosion-resistant properties.

Titanium-based alloys in general show excellent biocompatibility and are hence preferred in many biomedical applications. In comparison to conventional stainless steels and cobalt-based alloys, Ti-based alloys have a high strength-to-weight ratio,

**FIGURE 11.11**    Final shape of bone scaffold after sintering [37].

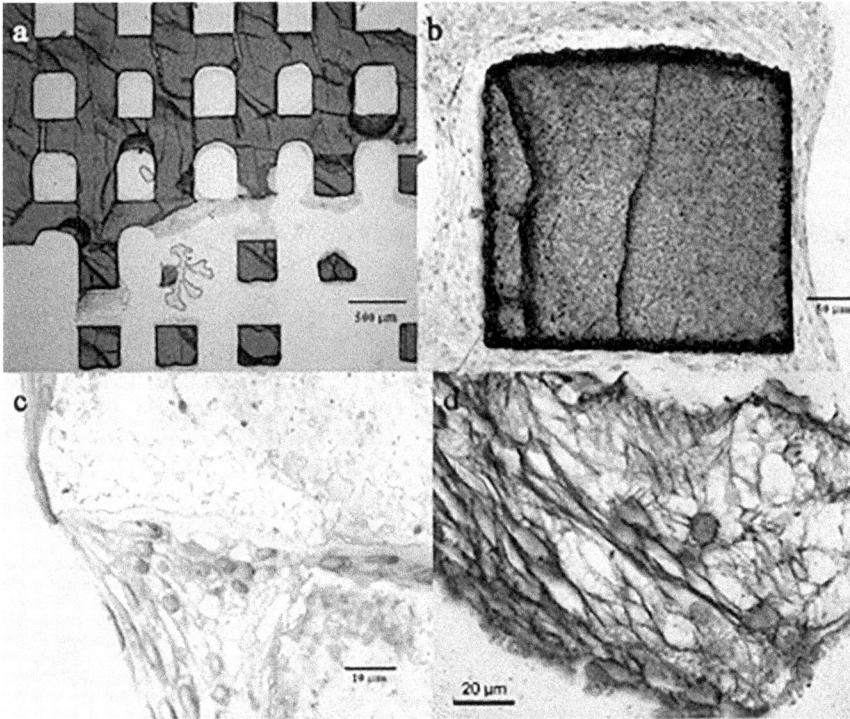

**FIGURE 11.12**    (a) HA scaffold after culturing with MC3T3-E1 cells for two weeks. (b) A single strut of a scaffold (gray), completely covered by cells (blue/pink). (c) Crack between two struts of the scaffold completely filled by MC3T3 cells (blue) and matrix generated by the cells (pink). (d) Collagen in the matrix formed by the cells [36].

and mechanical and corrosion properties which makes them a suitable material for load-bearing implants [38].

Kobryn and Semiatin [39] fabricated Ti64 alloy on a hot rolled Ti64 substrate plate material using laser additive manufacturing (LAM). They investigated the effect of variation of LAM parameters on the room temperature microstructure in Ti-6Al-4V deposits. For their experiments, they used different LAM systems, one with low-power ND:YAG-based systems and the other with high-power $CO_2$-based systems. They found that for both systems, columnar grains were deposited on the substrate. For the ND:YAG-based systems, columnar grains grew at a fixed acute angle to the substrate, whereas for the $CO_2$-based system columnar grains grew perpendicular to the substrate. The microstructure of the Ti64 deposits formed by ND:YAG-based systems showed a fine Widmanstätten pattern with fine equiaxed α particles distributed within the grains and along the grain boundaries. In contrast, the $CO_2$-based systems formed deposits with much coarser microstructure. The internal grain structure showed a Widmanstätten pattern with small α colonies along the grain boundaries. They also observed that the LAM parameters affect the grain width of the β phase. The grain width increased with the laser energy due to slower cooling rates. The average grain width was between 94 and 165 mm (Figure 11.13).

Heinl et al. [40] fabricated cellular titanium from Ti64 gas-atomized powder using selective electron beam melting (SEBM). In this technique, the powder particles are melted and deposited layer by layer with an electron gun. The main advantage of this technique is that the entire process is in a vacuum which prevents contamination from oxygen and nitrogen. They fabricated three different cellular titanium with interconnected pores and average porosities of about 60%, 25%, and 38%. Figure 11.14 shows the SEM images of the microstructure of the lateral and

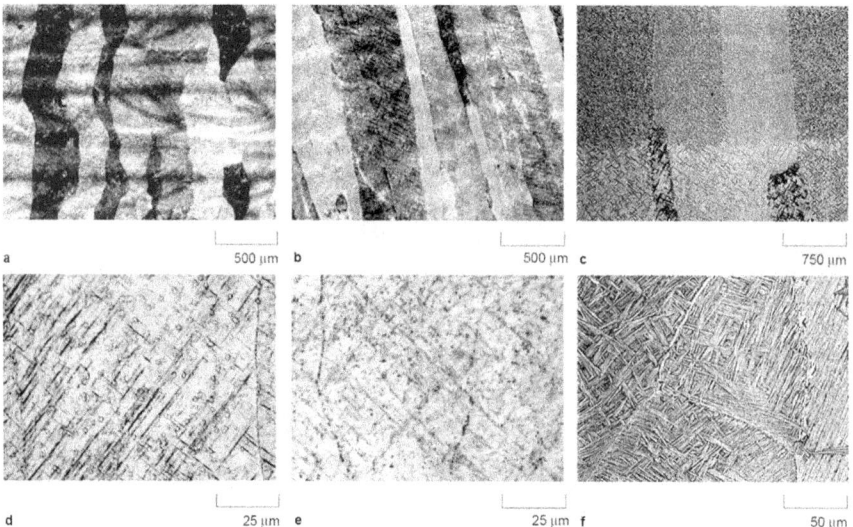

**FIGURE 11.13**  Microstructure of Ti-6Al-4V deposits from Nd:YAG system 1,2 and the $CO_2$ system [39].

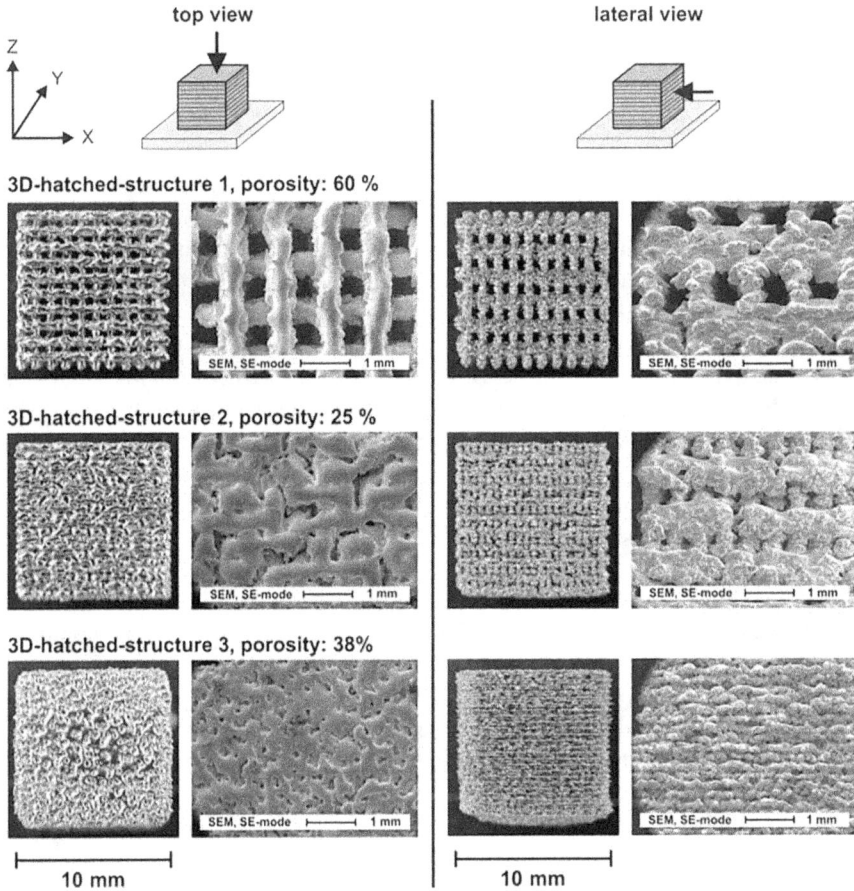

**FIGURE 11.14** SEM microstructure of the cellular titanium with different porosities [40].

top views of the three cellular titanium. The pore sizes vary from 100 to 400 μm, which is considered ideal for bone ingrowth leading to good fixation of the implant [40]. They also tested their mechanical properties and observed that their elastic moduli varied between 1 GPa and 30 GPa. These values are quite comparable to that of the elastic modulus of human bone [41].

### 11.3.3 POLYETHER ETHER KETONE (PEEK)

Polyether ether ketone, widely recognized as PEEK, is a semi-crystalline, high-performance thermoplastic belonging to the polyaryletherketone (PAEK) family. Known for its excellent mechanical properties, thermal stability, and resistance to various chemical interactions, PEEK is frequently employed in several high-performance applications [42]. Moreover, PEEK exhibits natural radiolucency and biocompatibility, making it particularly advantageous in the biomedical field, where

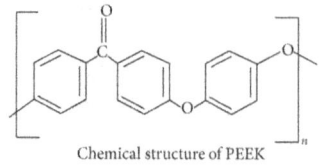

**FIGURE 11.15** The chemical structural formula of polyetheretherketone (PEEK) [44].

Chemical structure of PEEK

it is commonly used in the production of dental prosthetics, spinal implants, and cranial plates [43] (Figure 11.15).

PEEK's processability permits the use of several additive manufacturing techniques, including fused deposition modeling (FDM), selective laser sintering (SLS), and even bioprinting. In FDM, PEEK filaments are heated until they reach a semi-fluid state, after which they are extruded layer by layer to create the desired structure. The outcome is highly precise constructs that can be tailored to the patient's unique needs [45]. However, due to PEEK's high processing temperatures, a challenge lies in balancing printability and mechanical properties, which might require the development of modified formulations or specific printing environments [43].

In SLS, PEEK powders are sintered layer by layer with a laser to form three-dimensional structures. This method has been utilized to generate complex, porous constructs for bone tissue engineering, promoting bone ingrowth and enhancing implant fixation [46]. Nonetheless, the use of SLS in processing PEEK is limited by the high cost of PEEK powder and the need for stringent control of sintering parameters to maintain the integrity and performance of the final products [46].

In a study by Jahani et al. [47], PEEK was incorporated into a bio-ink for bioprinting. The composite bio-ink, consisting of PEEK and hydrogel, was used to manufacture cartilage constructs. They found that the inclusion of PEEK enhanced the mechanical properties of the printed structures while supporting chondrocyte survival and proliferation, demonstrating its potential for cartilage tissue engineering applications.

Expanding on this work, Chen et al. [48] developed a novel approach to create PEEK-based constructs with a high degree of vascularization, a critical aspect for the success of any tissue engineering application. They used bioprinting to incorporate endothelial cells into PEEK-hydrogel bio-inks, creating a vascularized 3D network within the PEEK structure. After several weeks of in vitro culture, the bio-printed constructs exhibited well-formed capillary networks, indicating the potential of this approach for fabricating vascularized tissue engineering scaffolds (Figure 11.16).

Furthermore, PEEK's inherent radiolucency, which allows for easy post-operative monitoring of the implant, has been exploited in the creation of patient-specific orthopedic implants. Nouri et al. [50] used SLS to fabricate PEEK implants for total knee arthroplasty, demonstrating superior fit and decreased wear compared to conventional metal implants. In another study by Adil and Lazoglu [51], SLS-manufactured PEEK spinal implants showed promising results in terms of maintaining intervertebral disc height and promoting bone fusion.

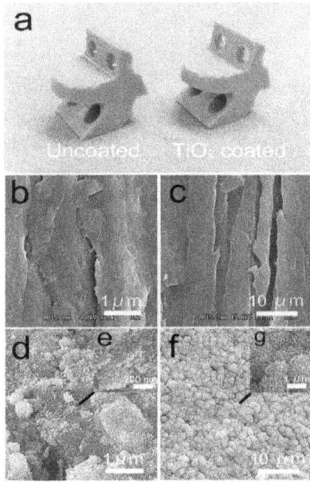

FIGURE 11.16 (a) View of the PEEK implants. (b) SEM image of the uncoated PEEK implant. (c) SEM image of the uncoated PEEK implant after soaking in simulated body fluid (SBF). (d) SEM image of the TiO2-coated PEEK implant. (e) View of the TiO2-coated PEEK implant. (f) SEM image of the TiO2-coated PEEK implant after soaking in SBF. (g) View of the TiO2-coated PEEK implant [49].

## 11.3.4 Poly-Lactic Acid (PLA)

Poly-lactic acid, or PLA, is a biodegradable and biocompatible polyester derived from renewable resources such as corn starch or sugar cane. Because of its properties, it is a wide choice in the biomedical field for applications ranging from drug delivery systems to tissue engineering scaffolds [52]. Its ability to safely degrade within the body over time into innocuous by-products makes it an ideal material for temporary implants that support tissue regeneration and subsequently dissolve, avoiding the need for secondary surgeries (Figure 11.17).

Like PEEK, PLA is often processed through FDM, allowing the fabrication of patient-specific implants. However, PLA's relatively low melting temperature compared to PEEK enables a more accessible and less energy-intensive manufacturing process. In a study by Hu et al. [54], the authors exploited FDM's capabilities to create complex, personalized, and degradable PLA-based nerve guides that could accurately replicate the anatomy of the patient's injured nerve.

Moreover, the SLS technique has also been employed to process PLA into highly porous and customizable constructs. This method creates more robust and denser structures, beneficial for load-bearing applications such as bone tissue engineering [55].

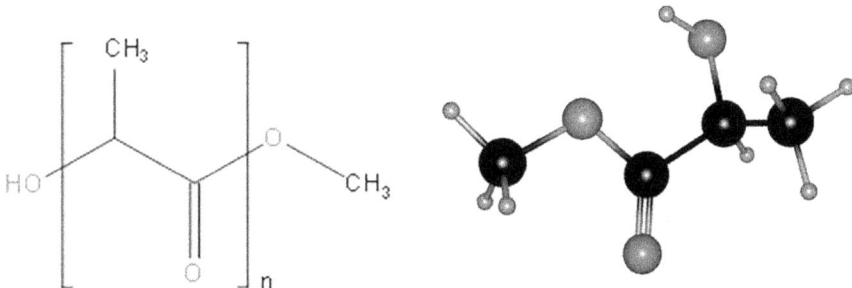

FIGURE 11.17 Chemical structure of poly (lactic acid) (PLA) [53].

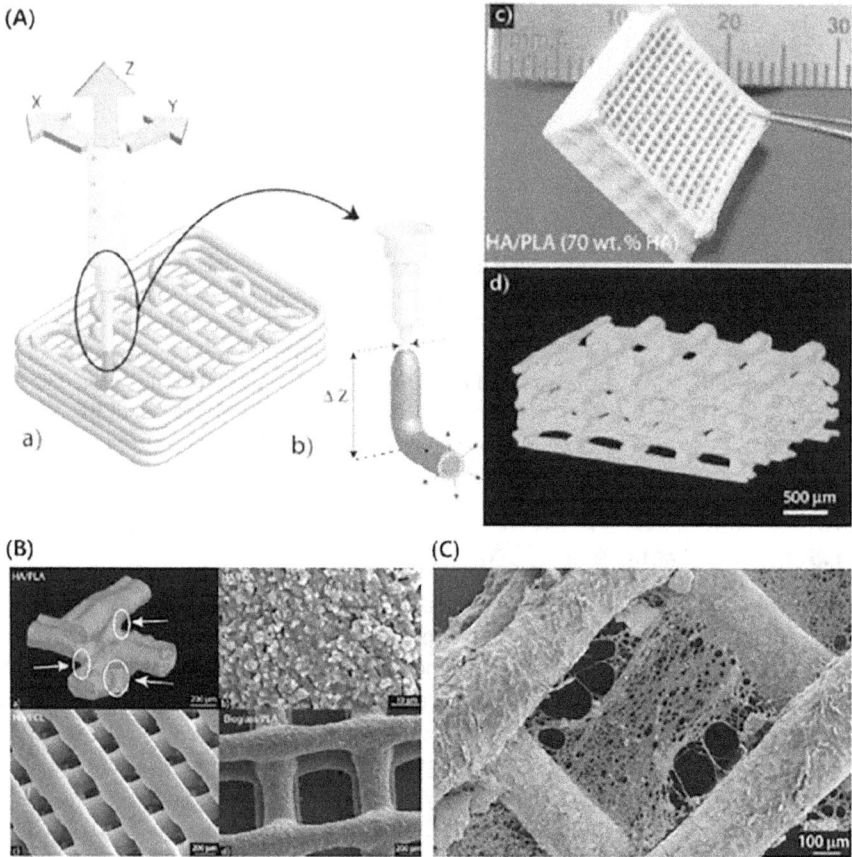

**FIGURE 11.18** (A(a)) Flow of liquid through the nozzle during robocasting. (A(b)) Formation of solid skin lines. (A(c)) A 17-layer scaffold made of PLA/HAp (70 wt-%). (A(d)) Synchrotron X-ray computed tomography reveals the intricate grid structure of PLA/ HAp (70 wt-%) scaffold. (B(a)) Synchrotron X-ray computed tomography visualizes the junction between PLA/HAp (70 wt-%) printed filaments. (B(b)) Homogeneous distribution of HAp particles in PLA/HAp (70 wt-%) printed filament. (B(c)) Homogenous filaments of PCL/ HAp (70 wt-%) and PLA/6P53B glass (70 wt-%). (C) SEM image demonstrates the surface degradation of a PCL/HAp (70 wt-%) printed scaffold after 20 days of incubation [58].

For instance, Tappa et al. [56] have developed a novel method for the fabrication of PLA-based bioresorbable vascular stents using SLS. The produced stents showed excellent mechanical properties, radial strength, and flexibility, proving the potential of this manufacturing process for the development of patient-specific vascular grafts.

The versatility of PLA as a biomaterial for additive manufacturing extends beyond FDM and SLS. For example, 3D Bioprinting has also been used to produce PLA-based bio-inks for the development of tissue engineering scaffolds. In a study by Piao et al. [57], they fabricated 3D bio-printed PLA scaffolds incorporated with graphene oxide for enhanced mechanical properties. The cell-laden structures demonstrated improved cell viability and proliferation, hinting at the feasibility of creating composite materials with PLA for tissue engineering applications (Figure 11.18).

Given its biocompatibility and biodegradability, PLA presents an attractive option for temporary implants and tissue engineering scaffolds. The application of various additive manufacturing techniques to PLA further broadens its potential, enabling the creation of complex, patient-specific constructs.

## 11.4 CONCLUSION

Additive manufacturing is widely being used in numerous industries today. One of the most significant aspects of AM is the ease of producing parts with very complex geometric designs which are generally used for biomedical or automobile applications. Through conventional mediums, it is either challenging to fabricate such parts or the process becomes uneconomical due to the high material losses associated. Biomaterials have the potential to repair or replace any injured part of the body. AM of biomaterials has the potential to make complex treatments accessible and affordable to a wider public.

Hydroxyapatite or HA is a naturally occurring mineral form of calcium apatite. The potential use of this as a replacement for bone material is being widely researched. We discussed some work on the fabrication of HA through vat polymerization techniques. It was seen that these additive-manufactured HA implants showed good biocompatibility and mechanical stability. Pores in these HA implants were seen to be important to facilitate cell growth inside the implants forming a tissue-like structure.

For load-bearing applications, Ti-based alloys are an ideal choice as they show excellent biocompatibility and have a high strength-to-weight ratio. Ti64 is one such Ti-based alloy which shows high specific strength and good corrosion properties. The fabrication of Ti64 alloy implants using laser additive manufacturing (LAM) and selective electron beam melting (SEBM) was discussed. The implants were reported to have good porosity and comparable elastic modulus to that of the human bone.

Polyetheretherketone (PEEK) and poly-lactic acid (PLA), two vital biomaterials in this sphere, have been extensively studied and utilized, showing remarkable versatility in application.PEEK is a high-performance thermoplastic with excellent mechanical and chemical resistance, making it a top choice for implants, particularly in the orthopedic sector. With additive manufacturing techniques such as FDM, SLS, and 3D bioprinting, the manufacturing of PEEK has transcended traditional boundaries, allowing for intricate, personalized constructs with incorporated bioactive agents, aimed at improving osseointegration and cellular response.

On the other hand, PLA, a biodegradable polymer, has become a cornerstone in tissue engineering and drug delivery systems. Its biocompatibility and ability to degrade into harmless by-products have proven invaluable for temporary implants. Using manufacturing methods such as FDM, SLS, and 3D bioprinting, PLA can be processed into complex, degradable constructs mirroring the patient's anatomy.

Recent developments even see PLA incorporated with other materials like graphene oxide, to enhance its properties.

In conclusion, PEEK and PLA, through additive manufacturing, demonstrate tremendous promise in improving the quality of life for patients, fostering a new era in biomaterial engineering. Their surfaces, tailor-made by precise manufacturing techniques, offer improved biocompatibility and biomechanical behavior, echoing the true potential of additive manufacturing in the realm of biomaterials.

## REFERENCES

1. Berman, B. (2012). 3-D printing: The new industrial revolution. *Business Horizons*, 55(2), 155–162. 10.1016/j.bushor.2011.11.003
2. Bose, S., Ke, D., Sahasrabudhe, H., & Bandyopadhyay, A. (2018). Additive manufacturing of biomaterials. *Progress in Materials Science*, 93, 45–111. 10.101 6/j.pmatsci.2017.08.003
3. Gebler, M., Uiterkamp, A. J. S., & Visser, C. (2014). A global sustainability perspective on 3D printing technologies. *Energy Policy*, 74, 158–167. 10.1016/ j.enpol.2014.08.033
4. Attaran, M. (2017). The rise of 3-D printing: The advantages of additive manufacturing over traditional manufacturing. *Business Horizons*, 60(5), 677–688. 10.1016/j.bushor.2017.05.011
5. Najmon, J. C., Raeisi, S., & Tovar, A. (2019). *Review of additive manufacturing technologies and applications in the aerospace industry*. Elsevier eBooks (pp. 7–31). 10.1016/b978-0-12-814062-8.00002-9
6. Sahoo, P., & Behera, A. (2022). *Plasma technology in antimicrobial nanocoatings*. Elsevier eBooks (pp. 207–219). 10.1016/b978-0-323-89930-7.00002-9
7. Biomaterials - R&D - Outsourced Research and Development - Indivenire. (2019, September 16). R&D - Outsourced Research and Development - Indivenire. https:// indiveni.re/biomaterials/
8. Zadpoor, A. A., & Malda, J. (2016a). Additive manufacturing of biomaterials, tissues, and organs. *Annals of Biomedical Engineering*, 45(1), 1–11. 10.1007/s10439-016-1719-y
9. Fei, F., Yao, H., Wang, Y., & Wei, J. (2023). Graphene oxide/RhPTH(1-34)/ polylactide composite nanofibrous scaffold for bone tissue engineering. *International Journal of Molecular Sciences*, 24(6), 5799. 10.3390/ijms24065799
10. Li, J., Chen, M., Fan, X., & Zhou, H. (2016). Recent advances in bioprinting techniques: approaches, applications and future prospects. *Journal of Translational Medicine*, 14(1). 10.1186/s12967-016-1028-0
11. Bathula I. S. R., Virupakshi, Mali H. (2017). 3D printing for foot. *MOJ Proteomics Bioinform* 5(6): 00176. 10.15406/mojpb.2017.05.00176
12. Melchels, F. P., Feijen, J., & Grijpma, D. W. (2010). A review on stereolithography and its applications in biomedical engineering. *Biomaterials*, 31(24), 6121–6130. 10.1016/j.biomaterials.2010.04.050
13. Criscenti, Giuseppe, De Maria, Carmelo, Vozzi, Giovanni, & Moroni, Lorenzo. (2018). Characterization of additive manufactured scaffolds. 10.1007/978-3-31 9-45444-3_4.
14. Yap, C. Y., Chua, C. K., Dong, Z., Liu, Z., Zhang, D., Loh, L. E., & Sing, S. L. (2015). Review of selective laser melting: Materials and applications. *Applied Physics Reviews*, 2(4), 041101. 10.1063/1.4935926

15. Hribar, Kolin, Soman, Pranav, Warner, John, Chung, Peter, & Chen, Shaochen. (2013). Light-assisted direct-write of 3D functional biomaterials. *Lab on a Chip*. 14. 10.1039/c3lc50634g.

16. Kruth, J., Leu, M., & Nakagawa, T. (1998). Progress in additive manufacturing and rapid prototyping. *CIRP Annals*, 47(2), 525–540. 10.1016/s0007-8506(07)63240-5 Kurtz, S. M., & DeVine, J. G. (2007). PEEK biomaterials in trauma, orthopedic, and spinal implants. Biomaterials, 28(32), 4845–4869. 10.1016/j.biomaterials.2007.07.013

17. Frazier, W. A. (2014). Metal additive manufacturing: A review. *Journal of Materials Engineering and Performance*, 23(6), 1917–1928. 10.1007/s11665-014-0958-z

18. Ligon, S. C., Liska, R., Stampfl, J., Gurr, M., & Mülhaupt, R. (2017). Polymers for 3D printing and customized additive manufacturing. *Chemical Reviews*, 117(15), 10212–10290. 10.1021/acs.chemrev.7b00074

19. Razavykia, A., Brusa E., Delprete C., and Yavari R. 2020. An overview of additive manufacturing technologies—A review to technical synthesis in numerical study of selective laser melting. *Materials*, 13(17), 3895. 10.3390/ma13173895

20. Murphy, S. D., & Atala, A. (2014). 3D bioprinting of tissues and organs. *Nature Biotechnology*, 32(8), 773–785. 10.1038/nbt.2958

21. Xu, T., Baicu, C. F., Aho, M., Zile, M. R., & Boland, T. (2009). Fabrication and characterization of bio-engineered cardiac pseudo tissues. *Biofabrication*, 1(3), 035001. 10.1088/1758-5082/1/3/035001

22. https://www.sigmaaldrich.com/US/en/technical-documents/technical-article/cell-culture-and-cell-culture-analysis/3d-cell-culture/3d-bioprinting-bioinks

23. Zheng, X., Deotte, J. R., Alonso, M. P., Farquar, G. R., Weisgraber, T. H., Gemberling, S., Lee, H., Fang, N. X., & Spadaccini, C. M. (2012). Design and optimization of a light-emitting diode projection micro-stereolithography three-dimensional manufacturing system. *Review of Scientific Instruments*, 83(12), 125001. 10.1063/1.4769050

24. Chae, M., Rozen, W. M., McMenamin, P. G., Findlay, M., Spychal, R. T., & Hunter-Smith, D. J. (2015). Emerging applications of bedside 3D printing in plastic surgery. *Frontiers in Surgery*, 2. 10.3389/fsurg.2015.00025

25. Zhang, J., Hu, Q., Wang, S., Tao, J., & Gou, M. (2019). Digital light processing based three-dimensional printing for medical applications. *International Journal of Bioprinting*, 6(1), 1. 10.18063/ijb.v6i1.242

26. Malinauskas, M., Žukauskas, A., Hasegawa, S., Hayasaki, Y., Mizeikis, V., Buividas, R., & Juodkazis, S. (2016). Ultrafast laser processing of materials: from science to industry. *Light-Science & Applications*, 5(8), e16133. 10.1038/lsa.2016.133

27. Tumbleston, J. R., Shirvanyants, D., Ermoshkin, N., Janusziewicz, R., Johnson, A. C., Kelly, D. J., Chen, K., Pinschmidt, R., Rolland, J. P., Ermoshkin, A., Samulski, E. T., & DeSimone, J. M. (2015). Continuous liquid interface production of 3D objects. *Science*, 347(6228), 1349–1352. 10.1126/science.aaa2397

28. Espera, A. H., Dizon, J. R. C., Chen, Q., & Advincula, R. C. (2019). 3D-printing and advanced manufacturing for electronics. *Progress in Additive Manufacturing*, 4(3), 245–267. 10.1007/s40964-019-00077-7

29. Murr, L. E., Gaytan, S. M., Martinez, E., Medina, F., & Wicker, R. B. (2012). Next generation orthopaedic implants by additive manufacturing using electron beam melting. *International Journal of Biomaterials*, 2012, 1–14. 10.1155/2012/245727

30. Ponche, R., Hascoet, J., Kerbrat, O., & Mognol, P. (2012). A new global approach to design for additive manufacturing. *Virtual and Physical Prototyping*, 7(2), 93–105. 10.1080/17452759.2012.679499

31. Zhang, L., Liu, Y., Li, S., & Hao, Y. (2017). Additive manufacturing of titanium alloys by electron beam melting: a review. *Advanced Engineering Materials*, 20(5), 1700842. 10.1002/adem.201700842

32. Singh, A., Tiwari, A., Bajpai, A. K., & Bajpai, A. K. (2018). Polymer-based antimicrobial coatings as potential biomaterials. In *Handbook of antimicrobial coatings*. 10.1016/b978-0-12-811982-2.00003-2

33. Rey, C., Combes, C., Drouet, C., & Grossin, D. (2011). *Bioactive ceramics: Physical chemistry*. Elsevier eBooks (pp. 187–221). 10.1016/b978-0-08-055294-1.00178-1

34. Junqueira, L. C. U., & Carneiro, J. (2003). Basic histology: Text & atlas.

35. Chu, T. G., Halloran, J. W., Hollister, S. J., & Feinberg, S. E. (2001). Hydroxyapatite implants with designed internal architecture. *Journal of Materials Science: Materials in Medicine*, 12(6), 471–478. 10.1023/a:1011203226053

36. Woesz, A., Rumpler, M., Stampfl, J., Varga, F., Fratzl-Zelman, N., Roschger, P., Klaushofer, K., & Fratzl, P. (2005). Towards bone replacement materials from calcium phosphates via rapid prototyping and ceramic gel casting. *Materials Science and Engineering: C*, 25(2), 181–186. 10.1016/j.msec.2005.01.014

37. Kim, J. M., Lee, J. Y., Lee, S., Park, E. K., Kim, S., & Cho, D. (2007). Development of a bone scaffold using HA nanopowder and micro-stereolithography technology. *Microelectronic Engineering*, 84(5–8), 1762–1765. 10.1016/j.mee.2007.01.204

38. Long, M., & Rack, H. J. (1998). Titanium alloys in total joint replacement—a materials science perspective. *Biomaterials*, 19(18), 1621–1639. 10.1016/s0142-9612(97)00146-4

39. Kobryn, P., & Semiatin, S. L. (2001). The laser additive manufacture of Ti-6Al-4V. *JOM*, 53(9), 40–42. 10.1007/s11837-001-0068-x

40. Heinl, P., Rottmair, A., Körner, C., & Singer, R. H. (2007). Cellular titanium by selective electron beam melting. *Advanced Engineering Materials*, 9(5), 360–364. 10.1002/adem.200700025

41. Rho, J. Y., Kuhn-Spearing, L., & Zioupos, P. (1998). Mechanical properties and the hierarchical structure of bone. *Medical Engineering & Physics*, 20(2), 92–102. 10.1016/s1350-4533(98)00007-1

42. Jiang, Z., Zhu, Z., Zhao, M., Chen, H., & Sue, H. (2022). Well-dispersed poly (ether-ether-ketone)/multi-walled carbon nanotubes nanocomposite for harsh environment applications. *Journal of Applied Polymer Science*, 139(33). 10.1002/app.52784

43. Vaezi, M., & Yang, S. (2015). Extrusion-based additive manufacturing of PEEK for biomedical applications. *Virtual and Physical Prototyping*, 10(3), 123–135. 10.1080/17452759.2015.1097053

44. Najeeb, S., Khurshid, Z., Matinlinna, J. P., Siddiqui, F. J., Nassani, M. Z., & Baroudi, K. (2015). Nanomodified peek dental implants: Bioactive composites and surface modification—A review. *International Journal of Dentistry*, 2015, 1–7. 10.1155/2015/381759

45. Boparai, K. S., Singh, R., & Singh, H. (2016). Development of rapid tooling using fused deposition modeling: a review. *Rapid Prototyping Journal*, 22(2), 281–299. 10.1108/rpj-04-2014-0048

46. Schmidt, M. P., Pohle, D., & Rechtenwald, T. (2007). Selective laser sintering of PEEK. *CIRP Annals*, 56(1), 205–208. 10.1016/j.cirp.2007.05.097

47. Jahani, B., Wang, X., & Brooks, A. (2020). Additive manufacturing techniques for fabrication of bone scaffolds for tissue engineering applications. *Recent Progress in Materials*, 2(3), 1–41. 10.21926/rpm.2003021

48. Chen, E. P., Toksoy, Z., Davis, B. H., & Geibel, J. P. (2021). 3D bioprinting of vascularized tissues for in vitro and in vivo applications. *Frontiers in Bioengineering and Biotechnology*, 9. 10.3389/fbioe.2021.664188

49. Shimizu, T., Fujibayashi, S., Yamaguchi, S., Otsuki, B., Okuzu, Y., Matsushita, T., Kokubo, T., & Matsuda, S. (2017). In vivo experimental study of anterior cervical

fusion using bioactive polyetheretherketone in a canine model. *PLoS One*, 12(9), e0184495. 10.1371/journal.pone.0184495

50. Nouri, A., Shirvan, A. R., Li, Y., & Wen, C. (2021). Additive manufacturing of metallic and polymeric load-bearing biomaterials using laser powder bed fusion: A review. *Journal of Materials Science & Technology*, 94, 196–215. 10.1016/j.jmst.2021.03.058

51. Adil, S., & Lazoglu, I. (2022). A review on additive manufacturing of carbon fiber-reinforced polymers: Current methods, materials, mechanical properties, applications and challenges. *Journal of Applied Polymer Science*, 140(7). 10.1002/app.53476

52. Farah, S., Anderson, D. G., & Langer, R. (2016). Physical and mechanical properties of PLA, and their functions in widespread applications — A comprehensive review. *Advanced Drug Delivery Reviews*, 107, 367–392. 10.1016/j.addr.2016.06.012

53. Villadiego, K. M., Tapia, M. L., Useche, J., & Macías, D. L. (2021). Thermoplastic starch (TPS)/polylactic acid (PLA) blending methodologies: A review. *Journal of Polymers and the Environment*, 30(1), 75–91. 10.1007/s10924-021-02207-1

54. Hu, Y., Wu, Y., Gou, Z., Tao, J., Zhang, J., Liu, Q., Kang, T., Jiang, S., Huang, S., He, J., Chen, S., Du, Y., & Gou, M. (2016). 3D-engineering of cellularized conduits for peripheral nerve regeneration. *Scientific Reports*, 6(1). 10.1038/srep32184

55. Wiria, F. E., Leong, K. F., Chua, C. K., & Liu, Y. (2007). Poly-ε-caprolactone/hydroxyapatite for tissue engineering scaffold fabrication via selective laser sintering. *Acta Biomaterialia*, 3(1), 1–12. 10.1016/j.actbio.2006.07.008

56. Tappa, K., Jammalamadaka, U., Ballard, D. J., Bruno, T., Israel, M. R., Vemula, H., Meacham, J. M., Mills, D. A., Woodard, P. K., & Weisman, J. A. (2017). Medication eluting devices for the field of OBGYN (MEDOBGYN): 3D printed biodegradable hormone eluting constructs, a proof of concept study. *PLoS One*, 12(8), e0182929. 10.1371/journal.pone.0182929

57. Piao, Y., You, H., Xu, T., Bei, H. P., Piwko, I. Z., Kwan, Y. H., & Zhao, X. (2021). Biomedical applications of gelatin methacryloyl hydrogels. *Engineered Regeneration*, 2, 47–56. 10.1016/j.engreg.2021.03.002

58. Kumar, A., Kargozar, S., Baino, F., & Han, S. K. (2019). Additive manufacturing methods for producing hydroxyapatite and hydroxyapatite-based composite scaffolds: A review. *Frontiers in Materials*, 6. 10.3389/fmats.2019.00313

# 12 Surface Modification

## Carbide-, Silicide-, Nitride-Based Surface

*Rojaleen Lenka and Subhasmita Swain*
Biomaterials and Tissue Regeneration Lab, CETMS, Institute
of Technical Education and Research, Siksha 'O'Anusandhan
(Deemed to be University), Bhubaneswar, Odisha, India

*Tae Yub Kwon*
Dept of Dental Biomaterials, Kyungpook National University,
Samduk-dong, Jung-gu, Daegu, Republic of Korea

*Tapash R. Rautray*
Biomaterials and Tissue Regeneration Lab, CETMS, Institute
of Technical Education and Research, Siksha 'O'Anusandhan
(Deemed to be University), Bhubaneswar, Odisha, India

## 12.1 INTRODUCTION

To build future generations of biomaterials, surface engineering is essential since the interactions between their biological surroundings and artificial materials take place at the surface [1–7]. Surface modification technologies are developing quickly [8]. Hard implants frequently experience osteointegration failure and dissociation from the host tissue as a result of poor biocompatibility [9]. Changes in topography and surface chemistry have a great impact on biocompatibility [10,10]. The addition of material, removal of material, and alteration of the existing material are the primary approaches that are performed to alter the surface properties. Plasma spraying and sol-gel are two novel techniques that can produce coatings that are a bit thicker. The optical and electronics sectors have contributed several novel technologies for producing thin film coatings through vapor deposition procedures, in which a coating substance evaporates into vapor and then condenses on the surface of the substrate [11]. Sputter-etching and glow discharge treatment are the methods for removing particles from surfaces for cleaning objectives. By using electron beam and laser treatments, surface characteristics may also be altered without the inclusion or elimination of material.

DOI: 10.1201/9781003429920-14

Also, incorporation of required substance, modifying coatings, and altering microstructures are all possible with ion implantation.

## 12.2 SURFACE MODIFICATION OF BIOMATERIALS

The bulk features of a material may first establish its acceptability for a certain use, but several biomedical devices perform effectively due to the surface morphology and chemistry. The objective of surface engineering is to enhance biocompatibility while maintaining the essential bulk characteristics of the material. The surface of a biomaterial can be modified physically and chemically. When a surface is physically altered, such as when it is etched, grit-blasted, or machined, the morphology (topography) of the surface changes, whereas the chemistry of the surface mostly remains unchanged. Atomic layer deposition, plasma-enhanced chemical vapor deposition, and electrochemical deposition are instances of well-known chemical processes [12–14]. Ion infusion, functionalization of surface, single-layer coatings, and coatings with several layers of various compositions are all possible outcomes of chemical treatment. A surface may also be oxidized, carbided, or nitrided [15,16]. To promote a positive cellular reaction in soft or hard tissue, it is necessary to change the surface of a material in a certain physical and chemical manner. When tissue integration is required, the physical surroundings contain elements on the micro, macro, and even nanoscale that permit cell adhesion, proliferation, as well as migration [17–25]. It is crucial to remember that textured surfaces might sometimes make the functionality of a device even worse, such as cardiovascular apparatus or articulating surfaces [26].

## 12.3 OBJECTIVES OF SURFACE MODIFICATION

Elements of biological systems like tissue, blood, body fluid, and bone are primarily communicating with the implant surface. Researchers are working on developing new techniques for altering surface properties, to attain the features that surface topography and surface properties influence over. The major goal of surface treatment for biomaterials is to increase the efficacy of the materials by enhancing their interfacial qualities, such as roughness, wettability, and protein-ligand adsorption [27].

Figure 12.1 illustrates the different purposes of surface engineering used to change the surface characteristics of the implants. These purposes involve changing the topography of the surface to minimize adhesion of bacterial cells, which lowers the incidence of implant malfunctions due to biofilm development, changing the surface topography as well as roughness to optimize tissue growth on the surface of the implant, improving the properties of lubrication in the implant (joint implants), altering the properties of hydrophobicity and hydrophilicity, and improving blood compatibility, etc. (Figures 12.2 and 12.3).

## 12.4 TECHNIQUES OF SURFACE MODIFICATION

The modification of the surface morphology or the binding of specific molecules or ligands to produce certain biological, chemical, or physical characteristics is

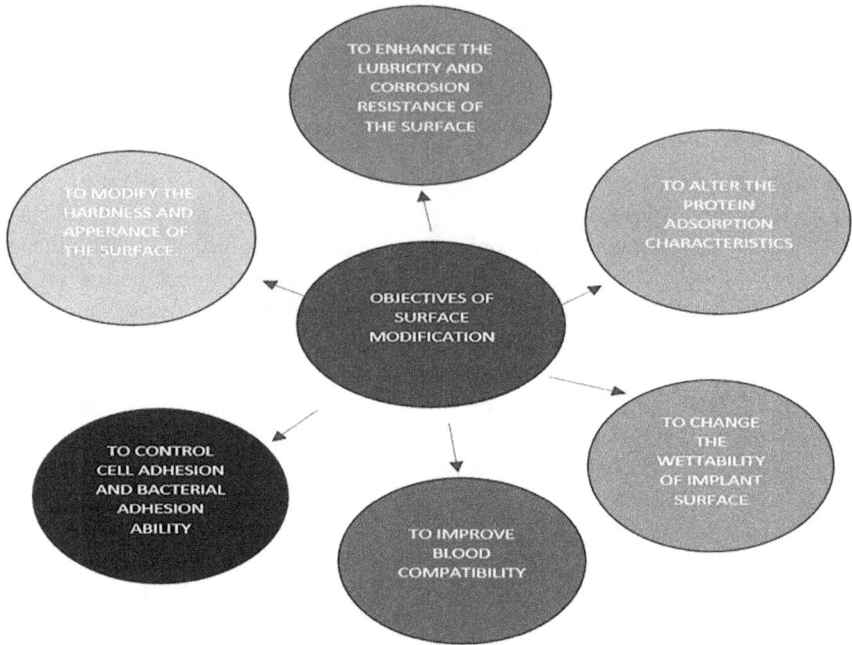

**FIGURE 12.1**  The objectives of surface modification.

associated with surface modification. Some of the key surface characteristics of implants utilized in the application of the biomedical field include cell adhesion capability, protein adsorption capacity, biomimetics, biocompatibility, hydrophobicity or hydrophilicity, biodegradation, etc. According to Alam et al., modification of any surface can be accomplished by layering (coating) the surface or by altering the molecules or atoms of the surface physically or chemically [27]. Several surface engineering methods are now being used to improve surface characteristics (by treating the untreated substrate surface through surface grafting, roughening, patterning, single-layer coating, multilayer coating, etc.). These approaches may be roughly divided into three classes.

## 12.4.1  MECHANICAL SURFACE MODIFICATION METHOD

The most popular methods for surface treatment are mechanical procedures that are being utilized to improve surface characteristics by modifying the morphology and roughness of the surface. Micro-machining, polishing, sand blasting, and grinding are some mechanical surface engineering techniques that entail physical modification, shaping, or elimination of materials from the surface [28]. Obtaining certain surface properties, removing surface contaminants, as well as enhancing surface adhesion are the basic goals of mechanical modification.

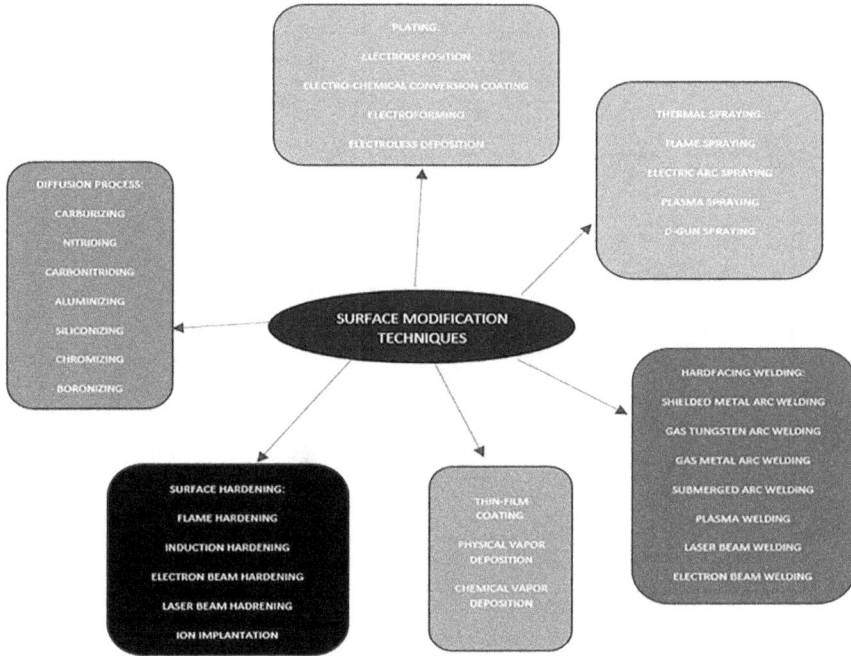

**FIGURE 12.2** Techniques of surface modification.

## 12.4.2 Physicochemical Surface Modification

Key factors that affect osteoblast attachment comprise physicochemical properties including surface charge, surface free energy, surface wettability, and chemical composition. Manufacturing procedures are the main means of changing physicochemical characteristics. Sol-gel, acid etching, alkaline treatment, anodic oxidation, chemical vapor deposition (CVD), electrochemical treatment, and biochemical modification, are a few of the chemical techniques. At the surface between the solution and biomaterial, processes such as electrochemical, biochemical, or chemical changes take place throughout the electrochemical treatment, biochemical treatment, and chemical treatment, respectively [28].

## 12.4.3 Physical Surface Modification

Certain surface engineering techniques, such as physical vapor deposition and thermal spraying, do not involve any chemical interactions. In this instance, kinetic, electrical, and thermal energy is primarily liable for the development of the coatings, modified layers, or films on the surface of biomaterials [28]. Thermal spray, ion implantation, glow discharge plasma treatment, physical vapor deposition (PVD), etc. are included in this technique.

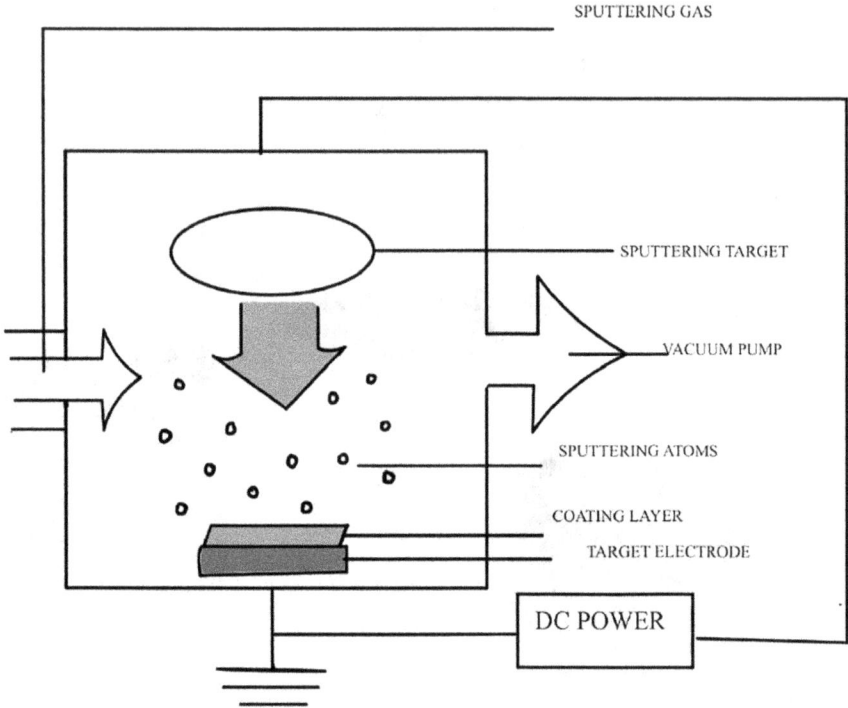

**FIGURE 12.3** Showing schematic representation of PVD.

## 12.5 BRIEF DISCUSSION ON CARBIDE, SILICIDE, NITRIDE

Carbon (C) is a chemical inorganic element having an atomic number (Z) "6" and belongs to the 14th group of the periodic table. This nonmetallic element makes covalent bonding due to its tetravalency property. When chemical bonds are formed between carbon and less electronegative elements, it results in the formation of carbides. Several bonding types occur between the constituent components based on their varying valence states and electronegativity. Thus, they are categorized as covalent compounds such as $B_4C$ and SiC, interstitial compounds that are formed with fourth, fifth, and sixth groups of transition metals (except chromium), intermediate transition metal carbides (composition of carbide ion and transition metal), and salt like compounds, where carbon behaves as a pure anion while the remaining elements behave as adequately electropositive. Carbides are generally refractory, have high melting points, and exhibit metallic characteristics. The constituent elements determine their distinct characteristics [29]. Carbides can be created from intricate iron-based mixtures either through the solid-state precipitation while cooling or directly through the solidification process. It is noticed that lower chances of re-dissolution are associated with simpler crystalline structures [30]. Iron carbide (cementite) is created when a combination of iron and carbon solidifies from the melt at certain temperatures. As powerful carbide precursors, alloying elements such as chromium, vanadium, tungsten, and molybdenum raise the wear resistance and hardness of the steel. On the other hand, silicon encourages the precipitation of graphite, which lowers the production of carbides.

Silicon (Si), having atomic number 14, is a nonmetallic element belonging to the carbon family, i.e., group IV-A (14th group) of the periodic table, and is found abundantly (27.7%) in the Earth's crust. A silicide is a sort of compound that is formed through the chemical bonding of silicon and an element that is often highly electropositive. Relative to carbon, silicon is highly electropositive and from a structural perspective, silicides resemble borides more than carbides. Based on the electronegativity of the constituent elements, the bonds in silicides can be mostly ionic or covalent. When compared to several alloys at high temperatures, silicides show superior stiffness and strength. $TiSi_2$, $Ti5Si_3$, and $MoSi_2$ are some of the commonly used silicides [31]. Transition metal silicides constitute a huge class of refractory materials that are used for various applications such as thin film coatings, electrical heating elements, thermoelectric, photovoltaics, and CMOS devices, etc.

Nitrogen (N), having atomic number 7, is a nonmetal, which belongs to group 15 [V-A] of the periodic table. This odorless, colorless, and flavorless gas is a component of all living organisms and is found abundantly in the atmosphere. When nitrogen is combined with other elements that have lower electronegativity as compared to nitrogen, nitrides are formed (the oxidation state of nitrogen is −3, i.e., $N^{3-}$). Nitrides are categorized into three classes: ionic ($Mg_3N_2$, $Be_3N_2$), covalent (BN, $P_3N_5$), and interstitial ($V_2(SO_4)_3$). Nitrides can be prepared using one of the two main techniques. One process involves the elements reacting directly (often at high temperatures), while the other involves a metal amide thermally decomposing, which results in the release of ammonia. In addition, nitrides are synthesized when ammonia is treated to temperatures generally between 1,050°F for 5–10 hours, (based on the required depth of the toughened structure) and in another process by reducing a metal oxide or halide in the availability of nitrogen gas. Some metal nitrides are not stable and produce ammonia, and metal hydroxide or metal oxide by reacting with water. Titanium, boron, vanadium, tantalum, and silicon nitrides are extremely hard, refractory, and resistant to chemicals. Hence, they are used as effective abrasives as well as for the fabrication of crucibles. Several forms of nitride coatings, such as TiN (Titanium nitride) and TiNbN (Titanium niobium nitride), were produced by the physical vapor deposition method to enhance the wear properties of knee or hip transplant bearings [32].

## 12.6   SURFACE TREATMENT: CARBIDE-BASED SURFACE

Some widely used carbides are described below.

### 12.6.1   TITANIUM CARBIDE (TiC)

This transition metal carbide exhibits a face-centered cubic (FCC) crystal structure (similar to NaCl structure). TiC is extremely hard metallic ceramic (hardness value > 10 GPa), having an elastic modulus of 400 GPa and a shear modulus of 188 GPa. With low density, it shows properties like good thermal conductivity, chemical stability, and wear resistance. TiC coatings are useful in tribological applications due to their shielding ability against oxidation, corrosion, wear, and abrasion [33].

## 12.6.2  SILICON CARBIDE (SiC)

The chemical combination of silicon and carbon gives rise to silicon carbide (carborundum). For more than a century, silicon carbide has been manufactured and used in the production of abrasive materials. Currently, premium grade technical class SiC ceramic featuring excellent mechanical characteristics has been manufactured. Being a covalent solid, its alpha form exists as a hexagonal crystal structure whereas its beta form has a face-centered cubic structure. This material possesses superior wear resistance, excellent chemical inertness, high hardness (25 Gpa), low density (3.21 gm/cm$^3$), good tensile strength (240 Ksi), and high elastic modulus (440 Gpa). SiC is employed in coatings for coronary heart stents, BMI devices, bone prosthetics, and dental applications [34].

## 12.6.3  NIOBIUM CARBIDE (NbC)

This refractory ceramic material possesses high hardness, superior ultimate strength (244 Mpa), high elastic modulus (330–537 Gpa), and high compressive strength (2374 Mpa). NbC has excellent biocompatibility, the capability to develop strong associations with metallic surfaces, as well as anticorrosive characteristics [35]. All these properties make NbC a suitable coating biomaterial.

An overview of some experiments performed by researchers on carbide coatings follows.

In a unique layering technique based on IPPA (Ion Plating Plasma Assisted) technology, Longo et al. [36] coated titanium implants using hard and thin nanostructured layer made of titanium oxide and titanium carbide arranged in such a way that surrounds the graphitic carbon. This layer serves two functions: it shields the titanium surface from the extreme circumstances of living tissue and, simultaneously, stimulates activity on osteoblasts. So, a robust nanostructured TiC (titanium carbide) coating with good morphological, mechanical, and chemical characteristics (synthesized by IPPA treatment) is coated to titanium surfaces, producing superior *in vitro* biological outcomes. The osseointegration process is accelerated by such altered surface and thus promotes osteoblastic activity, cell adhesion, and proliferation. The distinct mechanical and physical characteristics of the coating can shield the titanium surface from the abrasive impacts of body fluids and this in turn will enhance biocompatibility. Lastly, the coating cycle may be implemented on any commercial procedure without altering the blueprints since the IPPA deposition process does not involve excessively costly equipment. All these crucial characteristics suggest that the IPPA deposition with the TiC coating can prove to be a significant move in the creation of bone and dental implants. The probability of implant failure must be minimized if the titanium implant is shielded by a robust nanostructured TiC layer. Furthermore, the components of the coating promote osteoblastic activities and this is an indication of effective and successful osseointegration on the implant surface. Also, this layer will be the appropriate base for other stimulatory substances due to its lack of toxicity and exceptional qualities, which will have a significant impact on cell proliferation and bone growth.

Algodi et al. [37] have studied the electrical discharge coating (EDC) technique as it is employed to produce TiC-Fe coatings on 304stainless steel, as a function of rising current (from 2 to 19 Ampere) and pulse-on time 2 to 64 s. By using methods of EDS (energy dispersive spectroscopy), cross-sectional TEM (transmission electron microscopy), XRD (X-ray diffractometry), and SEM (scanning electron microscopy) coating morphology made up of a combination of TiC, amorphous carbon, ɑ-Fe, and γ-Fe, was analyzed. Relying on the distribution and the content of TiC (titanium carbide) nanoparticles inside the matrix of Fe, the produced coatings displayed varying values of hardness that may exhibit greater order of magnitude than the substrate. The quantity of TiC introduced into the coatings varied, as evidenced by the fact that the hardness of coatings was observed to rise with rising current but fall under circumstances of prolonged pulse-on durations. Having high wear resistance, excellent thermal and chemical stability, low coefficient of friction and high hardness TiC ceramic coating can be used as a protective layer to enhance surface morphology.

An experiment conducted by Prakash et al. [38] demonstrates a unique method for surface modification of beta titanium (β-Ti) implants utilizing SMP-ESA (silicon powder mixed electro spark alloying) for orthopedic treatments. At a pulse current of 15 Ampere, a larger pulse duration with powder of Si particles (having a concentration of 8 g/l), a dense coating of bioceramic carbides and oxides was developed on the surface. It was verified by analyzing EDS (energy dispersive X-ray spectroscopy) that SMP-ESA produced oxide and carbide loaded surface, which provided suitable surface features to promote the biocompatibility of β-Ti alloy. Coating the surface with SPM-ESA has boosted its microhardness by two times as compared to the original (unmodified) surface. Also, the tribological studies found that the SPM-ESA enhanced layer provides superior wear resistance and great friction minimization properties as compared to unaltered Ti specimens. TiC, NbC, SiC $Nb_2O_5$, $TiO_2$, $SiO_2$, and $ZrO_2$ coatings (composite layers of bioceramic carbides and oxides) on the implant surface provide superior bioactivity as well as corrosion resistance.

By using PLD (pulsed laser deposition) technology, Brama et al. [39] carried out an experiment demonstrating a coating method to yield titanium carbide (TiC) layers on titanium surface and assess the biological responses (*in vivo* and *in vitro*). According to X-ray photoelectron spectroscopy (XPS) study, the surface level comprises 18.6–21.5% of TiC along with titanium oxides (Ti2O3- 11.1–13.0%, TiO2- 50.8–55.8%, TiO- 14.5–14.7%). The gene expressions for alkaline phosphatase, osteocalcin, A2 pro-collagen type 1, BMP-4, Cbfa-1, and TGF β were stimulated in primary human osteoblasts, ROS.MER#14, and hFOB1.19 cell lines, grown on TiC, especially in comparison with non-layered titanium surface (measured through real-time PCR and quantitative PCR). While M-CSF and IL-6, which regulate osteoclastic behavior and osteoclastogenesis, remained unchanged. In comparison to untreated titanium, bone density surrounding TiC-coated implant was shown to be higher (after the second and fourth weeks in sheep, and fourth and eighth weeks in rabbits). Using intravital staining, accelerated bone growth was visible in the experimental rabbit model after two weeks. It is confirmed from this experiment that TiC coating can enhance osseointegration, hardness, and biocompatibility of the implant as compared to untreated surfaces.

Bio-SiC (Bio-inspired silicon carbide) might be regarded as a promising material due to its favorable mechanical, chemical, and biological characteristics as well as its distinctive interconnected porosity. This ceramic has the potential for use in biomedical applications, such as tissue engineering scaffolds and regulated drug delivery systems. In 2010, López-Álvarez et al. [40] developed this novel substance by infiltrating carbon templates with molten silicon after carefully pyrolyzing vegetable precursors. The resulting SiC ceramic has a porous, interconnecting microstructure that resembles the structure of natural bone tissue and this promotes angiogenesis along with internal growth of bone tissue. The *in vitro* study of the prepared bio-SiC ceramics (obtained from the sapelli tree, *Entandrophragma cylindricum*) was assessed. Scanning electron microscopy, confocal laser scanning microscopy, interferometric profilometry, red alizarin staining, and MTT assay were used to examine the cytoskeleton organization, mineralization, adherence, migration, and proliferation of the preosteoblastic cell line MC3T3-E1 for up to seven weeks of culture. The preosteoblastic shape of the cells was maintained during the growth period and the cells could appropriately adhere, migrate, and multiply rapidly. Also, the surface of bio-SiC showed a considerable degree of mineralization.

By coupling hydrophilicity and appropriate surface topography, a biomaterial can stimulate osteogenesis. To examine cell activity, in 2021, by using an osteoblast model (*in vitro* study), Ghezzi et al. [41] synthesized SiC (silicon carbide) as well as SiC/SiOx (core silicon carbide/silicon dioxide nanowires) with variable hydrophilicity. Initially, on silicon substrates, CVD (chemical vapor deposition) was used to produce SiC along with SiC/SiOx nanowires. After that, by chemically etching the core/shell nanowires, SiC nanowires (cubic silicon carbide nanowires) were produced. Later, self-made equipment (Lintes Labs) was used to change their hydrophilicity via LPPE (low-pressure plasma enhanced) treatment in hydrogen with RF-ICP (plasma inductive radio frequency). The contact angle and morphology of the four different kinds of nanowires, as well as their activities were assessed in the environment of MC3T3-E1 murine osteoblast cells. The development of focal adhesions of the cells, cell morphology, viability, metabolic activity, etc. was studied. SiC and SiC/SiOx nanowires have various dimensions, according to morphological data. Before the hydrogen plasma treatment, SiC nanowires had a very low contact angle, which was eliminated following the treatment. For each sample, osteoblastic cells seemed to be in an excellent state. Surprisingly, SEM (scanning electron microscopy) data showed that both hydrophilic SiC/SiOx nanowires and SiC nanowires produced desirable dispersion of focal adhesions all around the cell surface. Moreover, osteoblasts developed on SiC/SiOx nanowires treated with hydrogen plasma showed a rise in the cell count, which indicates greater metabolic activity. In summary, hydrogen plasma processing of SiC/SiOx nanowires and SiC nanowires improves the cellular adhesion of osteoblasts by enhancing the wettability of the nanowires.

In 2021, Hu et al. [42] experimented by combining UV (ultraviolet) light cross-linking and semi-IPN (semi-interpenetrating network) technology mechanisms to synthesize composite hydrogels based on salecan/poly (DEAEMA-co-HEAA), i.e., poly(2-(diethylamino)ethyl methacrylate-co-N-hydroxyethyl acrylamide) network incorporated with NbC(niobium carbide) nanoparticles. By adjusting the salecan concentration, the swelling and morphology of the hydrogels could be precisely controlled. The patterns shown by photo-cross-linked composite hydrogels indicated peaks of XRD diffraction associated with NbC. Results from thermal analysis and compressive strength test, distinctively showed that the incorporation of NbC nanoparticles into the network structure improved thermal stability and mechanical strength.

## 12.7 SURFACE TREATMENT: SILICIDE-BASED SURFACE

Some widely used silicides are discussed below.

### 12.7.1 TITANIUM SILICIDE

Having orthorhombic C54 and complex hexagonal $D_8$ lattice structures, the refractory titanium silicides ($TiSi_2$ and $Ti_5Si_3$) possess excellent mechanical and physical characteristics, like elastic stiffness, flow stresses, high hardness, low densities, and exceptional oxidation and creep resistance [43]. The titanium silicide layer significantly boosts the oxidation and wear resistance of the implant surface.

### 12.7.2 CALCIUM SILICIDE

Calcium mono and di silicide ($CaSi$, $Casi_2$) having low densities and high melting points exhibit orthorhombic and trigonal crystal structures respectively. Buga et al. [44] showed that for orthopedic and dental implant procedures, the electro-sprayed CaSi nanocrystal coating prepared at an annealing temperature of 750°C can be served as a promising coating as it exhibits enhanced antibacterial properties and osteogenic activities.

An overview of some experiments performed by researchers on silicide coatings follows.

In 2011, Antonova et al. [45] developed borate-silicate glasses. They were coated using an enameling technique on a titanium surface. An intermediate layer made of titanium silicide ($Ti_5Si_3$) was used to strengthen the adhesion between the glass coating and the substrate. It was discovered that $Ti_5Si_3$ was formed as a consequence of a reaction between Si and metallic Ti that began at a temperature over 800°C. Depending on the glass content, the coating composite system showed rupture tensile strength ranging from 15 to 20 MPa.

In an experiment conducted in 2021 by Buga et al. [44], a layer of calcium silicate (CaSi) nanostructure having both bioactive and antibacterial properties was developed by electrospray deposition on a titanium surface. The adhesive strength of CaSi coating was then enhanced by annealing at temperatures of 700°C, 750°C, and 800°C, respectively. Investigations were carried out on the phase composition, bonding strength, and microstructure of the CaSi coating layers. The antibacterial efficacy and osteogenicity of the coatings were examined using gram-negative *Escherichia coli* (*E. coli*), gram-positive *Staphylococcus aureus* (*S. aureus*) species, and human mesenchymal stem cells (hMSCs). According to experimental findings, the prepared CaSi layer was primarily composed of a phase of dicalcium silicate having a particle size of 300 nm. Upon annealing, the elemental depth profile and the cross-sectional morphology clearly showed that the width of the oxidation layer grew from 0.3 μm to 1 μm with temperature rise. Notably, the adhesive strength of the annealed coating (at 750°C) was 19 MPa, i.e., more than 15 MPa, which meets the minimum requirement of ISO 13,779 standard. For orthopedic and dental implant procedures, the electro-sprayed CaSi nanocrystal coating prepared at an annealing temperature of 750°C can serve as a promising coating in terms of enhanced antibacterial properties and osteogenic activities.

On Ti-6Al-4V alloy, both pure silicide coating layers and Y-Ce incorporated silicide coating layers were developed by Tian et al. [46], using the pack cementation procedure. Comparative studies were conducted on the coating structures and wear behaviors. The findings demonstrated that the $Ti_5Si_4$ inner layers, TiSi middle layers, and $TiSi_2$ outer layers were the primary components of the coating structures for both silicide coating and the Y-Ce augmented silicide coating (both are synthesized at 1,080°C for 4 h). Because of the favorable impacts of Ce and Y on the durability (wear resistance) of silicide coating, the wear tests showed that the mass reduction of the unadulterated silicide coating was approximately two times greater than that of the Y-Ce augmented silicide coating.

Ma et al. [47] experimented to inhibit post-surgical skin tumor reappearance and to heal skin wounds. Hollow nanospheres of glucose oxidase-blended manganese silicate augmented with alginate hydrogel were developed for skin tissue regeneration and starvation-photothermal treatment. With a significant photothermal effect, these nanospheres demonstrated a photothermal conversion efficacy of 38.5% in composite hydrogels. Hydrogen peroxide ($H_2O_2$) was degraded to release oxygen ($O_2$) by the action of the composite hydrogels (by availing the catalytic property of Mn ions). This helped with the issue of tumor hypoxia in the microenvironment and provided glucose oxidase the ability to absorb glucose in the oxygen-rich surroundings thus starvation of the tumor took place. Along with accelerating the rate of $H_2O_2$ breakdown by these hollow nanospheres and glucose intake by glucose oxidase, near-infrared-induced hyperthermia can also kill tumor cells. The anti-tumor data demonstrated that the combination of starvation and photothermal therapy caused the maximum rate of tumor cell destruction across all groups, and the anti-tumor effect shown by this combination was superior to that of either starvation alone or photothermal therapy alone. It is interesting to note that, in comparison to hydrogels without manganese-silicate hollow nanospheres, the addition of manganese-silicate hollow nanospheres might significantly stimulate the epithelialization of the lesion by delivering Mn ions. In the future, it is anticipated that a multipurpose system that includes starvation-photothermal treatment will be beneficial for curing skin abnormalities originating from tumors in association with its regenerative bioactivity.

## 12.8   SURFACE TREATMENT: NITRIDE-BASED SURFACE

Some widely used nitrides are discussed below.

### 12.8.1   CARBON NITRIDE (CN)

Having low density (2.336 g/cm$^3$), CN has a graphite-like structure. Nanostructures manufactured using graphitic carbon nitrides are promising substances with considerable surface area, exceptional chemical and thermal resilience, and outstanding band structure. They also have distinctive structural, compositional, electronic, and optical features. These characteristics have led to the development of graphitic-CN-based nanomaterials with potential biomedical implementation and improved productivity [48].

### 12.8.2   SILICON NITRIDE

The chemical combination of nitrogen and silicon is termed silicon nitride. Out of all the silicon nitrides, $Si_3N_4$ is commercially significant and the most thermally stable one. $Si_3N_4$ is categorized into trigonal ($\alpha$-$Si_3N_4$), hexagonal ($\beta$-$Si_3N_4$), and cubic ($\gamma$-$Si_3N_4$) crystallographic structures. $Si_3N_4$ is a very hard and chemically inert ceramic having a high melting point, and great thermal stability. This ceramic possesses high flexural and compressive strength, reduced friction coefficient, superior corrosion resistance, improved osteoconductivity and biocompatibility, excellent fracture toughness, superior antibacterial activity, and microengineering capabilities that enable the integration of computing, optical, and chemical aspects with mechanics and electronics. Having these excellent characteristics this material is used in designing of dental implants, scaffolds, antiviral and antibacterial coatings, knee and hip endoprosthesis grafts, photonic ICs, micro-spectroscopic imaging devices, waveguides (for medical diagnosis), and micro-tubes (for intelligent neural circuits) [49]. Recently, it was found that silicon nitride can deactivate the ss-RNA (single-stranded RNA) along with the severe acute respiratory syndrome coronavirus 2 (SARS-CoV-2) [50].

### 12.8.3   TITANIUM NITRIDE

Extremely hard titanium nitride (TiN) ceramic having a density of 5.48 gm/cm$^3$ exhibits a face-centered-cubic crystal structure. It is strong, durable, heat and wear-resistant, and possesses high mechanical strength, chemical stability, as well as a high melting point [51–58]. Because of its superior corrosion and wear resistance, biocompatibility, and chemical stability titanium nitride (TiN) films have become a common protective layer for implants [59].

### 12.8.4   OVERVIEW OF SOME EXPERIMENTS PERFORMED BY RESEARCHERS ON NITRIDE COATINGS

Guo et al. [60], in 2023, conducted an experiment that holds a special opportunity for bone grafts with antibacterial and osteogenic properties. Utilizing MAO (micro-arc oxidation) technique, coating layers of $Si_3N_4$ (silicon nitride) with $TiO_2$ (titanium dioxide) were produced on the Ti implant surface by spreading nanoparticles of $Si_3N_4$ in various concentrations inside the electrolyte. When micro-arc oxidized silicon nitride-coated Ti implant was compared with pure Ti, relevant changes took place in surface properties like hydrophilicity, surface energy, and nanotopography and these changes were dependent on the concentration of the silicon nitride nanoparticles. It was noticed that with an increasing concentration of $Si_3N_4$ nanoparticles, the enhancement in cellular response like osteoblastic differentiation and proliferation took place. Significant improvement in antibacterial properties was also observed. Hence, the introduction of $Si_3N_4$ into the micro-arc oxidized coating showed excellent antibacterial as well as osteogenic effects.

Mishra et al. [61] demonstrated the implementation of gallium nitride (GaN) surface by using a three-step procedure to improve cell adhesion and its productive use *in vitro* and *in vivo* applications. Being chemically stable, GaN offers promising potential in biomedical research. It has displayed biocompatibility, and there are several ways to modify its natural surface features to encourage effective biomolecule attachment with specific functions. Sodium hydroxide along with 3-aminopropyltriethoxysilane, followed by an amalgamation of T1HC (type 1-human collagen) was used to alter the surface chemistry and morphological features of the GaN substrate surface. Chemical and morphological changes along with changes in aqueous stability, wettability, and pH were then examined. Using HPdlF (human periodontal ligament fibroblast) further studies of cell culture were carried out. On the functionalized surface of GaN, the cell adhesion and proliferation were conducted and then assessed by using SEM and MTT assay analysis at various intervals of time. The adhesion of human collagen and silane binding agent was noticed to be promoted by protonated amines and OH⁻groups. The modified surfaces showed superior cell viability and cellular adhesion (near about 95%). It was revealed that biochemically tailored GaN substrate surfaces having excellent soft tissue amalgamation can be employed in the field of biomedical as well as in dental applications.

Chukwuike et al. [62] carried out an experiment by sputtering TiBN (titanium boron nitride)-thin films on SS316L surfaces (by using pulsed DC magnetron sputtering) and their biocompatibility and microstructures were analyzed. XRD analysis revealed the TiN-containing phases only. Ion sputtering spectroscopy (ISS), high-resolution transmission electron microscopy (HRTEM), and X-ray photoelectron spectroscopy (XPS) indicated the existence of boron. Block on ring wear tester and nanoindentation were used to characterize the mechanical characteristics of the produced thin films and it was revealed that these modified surfaces possessed coefficient of friction and hardness of 0.27 and 13 GPa, respectively. From the biocompatibility test (by utilizing hemocompatibility studies, cytotoxicity assay, and mineralization assay), the non-thrombogenic and non-cytotoxic nature of the produced films were assessed. Using MC3T3-E1 cells, it was observed that the mineralization of calcium was eased by the coated films. As compared to the unmodified surface, the TiBN-coated surface showed superior corrosion behavior.

Fu et al. [63], in 2023, developed a cost-effective and simple nitriding technique to improve the antibacterial efficacy of Ti-based alloy surfaces. This nitriding technique was implemented through thermal treatment and wet chemical etching under well-regulated circumstances and analyzed by using SEM, GIXRD, and XPS. These results demonstrated that this modification technique had a substantial impact on the morphology and phase composition of the newly generated phases on the Ti surfaces. Crystalline $TiO_2$ and TiN were produced and amorphous oxynitrides and nitrides were observed on the substrate surface. In comparison to the untreated Ti surface, the modified surface exhibited inhibitory effects against bacterial infection. Due to the release of the ammonia group from the treated Ti surface, this inhibitory response was noticed. Nitriding of the Ti implant surface using the thermal treatment and wet chemical etching has the potential to improve the antibacterial properties in orthopedic and dental applications.

Starosvetsky et al. [64], in 2001, conducted an experiment by modifying Ni-Ti (nitinol) shape memory alloy surface by using the PIRAC (powder immersion reaction assisted coating) nitriding technique (grounded based on annealing of the material in the environment of extremely reactive atmospheric nitrogen). Following annealing at 900°C for 1.5 h or annealing at 1,000°C for 1 h, the modified surface was found to be composed of an upper thin TiN layer and below this a denser $Ti_2Ni$ layer. The treated surface exhibited high hardness, superior wear resistance, and excellent corrosion resistance in Ringer's solution.

## 12.9 SOME COMMONLY USED COATING TECHNIQUES TO MODIFY THE REQUIRED BIOMATERIAL SURFACE

### 12.9.1 PHYSICAL VAPOR DEPOSITION

Physical vapor deposition (PVD) procedures involve the vaporization of atoms or molecules of material from a liquid or solid source, transport of the vapor via a gaseous environment having low pressure, or via vacuum and condensation on the surface of a substrate. Various PVD techniques are employed for film deposition of elemental, compound, and alloy materials along with some polymeric substances [65]. Ion plating, sputtering, electron beam sputtering, and magnetron sputtering are various types of physical vapor deposition. Material coating layers may vary in thickness from angstroms to millimeters during the sputtering technique [66]. In the initial stage of PVD, atoms are ejected from the target object by using an ion source having high energy in a vacuum and the presence of an inert gas (usually Argon). The target object is supplied with a source having high energy; hence, vaporization of atoms takes place from the surface of the target material. The atoms are then deposited in the chamber after moving towards the surface of the material. When a metal oxide, nitride, or carbide is deposited, the reaction takes place. When the vaporized atoms hit the surface of the substrate, they build up a thin deposition layer there. On the surface of the substrate, the PVD process creates a homogeneous coating layer that varies in size from nano to visible scale. By using heterogeneous nucleation, a compact layer is formed with superior mechanical qualities including wear resistance and hardness as less resistance is generated during the PVD process [66].

## 12.9.2 Chemical Vapor Deposition

Another method for producing films is chemical vapor deposition (CVD). PVD and CVD vary in that the target material in PVD is solid, while the target material in CVD is in a gaseous state. A chemical reaction that takes place on or in proximity to a typically heated surface produces a solid deposition from vapor in the process of CVD. The resultant solid substance might be a single crystal, powder, or thin layer. Materials with a variety of chemical, physical, and tribological characteristics can be developed by modifying experimental circumstances, such as substrate material, total pressure gas flow, reaction gas mixture composition, substrate temperature, etc. The possibility of selective or localized deposition on patterned surfaces is another crucial aspect. The CVD process stands out for its superior throwing ability, which makes it possible to produce coatings with minimal porosity and homogeneous thickness, even on surfaces with complex geometry. The possibility of selective or localized deposition on patterned surfaces is another crucial aspect. Highly pure, improved strength and dense materials are generated using the CVD technique. In the semiconductor and ceramic industries, it is utilized to make material used in semiconductors. This technique is heat sensitive and is used to produce nanomaterials when a metal catalyst is involved [66]. Different kinds of CVD include APCVD (atmospheric pressure CVD), UHVCVD (ultrahigh vacuum CVD), LPCVD (low-pressure CVD), DLICVD (direct liquid injection CVD), AACVD (aerosol-assisted CVD), MPCVD (microwave plasma-assisted CVD), atomic layer CVD ALCVD (atomic layer CVD), RPECVD (remote plasma-enhanced CVD), MOCVD (metal-organic chemical vapor deposition), hot wire CVD HWCVD (hot wire CVD), HPCVD (hybrid physical–chemical vapor deposition), VPE (vapor-phase epitaxy), and RTCVD (rapid thermal CVD). The deposition begins to occur and is sustained by heat in TACVD (thermally activated CVD). However, the induction and maintenance of CVD processes may be aided by electrons, photons, ions, or a mixture of these (plasma-triggered CVD)[67]. Reactants are introduced into a reactor in all the CVD techniques. The following chemical reaction takes place on or in the proximity of a heated surface:

*Gaseous reactants → Solid material + Gaseous by − products*

Figure 12.4 illustrates the five crucial reaction phases that form throughout the CVD with temperature and gas fluxes. The interactions taking place in the reaction domains have an impact on the characteristics of CVD materials. The surface of the coating is traversed by the flow of the reaction gas combination during CVD. A boundary layer (stagnant in nature) forms in the vapor next to the coating as a result of the fluid dynamics. The gaseous reactants and the products are moved through the layer of this boundary amid the deposition procedure. Uniform reactions may take place in the vapor in both the reaction gas mixture (main gas stream) and in zone 1. This might result in an unfavorable uniform (homogenous) nucleation that has a non-adherent and flaky covering. However, in other circumstances, when they are not associated with uniform nucleation, these reactions are advantageous to the CVD techniques (CVD of $B_{13}C_2$, Si, and $Al_2O_3$). In zone 2 (the phase boundary

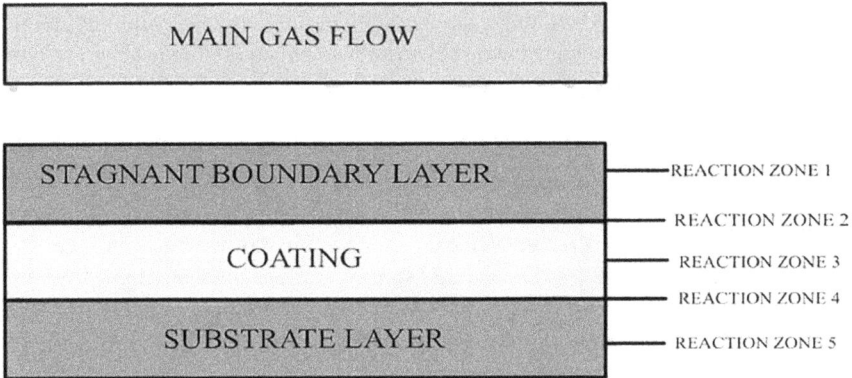

FIGURE 12.4   Depicting reaction zones involved in CVD.

vapor/coating), heterogeneous reactions take place. The rate of deposition and the characteristics of coatings are generally determined by these processes. The solid-state processes that can occur in zones 3–5 of the CVD process include precipitation, phase transitions, recrystallization, and grain growth (due to the high-temperature treatment during CVD). Several intermediary phases occur in zone 4 (diffusion zone). Reactions occurring in this region are crucial as they impact the ability of the coating to adhere to the surface of the substrate (Figure 12.5).

## 12.9.3 THERMAL SPRAYING

To shield the surface, to enhance the surface characteristics, and to meet the precise surface specifications thermal spraying is an effective technique. This process involves heating the feedstock material to a molten or semi-molten state, then

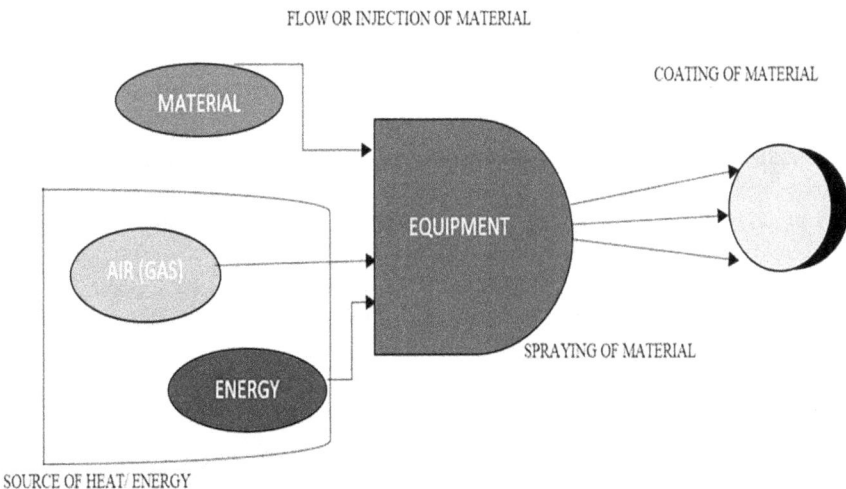

FIGURE 12.5   Shows a schematic presentation of thermal spraying.

accelerating it to a rapid speed using an energetic motion, and then spraying it onto the suitable surface of the substrate [68]. Major applications of thermal spraying include corrosion protection, restoration, and repair; various forms of wear such as erosion, abrasion, and scuffing; oxidation; heat conduction or insulation; electrical conductors or insulators; abradable coatings; near-net-shape manufacturing; decorative purpose; engineered emissivity; and more. Nearly all sectors may benefit from the employment of thermal spray techniques since they can be used effortlessly, prolong component life, offer greater execution, are inexpensive to run, and have other favorable characteristics. The thermal spray includes a detonation gun (D-Gun), plasma spray, twin wire arc process, low-velocity combustion, flame spray, cold spray, low-pressure plasma spraying, high-velocity oxy-fuel processes, etc.

Consumable substances a heat or energy source are required to carry out thermal spraying. In some circumstances, gases are required in addition to air to pump materials into the gun and produce the heat required for melting. Material is ejected as tiny molten particles which hit the structure undergo solidification, and then adhere to the surface due to the elevated gas velocities that occur during these procedures. The adhesion mechanism is sometimes metallurgical but primarily mechanical. The interaction of thermal and kinetic energy directly affects the characteristics of coatings. Three of the fundamental processes involved in combustion include high-velocity oxygen fuel (HVOF), detonation, and flame spray powder or wire. The methods of plasma and wire arc also use electric energy to assist melt consumable components. Exceptionally strong bond strength and highly dense microstructures are produced via HVOF and detonation spraying. Additionally, whether applied using either vacuum plasma spray (VPS) or low-pressure plasma spray (LPPS), plasma coatings have strong binding strengths with reasonably thick oxide-free microstructures. A new technique called cold spray depends more on kinetic energy and high velocity than it does on thermal energy [69].

### 12.9.4 Ion Implantation

Ion implantation involves a mechanism that includes the acceleration of ions from the surface of a material using an external electric field so that the bombardment of these ions takes place at the solid surface [70]. This is the process of bombarding a surface with highly energetic ions that have enough energy to significantly penetrate the surface layer [71]. Ions with energies ranging between 100 keV and 2 MeV are often used in ion implantation. The elemental content of the target material will be altered, and the electrical, physical, and chemical characteristics of the specimen might vary if the ions are different in composition as compared to the target [70]. The ion implantation apparatus comprises a target chamber, an accelerator (for accelerating the ions to high energies), and an ion source (for producing ions of the required element). Typical beam-line ion implantation uses a line of vision method in which ions are drawn from the source of the ion, then undergo high-energy acceleration, and finally allowed to strike the substrate [73,72]. The schematic design of HEMII-80 (a conventional high-energy metal ion implantation machine) from Hong Kong City University is shown in Figure 12.6. Since the ion beam is often extremely small, raster scanning is used to accomplish homogeneous implantation across a broad region. It

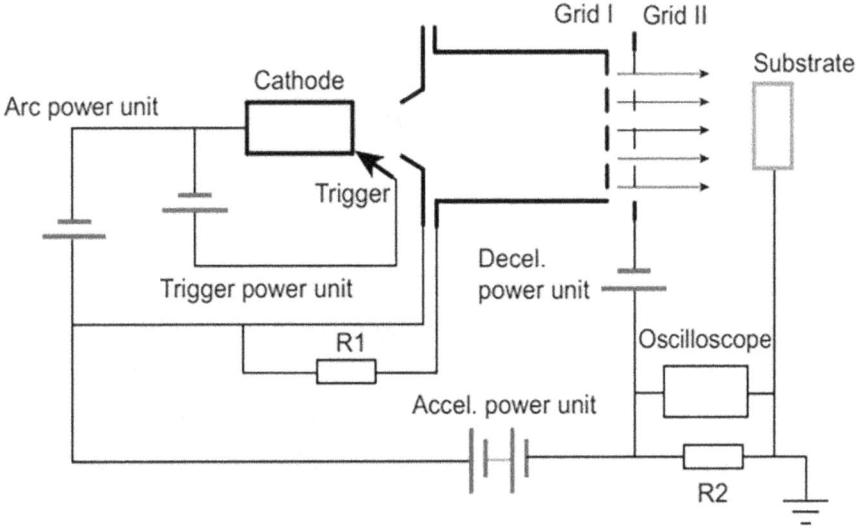

**FIGURE 12.6** Schematic presentation of HEMII-80 (high-energy metal ion implanter) [74].

could be challenging to achieve conformal ion implantation using this technique for materials exhibiting complex structures [73]. PIII (plasma immersion ion implantation), is a cutting-edge method that is more efficient and can better handle samples with a complicated geometry rather than traditional ion implantation. Several plasma generators, such as radio frequency (RF) or electron cyclotron resonance (ECR), generate plasma inside the vacuum chamber. To produce an intermixed atomic layer between the coating and substrate, along with ion implantation and coating, a hybridPIII&D (plasma immersion ion implantation and deposition) technique that includes ion implantation and deposition, can be carried out when metal plasmas are present [70] (Figure 12.7).

## 12.9.5 Diffusion

By developing boride, carbide, or nitride-dispersed phases in the vicinity of the surface by thermal diffusion of a reactive species, substrate surfaces may be reinforced and their distribution can be improved. Nitrogen can be thermally diffused into the surface of steels that include chromium, aluminum, molybdenum, tungsten, or vanadium to harden them. Normally, the nitriding procedure is executed at 500°C to 550°C for about 48 h in a gaseous environment, producing a hardened thickness that ranges about several hundred microns. Through diffusion via carbon-carrying vapor at 900°C, carburizing raises the carbon percentage of low-carbon steel (from 0.1%–0.2% to 0.65%–0.8%) [71]. By introducing carbon and nitrogen onto the surface of a ferrous substance, carbonitriding can be accomplished. Since nitrogen can diffuse more quickly than carbon, it forms a layer underneath the carbonitrided layer that, when cooled down, gives a boost to its fatigue strength. Any substance that has a

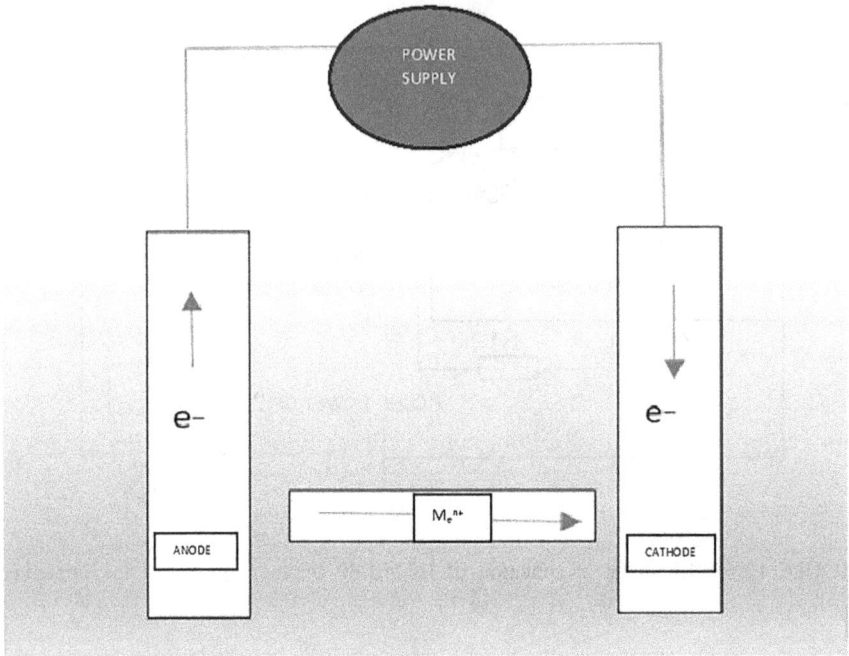

**FIGURE 12.7**   Schematic representation of electrodeposition.

component that creates a stable boride ($CrB_2$, $Fe_2B$, $NiB_2$, MoB, etc.), can be hardened by boronizing procedure. Siliconizing, aluminizing, and chromizing procedures are used to carry out the pack cementation procedure. Pack cementation can be achieved through solid-state diffusion or by methods similar to chemical vapor deposition. In situ, cleaning of the surface by hydrogen reduction and sputtering is made possible by the employment of plasma for ion bombardment, which also improves diffusion and chemical reactions. The substance under treatment may also be heated by the bombardment. Usually, for nitriding of steel, a plasma comprising of $N_2$, $NH_3$, or $N_2$-$H_2$ is employed combined with surface heating to 500°C–600°C. The procedure of plasma nitriding is referred to as ionitriding. Gears for robust machinery applications are hardened using this procedure in the industrial sector. Before the deposition of TiN coating, a steel surface can be plasma-nitrided by being bombarded using nitrogen plasma. Steel has been nitrided using nitrogen ion beams, and the modifications to the structure resulting from this method are comparable to those produced by nitriding. In a carbon-rich atmosphere, plasma carburizing is carried out. Additionally, plasma boronizing at low temperatures can be achieved [71].

### 12.9.6  PLATING

Electrodeposition, electrochemical conversion coating, electroforming, and electroless deposition come under the plating technique.

### 12.9.6.1 Electrochemical Deposition

An outdated yet effective method to produce metallic coatings is the peculiarity of electrochemical plating [75]. Electroplating and electroless plating are the two main categories of electrochemical deposition. In both instances, the metal salt (inside a solution) gets reduced to a metallic form on the substrate. The difference between electroless plating and electroplating is that in electroless plating, the electrons come from a chemical reducing agent present inside the solution, or the substrate itself generates reducing electrons in case of immersion deposition, whereas in electroplating the reducing electrons are received from an outside source [76].

### 12.9.6.2 Electrodeposition

The surface is altered in the electroplating or electrodeposition process, with the application of an external source of power in an electrolytic environment that is either aqueous or nonaqueous. The cathode of the electrolysis cell, which is coated, is submerged in a solution having the necessary metal in an oxidized condition. The metal that will be coated on the substrate serves as the anode. When the power is switched on, the metal atoms undergo the following process, which leads to their oxidation and dissolution in the solution.

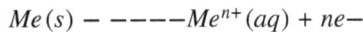

$$Me(s) - ----Me^{n+}(aq) + ne-$$

Following the reaction given below, the immersed metal ions present in the solution get reduced at the cathode-solution interface:

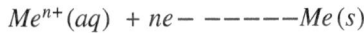

$$Me^{n+}(aq) + ne- ------Me(s)$$

In most cases, the dissolution rate of the anode and the plating rate of the cathode are identical. As a result, the anode constantly replenishes the ions in the solution. Additionally, electrodeposition can make use of a non-consumable anode like carbon or lead [77–80]. In this scenario, after being drawn out of the solution, the ions of the metal to be coated must be regularly supplied in the solution.

### 12.9.6.3 Electroless Deposition

Electroless deposition employs a single electrode and does not require a supply of any external power, unlike electrodeposition. Galvanic displacement and autocatalytic deposition are the two main categories of electroless deposition. A reducing agent (can be oxidized on its own to bring about reduction) can be used to carry out the reduction of metallic ions (inside the solution) along with deposition of film during the process of autocatalytic deposition. This reducing agent can carry out spontaneous oxidation and release electrons for the reduction of metallic ions at a certain temperature that relies on the bath composition of the reducing agent. Since the oxidation of the reducing component can only begin or end up self-sustaining on the surface of the coated metal, it is called an autocatalytic reaction [77–80]. A schematic representation of this deposition technique involving the reducing agent as the donor of electrons is shown in Figure 12.8. The mechanism of immersion plating or galvanic

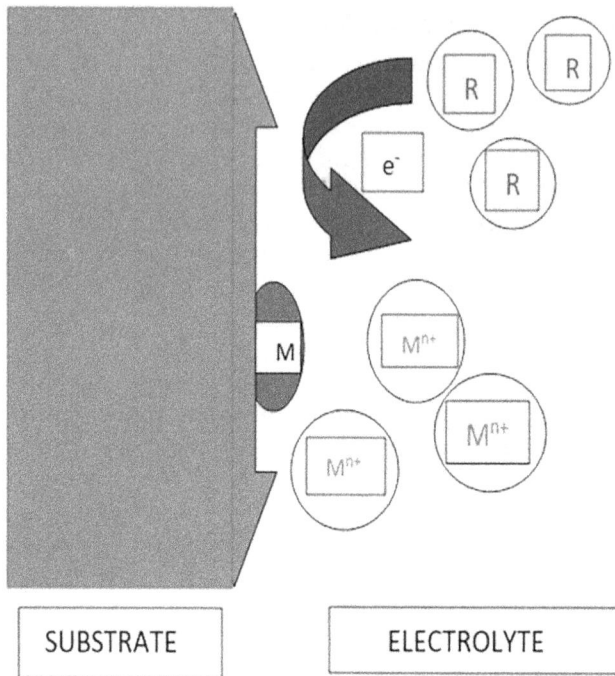

**FIGURE 12.8** Depicts reducing agent "R" (the source of electrons) executing electroless deposition.

displacement is not identical to the method of autocatalytic deposition. There is no requirement for reducing agents in this technique as the base components can perform this function. When the metallic ion (in the solution) with a lesser oxidation potential (less than the replaced base metal ion) displaces the base material, galvanic displacement occurs. The base material is disintegrated in the solution, after which the metallic ions become reduced on the substrate (base material).

A schematic representation for electroless nickel plating is shown in Figure 12.9 in simplified form. A plating tank, a specimen rotator, and a thermostat for controlling temperature comprise this device. Within a glass tank, the samples are put with a Teflon top over it that also has a thermostat [81]. As compared to electroplating, electroless plating shows superiority as there is no need for power sources. Additionally, autocatalytic deposition can improve thickness uniformity by avoiding the impacts of current distribution. As a result, plating components with complicated shapes are better suited to it. The common drawbacks of this coating technique include its poor speed and inability to produce thick coating layers [70].

## 12.10 CONCLUSION

Specifically in biomedical, aerospace, engineering, and automotive applications, the science of surface engineering has grown significantly in the domain of materials

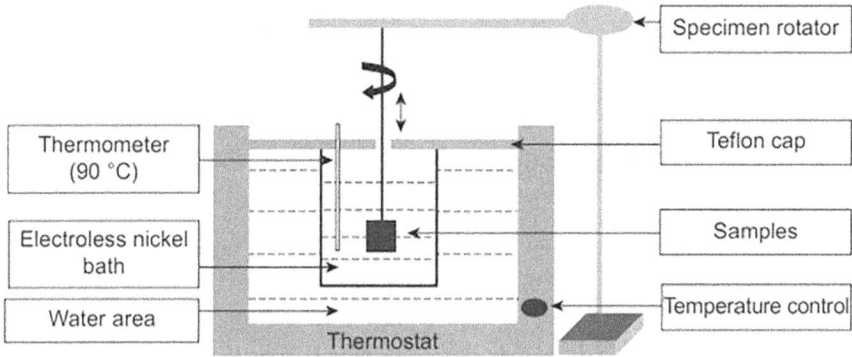

**FIGURE 12.9**   The experimental setup of electroless nickel plating [77].

technology in the past few years. To introduce the functional characteristics that the base material is unable to supply, the substrate surface undergoes surface engineering (modification), which improves the efficacy of the base substance at a reasonable cost. Based on the estimation, it has been figured out that corrosion and wear significantly impair the characteristics of materials. Surface modification can solve a wide range of technical challenges as it can improve the efficiency of the provided materials at a lower cost. Numerous techniques of surface modification, including physical and chemical vapor deposition, ultrasonic nanocrystal surface modification, ion implantation, glow discharge plasma, thermal spraying, electrophoretic deposition, coatings, sol-gel, diffusion processes, etc., are widely used and aid in preventing the corrosion of biomedical implants while meeting the highly precise norms of biomaterial surfaces for a specific biomedical objective.

# REFERENCES

1. Swain, Subhasmita, and Tapash R. Rautray. "Estimation of trace elements, antioxidants, and antibacterial agents of regularly consumed Indian medicinal plants." *Biological Trace Element Research* 199, no. 3 (2021): 1185–1193.
2. Swain, S., and T. R. Rautray. "Silver doped hydroxyapatite coatings by sacrificial anode deposition under magnetic field." *Journal of Materials Science: Materials in Medicine* 28 (2017): 1–5.
3. Rautray, Tapash R., Bijayinee Mohapatra, and Kyo-Han Kim. "Fabrication of strontium–hydroxyapatite scaffolds for biomedical applications." *Advanced Science Letters* 20, no. 3-4 (2014): 879–881.
4. Rautray, Tapash R., and Kyo Han Kim. "Synthesis of silver incorporated hydroxyapatite under magnetic field." In *Key engineering materials*, vol. 493, pp. 181–185. Trans Tech Publications Ltd, 2012.
5. Praharaj, Rinmayee, Snigdha Mishra, and Tapash R. Rautray. "The structural and bioactive behaviour of strontium-doped titanium dioxide nanorods." *Journal of the Korean Ceramic Society* 57 (2020): 271–280.
6. Mishra, Saswati, and Tapash R. Rautray. "Silver-incorporated hydroxyapatite–albumin microspheres with bactericidal effects." *Journal of the Korean Ceramic Society* 57 (2020): 175–183.

7. Mohapatra, Bijayinee, and Tapash R. Rautray. "Strontium-substituted biphasic calcium phosphate scaffold for orthopedic applications." *Journal of the Korean Ceramic Society* 57 (2020): 392–400.

8. Lee, Ki-Won, Cheol-Min Bae, Jae-Young Jung, Gi-Bong Sim, Tapash Ranjan Rautray, Hyo-Jin Lee, Tae-Yub Kwon, and Kyo-Han Kim. "Surface characteristics and biological studies of hydroxyapatite coating by a new method." *Journal of Biomedical Materials Research Part B: Applied Biomaterials* 98, no. 2 (2011): 395–407.

9. Kurella, Anil, and Narendra B. Dahotre. "Surface modification for bioimplants: the role of laser surface engineering." *Journal of Biomaterials Applications* 20, no. 1 (2005): 5–50.

10. Satpathy, Anurag, Rinkee Mohanty, and Tapash R. Rautray. "Bio-mimicked guided tissue regeneration/guided bone regeneration membranes with hierarchical structured surfaces replicated from teak leaf exhibits enhanced bioactivity." *Journal of Biomedical Materials Research Part B: Applied Biomaterials* 110, no. 1 (2022): 144–156.

11. Wagner, Warren C. "A brief introduction to advanced surface modification technologies." *The Journal of Oral Implantology* 18, no. 3 (1992): 231–235.

12. Stallard, Charlie P., Pavel Solar, Hynek Biederman, and Denis P. Dowling. "Deposition of non-fouling PEO-like coatings using a low temperature atmospheric pressure plasma jet." *Plasma Processes and Polymers* 13, no. 2 (2016): 241–252.

13. Roy, Mangal, Amit Bandyopadhyay, and Susmita Bose. "Induction plasma sprayed nano hydroxyapatite coatings on titanium for orthopaedic and dental implants." *Surface and Coatings Technology* 205, no. 8–9 (2011): 2785–2792.

14. O'Shaughnessy, W. Shannan, Meiling Gao, and Karen K. Gleason. "Initiated chemical vapor deposition of trivinyltrimethylcyclotrisiloxane for biomaterial coatings." *Langmuir* 22, no. 16 (2006): 7021–7026.

15. Roy, Mangal, Gary A. Fielding, Haluk Beyenal, Amit Bandyopadhyay, and Susmita Bose. "Mechanical, in vitro antimicrobial, and biological properties of plasma-sprayed silver-doped hydroxyapatite coating." *ACS Applied Materials & Interfaces* 4, no. 3 (2012): 1341–1349.

16. Das, Kakoli, Susmita Bose, Amit Bandyopadhyay, Balu Karandikar, and Bruce L. Gibbins. "Influence of silver on antibacterial activities of surface modified Ti." *Journal of Biomedical Materials Research Part B* 87, no. 2 (2008): 455–460.

17. Mohapatra, Bijayinee, and Tapash R. Rautray. "Facile fabrication of Luffa cylindrica-assisted 3D hydroxyapatite scaffolds." *Bioinspired, Biomimetic and Nanobiomaterials* 10, no. 2 (2021): 37–44.

18. Swain, Subhasmita, and Tapash Ranjan Rautray. "Assessment of polarized piezo-electric SrBi4Ti4O15 nanoparticles as an alternative antibacterial agent." *bioRxiv* (2021): 2021-01.

19. Swain, Subhasmita, Tae Yub Kwon, and Tapash R. Rautray. "Fabrication of silver doped nano hydroxyapatite-carrageenan hydrogels for articular cartilage applications." *bioRxiv* (2021): 2020-12.

20. Rautray, Tapash Ranjan, and Kyo Han Kim. "Synthesis of controlled release Sr-hydroxyapatite microspheres." In *Bioceramics-24*. 2012.

21. Swain, Subhasmita, Sapna Mishra, Abhishek Patra, Rinmayee Praharaj, and Tapash Rautray. "Dual action of polarised zinc hydroxyapatite-guar gum composite as a next generation bone filler material." *Materials Today: Proceedings* 62 (2022): 6125–6130.

22. Swain, Subhasmita, Abhishek Patra, Shubha Kumari, Rinmayee Praharaj, Satrujit Mishra, and Tapash Rautray. "Corona poled gelatin-magnesium hydroxyapatite composite demonstrates osteogenicity." *Materials Today: Proceedings* 62 (2022): 6131–6135.

23. Priyadarshini, Itishree, Subhasmita Swain, Janardhan Reddy Koduru, and Tapash Ranjan Rautray. "Electrically polarized Withaferin A and Alginate-incorporated biphasic calcium phosphate microspheres exhibit osteogenicity and antibacterial activity in vitro." *Molecules* 28, no. 1 (2022): 86.
24. Swain, Subhasmita, Shubha Kumari, Priyabrata Swain, and Tapash Rautray. "Polarised strontium hydroxyapatite–xanthan gum composite exhibits osteogenicity in vitro." *Materials Today: Proceedings* 62 (2022): 6143–6147.
25. Swain, Subhasmita, Janardhan Reddy Koduru, and Tapash Ranjan Rautray. "Mangiferin-enriched Mn–Hydroxyapatite coupled with β-TCP scaffolds simultaneously exhibit osteogenicity and anti-bacterial efficacy." *Materials* 16, no. 6 (2023): 2206.
26. Bose, Susmita, Samuel Ford Robertson, and Amit Bandyopadhyay. "Surface modification of biomaterials and biomedical devices using additive manufacturing." *Actabiomaterialia* 66 (2018): 6–22.
27. Alam, Fahad, Vivek Verma, and Kantesh Balani. "Fundamentals of surface modification." *Biosurfaces: A Materials Science and Engineering Perspective* 2 (2015): 126.
28. Subramani, Karthikeyan, Reji T. Mathew, and Preeti Pachauri. "Titanium surface modification techniques for dental implants—from microscale to nanoscale." *Emerging Nanotechnologies in Dentistry* (2018): 99–124.
29. Manara, Dario, Bruycker Franck De, A. K. Sengupta, Renu Agarwal, and H. S. Kamath. "Thermodynamic and thermophysical properties of the actinide carbides." (2012).
30. Hernández, María José Quintana, Jose Antonio Pero-Sanz, and Luis Felipe Verdeja. *Solidification and solid-state transformations of metals and alloys.* Elsevier, 2017.
31. Gupta, N., and B. Basu. "Hot pressing and spark plasma sintering techniques of intermetallic matrix composites." In *Intermetallic matrix composites*, pp. 243–302. Woodhead Publishing, 2018.
32. Gotman, Irena, and E. Y. Gutmanas. "Titanium nitride-based coatings on implantable medical devices." *Advanced Biomaterials and Devices in Medicine* 1, no. 1 (2014): 53–73.
33. Larhlimi, Hicham, Anas Ghailane, Mohammed Makha, and Jones Alami. "Magnetron sputtered titanium carbide-based coatings: A review of science and technology." *Vacuum* 197 (2022): 110853.
34. Saddow, Stephen E. "Silicon carbide materials for biomedical applications." In *Silicon carbide biotechnology*, pp. 1–25. Elsevier, 2016.
35. Vladescu, Alina, Vasile Pruna, Sawomir Kulesza, Viorel Braic, Irina Titorencu, Miroslaw Bramowicz, Anna Gozdziejewska et al. "Influence of Ti, Zr or Nb carbide adhesion layers on the adhesion, corrosion resistance and cell proliferation of titania doped hydroxyapatite to the Ti6Al4V alloy substrate, utilizable for orthopaedic implants." *Ceramics International* 45, no. 2 (2019): 1710–1723.
36. Longo, Giovanni, Caterina Alexandra Ioannidu, Anna Scotto d'Abusco, Fabiana Superti, Carlo Misiano, Robertino Zanoni, Laura Politi et al. "Improving osteoblast response in vitro by a nanostructured thin film with titanium carbide and titanium oxides clustered around graphitic carbon." *PLoS One* 11, no. 3 (2016): e0152566.
37. Algodi, Samer J., James W. Murray, Michael W. Fay, Adam T. Clare, and Paul D. Brown. "Electrical discharge coating of nanostructured TiC-Fe cermets on 304 stainless steel." *Surface and Coatings Technology* 307 (2016): 639–649.
38. Prakash, Chander, H. K. Kansal, B. S. Pabla, and Sanjeev Puri. "Potential of silicon powder-mixed electro spark alloying for surface modification of β-phase titanium alloy for orthopedic applications." *Materials Today: Proceedings* 4, no. 9 (2017): 10080–10083.

39. Brama, Marina, Nicholas Rhodes, John Hunt, Andrea Ricci, Roberto Teghil, Silvia Migliaccio, Carlo Della Rocca et al. "Effect of titanium carbide coating on the osseointegration response in vitro and in vivo." *Biomaterials* 28, no. 4 (2007): 595–608.

40. López-Álvarez, M., A. de Carlos, P. González, J. Serra, and B. León. "Cytocompatibility of bio-inspired silicon carbide ceramics." *Journal of Biomedical Materials Research Part B: Applied Biomaterials* 95, no. 1 (2010): 177–183.

41. Ghezzi, Benedetta, Paola Lagonegro, Giovanni Attolini, Pasquale Mario Rotonda, Christine Cornelissen, Joice Sophia Ponraj, Ludovica Parisi, Giovanni Passeri, Francesca Rossi, and Guido Maria Macaluso. "Hydrogen plasma treatment confers enhanced bioactivity to silicon carbide-based nanowires promoting osteoblast adhesion." *Materials Science and Engineering: C* 121 (2021): 111772.

42. Hu, Xinyu, Yongmei Wang, Liangliang Zhang, Man Xu, Jianfa Zhang, and Wei Dong. "Mechanical and thermal reinforcement of photocrosslinked salecan composite hydrogel incorporating niobium carbide nanoparticles for cell adhesion." *Polymer Testing* 69 (2018): 396–404.

43. Frommeyer, G., and R. Rosenkranz. "Structures and properties of the refractory silicides Ti 5 Si 3 and TiSi 2 and Ti-Si-(Al) eutectic alloys." In *Metallic materials with high structural efficiency*, pp. 287–308. Springer Netherlands, 2004.

44. Buga, Csaba, Chun-Cheng Chen, Mátyás Hunyadi, Attila Csík, Csaba Hegedűs, and Shinn-Jyh Ding. "Electrosprayed calcium silicate nanoparticle-coated titanium implant with improved antibacterial activity and osteogenesis." *Colloids and Surfaces B: Biointerfaces* 202 (2021): 111699.

45. Antonova, O. S., V. V. Smirnov, S. M. Barinov, N. V. Bakunova, L. Medvecky, and J. Durisin. "Bioactive silicium-containing coatings on titanium substrate." *Powder Metallurgy Progress* 11, no. 3-4 (2011): 271–276.

46. Tian, J., W. H. Yu, W. Tian, J. Zhao, Y. Q. Li, and Y. Z. Liu. "Effects of Y–Ce on wear behaviours of silicide coatings." *Surface Engineering* 31, no. 4 (2015): 289–294.

47. Ma, Hongshi, Qingqing Yu, Yu Qu, Yufang Zhu, and Chengtie Wu. "Manganese silicate nanospheres-incorporated hydrogels: starvation therapy and tissue regeneration." *Bioactive Materials* 6, no. 12 (2021): 4558–4567.

48. Deshmukh, Shamkumar, Krishna Pawar, Valmiki Koli, and Pradip Pachfule. "Emerging graphitic carbon nitride-based nanobiomaterials for biological applications." *ACS Applied Bio Materials* (2023).

49. Heimann, Robert B. "Silicon nitride, a close to ideal ceramic material for medical application." *Ceramics* 4, no. 2 (2021): 208–223.

50. Rautray, Tapash R., R. Narayanan, and Kyo-Han Kim. "Ion implantation of titanium based biomaterials." *Progress in Materials Science* 56, no. 8 (2011): 1137–1177.

51. Swain, Subhasmita, and Tapash Ranjan Rautray. "Effect of surface roughness on titanium medical implants." *Nanostructured Materials and Their Applications* 1 (2021): 55–80.

52. Rautray, T. R., and K-H. Kim. "Nanoelectrochemical coatings on titanium for bioimplant applications." *Materials Technology* 25, no. 3–4 (2010): 143–148.

53. Behera, Dipti Rani, Pratibindhya Nayak, and Tapash Ranjan Rautray. "Phosphatidylethanolamine impregnated Zn-HA coated on titanium for enhanced bone growth with antibacterial properties." *Journal of King Saud University-Science* 32, no. 1 (2020): 848–852.

54. Swain, Subhasmita, R. D. K. Misra, C. K. You, and Tapash R. Rautray. "TiO2 nanotubes synthesised on Ti-6Al-4V ELI exhibits enhanced osteogenic activity: A potential next-generation material to be used as medical implants." *Materials Technology* 36, no. 7 (2021): 393–399.

55. Praharaj, Rinmayee, Snigdha Mishra, R. D. K. Misra, and Tapash R. Rautray. "Biocompatibility and adhesion response of magnesium-hydroxyapatite/strontium-titania(Mg-HAp/Sr-TiO2) bilayer coating on titanium." *Materials Technology* 37, no. 4 (2022): 230–239.

56. Rautray, Tapash R., R. Narayanan, Tae-Yub Kwon, and Kyo-Han Kim. "Surface modification of titanium and titanium alloys by ion implantation." *Journal of Biomedical Materials Research Part B: Applied Biomaterials* 93, no. 2 (2010): 581–591.

57. Rautray, Tapash R., Subhasmita Swain, and Kyo-Han Kim. "Formation of anodic TiO2 nanotubes under magnetic field." *Advanced Science Letters* 20, no. 3-4 (2014): 801–803.

58. Pezzotti, Giuseppe, Francesco Boschetto, Eriko Ohgitani, Yuki Fujita, Wenliang Zhu, Elia Marin, Bryan J. McEntire, B. Sonny Bal, and Osam Mazda. "Silicon nitride: a potent solid-state bioceramic inactivator of ssRNA viruses." *Scientific Reports* 11, no. 1 (2021): 1–18.

59. Liu, Hui, Xiyue Zhang, Shujing Jin, Yanhui Zhao, Ling Ren, and Ke Yang. "Effect of copper-doped titanium nitride coating on angiogenesis." *Materials Letters* 269 (2020): 127634.

60. Guo, Lingyun, Chunna Gao, Fan Wang, Jie Wei, Jun Hu, and Yubo Xu. "Influence of content of silicon nitride nanoparticles into micro-arc oxidation coating of titanium on bactericidal capability and osteoblastic differentiation." *Surface and Coatings Technology* 458 (2023): 129346.

61. Mishra, Monu, Jitendra Sharan, Veena Koul, Om P. Kharbanda, Ashish Kumar, Ashok Sharma, Timothy A. Hackett, Ram Sagar, Manish K. Kashyap, and Govind Gupta. "Surface functionalization of gallium nitride for biomedical implant applications." *Applied Surface Science* 612 (2023): 155858.

62. Chukwuike, V. I., Dmitry V. Shtansky, and B. Subramanian. "Biocompatibility study of nanocomposite titanium boron nitride (TiBN) thin films for orthopedic implant applications." *Surface and Coatings Technology* 410 (2021): 126968.

63. Fu, Le, Karthik Rajaseka, Ioannis Katsaros, Yihong Liu, Helen Wang, Håkan Engqvist, and Wei Xia. "Enhanced bacteriostatic properties of Ti alloys by surface nitriding." *Biomedical Materials & Devices* (2023): 1–12.

64. Starosvetsky, D., and I. Gotman. "Corrosion behavior of titanium nitride coated Ni–Ti shape memory surgical alloy." *Biomaterials* 22, no. 13 (2001): 1853–1859.

65. Mattox, Donald M. "Physical vapor deposition (PVD) processes." *Metal Finishing* 100 (2002): 394–408.

66. Rafique, Muhammad Shahid, Muhammad Rafique, Muhammad Bilal Tahir, Syeda Hajra, Tasmia Nawaz, and Falak Shafiq. "Synthesis methods of nanostructures." In *Nanotechnology and photocatalysis for environmental applications*, pp. 45–56. Elsevier, 2020.

67. Carlsson, Jan Otto, and Peter M. Martin. "Chemical vapor deposition handbook of deposition technologies for films and coatings." *Science, Technology and Applications* 1 (2010): 444–445.

68. Nouri, Alireza, and Antonella Sola. "Powder morphology in thermal spraying." *Journal of Advanced Manufacturing and Processing* 1, no. 3 (2019): e10020.

69. Dorfman, Mitchell R. "Thermal spray coatings." In *Handbook of environmental degradation of materials*, pp. 469–488. William Andrew Publishing, 2018.

70. Chu, P. K., and G. S. Wu. "Surface design of biodegradable magnesium alloys for biomedical applications." In *Surface modification of magnesium and its alloys for biomedical applications*, pp. 89–119. Woodhead Publishing, 2015.

71. Mattox, Donald M. "Substrate ("Real") surfaces and surface modification." In *Handbook of physical vapor deposition (PVD) processing*, pp. 25–72. 2010.

72. Rautray, Tapash R., and Kyo Han Kim. "Synthesis of Mg2+ incorporated hydroxyapatite by ion implantation." In *Key engineering materials*, vol. 529, pp. 114–118. Trans Tech Publications Ltd, 2013.

73. Chu, Paul K., Shu Qin, Chung Chan, Nathan W. Cheung, and Lawrence A. Larson. "Plasma immersion ion implantation—A fledgling technique for semiconductor processing." *Materials Science and Engineering: R: Reports* 17, no. 6-7 (1996): 207–280.

74. Feng, Kai, Guosong Wu, Tao Hu, Zhuguo Li, Xun Cai, and Paul K. Chu. "Dual Ti and C ion-implanted stainless steel bipolar plates in polymer electrolyte membrane fuel cells." *Surface and Coatings Technology* 206, no. 11-12 (2012): 2914–2921.

75. Swain, Subhasmita, Tapash Ranjan Rautray, and Ramaswamy Narayanan. "Sr, Mg, and Co substituted hydroxyapatite coating on TiO2 nanotubes formed by electrochemical methods." *Advanced Science Letters* 22, no. 2 (2016): 482–487.

76. Gray, JEl, and Ben Luan. "Protective coatings on magnesium and its alloys—a critical review." *Journal of Alloys and Compounds* 336, no. 1-2 (2002): 88–113.

77. Sudagar, Jothi, Jianshe Lian, and Wei Sha. "Electroless nickel, alloy, composite and nano coatings–A critical review." *Journal of Alloys and Compounds* 571 (2013): 183–204.

78. Carraro, Carlo, Roya Maboudian, and Luca Magagnin. "Metallization and nanostructuring of semiconductor surfaces by galvanic displacement processes." *Surface Science Reports* 62, no. 12 (2007): 499–525.

79. Kanani, N. "Electroplating: Basic principles and practice." (2005).

80. Schlesinger, M., and M. Paunovic. Modern electroplating, pp. 285–308. John Wiley & Sons, Inc., 2010.

81. Li, Libo, Maozhong An, and Gaohui Wu. "A new electroless nickel deposition technique to metalliseSiCp/Al composites." *Surface and Coatings Technology* 200, no. 16–17 (2006): 5102–5112.

# 13 Nanoengineering in Biomaterials

*Neeraj Mehta*
Physics Department, Institute of Science, Banaras Hindu University, Varanasi, India

## 13.1 INTRODUCTION

Biomaterials are defined as natural or synthetic materials, whether natural or manufactured, interacting with natural environments for medicinal uses such as supporting, enhancing, repairing, and regulating biological activities. Both the bulk and surface qualities of these materials affect their usability and applicability. The science of nanomaterials provides information about their structural properties at a nanosized level where there is a drastic modification in their physical properties. By using the nanotechnologies we can design the nanomaterials of different sizes and shapes and can control both for the possible applications in different scientific fields. In recent decades, the search for novel multifunctional biomaterials has become a new movement for their applications in nanoengineering in continuation to the fast-growing field of biotechnology. Thus, the field of biomaterials has also not remained untouched by the influence of nanotechnology. Nanotechnology has proved important in developing biomaterials [1–5].

Recent developments in nanoengineering are significantly responsible for recent advancements in biomedicine [6]. Because traditional methods have a difficult time gaining access to biomaterials' distinctive qualities, research and development in nanoengineering has shown to be an effective strategy. As a result, the functionality and performance of biomaterials have been considerably enhanced. Based on the idea of nanoengineering, the simulation of biomaterials is also an effective approach that reveals molecular interactions and nanoscale conformational information during physical and chemical processes. This helps in the identification of primary symptoms of diseases in the human body and controls the macroscopic properties of nanoengineered biomaterials accordingly for providing medical solutions.

Nanoengineering facilitates unique ways of novel designs of biomaterials and has the capability to modify their characteristics and improve their performance in the desired directions [1–5]. Thus, biomaterials with unique properties become the potential candidates for future perspective and advanced prospects and different applications in the fields of bioengineering, biomedical industry, biomedical, and biotechnological. Nanoengineering permits the fabrication of nanomaterials with preprogrammed functionality and the handling of nano-designing at the atomic and

DOI: 10.1201/9781003429920-15

molecular levels. Nanoengineering is important because it can be used to discover the micro- and nanoscale clues hidden behind the performance of materials at the macroscale. Researchers in the biomedical, biophysics, and bioengineering sectors are increasingly realizing the special characteristics displayed by materials having specific nanoscale configurations (identical dimension range of biomolecules). In the multiscale investigation, Figure 13.1(a) presents an overview of biomaterials from macroscale observation to nanoscale modeling.

The characteristics of the constituent materials have a significant impact on the performance and functionality of medicines and medical devices. The four most prevalent types of synthetic biomaterials employed in biomedical applications are metallic materials, polymers, ceramics, and composites, as shown in Figure 13.1(b). The development of advanced computational techniques in nanoengineering has made it possible to do molecular simulations of aforesaid biomaterials for acquiring essential information about their structure and functionality for modeling and designing them [7].

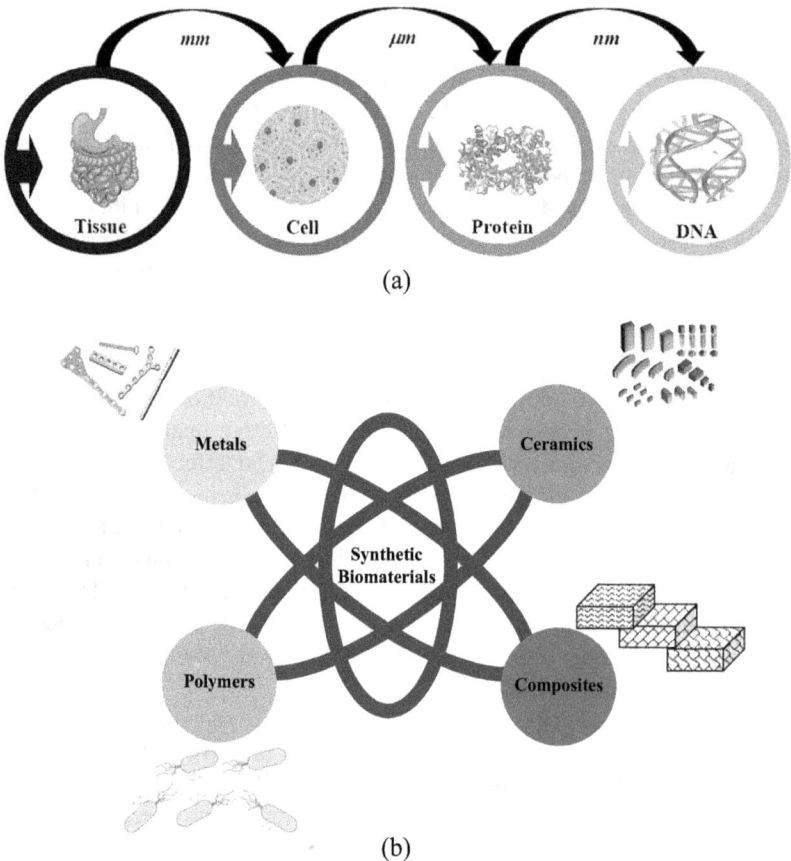

(a)

(b)

**FIGURE 13.1** (a) Nanoengineered biomaterials from macro-scale to nano-scale, and (b) different kinds of materials used for the fabrication of synthetic biomaterials.

Metals are frequently employed in a range of biomedical applications due to their excellent biocompatibility, strong corrosion resistance, exceptional mechanical features, and good implantation capabilities. To better understand industrial uses of metallic materials used in biomedical devices and therapies, the investigation of metallic materials at the nanoscale has grown thanks to nanoengineering. By offering the best molecular selection, accurate estimation of the desired properties, and effective processing guidelines, the manipulation of polymeric materials in the desired way is possible in nanoengineering. The structural adaptability of polymers allows for many different uses in biomedical engineering like immunological treatments, medication, and drug delivery. The structure-property connection has received a great deal of attention since artificial polymers are essential parts for non-natural tissues, stents, and transplants. The engineering of polymers has helped in the improvement of drug delivery by introducing systems of polymers that can detect signals and react to temperature drift, pH variation, and the existence of biological specimens. The goal of integrating unique structural designs from metallic and polymeric substances into nanocomposites is to increase functionality and biocompatibility for biomedical applications.

## 13.2  NANOENGINEERING TECHNIQUES FOR SYNTHESIS AND CHARACTERIZATION OF BIOMATERIALS

Biomaterials for nano-engineering have aided in the creation of several systems with different applications in nanomedicine and nano-biotechnology. Each nanostructure is distinct due to its structural and physicochemical characteristics, including composition, structure, surface, hydrophobicity, and charge. The applications of tissue engineering (where biomaterials, particularly polymers, metals, and their alloys are considered for medical use in medical situations dealing with the bones) experience unanticipated degradation that results in the removal of metal ions and other suspended waste from bulk substances, surface changes are desirable [8]. The interactions that occur at the boundary between the surface of a biomaterial and its immediate living environment are controlled by the biomaterial's surface. Therefore, a biomaterial's surface characteristics are crucial for its effective performance. Many biomaterials lack the proper surface functionalities and characteristics to meet particular requirements and applications. Modifying specific surface qualities would also be more efficient and take less time. Therefore, it would be extremely desirable and sought after for proper surface modification and functionalization to result in the greatly increased performance of biomaterials. Nanotechnology is a powerful method for surface functionalization in contemporary surface science and engineering, introducing thin coatings with nanometer-scale characteristics into materials. To nano-functionalize the surfaces of biomaterials, functional metal oxides, and metals in thin films and coatings can be employed.

The consequences of such implanted materials on human health are possible. Therefore, surface modification techniques have drawn a lot of attention in an attempt to safeguard against such undesirable and harmful processes occurring in vitro and in vivo [9]. The bulk qualities of the materials dictate their suitability for application, while the physical and chemical features of these surfaces specify their functional characteristics and behavior within the biological environment to elicit an acceptable and optimal host reprisal. Thus, for their application to be successful, materials must offer cutting-edge surface qualities like biocompatibility. During host-material interactions, these surface features control and influence biophysical and biochemical signals.

Due to their similar mineral composition to that of teeth and bones, coatings applied to implantation surfaces using physical vapor deposition methods are of excessive importance [10]. This makes it simpler for the coated regions to combine with the surrounding native tissues and avoid corrosion. Chemical vapor deposition techniques have several advantages [11] over solvent-based processes because they easily generate lean mono- or multilayers on surfaces, eliminating surface flaws like surface dewetting, congregate creation, surface nonuniformity, and toxicity risks to tissues and cells resulting from leftover solvents. Atomic layer deposition is another method [12] that has been extensively employed in applications such as bioelectronic products, implantable tissue engineering and medication delivery systems, and biosensors due to its uniform surface functionalization capabilities. For several different purposes, including bioelectronics, implant surfaces, biosensors, and diagnostic tools, self-assembled monolayers (SAMs) of biomaterials are being employed frequently for nanoscale modification of the surface characteristics of many substrates, including ceramic, metal-based, and polymer substrates [13].

Numerous techniques, for instance, attenuated total reflection-Fourier transform infrared spectroscopy, calorimetry, ellipsometry, atomic force microscopy, and X-ray photoelectron spectroscopy have been utilized to evaluate the surface quality of biomaterials in recent reports [14–19], as shown in the flow chart of Figure 13.2.

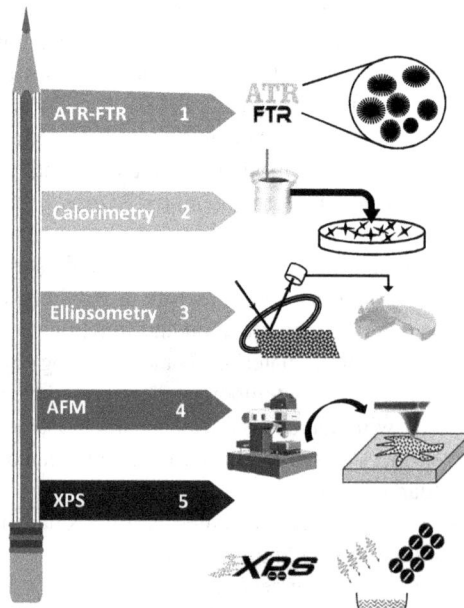

**FIGURE 13.2** List of some characterization techniques employed for the analysis of nanoengineered biomaterials.

## 13.3  APPLICATION OF NANOENGINEERED BIOMATERIALS

In the last three decades, various research papers on the fabrication of novel nanoengineered biomaterials have been published. Some of them report the foremost developments of such biomaterials in biomedical fields [such as cancer research and regenerative medicine]. In bioinspired robotics, nanoengineered biomaterials [20] have also been used to mimic particular actuation and sensing qualities; for instance, the perceptible properties of human skin or some animals' ability to feel tiny vibrations in their surroundings.

A conductive substance that converts biological reactions into audible electronic impulses is used to interface biological materials with electronic equipment in the development of bioelectronics. Enzyme-based bioelectronics, which is the core of wearable biodevices and is based on various biocatalytic reactions, in particular, offers specific advantages [20]. These on-body applications, which range from extremely selective nonsurgical biomarker tracking to epidermis storage of energy, have already shown immense potential for wearable bioelectronic devices that primarily rely on oxidase enzymes. Thus, such devices have the potential to significantly improve wearable technological advances and analytical capabilities, moving them away from common monitoring of movement and vital signs and towards a nonsurgical analysis of significant chemically based biomarkers. Exciting prospectives are presented by wearable enzyme electrodes in a range of industries, including medical services, games, ecological preservation, and safeguards. These gadgets can be set up to function as self-sustaining biosensors as well since they frequently come into contact with human tissues.

The surface and interface (where cells typically interact with biomaterials) of the biomaterials often determine the biocompatibility and subsequent uses of medical devices. Cardiovascular disease (CVD) is a group of serious health concerns that are typically brought on by coronary artery disease and have remained the primary contributor to disease and mortality in the world for many years. The serious conditions of cardiovascular diseases are frequently treated using stents used in long-term interventional therapy, with a significant number of these stents comprised of metallic biomaterials [21]. Conductive polymers are bioengineered materials that provide effective solutions for cardiovascular applications. Technologies based on conductive polymers are being developed to improve the healing process by increasing angiogenesis, detecting indicators of cardiac failure, monitoring and boosting cardiac activity, and delivering electrical commands through implanted devices.

Tissue engineering gained momentum in the 1990s when the more influential discipline of regenerative medicine was formed by merging it with stem cell transplantation [22–24]. The development of biomaterials has been done in a variety of forms and tested as a result of the private sector's growing interest as goods started to be successfully launched [21]. In this context, when a material is referred to as biomaterial, it is meant to refer to a system that has been engineered to be so complex that, when utilized in direct contact with healthy tissue, it is capable of interacting with the biological system in a controlled way. Consequently, any medicinal or diagnostic procedure may employ it.

These applications include control of the microbial environment to improve the membrane potential in cardiac cells and interactions between them, restoring nerve function and repair using the electrode-tissue interface of prosthetic devices, and healing wounds by triggering vascular development. Myocardial infarction has been treated using stem cell therapy and cardiac tissue engineering. The conductive polymers are intelligent materials that have the potential to expand this research's understanding of how to support and stimulate stem cells.

The nanoengineered biomaterials have been developed with a specific application of regenerative medicine in different organ tissues. Due to their significant advantages over traditional approaches for tissue repair, nanoengineered biomaterials have attracted considerable attention. Figure 13.3 highlights the flow chart of different human organs and body parts in which nanoengineered biomaterials are used for tissue regeneration.

Numerous researchers have been looking into the potential applications of cardiac tissue engineering, particularly biomimetic scaffolding, in the past few years [25–32]. The fibroin scaffolds have appealing possibilities for regenerating bones and the engineering of tissues so various research groups fabricated different designs of scaffolds by using biomaterials [25–32]. As a result, conducting scaffold processing and tissue engineering technologies have a big impact on tissue engineering and other bio-related fields. Conductive polymer substrates have the ability to trigger the scaffolds electrically and also greatly speed up the healing of electrically charged tissue, such as nervous and muscular tissues. Solvent casting,

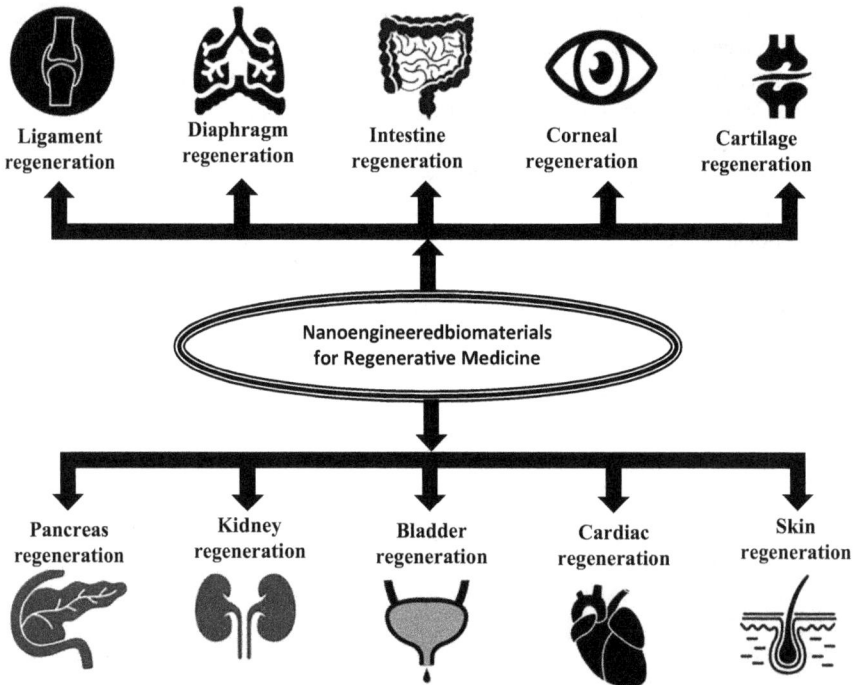

**FIGURE 13.3** List of various organ tissues where nanoengineered biomaterials are employed for tissue regeneration.

particle discharge, electrically spinning polymeric conductive mixtures with additional polymers, and polymer layering on the scaffold framework are methods used to create electrically conductive scaffolds [25–32]. Bioactive compounds are also immobilized on conducting scaffolds, enabling them to serve as multipurpose tissue engineering applications [33–35]. Further, the synthetic polymeric scaffolds are easily adapted to different tissue requirements and so they are also excellent candidates for creating cardiac tissue patches. These polymers are easily manipulable and have great biocompatibility and mechanical strength. Additionally, polymers made from synthetic materials can be engineered to have properties that are similar to those of natural cardiac tissue. However, there are still several difficulties in developing scaffolds that are biocompatible and possess sufficient electrical conduction and mechanical features for the regeneration of heart tissue [36]. A notable disadvantage of biomaterials in cardiovascular bioengineering is the low accessibility to host blood. In significantly surrounded and contracted organs, like the heart muscle, the degree of complications for engineering tissues increases because substances must also contain electroconductive structures.

For food and pharmaceutical items to be contained, stored, shipped, delivered, and used in their final state, packaging is a significant procedure [37]. Protection from mechanical harm, as well as the avoidance or inhibition of chemical alterations, metabolic changes, and microbial decomposition, are the primary goals of packaging. Bioengineering of materials has potential applications in the food packaging industry where biomaterials are deliberately used as supplemental components in the wrapping method to enhance the shelf life of consumable items, communicate freshness, display quality information, and enhance safety [38]. In the coming years, modern agrochemicals are going to be employed to supplement existing agricultural and intensive food processing. The application of biological fertilizers to the systems of both crops and food storage will be ensured by the introduction of organic agrochemical nano-formulation [39]. The use of genetically engineered organisms in agriculture without doing proper risk assessments is one of the alternative food production techniques that must be stopped. Farming organically and using smart packaging for meals should adopt more efficient ecological methods to synthesize the nanoscale components for more effective crop safety and secure food packaging techniques. Most of the environmentally friendly methods that have been studied in-depth at the laboratory level show decreased toxicity and offer excellent mechanical packaging properties. The next stage belongs to the introduction of these antimicrobial biomaterials into the commercial-scale packaging industries. With lower prices and more effectiveness, the commercialization of innovative biomaterials will confirm the protection, taste, and freshness of food items. It is clear that advances in nano-biotechnology have led to packaging and bioactive food additives that significantly increase the commodity's health needs.

By preventing moisture loss, bioengineered packaging materials can be utilized to enhance the standard of frozen, fresh, and processed animal products, chicken, and marine items as well as to increase the storage duration of products (e.g., vegetables and fruits). By adding a layer of batter or bread, they can also block the oxidation of lipids reaction, which will finally stop discoloration, retain the appearance of the product, and increase oil absorption throughout frying.

Therefore, in addition to various types of known packaging made of recyclable, ecologically suitable materials, it is crucial to plan bioengineered wrapping materials for food items that can be the best choice [37].

One element that is crucial to the survival of living things is water. Thus, it is a key goal to improve water quality because polluted water could have serious consequences [40]. The quality of drinkable water is threatened by industrial pollution, agricultural waste, and untreated wastewater. For refineries and other machinery operations, water has been used as a supply of kinematic power throughout the history of human civilization. It has also served as a fuel for biological processes like crop growth and food digestion. For the aquatic ecology and the health of society as a whole, it is crucial to remove organic and inorganic pollutants and treat wastewater effectively. To enhance the quantity of harmless and usable water and avoid contaminating naturally available sources of water, traditional water remediation techniques are being reinvented by nanoengineering in biomaterials. Recent studies demonstrate the usefulness of synthetic biomaterials for wastewater treatment [40,41]. Due to recyclable and environmentally friendly characteristics, bioengineered nanomaterials offer a sustainable technological approach to water treatment [42]. For effective water cleanup, various promising nanofiltration membranes have been developed over the last five years [43]. These bio-nanofilters have improved durability, water flow, and rejection efficiency in addition to being environmentally caring.

Interest in renewable sources of energy has increased as a result of climate change and the depletion of fossil fuels. Because of its benefits over traditional fossil fuels, for instance, lowered greenhouse gas emissions and sustainable development over time, bioenergy can replace them. Algal biofuels, solid biofuels, biobutanol, hydrogen, biogas, syngas, biodiesel, bioethanol, and several other kinds of bioenergy are among them [44]. Depending on their nature, several kinds of nanomaterials are frequently utilized for the generation of biological energy [45]. The electron transport from a bio-electrode to an enzyme is intended to be streamlined. Consequently, various studies have been carried out using manufacturing techniques for advanced engineered nanomaterials and their composite counterparts [46]. Additionally, an appropriate microenvironment can be offered by these substances for the enzymes' typical catalytic activities [47]. To provide effective electron transfer, the size, shape, and measurements of nanomaterials are intended for the reduction of the distance between blocked energetic sites of enzymes and the bioelectrodes. Therefore, it is essential to have a thorough understanding of how the enzyme causes bio-catalysis and efficient immobilization techniques to enhance the enzyme's activity and stability at various pH levels and temperatures. Due to this, high-performance bio-anodes are being engineered, and substantial advancements in biofuel cells based on glucose oxidizing enzymes have been made [48]. Furthermore, chemically altered electrodes with nanoengineering of advanced functional materials, particularly nanostructured substances, and innovative conducting polymeric materials, could demonstrate electron-mediating functions and facilitate the enzyme's direct charge transfer progressions.

Most nanoengineered biomaterials utilized in medical applications, whether synthetic or natural, are environment-friendly, biocompatible, and safe. They are frequently utilized in the targeted delivery of a variety of therapeutic agents, including RNA, DNA, antibacterial agents, enzymes, and cancer-preventive drugs. Some of them have the potential to influence the body's immunogenicity [49]. Nowadays, drug delivery uses nanoengineering extensively. The integration of nanoengineering technology has improved medicine delivery systems. Nanoengineering is used to create biomaterials facilitating the delivery of drugs, which improves the physico-chemical, and biological characteristics of the drug molecule [49]. The number of choices is increasing continuously for the use of different nanoengineered materials in numerous biomedical fields for drug delivery because of their small size, which enables them to permeate tissues and be absorbed by cells. The drug molecule is transported using the modified biomaterials. Nanoengineering is used to create nanosized compounds that aid in drug distribution from modified biomaterials. By combining medicine with nanoparticles that have superior qualities, the properties of the drug are enhanced. Sometimes a drug has active biological functionality but is possibly not fully soluble or has a limited life span or the opposite. Sometimes, a drug has aforesaid features but lacks effective biological functionality. In both cases, its properties can be improved by coupling the enhanced nanoparticles with them [50]. An effective medicine delivery system can now be created thanks to nanoengineering.

Degradable polymers for cancer drug delivery and implantable stents for treating blocked blood arteries are only two examples of how nanoengineering in biomaterials has transformed medicine [50]. It is now understood that the engineering of these materials also plays a crucial impact in determining how well they function biologically [49]. Therefore, understanding the unique characteristics of materials engineering and fusing these benefits with developments in materials chemistry may spark future advancements in the biomedical industry. Figure 13.4 summarizes the unique applications of nanoengineered biomaterials in different fields.

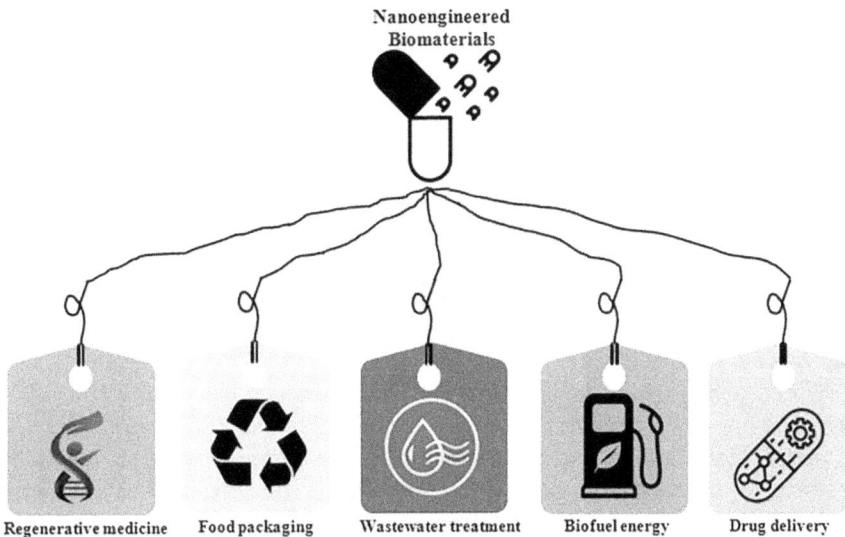

**FIGURE 13.4** A glimpse of exclusive applications of nanoengineered biomaterials in diverse fields.

## 13.4 SUMMARY

The substances we acquire from biological sources are known as biomaterials. Numerous biomaterials are altered using nanoengineered methods so they can be used effectively in various applications. Until now, the interference of nanoengineering in biomaterials has provided several products for therapy, bioenergy, water treatment, diagnosis, etc. Some of them are in the stage of clinical trials, while others are available in the market with applications. Thus, nanoengineering in biomaterials is fulfilling continuously its promises in biomedical fields. This chapter sheds light on the new developments in the arena of nanoengineered biomaterials for utilization in different applications (regeneration, drug delivery, implantation, biofuel cells, etc).

In order to engage with biological systems in clinical settings for therapeutic and diagnostic purposes, biomaterials are designed. Modern medicines and medical equipment both contain these elements. A biomedical device or living structure that can be used to carry out, improve, augment, or restore a natural function can be made entirely of biomaterials, which can be created in a lab or directly from natural resources. Natural polymers (e.g., proteins and polysaccharides) are generally derived from algae, plant-based animals, or human-made sources. Because of their excellent biological compatibility, they are believed to be advantageous elements for tissue science.

The idea of biomaterials has changed from solely mechanical replacement devices to actual biological solutions with the advancement of biodegradable implants. Though corrosion is typically regarded as a serious issue in metallurgy, it can be advantageous when used as a component of biodegradable implants. Examples of these metals include magnesium and iron. These substances function as necessary for healing and the growth of new tissue, then they disintegrate. Including recyclable polymer-based drugs availability, nanopatterned superficial surfaces, and polymers having electroactive properties for integrated detection to the tissue-engineered approach may result in innovative solutions for various materials. By integrating engineering and medical needs for implants, these metals have revolutionized orthopedic and cardiovascular surgery. In recent years, a lot of research groups investigated the uses of cardiac tissue engineering, particularly biomimetic scaffolding. Thus, the purpose of nanoengineering of biomaterials is to address unmet medical needs. With the aging population and rising health issues and medical needs, the engineering of biomaterials has the potential to revolutionize tissue engineering and regenerative medicine by modifying the biophysical characteristics of biomaterials, controlling cell-matrix interactions, and much more.

With better barrier qualities and robustness compared to biopolymer materials, a new type of next-generation material for packaging in the meals packaging sector comprises bioengineered packaging materials. Since they are made from renewable resources, they are biodegradable, environmentally friendly, and have superior mechanical strength to synthetic polymeric films. They also have better barrier properties. Thus, the storage of food goods can be improved with bioengineered packaging materials.

Recent developments in engineered nanomaterials and innovative polymers are offering fresh perspectives on how to create enzyme-based biofuel cells with greater efficiency. The fabrication of enzyme-based biofuel cells used advanced engineered materials, which helped with the charge transfer between electrodes and biocatalysts. According to current research, a variety of nanomaterials can be used and are proven to be effective replacements for electrode materials to improve the efficiency of enzyme-based biofuel cells too.

## REFERENCES

1. J. C. Scimeca, E. Verron, Nano-engineered biomaterials: Safety matters and toxicity evaluation. *Materials Todays Advances*, 15 (2022) 100260.
2. S. Mclaughlin, J. Podrebarac, M. Ruel, E. Suuronen, B. Mcneill, E. Alarcon, Nano-engineered biomaterials for tissue regeneration: What has been achieved so far? *Frontiers in Materials*, 3 (2016) 27.
3. Y. Shao, J. Fu, Integrated micro/nanoengineered functional biomaterials for cell mechanics and mechanobiology: A materials perspective. *Advanced Materials*, 26 (2014) 1494–1533.
4. L. M. Cross, A. Thakur, N. Jalili, M. Detamore, A. Gaharwar, Nanoengineered biomaterials for repair and regeneration of orthopedic tissue interfaces. *Acta Biomaterialia*, 42 (2016) 2–17.
5. Z. Jia, X. Xu, D. Zhu, Y. Zheng, Design, printing, and engineering of regenerative biomaterials for personalized bone healthcare. *Progress in Materials Science*, 134 (2023) 101072.
6. Q. Yan, H. Dong, Jin Su, J. Han, B. Song, Q. Wei, Y. Shi, A review of 3D printing technology for medical applications. *Engineering*, 4 (2018) 729–742.
7. D. Lau, W. Jian, Z. Yu, D. Hui, Nano-engineering of construction materials using molecular dynamics simulations: Prospects and challenges. *Composites Part B: Engineering*, 143 (2018) 282–291.
8. N. Eliaz, Corrosion of metallic biomaterials: A review. *Materials* 12 (2019) 407.
9. H. Amani, H. Arzaghi, M. Bayandori, A. S. Dezfuli, H. Pazoki-Toroudi, A. Shafiee, L. Morad, Controlling cell behavior through the design of biomaterial surfaces: A Focus on surface modification techniques. *Adv. Mater. Interf.*, 6 (2019) 1900572.
10. L. Geyao, D. Yang, C. Wanglin, W. Chengyong, Development and application of physical vapor deposited coatings for medical devices: A review. *Procedia CIRP*, 89 (2020) 250–262.
11. M. Vasudev, K. D. Anderson, T. Bunning, V. Tsukruk, R. Naik, Exploration of plasma-enhanced chemical vapor deposition as a method for thin-film fabrication with biological applications. *ACS Appl. Mater. Interfaces*, 5 (2013) 3983–3994.
12. A. K. Bishal, A. Butt, S. K. Selvaraj, B. Joshi, S. B. Patel, S. Huang, B. Yang, T. Shukohfar, C. Sukotjo, C. Takoudis, Atomic layer deposition in bio-nanotechnology: A brief overview. *Crit. Rev. Biomed. Eng.*, 43 (2015) 255–276.
13. M. Mrksich, Using self-assembled monolayers to understand the biomaterials interface. *Current Opinion in Colloid & Interface Science*, 2 (1997) 83–88.
14. R. Barbucci, M. Casolaro, A. Magnani, Characterization of biomaterial surfaces: ATR-FTIR, potentiometric and calorimetric analysis. *Clinical Materials*, 11 (1992) 37–51.
15. Krishnan K. Chittur, FTIR/ATR for protein adsorption to biomaterial surfaces. *Biomaterials*, 19 (1998) 357–369.
16. D. R. Miller, Nikolaos A. Peppas, The use of x-ray photoelectron spectroscopy for the analysis of the surface of biomaterials. *Journal of Macromolecular Science*, 26 (1986) 33–66.

17. K. D. Jandt, Atomic force microscopy of biomaterials surfaces and interfaces. *Surface Science*, 491 (2001) 303–332.
18. V. Samouillan, F. Delaunay, J. Dandurand, N. Merbahi, J. Gardou, M. Yousfi, A. Gandaglia, M. Spina, C. Lacabanne, The use of thermal techniques for the characterization and selection of natural biomaterials. *J. Funct. Biomater.*, 2 (2011) 230–248.
19. H. Arwin, Spectroscopic ellipsometry and biology: Recent developments and challenges. *Thin Solid Films*, 313–314 (1998) 764–774.
20. Jie-an Li, Ming Xin, Zhong Ma, Yi Shi, Lijia Pan, Nanomaterials and their applications on bio-inspired wearable electronics. *Nanotechnology*, 32 (2021) 472002.
21. F. Berthiaume, T. Maguire, M. Yarmush, Tissue engineering and regenerative medicine: History, progress, and challenges. *Annu Rev Chem Biomol Eng.*, 2 (2011) 403–430.
22. G. Sampogna, S. Guraya, A. Forgione, Regenerative medicine: Historical roots and potential strategies in modern medicine. *J. Microsc. Ultrastruct.*, 3 (2015) 101–107.
23. A. S. Slingerland, A. I. P. M. Smits, C. V. C. Bouten. Then and now: Hypes and hopes of regenerative medicine. *Trends Biotechnol.*, 31 (2013) 121–123.
24. H. Kaul, Y. Ventikos, On the genealogy of tissue engineering and regenerative medicine. *TissueEng. Part B*, 21 (2015) 203–217.
25. G. S. Karanasiou, M. I. Papafaklis, C. Conway, L. K. Michalis, R. Tzafriri, E. R. Edelman, D. I. Fotiadis, Stents: Biomechanics, biomaterials, and insights from computational modeling. *Annals Biomed. Engin.*, 45 (2017) 853–872.
26. H. Wu, Kaili Lin, C. Zhao, X. Wang, Silk fibroin scaffolds: A promising candidate for bone regeneration. *Front. Bioeng. Biotechnol.*, 25 (2022) 1054379.
27. Z.-H. Li, S. Ji, Y. Wang, X. Shen, H. Liang, Silk fibroin-based scaffolds for tissue engineering. *Front. Mater. Sci.*, 7 (2013) 237–247.
28. J. Nannan, X. Huang, Z. Li, L. Song, H. Wang, Yue-min Xu, H. Shao, Y. Zhang, Silk fibroin tissue engineering scaffolds with aligned electrospun fibers in multiple layers. *RSC Adv.*, 4 (2014) 47570–47575.
29. C. Scotti, D. Wirz, F. Wolf, D. Schaefer, Vivienne Burgin, A. U. Daniels, V. Valderrabano, C. Candrian, M. Jakob, I. Martin, A. Barbero, Engineering human cell-based, functionally integrated osteochondral grafts by biological bonding of engineered cartilage tissues to bony scaffolds. *Biomaterials*, 31 (2010) 2252–2259.
30. L. Liverani, J. Roether, P. Nooeaid, M. Trombetta, D. Schubert, A. Boccaccini, Simple fabrication technique for multilayered stratified composite scaffolds suitable for interface tissue engineering. *Materials Science and Engineering: A*, 557 (2012) 54–58
31. L.-P. Yan, J. Silva-Correia, M. B. Oliveira, C. Vilela, H. Pereira, R. A. Sousa, J. Mano, A. Oliveira, J. Oliveira, R. Reis, Bilayered silk/silk-nanoCaP scaffolds for osteochondral tissue engineering: In vitro and in vivo assessment of biological performance. *Acta Biomaterialia*, 12 (2015) 227–241
32. C. Li, C. Vepari, H. Jin, H.-J. Kim, D. Kaplan, Electrospun silk-BMP-2 scaffolds for bone tissue engineering. *Biomaterials*, 27, 16, (June 2006) 3115–3124.
33. Li Chen, J. Liu, M. Guan, T. Zhou, X. Duan, Z. Xiang, Growth factor and its polymer scaffold-based delivery system for cartilage tissue engineering. *Int. J. Nanomed.*, 15 (2020) 6097–6111.
34. R. Balint, N. Cassidy, S. Cartmell, Conductive polymers: Towards a smart biomaterial for tissue engineering. *Acta Biomater.*, 10 (2014) 2341–2353.
35. B. Guo, P. Ma, Conducting polymers for tissue engineering. *Biomacromolecules*, 19 (2018) 1764–1782.
36. E. S. Place, N. D. Evans, M. M. Stevens, Complexity in biomaterials for tissue engineering. *Nat. Mater.*, 8 (2009) 457–470.

37. R. Ahvenainen, E. Hurme, Active and smart packaging for meeting consumer demands for quality and safety. *Food Addit. Contam.* 14 (1997) 753–763.
38. D. Yuvaraj, J. Iyyappan, R. Gnanasekaran, G. Ishwarya, R. Harshini, V. Dhithya, M. Chandran, V. Kanishka, K. Gomathi, Advancements in bio food packaging – An overview. *Heliyon*, 7 (2021) e07998
39. R. R. Eapen Philip, A. Madhavan, R. Sindhu, A. Pugazhendhi, P. Binod, Ranjna Sirohi, M. Awasthi, A. Tarafdar, A. Pandey, Advanced biomaterials for sustainable applications in the food industry: Updates and challenges. *Environmental Pollution*, 283 (2021) 117071.
40. S. Lustenberger, R. Castro-Munoz, Advanced biomaterials and alternatives tailored as membranes for water treatment and the latest innovative European water remediation projects: A review. *Case Studies Chem. Environ.Engin.*, 5 (2022) 100205.
41. G. Amo-Duodu, E. K. Tetteh, S. Rathilal, E. Armah, J. A. Adedeji, M. N. Chollom, M. Chetty, Effect of engineered biomaterials and magnetite on wastewater treatment: Biogas and kinetic evaluation. *Polymers*, 13 (2021) 4323.
42. L. Muruganandam, M. P. S. Kumar, A. Jena, S. Gulla, B. Godhwani, Treatment of wastewater by coagulation and flocculation using biomaterials. *IOP Conf. Ser.: Mater. Sci. Eng.*, 263 (2017) 032006.
43. S. Elakkiya, G. Arthanareeswaran, A. F. Ismail, P. S. Goh, Y. L. Thuyavan, Review on characteristics of biomaterial and nanomaterials based polymeric nanocomposite membranes for seawater treatment application. *Environmental Research*, 197 (2021) 111177.
44. M.-H. Zhou, J. Yang, H. Wang, T. Jin, Dake Xu, T. Gu, Microbial fuel cells and microbial electrolysis cells for the production of bioelectricity and biomaterials. *Environmental Technology*, 34 (2013) 1915–1928.
45. A. Kumar, S. Sharma, L. Pandey, P. Chandra, Nanoengineered material based biosensing electrodes for enzymatic biofuel cells applications. *Mater. Sci. Energy Technol.*, 1 (2018) 38–48.
46. S.-J. Bao, C. M. Li, J.-F. Zang, X.-Q. Cui, Y. Qiao, J. Guo, New nanostructured TiO2 for direct electrochemistry and glucose sensor applications. *Adv. Funct. Mater.*, 18 (2008) 591–599.
47. K. Mahato, P. K. Maurya, P. Chandra, Fundamentals and commercial aspects of nanobiosensors in point-of-care clinical diagnostics. *3 Biotech*, 8 (2018), 149–152.
48. C. M. Moore, S. D. Minteer, R. S. Martin, Microchip-based ethanol/oxygen biofuel cell. *Lab Chip*, 5 (2005) 218–225.
49. A. Khademhosseini, N. A. Peppas, Micro-and nanoengineering of biomaterials for healthcare applications. *Adv. Healthcare Mater.*, 2 (2013) 10–12.
50. J. Shi, A. R. Votruba, O. C. Farokhzad, R. Langer, Nanotechnology in drug delivery and tissue engineering: From discovery to applications. *Nano Lett.*, 10 (2010) 3223–3230.

# 14 Smart Biomaterial Surface

*Sapna Mishra[#], Priyabrata Swain[#], and Subhasmita Swain*

Biomaterials and Tissue Regeneration Lab, CETMS, Institute of Technical Education and Research, Siksha 'O'Anusandhan (Deemed to be University), Bhubaneswar, Odisha, India

*V. John Kennedy*

National Isotope Center, GNS Science Limited, Lowerhutt, New Zealand

*Tapash R. Rautray*

Biomaterials and Tissue Regeneration Lab, CETMS, Institute of Technical Education and Research, Siksha 'O'Anusandhan (Deemed to be University), Bhubaneswar, Odisha, India

## 14.1 INTRODUCTION

### 14.1.1 Biomaterial Surface Science

In those primary days of medical approaches, the introduction of the surfaces along with the science, design, and modification played very important roles in medical approaches. The studies of surface modification and biological scopes are appreciated for their remarkable impact on biological utilization and drugs. In the biological as well as medical fields, surface science application is not very prominent as compared to surface application for catalysis [1]. The structure and surface chemistry of a material can be modified to regulate the bio-reactivity of cell-surface interaction. Biomaterials are classified basically into two groups: one is endogenous material, developed for curing human lives, and the other for material-biomolecule contact. The term *smart material* can be summarized as "materials, which alter their properties in correspondence with external stimuli" [2].

---

[#] authors contributed equally

DOI: 10.1201/9781003429920-16

### 14.1.1.1 Biocompatibility Biomaterials and Molecular Biorecognition Surfaces

Interaction of a material with a host body having zero risk of damage, infection, or rejection by natural immunity is known as biocompatibility. Modern biomaterial designs regulate the biocompatibility of materials in the primary stages of the medical field implementation. This ongoing chapter talks about the development of biocompatible and smart materials from synthetic polymers, metals, and ceramics with other biomaterials [3].

A few standardized procedures for developing (Figure 14.1) the ideas related to biomaterials and biological environment compatibility [4–6] are as follows.

*Hemocompatibility*: The property of the biomaterials having contact with blood to avoid thrombus formation by minimizing the active properties of platelets and coagulation of blood. This is known as hemocompatibility.

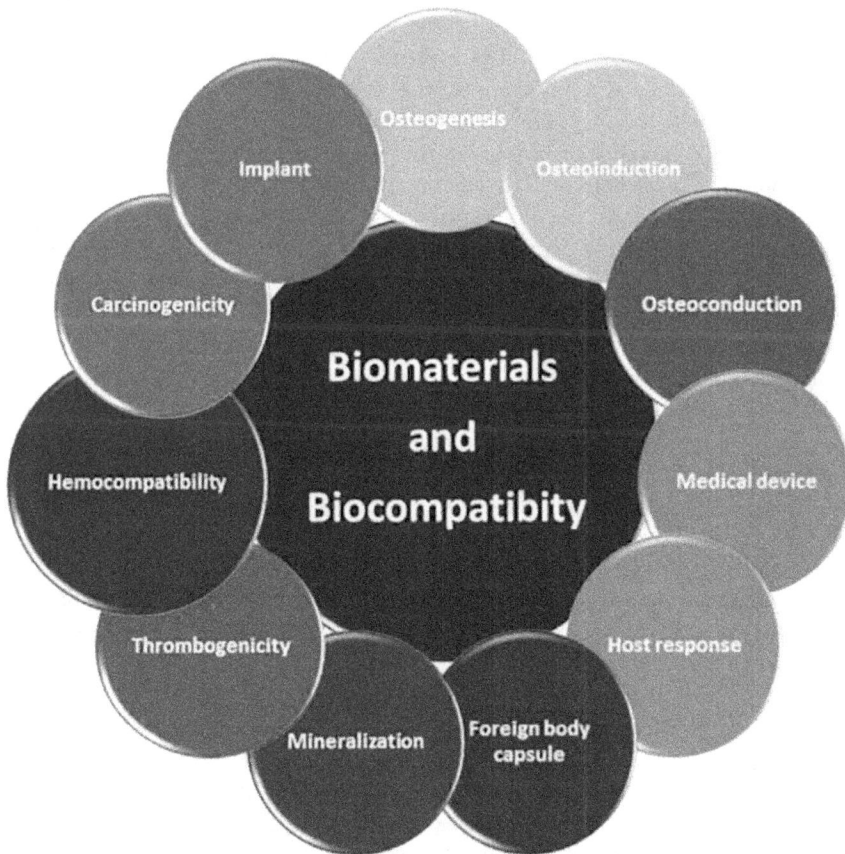

Current Opinion in Biomedical Engineering

**FIGURE 14.1** Schematic illustration of a few terms associated with biomaterials and biocompatibility.

*Carcinogenicity*: Either stimulating tumor formation, leading to cancer occurrence, or enhancing the rate of possibility of incidence, after the implantation, inhalation, injection, or ingestion.

*Mineralization*: Formation of a required structure (i.e., prosthetic devices or biomaterials), due to the time requisite mineral precipitation, is known as mineralization.

*Thrombogenicity*: The ability of a material to form a thrombus when in contact with blood, is called thrombogenicity.

*Osteoconduction*: The property of promoting bone tissue growth on the surface or inside of the porous spaces, and channels of an engineered implant or graft, is called osteoconductivity. Materials with osteoconduction ability help bone-forming cells in infiltration and proliferation and gradually new bone formation in the place of synthetic substitutes.

*Osteogenesis*: The procedure of emergence and development of bone is described by osteogenesis. An osteogenic material can produce novel bone from osteoblasts, which are previously presented in the material.

*Osteoinduction*: Osteoinduction is a process that stimulates or carries out the procedure of osteogenesis. Moreover, the process of stem cell division and maturation of bone-forming cells are stimulated by osteoinduction as well.

Antibody bio-recognition produces the structures for many clinical implementations like drug delivery, cancer imaging, diagnostic assay, etc. Artificial bio-recognition elements are the crucial phase of the contemporary bioreceptor assay that carries out selective and sensitive analysis. Artificial bio-binders have played the role of traditional biological recognition elements since the late 1970s [7].

Nowadays, where speed recovery is the key, along with the improvement of sensing characteristics such as stabilization, selection, and cost friendly, research on bio-recognition surfaces and their applications have been carried out. Atomic force microscopy (AFM) is used to study molecular recognition procedure [8,9]. The force of binding between two pair-like molecules (such as antigen-antibody) is measured by pico-newton sensitivity. It can be measured by breaking the covalent bond between those molecules onto the AFM nib and a substrate, respectively [10]. The AFM nib carried out with antibodies offers a nano-testbed, which is quite similar to an antibody conjugated nano ensemble, for examining the biorecognition-ability of the antibodies at a nanostructured surface [11].

### 14.1.1.2 Characterization of Complex Biological Surfaces

Implant materials or prosthetic devices in the biological environment come in contact with blood. For this reason, the implant surface should maintain electro-static potential. Moreover, the surface potential should be negative as blood cells do possess negative zeta potential. Of the negative potential, the blood cells and other compositions of blood could not attach to the repellent surface [12,13]. But the designed surfaces on implants result in blood-surface interaction and ultimately trigger thrombosis and embolism.

Henceforth, to counteract this event, the surface of the implant should be engineered to repel blood cells. This can be done by coating it with neointimal tissue. In these gradient surfaces, the progressively varying chemical compositions on the surface give rise in thickness, wettability, and di-electric constant along with many physiochemical properties [14].

However, the properties of smart surfaces or surface materials can be remarkably tailored by external agents (i.e., magnetic and electric fields in a regulated way, pH, moisture, stress, strain, temperature, etc). Different types of smart surface materials can be designed because of their altered properties. Many active biomaterials are chosen to combine the effect of multiple active additives within the same biomaterial for considerable and interdependent effects [15].

## 14.2 SMART BIOMATERIALS

### 14.2.1 Multiple Grades of Smart Biomaterials

INERT: By offering a mere biocompatible property and lesser risk to the host body, the bio-inert materials do not show bioreactive interactions. Thus, it is accepted by the host body without any absurd reaction.

ACTIVE: Particular and uncontrolled therapeutic effects are offered by active biomaterials. These are tailored and structured for the particular peculiar behavior in the biological environment and its surroundings.

RESPONSIVE: Biomaterials with responsive attributes receive and react to particular external/internal stimuli. Thus, the required biofunctions were executed by the materials.

AUTONOMOUS: These self-sufficient, futuristic biomaterials can adjust their properties according to the altered bio-environments and processes. These emergent biomaterials have a tremendous impact, can receive a stimulus, release a required payload, and can regulate their properties according to changing conditions.

### 14.2.2 Immune Modulatory Materials

A new group of biomaterials is introduced for bone regeneration purposes, contemporarily. Moreover, they can bring change in immunobiological behavior in the host body. Any biomaterials can manipulate the host immune system to induce hard tissue regeneration, systemically or locally. Immune-bioengineering overarching goals are to explore and enhance the immune system, numerous immune cells, like T and B lymphocytes, macrophages, dendritic cells, and polymorphonuclear leukocytes/neutrophils that participate in an appropriate manner in immune responses. There is plenty of evidence indicating the therapeutic effects of bio-glass in the orthopedic field; among all, immunomodulatory bio-glass has been remarkably utilized for bone tissue engineering. Moreover, bio-glass possesses wound healing as well as skin repair properties. According to immunomodulatory approaches, macrophage immune cells are the initial target. In vaccines, adjuvants help the host body for producing a better and enhanced immunological response. A physically and chemically engineered toxoid is a kind of toxin that poses less risk to the body, thereby ultimately helping the body to retain its immunogenicity. The nanostructured toxoid shows better immunogenicity towards methicillin-resistant *S. aureus* skin diseases [16,17].

## 14.3 CURRENT TRENDS

Rather than organ transplant, tissue engineering focuses in an enhancing way on the self-healing ability of damaged organs or tissues. Modern biomaterials and 3D printing are dedicated to building a platform that responds to neighbor exogenic and endogenic stimuli like ionic strength, temperature, and other environmental factors. With a high-precision 3D printer, it can be constructed by the addition of desired biomaterials [18]. By digital designs, in a very cost-effective way, one can design the required implants of any shape according to the damaged area. Selective laser sintering, inkjet-based 3D printing, fused deposition model, pressure-aided micro syringe, or stereolithography are used as different 3D printing procedures. But these 3D-printing scaffolds are stiff in nature, which offers less biological responses to the 4D printers, however, offer advantages over 3D printed implants. It can mimic both the tissue structure and its dynamic function ability [19]; further research and applications of 4D-printed structures are leading the way for promising success in bio-tissue engineering.

## 14.4 SURFACE ANALYSIS

The productivity and effectiveness of prosthetic devices and implants can be improved by doing alterations in surface properties. Moreover, in the future, surface modification can significantly affect the chemical composition, hydrophobicity, ionic group presence, morphology, and surface topography. But surface analysis also plays a very important role in surface modification. With the help of static secondary ion mass spectroscopy (SSIMS) and X-ray photoelectron spectroscopy (XPS), surface analysis (mainly for determining the chemical structure) can be executed. Along with these, scanning electron microscopy (SEM) is used to observe the surface morphology and cellular response to the synthetic biomaterial or implants. For elementary mapping, transmission electron microscopy (TEM), scanning auger microscopy, and SIMS are other widely used techniques [20].

## 14.5 CLASSIFICATION BASED ON SURFACE MODIFICATION

We cannot modify the implant surface of one size and expect that to fit all purposes and requirements. The shape and size of the required model may differ from site to site or depend on the degree of damage. Differences may be seen in surface modification according to the implementation and type of materials used to construct the implant (Figure 14.2). Corrosion and erosion resistance, along with biological and physio-chemical modifications are doable to improvise the implant surface properties.

### 14.5.1 CHEMICAL SURFACE MODIFICATION

For immobilizing the surface biomolecules, through chemical persevere of functioning moieties (silanization, florination, sulphonate substitution, and acetylation) and operationalization or implementation of pre-existing functional moieties like oxidation and reduction can be done.

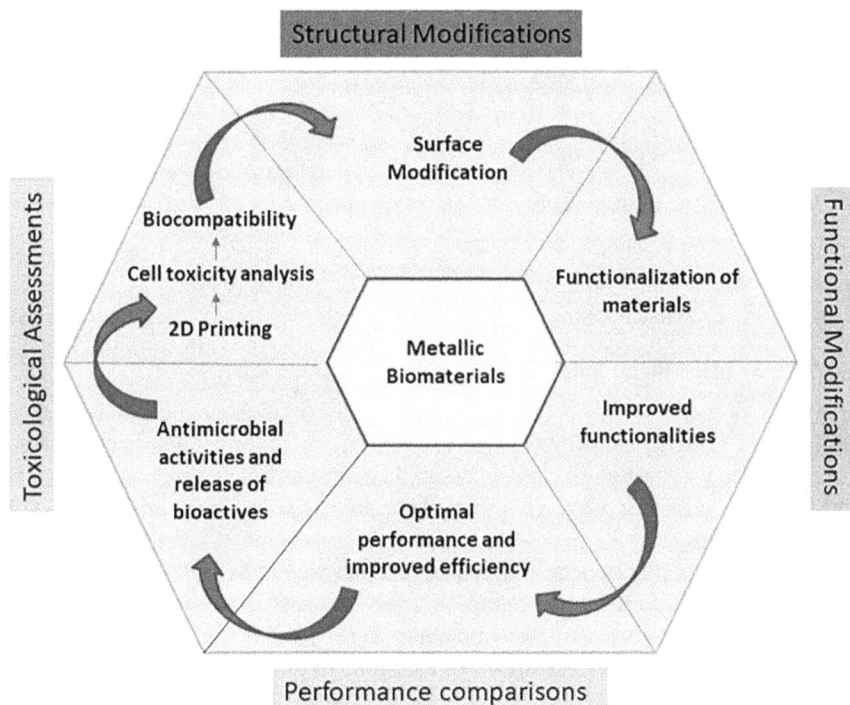

**FIGURE 14.2** Schematic illustration of processing and functional modifications of metallic biomaterials.

By forming a covalent bond, biomolecular adsorption onto the implant surface takes place. For a better crystal phase and surface morphology (at the nano and microscopic level), wet chemical procedures are convenient. By the introduction of multifunctional groups, prolonged chemical stability can be offered through surface grafting polymerization. As the metallic bio-implants lack an undeviating line of sight, the chemical modifications precisely depend upon the chemical reaction that is seen at the affiliation between the reactive fluid and implants. For this vast research, modern 3D structures are used for generating the hierarchical topography on a larger surface, ultimately on an implant. Obviously, a single implant surface alteration procedure cannot cover all the medical domains. For example, doing a surface modification against antimicrobial colonization does not cover osseointegration augmentation. Many researchers are in search of the achievement of multifunctional objectives and address most of the medical domain [21,22].

## 14.5.2 BIOCHEMICAL SURFACE MODIFICATION

To change the biochemical composition and the properties of the biomaterial, a biochemical surface modification procedure is adopted. Three types of adsorptive interactions are there in nature, i.e., adsorption, trapping, and covalent bond formation. Involvement of proteins results in better bone formation. These proteins remain immobilized and inactive on the implant surface [23]. Corrosion can cause toxicity and show other side effects like tissue infection, which ultimately leads to implant failure.

### 14.5.3 PHYSIOCHEMICAL SURFACE MODIFICATION

Development in a thermal spraying process is trending nowadays. These layering-based procedures are very handy in the extraction of energy like chemical decomposition, self-cleaning of over-layering (photocatalysis based), materials with biocompatibility, electronic-based function abilities, and many other applications. However, other than the thermal spaying procedure, the pulsed-laser deposition procedure (PLD) [24,25] helps in the ablation and condensation of work pieces onto the substrate surface. With the help of an ion-beam casting procedure, the preparation of pure epitaxial thin films at lower temperatures can be executed in a smoother way. This thin film production is nearly impossible with other techniques.

Surface topography can be tailored as per our needs by a variety of surface modifications. Based on the titanium proportion and surface topography, there exist different commercially available implants or prosthetic devices [26]. So, the clinicians can make a distinguishable selection of implants for the patients.

## 14.6 VARIOUS APPROACHES OF MODIFICATION

The fundamental ideas underlying many of the surface modification techniques under investigation have been around for a while. Although earlier research focused on investigating how protein concentrations and cells react to uniform as well as 2D surfaces with micro- and nano-topological features, there has recently been more focus on patterned surfaces to examine how geometric confinements affect cell spreading, in surfaces with gradients of chemical compositions, or in surfaces with varying protein surface densities [27] to investigate how proteins react to asymmetric surfaces, or else for high-throughput examination. The methods for microwell surfaces, gradient surfaces, and surface patterning typically make use of conventional surface modification techniques (typically thin coatings) or slight variations for the production of desired chemistries, in addition to the methods for producing specific gradients, patterns, and prior microfabrication of microwell units. Utilization of thiol-SAMs in early investigations for their capacity to produce well-controlled surface chemistries is one example and it continues to be widely used, even in recent studies. Since the last few decades, gas plasma techniques have seen promising requirements in biomaterials; improved apparatus and diagnostics have led to advancements in surface chemistry customization, application to patterning [28,29], and the development of gradient surfaces [30]. It is possible to generate precise patterns and deposit plasma polymer layers along with surface interactions useful for biomedical applications (like amine groups), with the help of atmospheric pressure microplasma jets. For a variety of applications, including tissue engineering, bioreactors, and biosensors, specific control of chemical activities and assignment of biomolecules, including growth factors, peptides, extracellular matrix proteins, and antibodies, is likely to be very helpful [31]. Future developments in the screening of smart biomaterial surfaces in 1, 2, and 3D are projected to include a variety of integrated skills, such as various manufacturing techniques, cutting-edge surface, and imaging analysis systems, etc.

## 14.7 APPLICATIONS OF SMART BIOMATERIALS

Various developed biomaterials with enhanced stimuli responsive and regenerative properties were engineered. These materials can perform irreversible and reversible modification in both the chemical composition and physical ability according to the physical stimuli (external electric and magnetic fields, light radiation, mechanical stimuli, ultrasound, etc), endogenic disease micro-environment (exposed ROS, medium acidic environment, endogenic electric field, proper ionic proportion, immunity probe environment, etc.) or mixing peripheral of above-mentioned phenomena [32].

### 14.7.1 RESPONDING TO INTERNAL STIMULI

Implant-cells or implant-tissue interactions in the presence of many biological processes related to osteoinduction and osteoconduction are regulated by material-istic properties such as surface topography and chemistry, mechanical properties, and charge. Various inner-body environment stimulus response phenomena are outlined and elaborated as follows [33].

#### 14.7.1.1 Oxidative Species Responsive Strategy

ROS (e.g., hydroxyl-radicals, peroxide, superoxide, nascent oxygen) are oxygen-containing active chemical agents. In bone tissue engineering, this ROS plays an essential role. Recent observations and studies are mainly focusing on the development of synthetic multifunctional biomaterials. Butyrate incorporated Ni-Ti LDH films can prevent bacteria colonization. Also, it can exhibit tumor formation and growth by expressing $H_2O_2$ into the tumor and surrounding environment, resulting in the release of butyral (cytotoxic) to suppress the metastasis and promote osteogenesis. Limited effective area and shorter life span of ROS, however, significantly hinder the stimuli responsive effect [34].

#### 14.7.1.2 Acidic Environment Responsive Strategy

A normal human body has a mildly alkaline micro-environment. However, for a few adverse conditions, it may be disturbed. By incorporation of calcium peroxide ($CaO_2$) and iron oxide ($Fe_3O_4$), a composite scaffold of akermanite (AKT)- $Fe_3O_4$-$CaO_2$ was structured. AKT here is to promote bone induction ability in the injury sites. The $CaO_2$ incorporation leads to adequate $H_2O_2$ production in a mild acidic tumor environment. Henceforth, $H_2O_2$ is catalyzed with $Fe_3O_4$ NPs for the production of -OH (hydroxyl radicals). This hydroxyl radical is a ROS that shows higher antibacterial efficacy. Therefore, ROS complements the therapeutic approaches of magnetic hyperthermia and bone remodeling [35,36]. For enhanced bone tissue growth and bifunctional bone dysfunction treatments, a few acidic environment responsive biomaterials have been designed. The construction of nano sacrificial layering is a fine example of it [37].

### 14.7.1.3 Endogenous Electric Field Responsive Strategy

By different studies, we have reached the conclusion that an external electric field promotes osteogenesis. The endogenic negatively charged electric field is supplied to the injured sites for the fabrication of electropositive nanofilms. The endogenic electronegative potential in injured sites for fabrication of nanofilms. It helps in osseointegration and biological self-healing ability. However, the protracted toxicity of the selected biomaterial and its prolonged stimuli-responsive nature should be examined thoroughly for further secure applications [38,39].

### 14.7.1.4 Specific Ionic Concentration Responsive Strategy

Ionic bond strength often varies from one kind of biofluid to another kind. For a given pathological environment, there exist different ion releases. A CS-polyelectrolyte and HA (CS- HA) develop scaffoldings and fill them with bovine serum albumin for studying the ion release ability. This kind of biomaterial possesses lower cytotoxicity and a higher drug capsulate property and is more biocompatible in nature [40].

### 14.7.1.5 Specific Enzyme Responsive Strategy

Enzymes are responsible for activation of various biological processes i.e., cell adhesion procedure and protein expression. Bio-reorganization and catalyzing physio-chemical material exchange are the major abilities for which enzymes are used for redesigning smart biomaterials. As per the substrate selection, the enzymes also exhibit astonishing selectivity, for which they activate biochemical reactions [41]. A construction of an enzyme-responsive base (LBL@MSN-Ag NPs) can be taken by enriching Ag-NPs in mesoporous silica NPs (MSNs) and combining poly-L-glutamic acid (PG) and polyallylamine hydrochloride [42]. This kind of exceptional enzyme responsive base shows good antibacterial efficacy and also promotes the tissue regeneration process [43]. Although this approach shows optimistic and excellent outcomes, certain challenging problems need to be resolved for future applications.

### 14.7.1.6 Specific Immune Environment Responsive Strategy

The cooperative activities of different immunological cells comprise the complicated living body immunological system. This cooperative activity is responsible for different cytokine formation. After implantation, an immune system may detect the implant as external or foreign material and immediately show immunological responses against it. Complete encapsulation was performed on the implant material by the immunological cells, assisted by the secretion of inflammatory cytokines. Formation of a neighboring bone microenvironment is obviously controlled by immune cells like microphages, T-lymphocytes, B- lymphocytes, mast and dendrite cells, and neutrophils [44,45]. Besides chronic diseases, patients with acute inflammatory reactions and cancer could get an adverse response from the effectiveness of smart specific immune environment responsive biomaterials. For obvious reasons and results, it is clever and efficient to initiate bone tissue engineering and regeneration based on the site-specific immunological environment of lesions; a number of problems need to be overlooked [46].

## 14.7.2 Responding to External Stimuli

Intensity of light, ultrasound, mechanical stimuli, external electrical as well as magnetic field stimulation, and many other external stimuli result in heat production and help in cellular adhesion, proliferation, and division in scaffolds, which ultimately result in osteoinduction and osteoconduction. Based on these studies, smart biomaterials are being fabricated [47]. Different varieties of external stimuli responsive methodologies are outlined and described below.

### 14.7.2.1 Piezoelectric Responsiveness

Piezoelectric phenomena are referred as the transformation of mechanical strength or energy into electrical energy. Such electrical signals/responses eventually serve to stimulate bone tissue growth and regeneration [48]. Modern synthetic redesign smart nanostructured biomaterials are capable of producing electric responses under stress or strain. These are identified as piezoceramics and belong to the class of superior mechanoelectrical transducers. The basic biochemical stimuli that affect osteoinduction, osteoconduction, regeneration, and self-healing ability of injured sites are developed from an endogenic direct electric current [49]. The piezoelectric biomaterials i.e., piezo-bioceramics (such as barium titanate, zinc oxide, magnesium silicate, etc.) and piezopolymers (e.g., polyvinylidene fluoride, poly hydroxybutyrate, etc.) have shown promising approaches in clinical domains. A non-conductive material such as quartz ($SiO_2$) has been the most dominant material that can give us a quantitative and qualitative overview of piezoelectric interaction. Moreover, the piezoelectric properties of bioceramics are quite requisite. Lead zirconium titanate is a widely used ceramic piezoelectric material. In addition to its piezoelectric property of ceramics, potassium sodium niobate (KNN) is being kept in sight for its effectiveness [50,51]. For better results in bone tissue regeneration, a novel composite of HA and $BaTiO_3$ via slip casting was developed. The composites gained better piezoelectric characteristics through consequent polarization [52]. Later, the development of a periosteum structure mimicking scaffoldings took a far-forward step in the bone regeneration of critical-sized bone defects by using piezoelectric property-based smart biomaterials. Nanosized piezoelectric interfaces play an essential role in the preparation of nanomedicines, which are used to stimulate the growth of cells and tissues as well [53]. The piezoelectric responsive biomaterial provides a biologically sound environment that smoothly regulates stem cell function ability without any external growth factors and effective drugs. Associated to it, a few limitations like higher temperature, effect of densification, and volatilization of alkali are there in synthesis processes, which need to be addressed.

### 14.7.2.2 pH Responsiveness

The materials that possess responses against pH level change and show new functions are categorized into pH stimuli smart materials. Generally, the pH levels of different parts of the human body vary accordingly [54]. On the basis of these pH differences, a wide range of materials have so far been developed, with regards to their ability to respond. Fabrication and investigation of pH-responsive materials have been carried out in a variety of forms, such as hollowed particles, three-dimensional porous structures, nano beads, and hydrogels. Studies on N-carboxymethyl chitosan and injectable hydrogel were carried out [55]. This study described that the changes in pH promote physical and chemical modification of the polymeric system to release the capsulated drug/antibiotics. Later, the application of sodium bicarbonate PLGA (polylactic-co-glycolic acid) hollow microspheres was studied. Water and carbon dioxide gas formation by carbonic acid and bicarbonate, respectively, takes place, for which the PLGA shell was ruptured, by resulting in the capsulated drug release. According to a study report, the dissolution and drug release at pH 8.0 have been observed for a hydrogel comprised of disintegrated hollowed particles, embedded into the poly(acrylamide) network [56]. In the acidic environment of the stomach, these pH-responsive materials were able to preserve the original morphological properties and drug release due to the breakdown of the disintegrated hollowed particles in a higher pH microenvironment. Another way for pH-responsive material production is by introducing acid-labile chemical bonds into a polymer. Although there is considerable progress in the field of pH-responsive material, certain challenges still are intact, among which, ineffective penetration depth into the deep and critical tumors and aggregation of nano-structured carriers within the tumor environment are very impactful [57–59].

### 14.7.2.3   Photo Responsiveness

Smart materials that can react to external light are termed *photo-responsive materials*. To accomplish effective therapies, regulated characteristics and non-intrusiveness of photo-triggered smart biomaterials are particularly crucial. Light is one of the most alluring environmental stimuli because of its adaptable and manageable characteristics. Besides, light-responsive stimuli can be utilized as a biomarker to monitor drug placement and targeting, as well as to visualize tumors using optical imaging techniques. Due to promising findings, this technique has been the subject of numerous studies, leading to the synthesis of various multifunctional materials. Photothermal activation, photodynamic, and photoactivation of chemotherapeutics are three prime categories of photo-responsive therapeutics. Azobenzenes (Azo) and spiropyrans (SP) are the two most often employed light-responsive substances. Since light-responsive biomaterials typically respond to shorter wavelengths of light, UV light has been frequently employed to stimulate drug release in these materials [60–64]. Fabricated light-responsive glycopolymer micelles built of Azo and β-galactose units, in order to generate targeted cancer therapeutics, were studied [65]. Recently, a group of researchers developed a multifunctional scaffold for effective osteogenesis by directly integrating biopolymer nanosheets within poly(lactic-co-glycolic acid) (PLGA) [66]. Apart from persisting tremendous outcomes in the treatment of osseous tumors, a photo-responsive approach was developed to address the challenging problems of infectious surgical sites as well. NIR light-responsive smart biomaterials were further fabricated for drug delivery applications; these materials are harmless for tissues as well as undergoing deep tissue penetration because of lower attenuation. In spite of recent advancements, the ability to manufacture numerous light-responsive smart biomaterials in bulk quantities is constrained by their highly difficult and technique-sensitive production processes [67,68].

### 14.7.2.4 Magnetic Responsiveness

Magnetic nanomaterial research has proven to possess numerous advantageous effects in a wide range of industrial along commercial applications, including magnetic storage, electrical, photonic, electrical, and biomedical therapeutics. "Magnetoresponsive materials" are defined as "materials that react as a stimuli carrier when a magnetic field is applied on it." Numerous methods, such as electrochemical synthesis, high-temperature thermal decomposition, laser pyrolysis, carbon arc, combustion, chemical vapor deposition, co-precipitation method, microbial synthesis, etc. have been successfully utilized to fabricate magneto-responsive biomaterials [69,70]. The magnetic 10Fe5Ca MBG scaffolds ($Fe_3O_4$-$CaO$-$SiO_2$-$P_2O_5$ system) synthesized may generate heat when exposed to an externally applied magnetic field. Additionally, the decreased ion dissolution rate and potential for maintaining a favorable pH level could also increase osteoblast cell proliferation, osteogenic expression, and alkaline phosphatase (ALP) activities as well [71]. In order to incorporate magnetic hyperthermia with tissue regeneration, combined calcium peroxide ($CaO_2$) and iron oxide ($Fe_3O_4$) NPs with an AKT scaffold (designated AKT-Fe$_3$O4-CaO$_2$) with the help of 3D-printing methodology shows better results. To cure particular types of cancers, magnetic nanoparticle-mediated hyperthermia has been fabricated [72]. Due to the magnetic traits of magneto nanoparticles in fluid, the technique effectively transforms the applied magnetic field into heat [73]. It is advantageous to be able to externally control the drug release since magnetic field generators can raise drug contents for sudden symptom flares. Additionally, as the agent approaches exhaustion, magnetic stimulation can be employed to keep medication elution rates constant. Even though all of the research on magneto-responsive bone repair and regeneration is still in its early stages, the positive outcomes are apprehending greater considerations [74].

### 14.7.2.5 Thermo Responsiveness

Those smart biomaterials that are responsive to temperature change are termed *thermo-responsive biomaterials*. A lot of research and applications have been carried out by scientists using these biomaterials. Thermo-responsive hydrogels are among those stimulus-responsive that display a sol-to-gel change when exposed to the appropriate physiological stimuli, like interaction with body pH (7.4) or temperature (37°C). The lower critical solution temperature (LCST) and the upper critical solution temperature (UCST) are the two primary types of polymers, respectively. Thermally responsive nanoparticle composites enable external control of polymer characteristics by combining the capacity of some metallic nanoparticles to transform external stimuli into heat with polymers that exhibit drastic characteristic changes in response to temperature alterations [75–77]. Numerous synthetic/artificial techniques, such as free radical copolymerization, photopolymerization, self-assembly method, copolymerization, sol-gel transition phase method, and end-group functionalization, have been effectively used to deliver various types of thermos-responsive smart biomaterials. Furthermore, medications are also delivered using certain features of thermo-responsive smart biomaterial. Controlled external stimuli can be taken into consideration by these biomaterials in order to administer the medications at a proper time and concentration. To promote cell growth and proliferation, thermo-responsive smart materials are frequently employed in tissue engineering [78,79]. In particular, by using stacking techniques to create thick tissues, these thermo-responsive polymers can improve cell-to-cell interactions, thereby further facilitating their usage for regenerative therapies. Despite having their own benefits, these polymers are only useful for a restricted number of practical applications due to major stability concerns [80].

### 14.7.2.6    Electro Responsiveness

Both in animal studies as well as in clinical trials, electrical stimulation has been shown to accelerate bone repair and help sustain the stemness of mesenchymal stem cells (BMSC). Electroactive polymers (EAPs), such as polythiophene, polypyrrole, and polyaniline, are the building blocks of electrical-responsive stimuli [81,82]. Because of their potential use in numerous sectors, including medication administration, robotics, artificial muscles, actuators, sensors, optical systems, as well as energy harvesting, this novel family of materials drew attention rapidly. Essentially, there are two types of electro-responsive materials: The first type is called ionic electro-responsive materials, those in which an electric field causes an increase or/and decrease in the local concentration of ions in a substance or solution. Electrostrictive polymers and dielectric elastomers are part of a second group. Since redox reactions are often reversible, various electro-responsive devices respond to repeated stimulation in a pulsatile on-off manner. In order to respond to a localized electrical stimulation, the loading of electroactive biomaterials like conductive polymers, inorganic electroactive materials, graphene, and carbon nanotubes, is a direct method. This results in better regulation of cell activities and enhanced regeneration effects. Apart from directly enhancing the drug release from conductive materials, electric stimuli are also utilized in order to upgrade the cellular intake of nanoparticles or drugs, and this technique is termed *electroporation*. Electric stimulation offers a quick and affordable way to modulate drug release as estimated by clinicians. Despite having their own advantages, their applications are restricted by relatively higher voltage and lengthy electrical potential activities [83–87].

## 14.8    CONCLUSION

This book chapter deals with the role of smart biomaterial surfaces, their technical applicability, their pros and cons, and an idea to further enhance their applications in bone tissue engineering. Today, at the existing time, no such implants have been developed that can endure the issues related to human lives, although novel materials with enhanced properties have the ability to tackle the problems such as reducing medical complexity, thereby inducing more sophisticated and realistic results to the patients. Since the last few decades, plenty of research has been consumed to resolve the issues like biocompatibility, which is the key factor in biomaterial applications. Apart from biocompatibility, various biomedical applications, including targeted cancer imaging and drug delivery, and diagnostic assays, are based on antibody recognition. Antibodies, particularly those that have been conjugated to interfaces, experience stability problems when kept in a non-refrigerated environment. Hence, it is of great importance to increase the stability of antibodies on interfaces as well as nanostructures for nano-biotechnology applications, under both elevated and ambient temperatures. There should be a thorough knowledge of structure together with the chemistry of a solid-liquid interface in order to develop an appropriate biomaterial that can be placed in given biological circumstances for bone tissue applications. Metallic biomaterial surfaces are frequently chemically altered to enhance their biological capabilities for therapeutic purposes. Moreover, it can be observed that research into surface modification is expanding beyond tribological issues like hardness, corrosion, and wear resistance of the modified layer to take into account the therapeutic factors including cell adhesion, cell proliferation, and antibacterial efficacy. There are numerous ways that surface science technology can be utilized in biomaterial applications. To assess particular surface interactions, systematic modifications in surface structure, chemical composition, and characteristics can be taken into consideration. Many of the procedures that are taken into consideration for surface modification and biomaterials analysis are neither novel nor original; rather, they are techniques that have been utilized for years in other fields of technology. It is necessary to develop a properly engineered surfaces in order to perform them effectively for a variety of applications; therefore, it is crucial to promote interdisciplinary research in order to better understand the surface properties and their co-relation with the complex bio-environment. All biomaterials must have certain surface characteristics; hence, consequently, the grafting of stimuli-responsive polymers, like chemical ($CO_2$, pH, etc.), physical (magnetic field, electric field, light, humidity, temperature, pressure), and biological stimuli can possess various biomaterial applications. To date, numerous practices have been carried out in order to encounter efficient methods for the immobilization of nanoparticles onto the interface as well as for the preparation of particular scaffolds with desired properties. For regenerative applications, stimuli-responsive biomaterials that respond to environmental changes have become a topic of discussion. When there exist specific stimuli, modified biomaterials can be utilized in order to alter the polymers so as to discharge a drug in response to it. There are three types of stimuli that can be brought into applications: internal (includes those who respond to variation in microenvironment pH or specific enzymes), external (includes acoustic energy, light, or electromagnetic tec), and multiple stimuli. These advancements give clinicians and researchers great opportunities to enhance the recent biomaterials along with a better future scope for the same. Until the recommended stimulation is achieved, the payload is a major drawback that needs to be further modified in order to achieve a smart biomaterial surface. From *in vitro* preclinical treatment, it is evident that some polymeric biomaterials have exceeded in clinical trials in terms of safety and efficacy. Moreover, to further enhance the field of drug delivery, clinical assessments of safety need to be improved. Recent studies of smart biomaterials offer peculiar capabilities and therapeutic effects in cooperation with particular stimuli, but to further enhance the diagnosis, multi-responsive nanomaterials are of greater concern where individual smart biomaterials can perform multiple functions.

## 14.9 FUTURE SCOPE

More than 30 years ago, when surface science first gained prominence as a distinct field of research, there was barely any knowledge about even elementary model systems; moreover, there were limited number and complexity in experimental as well as theoretical tools available than they are now. Surface electrical structure was mainly undiscovered. The situation is now reversed for simple systems; total energies can be estimated, electronic and atomic structures can be determined, and lattice dynamics of surfaces are well understood. The use of progressive tissue engineering methods offers a promising ability for the researchers to identify the COVID-19/host interaction; further, it also provides a great means of regenerating the lung tissues that have been destroyed. By evoking a significant inflammatory response in the lungs, the COVID-19 infection severely damages the lower respiratory system and, in some cases, there is a necessity for tissue regeneration. Due to their anti-inflammatory along with tissue-regenerating properties, exogenous mesenchymal stem cells have been successfully used in research to heal damaged lung tissues and it has improved the survival rate of animal models. Significant progress has been made in the design and characterization of various smart material surfaces to a particular stimuli response, as detailed in the chapter, although there are still a few issues that need to be resolved for more effective future applications. For better future applications, well-developed piezoelectric materials must be carefully taken into account. Future thermo-responsive materials may also be used in the treatment of chronic illnesses like diabetes, which require regular medication. Dual or multi-stimuli-response smart materials were used with successful therapeutic outcomes in a variety of biomedical applications. While engineered smart biomaterials have played a transformative role in the last 100 years, from inert biomaterials used as a temporary replacement for missing teeth to complex autonomous materials that are able to detect and react to various environmental stimuli, on the other hand naturally occurring materials have occasionally been used for medication since the last thousands of years. As scientists have become more familiar with their capabilities, we predict that demand for biomaterials with immunomodulatory properties will increase over the coming decades. Because they can control the immune system of the host to foster an environment where healing, regeneration, and repair are encouraged as well as controlled, these immunomodulatory biomaterials will represent the future generation of more adaptive and smarter biomaterials. An intensive effort must be made to amplify the smart biomaterials from laboratory use to clinical purposes, once their research and development in the lab are finished so that human life can benefit from the unprecedented capabilities like repair and regeneration of smart biomaterials (Figure 14.3).

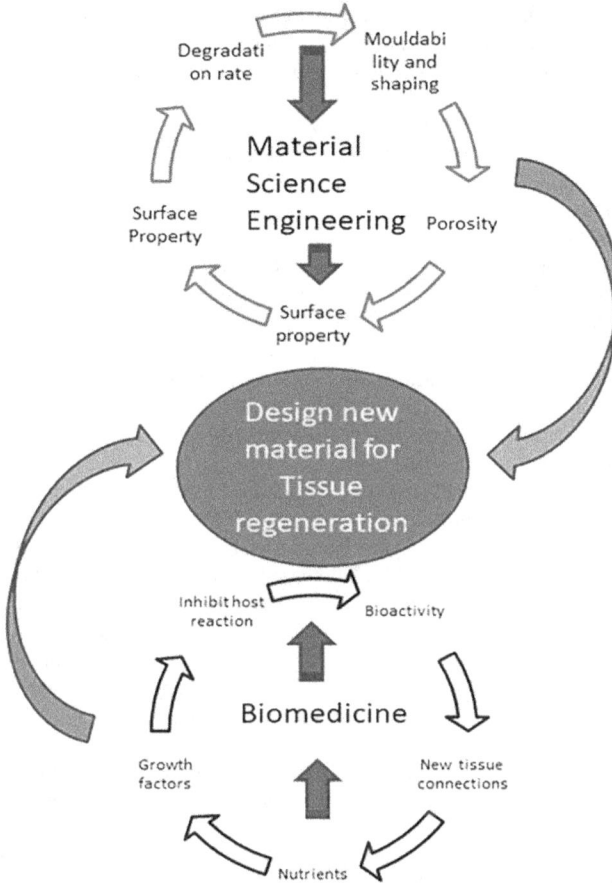

**FIGURE 14.3** Necessities for the optimal designing of smart biomaterial.

## REFERENCES

1. Kasemo, B., and J. Lausmaa. "Surface science aspects on inorganic biomaterials." *CRC Crit. Rev. Clin. Neurobiol.; (United States)* 4 (1986).
2. Rautray, Tapash R., R. Narayanan, and Kyo-Han Kim. "Ion implantation of titanium based biomaterials." *Progress in Materials Science* 56, no. 8 (2011): 1137–1177.
3. Swain, Subhasmita, and Tapash Ranjan Rautray. "Effect of surface roughness on titanium medical implants." *Nanostructured Materials and their Applications* (2021): 55–80.
4. Hsu, Robert Wen-Wei, Chun-Chen Yang, Ching-An Huang, and Yi-Sui Chen. "Investigation on the corrosion behavior of Ti–6Al–4V implant alloy by electro-chemical techniques." *Materials Chemistry and Physics* 86, no. 2-3 (2004): 269–278.
5. Williams, David F. "Biocompatibility pathways: Biomaterials-induced sterile inflammation, mechanotransduction, and principles of biocompatibility control." *ACS Biomaterials Science & Engineering* 3, no. 1 (2017): 2–35.
6. Dumitriu, Severian, and Valentin Popa, eds. *Polymeric Biomaterials: Medicinal and Pharmaceutical Applications, Volume 2.* Vol. 2. CRC Press, 2013.

7. Rautray, Tapash R., R. Narayanan, Tae-Yub Kwon, and Kyo-Han Kim. "Surface modification of titanium and titanium alloys by ion implantation." *Journal of Biomedical Materials Research Part B: Applied Biomaterials* 93, no. 2 (2010): 581–591.

8. Rich, Rebecca L., and David G. Myszka. "Advances in surface plasmon resonance biosensor analysis." *Current Opinion in Biotechnology* 11, no. 1 (2000): 54–61.

9. Bazak, Remon, Mohamad Houri, Samar El Achy, Serag Kamel, and Tamer Refaat. "Cancer active targeting by nanoparticles: a comprehensive review of literature." *Journal of Cancer Research and Clinical Oncology* 141 (2015): 769–784.

10. Bedabrata, Saha, and Prins Menno WJ. "How antibody surface coverage on nanoparticles determines the activity and kinetics of antigen capturing for biosensing." (2014).

11. Rautray, T. R., and K-H. Kim. "Nanoelectrochemical coatings on titanium for bioimplant applications." *Materials Technology* 25, no. 3-4 (2010): 143–148.

12. Heinz, William F., and Jan H. Hoh. "Spatially resolved force spectroscopy of biological surfaces using the atomic force microscope." *Trends in Biotechnology* 17, no. 4 (1999): 143–150.

13. Rautray, T. R., V. Vijayan, and S. Panigrahi. "Synthesis of hydroxyapatite at low temperature." *Indian Journal of Physics* 81 (2007): 95–98.

14. Dammer, Ulrich, Martin Hegner, Dario Anselmetti, Peter Wagner, Markus Dreier, Walter Huber, and Hans-Joachim Güntherodt. "Specific antigen/antibody interactions measured by force microscopy." *Biophysical Journal* 70, no. 5 (1996): 2437–2441.

15. Swain, Subhasmita, Chris Bowen, and Tapash Rautray. "Dual response of osteoblast activity and antibacterial properties of polarized strontium substituted hydroxyapatite—Barium strontium titanate composites with controlled strontium substitution." *Journal of Biomedical Materials Research Part A* 109, no. 10 (2021): 2027–2035.

16. Takagi, Toshinori. "A concept of intelligent materials and the current activities of intelligent materials in Japan." In *First European Conference on Smart Structures and Materials*, vol. 1777, pp. 23–28. SPIE, 1992.

17. Nakkala, Jayachandra Reddy, Ziming Li, Wajiha Ahmad, Kai Wang, and Changyou Gao. "Immunomodulatory biomaterials and their application in therapies for chronic inflammation-related diseases." *Acta Biomaterialia* 123 (2021): 1–30.

18. Swain, Subhasmita, Rabindra Nath Padhy, and Tapash Ranjan Rautray. "Polarized piezoelectric bioceramic composites exhibit antibacterial activity." *Materials Chemistry and Physics* 239 (2020): 122002.

19. Castro, Nathan J., Christoph Meinert, Peter Levett, and Dietmar W. Hutmacher. "Current developments in multifunctional smart materials for 3D/4D bioprinting." *Current Opinion in Biomedical Engineering* 2 (2017): 67–75.

20. Behera, Dipti Rani, Pratibindhya Nayak, and Tapash Ranjan Rautray. "Phosphatidylethanolamine impregnated Zn-HA coated on titanium for enhanced bone growth with antibacterial properties." *Journal of King Saud University-Science* 32, no. 1 (2020): 848–852.

21. Bagno, Andrea, Monica Dettin, and G. Santoro. "Characterization of Ti and Ti6Al4V surfaces after mechanical and chemical treatments: a rational approach to the design of biomedical devices." *J. Biotechnol. Biomater* 2, no. 7 (2012): 151–157.

22. Swain, Subhasmita, Rabindra Nath Padhy, and Tapash Ranjan Rautray. "Electrically stimulated hydroxyapatite–barium titanate composites demonstrate immunocompatibility in vitro." *Journal of the Korean Ceramic Society* 57, no. 5 (2020): 495–502.

23. Ratner, B. D., A. Chilkoti, and D. G. Castner. "Contemporary methods for characterizing complex biomaterial surfaces." *Biologically Modified Polymeric Biomaterial Surfaces* (1992): 25–36.

24. King, William J., and Paul H. Krebsbach. "Growth factor delivery: how surface interactions modulate release in vitro and in vivo." *Advanced Drug Delivery Reviews* 64, no. 12 (2012): 1239–1256.

25. Zhou, Jing, Xiaodong Guo, Qixin Zheng, Yongchao Wu, Fuzai Cui, and Bin Wu. "Improving osteogenesis of three-dimensional porous scaffold based on mineralized recombinant human-like collagen via mussel-inspired polydopamine and effective immobilization of BMP-2-derived peptide." *Colloids and Surfaces B: Biointerfaces* 152 (2017): 124–132.

26. Fleer, G. J., J. M. H. M. Scheujtens, M. A. Cohen-Stuart, B. Vincent, and T. Cosgrove. "Polymers at interfaces, chapman hall." *London-Glasgow-New York* (1993).

27. Rautray, Tapash R., R. Narayanan, Tae-Yub Kwon, and Kyo-Han Kim. "Surface modification of titanium and titanium alloys by ion implantation." *Journal of Biomedical Materials Research Part B: Applied Biomaterials* 93, no. 2 (2010): 581–591.

28. Lee, Ki-Won, Cheol-Min Bae, Jae-Young Jung, Gi-Bong Sim, Tapash Ranjan Rautray, Hyo-Jin Lee, Tae-Yub Kwon, and Kyo-Han Kim. "Surface characteristics and biological studies of hydroxyapatite coating by a new method." *Journal of Biomedical Materials Research Part B: Applied Biomaterials* 98, no. 2 (2011): 395–407.

29. Culver, Heidi R., Adam M. Daily, Ali Khademhosseini, and Nicholas A. Peppas. "Intelligent recognitive systems in nanomedicine." *Current Opinion in Chemical Engineering* 4 (2014): 105–113.

30. Ogaki, Ryosuke, Folmer Lyckegaard, and Peter Kingshott. "High-resolution surface chemical analysis of a trifunctional pattern made by sequential colloidal shadowing." *ChemPhysChem* 11, no. 17 (2010): 3609–3616.

31. Foley, Jennifer O., Elain Fu, Lara J. Gamble, and Paul Yager. "Microcontact printed antibodies on gold surfaces: function, uniformity, and silicone contamination." *Langmuir* 24, no. 7 (2008): 3628–3635.

32. Kamila, Susmita. "Introduction, classification and applications of smart materials: an overview." *American Journal of Applied Sciences* 10, no. 8 (2013): 876.

33. Shamsoddin, Erfan, Behzad Houshmand, and Mehdi Golabgiran. "Biomaterial selection for bone augmentation in implant dentistry: a systematic review." *Journal of Advanced Pharmaceutical Technology & Research* 10, no. 2 (2019): 46.

34. Tan, ShihJye, Josephine Y. Fang, Zhi Yang, Marcel E. Nimni, and Bo Han. "The synergetic effect of hydrogel stiffness and growth factor on osteogenic differentiation." *Biomaterials* 35, no. 20 (2014): 5294–5306.

35. Jalili, Nima A., Manish K. Jaiswal, Charles W. Peak, Lauren M. Cross, and Akhilesh K. Gaharwar. "Injectable nanoengineered stimuli-responsive hydrogels for on-demand and localized therapeutic delivery." *Nanoscale* 9, no. 40 (2017): 15379–15389.

36. Satpathy, Anurag, Rinkee Mohanty, and Tapash R. Rautray. "Bio-mimicked guided tissue regeneration/guided bone regeneration membranes with hierarchical structured surfaces replicated from teak leaf exhibits enhanced bioactivity." *Journal of Biomedical Materials Research Part B: Applied Biomaterials* 110, no. 1 (2022): 144–156.

37. Praharaj, Rinmayee, Snigdha Mishra, and Tapash R. Rautray. "Growth mechanism of aligned porous oxide layers on titanium by anodization in electrolyte containing Cl." *Materials Today: Proceedings* 62 (2022): 6216–6220.

38. Xie, Feiyu, Mina Wang, Qishuang Chen, Tiange Chi, Shijie Zhu, Peng Wei, Yingying Yang, Le Zhang, Xuexin Li, and Zehuan Liao. "Endogenous stimuli-responsive nanoparticles for cancer therapy: from bench to bedside." *Pharmacological Research* (2022): 106522.

39. Liu, Mengrui, Hongliang Du, Wenjia Zhang, and Guangxi Zhai. "Internal stimuli-responsive nanocarriers for drug delivery: design strategies and applications." *Materials Science and Engineering: C* 71 (2017): 1267–1280.

40. Khatoon, Shakera, Hwa Seung Han, Minchang Lee, Hansang Lee, Dae-Woong Jung, Thavasyappan Thambi, M. Ikram, Young Mo Kang, Gi-Ra Yi, and Jae Hyung Park. "Zwitterionic mesoporous nanoparticles with a bioresponsive gatekeeper for cancer therapy." *Actabiomaterialia* 40 (2016): 282–292.

41. Swain, Subhasmita, Tae Yub Kwon, and Tapash R. Rautray. "Fabrication of silver doped nano hydroxyapatite-carrageenan hydrogels for articular cartilage applications." *bioRxiv* (2021): 2020-12.

42. Rautray, Tapash Ranjan, and Kyo Han Kim. "Synthesis of controlled release sr-hydroxyapatite microspheres." In *Bioceramics-24*. 2012.

43. Guan, Qian, Yilei Fang, Xu Wu, Ranwen Ou, Xinyu Zhang, Hao Xie, Mengyu Tang, and Guisheng Zeng. "Stimuli responsive metal organic framework materials towards advanced smart application." *Materials Today* (2023).

44. Wang, Yuxiao, Jiaxin Li, Maomao Tang, Chengjun Peng, Guichun Wang, Jingjing Wang, Xinrui Wang, Xiangwei Chang, Jian Guo, and Shuangying Gui. "Smart stimuli-responsive hydrogels for drug delivery in periodontitis treatment." *Biomedicine & Pharmacotherapy* 162 (2023): 114688.

45. Abedi, F., P. Ghandforoushan, F. Adeli, M. Yousefnezhad, A. Mohammadi, S. V. Moghaddam, and S. Davaran. "Development of stimuli-responsive nanogels as drug carriers and their biomedical application in 3D printing." *Materials Today Chemistry* 29 (2023): 101372.

46. Swain, Subhasmita, Janardhan Reddy Koduru, and Tapash Ranjan Rautray. "Mangiferin-enriched Mn–hydroxyapatite coupled with β-TCP scaffolds simultaneously exhibit osteogenicity and anti-bacterial efficacy." *Materials* 16, no. 6 (2023): 2206.

47. Montoya, Carolina, Yu Du, Anthony L. Gianforcaro, Santiago Orrego, Maobin Yang, and Peter I. Lelkes. "On the road to smart biomaterials for bone research: Definitions, concepts, advances, and outlook." *Bone Research* 9, no. 1 (2021): 12.

48. Jacob, Jaicy, Namdev More, Kiran Kalia, and Govinda Kapusetti. "Piezoelectric smart biomaterials for bone and cartilage tissue engineering." *Inflammation and regeneration* 38, no. 1 (2018): 2.

49. Ihlefeld, Jon F. "Fundamentals of Ferroelectric and Piezoelectric Properties." In *Ferroelectricity in Doped Hafnium Oxide: Materials, Properties and Devices*, pp. 1–24. Woodhead Publishing, 2019.

50. Rautray, Tapash R., Subhasmita Swain, and Kyo-Han Kim. "Formation of anodic TiO2 nanotubes under magnetic field." *Advanced Science Letters* 20, no. 3-4 (2014): 801–803.

51. Swain, Subhasmita, Tapash Ranjan Rautray, and Ramaswamy Narayanan. "Sr, Mg, and Co substituted hydroxyapatite coating on TiO2 nanotubes formed by electrochemical methods." *Advanced Science Letters* 22, no. 2 (2016): 482–487.

52. Minary-Jolandan, Majid, and Min-Feng Yu. "Nanoscale characterization of isolated individual type I collagen fibrils: polarization and piezoelectricity." *Nanotechnology* 20, no. 8 (2009): 085706.

53. Kitsara, Maria, Andreu Blanquer, Gonzalo Murillo, Vincent Humblot, Sara De Bragança Vieira, Carme Nogués, Elena Ibáñez, Jaume Esteve, and Leonardo Barrios. "Permanently hydrophilic, piezoelectric PVDF nanofibrous scaffolds promoting unaided electromechanical stimulation on osteoblasts." *Nanoscale* 11, no. 18 (2019): 8906–8917.

54. Li, Shengliang, Kelei Hu, Weipeng Cao, Yun Sun, Wang Sheng, Feng Li, Yan Wu, and Xing-Jie Liang. "pH-responsive biocompatible fluorescent polymer nanoparticles

based on phenylboronic acid for intracellular imaging and drug delivery." *Nanoscale* 6, no. 22 (2014): 13701–13709.

55. Wei, Menglian, Yongfeng Gao, Xue Li, and Michael J. Serpe. "Stimuli-responsive polymers and their applications." *Polymer Chemistry* 8, no. 1 (2017): 127–143.

56. Qu, Jin, Xin Zhao, Peter X. Ma, and Baolin Guo. "pH-responsive self-healing injectable hydrogel based on N-carboxyethyl chitosan for hepatocellular carcinoma therapy." *Acta Biomaterialia* 58 (2017): 168–180.

57. Swain, Subhasmita, R. D. K. Misra, C. K. You, and Tapash R. Rautray. "TiO2 nanotubes synthesised on Ti-6Al-4V ELI exhibits enhanced osteogenic activity: A potential next-generation material to be used as medical implants." *Materials Technology* 36, no. 7 (2021): 393–399.

58. Rautray, Tapash R., and Kyo Han Kim. "Synthesis of Mg2+ incorporated hydroxy-apatite by ion implantation." In *Key Engineering Materials*, vol. 529, pp. 114–118. Trans Tech Publications Ltd, 2013.

59. Shi, Shengyu, Yajing Liu, Yu Chen, Zhihuang Zhang, Yunsheng Ding, Zongquan Wu, Jun Yin, and Liming Nie. "Versatile pH-response micelles with high cell-penetrating helical diblock copolymers for photoacoustic imaging guided synergistic chemo-photothermal therapy." *Theranostics* 6, no. 12 (2016): 2170.

60. Park, Wooram, Byoung-chan Bae, and Kun Na. "A highly tumor-specific light-triggerable drug carrier responds to hypoxic tumor conditions for effective tumor treatment." *Biomaterials* 77 (2016): 227–234.

61. Alonso-Cristobal, Paulino, Olalla Oton-Fernandez, Diego Mendez-Gonzalez, J. Fernando Díaz, Enrique Lopez-Cabarcos, Isabel Barasoain, and Jorge Rubio-Retama. "Synthesis, characterization, and application in HeLa cells of an NIR light responsive doxorubicin delivery system based on NaYF4: Yb, Tm@ SiO2-PEG nanoparticles." *ACS Applied Materials & Interfaces* 7, no. 27 (2015): 14992–14999.

62. Hui, Liwei, Shuai Qin, and Lihua Yang. "Upper critical solution temperature polymer, photothermal agent, and erythrocyte membrane coating: an unexplored recipe for making drug carriers with spatiotemporally controlled cargo release." *ACS Biomaterials Science & Engineering* 2, no. 12 (2016): 2127–2132.

63. Huang, Peng, Zhiming Li, Jing Lin, Dapeng Yang, Guo Gao, Cheng Xu, Le Bao et al. "Photosensitizer-conjugated magnetic nanoparticles for in vivo simultaneous magneto-fluorescent imaging and targeting therapy." *Biomaterials* 32, no. 13 (2011): 3447–3458.

64. Mohapatra, Bijayinee, and Tapash R. Rautray. "Facile fabrication of Luffa cylindrica-assisted 3D hydroxyapatite scaffolds." *Bioinspired, Biomimetic and Nanobiomaterials* 10, no. 2 (2021): 37–44.

65. Lan, M., S. Zhao, W. Liu, C. S. Lee, W. Zhang, and P. Wang. "Photosensitizers for photodynamic therapy". Adv Healthc Mater 8, no. 13 (2019): e1900132.

66. Pierini, Filippo, Paweł Nakielski, Olga Urbanek, Sylwia Pawłowska, Massimiliano Lanzi, Luciano De Sio, and Tomasz Aleksander Kowalewski. "Polymer-based nanomaterials for photothermal therapy: from light-responsive to multifunctional nanoplatforms for synergistically combined technologies." *Biomacromolecules* 19, no. 11 (2018): 4147–4167.

67. Mohapatra, Bijayinee, and Tapash R. Rautray. "Strontium-substituted biphasic calcium phosphate scaffold for orthopedic applications." *Journal of the Korean Ceramic Society* 57 (2020): 392–400.

68. Shell, Thomas A., and David S. Lawrence. "Vitamin B12: a tunable, long wavelength, light-responsive platform for launching therapeutic agents." *Accounts of Chemical Research* 48, no. 11 (2015): 2866–2874.

69. Mishra, Saswati, and Tapash R. Rautray. "Fabrication of Xanthan gum-assisted hydroxyapatite microspheres for bone regeneration." *Materials Technology* 35, no. 6 (2020): 364–371.

70. Laurent, Sophie, Silvio Dutz, Urs O. Häfeli, and Morteza Mahmoudi. "Magnetic fluid hyperthermia: focus on superparamagnetic iron oxide nanoparticles." *Advances in Colloid and Interface Science* 166, no. 1-2 (2011): 8–23.

71. Guisasola, Eduardo, Alejandro Baeza, Marina Talelli, Daniel Arcos, María Moros, Jesus M. de la Fuente, and María Vallet-Regí. "Magnetic-responsive release controlled by hot spot effect." *Langmuir* 31, no. 46 (2015): 12777–12782.

72. Swain, Subhasmita, Joo L. Ong, Ramaswamy Narayanan, and Tapash R. Rautray. "Ti-9Mn β-type alloy exhibits better osteogenicity than Ti-15Mn alloy in vitro." *Journal of Biomedical Materials Research Part B: Applied Biomaterials* 109, no. 12 (2021): 2154–2161.

73. Praharaj, Rinmayee, Snigdha Mishra, R. D. K. Misra, and Tapash R. Rautray. "Biocompatibility and adhesion response of magnesium-hydroxyapatite/strontium-titania (Mg-HAp/Sr-TiO2) bilayer coating on titanium." *Materials Technology* 37, no. 4 (2022): 230–239.

74. Rautray, Tapash R., and Kyo Han Kim. "Synthesis of Mg2+ incorporated hydroxy-apatite by ion implantation." In *Key Engineering Materials*, vol. 529, pp. 114–118. Trans Tech Publications Ltd, 2013.

75. Ren, Zhiwei, Yang Wang, Shiqing Ma, Shun Duan, Xiaoping Yang, Ping Gao, Xu Zhang, and Qing Cai. "Effective bone regeneration using thermosensitive poly (N-isopropylacrylamide) grafted gelatin as injectable carrier for bone mesenchymal stem cells." *ACS Applied Materials & Interfaces* 7, no. 34 (2015): 19006–19015.

76. Liao, Han-Tsung, Chien-Tzung Chen, and Jyh-Ping Chen. "Osteogenic differentiation and ectopic bone formation of canine bone marrow-derived mesenchymal stem cells in injectable thermo-responsive polymer hydrogel." *Tissue Engineering Part C: Methods* 17, no. 11 (2011): 1139–1149.

77. Pentlavalli, Sreekanth, Philip Chambers, Binulal N. Sathy, Michelle O'Doherty, Marine Chalanqui, Daniel J. Kelly, Tammy Haut-Donahue, Helen O. McCarthy, and Nicholas J. Dunne. "Simple radical polymerization of poly (alginate-graft-N-isopropylacrylamide) injectable thermoresponsive hydrogel with the potential for localized and sustained delivery of stem cells and bioactive molecules." *Macromolecular Bioscience* 17, no. 11 (2017): 1700118.

78. Swain, Subhasmita, and Tapash R. Rautray. "Estimation of trace elements, antioxidants, and antibacterial agents of regularly consumed Indian medicinal plants." *Biological Trace Element Research* 199, no. 3 (2021): 1185–1193.

79. Swain, S., and T. R. Rautray. "Silver doped hydroxyapatite coatings by sacrificial anode deposition under magnetic field." *Journal of Materials Science: Materials in Medicine* 28 (2017): 1–5.

80. Rautray, Tapash R., Bijayinee Mohapatra, and Kyo-Han Kim. "Fabrication of strontium–hydroxyapatite scaffolds for biomedical applications." *Advanced Science Letters* 20, no. 3–4 (2014): 879–881.

81. Pillay, Viness, Tong-Sheng Tsai, Yahya E. Choonara, Lisa C. du Toit, Pradeep Kumar, Girish Modi, Dinesh Naidoo, Lomas K. Tomar, Charu Tyagi, and Valence M. K. Ndesendo. "A review of integrating electroactive polymers as responsive systems for specialized drug delivery applications." *Journal of Biomedical Materials Research Part A* 102, no. 6 (2014): 2039–2054.

82. Atoufi, Zhale, Payam Zarrintaj, Ghodratollah Hashemi Motlagh, Anahita Amiri, ZohrehBagher, and Seyed Kamran Kamrava. "A novel bio electro active alginate-aniline tetramer/agarose scaffold for tissue engineering: synthesis, characterization, drug release and cell culture study." *Journal of Biomaterials Science, Polymer Edition* 28, no. 15 (2017): 1617–1638.

83. Wang, Jinmin, Xiao Wei Sun, and Zhihui Jiao. "Application of nanostructures in electrochromic materials and devices: recent progress." *Materials* 3, no. 12 (2010): 5029–5053.
84. Mishra, Saswati, and Tapash R. Rautray. "Silver-incorporated hydroxyapatite–albumin microspheres with bactericidal effects." *Journal of the Korean Ceramic Society* 57 (2020): 175–183.
85. Praharaj, Rinmayee, Snigdha Mishra, and Tapash R. Rautray. "The structural and bioactive behaviour of strontium-doped titanium dioxide nanorods." *Journal of the Korean Ceramic Society* 57 (2020): 271–280.
86. Sun, Yi-xin, Ke-feng Ren, Yi-xiu Zhao, Xiang-sheng Liu, Guo-xun Chang, and Jian Ji. "Construction of redox-active multilayer film for electrochemically controlled release." *Langmuir* 29, no. 35 (2013): 11163–11168.
87. Rautray, Tapash R., and Kyo Han Kim. "Synthesis of silver incorporated hydroxyapatite under magnetic field." In *Key Engineering Materials*, vol. 493, pp. 181–185. Trans Tech Publications Ltd, 2012.

# 15 Energy Biomaterial Surface

*Debasis Nayak*
Department of Materials Science and Metallurgy, University of Cambridge, Cambridge, UK

*Ajit Behera*
Department of Metallurgical & Materials Engineering, National Institute of Technology, Rourkela, Odisha, India

## 15.1  INTRODUCTION

Transplant and repair of tissue have become an easy affair with continuous progress in cell biology, medicine for tissue healing, and scaffolding of tissues, which facilitate the artificial regeneration of several biological systems [1–3]. Thus, an in-depth knowledge of cell characteristics that include cell adhesion, growth, proliferation, and migration in vitro on an artificial surface are important aspects to understand in life science and materials research [4]. Several researchers tried to understand the effect of surface properties on cellular behaviors [5–8]. When a metal/polymeric substrate and biological cells are in contact while suspended in a fluid system, the surface properties, namely surface energy, charge, etc. play a vital role in several factors, such as adsorption of serum protein, cell attachment, and cell proliferation [9–11]. Previous studies suggest that cells tend to develop on higher energy surfaces, discounting the fact that they have been functionalized with proteins or not [12–15]. In the case of polymer surfaces, polar groups react with cell surface groups and form chemical bonds, and non-polar groups tend to form short-range van der Waals interactions. On the other hand, a few studies revealed that low-energy surfaces (superhydrophobic) provide a preferential site for cell growth [8,16,17]. Moreover, chemical influence and surface topographical factors (roughness and porosity) are other factors to influence cell growth [6,8,17].

Like metal surfaces, polymer surfaces also significantly influence cell-substrate interactions. On a similar note, surface topology on polymers also impart significantly the chemical interactions and cellular behaviors [13]. Surface properties such as roughness promote mechanical anchoring to the cells. The cells must adhere to the surface and proliferate afterward. Thus, surface topography also defines cell growth parameters and protein adsorption. However, a few results show a contrasting result of a decrease in cell adhesion and growth with an upsurge in surface micro/nano roughness [1,18,19]. As mentioned, the surface properties influence highly the surface adhesion of biomaterials with tissues. Thus, several researchers tried to understand

DOI: 10.1201/9781003429920-17

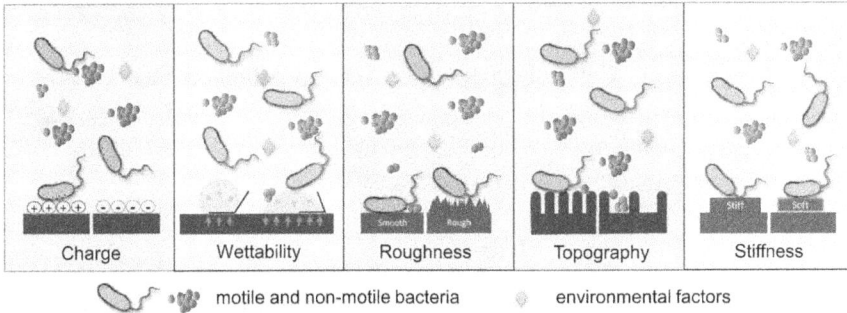

| Charge | Wettability | Roughness | Topography | Stiffness |

🦠🌀 •🦠 motile and non-motile bacteria  🌀 environmental factors

**FIGURE 15.1** Schematic diagram of several surface parameters that affect bacterial adhesion. Bacterial adhesion and growth is influenced by several surface properties, such as surface charge density, wettability, roughness, topography, and stiffness. This schematic reflects the significant aspect of bacterial response owing to each surface parameter [3].

surface properties and their effect on biocompatibility. Moreover, many tried to modify the biomaterial surface properties to enhance their biocompatibility. A schematic description of several surface parameters and their influence on bacterial adhesion on cell growth is demonstrated in Figure 15.1.

Various research demonstrated the formation of biofilm is instigated by bacterial adhesion to a surface. Several factors, such as bacteria exposure time, initial population, bacterial features (e.g., cell wall constituents, periphery, and motility), and type of nutrients, play a huge role in bacterial adhesion to a surface. Moreover, other surface properties discussed above also play important roles in influencing beginning bacterial adhesion to surfaces (Figure 15.1) [3]. A comprehensive list of such approaches to studies of surface properties and enhancement methods and effects are mentioned below.

## 15.2 SURFACE ROUGHNESS

Cellular adhesion is an important aspect to be considered in the growth of scaffolds for tissue engineering. This involves the design of biosensors and in making antibacterial substrates [20]. Moreover, bacterial infection and biofilm growth can affect several facets of society, such as bio-induced corrosion of piping in industries and other materials and serious health hazards in contaminated individuals [6]. The physico-chemical characteristics of the bacterium, sub-stratum, and the surrounding environment increase the propensity of bacterial cells to a substratum surface. Such physico-chemical contacts are van der Waals (LW), Lewis acid-base (AB), and electrostatic (EL) interactions relying on the surrounding fluid, as well as by the material surfaces and the bacterial membranes. However, as mentioned, surface topography and sanitation efficiency's role in bio-adhesion may sometimes act as safe zones for bacteria. On the other hand, a very distinct topography decreases the cell adhesion [21].

The growth of bacterial cells on several material surfaces such as glass, silica, metals, and polymers has been investigated to a greater extent. The reduction of the

**TABLE 15.1**

**Surface Properties of Cell-Substrate Using Several Attached Media [24]**

| Contact angle | Surface energy $(mJ/m^2)$ | Attachment media | Attached biomass colonization (g/L) |
|---|---|---|---|
| 72.1 | 40-50 | Cellulose acetate–nitrate membranes | – |
| 92.9 | 36.4 | Latex rubber | 40 |
| 84.9 | 30.9 | Plexiglass | 31.99 |
| 96.34 | 49.28 | CA-CN membrane | 10.67 |
| 81.5 | 42.64 | PTFE membrane | 0.05 |
| 33.8 | 36.8 | Nitrocellulose membrane | 1.2 |

initial adhesion of bacterial cells on the surfaces has been looked into for several approaches, such as chemical modification/functionalization of surfaces, development of self-assembled monolayers, and manipulating the surface topography, etc. [6].

Study of the influence of random surface topology on bacterial adhesion and proliferation have been conducted for at least the last three decades [6,21,22]. However, a clear understanding of how these random topological features can influence colonization has not been understood fully. Gentile et al. [23] showed that moderate surface roughness can enhance stable cell adhesion and proliferation. Although they observed it for four distinct cell lines, corresponding to two dissimilar species, proliferating over seven electrochemically etched silicon substrates having distinct surface features for three days. They studied (i) substrate topography, by measuring the spectral density of power along with the root mean square (RMS) value of roughness; and (ii) by making a compelling micro/ nanofabrication protocol for developing substrates where both roughness and fractal dimension could be taken into consideration to be controlled independently. Table 15.1 shows how surface characteristics can impact biomass colonization growth.

## 15.3 SURFACE CHARGE

Surface charge density is one of the significant surface characteristics that influence bacterial adhesion quantity onto surfaces. Van der Waals forces and electrostatic interactions play vital roles in bacterial adhesion onto surfaces of different materials. Many have proposed the mechanism by which surface charge affects bacterial initial adhesion. Moreover, apart from initial bacterial attachment, surface charge density can also impact biofilm assembly on material surfaces [3]. Several researches imply that negatively charged surfaces can bring down bacterial adhesion on surface, but other reports give contradictory evidence [25–27]. Table 15.2 below describes a summary of such contradictions of charge on the surface of bacterial adhesion followed by biofilm growth.

**TABLE 15.2**
**Influence of Surface Charge on Bacterial Adhesion and Biofilm Growth** [3]

| Microorganism Type | Surface Material | Consequences on Biofilm Growth and Adhesion |
|---|---|---|
| *Escherichia coli (E. coli)* | Polythylene sheets properties are improved by radiation-induced graft polymerization (RIGP) of an epoxy-group that has monomer glycidyl methacrylate (GMA) | Bacterial adhesion increases on positively charged surface, biofilm was dense, homogenous, and uniform |
| Pseudomonas, *E. coli,* and *Staphylococcus aureus* (*S. aureus*) | Positively charged poly(acrylic acid) (PAA) and poly(diallyl dimethyl ammonium chloride) (PDADMAC) | Increased bacterial adhesion on a positively charged surface |
| *Pseudomonas aeruginosa* (*P. aeruginosa*) | Poly(allylamine hydrochloride)/sodium poly(4-styrenesulfonate) (PAH/PSS) polyelectrolyte multilayers | Bacterial adhesion increases on a positively charged surface |
| *S. aureus* and *E. coli* | Gold coated plates with thin thiol layers of 1-octanethiol, 1-decanethiol, 1-octadecanethiol, 16-mercaptohexadecanoic acid, and 2-amino ethanethiol hydrochloride | Hydrophilic substrates having a positive charge promote bacterial adhesion and biofilm thickness |
| *E. coli* | Cationicpolyvinylamine (PVAm)/ anionic cellulose nanofibril/PVAm fabricated layer-by-layer (LbL) | Increased bacterial adhesion and proliferation as surface charge increases |
| *S. aureus* and *E. coli* | Several layers of polyethylenimine | Negatively charged surface decreases bacterial adhesion |
| *S. mutans* | Chimaeric peptide-mediated nanocomplexes of carboxymethyl chitosan/amorphous calcium phosphate (CMC/ACP) | Bacterial adhesion decreases on a positively charged surface |

## 15.4 SURFACE POLARITY

Surface polarity also influences a lot the adhesion of tissues and bacteria growth. In order to comprehend the effect of surface polarity, Allion et al. [21] used two coatings on stainless steel: silicon oxide (hydrophilic) and polysiloxane (hydrophobic). They found that cell adhesion decreases on a smooth surface with a polar coating, and increases with surface polarity. On a similar note, Alavarsa et al. [15] deposited a thin (~ 60 nm) xanthan gum (XG)-amino acid coating on different substrates, such as Si/SiO$_2$ wafers, quartz cuvettes, and glass slides. Apart from XG, the grafting of glutamic acid (Glu), cysteine (Cys), histidine (His), or tryptophan (Trp) to the XG chains decreases polar surface energy ($\gamma$pS), thereby decreasing the adhesion of SH-SY5Y cells from 30% or 60% of that obtained for the control (plastic plate) following 3 h or 24 h of the incubation period, respectively. On the contrary, the dispersive surface energy ($\gamma$dS) component was not affected by the process.

Again, in another study, Ratheesh et al. [12] developed a polar-functionalized biochar in PVA/CBC (polyvinyl alcohol/chemically engineered biochar) to get better bacterial interaction. These developed redox-active multi-functionalized and perforated biochar-based electrodes work as a bioanode and thus improve microbial activities to achieve a current density of up to $2.53 \times 10^2$ mA m$^{-2}$. Moreover, they got moderate hydrophilicity (64.8°) and effective surface area (11.82 m$^{-2}$ g$^{-1}$) that increased bacteria adhesion and subsequent colonization. Thus, an effective charge interaction was achieved, leading to increased electron transfer.

## 15.5  BIOMATERIAL SURFACE WETTABILITY

Surface wettability highly impacts cell adhesion. Moreover, the surface functional group and its density also affect cell growth. Arima et al. [28] studied the impact of surface wettability on human umbilical vein endothelial cells (HUVECs) and human cervical carcinoma (HeLa) cells. Both HUVECs and HeLa cells adhere well on moderately wettable surfaces (contact angles of 40–60°) of polymers. They found that HUVECs could not grow properly on adsorbed hydrophobic self-assembled monolayers (SAMs) when pretreated with albumin. Thus, albumin can adsorb on the surface and create a hindrance towards replacement by adhesion of cell proteins on hydrophobic SAMs. However, cells were easy to adhere to albumin-adsorbed hydrophilic SAMs. Moreover, the hydrophilic SAMs are effective in displacing the preadsorbed albumin with cell adhesive proteins. This was possible due to moderate wettability that gives suitable surfaces for cell adhesion.

The surface wettability can be modulated with surface acoustic waves (SAWs), thus giving it an advantage of manipulating cell adhesion. Surface acoustic waves (SAWs) are an impactful technique due to low power requisite, facile fabrication methods, and non-invasive nature. Nampoothiri et al. [29] carried out an experimental and theoretical model to understand the effect of SAW-based generation and the effect of micron-sized droplets through atomization and the role of substrate wettability on it. They studied the effect of these water droplets on wettable substrates with water contact angles (WAC) varying from 5° to 145°. The contact line diameter and polydispersity index describe the interaction between wettable substrates and droplets. They found out that polydispersity decreases when WAC increases; simultaneously, the inheritance between droplet–droplet increases, i.e., the number density of droplets decreases for an increased WCA. Hence, one can choose a substrate with an appropriate WAC depending on the application requirement.

Liu et al. [30] prepared flexible LIG/PPy (laser-induced graphene/polypyrrole) composite electrodes employing electrochemical deposition of PPy on LIG through laser etching polyimide (PI) film. They studied the growth, proliferation, and differentiation of PC-12 cells on the surface of composite electrodes under electrical stimulation (ES) of 400 mV/cm with different conductivity of the composite films and ES time. In their study, they found that ES induces a field effect that does not depend on the conductivity or internal resistance of the composite electrodes. Moreover, ES time increases the growth, proliferation, and differentiation of PC-12 cells. Again, protein adsorption capacity is enhanced with a different ES time.

However, the cell proliferation was sluggish up to 4 h of ES, whereas neural phenotype increase was not evident up to 8 h of ES.

Patel et al. [31] measured the water contact angle to understand surface wettability and cell behavior. They drew a correlation between contact angles ($\theta$) before and after poly-L-lysine (PLL) modification. They used Pt electrode as a substrate for the deposition of commercially available para substituted nitro- and methoxybenzene diazonium salts, as well as meta di-substituted chlorobenzene diazonium salt. It is interesting to note that the contact angle of different diazonium salts (DAZ)-coated electrodes before a PLL coating was similar in the case of DAZ 1 (4-nitrobenzenediazonium), DAZ 2 (4-methoxybenzenediazonium), and (3,5-dichlorophenyldia-zonium) DAZ 3 ($\theta$ between 29° and 41°). Although these surfaces have different functional groups, their chemical character did not affect surface wettability. In the case of DAZ 4, surface wettability; ($\theta = 65°$) decreased to a greater extent. PLL increases the hydrophobicity of the surface in the case of all types of DAZ-coated electrodes. Thus, these surface coatings change surface wettability and provide a moderate hydrophilicity that can enhance cell culture.

Ashi et al. [32] used bioceramic cement as retrograde material to control bacterial growth and calcium phosphate surface layer to support healing. Premixed cement is easier to use than powder–liquid cement.

Oliveira et al. [26] used a chemical modification technique in order to superhydrophobic polystyrene surfaces that controlled cell attachment/proliferation. They found that employing random micro/nano roughness and introducing the surface to UVO irradiation is the best approach to control contiguously the attachment/proliferation of cells in several materials with possible uses in high-throughput analysis, microfluidic systems, or even in 3D systems.

## 15.6 PLASTICITY OF BIOCOMPARTIBILITY

The concept of plasticity in biocompartibility has been borrowed from the concept of metallurgy. In metallurgy, plasticity defines the permanent change in the shape of a solid under stress; thus, an irreversible change. Compared with biological application, it is defined as biological pathways that are not reversible. For our recent chapter, plasticity-driven processes follow substitute biocompatibility pathways; often the variance in consequence with alike technologies is due to biological plasticity instead of material or device deficiency. The above surface condition discussed promotes irreversible bacterial formation pathways that are not reversible [33]. In such a scenario, surface modification is required to enhance the surface biocompartibility properties, which are discussed in the below section.

## 15.7 SURFACE MODIFICATION

Considering all the above concerns, the surface of the biomaterials can be enhanced by modifying their properties. By enhancing the surface properties, the growth of bacteria can be reduced, biocompatibility can be enhanced, and the adhesion of tissue with material can be improved. Such surface modification techniques are coating thin films [25,30,34,35], bioactive retrograde filling [32], laser-assisted surface modification [36–38], and surface patterning [19,26,39,40], etc. There are several coating methods to form thin film, such as pulsed laser deposition [41–43] sputtering [44–46], spin coating [29,30,34], etc. Experiments suggest that coating TiAlN on Ti6Al4V surfaces is beneficial in improving biocorossion properties with an increase in corrosion potential by 150 mV [47]. The wetting angle increased from 70.61° to 86.27° by TiAlN coating. Thus, the coating benefits in reducing corrosion current density and thereby charge transfer resistance. Moreover, coating improves surface mechanical properties.

Nazarov et al. [48] used titania nanofilms and silver nanoparticles on titanium disks using atomic layer deposition (ALD) to improve antibacterial properties. Such films improved hydrophobicity and thereby improved surface antibacterial properties. Patel et al. [31] modified platinum electrodes with selected diazonium salts and poly-L-lysine. By doing so, they improved the number of cell adhesion sites. These characteristics can be assessed through the surface chemical and morphological properties with increased wettability. They used biofunctionalized electrodes as substrates to monitor the cell attachment process. These substrates helped in culturing human neuroblastoma SH-SY5Y cells. Thus, cell adhesion is improved on the surface of diazonium-modified and poly-L-lysine coated electrodes, signifying their viability in strengthening the incorporation between bioelectronic devices and neural cells.

On a similar note, the acid-etching process can be another method to change the surface morphology. Bloise et al. [49] used this technique in both the solid and porous titanium plates and were able to create a macroscopic 3D pores/rough surface. Thus, it produced nanoscale surface features on the porous polyetheretherketone titanium composite (PTC) implant endplates. The micro- and nanosurface morphology is vital in determining stem cell growth. Moreover, acid-treated surfaces influence the response of human bone-marrow-derived mesenchymal stem cells, and micro- and nanoscale features play a vital role in this process as well. Hence, acid-treated PTC implant endplates help in human stem cell proliferation, followed by the expression of early osteogenic markers. Use of bioactive peptides can improve wound dressing by surface functionalization. Such properties have been verified by in vivo studies to enhance wound healing of the pig skin. This innovative method transforms an inert dressing into a bioactive dressing, which eventually leads to wound healing [50]. Lu et al. [51] came up with an innovative solution to improve the grafting efficiency of dopamine through $CO_2$ plasma-assisted treatment. They saw that by doing so, many polar groups were introduced to the surface of polydimethylsiloxane without any cracks. Thus, it will boost the development of poly(dopamine) coating through an amidation reaction. Moreover, contact angle and adhesion results suggest that wettability and shearing adhesion of polydimethylsiloxane are improved by this method.

Jun et al. [52] developed an innovative method to promote muscle regeneration by both enhancing surface topography and biphasic electric current. They found that the surface topology impacts significantly on deciding the growth direction of human myogenic precursor cell (hMPC) compared to the applied electrical field direction. However, electrical field simulation impacts myogenic differentiation. They made surface groove patterns with a laser and found that the growth and elongation directions were aligned with these groove directions. However, electrical stimulation was not able to reverse the cell growth direction.

## 15.8 SUMMARY

Over the last decade, biomaterials and their surface properties have been a focus of study. Tuning the surface properties can increase the adhesion of cells and reduce bacteria growth. Thus, the biological processes are highly mechanosensitive. We concentrated on surface charge, wettability, roughness/topology, and polarity to understand the behaviors. Thus, proper consideration of several surface modification techniques can improve cell adhesion and lessen bacterial growth for biomaterials.

## REFERENCES

1. E. Ellermann, N. Meyer, R. E. Cameron, and S. M. Best, "In vitro angiogenesis in response to biomaterial properties for bone tissue engineering: A review of the state of the art," *Regen. Biomater.*, vol. 10, no. March, 2023, doi: 10.1093/rb/rbad027.
2. A. F. Kanaan and A. P. Piedade, "Electro-responsive polymer-based platforms for electrostimulation of cells," *Mater. Adv.*, vol. 3, no. 5, pp. 2337–2353, 2022, doi: 10. 1039/d1ma01012c.
3. S. Zheng et al., "Implication of surface properties, bacterial motility, and hydro-dynamic conditions on bacterial surface sensing and their initial adhesion," *Front. Bioeng. Biotechnol.*, vol. 9, no. February, pp. 1–22, 2021, doi: 10.3389/fbioe.2021 .643722.
4. R. G. Wells, "The role of matrix stiffness in regulating cell behavior," *Hepatology*, vol. 47, no. 4, pp. 1394–1400, 2008, doi: 10.1002/hep.22193.
5. T. Cardoso, C. N. Elias, A. B. Lemos, A. B. Soares, A. A. Pelegrine, and L. N. Teixeira, "In vitro evaluation of two tissue substitutes for gingival augmentation," pp. 23–34, 2023, doi: 10.4236/jbnb.2023.142002.
6. R. J. Crawford, H. K. Webb, V. K. Truong, J. Hasan, and E. P. Ivanova, "Surface topographical factors influencing bacterial attachment," *Adv. Colloid Interface Sci.*, vol. 179–182, pp. 142–149, 2012, doi: 10.1016/j.cis.2012.06.015.
7. W. Cho, S. H. Yoon, and T. D. Chung, "Streamlining the interface between electronics and neural systems for bidirectional electrochemical communication," *Chem. Sci.*, pp. 4463–4479, 2023, doi: 10.1039/d3sc00338h.
8. L. Chen, C. Yan, and Z. Zheng, "Functional polymer surfaces for controlling cell behaviors," *Mater. Today*, vol. 21, no. 1, pp. 38–59, 2018, doi: 10.1016/j.mattod. 2017.07.002.
9. D. H. Kim, P. P. Provenzano, C. L. Smith, and A. Levchenko, "Matrix nanotopo-graphy as a regulator of cell function," *J. Cell Biol.*, vol. 197, no. 3, pp. 351–360, 2012, doi: 10.1083/jcb.201108062.
10. A. Bandzerewicz and A. Gadomska-Gajadhur, "Into the tissues: Extracellular matrix and its artificial substitutes: Cell signalling mechanisms," *Cells*, vol. 11, no. 5, 2022, doi: 10.3390/cells11050914.
11. B. Galateanu et al., "Applications of polymers for organ-on-chip technology in urology," *Polymers (Basel)*, vol. 14, no. 9, 2022, doi: 10.3390/polym14091668.
12. A. Ratheesh, B. R. Sreelekshmy, and S. M. A. Shibli, "Multi-functionalization of woody biochar tuned for sustainable surface microbiological processes: A case study for energy applications," *Sustain. Energy Fuels*, pp. 1454–1465, 2023, doi: 10.1039/ d2se01701f.
13. N. J. Hallab, K. J. Bundy, K. O'Connor, R. L. Moses, and J. J. Jacobs, "Evaluation of metallic and polymeric biomaterial surface energy and surface roughness character-istics for directed cell adhesion," *Tissue Eng.*, vol. 7, no. 1, pp. 55–70, 2001, doi: 10. 1089/107632700300003297.

14. Y. Inoue, T. Nakanishi, and K. Ishihara, "Elastic repulsion from polymer brush layers exhibiting high protein repellency," *Langmuir*, vol. 29, no. 34, pp. 10752–10758, 2013, doi: 10.1021/la4021492.

15. A. C. Alavarse, E. C. G. Frachini, J. B. Silva, R. dos S. Pereira, H. Ulrich, and D. F. S. Petri, "Amino acid decorated xanthan gum coatings: Molecular arrangement and cell adhesion," *Carbohydr. Polym. Technol. Appl.*, vol. 4, no. June, p. 100227, 2022, doi: 10.1016/j.carpta.2022.100227.

16. R. P. Trueman et al., "Improving the biological interfacing capability of diketopyr-rolopyrrole polymers via p-type doping," *J. Mater. Chem. C*, vol. 11, no. 21, pp. 6943–6950, 2023, doi: 10.1039/D3TC01148H.

17. M. Wytrwal-Sarna et al., "The effect of the topmost layer and the type of bone morphogenetic protein-2 immobilization on the mesenchymal stem cell response," *Int. J. Mol. Sci.*, vol. 23, no. 16, 2022, doi: 10.3390/ijms23169287.

18. B. S. Calin and I. A. Paun, "A review on stimuli-actuated 3D micro/nanostructures for tissue engineering and the potential of laser-direct writing via two-photon polymeri-zation for structure fabrication," *Int. J. Mol. Sci.*, vol. 23, no. 22. 2022, doi: 10.3390/ijms232214270.

19. N. M. Alves, I. Pashkuleva, R. L. Reis, and J. F. Mano, "Controlling cell behavior through the design of polymer surfaces," *Small*, vol. 6, no. 20, pp. 2208–2220, 2010, doi: 10.1002/smll.201000233.

20. P. Decuzzi and M. Ferrari, "Modulating cellular adhesion through nanotopography," *Biomaterials*, vol. 31, no. 1, pp. 173–179, 2010, doi: 10.1016/j.biomaterials.2009.09.018.

21. A. Allion, J. P. Baron, and L. Boulange-Petermann, "Impact of surface energy and roughness on cell distribution and viability," *Biofouling*, vol. 22, no. 5, pp. 269–278, 2006, doi: 10.1080/08927010600902789.

22. R. Iwata, P. Suk-In, V. P. Hoven, A. Takahara, K. Akiyoshi, and Y. Iwasaki, "Control of nanobiointerfaces generated from well-defined biomimetic polymer brushes for protein and cell manipulations," *Biomacromolecules*, vol. 5, no. 6, pp. 2308–2314, 2004, doi: 10.1021/bm049613k.

23. F. Gentile et al., "Cells preferentially grow on rough substrates," *Biomaterials*, vol. 31, no. 28, pp. 7205–7212, 2010, doi: 10.1016/j.biomaterials.2010.06.016.

24. H. Rawindran et al., "Mechanistic behaviour of Chlorella vulgaris biofilm formation onto waste organic solid support used to treat palm kernel expeller in the recent Anthropocene," *Environ. Res.*, vol. 222, no. January, p. 115352, 2023, doi: 10.1016/j.envres.2023.115352.

25. H. Du, Z. Chen, X. Gong, M. Jiang, G. Chen, and F. Wang, "Surface grafting of sericin onto thermoplastic polyurethanes to improve cell adhesion and function," *J. Biomater. Sci. Polym. Ed.*, vol. 0, no. 0, pp. 1–16, 2023, doi: 10.1080/09205063.2023.2166339.

26. S. M. Oliveira, W. Song, N. M. Alves, and J. F. Mano, "Chemical modification of bioinspired superhydrophobic polystyrene surfaces to control cell attachment/prolif-eration," *Soft Matter*, vol. 7, no. 19, pp. 8932–8941, 2011, doi: 10.1039/c1sm05943b.

27. P. Nana, A. Plange, A. R. Aikins, K. J. Brobbey, and E. E. Kaufmann, "Cassava microfiber – reinforced gelatin scaffold holds promise for tissue engineering by exhibiting cytocompatibility with," pp. 1–12, 2023, doi: 10.1177/15353702231168143.

28. Y. Arima and H. Iwata, "Effect of wettability and surface functional groups on protein adsorption and cell adhesion using well-defined mixed self-assembled monolayers," *Biomaterials*, vol. 28, no. 20, pp. 3074–3082, 2007, doi: 10.1016/j.biomaterials.2007.03.013.

29. K. N. Nampoothiri, N. S. Satpathi, and A. K. Sen, "Surface acoustic wave-based generation and transfer of droplets onto wettable substrates," *RSC Adv.*, vol. 12, no. 36, pp. 23400–23410, 2022, doi: 10.1039/d2ra04089a.

30. R. Liu et al., "Mediate neurite outgrowth of PC-12 cells using polypyrrole-assisted laser-induced graphene flexible composite electrodes combined with electrical stimulation," *Eur. Polym. J.*, vol. 181, no. October, p. 111634, 2022, doi: 10.1016/j.eurpolymj.2022.111634.

31. T. Patel, M. Skorupa, M. Skonieczna, R. Turczyn, and K. Krukiewicz, "Surface grafting of poly-L-lysine via diazonium chemistry to enhance cell adhesion to biomedical electrodes," *Bioelectrochemistry*, vol. 152, no. October 2022, 2023, doi: 10.1016/j.bioelechem.2023.108465.

32. T. Ashi et al., "Physicochemical and antibacterial properties of bioactive retrograde filling materials," *Bioengineering*, vol. 9, no. 11, 2022, doi: 10.3390/bioengineering9110624.

33. D. F. Williams, "The plasticity of biocompatibility," *Biomaterials*, vol. 296, no. October 2022, p. 122077, May 2023, doi: 10.1016/j.biomaterials.2023.122077.

34. N. Marcuz, R. P. Ribeiro, E. C. Rangel, N. C. da Cruz, and D. R. N. Correa, "The effect of PEO treatment in a Ta-rich electrolyte on the surface and corrosion properties of low-carbon steel for potential use as a biomedical material," *Metals (Basel)*, vol. 13, no. 3, p. 520, 2023, doi: 10.3390/met13030520.

35. M. J. Jensen et al., "Cochlear implant material effects on inflammatory cell function and foreign body response," *Hear. Res.*, vol. 426, p. 108597, 2022, doi: 10.1016/j.heares.2022.108597.

36. V. Selvamani et al., "Laser-assisted nanotexturing and silver immobilization on titanium implant surfaces to enhance bone cell mineralization and antimicrobial properties," *Langmuir*, vol. 38, no. 13, pp. 4014–4027, 2022, doi: 10.1021/acs.langmuir.2c00008.

37. J. Vishnu et al., "Insights into the surface and biocompatibility aspects of laser shock peened Ti-22Nb alloy for orthopedic implant applications," *Appl. Surf. Sci.*, vol. 586, no. February, p. 152816, 2022, doi: 10.1016/j.apsusc.2022.152816.

38. B. Das, S. V. Seesala, P. Pal, T. Roy, P. G. Roy, and S. Dhara, "A vascularized bone-on-a-chip model development via exploring mechanical stimulation for evaluation of fracture healing therapeutics," *Vitr. Model.*, vol. 1, no. 1, pp. 73–83, 2022, doi: 10.1007/s44164-021-00004-7.

39. D. Blaschke, L. Rebohle, I. Skorupa, and H. Schmidt, "Local tuning of the surface potential in silicon carriers by ion beam induced intrinsic defects," *Adv. Mater. Phys. Chem.*, vol. 12, no. 11, pp. 289–305, 2022, doi: 10.4236/ampc.2022.1211019.

40. Y. Yucheng, S. Glubay, R. Stirling, Q. Ma, and J. McKenzie, "Improved fiber control through ohmic/convective flow behavior," *J. Mater. Sci.*, vol. 57, no. 22, pp. 10457–10469, 2022, doi: 10.1007/s10853-022-07304-7.

41. D. Nayak, S. Ghosh, and V. Adyam, "Thin film manganese oxide polymorphs as anode for sodium-ion batteries: An electrochemical and DFT based study," *Mater. Chem. Phys.*, vol. 217, pp. 82–89, Sep. 2018, doi: 10.1016/j.matchemphys.2018.06.065.

42. D. Nayak, S. Puravankar, S. Ghosh, and V. Adyam, "Asymmetric reaction pathway of Na+-ion during fast cycling in α- and γ-Fe2O3 thin film anode for sodium-ion battery," *Ionics (Kiel.*, vol. 25, no. 12, pp. 5857–5868, Dec. 2019, doi: 10.1007/s11581-019-03112-3.

43. A. Carradò, H. Pelletier, J. Faerber, G. Versini, and I. N. Mihailescu, "Pulsed laser deposition of thin coatings: Applications on biomaterials," *Mater. Sci. Forum*, vol. 638–642, pp. 530–535, 2010, doi: 10.4028/www.scientific.net/MSF.638-642.530.

44. A. A. Alamdari et al., "In vitro antibacterial and cytotoxicity assessment of magnetron sputtered Ti1.5ZrTa0.5Nb0.5W0.5 refractory high-entropy alloy doped with Ag nanoparticles," *Vacuum*, vol. 203, no. April, p. 111286, 2022, doi: 10.1016/j.vacuum.2022.111286.

45. M. Fernández-Lizárraga, J. García-López, S. E. Rodil, R. M. Ribas-Aparicio, and P. Silva-Bermudez, "Evaluation of the biocompatibility and osteogenic properties of metal oxide coatings applied by magnetron sputtering as potential biofunctional surface modifications for orthopedic implants," *Materials*, vol. 15, no. 15, 2022, doi: 10.3390/ma15155240.

46. E. S. Marchenko, K. M. Dubovikov, G. A. Baigonakova, I. I. Gordienko, and A. A. Volinsky, "Surface structure and properties of hydroxyapatite coatings on NiTi substrates," *Coatings*, vol. 13, no. 4, p. 722, Mar. 2023, doi: 10.3390/coatings13040722.

47. M. Hussein, M. Kumar, N. Ankah, and A. Abdelaal, "Surface, mechanical, and in vitro corrosion properties of arc-deposited TiAlN ceramic coating on biomedical Ti6Al4V alloy," *Trans. Nonferrous Met. Soc. China*, vol. 33, no. 2, pp. 494–506, 2023, doi: 10.1016/s1003-6326(22)66122-3.

48. D. Nazarov et al., "Antibacterial and osteogenic properties of Ag nanoparticles and Ag/TiO2 nanostructures prepared by atomic layer deposition," *J. Funct. Biomater.*, vol. 13, no. 2, 2022, doi: 10.3390/jfb13020062.

49. N. Bloise et al., "Early osteogenic marker expression in hMSCs cultured onto acid etching-derived micro-and nanotopography 3D-printed titanium surfaces," *Int. J. Mol. Sci.*, vol. 23, no. 13, pp. 1–18, 2022, doi: 10.3390/ijms23137070.

50. C. Pinese et al., "Bioactive peptides grafted silicone dressings: A simple and specific method," *Mater. Today Chem.*, vol. 4, pp. 73–83, 2017, doi: 10.1016/j.mtchem.2017.02.007.

51. M. Lu, L. Ding, T. Zhong, and Z. Dai, "Improving hydrophilicity and adhesion of PDMS through dopamine modification assisted by carbon dioxide plasma," *Coatings*, vol. 13, no. 1, pp. 1–11, 2023, doi: 10.3390/coatings13010126.

52. I. Jun et al., "Synergistic stimulation of surface topography and biphasic electric current promotes muscle regeneration," *Bioact. Mater.*, vol. 11, no. July 2021, pp. 118–129, 2022, doi: 10.1016/j.bioactmat.2021.10.015.

# Section 3

## Post-Modification of Biomaterial Surface

# 16 Surface Treatment of Polymeric, Ceramic, Metallic, and Composite Biomaterials for Bioimplants and Medical Device Applications

*Garima Mittal*
Independent researcher, Omaha, Nebraska, USA

*Shiladitya Paul*
Materials Innovation Centre, School of Engineering,
University of Leicester, UK

Materials Performance and Integrity Group, TWI,
Cambridge, UK

## 16.1 INTRODUCTION

Bioimplants and medical devices are designed to support, replace, or mimic the structure or function of natural tissues or organs such as bones, blood vessels, heart valves, joints, and cartilage. There are numerous biomedical implants and devices developed for various applications (Table 16.1).

Thoughtful consideration of material selection and design is paramount for the potential use of biomedical implants/devices due to many factors that influence their performance, including biocompatibility, bioactivity, strength, functionality, corrosion and wear resistance, ease of implantation, and cost-effectiveness. Typically, polymer, metal, ceramic, or composite biomaterials are utilized for designing these biomedical implants and devices, and specific materials and designs are selected based on the placement site and duration requirements. However, finding a material that satisfies all the needs for a specific biomedical implant or device is challenging. Therefore, surface treatment is carried out, which aids in optimizing the overall performance of the biomedical implant or device. Through surface treatment of biomaterials, surface texture, chemistry, surface energy, wettability, etc. can be

DOI: 10.1201/9781003429920-19

**TABLE 16.1**

**Examples of Biomedical Implants and Devices**

| Application | Function | Common Implant/Device |
|---|---|---|
| Orthopedic | Replace or repair bones and joints | Hip, knee, spinal, shoulder, and ankle implants, bone plate, and screws |
| Dental | Restore dental function or replace damaged/missing teeth | Dentures, dental (endosteal, subperiosteal, and zygomatic) implants and bridges |
| Ophthalmic | Treat conditions affecting eyes such as corneal dystrophies, iris defects, or retinitis pigmentosa | Intraocular lenses (IOLs), glaucoma implants, corneal implants, retinal implants, Iris implants |
| Cardiovascular | Treat heart and blood vessel conditions | Artificial valves, blood vessels and pacemakers, catheters, implantable cardioverter defibrillators (ICDs), and left ventricular assist devices (LVADs) |
| Neurology | Treat conditions affecting the brain and nervous system such as chronic pain, Parkinson's disease | Deep brain stimulation (DBS) devices, spinal cord stimulators, neuro and cognitive prosthetics, and catheters |

altered while keeping bulk properties of biomedical implant or devices.[1] It ultimately helps in improving tissue compatibility as well as chemical and mechanical durability. Since there is no common surface treatment method that is applicable to all biomaterials, the selection of a suitable technique depends on the substrate material and application of the implant. For a better understanding, in sections 16.2 and 16.3, different biomaterials and surface treatment methods are discussed in detail, respectively.

## 16.2   BIOMATERIALS

### 16.2.1   Polymeric Biomaterials

Polymer biomaterials, either solids or hydrogels, are gaining popularity in various biomedical applications such as in neural, cardiovascular, and dermal tissues due to their inert and biocompatible nature, as well as their broad range of customizable chemical and physical properties.[2] However, because of their inadequate mechanical strength and potential to cause immunogenic reactions, it is uncommon to use polymer biomaterials alone for bone replacement. Instead, they are typically employed as suture materials, ocular devices, biosensors, tissue substitutes, and bioadhesives. Solid polymeric biomaterials are generally used for applications that involve mechanical strength, such as polyetheretherketone (PEEK; properties close to bone) for bone replacement and hydrogels loaded with water soluble drugs, growth factors, and proteins and even live cells are appropriate for drug delivery and tissue engineering applications, where high-water content and biocompatibility is needed. Polymer biomaterial can be categorized into two types (Table 16.2):

- Bioinert polymers: Nonbiodegradable, poor surface interactions, good mechanical properties, non-bioactive, and expensive
- Biodegradable polymers: Biocompatible, biodegradable, easy processing, adjustable degradation properties, poor mechanical properties, and limited temperature resistance

While polymers exhibit unique properties, they can still show cytotoxicity, leading to bone degeneration, abnormalities, rapid corrosion, and implant degradation over time.[3] Additionally, because of high surface energy, polymers may exhibit poor wettability, resulting in inadequate protein or cell adhesion, which can induce fibrotic (scar) tissue formation. These issues can be addressed through surface modification (discussed in Section 16.3).

### 16.2.2 METALLIC BIOMATERIALS

Metallic biomaterials are typically used in dental and orthopedic implants and fixation, joint replacements, and stents because of their outstanding mechanical performance and corrosion resistance. The outstanding combination of strength and ductility is required for static or repetitive load-bearing applications of bioimplants, which is why they are often preferred over polymeric and ceramic implants. Moreover, metallic bioimplants need to be non-magnetic and highly dense to be compatible with magnetic resonance imaging (MRI) and X-ray imaging. Table 16.3 presents some examples of commonly used metallic biomaterials.

The selection of metallic biomaterials is determined by the specific application of the implant/device.[4] For instance, orthopedic implants require materials with high strength, toughness, and elasticity, while for joint replacement, materials with good wear resistance are needed. Similarly, artificial stents need materials with good plasticity for expansion and rigidity for maintaining dilation. Titanium is often selected over conventional steel or CoCr alloys owing to its superior rcorrosion resistance, mechanical performance, biocompatibility, and comparatively low weight and density.[5] However, while implanted, metallic biomaterials show some adverse effects such as inflammation, premature loss of mechanical strength, necrosis, and infections, undermining their performance.[6] To overcome these challenges and to improve bioactivity such as blood compatibility and bone conductivity, various bulk and surface modification techniques are implemented (see Section 16.3).

### 16.2.3 CERAMIC BIOMATERIALS

Ceramics biomaterials have been used in dentistry for centuries, but current technological advancements and extensive research have expanded their use in numerous biomedical applications, including biomimetic scaffolds, fillers, joint

**TABLE 16.2**

**Commonly Used Bioinert and Biodegradable Polymers in Biomedical Implants/Devices**

| Type | Polymer | Properties | Limitations | Applications |
|------|---------|-----------|-------------|--------------|
| Bioinert | Polymethyl methacrylate (PMMA) | Optical clarity, easy processing, cost-efficient | Low fracture toughness, poor wear resistance, non-biodegradable, long-term complications | Contact and intraocular lens, screw fixation in bone, as a filler for bone cavities and skull defects and vertebrae stabilization in osteoporotic patients |
| | Polyethylene (PE) | High impact strength, modifiable density | Long-term complications, not compatible with some sterilization methods | Craniomaxillo facial surgery, acetabular cup of hip implant, knee and shoulder joint replacements |
| | PTFE | Outstanding lubricity, hydrophobic, chemical resistance, high strength, no inflammation | Poor bonding properties, limited machinability, sometimes causes compression syndrome | Cardiovascular devices such as artificial heart valves, vascular grafts (bypass grafts and arteriovenous fistulas), self-expanding stents and catheters (central venous catheters, peripherally inserted central catheters (PICC), and dialysis catheters) |
| | Silicone | Highly flexible, elastic, hydrophobic, soft, good electrical properties, flame resistance, versatile, good electrical insulation | | Breast implants, oxygenator membrane, tubing, shunts, prostheses, heart peacemaker leads, heart valve structures, burn dressing |

| Category | Material | Advantages | Disadvantages | Applications |
|---|---|---|---|---|
| Biodegradable (natural) | Collagen | Versatile, promotes tissue growth, good mechanical strength and elasticity | Limited life span, sometimes triggers allergic reactions | Bone, cartilage, skin tissue regeneration, and cardiovascular repair |
| | Chitosan | Biocompatibility, promotes tissue growth, antimicrobial properties | Limited mechanical strength and stability, might trigger allergic reactions | Wound healing, drug delivery, gene therapy, and tissue engineering |
| | Gelatin | Promotes cell adhesion and proliferation | Prone to bacterial infections, limited life span | Surgical adhesives, bone, cartilage and skin regeneration, wound healing, and drug delivery |
| Biodegradable (synthetic) | Polycaprolactone (PCL) | Good mechanical strength, ease of processing, slow degradation rate | Do not stimulate tissue growth, highly hydrophobic | Sutures, absorbable surgical mesh, bone, cartilage, and nerve regeneration, orthopedic implants, and cardiovascular stents |
| | Polylactic acid (PLA) | Good mechanical properties (strength and stiffness), ease of processing | Brittleness, delayed crystallization, poor thermal stability, might spread communicable pathologies | Surgical tools, sutures, screws, pins, and plates for fracture fixation and bone healing, tissue engineering, cardiovascular implants, skin and tendon healing |

**TABLE 16.3**
**Commonly Used Metallic Biomaterials in Biomedical Implants/Devices**

| Type | Metal | Properties | Limitations | Application |
|---|---|---|---|---|
| Permanent | Stainless steel | Cost-efficient, good wear resistance | Inflammation, stress shielding | Joint replacements, screws, nails, fracture plates, dental implants, heart valves, stents, pacemaker, and surgical devices |
| | Titanium and its alloys | Excellent corrosion, good mechanical performance, biocompatibility, low density | High cost, poor tribological performance, may induce allergy/inflammation due to the prolonged use | Joint replacement, cochlear implant, dental implant, bone plate, screw, pacemaker, and surgical devices |
| | Cobalt-chromium alloys | Better fatigue strength and corrosion resistance than Ti-alloys, chemically inert | Low biocompatibility | Dental and orthopedic implants, hip and knee joint replacements, screws, rods in spinal fixation, and sutures |
| | Tantalum | Biocompatibility and outstanding corrosion resistance | High elastic modulus, difficult precision manufacturing, high temperature conductivity may induce temperature-dependent headaches | Artificial joints |
| Biodegradable | Magnesium and its alloys | Biocompatibility, nontoxic, low elastic modulus and low density, controlled degradation profile | Rapid corrosion rate and hydrogen desorption, resorption issues | Fracture fixation, stents, and bone repair |

| Emerging biometals | Nickel-titanium (NiTi; Nitinol) shape memory alloy (SMA) | Lightweight, high strength, good corrosion resistance, superelastic, no cytotoxicity | Carcinogenic and allergic reactions may arise due to the high quantity of nickel, expensive | Wires, palatal arches, intraspinal implants, intramedullary nails, staples, and devices for correcting scoliosis, spinal vertebrae spacer, self-expanding vascular stents |
|---|---|---|---|---|
| | Bulk metallic glass (BMG) | High strength, low elastic modulus, and a high elastic strain limit, low corrosion rate | Poor ductility, insufficient crack propagation, and mechanical behavior knowledge/data | Orthopedic prostheses, surgical scalpels, and flexible vascular stents Bioresorbable BMGs: bone screw or surgical plate, intramedullary nails or a temporary vascular stent |

replacements, and load-bearing parts.[7] The distinctive properties of ceramic biomaterials make them apt for different applications, for instance, high-stress applications like joints or dental implants requiring materials with high wear resistance and low friction coefficient. For tissue regeneration and osteointegration, ceramics with biocompatibility like hydroxyapatite or bioactive glass are used.[8] Moreover, ceramics have high melting temperature, corrosion resistance, and better mechanical properties with poor plasticity than metallic biomaterials. However, ceramic biomaterials possess limited sinterability and ductility, which can be resolved through doping with different materials, improving their performance as well as providing multifunctionality. However, it is very important to keep a balance among selected materials, their concentration and constituents, process parameters, and desired properties to avoid adverse effects on the host. Based on their interactions with the physiological environment during biomedical applications, ceramic biomaterials can be divided into three categories (Table 16.4):

- Bioinert: No interaction with surrounding tissues; resists corrosion
- Biodegradable/bioresorbable: Degrades after being implanted in the host and substituted by advancing tissues; no inflammation, used to treat fractured bone
- Bioactive: Osteoconductive (support bone cells to attach, migrate, grow and/or divide onto the implant/scaffold surface), possesses properties of both bioinert and biodegradable ceramics; used as coating to improve mechanical and anti-corrosive performance of implants

### 16.2.4 COMPOSITE BIOMATERIALS

Polymeric, metallic, and ceramic biomaterials have unique properties that are favorable to biomedical implants/devices, but these biomaterials alone may not meet all the requirements considered necessary for certain applications. Therefore, composites that consist of two or more chemically different materials and converging their properties can be helpful in designing biomedical implants/devices with improved properties, meeting specific needs.[9] Bone, which is made up of hard but brittle bone material and natural polymer collagen, is a good example of composite biomaterial that has excellent strength and high load-bearing capacity.

For biomedical applications, composites can be broadly grouped as bioinert composites, bioresorbable/biodegradable composites, and bioactive composites. The biocompatibility of bioactive composites is quite challenging, due to the presence of two or more different phases of materials (bioactive and bioresorable) at the interface, resulting in dynamic implant-tissue interfacial conditions.[10] Hence, a thorough evaluation of biocompatibility of the composite for desired application is essential. Examples of biomaterial composites that are used or being studied for different biomedical implants and devices are mentioned in Table 16.5.

Composites have become popular in tissue engineering attributed to their potential to improve tissue repair/regeneration and to provide multifunctionality. For example, incorporation of antibiotics or nanomaterials with antibacterial properties into the matrix resolves the infection issue, while the use of graphene

**TABLE 16.4**

**Examples of Ceramic Biomaterials Used in Biomedical Implants and Devices**

| Type of Ceramic | Ceramic | Properties | Limitations | Applications |
|---|---|---|---|---|
| Bioinert | Alumina | Good mechanical strength, low friction coefficient, and biocompatibility | Low fracture toughness, squeaking, high cost | Orthopedic, load bearing applications, and dental implants |
| | Zirconia | Same color as natural teeth, good wear resistance, biocompatibility | Crack generation with time, high cost, hydrothermal aging | Orthopedic and dental implants (pins, crowns, bridges, veneers, and orthodontic brackets) |
| | Non-oxide ceramics (silicon nitride) | Low friction of coefficient, wear resistant, biocompatible | Expensive, instable, electrical non-conductive | Spinal surrey and joint replacements |
| Biodegradable/ bioresorable | Calcium Phosphate | Good compressive strength (similar to the bone), resorbability and bioactivity can be controlled with Ca/P ratio | Lower mechanical strength than bioinert ceramics, poor metal-ceramic interface when used as coatings due to thermal coefficient difference | Bone grafting, mostly used as coating over implants/devices |
| | Calcium sulphate | Compressive strength similar to bone | Fragile and rapidly loses its strength when it is hydrolyzed, rapid degradability | Bone grafting, used as coating over implants/devices |

*(Continued)*

**TABLE 16.4 (*Continued*)**
**Examples of Ceramic Biomaterials Used in Biomedical Implants and Devices**

| Type of Ceramic | Ceramic | Properties | Limitations | Applications |
|---|---|---|---|---|
| Bioactive | Bioactive glass | Rapid healing rates, higher osteoconductivity, better bone fixation | Poor fracture toughness, manufacturing limitations, and sometimes relatively fast dissolution and resorption rates might lead to gap formation between implant/device and tissue formation | Bone grafting, used as coating over implants/devices |

**TABLE 16.5**

**Examples of Biocomposites Used or Being Studied in Different Biomedical Implants/Devices**

| Application | Biomaterial Composites |
|---|---|
| Fracture | Carbon fiber (CF) or glass fiber (GF)/epoxy, CF with PEEK, PLA, PLGA, or PCL and CF with bioresorbable ceramic |
| Joint replacement | CF/polysulfone and CF/PEEK (stem of hip prosthesis) and CF/UHMWPE (acetabulum cup) |
| Tendon/ligaments | Aramid fiber/PE, CF/PTFE (nondegradable) or CF/PLA, CF/PU (partially degradable) |
| Middle ear implant | HA/HDPE |
| Soft tissue repair | Bioglass®/HDPE and A-W glass-ceramic/HDPE |
| Bone tissue repair | HA/PSU and HA/PEEK |
| Dental posts | Dental posts: CF or GF/epoxy |
| Artificial tooth root | CF/carbon, SiC/carbon, CF/PMMA |

or carbon nanotubes (CNTs) provide improved mechanical strength or electrical conductivity (for neural tissue regeneration).[11] Despite this, there are many challenges that need to be resolved through multidisciplinary knowledge, such as bonding to bone and effect of body's environment on the different components of composites.

## 16.3  SURFACE TREATMENT

The properties of bulk biomaterial might be helpful in selecting the suitable material; however, it is surface properties that are responsible for the optimal function of the biomedical implant or device because interactions with host body occur at the surface. If a surface is not biocompatible, scar tissue formation takes place, ultimately causing the implant failure. This scenario can be avoided by altering the morphology and chemistry of the implant/device surface. Surface treatment of biomaterial assists in offering suitable conditions for cell adhesion, proliferation, and migration by improving the wettability, hydrophilicity, and protein adsorption of the surface.[12] There are several surface treatment/modification techniques available (Figure 16.1). Some common surface treatment methods are discussed in this section.

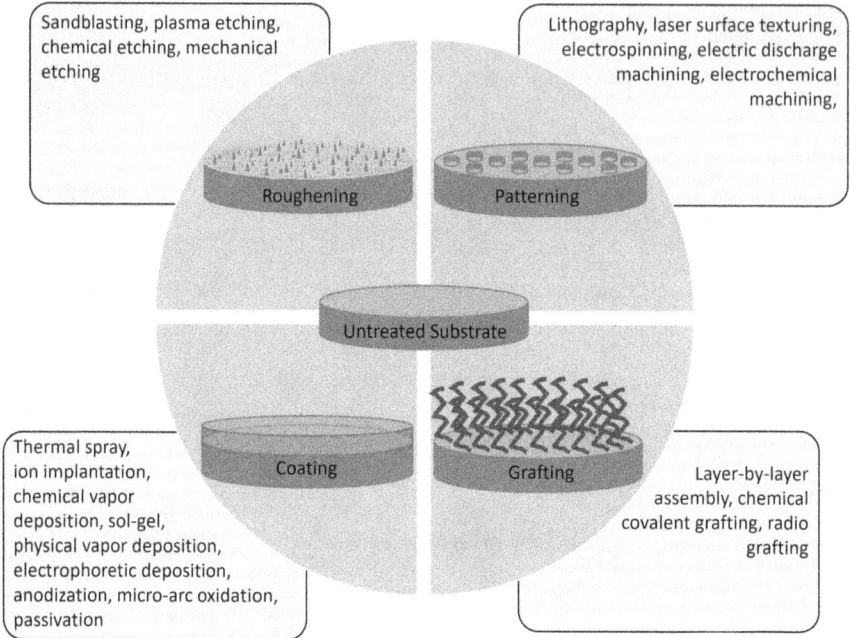

**FIGURE 16.1** Different methods of surface treatment.

## 16.3.1 ROUGHENING

### 16.3.1.1 Blasting

In the blasting surface treatment process, the surface of the implant is treated with a blasting media that is fired through a nozzle and propelled towards the implant surface using compressed air or steam. The generated macro-roughness varies with numerous variables such as the size, shape, and type of blasting particles, projection pressure, nozzle diameter, and distance between projection nozzle and the surface. Commonly used blasting particles include alumina, titania, silica, or hydroxyapatite that can help in boosting the surface properties of the implant/device for osteointegration.[13] Sometimes steel slag is used to treat medical grade 316 L stainless steel, which releases bioactive elements such as calcium, silicon, and magnesium into the host body, enhancing both mechanical properties and bioactivity of the material. Sandblasting also helps in removing surface contamination.[14] Blasting is an easy and industrially approved surface treatment method that is commonly used for metallic biomaterials, where surface texture is generated due to plastic deformation. However, it could be detrimental to use non-biocompatible blasting particles or to blast with particles made of a different material than implant/device, as this can alter the surface composition and releases ions into the host body, leading to inflammatory responses.[15]

### 16.3.1.2   Chemical Etching

Chemical etching involves the generation of a desired surface texture by removing material from a specific portion of the biomaterial using chemical reagents. Commonly, strong acids such as $H_2SO_4$, HCl, $HNO_3$, HF, and piranha solution ($H_2SO_4$ (70%vol):$H_2O_2$ (30%vol.)) are used to create a micro-pit-like texture (0.5 μm–2 μm). This quick surface treatment method is frequently used for metallic biomaterials to promote cell adhesion and osseointegration to the implant/device.[16] For instance, HF treatment supports osseointegration in titanium biomaterials due to the formation of soluble $TiF_4$.[17] The surface texture can be varied with the type and concentration of chemical reagent, exposure time, and working temperature. However, the degree of chemical etching influences the corrosion resistance along with the mechanical and biological properties of the surface.[18] Sometimes treated metallic biomaterials exhibit weakened mechanical properties such as fatigue resistance due to hydrogen embrittlement and inflammation in the host due to ion toxicity.[17] Etching is usually employed after sandblasting to eliminate surface damage caused by blasting and concurrently improve surface roughness characteristics. Selective infiltration etching (SIE) is a variation of this technique, where specific surface topography is achieved by diffusing molten glass between the grain boundaries of the metal, which rearranges the surface grains.[19] Later, the diffused glass is dissolved in an acidic bath, exposing a smooth and highly retentive surface that promotes cell adhesion and osseointegration.

### 16.3.1.3   Plasma Etching

Plasma etching is a surface modification technique that involves exposing biomaterial implant/device to low-power plasma gases such as water vapor, oxygen/argon, and ammonia to remove contaminants, including oxides or organic residues and to introduce hydrophilic functional groups such as carboxylic or hydroxyl groups on biomaterial implants/devices surface. This technique enhances tissue integration, reduces the risk of inflammation and develops nanoscale roughness that supports cell attachment and osseointegration.[20] The depth of the etched surface varies with the type of biomaterial and power and duration of the plasma treatment.[21] Highly crystalline polymers exhibit a low etching rate compared to amorphous polymers due to better structural integrity. Plasma etching also alters the zeta potential of the material by altering its polarity. The longer the sample is treated, the more the zeta potential increases, which can influence the cell behavior and protein formation on the biomaterial surface. Surface treatment of intra-ocular lenses to make them hydrophilic, ultra-fine cleaning of medical implants, and surface activation of balloon catheters to improve adhesion are the common examples of plasma etching used for biomedical devices.

## 16.3.2   PATTERNING

### 16.3.2.1   Laser Surface Texturing

Laser surface texturing (LST) technique is a highly efficient and inexpensive method to generate reproducible micro/nano patterns on the surface of biomaterial implants and devices, which are critical for promoting cell adhesion and osseointe-gration. In this method, a high-energy beam is used to melt and vaporize the material, resulting in groove-like patterns on the surface that exhibit superior cell/surface interactions compared to etched and blasted surfaces. With the adjustment of processing parameters such as wavelength, scan speed, and pulse rate, these groove pattern/geometry can be customized.[1] This technique improves the tribological performance of the biomaterial by providing lower friction coefficient. In titanium and its alloys, higher groove density provides low friction coefficient because of the entrapment of the wear debris from the passive oxide layer into the grooves[22]. These grooves also play a role of fluid reservoir in the host body, providing lubricity and, hence, less wear. Laser treatment is also utilised to change the surface energy and surface chemistry of polymeric biomaterialsby breaking existing chemical bonds and forming new functional groups, which support cell adhesion and proliferation.[23] However, if the surface is over-treated, an excessive number of new functional group might generate on the surface and become contra-productive for cell adhesion and proliferation.[24] Moreover, above ablation threshold treatment can be responsible for the loss of material from the surface. There are many variations of this technique, such as ultra-violet laser treatment, diode-pumped solid-state (DPSS) laser treatment, and laser shock peening. Higher photon energy associated with UV laser treatment might deplete the material properties by introducing micro-cracks and heat-affected areas in the grooves.

### 16.3.2.2 Lithography

Surface treatment through lithography techniques is a highly efficient and reproduc-ible method that allows the precise generation of patterns at nano- and micro-scale without altering the surface composition. Photolithography, electron beam lithog-raphy, and nano-imprint lithography are the examples of different types of lithography. In lithography, a selective area of the surface is specifically exposed to irradiation through a mask or template, which creates complex patterns with a high degree of accuracy and precision. Photolithography involves coating a photosensitive material onto the masked surface, which is then exposed to light, followed by a transfer of pattern through etching. Meanwhile, electron beam lithography utilizes a beam of electrons to create patterns (high resolution) and nanoimprint lithography deals with a template or stamp that generates nanoscale features on the surface. The nano and micro topographies generated through lithography help in promoting cell adhesion, proliferation, and differentiation by influencing cells behavior, including their alignment and migration, which ultimately enhance tissue regeneration and integration with the surrounding tissue.[25,26] Moreover, lithography can be helpful in functionalizing implant surfaces with biomolecules such as proteins, peptides, growth factors, or enzymes, which introduce multifunctionality into implants/devices.[27]

### 16.3.2.3 Machining

There are several machining techniques, including electric discharge machining (EDM), electrochemical machining (ECM), and ultrasonic-assisted machining (UAM).

EDM involves using a series of high-frequency electrical sparks to remove material from the surface. This technique does not compromise the material integrity as the electrode and sample both are in non-contact mode. Also, it is particularly useful for introducing carbides onto the surface, which enhances corrosion and wear resistance as well as hardness. In addition, EDM can generate a porous nanostructured oxide layer, which is beneficial for cell adhesion and growth. Pulse duration can be used to control the layer thickness. However, a drawback of EDM-functionalized surfaces is that it may compromise the fatigue performance of the material due to a recast layer formation, which can be addressed by sandblasting.[28]

ECM, on the other hand, selectively removes the material from the surface through an electrochemical reaction at the anode. The machine rate can be controlled by adjusting the electric current. This technique is particularly suitable for metallic bioimplants and devices and can achieve micro-machining with high aspect ratio features even on curved surfaces. Finally, UAM can produce high-precision textures on a variety of surfaces, including polymers, glasses, and stainless steels. This technique uses ultrasonic vibrations to induce material deformation and surface texturing. UAM has various advantages, including low tool wear and the capability to generate intricate patterns and features with great precision and accuracy. Overall, the patterns and chemistry generated through various machining techniques help in improving biocompatibility, functionality, and performance.[28–30]

### 16.3.2.4   Electrospinning

Electrospinning is a highly flexible, simple, and effective method that develops mesh-like fibrous patterns onto the surface of implants. This technique involves using a high volage to stretch and whip a polymer solution, resulting in the generation of continuous fibers with diameter ranging from a few nanometers to several micrometers. The resulting surface of electrospinning-treated implant/device exhibits a high surface-to-volume ratio, which is desirable for promoting cell adhesion and proliferation. Furthermore, thesurface properties of the electrospinning-treated surface can be customized by altering the polymer type, concentration, and electrospinning parameters that creates fibrous patterns with different morphologies, orientations, and mechanical properties, give the freedom to design surfaces for specific biomedical applications.[31] Electrospinning is also used for incorporation of various bioactive molecules or drugs into the fibers, which introduce multifunctionality to the implant surface.[32] Through electrospinning, it is possible to simulate extracellular matrix (ECM) structure of almost all connective tissues, including its porosity, surface-to-volume ratio, and mechanical performance as ECM has a similar structure to nanofibers.[11] Figure 16.2 summarizes different uses of electrospinning in biomedical applications.

## 16.4   COATING DEPOSITION

Nanocoatings have been found to promote osteoblast activities by mimicking the nanoscale features present in the structural components of naturally occurring body proteins at the coating-organ interface. Most commonly, titania, zirconia, silica, hydroxyapatite, carbon-based coatings, and biomimetic coatings are deposited on implants/devices. For further enhancement of performance and properties and to achieve multifunctionality, these coatings are often used in

**FIGURE 16.2**   Use of electrospinning in biomedical applications.

combination with other biomaterials. Among them, carbon-based and biomimetic coatings are discussed below.

- Carbon-based coatings, including pyrolytic carbon (PyC), nanocrystalline diamond (NCD), and diamond-like carbon (DLC), are frequently employed in biomedical applications due to their bioinert but biocompatible nature. These coatings are particularly useful for load-bearing and corrosion protection against body fluid. PyC is often used for lining artificial heart valves and in replacing small joints such as wrist joints, knuckles, and arthroplasty of proximal interphalangeal joints. NCD is commonly used for the implants where protection against wear is needed, cardiovascular devices, and antibacterial coatings, while DLC finds applications in many implant/devices, including knee replacement, cardiovascular stents, heart valves, surgery needles, medical wires, and contact lenses. Delamination of DLC from the substrate can be a concern, which can be addressed using buffer layers between DLC and the substrate or through doping.

| Physical method | Advantages | Limitations | Effects on implants/devices |
|---|---|---|---|
| Thermal spray | Good bonding strength, dense coatings, quick, wide selection of materials | Line-of-sight process, porosity might lead to delamination due to crack formation | Improves wear and corrosion resistance, and biocompatibility |
| Ion implantation | Strong adhesion, intact structural integrity of material, good biocompatibility | Line-of-sight process, low deposition rate, size limitation, inhomogeneity over large area | Improves wear and corrosion resistance and biocompatibility |
| PVD (Evaporation) | Quick, uniform coating, good adhesion, deposits wide variety of inorganic and some organic materials | Line-of-sight process, low deposition rate | Improves hardness, strength, corrosion resistance and biocompatibility |
| Electrophoretic deposition | Uniform coating, low-cost, deposits wide variety of materials, non-line-of-sight process | Needs post-sintering, comparatively poor adhesion | Improves biocompatibility, bioactivity, antibacterial properties, and introduce multifunctionality |

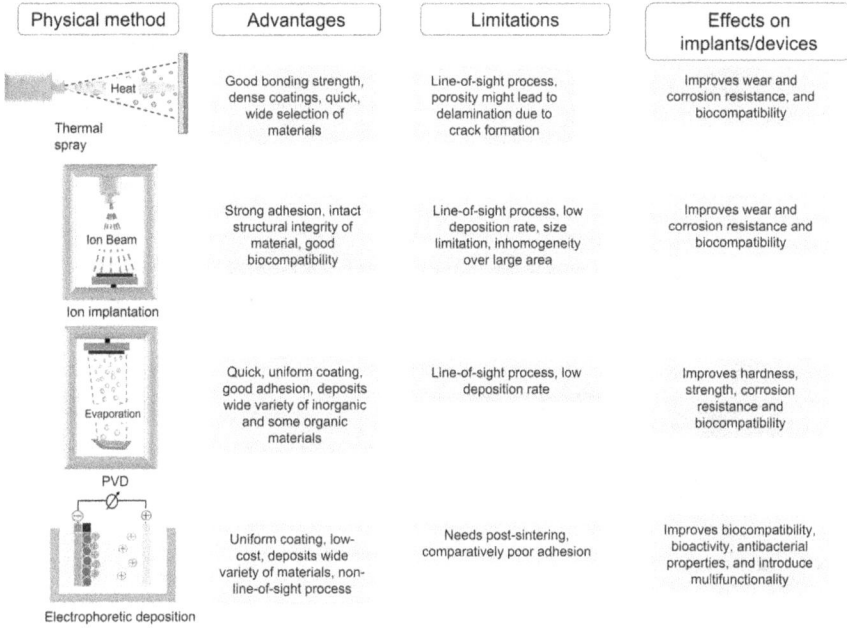

**FIGURE 16.3**  Examples of physical coating deposition methods.

- Biomimetic coatings are intended to replicate the structure and function of natural materials, enhancing the integration of implants/devices with surrounding tissues. Immobilization of cell proteins such as collagen, tropoelastin, silk fibronin, and fibronectin onto the implant/device surface via physical or covalent immobilization is a common strategy to stimulate osseointegration and bone cell formation.[33,34] Similarly, the formation of Ca/P-based coatings on orthopedic implants mimics the mineral composition of bone and improves the implant/surrounding tissue interactions. In cardiovascular applications, biomimetic coatings can imitate the structure and characteristics of the ECM along with being reinforced with biological factors, such as heparin, transforming growth factor-β, fibroblast, platelet-derived, and vascular endothelial growth factors, to stimulate tissue formation and cell attachment. These coatings have shown promising results in improving biocompatibility and bioactivity of implants and devices, reducing thrombosis and thromboembolic complications, and enhancing bone regeneration.[33,35]

There are several coating deposition methods (Figures 16.3 and 16.4) and some of them are discussed in the following sections.

## 16.4.1 THERMAL SPRAY

Thermal spray is a versatile and efficient technique to deposit coatings onto bioimplants and devices where a heat source is used to deposit a variety of

| Chemical method | Advantages | Limitations | Effects on implants/devices |
|---|---|---|---|
| Chemical vapor deposition | Good bonding strength, homogeneous coating, good control over composition | Low deposition rate, release toxic gases, expensive | Improves wear and corrosion resistance, and biocompatibility |
| Anodization | Strong adhesion, low-cost, homogeneous coating, nanotubular array can be formed | High energy consumption, high-cost, limited versatility | Improves wear and corrosion resistance and biocompatibility |
| Micro-arc oxidation | Strong adhesion, porous coatings, good hardness and corrosion resistance | High energy consumption, high-cost, limited versatility | Improves hardness, strength, corrosion resistance and biocompatibility |
| Sol-gel | Uniform coating, can be applied to complex geometries, low temperature required, high purity | Needs post-sintering, expensive precursors, poor tribological properties, a better control over reaction environment is required | Improves biocompatibility and bioactivity |
| Layer-by-layer assembly | Often used for multilayer biological coatings, good control over coating thickness, can be applied to variety of materials, strong adhesion | Time consuming, can cause cross contamination of stock material | Introduce multifunctionality, antimicrobial properties, and biomimic properties |

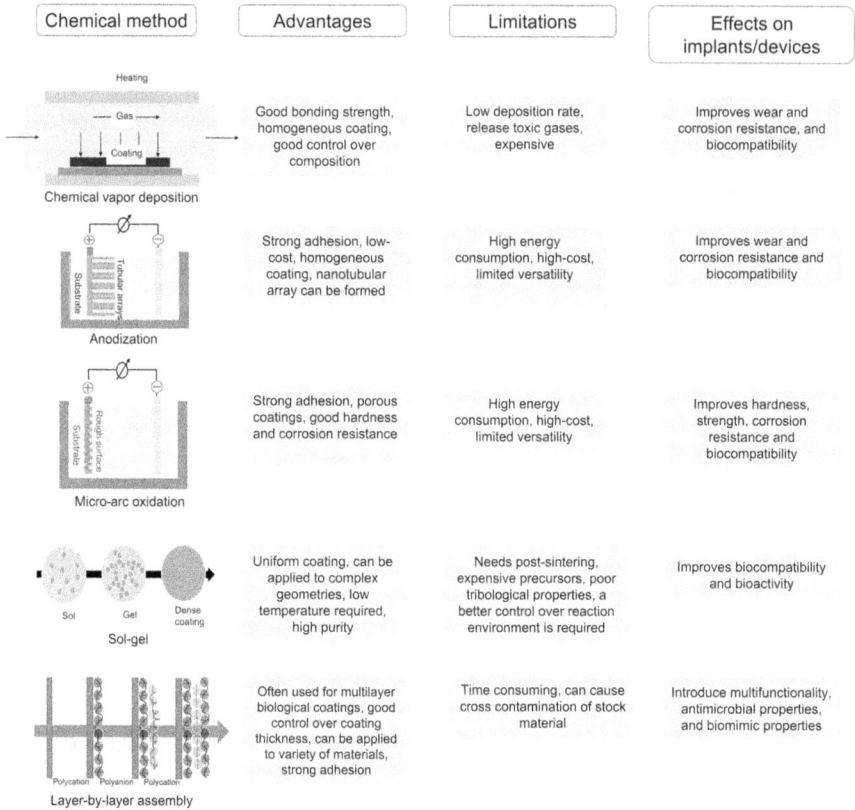

**FIGURE 16.4**    Examples of chemical coating deposition methods.

materials, containing ceramic, metals, and alloys are deposited onto various biomaterials, forming a functional layer.[36] Thermal spray coatings typically possess good corrosion and wear resistance. Based on the heat source, thermal spray can be categorized into different processes, such as flame spray, cold spray, plasma spray, and high-velocity oxygen fuel (HVOF) spray, where the processing temperature and coating properties vary accordingly. Plasma spray and HVOF are commonly used techniques to deposit robust and compact coatings with nano- and micro-features onto bioimplants/devices. Coating morphology and properties vary with the property of the deposited material, its state (solid or liquid) and processing parameters.[37] HA and Ca-P coatings on Ti or Ti alloys are often deposited using this technique as they are favorable for osseointegration and quick recovery of the host.[38] However, the structural alteration in HA due to high thermal energy and different thermal coefficient than the (metallic) substrate might lead to cracking and delamination of coating, leading to implant failure.[39] Sandblasting, pre-treatment, and post-treatment of the substrate could facilitate the coating-substrate interface.[40] Some studies use doped or drug loaded HA coatings for further improvement of the biocompatibility, mechanical performance, and antibacterial properties of the coated implant.[12,41]

## 16.4.2 ION IMPLANTATION

Ion implantation is a precise and highly controllable technique used for depositing high-purity coatings with outstanding adhesion to biomaterial substrates. During this process, energetic ions are accelerated towards the biomaterial surface in the presence of high potential difference, leading to ion-solid interactions and the deposition of the ions onto the surface. The penetration depth of ions is dictated by the energy possessed by ions, and for biocompatibility, near-surface penetration is desirable. Other factors that affect the penetration profile are ion species, ion beam current density or flux, and number of ions bombarded onto the substrate.[42] Compared to plasma spray, ion-implanted coatings exhibit better bonding strength due to the formation of atomic intermixing interfacial layer.[42] This technique also helps in avoiding stress shielding and improving fracture toughness of bioimplants. Moreover, ion implantation is helpful in depositing phosphorus, fluoride, and calcium ions onto Ti or Ti alloys. For instance, the formation of TiP phase improves corrosion resistance of Ti or Ti alloy substrates.[43] For polymers, ion implantation generates different chemical processes such as oxidation, reduction, cross-linking, ring formation, and hetero-atoms removal, making them suitable for cell adhesion and proliferation.[44] However, because of the use of high vacuum, the high-cost and size limitations are factors that limit the potential use of this technique.[45]

## 16.4.3 PHYSICAL VAPOR DEPOSITION (PVD)

PVD is a process that includes the ejection of material from a negatively charged target and its deposition onto a substrate in the presence of a gas plasma. This process provides a homogeneous coating with good control over microstructures. Since PVD can regulate calcium-phosphate ratios in the coating, it is often used to deposit HA coatings.[46,47] PVD coatings provide enhanced hardness, strength, corrosion resistance, and biocompatibility to biomaterials used in implants/devices.[47] However, the poor dissolution rate of HA coatings in the human body is a drawback that can be resolved using magnetron sputtering and vacuum evaporation techniques to provide comparatively thicker coatings.[48] Despite the ability to develop homogeneous and dense coatings with good adhesion strength, the high cost of PVD coatings has limited their commercial use.

## 16.4.4 CHEMICAL VAPOR DEPOSITION (CVD)

CVD is a method used to deposit a thin and uniform film onto a heated substrate through the reaction of precursor gases in a controlled environment. It differs from PVD in that chemical bonding occurs during coating formation rather than physical forces in PVD. It is a widely used method in the semiconductor industry, but recently, it has become popular for modifying the biomaterials surfaces. It is a non-line-of-sight process that provides better regulation over the quantity and quality of the product. Common applications of CVD in biomedical implants/devices include the deposition of DLC for improved corrosion resistance and tribological properties and HA coatings for osseointegration and bone formation on biomaterials' surface.[49,50] It is also used to deposit a functionalized polymer on Ti implants to improve the protein adsorption and facilitate better calcium deposition.[51] Despite having good quality, density, and purity, CVD-based coatings are not extensively used for bioimplants/devices due to associated high-cost, high processing temperature, potentially toxic gas source, and exhaust gas, which may have negative impacts on implantation process.[52]

## 16.4.5 CONVERSION COATINGS

Conversion coatings are coatings that are formed through specific interactions between materials and surroundings, such as passivation, anodization, and micro-arc oxidation. These coatings are well adhered to the substrate. Passivation is a straightforward method that involves immersing a metal implant/device in a solution with specific pH, promoting development of a passive oxide layer, which reduces the risk of inflammation or allergic reactions due to creation of a barrier against corrosion that restricts metal ions release into surrounding tissues.[53] Coating properties can be altered based on the type of metal, duration, and conditions of the passivation process.[54] Anodization (anodic oxidation; AO) involves a repeated formation and disappearance of a stable oxide layer with uniform micro/nanoscale porous or tubular arrays on metallic biomaterials placed in an electrolyte solution with applied electrical potential. The most frequently deposited coatings are $TiO_2$ coatings on Ti and its alloys, enhancing their biocompatibility by restricting the discharge of metal ions into the host body and promoting osseointegration.[55] Other coatings that are formed using AO are $Al_2O_3$, $ZrO_2$, MgO, and Ca/P coatings.[56] The coating properties can be controlled by adjusting the process parameters such as applied potential, treatment time, and electrolyte composition.[12] Micro-arc oxidation (MAO) is a variation of anodization that occurs above the breakdown voltage, forming a relatively rough and hard oxide film with better corrosion resistance. For further biocompatibility and bioactivity enhancement, elements that support bone formation, such as Si, Ca, Mg, and Sr, are incorporated into MAO coatings.[57] HA-modified MAO $TiO_2$ coatings on Ti implants are also widely used for bone formation and angiogenesis(new blood vessels formation) promotion.[58] Secondary coatings can be grown on MAO-treated surfaces for further improvement in cell adhesion and osteogenic differentiation due to nano-features formation.[59]

## 16.4.6 SOL-GEL COATINGS

Sol-gel is an extensively used coating method due to its simplicity, coating homogeneity, and non-line-of-sight approach, which provides a precise control over coating properties, including composition via adjusting factors, such as the pH of the solution, processing time, and surface-solution interface. The process involves hydrolysis polymerisation of inorganic or organic compounds in a solution, which forms a colloidal suspension that gradually transforms into a gel-like formation.

After thermal treatment, this gel-like formation solidifies into a coating due to polycondensation. Usually, ceramic coatings, such as $TiO_2$, $SiO_2$, $ZrO_2$, and HA-based and inorganic-organic composite coatings, such as forsterite/poly-3-hydroxybutyrate, chitosan/tetraethyl orthosilicate (TEOS), gentamicin/chitosan gelatin/silica, and hydroxyapatite/sorbitol sebacate glutamate coatings, are deposited on implants using this method, providing them better corrosion resistance, bio-compatibility, and bioactivity.[60–62]

### 16.4.7 ELECTROPHORETIC DEPOSITION (EPD)

EPD is a cost-efficient method used to coat a variety of materials, including HA, $TiO_2$, chitosan, collagen, alginate, different composites, peptides, proteinsontome-tallic, and ceramic and polymeric implants/devices to improve their properties and biocompatibility.[63–65] This technique works by applying an electric field to the suspension of charged particles in a liquid medium, causing the particles to move towards the substrate and deposit on it. The coating thickness can be controlled by adjusting the electric field strength, suspension concentration, and deposition time play a critical role in controlling the coating thickness.[66] EPD is used in developing multi-architectural coatings with improved surface properties, such as corrosion resistance, hardness and wettability, better bioactivity through supporting osseoin-tegration, and controlled drug release of metallic ions or drugs.[67,68] However, EPD coatings face a challenge in attaining strong adhesion to the biomaterial substrate, which can hinder the production of high-quality coatings for bioimplants/devices.[48]

### 16.4.8 LAYER-BY-LAYER ASSEMBLY (LBL) METHOD

LBL is a versatile method that involves the alternative deposition of oppositely charged polyelectrolytes on a charged substrate, forming an ordered structure, spontaneously. LBL coating method has several advantages, including not compromising the substrate's structural/chemical integrity, large area coverage, nanoscale resolution with a precise control over biological surface interactions, and aqueous assembly conditions, favorable for biological agents and small molecules incorporation.[69] This technique is often used to coat biomimicking and bioactive coatings on implants/devices for tissue engineering, drug delivery, enhancing antifouling and antimicrobial properties, and many other applications.[70,71]

## 16.5 CONCLUSIONS

The choice of materials and design for implants and devices depends mainly on bulk biomaterials, with metals and ceramics typically used for hard tissue healing or replacement and polymers for soft tissue repair. However, the surface of a biomaterial is crucial for determining the success or failure of an implant or device within the host body. Inadequate surface properties can cause scar tissue formation and implant/device failure. Surface modification techniques provide various options to alter the surface properties of biomaterials to enhance their biocompatibility, bioactivity, mechanical properties, wear resistance, and corrosion resistance. These techniques also facilitate the development of innovative biomaterials with specific properties. Future research should focus on designing novel surface modification techniques, developing new biomaterials, and extensive clinical testing to ensure safety and efficacy, ultimately affecting patients' lives positively.

## REFERENCES

1. Neděla, O., Slepička, P. & Švorčík, V. Surface Modification of Polymer Substrates for Biomedical Applications. *Materials* **10**, 1115 (2017).
2. Teo, A. J. T. *et al.* Polymeric Biomaterials for Medical Implants and Devices. *ACS Biomater. Sci. Eng.* **2**, 454–472 (2016).
3. Kalirajan, C., Dukle, A., Nathanael, A. J., Oh, T.-H. & Manivasagam, G. A Critical Review on Polymeric Biomaterials for Biomedical Applications. *Polymers* **13**, 3015 (2021).
4. Santos, G.A.D., The Importance of Metallic Materials as Biomaterials. *Adv. Tissue Eng. Regen. Med. Open Access* **3**, 300–302 (2017).
5. Prasad, K., Bazaka, O., Chua, M., Rochford, M., Fedrick, L., Spoor, J., Symes, R., Tieppo, M., Collins, C., Cao, A., Markwell, D., Ostrikov, K.K., Bazaka, K.. Metallic Biomaterials: Current Challenges and Opportunities. *Materials* **10**(8), 884 (2017). doi:10.3390/ma10080884. PMID: 28773240; PMCID: PMC5578250. https://www.mdpi.com/1996-1944/10/8/884
6. Jambagi, S. C. & Malik, V. R. A Review on Surface Engineering Perspective of Metallic Implants for Orthopaedic Applications. *JOM* **73**, 4349–4364 (2021).
7. Vaiani, L., Boccaccio, A., Uva, A.E., Palumbo, G., Piccininni, A., Guglielmi, P., Cantore, C., Santacroce, L., Charitos, I.A., & Ballini, A. Ceramic Materials for Biomedical Applications: An Overview on Properties and Fabrication Processes. *J. Funct. Biomater.* **14**(3), 146. doi:10.3390/jfb14030146
8. Punj, S., Singh, J. & Singh, K. Ceramic Biomaterials: Properties, State of the Art and Future Prospectives. *Ceram. Int.* **47**, 28059–28074 (2021).
9. Zagho, M. M., Hussein, E. A. & Elzatahry, A. A. Recent Overviews in Functional Polymer Composites for Biomedical Applications. *Polymers* **10**, 739 (2018).
10. Wang, M. & Zhao, Q. Biomedical Composites. In *Encyclopedia of Biomedical Engineering* (ed. Narayan, R.) 34–52 (Elsevier, 2019). doi:10.1016/B978-0-12-801238-3.99868-4
11. Zheng, S. *et al.* Carbon Nanomaterials for Drug Delivery and Tissue Engineering. *Front. Chem.* **10**, 990362 (2022).
12. Zhu, G., Wang, G. & Li, J. J. Advances in Implant Surface Modifications to Improve Osseointegration. *Mater. Adv.* **2**, 6901–6927 (2021).
13. Yurttutan, M. E. & Keskin, A. Evaluation of the Effects of Different Sand Particles that Used in Dental Implant Roughened for Osseointegration. *BMC Oral Health* **18**, 47 (2018).
14. Arifvianto, B., Wibisono, S., K. A. & Mahardika, M. Influence of Grit Blasting Treatment Using Steel Slag Balls on the Subsurface Microhardness, Surface Characteristics and Chemical Composition of Medical Grade 316l Stainless Steel. *Surf. Coat. Technol.* **210**, 176–182 (2012).
15. Bauer, S., Schmuki, P., von der Mark, K. & Park, J. Engineering Biocompatible Implant Surfaces: Part I: Materials and Surfaces. *Prog. Mater. Sci.* **58**, 261–326 (2013).
16. Velasco-Ortega, E. *et al.* Long-Term Clinical Outcomes of Treatment with Dental Implants with Acid Etched Surface. *Materials* **13**, 1553 (2020).

17. Le Guéhennec, L., Soueidan, A., Layrolle, P. & Amouriq, Y. Surface Treatments of Titanium Dental Implants for Rapid Osseointegration. *Dent. Mater.* **23**, 844–854 (2007).
18. Chauhan, P., Koul, V. , & Bhatnagar, N. Critical Role of Etching Parameters in the Evolution of Nano Micro SLA Surface on the Ti6Al4V Alloy Dental Implants. *Materials* **14**(21), 6344 (2021). doi:10.3390/ma14216344. PMID: 34771869; PMCID: PMC8585160.
19. Aboushelib, M. N., Feilzer, A. J. & Kleverlaan, C. J. Bonding to Zirconia Using a New Surface Treatment. *J. Prosthodont.* **19**, 340–346 (2010).
20. Liu, P., Wang, G., Ruan, Q., Tang, K. & Chu, P. K. Plasma-Activated Interfaces for Biomedical Engineering. *Bioact. Mater.* **6**, 2134–2143 (2021).
21. Donnelly, V.M. & Kornblit, A. Plasma Etching: Yesterday, Today, and Tomorrow. *J. Vac. Sci. Technol. A* **31**, 050825 (2013). doi:10.1116/1.4819316
22. Shivakoti, I., Kibria, G., Cep, R., Pradhan, B. B. & Sharma, A. Laser Surface Texturing for Biomedical Applications: A Review. *Coatings* **11**, 124 (2021).
23. Riveiro, A., Maçon, A. L. B., del Val, J., Comesaña, R. & Pou, J. Laser Surface Texturing of Polymers for Biomedical Applications. *Front. Phys.* **6**, (2018).
24. Jäger, M., Sonntag, F., Pietzsch, M., Poll, R. & Rabenau, M. Surface Modification of Polymers by using Excimer Laser for Biomedical Applications. *Plasma Process. Polym.* **4**, S416–S418 (2007).
25. Greant, C., Van Durme, B., Van Hoorick, J. & Van Vlierberghe, S. Multiphoton Lithography as a Promising Tool for Biomedical Applications. *Adv. Funct. Mater.* **n/a**, 2212641.
26. Tran, Khanh T.M. & Nguyen, Thanh D. Lithography-Based Methods to Manufacture Biomaterials at Small Scales. *J. Sci. Adv. Mater. Dev.* **2**(1), 1–14 (2017). doi:10.1016/j.jsamd.2016.12.001
27. Nowduri, B., Britz-Grell, A., Saumer, M. & Decker, D. Nanoimprint Lithography-Based Replication Techniques for Fabrication of Metal and Polymer Biomimetic Nanostructures for Biosensor Surface Functionalization. *Nanotechnology* **34**, 165301 (2023).
28. Prakash, C., Kansal, H. K., Pabla, B., Puri, S. & Aggarwal, A. Electric Discharge Machining – A Potential Choice for Surface Modification of Metallic Implants for Orthopedic Applications: A Review. *Proc. Inst. Mech. Eng. Part B J. Eng. Manuf.* **230**, 331–353 (2016).
29. Davis, R. *et al.* A Comprehensive Review on Metallic Implant Biomaterials and their Subtractive Manufacturing. *Int. J. Adv. Manuf. Technol.* **120**, 1473–1530 (2022).
30. Peng, P.-W., Ou, K.-L., Lin, H.-C., Pan, Y.-N. & Wang, C.-H. Effect of Electrical-discharging on Formation of Nanoporous Biocompatible Layer on Titanium. *J. Alloys Compd.* **492**, 625–630 (2010).
31. Uhljar, L. É. & Ambrus, R. Electrospinning of Potential Medical Devices (Wound Dressings, Tissue Engineering Scaffolds, Face Masks) and Their Regulatory Approach. *Pharmaceutics* **15**, 417 (2023).
32. Liu, Z., Ramakrishna, S. & Liu, X. Electrospinning and Emerging Healthcare and Medicine Possibilities. *APL Bioeng.* **4**, 030901 (2020).
33. Stewart, C., Akhavan, B., Wise, S. G. & Bilek, M. M. M. A Review of Biomimetic Surface Functionalization for Bone-integrating Orthopedic Implants: Mechanisms, Current Approaches, and Future Directions. *Prog. Mater. Sci.* **106**, 100588 (2019).
34. Vitale, F. *et al.* Biomimetic Extracellular Matrix Coatings Improve the Chronic Biocompatibility of Microfabricated Subdural Microelectrode Arrays. *PLoS ONE* **13**, e0206137 (2018).
35. Biran, R. & Pond, D. Heparin Coatings for Improving Blood Compatibility of Medical Devices. *Adv. Drug Deliv. Rev.* **112**, 12–23 (2017).

36. Prashar, G. & Vasudev, H. Thermal Sprayed Composite Coatings for Biomedical Implants: A Brief Review. *J. Therm. Spray Eng.* **2**, 50–55 (2020).

37. Mittal, G. & Paul, S. Suspension and Solution Precursor Plasma and HVOF Spray: A Review. *J. Therm. Spray Technol.* **31**, 1443–1475 (2022).

38. Sun, L. Thermal Spray Coatings on Orthopedic Devices: When and How the FDA Reviews Your Coatings. *J. Therm. Spray Technol.* **27**, 1280–1290 (2018).

39. Yang, Y. C. & Chang, E. Influence of Residual Stress on Bonding Strength and Fracture of Plasma-sprayed Hydroxyapatite Coatings on Ti–6Al–4V Substrate. *Biomaterials* **22**, 1827–1836 (2001).

40. Nimb, L., Gotfredsen, K. & Jensen, J. Mechanical Failure of Hydroxyapatite-coated Titanium And Cobalt-Chromium-Molybdenum Alloy Implants. An Animal Study. *Acta Orthop. Belg.* **59**, 333–338 (1993).

41. Sarkar, N. & Bose, S. Controlled Delivery of Curcumin and Vitamin K2 from Hydroxyapatite-Coated Titanium Implant for Enhanced in Vitro Chemoprevention, Osteogenesis, and in Vivo Osseointegration. *ACS Appl. Mater. Interfaces* **12**, 13644–13656 (2020).

42. Rautray, T. R., Narayanan, R. & Kim, K.-H. Ion Implantation of Titanium Based Biomaterials. *Prog. Mater. Sci.* **56**, 1137–1177 (2011).

43. Krupa, D. *et al.* Effect of Phosphorus-Ion Implantation on the Corrosion Resistance and Biocompatibility of Titanium. *Biomaterials* **23**, 3329–3340 (2002).

44. Wakelin, E. A., Yeo, G. C., McKenzie, D. R., Bilek, M. M. M. & Weiss, A. S. Plasma Ion Implantation Enabled Bio-Functionalization of Peek Improves Osteoblastic Activity. *APL Bioeng.* **2**, 026109 (2018).

45. Hornberger, H., Virtanen, S. & Boccaccini, A. R. Biomedical Coatings on Magnesium Alloys – A Review. *Acta Biomater.* **8**, 2442–2455 (2012).

46. Surmenev, R. A. *et al.* The Influence of the Deposition Parameters on the Properties of an Rf-Magnetron-Deposited Nanostructured Calcium Phosphate Coating and a Possible Growth Mechanism. *Surf. Coat. Technol.* **205**, 3600–3606 (2011).

47. Safavi, M. S., Surmeneva, M. A., Surmenev, R. A. & Khalil-Allafi, J. RF-Magnetron Sputter Deposited Hydroxyapatite-Based Composite & Multilayer Coatings: A Systematic Review From Mechanical, Corrosion, and Biological Points of View. *Ceram. Int.* **47**, 3031–3053 (2021).

48. Mohseni, E., Zalnezhad, E. & Bushroa, A. R. Comparative Investigation on the Adhesion of Hydroxyapatite Coating on Ti–6Al–4V Implant: A Review Paper. *Int. J. Adhes. Adhes.* **48**, 238–257 (2014).

49. Alakoski, E., Tiainen, V.-M., Soininen, A. & Konttinen, Y. T. Load-Bearing Biomedical Applications of Diamond-Like Carbon Coatings – Current Status. *Open Orthop. J.* **2**, 43–50 (2008).

50. Darr, J. A., Guo, Z. X., Raman, V., Bououdina, M. & Rehman, I. U. Metal Organic Chemical Vapour Deposition (MOCVD) of Bone Mineral Like Carbonated Hydroxyapatite Coatings. *Chem. Commun.* 696–697 (2004) doi:10.1039/B312855P.

51. Park, S. W. *et al.* Generation of Functionalized Polymer Nanolayer on Implant Surface Via Initiated Chemical Vapor Deposition (iCVD). *J. Colloid Interface Sci.* **439**, 34–41 (2015).

52. Xue, T. *et al.* Surface Modification Techniques of Titanium and its Alloys to Functionally Optimize Their Biomedical Properties: Thematic Review. *Front. Bioeng. Biotechnol.* **8**, (2020).

53. O'Brien, B., Carroll, W.M., Kelly, M.J. Passivation of Nitinol Wire for Vascular Implants—A Demonstration of the Benefits. *Biomaterials* **23**(8), 1739–1748 (2002). doi:10.1016/S0142-9612(01)00299-X

54. Asri, R.I.M., Harun, W.S.W., Samykano, M., Lah, N.A.C., Ghani, S.A.C., Tarlochan, F., and Raza, M.R. Corrosion and Surface Modification on Biocompatible Metals: A Review. *Mater. Sci. Eng. C* **77**, 1261–1274 (2017). doi:10.1016/j.msec.2017.04.102

55. Alipal, J., Lee, T. C., Koshy, P., Abdullah, H. Z. & Idris, M. I. Evolution of Anodised Titanium for Implant Applications. *Heliyon* **7**, e07408 (2021).

56. Zhao, X. *et al.* Fabrication of $Al_2O_3$ by Anodic Oxidation and Hydrothermal Synthesis of Strong-bonding Hydroxyapatite Coatings on Its Surface. *Appl. Surf. Sci.* **470**, 959–969 (2019).

57. Li, Y. *et al.* Characterization and Cytocompatibility of Hierarchical Porous $Tio_2$ Coatings Incorporated with Calcium and Strontium by One-step Micro-Arc Oxidation. *Mater. Sci. Eng. C* **109**, 110610 (2020).

58. Bai, L. *et al.* A Multifaceted Coating on Titanium Dictates Osteoimmunomodulation and Osteo/Angio-genesis Towards Ameliorative Osseointegration. *Biomaterials* **162**, 154–169 (2018).

59. Li, G. *et al.* Enhanced Osseointegration of Hierarchical Micro/Nanotopographic Titanium Fabricated by Microarc Oxidation and Electrochemical Treatment. *ACS Appl. Mater. Interfaces* **8**, 3840–3852 (2016).

60. Pan, J., Prabakaran, S. & Rajan, M. In-vivo Assessment of Minerals Substituted Hydroxyapatite / Poly Sorbitol Sebacate Glutamate (PSSG) Composite Coating on Titanium Metal Implant for Orthopedic Implantation. *Biomed. Pharmacother.* **119**, 109404 (2019).

61. Jaafar, A., Hecker, C., Árki, P. & Joseph, Y. Sol-Gel Derived Hydroxyapatite Coatings for Titanium Implants: A Review. *Bioengineering* **7**, 127 (2020).

62. Tranquillo, E. & Bollino, F. Surface Modifications for Implants Lifetime extension: An Overview of Sol-Gel Coatings. *Coatings* **10**, 589 (2020).

63. Sikkema, R., Baker, K. & Zhitomirsky, I. Electrophoretic Deposition of Polymers and Proteins for Biomedical Applications. *Adv. Colloid Interface Sci.* **284**, 102272 (2020).

64. Radice, S., Kern, P., Bürki, G., Michler, J. & Textor, M. Electrophoretic Deposition of Zirconia-Bioglass® Composite Coatings for Biomedical Implants. *J. Biomed. Mater. Res. A* **82A**, 436–444 (2007).

65. Sun, F., Pang, X. & Zhitomirsky, I. Electrophoretic Deposition of Composite Hydroxyapatite–Chitosan–Heparin Coatings. *J. Mater. Process. Technol.* **209**, 1597–1606 (2009).

66. Maciąg, F. *et al.* The Effect of Electrophoretic Deposition Parameters on the Microstructure and Adhesion of Zein Coatings to Titanium Substrates. *Materials* **14**, 312 (2021).

67. Djošić, M., Janković, A. & Mišković-Stanković, V. Electrophoretic Deposition of Biocompatible and Bioactive Hydroxyapatite-Based Coatings on Titanium. *Materials* **14**, 5391 (2021).

68. Boccaccini, A. R., Keim, S., Ma, R., Li, Y. & Zhitomirsky, I. Electrophoretic Deposition of Biomaterials. *J. R. Soc. Interface* **7**, S581–S613 (2010).

69. Zhang, S., Xing, M. & Li, B. Biomimetic Layer-by-Layer Self-Assembly of Nanofilms, Nanocoatings, and 3D Scaffolds for Tissue Engineering. *Int. J. Mol. Sci.* **19**, 1641 (2018).

70. Gentile, P., Carmagnola, I., Nardo, T. & Chiono, V. Layer-by-Layer Assembly for Biomedical Applications in the Last Decade. *Nanotechnology* **26**, 422001 (2015).

71. Chen, W. *et al.* Surface Functionalization of Titanium Implants with Chitosan-Catechol Conjugate for Suppression of ROS-Induced Cells Damage and Improvement of Osteogenesis. *Biomaterials* **114**, 82–96 (2017).

# 17 Surface Functionalization for Biomaterials

*Mona M. Agwa*
Department of Chemistry of Natural and Microbial Products, Pharmaceutical and Drug Industries Research Institute, National Research Centre, Giza, Egypt

*Hesham S.M. Soliman*
Department of Pharmacognosy, Faculty of Pharmacy, Helwan University, Ain Helwan, Cairo, Egypt

Pharm D Program, Egypt-Japan University of Science and Technology, New Borg El-Arab City, Egypt

*Heba Elmotasem*
Pharmaceutical Technology Department, Pharmaceutical and Drug Industries Research Institute, National Research Centre, Giza, Egypt

*Sally Sabra*
Department of Biotechnology, Institute of Graduate Studies and Research, Alexandria University, Alexandria, Egypt

## 17.1 INTRODUCTION

Biomaterials are natural or synthetic materials that interact with biological environments and are used for therapeutic purposes such as sustaining, augmenting, mending, and controlling physical activity [1]. They are assorted into organic, inorganic, and bio-based materials. Examples of organic core material include polymers approved by Food and Drug Administration (FDA) like poly(lactic-co-glycolic acid) (PLGA), and polyethylene glycol (PEG) [2]. Inorganic core biomaterials include mesoporous silica, gold, iron oxide, and upconversion nanoparticles, while the bio-based materials are fundamentally generated from natural sources like protein nanosystems, cell membranes, lipids, and exosomes [3,4]. Advances in materials science have resulted in the development of multifunctional biomaterials with inherent specialized functionalities and biological effects to satisfy the demands of many biomedical

DOI: 10.1201/9781003429920-20

applications [5]. The surface properties of biomaterials are crucial for influencing biomolecular and cellular responses, making surface modification essential for reliable drug delivery.

Therapeutic delivery systems aim to maximize the performance in diseased sites while minimizing adverse effects [6]. Biopharmaceutical hurdles like first-pass effect, clearance, and biological barriers limit therapeutic potential. Nanotechnology has been applied to design therapeutic nanocarriers with targeting functions, ensuring precise distribution and protection [7]. Research focuses on engineering nano-carriers with improved physiochemical characteristics, enhancing selectivity, stability, biocompatibility, and pharmacokinetics through bioinspired surface functionalization [8,9].

In the arsenal of developing advanced biomimetic drug delivery, the nano-systems based on cellular architecture have demonstrated significant benefits over synthetic counterparts. Nano-systems that mimic cellular architecture are signifi-cantly superior to synthetic alternatives in developing advanced biomimetic drug delivery systems. Harnessing of materials that the body does not identify as strange particles prevents quick detection through the reticuloendothelial system, resulting in longer clearance and the avoidance of an immunological response, and even targets a disease region via the surface cell membrane protein's homing tendency [10]. Various sources of cell membranes for camouflaging different therapeutic delivery systems have been exploited, involving platelets, red blood cells (RBCs), macrophages, cancer cells, and white blood cells (WBCs) (Figure 17.1). The intrinsic merits of each cell membrane are linked directly to the proteins integrated into the lipid bilayer and are harnessed successfully to bypass shortages faced by the original medications. The RBC membranes include proteins that extend the circulation period via immune escaping. In contrast, proteins located in WBCs

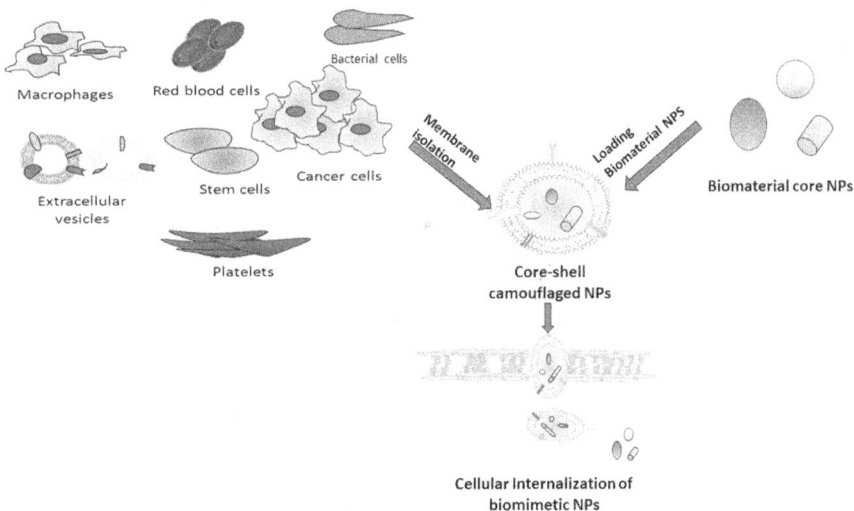

**FIGURE 17.1**  Simplified representation of biological membrane-coated core NPs.

favor passing the endothelial barriers to selectively accumulate inside the target cells. Additionally, camouflaging with cancer cells or platelet membranes manifested homologous cancer targeting through surface antigen and prolonged survival by minimizing macrophage uptake respectively via their unique proteins [11]. Thus, the cell membrane's natural properties can be harnessed to promote efficient targeting through the homing propensity of membrane proteins without sophisticated functionalization techniques [12]. Cutting-edge coated nano-drug carriers have enhanced drug delivery capabilities and have been exploited based on biomembranes of extracellular vesicles (EVs) and other natural cell membranes [13]. All of these have shown promising results as potential tools for targeted biomimetic delivery systems.

Cell membrane-wrapped NPs implicate a therapeutic NPs-based biomaterials inner core like inorganic, metallic, polymeric, and lipid bio-based materials surrounded by a cell membrane, making them ideal for targeted medication delivery, photothermal therapy, diagnostics, and imaging due to their inherent biological properties and outstanding biocompatibility [14]. These "core-shell" structures are more complex but more adaptable than conventional NP-based biomaterials. Organic biomaterials offer biocompatibility and biodegradability. In contrast, magnetic, optical, and electrical properties determine the selection of an inorganic core for cell membrane-coated NPs [15]. Researchers have demonstrated that cell membrane–wrapped NPs can skip biological barriers such as BBB impediments, prevent immune system damage, prolong the time that NPs remain in blood circulation, and have adequate biocompatibility and decreased cytotoxicity, thereby enhancing the drug delivery performance [16]. In developing biomimetic nano drug delivery carriers, the research has concentrated on employing entire whole cells, cell-derived membranes, and extracellular vesicles (EVs) [17]. However, utilizing whole cells was limited because they are more liable to be damaged and affected by the enclosed drugs. Therefore, cell-derived extracellular vesicles and cell-derived membranes are preferred and repurposed to carry therapeutic drugs or encase drug-loaded NPs. This chapter aims to discuss the diverse origins of biomimetic cell membranes camouflaged biomaterials to be harnessed as targeted delivery systems via homotopic or homologous tendencies.

## 17.2  EXTRACELLULAR VESICLES

Extracellular vesicles (EVs) are the subject of extensive research as delivery systems owing to their advantageous biomedical structures and therapeutic merits. It is anticipated to be a unique next-generation biomimetic pharmaceutical delivery system if properly advanced. The International Society for Extracellular Vesicles (ISEV) advocated adopting the word EVs as the distinctive descriptor for particles produced spontaneously from cells distinguished by a lipid bilayer and unable to reproduce [18]. They are tiny structures in animals, plants, and fungi. They contain a lipid bilayer, enclosed vesicles released by cells, comprising antigens, proteins, proteases, miRNA, and mRNA in the inner reservoir. They are assorted into exosomes, apoptotic bodies, and microvesicles (MVs). Exosomes and MVs are commonly used in delivery of lipophilic and hydrophilic therapeutics, owing to their nanoscale size and versatility [19].

They also act as therapeutics that promote tissue regeneration, modulate the immune system, and serve as substitutes for stem cell treatment [20]. EVs have the potency to skip biological barriers with a high targeting capacity while also being biocompatible and nearly non-immunogenic [21]. Also, EVs have been found to prolong systemic circulation [22]. Unlike other naturally based membranes, EVs equip unmet targeting potency linked to their role in cell–cell connection [21]. Different biofluids contain EVs, which are conclusive for cellular connection as they exchange bio-information between cells. These vesicles, decorated with unique biomolecules, carry specific markers proteins and lipids that and their structure stabilize their core cargo and help locate their targets both nearby and at far distances [23]. Thus, these EV-based nanoparticles as biomimetic approaches for surface coating enable the development of delivery systems targeting several receptors simultaneously for more effective absorption into target cells that express different receptors [24]. Different phases of clinical trials revealed that EVs might boost the immune response to many types of cancer [25,26].

## 17.2.1 EXOSOMES

Exosomes have been distinguished in many body fluids (including sweat, urine, saliva, milk, and blood). Also, a wide range of cell types can generate them involving red blood cells, platelets, tumors, brain, and stem cells [18]. They are an effective natural carrier to enclose material in their core as they have a natural lipid bilayer architecture that easily attaches to cellular membranes which express different surface proteins depending on their origin cell, altering their targeting potential to facilitate their cellular uptake and augment their therapeutic efficiency [27,28]. Exosomes exhibit potential as a viable biomimetic alternative for drug delivery due to their minimal side-effect profile and low proclivity to elicit immune reactions [29]. Researchers have successfully targeted and delivered therapies to tumor cells using biomimetic coats from cancer cells secreted by extracellular vesicles. Illes et al., for example, used HeLa cell exosomes to encapsulate metal organic frame nanoparticles, which were subsequently taken up by HeLa cells to deliver the anticancer medication [30]. Dumontel et al. wrapped zinc oxide nanocrystals with EVs generated from cancer cells, to mend their preferential internalization [31]. Another novel strategy was used by Sancho-Albero et al., who constructed palladium nanosheets directly into cancer cell exosomes, allowing the directed delivery of catalytic cargo to cancer cells [32]. Gao et al. wrapped PLGA nanoparticles with EVs from *Staph aureus* bacteria, which resulted in preferential absorption by macrophages exposed to this bacteria [33].

## 17.2.2 PLANT-DERIVED EXTRACELLULAR VESICLES

Plant extracellular vesicles (PEVs) were discovered in the 1960s, but their significance was overlooked until the early 2000s when researchers extracted them from apoplastic fluids [34–36]. Afterward, plant-derived exosome-like nanovesicles

(PELNs) with exosome-like morphology and composition were prepared for therapeutic encapsulation. Subsequently, lipids extracted from PELNs as a new class of PEVs drug delivery nanoparticles were prepared and known as PLNVs. This new class comprises the extracted lipids because of its high preparation method reproducibility and promising results when employed in biomimetic coated nano drug delivery systems [34,36]. Studies have stated the existence of exosome-like nanoparticles (ELNs) in plants, fruits, and mushrooms [37,38]. Due to their therapeutic capabilities, exosome-like nanoparticles produced from plants are projected to become viable therapeutic techniques for disease treatment or medication delivery. They can be employed as therapeutic agents due to the presence of phytoconstituents from the plant source. The biomimetic PELNs are minimally harmful to healthy tissues, demonstrating high biocompatibility, and are directed toward malignancies by targeting specific tissues when loaded with free drugs or drugs formulated within conventional nanocarriers to fine-tune their properties via distinct endocytosis pathways [39]. Plant-derived EVs have similar structures and functionalities to mammalian-derived counterparts [40]. In addition, they are characterized by multiple benefits over animal-derived counterparts; thus, they have been employed as an efficient therapy for various ailments [41]. PEVs promote regenerative tissue healing, improve antimicrobial immunity, originate from renewable resources, are biocompatibility, safety, easy to scale up, and offer multiple sources for targeting specific diseases [42]. On the other hand, the limited living sources of animal-based EVs complicated the scaling-up steps [43]. Plant-derived exosome-like nanovesicles are being studied for targeted drug delivery. Moreover, they are eligible for extra-surface modifications with ligands and biomimetic cell membrane fusion to improve organ or cell targeting, making them ideal for tailored hybrid biomimetic nanocarriers [44]. Wang et al. prepared a smart NP employing grapefruit-derived nanovesicles covered with an active leukocyte derived from the plasma membrane to improve the tumor's therapeutic potential and targeted specificity. This approach improved tumor permeability and decreased its size volume compared to the free drug and unmodified vesicles [45]. Niu et al. designed a biomimetic therapeutic system against glioma that combines grapefruit EVs and heparin-loaded doxorubicin (DOX) NPs. These nanoparticles bypassed the blood-brain barrier (BBB), promoting cellular intake, anti-proliferation, and prolonging their circulation period [46].

### 17.2.3 Isolation Techniques of EVs

Efficient extraction and isolation methods for EVs are crucial for developing biomimetic-coated NPs. Various methods were employed for EVs isolation from different sources, such as sonication, polymer precipitation, density-gradient ultracentrifugation, ultrafiltration, immunoaffinity separation, differential ultracentrifugation, and some innovative new isolation kits. The commercially available kits supplied an easy and proper procedure for EVs isolation as it can provide higher isolation yield than other isolation methods, but are utilized on a small scale due to weak purity [43]. Standardized methods for separating and purifying large amounts of exosomes are urgently demanded [47].

## 17.3 SOURCES OF CELL MEMBRANE FOR BIOMIMETIC COATING

Various sources of naturally inspired cell membranes for biomimetics surface functionalized therapeutic delivery systems have been exploited to fulfill extended circulation or targeting involving platelets, RBCs, macrophages, cancer cells, and WBCs. The innate merits of each cell membrane typeare linked directly to the proteins integrated into the lipid bilayer. They are harnessed successfully to bypass shortages faced by conventional nanocarriers [11].

### 17.3.1 RED BLOOD CELLS

Red blood cells (RBCs) are an important source of membrane coatings due to their long life span, reaching 120 days, and their abundance in the body. Also, mature RBCs don't exhibit a nucleus and several organelles, making their isolation very favorable. Coating nanoparticles with RBCs membranes enhance long-term circulation, pathogen elimination, and detoxification [48]. Employing RBCs membrane-camouflaged nanoparticles (NPs) in cancer treatment revealed extended blood half-life and retention better than PEGylated-PLGA owing to multiple proteins integrated with their membrane. Membrane proteins like CD47 hinder RES system macrophage phagocytosis. RBCs membrane-coated NPs improve cargo encapsulation and particle stability. RBCs functionalized NPs boost host anti-cancer immunity for cancer immunotherapy [12]. Arayal et al. developed polymeric NPs comprising poly(lactic acid) (PLA) core loaded in RBCs' membrane. Their results revealed the benefits of lengthy blood circulation and regulated medication release [49]. In another study, RBC membranes were employed by Liang et al. to coat the black-phosphorus quantum dot. The results showed that the biomimetic formulation with an increased in vivo NP circulation time and tumor internalization [50]. It was also reported that RBCs' membrane wrapping for AuNCs surfaces preserves the distinctive porous architectures, yielding RBC-membrane-coated AuNCs with good colloidal stability. These biomimetic surface-modified AuNCs exhibit photothermal effects and selective in vitro malignant cell ablation while also extending circulation lifetime in an animal study [51]. Also, the RBC-membrane stealth coating approach was anticipated to pave the path for enhanced PPT modulation by biomimetic metal NPs.

### 17.3.2 PLATELETS CELLS

Platelets (PLTs) are crucial, tiny cytoplasmic fragments in hemostasis and immune reactions that prompt the healing and inflammatory response in the injured areas. Encapsulating nanotherapeutics with PLT membranes containing p-selectin and other glycoproteins (integrins) that allow cohesion to the sub-endothelial matrix was employed in targeting various disorders like cancer, cardiovascular disease, inflammatory diseases, and bacterial infections. Fortunately, those membranes are implicated in targeting metastatic circulating cancer cells. Platelets membrane can successfully deliver siRNA to target cancer cells for a therapeutic gene approach [52]. Coating conventional NPs with a platelet membrane protects them from macrophage recognition, enhances preferential adhesion to malignant tissue or injured vessels, targets blood vessel abnormalities, promotes binding to circulating tumor cells, and helps pathogen clearance [53]. Zhou et al. modified the surface of PLGA coupled hyaluronic acid NPs (PLGA-SS-HA) with PLT membranes and rVAR2 peptide to construct redox-reactivedocetaxel-loaded NPs (rVAR2-PLTs/PLGA-SS-HA/DTX) for efficient primary and metastatic in vivo tumor homing [54]. The tumor homing was attributed to the intrinsic cellular adhesion feature of PLTs and rVAR2 peptide that can specifically recognize and bind the overexpressed oncofetal chondroitin sulfate. Refaat et al. developed immunoliposomes loaded with IR780, a photothermal dye, and finally enveloped by activated PLTs membranes for targeted photothermal thrombolysis in vivo via preferential aggregation in the thrombotic area and the generation of localized mild hyperthermia under laser exposure [55]. The fabricated targeted photo-therapy alone allows thrombolysis efficiently without demanding adjuvant fibrinolytic drugs. Moreover, it provides a higher potential for NIR (near infra-red) thrombosis fluorescence imaging. Recently, Zhou et al. inspireda porphyrinic metal-organic framework (MOF) loaded with anticancer drug and incorporated it inside sonosensitive PLT membranes for ultrasound-triggered sonodynamic tumor therapy [56]. The targeted nanoplatform exhibited excellent tumor homing via the innate TME (tumor micro-environment) tropism of PLTs. Upon reaching the tumor region, the porphyrinic MOF generates ROS under ultrasound stimulation, which acts synergistically with the released anticancer drug to eradicate the tumor in vivo.

### 17.3.3 WHITE BLOOD CELLS

Leukocytes, or white blood cells (WBCs), are a type of immune cells in charge of preserving the body against infections via engulfing external toxins and restoring the injured tissues. Many types of WBCs involving neutrophils, natural killer cells, macrophage, and lymphocytes share common qualities, making them a primary player in various diseases like endothelial barrier penetration and infiltrating other cells. Integrins, selectins, and G-protein-coupled receptors are the major membrane proteins in WBCs that confer the ability to pass the endothelial barriers and immune activation. WBCs have also been utilized for biomimetic nanoparticle coatings against major disorders such as cancer and infections [57]. The coated NPs operate to simulate original WBCs for bioactivity due to the inherited functions from the source cells. These benefits include a longer blood circulation period, improved cell-to-cell interaction, the capacity to detect antigens for better-targeting medication release, and higher biocompatibility. Moreover, transendothelial migration, tumor-targeting, and opsonization reduction are benefits that WBCs membrane coating NPs can achieve [58]. The chronic tumor inflammation is realized easily by leukocytes via recognizing and binding leukocyte-endothelial cell receptors at the inflamed tumor endothelium [59]. Parodi et al. fabricated leukocyte membrane-enveloped porous silica NPs incorporated chemotherapeutic, for targeted inflamed endothelial cells and avoid rapid hepatic clearance via evading lysosomal sequestration [60]. The fabricated biomimetic NPs exhibited higher in vivo tumor homing via cellular membrane recognition associated with efficient killing performance against breast cancer.

### 17.3.3.1   Neutrophil Cells

Neutrophils are the first defense line against pathogens. They manifested a pro-inflammatory attitude when accumulated at the inflammation area involving the generation of antimicrobial ROS, enzymes, peptides, and cytokines. Neutrophils migrate to infection or inflammation sites through cytokines and chemokines, targeting tumors, and inflammatory areas with integrin adhesion receptors. The presence of biomimetic neutrophil membrane coat NPs can target primary and metastatic inflammatory tumor regions via adhesion receptors, including L-selectin, and β1integrin [61]. Kang et al. designed neutrophil-mimicking drug delivery NPs based on surface functionalization medicated PLGA nanoparticles incorporated anticancer carfilzomib with neutrophil membranes to preserve cellular binding activity [62]. This system potentially inhibits the de novo metastatic progression and the already created metastasis.

### 17.3.3.2   Natural Killer Cells

Natural killer cells (NK) are components of the natural immune system that act as the initial line of defense in combating cancers and virus-pathogens. Unlike T and B cells, NK does not require prior activation. They feature abundant activating and inhibiting receptors on their exterior that particularly target tumor or virus-injured cells while leaving noninjured cells intact [63]. They can differentiate normal cells from malignant cells through exterior inhibitory receptors involving CD94-NKG2A, and killer cell immunoglobulin like receptors. The recognition of tumor cells is attributed to the absence of major histocompatibility complex (MHC) class I that is expressed in noninjured ones. They can stimulate tumor cell apoptosis via expressing death ligands on their surface to bind death receptors on the tumor exterior. Thus, NK cell-based immunotherapy against cancer is favorable due to its innate anti-tumor efficiency [59]. Pitchaimani et al. designed NK cell membranes fused into cationic liposomes loaded with anticancer doxorubicin (DOX) for efficient in vivo tumor killing via enhanced tumor homing [63]. The preferable tumor homing is attributed to the innate features of NK-92 cell surface receptors accompanied by a delayed blood retention period. In another survey, a biomimetic NPs incorporated NIR fluorescent dye and fused with NK cell membranes was constructed for deep tumor bio-imaging. The fabricated biomimetic NPs displayed efficient tumor homing ability accompanied by extended blood retention in vivo [64].

### 17.3.3.3   T-cell Membranes

T-cells (the name originated from thymus) are one of the remarkable sorts of WBCs that exert a master function in the immune response, range from eradicating both malignant and virus injured tissues to coordinating the immune response [65]. They can be discriminated from other lymphocytes by the expression of T-cell receptors on their exteriors to facilitate the recognition of different pathogens by the immune system. Stimulation of these receptors initiated via binding to their ligands on the APCs (antigen presenting cells) massively expressed in malignant and virus injured cells that trigger the intracellular receptor phosphorylation associated with recruitment of signal transducer proteins from the cytoplasm to the membrane-connected receptors, which affect the cytoskeleton and the nucleus [65]. Thus, T-cell membrane camouflaged NPs can target cancer and pathogens via binding antigens presented on their exteriors and destroy them by releasing cytotoxic molecules via providing the Fas ligand apoptotic signals [66]. Recent investigations have revealed that T-cell membrane-wrapped PLGA incorporated anticancer dacarbazine may target tumors via T-cell derived proteins to successfully eradicate tumor via releasing anticancer medicines and providing the Fas ligand apoptotic signals [67].

### 17.3.3.4   Dendritic Cells

Dendritic cells (DCs) bridge adaptive and innate immunity. Dendritic cells activate once a pathogen enters the body. They trigger T and B cells by presenting antigens. Dendritic cell membranes express co-stimulatory, CD86, and CD83, CD8. After connecting to T-cell receptors, dendritic cells also produce cytokines as interleukin. One mature DCs activates hundreds to thousands of T cells [67]. Thus, biomimetic coating with mature dendritic cell membranes can activate T cells to produce an efficient immune response to heal different malignancies and infections. Cheng et al. designed a biomimetic nanovaccine constructed from biodegradable PLGA-fused DCs membranes. The fabricated biomimetic nanovaccine displayed superior in vivo tumor homing via their plasma membrane proteins accompanied by augmented therapeutic, prophylactic, and anti-metastatic potential [68].

### 17.3.3.5   Macrophages Cells

Macrophages are innate immune cells that eliminate foreign undesired toxins from the body. They are drawn from circulating monocytes found in all tissues, where cytokines attract them during infections or tissue injury; chemokine receptors on the macrophage membrane increase cell adhesion and recruitment. The targeting strength of macrophages is mostly attributable to the proteins expressed on their membrane involving interleukin-1 receptors and toll-like receptors, which identify the tumor endothelium [59]. Furthermore, they recognize and phagocytose odd toxins and injurious cells like malignant cells. Coating the nanoparticles with macrophage membrane allows pathogens to attach and quickly circumvent macrophage identification, allowing active targeting in inflammatory and tumor locations [69]. Long et al. designed baicalin liposomes wrapped with macrophages to target brain cells and treat encephalopathy. The results showed improved brain targeting, prolonged circulation, and preferable neuroprotective effects on rats, compared to baicalin conventional liposomes [70].

### 17.3.4   Stem Cells

Stem cells multiply and develop into particular cell types. Mesenchymal (MSC)-based regenerative medicine has great potential [71]. MSCs target cancer and metastasis due to their tumor-tropic properties. They are stable and easily isolated. MSCs roll, adhere, and extravasate to inflammation or damaged sites through chemokines and cytokine receptors. Thus, a stem cell membrane on the core NPs targets tumors and degenerative disorders [71]. Gao et al. designed DOX-incorporated gelatin nanogels fused with the mesenchymal stem cell membrane to exploit MSC-specific tumor targeting power, which in turn augments the intratumoral DOX accumulation to improve antimalignant therapeutic efficiency in vivo [72]. In another study, biomimetic NPs constructed from PLGA NPs incorporated DOX and wrapped with MSC cell membranes were designed by Yang et al. The biomimetic NPs disclosed remarkable antimalignant power via exploiting specific tumor-homing merits [73]. Zhou et al. developed targetable NPs constructed from DOX-loaded MOFs NPs wrapped with MSC cell membranes for oral squamous cell cancer [74]. The NPs were found to be prospective selective targeted drug delivery vehicles for oral squamous cell carcinoma, causing cancer cells death in vitro and inhibiting tumor development in vivo.

## 17.3.5 CANCER CELLS

Several tumor-specific antigens and adhesion molecules are presented on the exterior of cancer cell membranes depending on their type, tumor stage, and location. Homotypic cancer adhesion is deemed to be a precious tool in directed therapeutics delivery to primary and metastatic circulating tumors. The homologous interaction is fundamentally assigned to the expression of glycoprotein adhesion molecules on cell membranes that vary among different cells but commonly comprise proteins like cadherins, selectins, lymphocyte-homing receptors resembling CD44-molecules, and integrins [11]. Employing cancer cells to create biomimetic nanoparticles has numerous benefits, including homologous targeting via homotypic adhesion to comparable tumor cells, immune evasion, and enhanced antitumor response. Nevertheless, there are constraints due to the unknown pathogenicity [75]. Cancer cell membrane-wrapped NPs aid in creating personalized cancer medication exhibiting superior cellular internalization, tumor targeting, therapeutics accumulation, and endogenous biomimetic "stealth" delivery in vivo [16]. Disguising biomimetic NPs constructed from glioma cell membranes coated liposome incorporated photothermal dye can improve their biocompatibility for directed phototherapeutic treatment of orthotropic glioma in vivo, efficient fluorescence tumor imaging, and allow them to traverse the blood-brain barrier and circulate long term [76]. Ji et al. designed modified biomimetic NPs constructed from cell penetrating peptide amalgamated melanoma tumor membranes modified paclitaxel (PTX)-loaded nanomicelles [77]. The fabricated biomimetic NPs exhibited outstanding tumor homing performance and simultaneous therapeutic delivery for superior melanoma immunotherapy. Li et al. manufactured self-amplified biomimetic NPs constructed from tumor cell membrane wrapped MOFs NPs incorporated epirubicin (EPI), hemin, and glucose oxidase (Gox) [78]. The biomimetic NPs can successfully penetrate the tumor cells via homotypic targeting, suppressing tumor growth and metastasis via triggering ROS generation that acts synergistically with the immunogenic cell death induced by EPI to maximize the therapeutic sensitivity of anti-PD-L1 antibody towards TNBC (triple negative cell).

## 17.3.6 CELLULAR ORGANELLES' PHOSPHOLIPID MEMBRANE

Biomimetic targeting based on intracellular membrane activity has applications in gene therapy, drug-resistant infections, and cancer treatment. Camouflaged-coated biomimetic nanoparticles constructed using intracellular membranes have recently been studied for their potential to target detoxification and molecular detection in induced thrombocytopenia [79].

### 17.3.7 BACTERIAL CELLS

Unlike human cells, bacteria have a peptidoglycan cell wall. Thus, covering the NPs with bacterial membranes induces an antibacterial immune response, bacterial infection vaccination, and tumor targeting [48,80]. Gao et al. investigated gold nanoparticle vesicles coated with the bacterial outer membrane [81]. The prepared biomimetic NPs successfully improve the NPs' stability and rapidly activate dendritic cells, boosting antibody response against the used membrane bacteria source. This shows that encapsulating synthetic nanoparticles with natural bacterial membranes may provide effective antibacterial vaccinations [10].

### 17.3.8 HYBRID CELL MEMBRANE

Hybrid cell membrane NP wrapping combines cell membranes from various cell types to imitate numerous cell-specific functional features. Designing coated NP amalgamated membranes depends on the requested purpose. These hybrid membrane-wrapped NPs provide a customized strategy for biocompatible NPs with merged source cell functions, surpassing the restrictions of single membrane coating techniques. Researchers have effectively fabricated RBC-PLT combination membrane-coated PLGA NPs attained new properties derived from both cells [82]. These include prolonged circulation, biological compatibility, active tumor site targeting, and notable internalization in breast cancer cells; thus, hybrid membrane camouflaging by fusing diverse cell types allows biomimetic NPs with several desirable capabilities to be developed, which has many therapeutic benefits. He et al. designed a novel hybridized biomimetic nanoplatform called Leutusome to prevent immune system detection [83]. This platform comprises PTX incorporated liposomal nanoparticles wrapped with fused plasma membranes constructed from leukocytes and tumor cells. They discovered that encapsulating chemotherapy within leutusome can effectively limit tumor development without deleterious effects to the healthy body regions in vivo. Gong et al. survey showed that the co-delivery of immuno-metabolic modulator metformin and siRNA could be achieved using a pH-responsive targeted platform [84]. This platform used hybrid biomimetic membranes constructed from cancer and macrophages cells coated poly(lactic-co-glycolic acid) NPs. The in vitro and in vivo studies demonstrates the promising therapeutic efficiency of the immunotherapy against breast cancer. Wang et al. reported a hybrid membrane constructed from an outer bacterial membrane and cancer cell membrane–wrapped hollow polydopamine NPs [85]. The biomimetic NPs displayed prime melanoma targeting and eminent synergistic antimalignant immune response combined photothermal therapy without prominent toxicity. Also, exosome-liposome complex hybrids were investigated. They efficiently delivered large payloads and extended circulation, combining exosomes' and liposomes' benefits. They shield contents, targeting specificity, and immune detection escaping, making them valuable biomedical tools [86].

## 17.4 TECHNIQUES FOR LOADING CORE CARGOES INTO BIOMIMETIC MEMBRANES

### 17.4.1 LOADING CARGOES INTO EVs

Drug loading into EVs can be accomplished through direct loading into exosomes or in the mother cell during exosome formation. Methods such as electroporation, incubation, sonication, and thawing are used to load pharmaceuticals into EVs, though success rates vary, and some cause deterioration to either the EVs or their payload. Sonication, pH gradient, and extracellular vesicles-liposome fusion methods are successful ways of loading cargoes into extracellular vesicles [87]. The sonication approach mixes exosomes and core cargo using ultrasonic force, resulting in excellent loading efficiency. The pH gradient method for loading uses a pH gradient to raise the acidity of extracellular vesicles, increasing the effectiveness of loading negatively charged payloads such as nucleic acids [88].

### 17.4.2 LOADING CARGOES INTO CELLULAR MEMBRANES

#### 17.4.2.1 Co-extrusion

Mechanical extrusion across a porous membrane is employed in enveloping NPs in cell membranes. Extrusion mechanically adjusts cell membrane structure, allowing core NPs to move across the phospholipid bilayer, for coated membrane nanoparticle production [89]. Coated NPs extrude uniformly due to constant pore diameter, preserving cell membrane proteins as it is free of chemical processing. However, co-extrusion often wastes raw resources because materials stick to the extrusion film [90].

#### 17.4.2.2 Sonication

Sonication prevents material degradation by mixing cell membranes and nanoparticles. It is more effective than extrusion in preventing material loss. Sonication creates an attraction between NPs and membranes, making it suitable for the smallest structures. However, it may not be ideal for certain nanocore structures due to size and durability [91].

#### 17.4.2.3 Electroporation

Cell membrane electroporation is based on exposing cell membranes to a strong external electric field, which forms numerous pores that produce membrane semi-permeability. These pores allow NPs to enter the cell membrane. This strategy is gaining popularity among scientists, since it does not impair NPs stability by prohibiting agglomeration [92].

#### 17.4.2.4 Microfluidic Electroporation

A microfluidic electroporation approach generates membrane-coated nanoparticles using a Y-shaped merging channel, S-shaped mixing channel, and electroporation zone. This superior method offers superior colloidal stability and in vivo effects. Microfluidic electroporation could potentially overcome the mechanical extrusion technique's limited throughput. Although this method has the benefit of minimizing membrane protein damage and raw material wastage, its technical expense is relatively high [93].

### 17.4.2.5 Cell Membrane Coating Using Graphene Nanoplatform
Researchers developed a single-step process for extracting and assembling cell membranes using leukocyte cells, focusing on targeting biomimetic-coated nanoparticles. The technology uses graphene nanosheets to extract phospholipids, integrating membrane extraction and coating [94].

### 17.4.2.6 In Situ Packaging Technique of Coating
In situ, packaging of cell-derived vesicles is another interesting method. NPs are left to be incubated alongside cells under a starvation condition; this promotes NPs to enter cells by endocytose and thus release coated NPs. This method offers versatile and multifunctional nanoplatforms, retaining membrane integrity and promoting a bio-inspired approach to membrane-coated NP technology [95].

### 17.4.2.7 Extra-Optimization Techniques for Camouflaging Cell Membrane
Additional functionalization techniques may employ targeting ligands and cell membrane fusion for homing to certain organs or cells for more localization. Optimization of cell membrane camouflaged nanoparticles is feasible through the use of enzyme-mediated techniques, as well as noncovalent and covalent modifications. In addition, the camouflaged membrane NPs can be modified at the cellular level by involving genetic engineering and membrane fusion [14].

## 17.5  CONCLUSION
Engineering the exteriors of the nano-delivery systems is crucial in the styling procedure, predominantly for bio-interfacing. Biomimetic delivery systems deduced from cell membranes are hopeful technology with abundant privileges as they authorize the easy fabrication of NPs with modified surfaces that imitate the susceptibility requisite for efficacious bio-interfacing. The intrinsic merits of each cell membrane are linked directly to the proteins integrated into the lipid bilayer and are harnessed successfully to bypass shortages faced by the original medications. These biomimetic NPs can interface with biological systems and reinforce the therapeutics aggregation and efficacy at intended sites owing to mimicking the characteristics of the parent cells. They can execute diverse roles like expanded circulation and escaping rapid immune clearance. It is likely that the application of EVs camouflaging NPs will expand in the future and could even substitute current targeting approaches like aptamers, antibodies, and peptides. Despite these privileges, there are hurdles to bypass when applying biomimetic cell-derived nanocarriers involving large-scale fabrication that require sterile conditions, standardized protocols for particular cell sorts, reproducibility, and well-settled characterization equipments. Due to their heterogeneity, long-term storage can also be problematic, and methods must be developed to validate their structural integrity and functionality.

## REFERENCES

1. Rahmati, B., et al., Development of tantalum oxide (Ta-O) thin film coating on biomedical Ti-6Al-4V alloy to enhance mechanical properties and biocompatibility. *Ceramics International*, 2016. **42**(1): p. 466–480.
2. Virlan, M.J.R., et al., Organic nanomaterials and their applications in the treatment of oral diseases. *Molecules*, 2016. **21**(2): p. 207.
3. Han, X., et al., Biomaterial-assisted biotherapy: A brief review of biomaterials used in drug delivery, vaccine development, gene therapy, and stem cell therapy. *Bioactive Materials*, 2022. **17**: p. 29–48.
4. Hamza, K. H., et al., Topically Applied biopolymer-based tri-layered hierarchically structured nanofibrous scaffold with a self-pumping effect for accelerated full-thickness wound healing in a rat model. *Pharmaceutics*, 2023. **15**(5): p. 1518.
5. Li, J., et al., Applications of 3D printing in tumor treatment. *Biomedical Technology*, 2024. **5**: p. 1–13.
6. Agwa, M.M., et al., Self-assembled lactoferrin-conjugated linoleic acid micelles as an orally active targeted nanoplatform for Alzheimer's disease. *International Journal of Biological Macromolecules*, 2020. **162**: p. 246–261.
7. Agwa, M.M., et al., Vitamin D3/phospholipid complex decorated caseinate nanomicelles for targeted delivery of synergistic combination therapy in breast cancer. *International Journal of Pharmaceutics*, 2021. **607**: p. 120965.
8. Agwa, M.M. and S. Sabra, Lactoferrin coated or conjugated nanomaterials as an active targeting approach in nanomedicine. *International Journal of Biological Macromolecules*, 2021. **167**: p. 1527–1543.
9. Agwa, M.M., et al., Carbohydrate ligands-directed active tumor targeting of combinatorial chemotherapy/phototherapy-based nanomedicine: A review. *International Journal of Biological Macromolecules*, 2023. **239**: p. 124294.
10. Soprano, E., et al., Biomimetic cell-derived nanocarriers in cancer research. *Journal of Nanobiotechnology*, 2022. **20**(1): p. 538.
11. Bigaj-Józefowska, M.J. and B.F. Grześkowiak, Polymeric nanoparticles wrapped in biological membranes for targeted anticancer treatment. *European Polymer Journal*, 2022. **176**: p. 111427.
12. Fang, R. H., et al., Cell membrane coating nanotechnology. *Advanced Materials*, 2018. **30**(23): p. 1706759.
13. Abesekara, M. S. and Y. Chau, Recent advances in surface modification of micro-and nano-scale biomaterials with biological membranes and biomolecules. *Frontiers in Bioengineering and Biotechnology*, 2022. **10**: p. 972790.
14. Zhang, M., et al., Membrane engineering of cell membrane biomimetic nanoparticles for nanoscale therapeutics. *Clinical and Translational Medicine*, 2021. **11**(2): p. e292.
15. Anselmo, A. C. and S. Mitragotri, A review of clinical translation of inorganic nanoparticles. *The AAPS Journal*, 2015. **17**: p. 1041–1054.
16. Zhong, X., et al., Cell membrane biomimetic nanoparticles with potential in treatment of Alzheimer's disease. *Molecules*, 2023. **28**(5): p. 2336.
17. Mitchell, M.J., et al., Engineering precision nanoparticles for drug delivery. *Nature Reviews Drug Discovery*, 2021. **20**(2): p. 101–124.
18. Ali, N.B., et al., Theragnostic applications of mammal and plant-derived extracellular vesicles: Latest findings, current technologies, and prospects. *Molecules*, 2022. **27**(12): p. 3941.
19. Meng, W., et al., Prospects and challenges of extracellular vesicle-based drug delivery system: Considering cell source. *Drug Delivery*, 2020. **27**(1): p. 585–598.

20. Wiklander, O.P., et al., Advances in therapeutic applications of extracellular vesicles. *Science Translational Medicine*, 2019. **11**(492): p. 8521.
21. Saari, H., et al., Microvesicle-and exosome-mediated drug delivery enhances the cytotoxicity of Paclitaxel in autologous prostate cancer cells. *Journal of Controlled Release*, 2015. **220**: p. 727–737.
22. Lai, C.P., et al., Dynamic biodistribution of extracellular vesicles in vivo using a multimodal imaging reporter. *ACS Nano*, 2014. **8**(1): p. 483–494.
23. Hassanpour, M., et al., Exosomal cargos modulate autophagy in recipient cells via different signaling pathways. *Cell & Bioscience*, 2020. **10**(1): p. 1–16.
24. Fathi, P., L. Rao, and X. Chen, Extracellular vesicle-coated nanoparticles. *View*, 2021. **2**(2): p. 20200187.
25. Escudier, B., et al., Vaccination of metastatic melanoma patients with autologous dendritic cell (DC) derived-exosomes: Results of the first phase I clinical trial. *Journal of Translational Medicine*, 2005. **3**: p. 1–13.
26. Besse, B., et al., Dendritic cell-derived exosomes as maintenance immunotherapy after first line chemotherapy in NSCLC. *Oncoimmunology*, 2016. **5**(4): p. e1071008.
27. Hussen, B.M., et al., Strategies to overcome the main challenges of the use of exosomes as drug carrier for cancer therapy. *Cancer Cell International*, 2022. **22**(1): p. 323.
28. Kim, M. S., et al., Engineering macrophage-derived exosomes for targeted paclitaxel delivery to pulmonary metastases: in vitro and in vivo evaluations. *Nanomedicine: Nanotechnology, Biology and Medicine*, 2018. **14**(1): p. 195–204.
29. Kooijmans, S.A., et al., Electroporation-induced siRNA precipitation obscures the efficiency of siRNA loading into extracellular vesicles. *Journal of Controlled Release*, 2013. **172**(1): p. 229–238.
30. Illes, B., et al., Exosome-coated metal–organic framework nanoparticles: An efficient drug delivery platform. *Chemistry of Materials*, 2017. **29**(19): p. 8042–8046.
31. Dumontel, B., et al., ZnO nanocrystals shuttled by extracellular vesicles as effective Trojan nano-horses against cancer cells. *Nanomedicine*, 2019. **14**(21): p. 2815–2833.
32. Sancho-Albero, M., et al., Cancer-derived exosomes loaded with ultrathin palladium nanosheets for targeted bioorthogonal catalysis. *Nature Catalysis*, 2019. **2**(10): p. 864–872.
33. Gao, F., et al., Kill the real with the fake: Eliminate intracellular Staphylococcus aureus using nanoparticle coated with its extracellular vesicle membrane as active-targeting drug carrier. *ACS Infectious Diseases*, 2018. **5**(2): p. 218–227.
34. Ly, N.P., et al., Plant-derived nanovesicles: Current understanding and applications for cancer therapy. *Bioactive Materials*, 2023. **22**: p. 365–383.
35. Halperin, W. and W.A. Jensen, Ultrastructural changes during growth and embryogenesis in carrot cell cultures. *Journal of Ultrastructure Research*, 1967. **18**(3-4): p. 428–443.
36. Ju, S., et al., Grape exosome-like nanoparticles induce intestinal stem cells and protect mice from DSS-induced colitis. *Molecular Therapy*, 2013. **21**(7): p. 1345–1357.
37. Liu, B., et al., Protective role of shiitake mushroom-derived exosome-like nanoparticles in D-galactosamine and lipopolysaccharide-induced acute liver injury in mice. *Nutrients*, 2020. **12**(2): p. 477.
38. De Robertis, M., et al., Blueberry-derived exosome-like nanoparticles counter the response to TNF-α-Induced change on gene expression in EA. hy926 cells. *Biomolecules*, 2020. **10**(5): p. 742.
39. Kim, J., et al., Plant-derived exosome-like nanoparticles and their therapeutic activities. *Asian Journal of Pharmaceutical Sciences*, 2022. **17**(1): p. 53–69.
40. Rome, S., Biological properties of plant-derived extracellular vesicles. *Food & Function*, 2019. **10**(2): p. 529–538.

41. Cai, Q., et al., Plants send small RNAs in extracellular vesicles to fungal pathogen to silence virulence genes. *Science*, 2018. **360**(6393): p. 1126–1129.
42. Teng, Y., et al., Plant-derived exosomal microRNAs shape the gut microbiota. *Cell host & microbe*, 2018. **24**(5): p. 637–652. e8.
43. Alzahrani, F.A., et al., Plant-derived extracellular vesicles and their exciting potential as the future of next-generation drug delivery. *Biomolecules*, 2023. **13**(5): p. 839.
44. Wang, Q., et al., Delivery of therapeutic agents by nanoparticles made of grapefruit-derived lipids. *Nature Communications*, 2013. **4**(1): p. 1867.
45. Wang, Q., et al., Grapefruit-derived nanovectors use an activated leukocyte trafficking pathway to deliver therapeutic agents to inflammatory tumor sites hijacked leukocyte pathway for targeted delivery. *Cancer Research*, 2015. **75**(12): p. 2520–2529.
46. Niu, W., et al., A biomimetic drug delivery system by integrating grapefruit extracellular vesicles and doxorubicin-loaded heparin-based nanoparticles for glioma therapy. *Nano Letters*, 2021. **21**(3): p. 1484–1492.
47. Chen, C., et al., Toward the next-generation phyto-nanomedicines: Cell-derived nanovesicles (CDNs) for natural product delivery. *Biomedicine & Pharmacotherapy*, 2022. **145**: p. 112416.
48. Chugh, V., K. Vijaya Krishna, and A. Pandit, Cell membrane-coated mimics: a methodological approach for fabrication, characterization for therapeutic applications, and challenges for clinical translation. *ACS Nano*, 2021. **15**(11): p. 17080–17123.
49. Aryal, S., et al., Erythrocyte membrane-cloaked polymeric nanoparticles for controlled drug loading and release. *Nanomedicine*, 2013. **8**(8): p. 1271–1280.
50. Liang, X., et al., Photothermal cancer immunotherapy by erythrocyte membrane-coated black phosphorus formulation. *Journal of Controlled Release*, 2019. **296**: p. 150–161.
51. Piao, J.-G., et al., Erythrocyte membrane is an alternative coating to polyethylene glycol for prolonging the circulation lifetime of gold nanocages for photothermal therapy. *ACS Nano*, 2014. **8**(10): p. 10414–10425.
52. Zhuang, J., et al., Targeted gene silencing in vivo by platelet membrane–coated metal-organic framework nanoparticles. *Science Advances*, 2020. **6**(13): p. eaaz6108.
53. Xu, L., et al., Platelet membrane coating coupled with solar irradiation endows a photodynamic nanosystem with both improved antitumor efficacy and undetectable skin damage. *Biomaterials*, 2018. **159**: p. 59–67.
54. Zhou, M., et al., Platelet membrane-coated and VAR2CSA malaria protein-functionalized nanoparticles for targeted treatment of primary and metastatic cancer. *ACS Applied Materials & Interfaces*, 2021. **13**(22): p. 25635–25648.
55. Refaat, A., et al., Activated platelet-targeted IR780 immunoliposomes for photo-thermal thrombolysis. *Advanced Functional Materials*, 2023. **33**(4): p. 2209019.
56. Zhou, L., et al., Nanoengineered sonosensitive platelets for synergistically augmented sonodynamic tumor therapy by glutamine deprivation and cascading thrombosis. *Bioactive Materials*, 2023. **24**: p. 26–36.
57. Wang, D., et al., White blood cell membrane-coated nanoparticles: recent develop-ment and medical applications. *Advanced Healthcare Materials*, 2022. **11**(7): p. 2101349.
58. Zeng, S., et al., Cell membrane-coated nanomaterials for cancer therapy. *Materials Today Bio*, 2023. **20**: p. 100633.
59. He, Z., Y. Zhang, and N. Feng, Cell membrane-coated nanosized active targeted drug delivery systems homing to tumor cells: a review. *Materials Science and Engineering: C*, 2020. **106**: p. 110298.
60. Parodi, A., et al., Synthetic nanoparticles functionalized with biomimetic leukocyte membranes possess cell-like functions. *Nature Nanotechnology*, 2013. **8**(1): p. 61–68.

61. Zhang, Q., et al., Neutrophil membrane-coated nanoparticles inhibit synovial inflammation and alleviate joint damage in inflammatory arthritis. *Nature Nanotechnology*, 2018. **13**(12): p. 1182–1190.

62. Kang, T., et al., Nanoparticles coated with neutrophil membranes can effectively treat cancer metastasis. *Acs Nano*, 2017. **11**(2): p. 1397–1411.

63. Pitchaimani, A., T.D. T. Nguyen, and S. Aryal, Natural killer cell membrane infused biomimetic liposomes for targeted tumor therapy. *Biomaterials*, 2018. **160**: p. 124–137.

64. Pitchaimani, A., et al., Biomimetic natural killer membrane camouflaged polymeric nanoparticle for targeted bioimaging. *Advanced Functional Materials*, 2019. **29**(4): p. 1806817.

65. Hui, E., Understanding T cell signaling using membrane reconstitution. *Immunological Reviews*, 2019. **291**(1): p. 44–56.

66. Sterner, R. C. and R.M. Sterner, CAR-T cell therapy: current limitations and potential strategies. *Blood Cancer Journal*, 2021. **11**(4): p. 69.

67. Kang, M., et al., T-Cell-mimicking nanoparticles for cancer immunotherapy. *Advanced Materials*, 2020. **32**(39): p. 2003368.

68. Cheng, S., et al., Artificial mini dendritic cells boost T cell-based immunotherapy for ovarian cancer. *Advanced Science*, 2020. **7**(7): p. 1903301.

69. Xuan, M., et al., Macrophage cell membrane camouflaged Au nanoshells for in vivo prolonged circulation life and enhanced cancer photothermal therapy. *ACS Applied Materials & Interfaces*, 2016. **8**(15): p. 9610–9618.

70. Long, Y., et al., Macrophage membrane modified baicalin liposomes improve brain targeting for alleviating cerebral ischemia reperfusion injury. *Nanomedicine: Nanotechnology, Biology and Medicine*, 2022. **43**: p. 102547.

71. Han, Y., et al., Mesenchymal stem cells for regenerative medicine. *Cells*, 2019. **8**(8): p. 886.

72. Gao, C., et al., Stem cell membrane-coated nanogels for highly efficient in vivo tumor targeted drug delivery. *Small*, 2016. **12**(30): p. 4056–4062.

73. Yang, N., et al., Surface functionalization of polymeric nanoparticles with umbilical cord-derived mesenchymal stem cell membrane for tumor-targeted therapy. *ACS Applied Materials & Interfaces*, 2018. **10**(27): p. 22963–22973.

74. Zhou, D., et al., Modification of metal-organic framework nanoparticles using dental pulp mesenchymal stem cell membranes to target oral squamous cell carcinoma. *Journal of Colloid and Interface Science*, 2021. **601**: p. 650–660.

75. Rao, L., et al., Cancer cell membrane-coated nanoparticles for personalized therapy in patient-derived xenograft models. *Advanced Functional Materials*, 2019. **29**(51): p. 1905671.

76. Jia, Y., et al., Phototheranostics: active targeting of orthotopic glioma using biomimetic proteolipid nanoparticles. *ACS Nano*, 2018. **13**(1): p. 386–398.

77. Ji, Z., et al., Multifunctional modified tumor cell membranes-coated adjuvant PTX against melanoma. *Biomolecules*, 2023. **13**(1): p. 179.

78. Li, Z., et al., A tumor cell membrane-coated self-amplified nanosystem as a nanovaccine to boost the therapeutic effect of anti-PD-L1 antibody. *Bioactive Materials*, 2023. **21**: p. 299–312.

79. Gong, H., et al., Nanomaterial biointerfacing via mitochondrial membrane coating for targeted detoxification and molecular detection. *Nano Letters*, 2021. **21**(6): p. 2603–2609.

80. Tu, Y., et al., Destructive extraction of phospholipids from Escherichia coli membranes by graphene nanosheets. *Nature Nanotechnology*, 2013. **8**(8): p. 594–601.

81. Gao, W., et al., Modulating antibacterial immunity via bacterial membrane-coated nanoparticles. *Nano Letters*, 2015. **15**(2): p. 1403–1409.

82. Dehaini, D., et al., Erythrocyte–platelet hybrid membrane coating for enhanced nanoparticle functionalization. *Advanced Materials*, 2017. **29**(16): p. 1606209.
83. He, H., et al., Leutusome: a biomimetic nanoplatform integrating plasma membrane components of leukocytes and tumor cells for remarkably enhanced solid tumor homing. *Nano Letters*, 2018. **18**(10): p. 6164–6174.
84. Gong, C., et al., Regulating the immunosuppressive tumor microenvironment to enhance breast cancer immunotherapy using pH-responsive hybrid membrane-coated nanoparticles. *Journal of Nanobiotechnology*, 2021. **19**(1): p. 1–20.
85. Wang, D., et al., Bacterial vesicle-cancer cell hybrid membrane-coated nanoparticles for tumor specific immune activation and photothermal therapy. *ACS Applied Materials & Interfaces*, 2020. **12**(37): p. 41138–41147.
86. Sato, Y. T., et al., Engineering hybrid exosomes by membrane fusion with liposomes. *Scientific Reports*, 2016. **6**(1): p. 21933.
87. Hussen, B.M., et al., Strategies to overcome the main challenges of the use of exosomes as drug carrier for cancer therapy. *Cancer Cell International*, 2022. **22**(1): p. 1–23.
88. Xi, X.-M., S.-J. Xia, and R. Lu, Drug loading techniques for exosome-based drug delivery systems. *Die Pharmazie-An International Journal of Pharmaceutical Sciences*, 2021. **76**(2-3): p. 61–67.
89. Song, W., et al., Cell membrane-camouflaged inorganic nanoparticles for cancer therapy. *Journal of Nanobiotechnology*, 2022. **20**(1): p. 1–29.
90. Liu, W. and Y. Huang, Cell membrane-engineered nanoparticles for cancer therapy. *Journal of Materials Chemistry B*, 2022. **10**(37): p. 7161–7172.
91. Hu, C.-M.J., et al., Nanoparticle biointerfacing by platelet membrane cloaking. *Nature*, 2015. **526**(7571): p. 118–121.
92. Rao, L., et al., Microfluidic electroporation-facilitated synthesis of erythrocyte membrane-coated magnetic nanoparticles for enhanced imaging-guided cancer therapy. *Acs Nano*, 2017. **11**(4): p. 3496–3505.
93. Han, Z., et al., Improving tumor targeting of exosomal membrane-coated polymeric nanoparticles by conjugation with aptamers. *ACS Applied Bio Materials*, 2020. **3**(5): p. 2666–2673.
94. Zhou, X., et al., Leukocyte-repelling biomimetic immunomagnetic nanoplatform for high-performance circulating tumor cells isolation. *Small*, 2019. **15**(17): p. 1900558.
95. Silva, A.K.A., et al., Cell-derived vesicles as a bioplatform for the encapsulation of theranostic nanomaterials. *Nanoscale*, 2013. **5**(23): p. 11374–11384.

# 18 Surface Modification Technologies and Methods of Biomaterials

*Mojtaba Najafizadeh*
Faculty of Chemical and Materials Engineering, Shahrood University of Technology, Shahrood, Iran

Department of Innovation Engineering, University of Salento, Lecce, Italy

*Payam Sarir*
College of Civil Engineering, Tongji University, Shanghai, China

*Sahar Yazdi*
School of Medical Sciences, Shahrood Branch, Islamic Azad University, Shahrood, Iran

*Ehsan Marzban Shirkharkolaei*
Faculty of Mechanical Engineering, Isfahan University of Technology, Isfahan, Iran

*Mansoor Bozorg*
Faculty of Chemical and Materials Engineering, Shahrood University of Technology, Shahrood, Iran

*Morteza Hosseinzadeh*
Department of Engineering, Ayatollah Amoli branch, Islamic Azad University, Amol, Iran

*Pasquale Cavaliere*
Department of Innovation Engineering, University of Salento, Lecce, Italy

## 18.1 INTRODUCTION

This chapter comprehensively scrutinizes the surface functionalization of metallic and polymeric biomaterials and presents recent empirical evidence on their

DOI: 10.1201/9781003429920-21

promising potential for medical applications. This study concentrates predominantly on metallic implant biomaterials, including titanium, Ti6Al4V, Gama-TiAl, and stainless steel, as well as synthetic polymeric biomaterials such as polyethylene, poly(ethylene terephthalate), polyurethane, polystyrene, polyimide, and parylene C. These biomaterials have been subjected to surface functionalization with renewable materials, specifically biopolymers like alginate, chitosan, starch, hyaluronate, and others. The present study discusses the potential applicability of surface-modified biomaterial composites that comprise both conventional medical biomaterials and renewable materials, to enhance their functionalization. The outcomes of this research indicate that such composites hold great promise for successful use across a multitude of medical domains.

The biocompatibility and physicochemical surface characteristics of a biomaterial are critical factors in determining its reaction. A comprehensive evaluation of potential implant applications and their corresponding insertion sites necessitates careful consideration of pivotal determinants, including but not limited to blood compatibility, osseointegration, non-toxic properties, immunogenicity, tumorigenicity, genotoxicity, and hemostatic responses. Furthermore, it is imperative to take into account a range of physicochemical properties, including but not limited to corrosion resistance, mechanical attributes, tribological parameters, surface tension, surface free energy, surface topography, and coating composition. Frequently, the enhancement of these parameters and the ability to be applied in vivo as biomaterials necessitates surface modification. Functional coatings employed for the purpose of surface modification are required to satisfy a multitude of criteria which include but are not limited to reproducibility, durability, ease of application, cost-effectiveness, tailored surface morphology, and environmentally sustainable production. Moreover, these coatings must possess distinctive properties that are specific to their intended application.

## 18.2 BIOMATERIALS AND SURFACE MODIFICATION

A biomaterial refers to an inanimate substance that interacts directly with bodily tissues to assess, remedy, enhancing, or substitute any tissue, organ, or function of the human body [1–3]. A variety of biomaterials have demonstrated efficacy in applications within the human body, including but not limited to titanium alloys and both natural and synthetic polymers such as alginate, chitosan, polyethylene, and polyurethane. The destiny of the implant material is predominantly influenced by its surface properties, comprising surface chemistry and structure, which are integral components in the biomaterial's interaction with the adjacent tissues.

Titanium and its derivatives are widely recognized as auspicious biomaterials for implantation in orthopedic procedures including joint replacement as well as mending fractured bones via the utilization of bone pins, plates, and screws [2]. The utilization of these materials in the manufacturing of biomedical devices on a broad scale hinge largely on their amalgamation of superior corrosion resistance, adequate mechanical properties, and biocompatibility. The remarkable corrosion resistance exhibited by titanium and its alloys in biological systems is largely ascribed to the development of an oxide layer on their surfaces that is distinguished by its chemical stability, continuity, and strong adhesion. Notwithstanding the spontaneous inception of the oxide layer, the liberation of metal-laden entities remains feasible. Moreover, the thickness of the oxide layer, ranging between 2–20 nm, is inadequate to sustain the potential motions that may occur between the implanted biomaterial and the surrounding tissue, as supported by previous studies [2,4,5].

Biomedical polymers are synthetic products that are considered the most commercially versatile class of biomaterials. Surgical devices, artificial organs, orthopedic and vascular prostheses, and other medical devices are some applications of biomedical polymers. Another application of biomedical polymers is in tissue engineering and regenerative medicine, which can be classified into the following four categories [6–8]:

1. Relatively hydrophobic materials such as alpha hydroxy acid (such as poly (lactic-co-glycolic acid), PLGA)
2. Polyanhydrides
3. Natural polymers (such as chitosan and hyaluronan)
4. Minerals (e.g., hydroxyapatite)

The four possible states after biomaterial implantation in the body include the following:

1. Successful osseointegration
2. Rejection
3. Micro implant movement
4. Bacterial infection

In particular, osseointegration is considered a key factor in successful implant implantation in the body, which is influenced by two main factors: mechanical properties and biological interactions of the implant surface with the surrounding tissues. After surgical implantation and as a result of bone proliferation and differentiation processes, the process of osteoconduction of healing will occur. The healing response typically consists of acute inflammation, chronic inflammation, granulation tissue formation, foreign body reaction, and fibrosis. After surgical implantation, we will always have inflammation, which is independent of the type of biomaterial used and the place of implantation, and it may also bring acute inflammation for a few days. The last stage of improvement of encapsulation of birch implants is fibrous tissue, which is called fibrosis. It is important to note that the human body recognizes the implant material as foreign and isolates it by encasing it in fibrous tissues if the surface properties are incapable of forming a stable bond between the implant surface and the surrounding tissue. As the result of an acute external body response caused by the inflammatory process at the implantation site, this may result in implant rejection. Moreover, it is noticeable that in the case of orthopedic and dental implants, since they cannot withstand the same physical stress as bones, the fibrosis process is undesirable and hence it has micromovements and implant failure [3,7,9,10].

The results of studies have shown that short-term or long-term implant failure can be a result of bacterial infection and biofilm formation on the surface of the implant, and therefore implantable biological materials such as bone or dental implants, stents, wound dressings, and catheters play a key role in maintaining health and they have human treatment. For example, one of the applications of catheters is to transfer liquids and drugs intravenously with a flexible silicone rubber structure and with a very low level of toxicity. Unfortunately, these materials are prone to bacterial or microbial colonization. Once settled, microbes form a biofilm that is resistant to antimicrobial agents and the body's immune response. These biofilms are the main threat to human life [11–15].

As described before, it is presented that there is a significant need to enhance the implant's surface (e.g., metal alloys and polymeric biomaterials) to prevent implant failure or rejection for four reasons:

i. increasing tissue adhesion,
ii. enhancing implant integration,
iii. reducing bacterial adhesion,
iv. and minimizing inflammatory and foreign body responses.

Based on mentioned information, it is possible to predict some relationships between the cell behavior and the physico-chemical properties of the material. Therefore, the interactions between the cells and the surface of the substrate should be considered as follows:

i. rigidity and deformability,
ii. surface roughness and morphology,
iii. surface chemistry,
iv. cell adhesion, proliferation, and differentiation

## 18.2.1 RIGIDITY AND DEFORMABILITY

The mechanical properties of cells on the surface are defined based on physical parameters such as their rigidity and deformation. The concept of elasticity refers to the ability of a given material to withstand stress without permanent deformation, and the slope of the stress-strain curve represents the elastic modulus, which is also called Young's modulus [4]. In biological systems, nonlinear elasticity is typically observed because the extracellular matrix (ECM) proteins that make up biological networks consist of long peptide chains that can rapidly and irreversibly rigidity [4].

One of the potential characteristics of cells is their ability to recognize the rigidity of their environment, which reacts with migration or cell differentiation under the influence of the degree of rigidity of the environment [4,16,17]. The substrate's rigidity and deformability are crucial determinants influencing several cellular processes, including the formation of cell surface adhesion complexes, the development of the actin cytoskeleton, cell survival, activity, differentiation, growth, and spreading, as well as the generation of fibrin matrices. This is especially critical in the case of scaffold components for tissue engineering. For example, coatings on the inner surface of polymeric vascular grafts could be used to accelerate endothelialization, or delivery systems for various types of cells (e.g., mesenchymal stem cells, chondrocytes) could be employed for tissue wound healing [18].

The impact of molecular properties on the determination of material stiffness can be effectively illustrated in the case of polymeric surfaces. Various material characteristics are capable of modifying and regulating the stiffness of a matrix. These characteristics include, but are not limited to, the concentration of the material, the size of the pores, the order of the network, the degree of crystallinity, the average molecular weight or chain length, as well as the degree of cross-linking.

It is crucial to consider that any alteration made to a surface yields an impact not solely on the material's chemistry and wettability, but also its physicochemical properties such as rigidity and flexibility. It is imperative to take into account that any modification of a surface has an impact not only on the material chemistry and wettability but also on the physicochemical properties, including the rigidity and flexibility of the treated surface. The prevailing development in materials with the ability to modify stiffness is moving away from synthetic 2D materials towards 3D materials that are more biomimetic and sustainable in nature. Furthermore, the incorporation of three-dimensional (3D) stiffness into emerging biomaterials or their surfaces can offer valuable insight into phenomena such as tissue growth and injury recovery, while also paving the way for advancements in the field of tissue engineering. Figure 18.1 illustrates various influential factors that impact Young's modulus.

## 18.2.2 BIOMATERIAL SURFACE ROUGHNESS

The harshness of a structure advances biomineralization by expanding the surface zone accessible for a mineral statement [3]. Large scale-, smaller scale-, and nano unpleasantness altogether influence the behavior of an embed after implantation, especially concerning cell attachment, development, and development. Expanded unpleasantness comes about in an expanded genuine surface region and surface vitality, which in turn leads to an increment in contact point on hydrophobic surfaces (directed by Young's condition) [19]. Be that as it may, the behavior of cells on an unpleasant surface depends on numerous physical and chemical variables such as surface vitality, the morphology of abnormalities, wettability, and sort of accessible functionalities. By and large, microroughness emphatically influences cell grip and development due to the moderately huge unlimited surface

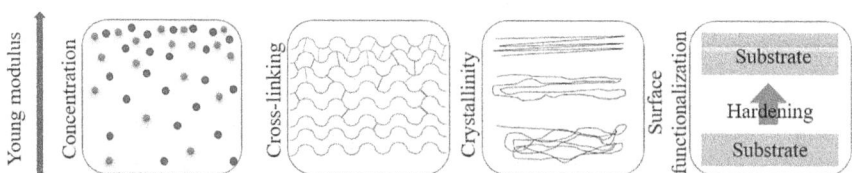

**FIGURE 18.1** Graphical representation depicts the various influential factors that impact Young's modulus.

range given by the abnormalities. Interests, various reports on the impact of microroughness on cell behavior at amalgam surfaces after implantation have appeared both positive (e.g., expanded spreading of rodent osteoblasts on micro-porous titanium dental inserts, human osteoblast-like MG-63 cell separation on Ti substrate with the microroughness surface) and negative impacts (e.g., the slower multiplication of MG-63 cells on a Ti-6Al-4V amalgam with the microroughness surface) [20–23]. Despite clashing coming about within the writing, it has been concluded that expanding surface harshness decreases cell movement and leads to slower expansion [18]. The component behind this double impact of surface microroughness remains hazy. Nanoroughness is accepted to promote cell grip and encompasses a positive impact on cell development and development. Also, materials with a nanostructured surface minimize the potential chance of immuno-genicity and provocative reactions [24]. Gradually things appeared that on nanostructured surfaces, extracellular network (ECM) proteins adsorb in a fitting geometrical introduction, empowering their amino corrosive grouping (RGD) to tie to cell adhesion–mediating receptors (see Sections 18.2.3 and 18.2.4) [18].

### 18.2.3  CHEMISTRY OF SUBSTRATE'S SURFACE

The biocompatibility of functionalized materials is emphatically impacted by surface chemistry properties, such as free vitality, hydrophilicity, contact point, extremity, electrostatic intuitive, chemical reactivity, and versatility of surface useful bunches.

The free vitality of a biomaterial surface depends on the accessible useful bunches and the electrical charges shown. For solids, surface vitality compares to adhesiveness and fondness for intelligence with other materials. A tall surface vitality guarantees wettability by most fluids, where any fluid with a lower surface pressure than the surface-free vitality of a biomaterial will dampen its surface. In outline, surface-free vitality controls wettability and hydrophilicity as well. A hydrophilic surface for the most part has polar functionalities and electrical charges, coming about in tall surface vitality. Such a surface does not bolster protein and cell adsorption, as the intelligence with encompassing water particles is weaker. On the other hand, a hydrophobic surface advances protein attachment, by driving theenergetically favorable hydrophobic intuitive. Be that as it may, this will initiate solid and irreversible intelligent, coming about in protein denaturation and misfortune of action [4,18]. The hydrophilicity of materials is essentially decided by water attachment pressure and is subordinate to the contact point. Biomaterials with a contact point more prominent than 65° are classified as hydrophobic, whereas those with a contact point less than 65° are considered hydrophilic [4]. Exceedingly hydrophilic surfaces restrain or restrain cell connection and development. Be that as it may, exceedingly hydrophilic surfaces empower the adsorption and authority of cell adhesion–mediating particles with moderately powerless powers. Within the case of an expansive number of bound cells, particularly amid longer hatching times, the separation of these atoms is conceivable. Subsequently, it is as of now postulated that surfaces with direct wettability with water are ideal and craved, as they don't adjust protein adaptation and encourage cell adsorption and development.

The surface functionalities most commonly displayed on biomaterials incorporate hydroxyl (–OH), amino (–NH2), carboxyl (–COOH), and methyl (–CH3) bunches. The nearness of these bunches on the surface and their intuition with cement proteins decide the ensuing cell connection. For the most part, a negative charge on the fabric surface diminishes cell-material intuitive and cell-to-cell grip, whereas a positive charge advances such intuitive. Oxygen-containing bunches influence the polar component of surface-free vitality, making the surface more wettable, stickier, and more inclined to the attachment of interceding proteins, such as ECM proteins (e.g., fibronectin, laminin, vitronectin, collagen). Then again, cell non-adhesive atoms, such as egg whites, lean toward less oxygenated and more hydrophobic surfaces, coming about in lessened adsorption [4,18,25].

Biomaterials with hydroxyl bunches show an unbiased hydrophilic character that encourages cell attachment, to the extent of the sum of oxygen-containing functionalities [26]. Biomaterials with amine bunches have a positive charge on the surface at physiological pH, driving expanded endothelial cell development and improved myoblast grip, separation, and mineralization of osteoblasts [27,28].

Biomaterials possessing carboxyl groups exhibit a negative charge which confers hydrophilicity to the surface. This property, in turn, exerts a favorable effect on the adsorption of fibrinogen and albumin, as well as the binding of various biomolecules that support cellular adhesion such as epidermal growth factor, collagen IV, and chondroitin sulfate. Additionally, the expression of integrin receptors is promoted which ultimately results in enhanced cellular proliferation and differentiation, specifically in the case of myoblasts [29–31]. Despite this fact, it has also been elucidated that the presence of carboxyl (-COOH) functionalities on the surface inhibits cellular attachment and subsequent proliferation (particularly in osteoblasts) [18,25,32].

Methyl or alkyl groups are commonly encountered constituents on biomaterial surfaces or coatings. These groups engender a hydrophobic surface with a net neutral charge, facilitating the adsorption of proteins through hydrophobic-hydrophobic interactions, such as fibrinogen, albumin, and IgG [26,29].

However, these strong interactions may also cause significant alterations in the conformations, leading to protein denaturation [26]. In conclusion, surfaces with mixed functionalities are the most desired, as they allow for diverse available interactions with proteins and cells.

Introducing a variety of additional functional groups, such as hydroxyl, carboxyl, and amine groups, to chemically inert biomaterials enables the covalent immobilization of proteins, including adhesive proteins or lytic enzymes. Adhesive proteins derived from ECM proteins, such as fibronectin, laminin, vitronectin, or collagen, can promote cell adhesion through ligand-acceptor interactions. Among a variety of adhesive peptides, the Arg-Gly-Asp (RGD) sequence is believed to be the binding domain for integrin receptors on the cell membrane. Integrins are a widely expressed family of heterodimeric transmembrane receptors that bind to these adhesive peptide motifs present in various ECM [33,34]. After ligand binding, integrins interact with elements of the cytoskeleton, forming focal adhesions of supramolecular assemblies composed of structural and signaling proteins. These changes trigger signaling cascades that regulate cell cycle progression and differentiation. Integrin-mediated cell attachment significantly affects other cellular processes, including cell migration and apoptosis [33,35]. From a functionalization process perspective, it is worth mentioning that introducing a variety of additional functional groups to chemically inert biomaterials enables the covalent immobilization of proteins, including adhesive proteins or lytic enzymes.

The second category of adhesive proteins pertains to immobilized growth factors (GF) that serve as signals for regulating cellular growth, proliferation, and differentiation.

Growth factors (GFs) refer to endogenous polypeptides with small molecular sizes, which are soluble and commonly secreted by cells to facilitate communication between cells in various biological contexts. As per Tallawi et al. [34], it can be inferred that insulin-like growth factor (IGF-1), hepatocyte growth factor (HGF), and vascular endothelial growth factor (VEGF) are commonly acknowledged as prime examples of growth factors (GFs). The feasibility of utilizing the solid-phase presentation of hepatocyte growth factor (HGF), a seminal protein integral to liver development and regeneration, as an alternative to conventional methodologies employing growth factors in a soluble form, has been explored. Jones et al. conducted a study, as per their research endeavor. The present study documented that the integration of hepatocyte growth factor (HGF) into extracellular matrix (ECM) protein microarrays, specifically collagen and laminin, exhibited a considerable influence on the morphology and phenotype of primary hepatocytes, for a prolonged duration. In contemporary discourse, there exists a noteworthy fascination with the implementation of biocompatible and biodegradable polymers, namely thermoplastic aliphatic polyesters such as poly-l-lactide (PLA) and poly(lactide-co-glycolide) (PLGA), as plausible conveyance mechanisms for growth factors (GFs) [34]. Concerning the process of functionalization, the second class of adhesive proteins is composed of growth factors that are immobilized and function as signaling agents to regulate cellular growth, proliferation, and differentiation.

There exists a pronounced exigency to substitute petroleum-derived polymeric substances and plasticized substances with resources that are both sustainable and environmentally beneficial.Thus, biodegradable or edible films and coatings are proposed as new types of surface modification for metallic and polymeric materials. In particular, increasing attention has been focused on incorporating antimicrobial agents and/or immobilizing lytic enzymes in edible packaging or medical devices [36–39]. Bacterial biofilms possess distinctive characteristics that lead to an elevated level of resilience towards antibiotics and disinfectants, coupled with the ability to resist phagocytosis and other defense mechanisms of the human body [13–15,40]. Lytic enzymes, including lysozyme, lysostaphin, amylase, cellulase, alginate lyase, and DNase, have demonstrated the potential to degrade bacterial biofilms, thereby enhancing their susceptibility to antibiotics and other antimicrobial agents [11,41–43]. Henceforth, the present surface functionalization is anticipated to impede the process of food spoilage resulting from microbial proliferation, prolong the span of food commodities, and avert the onset of bacterial infections linked to hospitalization. From an environmental and health standpoint, there exists an imminent necessity to substitute petrochemical-derived synthetic polymers and plastics with sustainable and eco-friendly materials, including biodegradable or food-grade films and coatings.

## 18.2.4  SURFACE PREPARATION

In preparing surfaces for metal alloys, polymeric biomaterials, medical devices, implants, or artificial organs, at least three aspects are crucial for supporting cell growth [4,10,18]:

1. Modifying surface tension by grafting chemical groups to the surface to increase hydrophilicity.
2. Introducing amine and carboxylic groups for bonding of proteins to allow cells to identify the surface as "own matter".
3. Binding growth factors (cytokines) to the surface to accelerate cell growth. From a cellular growth support perspective, the preparation of surfaces for metal alloys, polymeric biomaterials, medical devices, implants, or artificial organs must consider at least three critical aspects: modifying surface tension, introducing amine and carboxylic groups for protein binding, and binding growth factors to the surface to accelerate cell growth.

For adherent cells in the initial phase of cell spreading, the size of the adhesion substrate has a positive correlation with cell proliferation activity. Restriction of cell adhesion, which limits extension on the growth surface, can result in anoikis, a specific type of apoptosis [44]. It is now well established that cell spreading activates cell proliferation through two mechanisms: biochemical and mechanical. Both mechanisms begin with the adsorption of cell adhesion-mediated molecules (RGD) and the binding of their attachment sequence of peptides (Arg-Gly-Asp) to glycoprotein receptors (integrins) [33]. This triggers their clustering into specific domains ("adhesion sites"), where communication of structural and signaling proteins occurs. These molecules play a crucial role in determining cell fate, such as cell proliferation and differentiation, survival, apoptosis, and many other functions [18,45,46]. The biochemical process entails the activation of signaling constituents within the "adhesion site," including focal adhesion kinase (FAK), tyrosine kinases (such as Src, PYK2, Csk, and Abl), and additional intracellular signaling pathways linked to growth factor receptors, such as mitogen-activated protein (MAP) kinases (ERK-1 and -2) and serine/threonine kinases (ILK, PCK). The aforementioned phenomenon engenders cellular advancement across the G0, G1, and S phases of the cellular life cycle, the process of DNA synthesis, culminating in the phenomenon of cellular division. Meanwhile, the mechanical mechanism that controls cell growth is associated with the regulation of cytoskeletal assembly. The actin fibers are linked to the structural components of the "adhesion site," such as talin, vinculin, paxillin, alpha-actinin, and tensin. At the same time, these fibers are anchored to the nuclear membrane, the membranes of cellular organelles, and various enzymes. Increasing tension of the cytoskeleton stimulates cell proliferation through nuclear expansion, enlargement of the nuclear pores, synthesis, and transportation of various extracellular cell cycle–regulating factors, and DNA synthesis [18]. In the case of adherent cells in the initial phase of cell spreading, the size of the adhesion substrate is positively correlated with cell proliferation activity, while restriction of cell adhesion can result in anoikis. Cell spreading activates cell proliferation through both biochemical and mechanical mechanisms, involving the adsorption of cell adhesion–mediated molecules and the clustering of glycoprotein receptors. These mechanisms trigger intercellular signaling pathways and cytoskeletal assembly, leading to cell progression through the cell cycle and DNA synthesis.

The correlation between cell spreading and proliferation activity remains incompletely comprehended. The phenomenon of cell proliferation capacity and migration on various substrates, namely smooth muscle cells, endothelial cells, fibroblasts, and mesenchymal cells, has been observed to peak at intermediate adhesion strength. Subsequently, high levels of adhesion have been observed to stimulate cell maturation, as corroborated by numerous studies [18,46]. Additionally, the correlation between cellular proliferation and differentiation necessitates clarification, given that accurate regulation of these mechanisms is of paramount importance in the field of tissue engineering. For example, excessive migratory and proliferation activities of vascular smooth muscle cells (VSMC) can cause stenosis and occlusion of the lumen of vascular prostheses, mainly due to the expression of surface adhesion molecules that bind to immune cells such as leucocytes, lymphocytes, monocytes, and macrophages [45]. The question arises of how to exclude these undesirable components from artificial vascular grafts. One solution is to create a bioinert surface or appropriate surface functionalization that prevents the attachment, spreading, and growth of VSMC and other cells that stimulate an immune response. However, this solution provides the body with an artificial replacement that loses its bio-inertness over time. Currently, artificial blood vessel replacements are constructed or functionalized with hydrophobic materials on one side and extremely hydrophilic materials on the other to adsorb protein and cells over time. Modification of the surface of prostheses with albumin, heparin, or other drugs exerting anticoagulant, anti-migratory, and/or antiproliferative activity is one proposed solution for preventing restenosis of artificial vascular prostheses. Another more physiological approach involves functionalizing the inner surface of grafts with a confluent, mature, quiescent, and semipermeable endothelial cell layer to inhibit adhesion and VSMC growth, prevent thrombus formation, and activate the immune system response [47–50]. A different tactic in developing new materials for vascular tissue engineering is not to inhibit VSMC adhesion completely but to use their physiological nature to enhance the endothelialization of artificial vascular prostheses, perform a contractile function, and participate in the elimination of bioresorbable components of artificial grafts by replacing them with natural ECM [18,45,46].

In the contemporary field of biomaterials science and tissue engineering, a major obstacle pertains to the surface functionalization of biomaterials that can be replenished, wherein the strategies employed involve either imitating the behavior of stem cells or their utilization [51–53]. Significant objectives in this domain encompass emulating the stem cell niche using surface modification, devising strategies for governing processes akin to the differentiation of vascular smooth muscle cells (VSMC) from mesenchymal stem cells, preserving VSMC in an inactive and contractile phenotype, and reconstructing both the tunica media and the media of vascular tissue via an artificial polymeric substrate.

## 18.3  SURFACE MODIFICATION TECHNOLOGIES AND METHODS

In the field of biomedical engineering, the replication of various environments necessitates the utilization of specific biomaterials, such as metal alloys and polymer materials. These materials, however, demand innovative approaches for the chemical functionalization of their surfaces to achieve their intended purpose. An alternative methodology involves utilizing low molecular weight compounds that act as functional constituents of larger molecules, which subsequently become immobilized to elicit the recruitment of suitable cells present within the host tissue. The surface property requirements of each system are diverse and specific, thereby precluding the ability to propose a universal model. This contention underscores the need for tailored approaches to surface property optimizations by each system's distinct parameters. The interactions between biological systems and artificial biomaterials occur mainly on the materials' surface, and the biological response depends on surface properties such as chemical composition, roughness, surface energy, corrosion, and affinity for denaturation of nearby proteins. The latter is critical because proteins such as albumin, fibrinogen, fibronectin, and vitronectin adsorb onto the implant surface within seconds to minutes after implantation and influence the overall kinetics and thermodynamics of the binding events between cells and the implant surface [4,18]. A schematic overview of the sequence of events that occur at the surface material during contact with the biological medium is shown in Figure 18.2.

The choice of surface functionalization strategy for platforms is based on components such as the sort of substrate (e.g., metals, polymers), platform structure, and specific surface chemistry necessities. Framework structure is continuously a

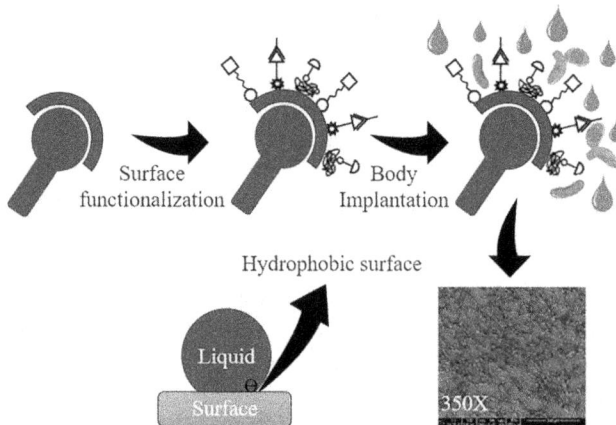

**FIGURE 18.2**   Schematic depicts multiple stages after implantation of the biomaterial.

vital figure to consider, but coating methods are chosen based on surface charge, dissolvable intuitive, or surface vitality. Surface alteration procedures can be categorized into five bunches: mechanical strategies, chemical strategies, physical strategies, organic strategies, and radiation strategies, as summarized in Table 18.1 [3,4,54,55].

The attainment of varying alterations in surface physico-chemical parameters is contingent upon the dictated implant location and intended functionality of a biomedical apparatus. The significance of possessing good hemocompatibility cannot be overstated for medical devices that are designed to make contact with

**TABLE 18.1**
**Surface Modification Methods**

Methods of Surface Modification

| Mechanical | Radiation | Physical | Chemical | Biological |
|---|---|---|---|---|
| Machining | Glow discharge (plasma) | Physical adsorption of active molecules | Chemical treatment: acidic, alkaline, hydrogen peroxide | Heparinization |
| Grinding | Corona discharge | Langmuir–Blodgett (LB) film deposition | Incorporation of functional groups | Hyaluronic acid coating |
| Polishing | Photoactivation (UV) | Surface micro- and nanopatterning | Salinization | Lipid immobilization |
| Blasting | Microwave activation | Inducing roughness and texture | Fluorination | Protein–enzyme immobilization |
| Attrition | Laser | Impinging etching | Sol-gel treatment | Antimicrobial activation, e.g., silver, antibiotic, or other antibacterial agents' coatings |
| | Ion beam | Physical vapor deposition (PVD) | Chemical vapor deposition (CVD) | Lytic enzyme immobilization |
| | Electron beam | Sputtering | Layer-by-layer (LbL) self-assembly | |
| | g irradiation | Thermal, flame, plasma spray | Ozone treatment | |
| | | Ion implantation and deposition | Hydrogel grafts | |
| | | Plasma immersion ion implantation (PIII) | | |

blood such as catheters, grafts, or stents. Similarly, excellent osseointegration is indispensable for materials intended for bone replacement purposes. Surface modification methodologies have the potential to enhance the hydrophilicity of materials, promote endothelialization, expedite osseointegration, mitigate thrombogenicity, function as drug delivery modalities, and/or provide dependable antibacterial properties. Surface treatments encompass several techniques such as surface roughening and patterning, surface films and coatings, and surface modification for biomolecule and pharmaceutical delivery, as depicted in Figure 18.3.

Although surface modification techniques have been extensively examined in numerous review articles and book chapters, the present article solely focuses on essential surface functionalization methods that significantly affect the interactions between biomaterials and surrounding tissues.

## 18.3.1  PHYSICAL PROPERTIES OF THE SURFACE

Mechanical techniques, such as grinding, polishing, machining, and blasting, are widely employed to alter the surface characteristics of titania and titanium alloys. These techniques impart designated surface **topography**, refine surface finish, induce surface roughness, and augment adhesive properties. The process of surface roughening primarily alters the topographical features of the surface, without inducing any chemical modifications [3,54,55]. This leads to a noteworthy amplification of the surface area and a concomitant impeding of cell mobility, ultimately culminating in augmented cell adhesion.

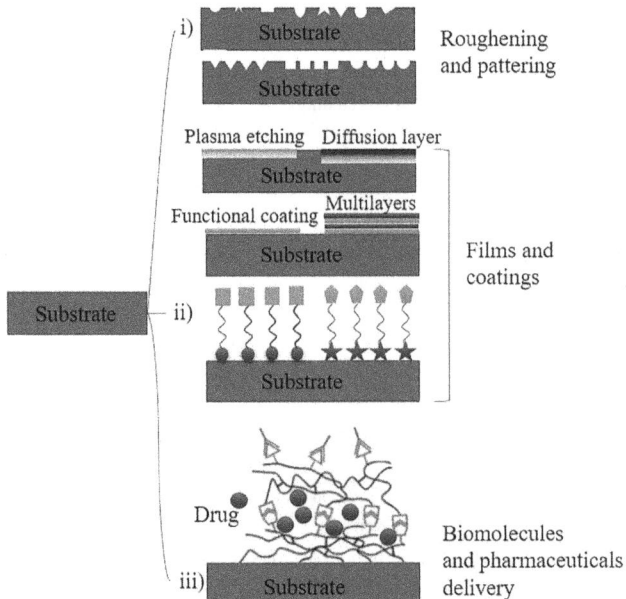

**FIGURE 18.3**   Schematic illustration of the most popular surface functionalization methods.

Oxygen or argon plasma deposition is a commonly used technique to alter surface topography by inducing processes such as melting and recrystallization. This leads to the formation of more ridges and a hydrophilic surface compared to the original surface [3,56].

Plasma and chemical-based etching are a phenomenon that transpires when a substrate is exposed to etching gas that frequently constitutes plasma comprising electrons, ions, radicals, and neutral molecules. This process is commonly employed in material science research, manufacturing industries, and surface modification applications. Polymeric materials undergo surface degradation and top layer alterations when exposed to etching gas, leading to the breaking of old bonds and the formation of new ones via chain scission. Plasma etching is a potential pre-treatment method to facilitate the coating process with a specific desired material, including polymeric hydrogels and bioactive molecules. Plasma pretreatment has been observed to have the capacity to modify surface roughness. Consequently, the implementation of biopolymer chains of varying lengths can be achieved subsequent to said treatment. The present empirical evidence has demonstrated that surface roughening at the nanoscale level can facilitate improved biocompatibility and cellular adherence [57]. In the graphical display denoted as Figure 18.4, our research team has presented a compilation of specific instances of plasma treatment that were scrutinized, and subsequently implemented in diverse medical substrates.

Compared to surface roughening, surface patterning provides a more organized type of surface alteration from the original material. This widely used technique enables the creation of surfaces patterned in both micro- and nanoscales with different topography, such as hollows or prominences with a variety of sizes,

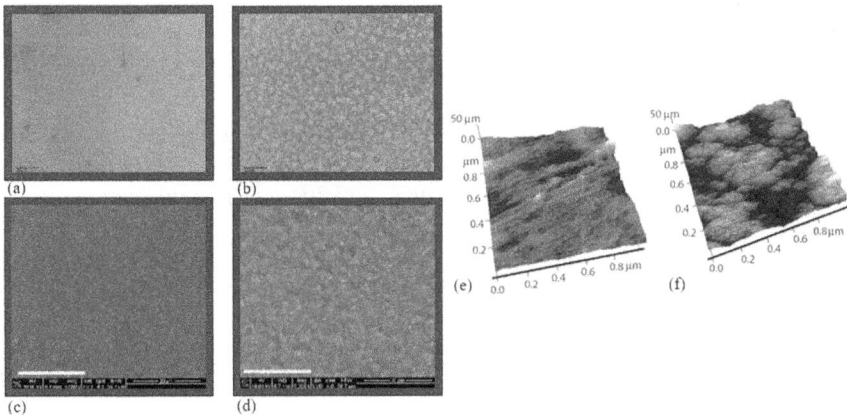

FIGURE 18.4 Plasma treatment of various medical substrates – optical images of Ti6Al7Nb alloy (magnification 40×): (a) unmodified and (b) treated with $O_2$ plasma and after deposition of CS coatings; SEM images of Ti6Al4V alloy: (c) unmodified and (d) treated with $O_2$ plasma and after deposition of CS coatings; AFM images of PE: (e) unmodified and (f) treated with $Ar/O_2$ plasma and after gradient coatings deposition (N-DLC/Si-DLC).

shapes, spacing, and distribution, including grooves, ridges, pits, pillars, boxes, cylinders, and honeycombs. Micro- and nanopatterned surfaces can contain domains with different physical and chemical properties such as chemical composition, chemical and biological reactivity, wettability, electrical charge, or topography [5,18,54,55].

Lithography constitutes a prevalent method for the patterning of polymer surfaces. The topographical features of hydrogel materials, employed as scaffolds in the field of tissue engineering, constitute a critical aspect. The implementation of lithography techniques presents an avenue to design, manipulate, and regulate the geometry and dimensions of hydrogel and various other biomaterials [58]. Photolithography is a widely employed technique in the manipulation of polymer surfaces, which involves the selective exposure of the polymer surface to photoirradiation, thus resulting in a patterned modification of the surface [55]. The aforementioned methodology was initially utilized for the development of surface features with physical variations measuring between 5 to 100 millimeters, specifically with a view toward research centered on stem cells [59].

## 18.3.2 Surface Coatings

Surface films and coatings represent promising methodologies for enhancing the biocompatibility of both metallic and polymeric surfaces. The functionalization techniques in question entail the attachment of secondary chemical groups, otherwise referred to as coatings, or surface modifications that follow familiar practices in conventional chemical modification methods, such as films. Spin and dip coatings, Langmuir-Blodgett (LB) and layer-by-layer (LbL) films, physical vapor deposition (PVD), and chemical vapor deposition (CVD) are among the most frequently employed methods for fabricating thin films and coatings in modern times. This phenomenon is governed by the interplay of viscous forces, surface tension, and gravity, which collectively regulate the thickness of the resultant layer or layers [4,55,58].

The process of spin coating is employed to apply thin films onto a level substrate in a rapid rotation, primarily in the case of semiconductors. Dip coating represents a straightforward technique for achieving the deposition of thin films through the immersion and subsequent withdrawal of the substrate from a coating solution. Regrettably, maintaining uniformity in a dip-coated film presents a challenging task, with a similar drawback also observed in dip casting. Both methods have demonstrated efficient functionalization of biomaterial surfaces, exemplified by the successful integration of synthetic peptide coatings onto orthopedic implants [60]. The aforementioned techniques are frequently implemented subsequent to the activation of the surface through methods such as plasma treatment.

Polymeric biomaterial surfaces may be functionalized using amphiphiles via the deposition of LB films. LB films comprise highly structured, densely arranged formations capable of being deposited and cross-linked onto the surface of the polymer. The LB technique facilitates the creation of deposited thin protein films on substrates through non-covalent and non-specific physical adhesion. The substrate is submersed within a solution intended for coating and then gradually extracted from the solution. The process by which molecules are deposited onto the surface of a substrate from a liquid occurs at the interface between the liquid and air during the dipping and removal phases. The assembly process is facilitated and regulated through the intricate interplay among a range of distinct forces, including inter-molecular, electrostatic, and capillary forces [4,61]. The deposition of LB films onto polymer surfaces can be accomplished through chemical treatment of the substrate. This process entails the attraction and formation of a monolayer of the LB film on the polymer surface. LB films have been found to improve cellular adhesion and hemocompatibility, with a simultaneous reduction in platelet adhesion [62].

The LbL technique facilitates the formation of nanostructured self-assembled films that exhibit a spectrum of geometries, encompassing straightforward two-dimensional multilayer films to intricate three-dimensional porous architectures. The present study reveals that the self-assembly process is predominantly governed by electrostatic interactions manifested between the constituent components as well as between the said components and the substrate surface. Nevertheless, alternative forms of intermolecular interactions can also be employed, including hydrophobic interactions and hydrogen bonding. To provide an instance, a polyanion possesses the ability to undergo adsorption onto a surface that is positively charged, thus resulting in the formation of a layer. The iterative procedure of generating alternating negative and positive layers is conducted until the desired thickness of the film is attained, encompassing dimensions that span from a few nanometers to tens of micrometers [4,61].

### 18.3.3 SURFACE MODIFICATION BY CHEMICAL TECHNIQUES

Chemical modification methodologies are employed for perturbing the surface attributes of a given substrate, without exerting considerable influence upon its inherent bulk characteristics, such as the desirable traits of high endurance and relatively lower modulus. The techniques under consideration are directed towards enhancing a variety of material properties, including wear and corrosion resistance, chemical reactivity (i.e., the potential for further modification), hydrophilicity/hydrophobicity, blood compatibility, endothelialization, cell adhesion and growth, antifouling characteristics, as well as antimicrobial properties. Various chemical modification approaches have been developed, including chemical vapor deposition (CVD), sol-gel treatment, grafting techniques, self-assembled monolayers (SAMs), and additional techniques [4,58,63].

CVD techniques are established on the fundamental principles of chemical reactions occurring between chemical reagents present in the gas phase and the surface of the substrate. This reaction leads to the formation of a nonvolatile coating that is subsequently deposited onto the substrate. Currently, a multitude of modifications to the chemical vapor deposition (CVD) technique exist, such as atmospheric pressure chemical vapor deposition (APCVD), low-pressure chemical vapor deposition (LPCVD), laser-enhanced chemical vapor deposition (LECVD), plasma-enhanced chemical vapor deposition (PECVD), and plasma-assisted chemical vapor deposition (PACVD).

The diverse modifications of the chemical vapor deposition (CVD) technique, encompassing atmospheric pressure chemical vapor deposition (APCVD), low-pressure chemical vapor deposition (LPCVD), laser-enhanced chemical vapor deposition (LECVD), plasma-enhanced chemical vapor deposition (PECVD), and plasma-assisted chemical vapor deposition (PACVD), have resulted in significant enhancements in the uniformity of the deposited coating, as well as augmented levels of hardness, wear and corrosion resistance, chemical stability, biocompatibility, and corrosion resistance [3,64–68].

The significance of precursor chemistry cannot be overstated as it is a vital processing parameter that has a substantial impact on the physical and chemical composition of the coating produced. This impact is further influenced by various other parameters such as plasma power, gas type, and process time. Accordingly, the PECVD methodology is currently in a nascent phase, primarily owing to the deficiency of suitable precursors that are amenable to gas-phase transport and plasma-induced activation [69].

The application of plasma techniques has proven to be efficacious in the alteration of the surface composition, and chemical reactivity, as well as the physico-chemical and biological properties of materials, thereby augmenting their interaction with proteins, cells, and living tissues within artificial biomaterials, joints and implants, and other medical devices. The utilization of plasma treatment on polymer materials has been demonstrated to lead to beneficial outcomes, particularly in terms of promoting cell adhesion and proliferation on the surface of said material. This effect is largely attributed to the enhancement of hydrophilicity and wettability. Moreover, the utilization of appropriate precursor materials within plasma deposition methodologies establishes the chemical landscape and surface chemistry characteristics of the resultant interfaces [70,71]. The adoption of plasma-enhanced methodologies has proven to be a valuable and dependable mechanism in the realm of surface engineering techniques. Such techniques aim to formulate and produce novel surfaces that enhance the metallic or polymeric surface properties which include but are not limited to, biocompatibility, blood compatibility, cell adhesion and proliferation, wear and corrosion resistance, as well as providing tailored drug delivery options [8,56,64,68,72–74].

Contemporary methodologies for addressing biocorrosion, the hastened erosion of metallic attributes resulting from the impact of microorganisms, primarily entail the implementation of safeguarding coatings, specifically antimicrobial polymer coatings [75,76]. The conventional approach for immobilizing antimicrobial polymers onto metal surfaces entails employing covalent coupling, layer-by-layer (LbL) sol-gel deposition, and classical free-radical graft polymerization [77]. Composite renewable materials employ these methods primarily to introduce linkers that facilitate coupling with other desired biomolecules. A novel bioinspired coating has been devised that comprises biomimetic linkers that enable the covalent grafting of natural bioagents onto biocompatible polymers. An environmentally beneficial technique for functionalizing metal substrates with potential biomedical applications involves the simultaneous integration of hydrophilic antifouling brushes and antibacterial enzymes or peptides, such as lysozyme, onto metal surfaces via catholic anchors [78].

The versatile and uncomplicated technique of layer-by-layer (LbL) electrostatic deposition has the capacity to be employed on diverse surfaces and geometries, including metallic and polymeric biomaterials. Chitosan (CS), owing to its cationic properties, has been identified as a promising candidate for the aforementioned technique, given its capacity to interact with the negatively charged bacterial cell wall, impairing the integrity of the cell membranes, and resulting in the demise of the bacterial cells [79,80]. Exemplars of efficacious surface functionalization with metal or polymer agents comprise the utilization of layer-by-layer (LbL) electro-static deposition of chitosan (CS) together with polyanionic compounds such as hyaluronic acid [81,82] or heparin [83], resulting in the establishment of stable multilayers. In addition, chitosan can be inserted into surfaces through the application of dopamine or 3-aminopropyltriethoxysilane (APTES) as linkers [6]. Illustrated in Figure 18.5 are several prevalent methods for surface modification.

The enhancement of bactericidal properties on metallic and polymeric implants, which are non-toxic to mammalian cells and possess the capability to hinder bacterial colonization, is an extremely desirable goal. In order to attain such a desired outcome, techniques such as plasma activation, sol-gel film synthesis, ion implantation, and grafting have emerged as widely adopted methods for generating refined functional antimicrobial arrangements upon the surfaces of biomaterials. The advantages of plasma-assisted techniques for the manufacturing and modification of biomaterials are of noteworthy import. The application of plasma-based techniques to modify the surface of biomaterials represents an economically viable, straightforward, and highly effective strategy for enhancing both biocompatibility

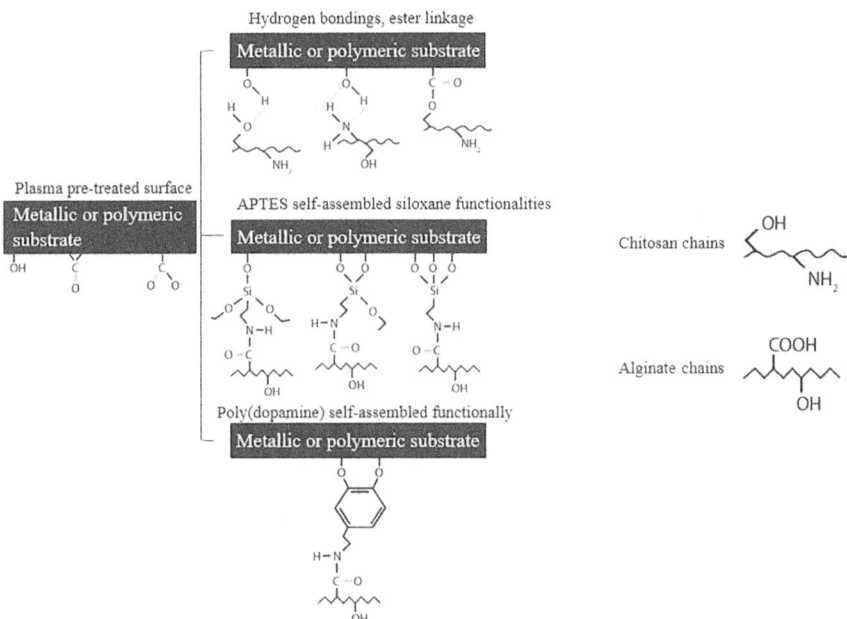

**FIGURE 18.5**   The attachment chemistry of the most common surface linkages.

and functionality, all while preserving the desirable bulk attributes of strength and inertness.

Currently, a pioneering methodology towards the functionalization of bio-material surfaces encompasses the utilization of hybrid multifunctional coatings that satisfy multiple needs of implantable materials simultaneously. Stents are frequently utilized as drug delivery systems to stimulate vessel restoration, thereby negating the requirement for further administration of oral anticoagulants. There exist two primary methods of drug delivery in the context of stents: direct attachment of the drug onto the surface of the stent or integration of the drug within the porous structure of the stent's surface. Surface modifications, including prolonged plasma etching, soft lithography, and photolithography, are commonly employed techniques for developing pores on the surface of polymers. These pores can be infused with drugs and utilized as advanced drug delivery systems [56]. Moreover, the utilization of acidic treatment on stainless-steel stents results in the development of a permeable surface that permits targeted drug delivery at the localized site [84].

The principal techniques for confining bioactive moieties onto polymeric substrates involve the use of adsorption through electrostatic forces, interactions between ligands and receptors, and covalent bonding. Non-covalent adsorption plays a crucial role in a diverse range of applications including drug delivery systems, renewable antimicrobial textiles, and systems utilizing the avidin-biotin tool. These applications span from research and diagnostics to medical devices and pharmaceuticals. The covalent immobilization process affords the most enduring bonds between the compound and the functionalized polymer surface, resulting in a prolonged half-life of biomolecules, prevention of metabolism, and the unimpeded maintenance of bioactivity, rendering it an optimal choice for functionalized medical devices such as vascular devices or catheters [73,85]. As evidenced by active food packaging, the utilization of a covalent bond effectively safeguards against the inadvertent contamination of food by bioactive compounds [11]. Moreover, the tethering of antimicrobial peptides (AMPs), which are considered the successors of antibiotics, serves as an exemplary case of state-of-the-art antimicrobial coatings developed for medical equipment.

Antimicrobial peptides, including but not limited to β-Defensins, indolicidin, cecropin A, and magainins, possess potent antimicrobial activities against a broad spectrum of microorganisms such as bacteria, fungi, parasites, and select types of viruses. The antimicrobial peptides (AMPs) in question may be rendered immobile upon solid surfaces via physical means, namely through adsorption or layer-by-layer (LbL) assembly, or by chemical means through covalent bonding [86].

### 18.3.3.1   Surface Functionalization of Metallic Biomaterials

Recently, hybrid materials composed of titanium support and a thin alginate hydrogel layer have been identified as potential scaffold materials for bone tissue engineering. The alginate chains are covalently attached to anchor layers made of self-assembled bisphosphonate etidronate, polymer films of APTES, or poly (dopamine), which are first immobilized on the activated titanium surface. The success of the titanium surface activation, anchoring coating formation, and alginate immobilization are proven by subjecting the materials to physiological-like conditions. It is expected that the suggested composites of thin alginate hydrogel can serve as a carrier of bioactive compounds (formed by ionic cross-linking) and enhance the adhesion, proliferation, and differentiation of osteoblasts. This can lead to better integration of titanium implants into bone tissue [87].

Additionally, CS, another biopolymer, is currently being investigated for its use as a coating suitable for functionalizing metallic implant surfaces. A CS-coated porous titanium alloy implant (CTI) has been proposed as a promising approach to enhance the osseointegration of pure porous titanium alloy implants. Since CS exhibits antioxidant activity, it is believed that CTI can reduce diabetes-mediated ROS accumulation, thus reversing poor osseointegration under diabetic conditions. Pathological ROS overproduction can lead to alterations in the PI3K/AKT signaling pathway, which is a major osteogenic signaling network for osteoblastic activities. In vitro and in vivo tests have shown that CTI stimulates AKT phosphorylation through ROS attenuation, reversing osteoblast dysfunction and improving osteoblast adhesion, proliferation, and alkaline phosphatase activity while decreasing cytotoxicity and apoptotic rate. The antioxidative activity of CS coating can provide an alternative therapeutic approach for increasing the integration of titanium and reducing implant failure in diabetic patients [88].

Xiao et al. successfully obtained a composite material for the controlled release of water-soluble drugs. They coated a Ti6Al4V implant with alginate using a dip-coating method and added dispersed gelatin particles loaded with gentamicin. This functionalization not only improved the surface properties of the implant but also ensured a slower initial burst release and prolonged retention time compared to a pure calcium alginate coating. The proposed functionalization exhibited better antibacterial activity against *Staphylococcus aureus* when compared to an unmodified Ti6Al4V implant [89]. It was concluded that such bioactive and biodegradable coatings can be used as drug-eluting systems with the advantage of controlled drug release to prevent bone infections. In addition, it was demonstrated that such coatings induce the formation of apatite in vitro and it was suggested that this could lead to bonding activation to the surrounding tissues in vivo.

Many scientists have suggested putting a CS layer on metal implants [90–94]. CS is a great material for medical use because it is safe for living things, can break down naturally, and can stop bacteria from growing [79,80,82,95–102]. CS is useful for carrying medicines and other helpful things that fight germs. This can stop germs from getting stronger and spreading more.

Greene et al. [94] demonstrated that CS coating can be sufficiently bonded to SS medical implants and deliver therapeutic agents. Biocompatibility, zone of bacterial inhibition, and antibiotic elution tests (gentamicin) unambiguously revealed that CS has the potential to be applied as a coating for orthopedic devices. The functional simulated bone study indicated that the coating strength is adequate for use in orthopedic applications. The use of antibiotic-loaded CS coatings on SS bone screws as internal fixation devices for contaminated bone fracture fixation is an innovative and promising strategy worth investigating and developing. However, further optimization of the efficiency of antibiotic loading and releasing expanded in vitro and in vivo investigations with various cell lines, microorganisms, and antibiotics, as well as detailed physico-chemical studies on the resulting coatings are still required before the widespread use in traumatology and translatology. Only stable fractures with internal fixation have a desirable influence on bone healing, while the implanted biomaterials used for these devices can serve as protection against contaminating bacteria.

The literature currently presents a versatile, suitable, and efficient method for incorporating growth factors (such as bone morphology protein-2, BMP-2) and antibacterial agents (Ag NPs) into coatings deposited on metallic implant surfaces. In this complex system, the biopolymer CS serves as the stabilizing agent to chelate and reduce Ag ions while simultaneously reducing Ag toxicity and retaining its antibacterial activity. A BMP/heparin solution is absorbed into the CS/Ag/HA coating, ensuring BMP-2 immobilization on the coating through electrostatic attraction between CS, heparin, and BMP-2. This hybrid system achieves sustained release of BMP-2 and Ag ions over a long period. The presented coating system shows excellent antibacterial activity in vitro against both *Staphylococcus epidermidis* and *Escherichia coli*. Additionally, this complex coating exhibits good biocompatibility with osteoblasts, enhances differentiation of bone marrow stromal cells, and shows good osteoinductivity after the implantation of Ti bars with BMP/CS/Ag/HA functionalization into the femur of rabbits [103].

A novel approach to combat bacterial colonization and prevent subsequent infection development on metallic implants is the grafting of lytic enzymes. Lysozyme is the most commonly proposed enzyme due to its ability to catalyze the hydrolysis of 1,4-beta-glycosidic linkages between N-acetylmuramic acid and N-acetylglucosamine, which are components of the peptidoglycan of the bacteria cell wall. Lysozyme is an efficient antibacterial agent for Gram(+) bacteria [41,42]. The main challenge in immobilizing enzymes onto solid substrates, such as metallic alloys, is to apply a functionalization method that preserves its activity. Direct adsorption of the enzyme may result in protein denaturation and loss of enzymatic activity. Additionally, enzyme release from the surface may occur in the long term. Therefore, only controlled covalent chemisorption of the enzyme can lead to a strongly attached enzyme, with its lytic activity maintained. Chemical grafting of lysozyme (covalent binding) has been demonstrated by various groups, resulting in surfaces that exhibit prevention of protein adsorption and bacterial adhesion, together with their biocidal properties [36,38].

Yuan et al. [78] proposed a surface functionalization idea based on environmentally benign modification, which imparts SS surfaces with antifouling and antibacterial functionalities. They performed surface-initiated atom transfer radical polymerization (ATRP) of poly(ethylene glycol) monomethacrylate (PEGMA) from the SS surface-coupled catholic L-3,4-dihydroxyphenylalanine (DOPA) with a terminal alkyl halide initiator. This was followed by the immobilization of lysozyme at the chain ends of the grafted PEGMA polymer brushes. This functionality, based on the incorporation of antifouling hydrophilic brushes and antibacterial enzymes (lysozyme), onto metal surfaces via catholic linkers, was effective in inhibiting bovine serum albumin (BSA) adsorption, preventing bacterial adhesion, and biofilm formation.

Kyzioł et al. [74] conducted plasmochemical surface modification of titanium alloys (Ti6Al4V and α-TiAl) using the plasma-assisted microwave chemical vapor deposition (PA MW CVD) method. This allowed for the formation of stable surface coatings and significant improvement in mechanical and surface parameters such as hardness, roughness, surface energy, and contact angle. Deposition of SiCNH coating on the α-TiAl alloy surface, without plasma nitriding process, resulted in the most hydrophobic structure with the largest surface area. The mechanical properties, such as surface hardness and Young's modulus, increased by approximately 30% and 10%, respectively, compared to the unmodified substrates. These alterations provided the excellent attachment of CT26 cells and promoted their growth in vitro. Plasmochemically modified γ-TiAl alloy can provide an alternative solution for titanium alloys with vanadium (i.e., Ti6Al4V), which is commonly used in implantology despite possessing undesirable properties, such as mainly vanadium carcinogenic, immunotoxin, and neurotoxic effects [104].

Furthermore, there is ongoing intensive research on new biomaterials, which is developing and changing significantly. All changes are aimed at mimicking the conditions in a living organism as closely as possible. Winkel et al. proposed a new concept of an experimental method to test the direct contact-based mechanisms of antibacterial effects. A new in vitro model with semi-coated titanium discs showed almost complete inhibition of the adhesion of several pathogenic bacteria strains, including *Streptococcus sanguinis, E. coli, S. aureus,* and *S. epidermidis.* Notably, soft tissue cells (human gingival or dermis fibroblasts) were less affected by the same coating. The viability and growth of human fibroblasts from peri-implant soft tissue were not modified at all, while a moderate influence on the initial adhesion of gingival fibroblasts was observed. The polymer coating composition consisting of 4-vinyl-N-hexyl pyridinium bromide (VP) and dimethyl-(2-methacryloyloxyethyl) phosphonate (DMMEP) in a ratio of VP:DMMEP 30:70 was proven to possess an equilibrium between biocompatibility and antimicrobial activity in vitro against selected bacteria strains relevant in different medical disciplines. It was concluded that the copolymer VP:DMMEP 30:70 could be considered a promising antimicrobial coating for some clinical applications [105].

Jesus Santamaria et al. [106] demonstrated a unique approach to combat bacterial growth using a new hybrid material. They formed the porous surface of a medical-grade stainless-steel pin, providing an innovative solution for the surface functionalization of implantable drug-eluting devices. The proposed concept of the drug release system for traumatology and orthopedic surgery applications was achieved by packing the interior of the hollow porous reservoir with mesoporous silica microparticles and adsorbed antibiotics (such as linezolid). The resulting system has the potential for satisfactory control of the rate of drug release, as the variables related to the filled drug type, solubility, and the porous wall reservoir's number and size of release orifices can be independently manipulated [107,108]. The proposed drug release devices offer new possibilities for applications in many medical fields, such as hollow screws used as maxillofacial implants or porous orthopedic implantable devices for localized delivery of therapeutic compounds.

### 18.3.3.2  Surface Functionalization of Polymeric Biomaterials: Selected Examples

Polymeric biomaterials with preserved bulk properties like elasticity, conductivity, mechanical strength, optical clarity, degradability, and biocompatibility remain extremely challenging to design and manufacture in the 21st century. To make the ideal polymeric biomaterial, special treatments are usually required to alter and improve the surface properties. One of the most intense and rapidly developing areas of biomaterials research is surface engineering, and multiple approaches are frequently required to meet all of its requirements. The selection of an appropriate biomaterial must consider not only its inherent properties, but also its reliability in terms of process dependability, reproducibility, and product yield.

Multiple natural (namely, CS, gelatin, alginate, collagen type I, hyaluronic acid, and fibrin glue) and synthetic (specifically, PEG, polyglycolic acid (PGA), PLA, PLGA, polyvinyl alcohol (PVA), polyglycerol sebacate (PGS), polycaprolactone, PURs, and poly(N-isopropyl acrylamide)) polymers have been evaluated as potential materials in the production of novel biodegradable biomaterials and biomaterial coatings [7,109]. Within the category of synthetic polymers, a significant amount of research has been conducted on the use of polyethylene glycol (PEG) in the realm of tissue engineering. The PEG hydrogels, known for their resistance to degradation, are commonly rendered degradable through the synthesis of diblock, triblock, and multiblock copolymers of PL(G)A/PEG. This is achieved via ring-opening polymerization of lactide and/or glycolide in the presence of PEG and catalysts. One alternative approach is to introduce enzyme-cleavable bonds within the polyethylene glycol (PEG) structure [4,6,18,110].

One of the simplest strategies for functionalizing polymeric surfaces to achieve antimicrobial activity is to introduce bactericidal polycationic groups. In the case of PET film, it was graft co-polymerized with 4-vinyl pyridine (4VP) and then functionalized with hexyl bromide through the quaternization of the grafted pyridine groups into pyridinium groups. The number of pyridinium groups on the film surface was controlled by varying the 4VP monomer concentration used for grafting. It was demonstrated that the higher the content of pyridinium groups introduced on the substrate surface, the more remarkable the antibacterial properties [111].

A more advanced strategy for the surface functionalization of polymers is their coating with bioactive molecules such as amino acid sequences of peptides, proteins, or other polymers, which improves cell adhesion to polymer scaffolds [112–114]. Polyurethanes (PURs) have been used in the production of various medical devices for over three decades, but there is now great interest in their functionalization with biopolymers (such as CS, collagen, starch, gelatin, etc.) to make them environmentally friendly and highly desirable in biomedical disciplines. Such PURs modified with natural polymers can be applied as wound dressings, scaffolds in tissue engineering, tissue implants, and vascular prostheses. In addition, a specific cell-recognition sequence RGD, present in ECR molecules such as vitronectin, fibronectin, laminin, collagen, and fibrillin, can be used for surface functionalization of polymeric biomaterials. To provide a stable RGD peptide linking to the polymer surface, a covalent bond between amine or hydroxyl functional groups of PURs and the carboxyl groups of RGD must be formed. In general, surface functionalization can be achieved by forming a strong covalent bond between biomolecule and biomaterial, which is required for cell adhesion. Most commercially available polymers are inert and require surface activation before biomolecule immobilization. Generated active groups may not be of the desired type or in the required quantity, necessitating the further introduction of reactive functional groups specific to desired biomolecules. Surface modification methods include ionized gas treatments, flame treatment, UV irradiation, chemical grafting, as well as various chemical addition reactions such as acetylation, fluorination, silanization, and incorporation of sulfonate groups. Additional functional group modifications such as oxidation and reduction are also possible [8,115,116]. It is noteworthy that biomolecule immobilization can often result in a loss of its bioactivity due to steric hindrance, and an introduction of an additional "spacer" is sometimes required. For instance, peptides are typically linked to spacers through stable amine bonds formed by surface carboxyl groups and the N-terminus of the peptide [86].

One common strategy is to activate the surface of PUR with ozone or plasma and modify it with poly(acrylic acid) to introduce carboxyl groups, which can be used for specific reactions with biopolymers [117].

Biofilm formation on biomedical polymeric implants, such as catheters, prosthetic devices, vascular grafts, or contact lenses, can lead to serious bacterial infections. Typical treatment methods for biofilm-mediated contamination of medical devices involve a surgical replacement of the implant, which is also associated with prolonged subsequent antibiotic therapy. This can result in long hospitalization times, life-threatening morbidity, severe functional impairments, increased healthcare costs, and can even lead to death due to biofilm-mediated complications. One promising strategy for fighting bacterial biofilm growth is lytic enzyme immobilization. In a different investigation, scholars explored the effective deployment of a PEG-conjugated enzyme, specifically α-amylase, for the purpose of developing PUR self-cleaning coatings [39]. Studies are currently being conducted to investigate antimicrobial coatings for various other applications, which may include food packaging. Other groups of researchers [37,118–120] proposed enzyme immobilization on the CS powder surface of the edible coating from a whey protein concentrate-CS, resulting in a bioactive coating that combines the bacteriolytic activity of the immobilized lysozyme and the bacteriostatic activity of CS, ensuring their synergy of action.

Theapsak et al. [121] have demonstrated the preparation of CS-coated PE packaging films with antibacterial activity against *E. coli* and *S. aureus*. The surface of the PE packaging polymer was functionalized with dielectric barrier discharge (DBD) plasma and subsequently coated with CS. Plasma treatment of the PE films increased the surface roughness and the number of oxygen-containing functional groups (i.e., C=O, C–O, –OH), indicating that DBD plasma pretreatment enhances the hydrophilicity of the PE films. This treatment also augments the interaction between the two applied polymers. The CS-coated PE films obtained could be a promising candidate for antibacterial packaging materials for food, small sterile implants (e.g., stents, endoprosthesis, dental screws), and disposable medical equipment (e.g., needles, syringes, catheters, wound dressing).

Goda-Cpa et al. [71,122] recently studied the surface functionalization of the polymer Parylene C (poly (chloro-para-xylylene)) by oxygen plasma treatment, which is a candidate for biocompatible antibacterial protective coatings for metal implants. The influence of the applied technology on nano-topography, chemical composition, hydrophilicity, and biocompatibility was examined. The chemical (oxygen insertion) and physical (nano-topography generation) changes induced by oxygen plasma treatment were found to have a significant impact on bio-compatibility (MG-63 human osteosarcoma) due to increased hydrophilicity and surface-free energy. Furthermore, no statistically significant effects on bacterial adhesion and biofilm formation of the selected reference strains and clinical isolates (*S. aureus, S. epidermidis, Pseudomonas aeruginosa*) were observed in vitro.

Therefore, it can be concluded that the proposed surface functionalization by the CVD method provides a polymeric surface with promising chemical properties and can be used as a helpful tool in the surface modification of bacteria-inert implant coating.

### 18.3.3.3 Conclusions and Research Perspective

Proper wound healing is one of the most important factors for the success of medical implantable devices, as the process begins at the surface of a material. The effectiveness of wound healing depends on various material properties, including surface topography, surface energy, surface chemistry, crystallinity, leachable content, and biocompatibility of degradation products, among others. Additionally, peri-implant infections resulting from bacterial biofilm development on artificial surfaces, which are mainly associated with antibiotic resistance, pose a common threat to all medical implants. The aforementioned infections pose a significant apprehension for healthcare practitioners and present an imposing obstacle for patients, as they have the potential to result in extensive hospital stays, potentially fatal complications, or even the failure of implants. Consequently, there exists a compelling necessity for the fabrication of novel biomaterials or surface coatings that possess the ability to impede or diminish the formation of biofilms. Such an enhancement can ensure the enduring utilization of medical materials or devices.

The need to develop new biomaterials in line with the structure design and increase application and improve usage conditions in medicine is an essential point. Current strategies for designing new implantable biomaterials mainly involve the preparation of hybrid multifunctional materials that provide the desired physico-chemical and biological properties, all at once. This does not always involve the creation of completely new materials from scratch but can also involve surface modification of existing materials that are widely used in medicine.

The process of surface functionalization of biomaterials presents a viable and cost-efficient approach towards achieving the desired bulk characteristics of the biomaterial, while simultaneously preserving its functionality and biocompatibility [123,124]. There is now a strong interest in using polymers derived from renewable resources for surface modification purposes. The scientific and technological communities are motivated to replace progressively disappearing fossil resources with sustainable counterparts, such as furans, vegetable oils, and polysaccharides [102,125,126].

## REFERENCES

1. Bauer, S., et al., Engineering biocompatible implant surfaces: Part I: Materials and surfaces. *Progress in Materials Science*, 2013. 58(3): p. 261–326.
2. Chen, Q. and G.A. Thouas, Metallic implant biomaterials. *Materials Science and Engineering: R: Reports*, 2015. 87: p. 1–57.
3. Kulkarni, M., et al., Biomaterial surface modification of titanium and titanium alloys for medical applications. *Nanomedicine*, 2014. 111(615): p. 111.
4. Taubert, A. and J.F. Mano, *Biomaterials Surface Science*. 2013: John Wiley & Sons.
5. Bhola, R., F. Su, and C.E. Krull, Retracted Article: Functionalization of titanium based metallic biomaterials for implant applications. *Journal of Materials Science: Materials in Medicine*, 2011. 22: p. 1147–1159.
6. Dumitriu, S. and V. Popa, *Polymeric Biomaterials: Medicinal and Pharmaceutical Applications, Volume 2*. Vol. 2. 2013: CRC Press.
7. Pradas, M.M. and M.J. Vicent, *Polymers in Regenerative Medicine*. 2015: Wiley.
8. Ma, Z., Z. Mao, and C. Gao, Surface modification and property analysis of biomedical polymers used for tissue engineering. *Colloids and Surfaces B: Biointerfaces*, 2007. 60(2): p. 137–157.
9. Broderick, N., Understanding chronic wound healing. *The Nurse Practitioner*, 2009. 34(10): p. 16–22.
10. Williams, D.F., On the mechanisms of biocompatibility. *Biomaterials*, 2008. 29(20): p. 2941–2953.
11. Banerjee, I., R.C. Pangule, and R.S. Kane, Antifouling coatings: recent developments in the design of surfaces that prevent fouling by proteins, bacteria, and marine organisms. *Advanced Materials*, 2011. 23(6): p. 690–718.
12. Olson, M.E., et al., Biofilm bacteria: formation and comparative susceptibility to antibiotics. *Canadian Journal of Veterinary Research*, 2002. 66(2): p. 86.
13. Pavithra, D. and M. Doble, Biofilm formation, bacterial adhesion and host response on polymeric implants—issues and prevention. *Biomedical Materials*, 2008. 3(3): p. 034003.
14. Wu, H., et al., Strategies for combating bacterial biofilm infections. *International Journal of Oral Science*, 2015. 7(1): p. 1–7.
15. Høiby, N., et al., Antibiotic resistance of bacterial biofilms. *International Journal of Antimicrobial Agents*, 2010. 35(4): p. 322–332.
16. Engler, A.J., et al., Matrix elasticity directs stem cell lineage specification. *Cell*, 2006. 126(4): p. 677–689.
17. Zaari, N., et al., Photopolymerization in microfluidic gradient generators: microscale control of substrate compliance to manipulate cell response. *Advanced Materials*, 2004. 16(23-24): p. 2133–2137.
18. Bacakova, L., et al., Modulation of cell adhesion, proliferation and differentiation on materials designed for body implants. *Biotechnology Advances*, 2011. 29(6): p. 739–767.
19. Shang, H.M., et al., Nanostructured superhydrophobic surfaces. *Journal of Materials Science*, 2005. 40(13): p. 3587–3591.
20. Kim, H., et al., Varying Ti-6Al-4V surface roughness induces different early morphologic and molecular responses in MG63 osteoblast-like cells. *Journal of Biomedical Materials Research Part A: An Official Journal of the Society for*

*Biomaterials, the Japanese Society for Biomaterials, and the Australian Society for Biomaterials and the Korean Society for Biomaterials*, 2005. 74(3): p. 366–373.

21. Sammons, R.L., et al., Comparison of osteoblast spreading on microstructured dental implant surfaces and cell behaviour in an explant model of osseointegration: a scanning electron microscopic study. *Clinical Oral Implants Research*, 2005. 16(6): p. 657–666.

22. Zhao, G., et al., High surface energy enhances cell response to titanium substrate microstructure. *Journal of Biomedical Materials Research Part A: An Official Journal of the Society for Biomaterials, the Japanese Society for Biomaterials, and the Australian Society for Biomaterials and the Korean Society for Biomaterials*, 2005. 74(1): p. 49–58.

23. Zhao, G., et al., Osteoblast-like cells are sensitive to submicron-scale surface structure. *Clinical Oral Implants Research*, 2006. 17(3): p. 258–264.

24. Saino, E., et al., Effect of electrospun fiber diameter and alignment on macrophage activation and secretion of proinflammatory cytokines and chemokines. *Biomacromolecules*, 2011. 12(5): p. 1900–1911.

25. Anselme, K., Osteoblast adhesion on biomaterials. *Biomaterials*, 2000. 21(7): p. 667–681.

26. Barbosa, J.N., et al., The influence of functional groups of self-assembled monolayers on fibrous capsule formation and cell recruitment. *Journal of Biomedical Materials Research Part A: An Official Journal of the Society for Biomaterials, the Japanese Society for Biomaterials, and the Australian Society for Biomaterials and the Korean Society for Biomaterials*, 2006. 76(4): p. 737–743.

27. Anderson, J., T. Bonfield, and N. Ziats, Protein adsorption and cellular adhesion and activation on biomedical polymers. *The International Journal of Artificial Organs*, 1990. 13(6): p. 375–382.

28. Lee, J.H., et al., Platelet adhesion onto chargeable functional group gradient surfaces. *Journal of Biomedical Materials Research: An Official Journal of the Society for Biomaterials, the Japanese Society for Biomaterials, and the Australian Society for Biomaterials*, 1998. 40(2): p. 180–186.

29. Hayward, J.A. and D. Chapman, Biomembrane surfaces as models for polymer design: the potential for haemocompatibility. *Biomaterials*, 1984. 5(3): p. 135–142.

30. Keselowsky, B.G., D.M. Collard, and A.J. García, Surface chemistry modulates focal adhesion composition and signaling through changes in integrin binding. *Biomaterials*, 2004. 25(28): p. 5947–5954.

31. Recum, H.V., et al., Growth factor and matrix molecules preserve cell function on thermally responsive culture surfaces. *Tissue Engineering*, 1999. 5(3): p. 251–265.

32. Lee, J.H., et al., Cell behaviour on polymer surfaces with different functional groups. *Biomaterials*, 1994. 15(9): p. 705–711.

33. Delon, I. and N.H. Brown, Integrins and the actin cytoskeleton. *Current Opinion in Cell Biology*, 2007. 19(1): p. 43–50.

34. Tallawi, M., et al., Strategies for the chemical and biological functionalization of scaffolds for cardiac tissue engineering: a review. *Journal of the Royal Society Interface*, 2015. 12(108): p. 20150254.

35. Ruoslahti, E., RGD and other recognition sequences for integrins. *Annual Review of Cell and Developmental Biology*, 1996. 12(1): p. 697–715.

36. Caro, A., et al., Grafting of lysozyme and/or poly (ethylene glycol) to prevent biofilm growth on stainless steel surfaces. *The Journal of Physical Chemistry B*, 2009. 113(7): p. 2101–2109.

37. Lian, Z.-X., et al., Preparation and characterization of immobilized lysozyme and evaluation of its application in edible coatings. *Process Biochemistry*, 2012. 47(2): p. 201–208.

38. Minier, M., et al., Covalent immobilization of lysozyme on stainless steel. Interface spectroscopic characterization and measurement of enzymatic activity. *Langmuir*, 2005. 21(13): p. 5957–5965.

39. Zhang, L., et al., Poly (ethylene glycol) conjugated enzyme with enhanced hydrophobic compatibility for self-cleaning coatings. *ACS Applied Materials & Interfaces*, 2012. 4(11): p. 5981–5987.

40. Penesyan, A., M. Gillings, and I.T. Paulsen, Antibiotic discovery: combatting bacterial resistance in cells and in biofilm communities. *Molecules*, 2015. 20(4): p. 5286–5298.

41. Cordeiro, A.L., C. Hippius, and C. Werner, Immobilized enzymes affect biofilm formation. *Biotechnology Letters*, 2011. 33: p. 1897–1904.

42. Thallinger, B., et al., Antimicrobial enzymes: an emerging strategy to fight microbes and microbial biofilms. *Biotechnology Journal*, 2013. 8(1): p. 97–109.

43. Yeroslavsky, G., et al., Antibacterial and antibiofilm surfaces through polydopamine-assisted immobilization of lysostaphin as an antibacterial enzyme. *Langmuir*, 2015. 31(3): p. 1064–1073.

44. Paoli, P., E. Giannoni, and P. Chiarugi, Anoikis molecular pathways and its role in cancer progression. *Biochimica et Biophysica Acta (BBA)-Molecular Cell Research*, 2013. 1833(12): p. 3481–3498.

45. Bacakova, L., et al., Adhesion and growth of vascular smooth muscle cells in cultures on bioactive RGD peptide-carrying polylactides. *Journal of Materials Science: Materials in Medicine*, 2007. 18: p. 1317–1323.

46. Bačáková, L., et al., Cell adhesion on artificial materials for tissue engineering. *Physiological Research*, 2004. 53(Suppl 1): p. S35–S45.

47. Chlupáč, J., E. Filova, and L. Bačáková, Blood vessel replacement: 50 years of development and tissue engineering paradigms in vascular surgery. *Physiological Research*, 2009. 58(Suppl 2): p. S119–S139.

48. Meinhart, J.G., et al., Enhanced endothelial cell retention on shear-stressed synthetic vascular grafts precoated with RGD-cross-linked fibrin. *Tissue Engineering*, 2005. 11(5-6): p. 887–895.

49. Punshon, G., Sales, K.M., Vara, D.S., Hamilton, G., Seifalian, A.M., Assessment of the potential of progenitor stem cells extracted from human peripheral blood for seeding a novel vascular graft material. *Cell Proliferation*, 2008. 41: p. 321–335.

50. Zilla, P., D. Bezuidenhout, and P. Human, Prosthetic vascular grafts: wrong models, wrong questions and no healing. *Biomaterials*, 2007. 28(34): p. 5009–5027.

51. Beckstead, B.L., D.M. Santosa, and C.M. Giachelli, Mimicking cell–cell interactions at the biomaterial–cell interface for control of stem cell differentiation. *Journal of Biomedical Materials Research Part A: An Official Journal of the Society for Biomaterials, the Japanese Society for Biomaterials, and the Australian Society for Biomaterials and the Korean Society for Biomaterials*, 2006. 79(1): p. 94–103.

52. Kolambkar, Y.M., et al., Nanofiber orientation and surface functionalization modulate human mesenchymal stem cell behavior in vitro. *Tissue Engineering Part A*, 2014. 20(1-2): p. 398–409.

53. Tung, J.C., et al., Engineered biomaterials control differentiation and proliferation of human-embryonic-stem-cell-derived cardiomyocytes via timed Notch activation. *Stem Cell Reports*, 2014. 2(3): p. 271–281.

54. Liu, X., P.K. Chu, and C. Ding, Surface modification of titanium, titanium alloys, and related materials for biomedical applications. *Materials Science and Engineering: R: Reports*, 2004. 47(3-4): p. 49–121.

55. Liu, X., P.K. Chu, and C. Ding, Surface nano-functionalization of biomaterials. *Materials Science and Engineering: R: Reports*, 2010. 70(3-6): p. 275–302.

56. Govindarajan, T. and R. Shandas, A survey of surface modification techniques for next-generation shape memory polymer stent devices. *Polymers*, 2014. 6: p. 2309–2331.

57. Chung, T.-W., et al., Enhancement of the growth of human endothelial cells by surface roughness at nanometer scale. *Biomaterials*, 2003. 24(25): p. 4655–4661.

58. Rashidi, H., Yang, J., Shakesheff, K.M., Surface engineering of synthetic polymeric materials for tissue engineering and regenerative medicine applications. *Biomaterials Science* 2, 2014: p. 1318–1331.

59. Curtis, A. and C. Wilkinson, Topographical control of cells. *Biomaterials*, 1997. 18(24): p. 1573–1583.

60. Reyes, C.D., et al., Biomolecular surface coating to enhance orthopaedic tissue healing and integration. *Biomaterials*, 2007. 28(21): p. 3228–3235.

61. Scholz, M., Biofunctionalized wound dressings for advanced wound care. *Biofunctional Surface Engineering*, 2014. 1: p. 251.

62. Tirrell, M., E. Kokkoli, and M. Biesalski, The role of surface science in bioengineered materials. *Surface Science*, 2002. 500(1-3): p. 61–83.

63. Govindarajan, T., Shandas, R., A survey of surface modification techniques for next-generation shape memory polymer stent devices. *Polymers*, 2014. 6: p. 2309–2331.

64. Desmet, T., et al., Nonthermal plasma technology as a versatile strategy for polymeric biomaterials surface modification: a review. *Biomacromolecules*, 2009. 10(9): p. 2351–2378.

65. Januś, M., et al., Plasma assisted chemical vapour deposition–technological design of functional coatings. *Archives of Metallurgy and Materials*, 2015. 60.

66. Jonas, S., et al. *Stability of a C: N: H layers deposited by RF plasma enhanced CVD.* In *Solid State Phenomena*. 2009: Trans Tech Publ.

67. Kyzioł, K., et al., A role of parameters in RF PA CVD technology of aC: N: H layers. *Vacuum*, 2008. 82(10): p. 998–1002.

68. Wu, S., et al., Plasma-modified biomaterials for self-antimicrobial applications. *ACS Applied Materials & Interfaces*, 2011. 3(8): p. 2851–2860.

69. Merche, D., N. Vandencasteele, and F. Reniers, Atmospheric plasmas for thin film deposition: A critical review. *Thin Solid Films*, 2012. 520(13): p. 4219–4236.

70. Bazaka, K., et al., Plasma-assisted surface modification of organic biopolymers to prevent bacterial attachment. *Acta Biomaterialia*, 2011. 7(5): p. 2015–2028.

71. Gołda, M., et al., Oxygen plasma functionalization of parylene C coating for implants surface: nanotopography and active sites for drug anchoring. *Materials Science & Engineering C-Materials for Biological Applications*, 2013. 33(7): p. 4221–4227.

72. Egitto, F.D. and L.J. Matienzo, Plasma modification of polymer surfaces for adhesion improvement. *IBM Journal of Research and Development*, 1994. 38(4): p. 423–439.

73. Kugel, A., S. Stafslien, and B.J. Chisholm, Antimicrobial coatings produced by "tethering" biocides to the coating matrix: A comprehensive review. *Progress in Organic Coatings*, 2011. 72(3): p. 222–252.

74. Kyzioł, K., et al., Structure, characterization and cytotoxicity study on plasma surface modified Ti–6Al–4V and γ-TiAl alloys. *Chemical Engineering Journal*, 2014. 240: p. 516–526.

75. Jain, A., et al., Antimicrobial polymers. *Advanced Healthcare Materials* 3: 1969–1985. 2014.

76. Siedenbiedel, F. and J.C. Tiller, Antimicrobial polymers in solution and on surfaces: overview and functional principles. *Polymers*, 2012. 4(1): p. 46–71.

77. Neoh, K.G. and E.-T. Kang, Combating bacterial colonization on metals via polymer coatings: relevance to marine and medical applications. *ACS Applied Materials & Interfaces*, 2011. 3(8): p. 2808–2819.

78. Yuan, S., et al., Lysozyme-coupled poly (poly (ethylene glycol) methacrylate)–stainless steel hybrids and their antifouling and antibacterial surfaces. *Langmuir*, 2011. 27(6): p. 2761–2774.
79. Krajewska, B., P. Wydro, and A. Kyzioł, Chitosan as a subphase disturbant of membrane lipid monolayers. The effect of temperature at varying pH: I. DPPG. *Colloids and Surfaces A: Physicochemical and Engineering Aspects*, 2013. 434: p. 349–358.
80. Krajewska, B., P. Wydro, and A. Jańczyk, Probing the modes of antibacterial activity of chitosan. Effects of pH and molecular weight on chitosan interactions with membrane lipids in Langmuir films. *Biomacromolecules*, 2011. 12(11): p. 4144–4152.
81. Chua, P.-H., et al., Surface functionalization of titanium with hyaluronic acid/chitosan polyelectrolyte multilayers and RGD for promoting osteoblast functions and inhibiting bacterial adhesion. *Biomaterials*, 2008. 29(10): p. 1412–1421.
82. Zhang, A., et al., Chitosan coupling makes microbial biofilms susceptible to antibiotics. *Scientific Reports*, 2013. 3(1): p. 3364.
83. Shu, Y., et al., Surface modification of titanium with heparin-chitosan multilayers via layer-by-layer self-assembly technique. *Journal of Nanomaterials*, 2011. 2011: p. 1–8.
84. Wieneke, H., et al., Synergistic effects of a novel nanoporous stent coating and tacrolimus on intima proliferation in rabbits. *Catheterization and Cardiovascular Interventions*, 2003. 60(3): p. 399–407.
85. Goddard, J.M. and J. Hotchkiss, Polymer surface modification for the attachment of bioactive compounds. *Progress in Polymer Science*, 2007. 32(7): p. 698–725.
86. Onaizi, S.A. and S.S. Leong, Tethering antimicrobial peptides: current status and potential challenges. *Biotechnology Advances*, 2011. 29(1): p. 67–74.
87. Pop-Georgievski, O., et al., Self-assembled anchor layers/polysaccharide coatings on titanium surfaces: a study of functionalization and stability. *Beilstein Journal of Nanotechnology*, 2015. 6: p. 617–631.
88. Li, X., et al., Osseointegration of chitosan coated porous titanium alloy implant by reactive oxygen species-mediated activation of the PI3K/AKT pathway under diabetic conditions. *Biomaterials*, 2015. 36: p. 44–54.
89. Xiao, J., et al., A composite coating of calcium alginate and gelatin particles on Ti6Al4V implant for the delivery of water soluble drug. *Journal of Biomedical Materials Research Part B: Applied Biomaterials: An Official Journal of the Society for Biomaterials, the Japanese Society for Biomaterials, and the Australian Society for Biomaterials and the Korean Society for Biomaterials*, 2009. 89(2): p. 543–550.
90. Aimin, C., et al., Antibiotic loaded chitosan bar: an in vitro, in vivo study of a possible treatment for osteomyelitis. *Clinical Orthopaedics and Related Research (1976–2007)*, 1999. 366: p. 239–247.
91. Bumgardner, J., et al., Contact angle, protein adsorption and osteoblast precursor cell attachment to chitosan coatings bonded to titanium. *Journal of Biomaterials Science, Polymer Edition*, 2003. 14(12): p. 1401–1409.
92. Bumgardner, J.D., et al., Chitosan: potential use as a bioactive coating for orthopaedic and craniofacial/dental implants. *Journal of Biomaterials Science, Polymer Edition*, 2003. 14(5): p. 423–438.
93. Di Martino, A., M. Sittinger, and M.V. Risbud, Chitosan: a versatile biopolymer for orthopaedic tissue-engineering. *Biomaterials*, 2005. 26(30): p. 5983–5990.
94. Greene, A.H., et al., Chitosan-coated stainless steel screws for fixation in contaminated fractures. *Clinical Orthopaedics and Related Research*, 2008. 466: p. 1699–1704.

95. Goy, R.C., D.d. Britto, and O.B. Assis, A review of the antimicrobial activity of chitosan. *Polímeros*, 2009. 19: p. 241–247.

96. Hamilton, V., et al., Bone cell attachment and growth on well-characterized chitosan films. *Polymer International*, 2007. 56(5): p. 641–647.

97. Prasitsilp, M., et al., Cellular responses to chitosan in vitro: the importance of deacetylation. *Journal of Materials Science: Materials in Medicine*, 2000. 11: p. 773–778.

98. Raafat, D., H.G. Sal, Chitosan and its antimicrobial potential – a critical literature survey. *Microbial Biotechnology*, 2009. 2: p. 186–201.

99. Raafat, D., K. von Bargen, A. Haas, G.H. Sahl, Insights into the mode of action of chitosan as an antibacterial compound. *Applied and Environmental Microbiology*, 2008. 74: p. 3764–3773.

100. Regiel-Futyra, A., M. Kus-Liśkiewicz, V. Sebastian, S. Irusta, M. Arruebo, G. Stochel, A. Kyzioł, Development of noncytotoxic chitosan–gold nanocomposites as efficient antibacterial materials. *ACS Applied Materials & Interfaces* 7, 2015: p. 1087–1099.

101. Regiel, A., et al., Preparation and characterization of chitosan–silver nanocomposite films and their antibacterial activity against Staphylococcus aureus. *Nanotechnology*, 2012. 24(1): p. 015101.

102. Thakur, V.K. and M.K. Thakur, Recent advances in graft copolymerization and applications of chitosan: a review. *ACS Sustainable Chemistry & Engineering*, 2014. 2(12): p. 2637–2652.

103. Xie, C.-M., et al., Silver nanoparticles and growth factors incorporated hydroxy-apatite coatings on metallic implant surfaces for enhancement of osteoinductivity and antibacterial properties. *ACS Applied Materials & Interfaces*, 2014. 6(11): p. 8580–8589.

104. Zwolak, I., Vanadium carcinogenic, immunotoxic and neurotoxic effects: a review of in vitro studies. *Toxicology Mechanisms and Methods*, 2014. 24(1): p. 1–12.

105. Winkel, A., et al., Introducing a semi-coated model to investigate antibacterial effects of biocompatible polymers on titanium surfaces. *International Journal of Molecular Sciences*, 2015. 16(2): p. 4327–4342.

106. Perez, L.M., et al., Hollow porous implants filled with mesoporous silica particles as a two-stage antibiotic-eluting device. *International Journal of Pharmaceutics*, 2011. 409(1-2): p. 1–8.

107. Gimeno, M., et al., A controlled antibiotic release system to prevent orthopedic-implant associated infections: An in vitro study. *European Journal of Pharmaceutics and Biopharmaceutics*, 2015. 96: p. 264–271.

108. Gimeno, M., et al., Porous orthopedic steel implant as an antibiotic eluting device: prevention of post-surgical infection on an ovine model. *International Journal of Pharmaceutics*, 2013. 452(1–2): p. 166–172.

109. Plackett, D., *Biopolymers: New Materials for Sustainable Films and Coatings*. 2011: John Wiley & Sons.

110. Dumitriu, S. and V. I. Popa, *Polymeric Biomaterials*. Vol. 2. 2013: CRC Press.

111. Cen, L., K. Neoh, and E. Kang, Surface functionalization technique for conferring antibacterial properties to polymeric and cellulosic surfaces. *Langmuir*, 2003. 19(24): p. 10295–10303.

112. Cen, L., et al., Assessment of in vitro bioactivity of hyaluronic acid and sulfated hyaluronic acid functionalized electroactive polymer. *Biomacromolecules*, 2004. 5(6): p. 2238–2246.

113. Huang, M.-N., Y.-L. Wang, and Y.-F. Luo, Biodegradable and bioactive porous polyurethanes scaffolds for bone tissue engineering. *Journal of Biomedical Science and Engineering*, 2009. 2(01): p. 36.

114. Jóźwiak, A.B., C.M. Kielty, and R.A. Black, Surface functionalization of poly-urethane for the immobilization of bioactive moieties on tissue scaffolds. *Journal of Materials Chemistry*, 2008. 18(19): p. 2240–2248.

115. Lee, H.Y. and S.B. Park, Surface modification for small-molecule microarrays and its application to the discovery of a tyrosinase inhibitor. *Molecular BioSystems*, 2011. 7(2): p. 304–310.

116. Vladkova, T.G., Surface engineered polymeric biomaterials with improved biocontact properties. *International Journal of Polymer Science*, 2010. 2010: 1–23.

117. Lee, C., et al., Bioinspired, calcium-free alginate hydrogels with tunable physical and mechanical properties and improved biocompatibility. *Biomacromolecules*, 2013. 14.

118. Dash, M., et al., Chitosan—A versatile semi-synthetic polymer in biomedical applications. *Progress in Polymer Science*, 2011. 36(8): p. 981–1014.

119. Martinez, L.R., et al., Demonstration of antibiofilm and antifungal efficacy of chitosan against candidal biofilms, using an in vivo central venous catheter model. *The Journal of Infectious Diseases*, 2010. 201(9): p. 1436–1440.

120. Orgaz, B., et al., Effectiveness of chitosan against mature biofilms formed by food related bacteria. *International Journal of Molecular Sciences*, 2011. 12(1): p. 817–828.

121. Theapsak, S., A. Watthanaphanit, and R. Rujiravanit, Preparation of chitosan-coated polyethylene packaging films by DBD plasma treatment. *ACS Applied Materials & Interfaces*, 2012. 4(5): p. 2474–2482.

122. Golda-Cepa, M., et al., Microbiological investigations of oxygen plasma treated parylene C surfaces for metal implant coating. *Materials Science and Engineering: C*, 2015. 52: p. 273–281.

123. Thakur, M.K., et al., Synthesis and applications of biodegradable soy based graft copolymers: a review. *ACS Sustainable Chemistry & Engineering*, 2016. 4(1): p. 1–17.

124. Voicu, S. I., et al., Sericin covalent immobilization onto cellulose acetate membrane for biomedical applications. *ACS Sustainable Chemistry & Engineering*, 2016. 4(3): p. 1765–1774.

125. Thakur, V.K. and M.K. Thakur, Recent trends in hydrogels based on psyllium polysaccharide: a review. *Journal of Cleaner Production*, 2014. 82: p. 1–15.

126. Thakur, V.K. and M.K. Thakur, Recent advances in green hydrogels from lignin: a review. *International Journal of Biological Macromolecules*, 2015. 72: p. 834–847.

# 19 Mechanical Surface Treatments of Biomaterials

*Shivani Chaudhary and Gautam Jaiswar*
Department of Chemistry, Institute of Basic Sciences, Dr.
Bhimrao Ambedkar University, Agra, India

## 19.1 INTRODUCTION

Biomaterial is defined as a material that is used in a specific form or structure to fabricate prostheses or biomedical devices that are intended to replace or recover an impaired body function in order to save or improve the patient's quality of life [1]. Over the past 50 years, biomaterials science has investigated different types of biomaterials and their applications to replace or restore the function of compromised or degenerated tissues or organs [2]. Every year, more than 13 million prostheses/medical devices are implanted solely in the U.S.A. Hundreds of different biomaterials have been investigated: synthetic and natural polymers (collagen, fibrin, chitosan, hyaluronan, heparin, cellulose, polyurethanes, polyesters, polytetrafluoroethylene, polymethylmethacrylate, hydrogels ... etc.), metals (steel, titanium, CrCoMoalloys .... etc.), and ceramics (calcium phosphates, hydroxyapatite, alumina, zirconia ... etc.). These materials are utilized in a wide class of implants like in ophthalmology as intraocular lenses, wound dressings, orthopedic implants, (cardio)-vascular surgery for blood vessels, heart valves or stents, and implants for dentistry [3].

Mechanical surface treatment comprises a number of processes from the main manufacturing group "Modification of material properties", which are used to improve the component behavior under operational loads. Mechanical surface treatments include, for example, shot peening, deep rolling, machine hammer peening, and some other processes that are used in customized industrial applications. The mechanical surface treatment of a component causes plastic deformation of its surface layer, resulting in local work hardening and the formation of residual compressive stresses [4]. There are several surface engineering methods that can be categorized into two main groups: physico-chemical and biological. Examples of physico-chemical methods are wet chemical processing such as acid etching/ oxidation, functional groups grafting, methods utilizing material irradiation such as cold plasmas, ion or electron beams, lasers, and photo-lithography [5]. In particular, the processes of deep rolling and machine hammer peening can also be used to smooth and structure surfaces due to their deterministic nature [6]. A combination of

DOI: 10.1201/9781003429920-22

smooth surface, work hardening, and residual compressive stresses is particularly advantageous for improving the service life properties in the fatigue stress frequently encountered in mechanical, automotive, and aircraft engineering [7]. Structured surfaces, such as bionic ones, can also be created to optimize wear behavior. The compatibility of a synthetic biomaterial in human body is very important for the successful implantation of artificial organs in body. It depends upon the chemical nature of the implant which must be non-reactive and the bond of biomaterial must be very integrated with the body. This would be possible with the modification of the biomaterial surface and modifications can be in the form of chemical inertness, avoid corrosion, degradation protection, etc. [8]. Mechanical properties determine the response of a biomaterial in different force conditions, which are always characterized by the stresses and strains in a material that results from displacements in defects. Systematical studies of the mechanical properties of biomaterials are obviously of great importance [9]. On one hand, biomaterials with insufficient mechanical properties cannot provide enough mechanical support for bone tissue regeneration, particularly in load-bearing applications, while overly stiff biomaterials could hinder bone tissue regeneration through the stress shielding phenomenon [10]. Therefore, it is important to carefully choose the mechanical properties of biomaterials aimed at bone tissue regeneration. Not only the static mechanical properties, but also the mechanical behavior of substituting biomaterials, need to be properly adjusted for optimum tissue regeneration performance.

Surface modifications of metallic biomaterials can be done using additive manufacturing. Addressing intrinsic material issues is critical for the success and endurance of all metallic devices. Modifying the surfaces of the implants can prevent corrosion, enhance biocompatibility, and improve osseointegration without compromising the bulk properties [11] with incorporating passivating elements (Al, Cr, Ti), and augmenting surface energy.

The modification of the surface can be categorized into several classes, but two important ones are surface coating and physico-chemical modifications. The former includes underlying support and the latter includes changes in atoms or molecules present on surface by the reactions like oxidation, reduction, or acetylation. The surface of a biomaterial has varying roughness and texture and this can be treated with different kinds of engineering techniques. Other methods like covalent and non-covalent coatings can also used for improving the surface texture depending on the type of polymer used for this purpose [12–15].

Surface integrity properties such as surface roughness (SR), cracks, impinged material, solid dissolution, and residual stress (RS) affect the corrosion performance of Mg alloys. Increased asperities increase surface area and contact with the salt solution, while impinged metallic fragments or impurities generate a galvanic cell with the surrounding $\alpha$-Mg. Beneficial compressive residual stress (CRS) impedes corrosion crack propagation. Moreover, the salt solution in the corrosive medium also affects the corrosion progression of the materials. For instance, the presence of Ca and P in solution produces hydroxyapatite (HA) as a corrosion product layer, thereby hindering $MgCl_2$ mobility and increasing corrosion resistance [16].

Since most mechanical failures are related to surface properties, surface severe plastic deformation (SSPD) has been an ideal solution to corrosion, cracking, fatigue, and wear damage problems [17]. The key underlying properties that can be controlled by surface SSPD are grain refinement, grain reorientation, lowered surface roughness (SR), increased surface hardness, and compressive residual stress (CRS). SSPD causes gradient microstructural change with depth up to a few hundreds of microns with nanograin refinement, followed by mild grain refinement and twinning layers. The major surface treatments that will be investigated are shot peening (SP), surface mechanical attrition treatment (SMAT), laser shock peening (LSP), ball burnishing (BB), and ultrasonic nanocrystalline surface modification (UNSM), as they have shown a great degree of potential in controlling the corrosion characteristics of Mg alloys. Due to the complexities of Mg alloy corrosion, the use of one surface treatment routine may not provide sufficient service life in physiological environments. Thus, researchers investigated the combined treatments. This surface treatment technique seems to be an effective method for industrial corrosion protection of lightweight structures made of Mg alloys, but its application in the biomedical industry as implants must be evaluated through an in vitro cell viability study. It was reported that higher aluminum content in implants causes neurological disorders or related diseases in patients [18].

Each surface treatment has its own unique advantage and process limitation, but the final outcome can be optimized by carefully controlling its process-specific key parameters.

Numerous factors are involved in the performance of a biomaterial in a biotechnology or medical device. These include its interaction with proteins and cells at the site of use; its stability, or engineered instability, to degradation in the site of use; its micro- and macromechanical properties relative to the properties of the tissue at the site of use; and its ability to be sterilized [19]. While advances in each of these issues have been made over the past few years, this review addresses exclusively approaches that have been developed to control the interaction between biomaterials and the components of the biological system, be they proteins, cells, systems, and networks of cells, and ultimately an entire organism.

Almost all biological interactions are mediated by specific biorecognition, i.e., high-affinity binding of receptors on cell surfaces to ligands on biomaterial surfaces. Such ligands could be present in proteins or other biomolecules that have spontaneously adsorbed to the material surface, or they could be present after intentional immobilization upon the material surface. Thus, an entirely nonbiological material implanted in the body interacts with the cells of the body through exclusively biological interactions; proteins from the body fluids rapidly adsorb to the nonbiological surface and, in effect, modify its surface such that is biological in nature [20].

The biocompatibility of an implant depends upon the material it is composed of, in addition to the prosthetic device's morphology, mechanical, and surface properties. Properties such as porosity and pore size should allow, when required, cell penetration and proliferation. Stiffness and strength, which depend on the bulk characteristics of the material, should match the mechanical requirements of the prosthetic applications [21]. Surface properties should allow integration in the surrounding tissues by activating proper communication pathways with the surrounding cells.

For biomedical applications, requirements are quite severe. Materials must be biocompatible; that is, they should not induce long-lasting or severe negative effects on the surrounding biological tissues [22], likely be bioactive to trigger a desired biological process (such e.g., those involved in the material integration), and should be durable or biodegradable depending on the application. In addition, they must fulfill the mechanical demand, as in prosthetic implants; likewise, when employed in other devices such as biosensors. Surface properties of biomaterial plays a dominant role in controlling biological phenomena that occur at the interface which determine the fate of the prosthesis [23]. In this respect, surface engineering has become crucial for biological and biomedical applications where biocompatibility and bioactive responses are required.

## 19.2 PURPOSE OF SURFACE TREATMENTS OF BIOMATERIALS

Mechanical properties quantify the response of a material under different force systems, which are almost characterized by the stresses and strains in a material that result from applied loads and displacements. Mechanical properties of the materials are measured by standardized mechanical tests, using specimens subjected to well-defined load conditions. To ensure a better comparison among mechanical tests carried out in different laboratories, specimens of standard dimensions are obtained from the material under evaluation, taking into account international standards (e.g., ISO, CEN, ASTM, DIN). In particular cases, some specific working conditions can be simulated (e.g., accelerated aging, wear, fatigue) to better study the real behavior of the material. The main classes of clinical biomaterials with respect to mechanical behavior are ceramics, metals, polymers, and composites thereof [24]. Reasons for the general differences in mechanical performance are different, and they actually are a primary subject of books on mechanical behavior, while several general points are to be figured out here. Differences in the elastic modulus and strength reflect differences in atomic bonding. In fact, mechanical properties such as the elastic modulus and fracture strength can be predicted from first principles through atomistic/molecular modeling incorporating bond strengths andatomic/molecular structure [25]. The great difference in electronegativity between metal and nonmetal atoms in ceramics results in strong covalent or ionic bonding, due to electrons being shared or donated, respectively, to fill the outer shell and a summary of physical technique is shown in Figure 19.1.

Studies have shown that degradation can significantly affect the mechanical properties of biomaterials and also cell functions, as a result of cell-material interactions, including cell proliferation, tissue synthesis, and host response. Degradation of biological materials such as collagen or hydroxyapatite usually proceeds significantly faster than synthetic materials with degradation products considered safe. Rapid and more unpredictable degradation kinetics, however, render biological materials unsuitable for long-term structural support and sustained particle delivery [26,27]. Such materials may be structurally reinforced by carefully choosing suitable fabrication methods or creating bio-synthetic composite materials. Possible mechanisms of mechanical properties influencing the degradability of biomaterials are demonstrated in Figure 19.2.

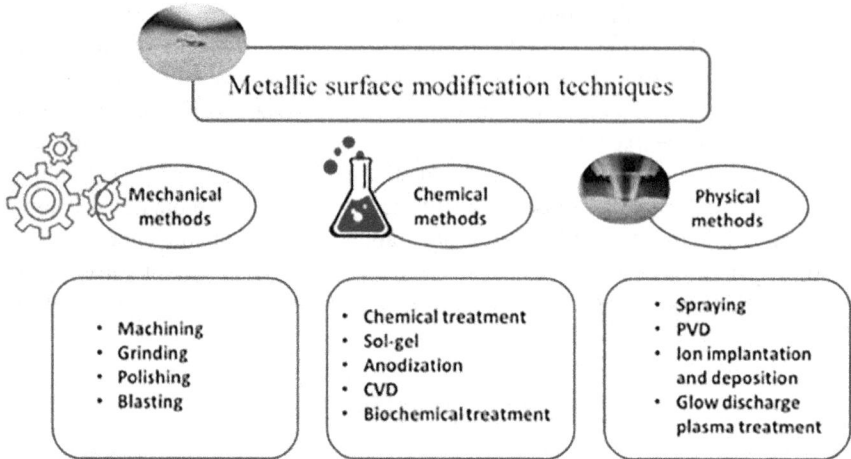

**FIGURE 19.1** A summary of different physical techniques for surface modifications of biomaterials.

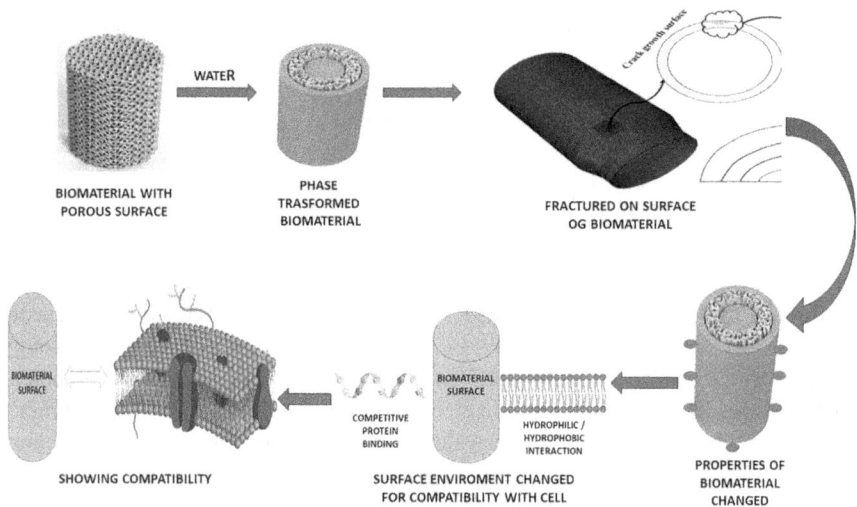

**FIGURE 19.2** Possible mechanisms of mechanical properties influencing degradability of biomaterials [28].

Biocompatibility is the basic and vital requirement of biomaterials for tissue repair. And bioadaptability would be an ideal benchmark for choosing the most suitable biomaterials in tissue engineering. Biomaterials with excellent bioadaptability should have one or several important characteristics, such as adaptable components that simulate cell response and tissue reconstruction, matched mechanical properties with natural tissues, suitable surfaces with beneficial tissue reaction, and bio-mimic multi-level structures [29].

In addition, in the field of tissue engineering, the mechanical properties of biomaterials in both macroscopic and microscopic scales could play crucial roles in the regulation of cell behavior. Furthermore, the mechanical signals from surroundings could be transduced into chemical and physical or other biological signals involved in a series of complicated physiological processes, which finally would direct the cell function, such as cell phenotype and genotype [30,31]. It is known that implanted products are typically exposed to different mechanical stresses. Thus, besides the original mechanical properties of the material, the mechanical influence from the deformation cannot be ignored, either.

## 19.3 DIFFERENT METHODS OF SURFACE TREATMENTS OF BIOMATERIALS

For better adaptation of the materials to the tissues being in contact with them, different techniques of surface modification can be employed. They are divided into two main types of modification: mechanical and physico-chemical. Mechanical processing methods are of little value as they frequently destroy and alter other material features. The methods of physico-chemical treatment can be classified as wet chemical methods based on chemical reactions in a liquid medium, and gas treatment such as a) plasma treatment andion, electron, photon (lasers) stream usage, or b) X-ray radiation [32]. To improve the biocompatibility of solid substrates, polymer coatings with other substances are often used to produce a so-called c) the antithrombotic layer. Different surface treatments are shown in Figure 19.3.

Nowadays, the development of materials has to involve the consideration of environmental responsibility and sustainability. Therefore, more environmentally friendly products, which are biodegradable are extensively desired. However, in widespread applications, most bio-based polymers when used alone present limitations on their functions and are more expensive [33]. There are two principal groups of biodegradable polymers.

The factors related primarily to the implant surface are of particular importance due to its close contact with the biological system. The phenomena in the outer layer of a biomaterial are strictly dependent on its properties and among others, these are the scheme in Figure 19.4:

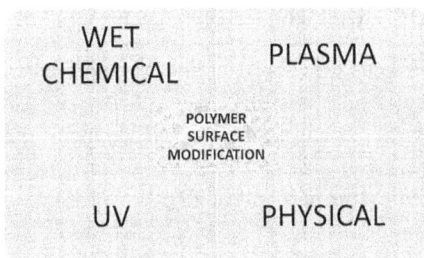

WET CHEMICAL PLASMA

POLYMER SURFACE MODIFICATION

UV PHYSICAL

FIGURE 19.3 Surface treatment methods usually applied for polymer materials by wet chemical, plasma, and UV methods.

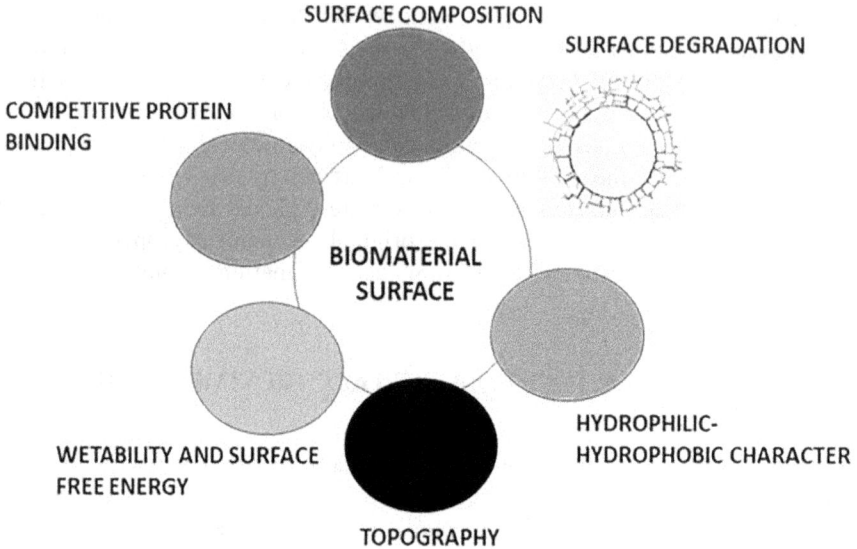

**FIGURE 19.4** Scheme of the key surface parameters and processes in directing biological responses to biomaterials.

surface composition (kind of functional groups and density/sign of charge),
surface degradation,
hydrophilic-hydrophobic character,
wettability and surface free energy (SFE),
topography (roughness, stiffness),
competitive protein binding

## 19.3.1 Mechanical Surface Modification Techniques

### 19.3.1.1 Simple Physical Adsorption

Safety, cost effectiveness, and biocompatibility are the main advantages of physical methods rather than chemical methods. Physical adsorption has been known as one of the simplest techniques to modify the surface of biomaterials via immersion of substrates in solutions containing the adhesive molecules [34]. Growth factors are commonly bound to the surface of the tissue-engineered scaffold via noncovalent modes, including ionic complexation and electrostatic interactions. Although interactions between surface groups and scaffolds are mainly governed via electrostatic force, other forces such as hydrophobic, van der Waals, and hydrogen bonds are also involved in the physical adsorption. Depending on the protein–surface interactions, the release kinetics of growth factors from the surface of substrates are different and can be affected by the surface roughness, charge, and biodegradation rate of substrate and surface energy. Physical adsorption is commonly employed in combination with other strategies such as covalent bonding.

Castellanos et al. used different types of surface modification techniques including surface hydrolysis, physical adsorption, and covalent bonding to immobilize adhesive peptides such as RGD, Arg–Glu–Asp–Val, and Tyr–Ile–Gly–Ser–Arg onto the surface of chromium carbide (CoCr) [35].

### 19.3.1.2 Layer-by-Layer Assembly

Layer-by-layer (LBL) multilayer assembly is able to provide a surface coating on precisely controlled scales (from a few nanometers to several micrometers). In the LBL technique, oppositely charged polymers (polyanions and polycations) are deposited on the charged substrates via an electrostatic force. Serial exposure of an inherently charged substrate to solutions containing oppositely charged species results in their electrostatic deposition and formation of ultrathin and uniform films. In addition to electrostatic force, hydrophobic, covalent, and hydrogen-bonding interactions are other forces that can be involved in assembling multilayer thin films through the LBL technique [36]. Precise control of the coating properties, capability of low-cost production, creation of a biomolecule-friendly environment with mild conditions, production of homogeneous layers with controlled thickness, the possibility of incorporation and controlled release of growth factors/biomolecules/therapeutic agents, and versatility for modification of all available surfaces are the most important advantages of the LBL multilayer assembly method.

## 19.4   EFFECT OF SURFACE TOPOGRAPHY ON BIOMATERIALS

In order to fabricate the biomaterials, it is relevant to consider the effect of material surface architecture on the reaction of a living organism. The researchers suggest that, beyond the surface composition and topography, the so-called protuberances and their shapes perform an essential function in cell attachment and ripening. The diverse surface texture causes different forms of contact between the cells and the adjacent environment. This influences cellular interactions and provokes changes in the cell-surface connections. The multistage and multidirectional in vitro research allows for promising and effective in vivo tests; as an example, the main stages of investigations in the development process of skeletal muscle tissue engineering [37].

Nanotechnology-based biomaterials are a better option to commonly used materials due to their size-dependent properties and improved biocompatibility. The nano-functionalized biomaterials simulating the properties of natural tissues have been shown to improve cell responses. They particularly affect the molecular, cellular, and subcellular functions, including controlling protein adsorption, cell adhesion, differentiation, proliferation, survival, and migration for implants [38]. Nanocomposites and nanofibrous scaffolds produced from synthetic to natural polymers such as chitosan, gelatin, collagen, or silk fibroin act as artificial extracellular matrices that can elevate the growth of tissues required for acceleration.

## 19.5   SURFACE MODIFICATION APPROACHES

The fundamental concepts for many of the surface modification strategies that are being investigated today often date back several decades. While earlier work concentrated on studying responses of protein solutions and cells to uniform flat surfaces and 2D surfaces with micro-topological and nano-topological features, more recently there has been considerable interest in patterned surfaces to study the effects of geometrical confinements on cell spreading. The modification and control of the surface chemistries of biomaterials continue to be very active areas of research [39]. The approaches for surface patterning, gradient surfaces, and microwell surfaces generally utilize established surface modification strategies (usually thin coatings) or minor variations for the generation of intended chemistries, together with approaches for creating specific patterns, gradients, or prior microfabrication of microwell structures. For instance, thiol-SAMs were used in early studies for their ability to provide well-controlled surface chemistries and are still being applied extensively. Gas plasma techniques have also been of interest for biomaterials applications for several decades; better equipment and diagnostics have brought advances in the tailoring of surface chemistries as well as application to patterning and the creation of gradient surfaces. Plasma surface treatments and plasma polymerization approaches with a wide variety of process vapors have been studied extensively and enable the generation of surfaces that support cell colonization as well as reactive surface chemistries for covalent immobilization of bioactive molecules. Chemical patterning to create complex surfaces with feature sizes ranging from those that accommodate single cells down to individual biomolecules is rapidly becoming an important tool in bio-interface science [40]. The scheme of chemical modification is shown in Figure 19.5.

Surfaces play an important role in a biological system for most biological reactions occurring at surfaces and interfaces. The development of biomaterials for tissue engineering is to create perfect surfaces that can provoke specific cellular responses and direct new tissue regeneration. The improvement in biocompatibility of biomaterials for tissue engineering by directed surface modification is an important contribution to biomaterials development and the series of biological events shown in

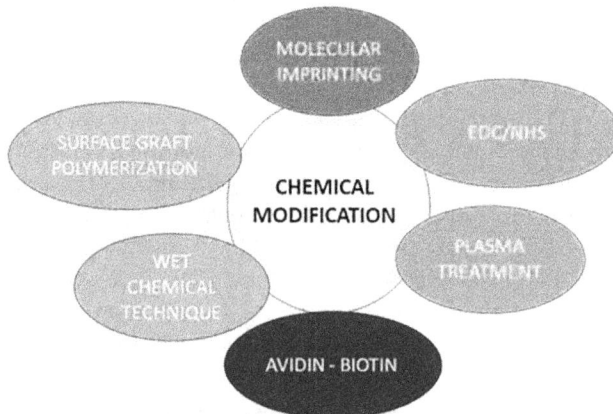

**FIGURE 19.5**   Schematic representation of different chemical surface modification techniques to improve the cell–material interactions [41].

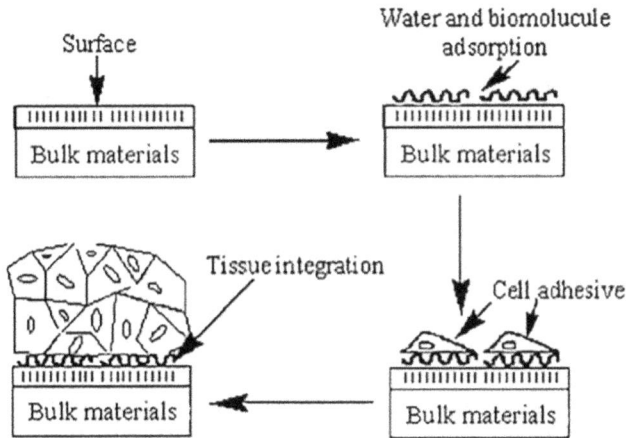

**FIGURE 19.6**   The sketch map of serial biological events evoked by the surface properties of the biomaterials implanted in the host.

Figure 19.6. Among many biomaterials used for tissue engineering, polyesters have been well documented for their excellent biodegradability, biocompatibility, and nontoxicity. However, poor hydrophilicity and the lack of natural recognition sites on the surface of polyesters have greatly limited their further application in the tissue engineering field [42].

The understanding of the interaction between the material surface and biological system is an important requirement for the development of biomaterials. When biomaterials are exposed to a living organism, there are many reactions occurring at the interface between the host and biomaterial surfaces [43].

## 19.6   SURFACE CHEMICAL GROUP/CHARGE MODIFICATIONS

The biocompatibility of a biodegradable polymer is affected by its chemical composition, molecular weight, and crystallinity. Surface characteristics such as free energy or hydrophilicity can affect cell adhesion, spreading, and signaling and hence regulate a wide variety of biological functions, including cell growth, cell migration, cell differentiation, synthesis of ECMs, and tissue morphogenesis [44].

### 19.6.1   CHEMICAL METHODS

In general, biodegradable aliphatic polyesters, such as poly(glycolic acid) (PGA) and poly(lactic acid) (PLA), are degraded by water via autocatalytic cleavage of main chain ester bonds. During modification of surface of the polymer carboxylic acid or hydroxyl groups can be free at the ends by breaking the ester bonds at the surface. Some investigators have reported that surface modifications of bio-degradable polymers in an alkaline solution could be used to generate a hydrophilic and rough surface for cell attachment [45].

The surface degradation of PLLA fibers was studied by incubating them in a concentrated NaOH solution. The results suggested that the PLLA fibers exhibited typical surface degradation in a concentrated alkaline solution. The surface degradation may be advantageous in the application where cell attachment is important [46].

### 19.6.2  PHOTO-INDUCED GRAFTING MODIFICATION METHODS

Photo-induced grafting is a useful technique well known for its advantages, such as low cost of operation, mild reaction conditions, selectivity of ultraviolet (UV) light absorption without affecting the bulk polymer, and permanent alteration of the surface chemistry.

Photo-induced grafting and polymerization of acrylic acid have been extensively used in the past. Their major objective was to modify the surface characteristics of PLLA by grafting a combination of hydrophilic polymers to produce a continuum of hydrophilicity. The results showed that acrylamide dominated the hydrophilicity of the film surface when copolymerized with vinyl acetate or acrylic acid, while the water contact angles of PLLA films grafted with the poly(vinyl acetate-co-acrylic acid) varied more gradually with feed composition [47].

### 19.6.3  PLASMA GRAFTING AND PLASMA TREATMENT

Plasma containing electrons, ions, radicals, and neutral molecules strongly interacts with polymer surfaces and, as a result, chemical and physical modifications occur on the surfaces. Plasma treatment is an effective and widely used method for modifying the surface of biomaterials. It has been reported that the surface of the cell culture in petri dish includes various membranes or microcarriers that can be modified by plasma treatment to improve cell adhesion and growth. Plasma treatment using non-polymerizing gases can create reactive sites such as amine group and carboxyl on polymer surfaces. It has been demonstrated that the plasma process can only proceed with the localized surface treatment (in depths from several hundred angstroms to 10 mm) without changing the bulk properties of the polymers [48].

## 19.7  APPLICATIONS OF SURFACE MODIFICATION OF BIOMATERIALS

Any biomaterial that is to be clinically used for the next generation should have excellent properties both in bulk and surface, but it is very rare that a material with good bulk properties also possesses the surface characteristics required for the biomaterial. This is the reason why surface modification is in many cases essential for a material to be applied in medicine. The surface of the materials used for medical application is modified for at least two purposes: one is to render the material surface biocompatible and the other to give it physiological activity [49].

For mechanical surface modification in the medical field, mainly very thin layers with a thickness of some ten to hundred nanometers are required. Exceptions are layers that have particular mechanical characteristics such as scratch resistance or, particularly, as effective barriers. Important for all these layers is that they possess a good adhesive strength onto the base material, and that their sterilization characteristics are satisfactory. Various methods are used to achieve such surface treatments [50].

### 19.7.1 LUBRICIOUS SURFACE

Tubular medical devices such as catheters, cannulae, endoscopes, and cytoscopes are inserted into the body orifices open to the outside. If there is a significantfrictional resistance between the device's surface and the mucous membrane of the body during the insertion or removal of the device, mechanical damage will occur on the mucous membrane. An effective method to reduce this friction is to make the device's surface very lubricious, similar to the surface of mucous membranes. This is readily accomplished by graft polymerization of non-ionic water-soluble monomers onto the device surface to a high level [51].

### 19.7.2 BLOOD-COMPATIBLE SURFACE

One of the most important biocompatibilities of biomaterials is blood compatibility. Since platelet aggregation and fibrin network formation take place sooner or later when a conventional foreign surface is brought into contact with fresh blood, a large number of methods have been proposed for rendering the material surface blood-compatible, as shown in Figure 19.7 [52].

### 19.7.3 MICROBIAL ADHESION

For a long time, research on antimicrobial adhesion has mainly focused on resisting bacterial biofilms, which are organic films composed of microorganisms, also called

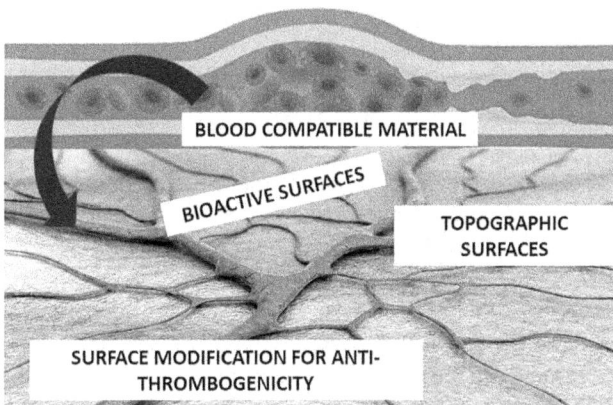

**FIGURE 19.7** Blood-compatible material for the surface modification of blood vessel and treatment of blood-related disease.

microbes, embedded in a polymer matrix of their own making. Microorganism adhesion to the surface of biomaterials and the subsequent formation of biofilms often leads to biomaterial-related infections. It has been shown that the biofilm formed by the mixture of microorganisms and organic material in the early stage is one of the important factors influencing the subsequent adhesion of fouling organisms. For a long time, research has focused on biomaterials modified by types of nanomaterials with antibacterial properties. However, many studies have shown that chemical surface modification can be used to improve the antibacterial properties of substrates [53].

### 19.7.4 Applications in Biocompatibility

According to the definition, biocompatibility refers to the concept of various biological, physical, chemical, and other reactions produced by the interaction between materials and organisms. It is foreseeable that the demand for medical implants and devices will increase in the biomedical field in the future. At the same time, with the development of chemical surface modification technology for biomaterials, an increasing number of studies have focused on enhancing bio-compatibility, which is a particularly important feature in the selection of medical implants. Moreover, some types of polymeric biomaterials have widespread biocompatible applications because of their excellent processability and physico-chemical and mechanical properties, such as PU, polyglycolic acid (PGA), polydioxanone (PDO), and polylactide (PL). At present, it is generally believed that biocompatibility can be simply divided into three categories: hemocompatibility, histocompatibility, and cytocompatibility [54].

#### 19.7.4.1 Hemocompatibility

Hemocompatibility is another feature of biocompatibility that describes the compatibility of the biomedical implant with circulating blood in the body; that is, a medical device that can come into direct contact with blood without causing unexpected adverse reactions. This property is governed not only by coagulation biochemistry and blood–material interactions, but also by the design and function of the device in the bloodstream. Biomaterial–blood compatibility is interpreted as a rather complex process and is generally divided into four important steps: the coagulation cascade, the complement system, platelets, and leukocytes. Considering these factors, it is possible, at least to an extent, that the surface modification strategies for addressing protein fouling and platelet adhesion mentioned above can also be used to improve hemocompatibility [55,56].

#### 19.7.4.2 Histocompatibility

Histocompatibility, also known as tissue compatibility, is the ability of tissues to coordinate with each other. This term is used in immunology to describe the genetic system that determines tissue and organ transplant rejection due to the immune recognition of histocompatibility antigens. Histocompatibility derives from alleles of a set of genes called human leukocyte antigens (HLAs) or the major histo-compatibility complex (MHC). This term also refers to the study of factors that determine the acceptance or rejection of grafted tissues or organs. In general, the higher the histocompatibility, the lower the degree of immune (or rejection) reaction; it is possible that even no immune/rejection reaction occurs. In recent decades, studies on surface modifications for histocompatibility have mainly focused on cell transplantation and medical implants to solve graft rejection; for example, immune isolation of pancreatic islets via thin-layer surface modification [57].

### 19.7.4.3 Drug Delivery

Over the past decades, a number of approaches to achieve site-specific and time-controlled drug delivery have been proposed to alleviate undesired side effects and enhance the efficacy of a given treatment. One of the most common methods of surface modification to achieve drug delivery is self-assembly. Zhu et al. reported a straightforward method for loading hydrophobic materials into commercially available polymer nano- or microparticles through surface modification via layer-by-layer (LbL) self-assembly. The findings suggested that this facile process could be useful in a wide range of applications for drug delivery. Recent studies on drug delivery systems for amphiphilic polymers have shown that these systems can deliver different types of drugs simultaneously in a controlled manner and thus have the potential to enhance the pharmacokinetics and therapeutic efficacy. Ho et al. proposed a simple and feasible method to synthesize a novel amphiphilic polymer; that is, farnesylated ethylene glycol chitosan, which was dispersed in an aqueous medium and self-assembled into nanoparticles. These existing examples of surface modification by self-assembly might provide a reference and comparison for future exploration of potential or feasible chemical surface modification approaches [58].

### 19.7.4.4 Bioactive Surfaces

Utilizing bioactive surfaces to realize switching interactions with biological systems is another recent research hotspot. "Bioactive surface" means the surface of an immobilized bioactive molecule capable of facilitating or supporting specific biological interactions. Bioactive surfaces are not only of great theoretical value but also of practical significance for biomedical applications. In a recent study, Zhan et al. developed a surface with switchable bioactivity in response to sugars based on dynamic covalent bonding between phenylboronic acid (PBA) and secondary hydroxyl groups on the "wide" rim of β-cyclodextrin (β-CD). The reported system consists of gold surfaces modified with PBA-containing polymer brushes and a series of functional β-CD derivatives conjugated to diverse bioactive ligands (CD-X) [59]. It was shown that the prepared surfaces, which can capture a specific protein and have switchable bacteria killing–release properties, can function under relatively simple, mild, and green chemical conditions without eliciting concerns about the negative effects of stimulation, such as high temperature, extreme pH, or harsh chemicals mentioned in the article. It is foreseeable that the reference significance of this work would not be limited to metal or polymeric substrates. As suggested in the above-mentioned work, considering the attributes of generality and versatility inherent in this system, this approach would have the potential as a biomolecule-responsive methodology for the design of dynamic bioactive surfaces for a variety of applications in the biomedical and biotechnology fields [60].

### 19.7.5 Plasma-Surface Modification for Blood-Contacting Devices

Biocompatibility is not an intrinsic property of a material. It is one of the essential parameters of any material's selection and application in biomedicine, which entails it having hydrophilicity and a low friction coefficient. The interactions between an implant and the surrounding tissues are very complicated. Most classes of polymeric materials have good mechanical strength and stability but they suffer due to the issues related to surface-induced thrombosis. When the device/implant comes in contact with blood, the deposition of plasma proteins (like fibrin) leads to the formation of thrombosis on the surface of implanted biomaterial followed by adhesion of platelets. The adhesion, activation, and spreading of platelets and protein deposition are determined by the interaction between the plasma proteins and the surface of the implant. The main problem associated with blood-contacting devices (like artificial heart valves, catheters, various joints, and implants) is that the surface of the artificial organ is not recognized by blood. In general, urgent replacement of artificial prostheses is indispensable for patients who suffer from heart-valve disease and arteriosclerosis. Unfortunately, most of the products do not provide stable long-term biofunctionality. Therefore, surface engineering is required in the biohybrid organ systems, which should possess a blood-contacting side that has an excellent hemocompatibility and a tissue-contacting side that should be cytocompatible. Materials used for such implants may have excellent resistance to wear, fatigue, and degradation [61].

However, the main problem of these materials is thrombogenicity (Goodman et al. 1996). For example, vascular prostheses used to replace large arteries work very well, but small-caliber artificial grafts having an internal diameter of less than 5 mm do not work well. This may happen due to the localized constriction (stenosis) in the vascular graft, which eventually restricts the blood flow through the affected vessel (Mustard and Packham 1975). Therefore, in order to minimize the risk of thromboembolic complications in patients, anticoagulation therapy is necessary. Hence, it is necessary to either find new biomaterials with better blood compatibility or suitably modify the existing blood contacting materials [62].

### 19.7.6 Application in Clinical Medical Fields

The application mentioned above has mainly remained in the stage of theoretical research, without apparent assessment in clinical trials. However, increasing numbers of recent studies have shown that surface modification strategies can also be applied to clinical medical fields [63]. This means that it is necessary to separately introduce clinical applications.

In a recent review from Janani et al., insights into silk-based products assessed in human clinical trials were presented. The paradigm of the silk structure–function relationship promoting the use of silk-based biomaterials in tissue engineering, drug delivery systems, and in vitro tissue models was explored. Meanwhile, the potential use of silk in drug delivery systems, in vitro models, and medical implants, as well as clinical evidence of silk-related products in the market, was introduced. As mentioned in the review, natural biopolymers, such as silk, have been used clinically for a long time and have received much impetus for a plethora of biomedical applications in the last two decades. Therefore, there is considerable application foreground in the field of surface-modified natural biopolymers for clinical treatment [64].

## 19.8 CONCLUSIONS AND FUTURE OUTLOOK

Recent advances in chemical and mechanical surface modification and polymeric biomaterials enable functionalization by grafting techniques to improve performance in biomedical applications at the interfaces of biomacromolecules, cells, tissues, and biomaterials. This chapter focuses on the surface modification of the biomaterials for their application in biomedical uses like tissue engineering, grafting, implants, etc. Many methods of functionalization by surface modification of biomaterials for biomedical applications have been introduced in recent years, such as chemical grafting, plasma-induced graft polymerization, radiation-induced graft polymerization, ozone graft polymerization, and photo-induced graft polymerization [65].

Although grafting strategies are promising tools and approaches to enhance the biomedical functions of surfaces such as antifouling properties or biocompatibility, additional studies are needed to determine the mechanisms and realize the full potential of these strategies. For example, future work could focus on achieving precise control of grafting onto different types of substrates with different shapes, elucidating the relationship between specific biomedical functions and the polymeric architectures grafted onto a surface, and achieving multifunctionalities by simultaneous grafting of different macromolecules or multifunctional polymers. However, it is again necessary to specify what is meant by "mechanical properties", as well as specific magnitudes of these properties, and determine defect boundary conditions for calculating stresses, although a number of different models have already been proposed for myocardium, blood vessels, and heart valves. This task is an enormous challenge mainly for two reasons: 1) there is a tremendous range of constitutive models defined for different cells and tissues, including different mechanical properties, and 2) as there has been no consensus model to use as a porous biomaterial design target yet; the same tissue is often modeled with a large number of different versions of these constitutive models considering the first reason. These issues both would make things difficult for defining mechanical-design requirements.

## ACKNOWLEDGMENT

This work was supported by the Department of Chemistry, Institute of Basic Sciences.

## REFERENCES

1. Gobbi, S. J., Reinke, G., Gobbi, V. J., Rocha, Y., Sousa, T. P., & Coutinho, M. M. (2020). Biomaterial: Concepts and basics properties. *European International Journal of Science and Technology*, 9, 23–42.
2. Harmon, M. D., Ramos, D. M., Nithyadevi, D., Bordett, R., Rudraiah, S., Nukavarapu, S. P., ... & Kumbar, S. G. (2020). Growing a backbone–functional biomaterials and

structures for intervertebral disc (ivd) repair and regeneration: Challenges, innovations, and future directions. *Biomaterials Science*, 8(5), 1216–1239.

3. Cvrček, L., & Horáková, M. (2019). Plasma modified polymeric materials for implant applications. In *Non-Thermal Plasma Technology for Polymeric Materials* (pp. 367–407). Elsevier.

4. Qin, Z., Li, B., Chen, T., Chen, C., Chen, R., Ma, H., & Xue, H. (2023). Comparative study of the effects of conventional shot peening and ultrasonic shot peening on very high cycle fatigue properties of GH4169 superalloy. *International Journal of Fatigue*, 175, 107799.

5. Vladkova, T. G. (2010). Surface engineered polymeric biomaterials with improved biocontact properties. *International Journal of Polymer Science*, 2010, 1–23.

6. Chan, W. L., & Cheng, H. K. F. (2022). Hammer peening technology—the past, present, and future. *The International Journal of Advanced Manufacturing Technology*, 118, 1–19.

7. He, Z., Shen, Y., Tao, J., Chen, H., Zeng, X., Huang, X., & Abd El-Aty, A. (2021). Laser shock peening regulating aluminum alloy surface residual stresses for enhancing the mechanical properties: Roles of shock number and energy. *Surface and Coatings Technology*, 421, 127481.

8. Nouri, A., & Wen, C. (2015). Introduction to surface coating and modification for metallic biomaterials. *Surface Coating and Modification of Metallic Biomaterials*, 1, 3–60.

9. Zadpoor, A. A. (2019). Mechanical performance of additively manufactured meta-biomaterials. *Acta Biomaterialia*, 85, 41–59.

10. Hedayati, R., Janbaz, S., Sadighi, M., Mohammadi-Aghdam, M., & Zadpoor, A. A. (2017). How does tissue regeneration influence the mechanical behavior of additively manufactured porous biomaterials? *Journal of the Mechanical Behavior of Biomedical Materials*, 65, 831–841.

11. Bandyopadhyay, A., Mitra, I., Goodman, S. B., Kumar, M., & Bose, S. (2022). Improving biocompatibility for next generation of metallic implants. *Progress in Materials Science*, 101053.

12. García, A. J. (2009). Surface modification of biomaterials. *Foundations of Regenerative Medicine: Clinical and Therapeutic Applications*, 368–378.

13. Crawford, R. J., Webb, H. K., Truong, V. K., Hasan, J., & Ivanova, E. P. (2012). Surface topographical factors influencing bacterial attachment. *Advances in Colloid and Interface Science*, 179, 142–149.

14. Coluccio, M. L., Gentile, F., Barbani, N., &Cristallini, C. (2023). Surface properties and treatments. In *Microfluidics for Cellular Applications* (pp. 189–222). Elsevier.

15. Yaseen, M. (2019). Study of morphology, physico-chemical properties and antibacterial effect of gallium oxide coated implantable substrates submitted to thermal oxidation treatment.

16. Frutos, G. (2020). Local surface environment and degradation processes of degradable magnesium biomaterials under simulated physiological conditions (Doctoral dissertation, University of Kiel).

17. Maurel, P., Weiss, L., Grosdidier, T., & Bocher, P. (2020). How does surface integrity of nanostructured surfaces induced by severe plastic deformation influence fatigue behaviors of Al alloys with enhanced precipitation?. *International Journal of Fatigue*, 140, 105792.

18. Chen, X. B., Birbilis, N., & Abbott, T. B. (2011). Review of corrosion-resistant conversion coatings for magnesium and its alloys. *Corrosion*, 67(3), 035005–1.

19. Wu, S., Liu, X., Yeung, K. W., Liu, C., & Yang, X. (2014). Biomimetic porous scaffolds for bone tissue engineering. *Materials Science and Engineering: R: Reports*, 80, 1–36.

20. Ramakrishnan, S. K., Zhu, J., & Gergely, C. (2017). Organic–inorganic interface simulation for new material discoveries. *Wiley Interdisciplinary Reviews: Computational Molecular Science*, 7(1), e1277.

21. Goonoo, N., Bhaw-Luximon, A., Bowlin, G. L., & Jhurry, D. (2013). An assessment of biopolymer-and synthetic polymer-based scaffolds for bone and vascular tissue engineering. *Polymer International*, 62(4), 523–533.

22. Cheong, W. F., Prahl, S. A., & Welch, A. J. (1990). A review of the optical properties of biological tissues. *IEEE Journal of Quantum Electronics*, 26(12), 2166–2185.

23. Bauer, S., Schmuki, P., Von Der Mark, K., & Park, J. (2013). Engineering biocompatible implant surfaces: Part I: Materials and surfaces. *Progress in Materials Science*, 58(3), 261–326.

24. Festas, A. J., Ramos, A., & Davim, J. P. (2020). Medical devices biomaterials–A review. *Proceedings of the Institution of Mechanical Engineers, Part L: Journal of Materials: Design and Applications*, 234(1), 218–228.

25. Kim, J. H., Jeong, J. H., Kim, N., Joshi, R., & Lee, G. H. (2018). Mechanical properties of two-dimensional materials and their applications. *Journal of Physics D: Applied Physics*, 52(8), 083001.

26. Adetunji, C. O., Oloke, J. K., Dwivedi, N., Ummalyma, S. B., Dwivedi, S., Hefft, D. I., & Adetunji, J. B. (Eds.). (2023). *Next-Generation Algae, Volume 1: Applications in Agriculture, Food and Environment*. John Wiley & Sons.

27. Wang, L., Wang, C., Wu, S., Fan, Y., & Li, X. (2020). Influence of the mechanical properties of biomaterials on degradability, cell behaviors and signaling pathways: Current progress and challenges. *Biomaterials Science*, 8(10), 2714–2733.

28. Wang, L., Wang, C., Wu, S., Fan, Y., & Li, X. (2020). Influence of the mechanical properties of biomaterials on degradability, cell behaviors and signaling pathways: Current progress and challenges. *Biomaterials Science*, 8(10), 2714–2733.

29. Venkatraman, S., Boey, F., & Lao, L. L. (2008). Implanted cardiovascular polymers: Natural, synthetic and bio-inspired. *Progress in Polymer Science*, 33(9), 853–874.

30. Kolahi, K. S., & Mofrad, M. R. (2010). Mechanotransduction: A major regulator of homeostasis and development. *Wiley Interdisciplinary Reviews: Systems Biology and Medicine*, 2(6), 625–639.

31. Nickerson, C. A., Ott, C. M., Wilson, J. W., Ramamurthy, R., & Pierson, D. L. (2004). Microbial responses to microgravity and other low-shear environments. *Microbiology and Molecular Biology Reviews*, 68(2), 345–361.

32. Samanta, K. K., Basak, S., & Chattopadhyay, S. K. (2014). Environment-friendly textile processing using plasma and UV treatment. *Roadmap to Sustainable Textiles and Clothing: Eco-friendly Raw Materials, Technologies, and Processing Methods*, 161–201.

33. Iwata, T. (2015). Biodegradable and bio-based polymers: Future prospects of eco-friendly plastics. *AngewandteChemie International Edition*, 54(11), 3210–3215.

34. Yang, K., Lee, J. S., Kim, J., Lee, Y. B., Shin, H., Um, S. H., … & Cho, S. W. (2012). Polydopamine-mediated surface modification of scaffold materials for human neural stem cell engineering. *Biomaterials*, 33(29), 6952–6964.

35. Castellanos, F. X., & Tannock, R. (2002). Neuroscience of attention-deficit/hyperactivity disorder: The search for endophenotypes. *Nature Reviews Neuroscience*, 3(8), 617–628.

36. Kharlampieva, E., Kozlovskaya, V., & Sukhishvili, S. A. (2009). Layer-by-layer hydrogen-bonded polymer films: From fundamentals to applications. *Advanced Materials*, 21(30), 3053–3065.

37. Niermeyer, W. L., Rodman, C., Li, M. M., & Chiang, T. (2020). Tissue engineering applications in otolaryngology—The state of translation. *Laryngoscope Investigative Otolaryngology*, 5(4), 630–648.

38. Le, X., Poinern, G. E. J., Ali, N., Berry, C. M., & Fawcett, D. (2013). Engineering a biocompatible scaffold with either micrometre or nanometre scale surface topography for promoting protein adsorption and cellular response. *International Journal of Biomaterials*, 2013.

39. Nouri, A., Shirvan, A. R., Li, Y., & Wen, C. (2023). Surface modification of additively manufactured metallic biomaterials with active antipathogenic properties. *Smart Materials in Manufacturing*, 1, 100001.

40. Lin, M. (2023). Multifunctional Nanoscale Surfaces for Biological Applications (Doctoral dissertation, Northwestern University).

41. Amani, H., Arzaghi, H., Bayandori, M., Dezfuli, A. S., Pazoki-Toroudi, H., Shafiee, A., & Moradi, L. (2019). Controlling cell behavior through the design of biomaterial surfaces: A focus on surface modification techniques. *Advanced Materials Interfaces*, 6(13), 1900572.

42. Abtahi, S., Chen, X., Shahabi, S., & Nasiri, N. (2023). Resorbable membranes for guided bone regeneration: Critical features, potentials, and limitations. *ACS Materials Au*, 3(5), 394–417.

43. Jiao, H., Sun, J., Shi, Y., Lu, X., Ali, S. S., Fu, Y., … & Liu, J. (2023). Recent advances in strategies of nanocellulose surface and/or interface engineering for potential biomedical applications as well as its ongoing challenges: A review. *Cellulose*, 1–31.

44. Zhang, Y., & Habibovic, P. (2022). Delivering mechanical stimulation to cells: State of the art in materials and devices design. *Advanced Materials*, 34(32), 2110267.

45. Tolabi, H., Bakhtiary, N., Sayadi, S., Tamaddon, M., Ghorbani, F., Boccaccini, A. R., & Liu, C. (2022). A critical review on polydopamine surface-modified scaffolds in musculoskeletal regeneration. *Frontiers in Bioengineering and Biotechnology*, 10, 1008360.

46. Javid-Naderi, M. J., Behravan, J., Karimi-Hajishohreh, N., & Toosi, S. (2023). Synthetic polymers as bone engineering scaffold. *Polymers for Advanced Technologies*.

47. Kikuchi, M., Saito, N., Ohke, M., Nagano, S., Nishitsuji, S., & Matsui, J. (2023). Order–order transitions in poly (N-octadecyl acrylamide-co-hydroxyethyl acrylamide) statistical copolymer films. *Soft Matter*, 19(17), 3058–3068.

48. Günther, R. (2023). Direct Joining and Debonding on-Demand of Polymers through Surface Modification and Metal Ion Interaction (Doctoral dissertation, ETH Zurich).

49. Williams, D. F. (2008). On the mechanisms of biocompatibility. *Biomaterials*, 29(20), 2941–2953.

50. Vijaya Kumar, P., & Velmurugan, C. (2022). Surface treatments and surface modification techniques for 3D built materials. In *Innovations in Additive Manufacturing* (pp. 189–220). Cham: Springer International Publishing.

51. Fan, Y. L. (1991). Hydrophilic lubricious coatings for medical applications. In *Cosmetic and Pharmaceutical Applications of Polymers* (pp. 311–319). Boston, MA: Springer US.

52. Gürbüzer, B., Pikdöken, L., Tunalı, M., Urhan, M., Küçükodacı, Z., & Ercan, F. (2010). Scintigraphic evaluation of osteoblastic activity in extraction sockets treated with platelet-rich fibrin. *Journal of Oral and Maxillofacial Surgery*, 68(5), 980–989.

53. Chenthamara, D., Subramaniam, S., Ramakrishnan, S. G., Krishnaswamy, S., Essa, M. M., Lin, F. H., & Qoronfleh, M. W. (2019). Therapeutic efficacy of nanoparticles and routes of administration. *Biomaterials Research*, 23(1), 1–29.

54. Bai, Q., Teng, L., Zhang, X., & Dong, C. M. (2022). Multifunctional single-component polypeptide hydrogels: The gelation mechanism, superior biocompatibility, high performance hemostasis, and scarless wound healing. *Advanced Healthcare Materials*, 11(6), 2101809.

55. Goddard, J. M., & Hotchkiss, J. H. (2007). Polymer surface modification for the attachment of bioactive compounds. *Progress in Polymer Science*, 32(7), 698–725.
56. Fan, X. L., Hu, M., Qin, Z. H., Wang, J., Chen, X. C., Lei, W. X., … & Ji, J. (2018). Bactericidal and hemocompatible coating via the mixed-charged copolymer. *ACS Applied Materials & Interfaces*, 10(12), 10428–10436.
57. Foster, G. A., & García, A. J. (2017). Bio-synthetic materials for immunomodulation of islet transplants. *Advanced Drug Delivery Reviews*, 114, 266–271.
58. Yu, H., Wang, Y., Jing, Y., Ma, J., Du, C. F., & Yan, Q. (2019). Surface modified MXene-based nanocomposites for electrochemical energy conversion and storage. *Small*, 15(25), 1901503.
59. Sun, W., Liu, W., Wu, Z., & Chen, H. (2020). Chemical surface modification of polymeric biomaterials for biomedical applications. *Macromolecular Rapid Communications*, 41(8), 1900430.
60. Han, T., Wang, X., Wang, D., & Tang, B. Z. (2021). Functional polymer systems with aggregation-induced emission and stimuli responses. *Aggregation-Induced Emission*, 287–333.
61. Wang, J., Dou, J., Wang, Z., Hu, C., Yu, H., & Chen, C. (2022). Research progress of biodegradable magnesium-based biomedical materials: A review. *Journal of Alloys and Compounds*, 923, 166377.
62. Özgüzar, H. F., Evren, E., Meydan, A. E., Kabay, G., Göçmen, J. S., Buyukserin, F., & Erogul, O. (2023). Plasma-assisted surface modification and heparin immobilization: dual-functionalized blood-contacting biomaterials with improved hemocompatibility and antibacterial features. *Advanced Materials Interfaces*, 10(6), 2202009.
63. Gautam, S., Bhatnagar, D., Bansal, D., Batra, H., & Goyal, N. (2022). Recent advancements in nanomaterials for biomedical implants. *Biomedical Engineering Advances*, 3, 100029.
64. Ghosh, S. K., Samanta, S., Hirani, H., & da Silva, C. R. V. Effective Waste Management and Circular Economy, 1.
65. Periasamy, K., Kandare, E., Das, R., Darouie, M., & Khatibi, A. A. (2023). Interfacial engineering methods in thermoplastic composites: An overview. *Polymers*, 15(2), 415.

# 20 Heat Treatment for Biomaterial Surface

*Akash Mishra, Amlan Prabhujyoti Sahu,
Priyabrata Mallick, and Ajit Behera*
Department of Metallurgical & Materials Engineering,
National Institute of Technology, Rourkela, Odisha, India

## 20.1 INTRODUCTION TO BIOMATERIALS

The initial phase of materials intended to be utilized within the human body emerged in the 1960s and 1970s. The discipline of biomaterials was founded on these innovations [1]. Prostheses are the name given to the biomaterial-based devices. Among the first people to grasp the significance of comprehending the mechanical characteristics of tissues, particularly bone, with the goal to produce trustworthy skeletal prostheses was Professor Bill Bonfield. One of the first attempts to comprehend how biomaterials behave with live tissues was his study. "To obtain an appropriate mix of physical characteristics that correspond to that of the substituted tissue while producing a minimum toxic reaction in the recipient" was the aim of all early biomaterials [1,2]. Over 50 implanted prostheses consisting of 40 distinct components were in use in hospitals by the year 1980. During that time, every year, patients all around the world received over 3 million implants of prosthetic components. The biological "inertness" of the majority of the 40 different materials constituted a prevalent characteristic. Nearly a majority of the substances utilized in our bodies were single-phase substances [2]. The majority of the implant components were improved versions of already-available commercial materials to prevent the production of hazardous by-products and reduce corrosion. The area of biomaterials has had sustained expansion due to the steady growth of fresh concepts and fruitful branches after its founding a little over 50 years ago [1]. The production of artificial substances with regulated characteristics for drug and cell carriers, the creation of biologically motivated substances that resemble natural processes, and the construction of advanced three-dimensional (3D) structures that generate established trends for diagnostics, such as biological mRNA, are all examples of the latest generation of biomaterials.

### 20.1.1 COMPATIBILITY OF BIOMATERIALS

The potential of biomaterials to function in harmony with biological systems, minimizing negative responses, and fostering effective integration is referred to

DOI: 10.1201/9781003429920-23

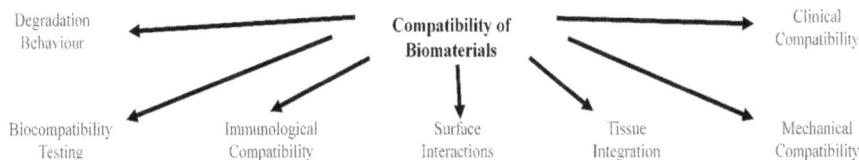

**FIGURE 20.1** Different variables that affect biocompatibility.

as biocompatibility [3]. Figure 20.1 shows various components that affect biocompatibility.

The components used for orthopedic applications must be able to perform in vivo without triggering any unfavorable local or systemic reactions, such as immunological, allergic, inflammatory, or carcinogenic ones. Biofunctionality and biostability have been parts of biocompatibility, in addition to bio-inertia [4]. This is a crucial idea that is greatly influenced by the characteristics of the material (charges, surface chemistry, wettability, surface chemistry, texture, breakdown products, crystallinity, stiffness), communication between it and the intended tissues' biological surroundings (inflammatory processes, blood contact, protein adsorption), application type, application period, and application properties. Various international standards that are used to check biocompatibility are given in Table 20.1.

Native immune cells are initially drawn to the area subsequently activate local cells in response to stimuli related to damage (and the placement of a scaffold), which is followed by the stimulation and polarisation of immune cells that are adaptive. Both tissue matching the initial host tissue as well as additional fibrous

**TABLE 20.1**

**International Standards for Biocompatibility Testing of Biomaterials [5]**

| | Standards | |
|---|---|---|
| **Biological Response** | **ISO** | **ASTM** |
| Cytotoxicity | 10993–5 | F813–07; F895–84; F1027–06 |
| Sensitization | 10993–10 | F720–81; F2147–01; F2148–07 |
| Irritation | 10993–10 | F719–81; F749–98 |
| Acute systemic toxicity | 10993–11 | F750–87 |
| Subacute toxicity | 10993–11 | – |
| Genotoxicity | 10993–3 | E1262–88 |
| Immunoresponsiveness | 10993–20 | F1906–98 |
| Hemocompatibility | 10993–4 | F756–08 |
| Chronic toxicity | 10993–11 | - |
| Carcinogenicity | 10993–3 | F1439–03 |
| Degradation | 10993–9, 10993–13, 10993–14, 10993–15 | F1983–14 |
| Implantation | 10993–6 | F1408–97; F763–04; F1904–98 F981–04; F1983–99 |

scar tissue may grow, depending upon the phenotype of the aforementioned immune cells, their relationship with diverse cell types, especially stem cells, and signals from the local micro surrounding (which includes the biomaterial scaffold) [5]. The creation of appropriate substitute tissue is supported by the establishment of a pro-regenerative immunity environment; nevertheless, imbalanced immune response activation can result in fibrosis or harmful inflammation. Growth-promoting agents have been delivered via biomaterial scaffolds frequently to promote the development of freshly formed bone [6]. The therapeutic success of growth regulators has been significantly influenced by the pharmacokinetics of their distribution. The exterior interactions that regulate the non-covalent integration of growth-promoting agents into scaffolds and the processes that regulate the expulsion of growth factors from therapeutically important biomaterials will be the main topics of our attention. We'll concentrate on the best way to administer synthetic human bone morphogenetic protein-2 using substances that are now in use in clinical settings, but we'll also make suggestions about how general processes that regulate growth factor integration and discharge that have been identified with this growth factor may also apply to various other systems. Even though treating illnesses using just cells has clearly had some success, it remains unknown if cells just encourage the growth of nearby cells or actually create fresh tissue. Considering both situations, the end effect is frequently out of sync with the surroundings and loses effectiveness after a short time. Complete integration may be attained by creating biomaterial-based smart patterns that serve as scaffolds for fresh tissue formation and cues and indicators for the linkages between functioning tissue. Scanning electron microscopy (SEM), X-ray diffraction (XRD), and transmission electron microscopy (TEM) studies are used to study the microstructures of materials [5,6]. Uniaxial tensile measurements are used to assess the mechanical characteristics, and the resonance vibration technique is used to quantify the dynamic elastic modulus. According to the experimental findings, following solution treatment, the Ti-25Ta alloy displays an orthorhombic martensite ($\alpha''$) paired structure. Among Ti-Ta alloys, the Ti-25Ta alloy possesses the smallest elastic modulus and the greatest strength to modulus ratio. The Ti-25Ta alloy is a viable contender for new metallic biomaterials since it has the highest mechanical compliance of any Ti-Ta alloy [7].

## 20.1.2 DIFFERENT TYPES OF BIOMATERIALS

Stainless steel, partly stabilized zirconia, titanium, alumina, and ultra-high molecular weight polyethylene are a few examples of materials that are considered bioinert because they have no effect on the tissue they are inserted in. The biofunctionality of bioinert implants often depends on tissue incorporation via the implant since a capsule of fibrous material may develop around them [8,9]. Bioactive materials react with the adjacent bone and, in certain circumstances, particularly soft tissue when they are introduced into the body of an individual. This happens as a result of their insertion into the living bone, which causes a time-dependent kinetic alteration on the outermost layer. A naturally active carbonate apatite (CHAp) coating that is chemically and crystallographically identical to the

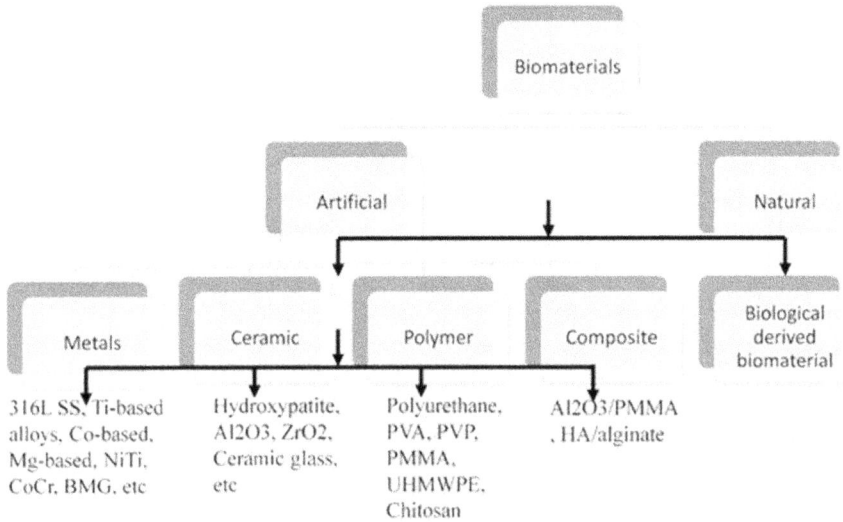

**FIGURE 20.2** Classification of biomaterials.

mineral phase within bone forms on an implant primarily as a result of an ion-exchange interaction among the bioactive implants and nearby bodily fluids [10]. Synthetic hydroxyapatite [Ca10(PO4)6(OH)2], glass ceramic A-W, and bioglass are exemplary instances of these types of materials. A substance is considered bioresorbable if it begins to disintegrate (be resorbed) after it is inserted into the body and is eventually replaced by growing tissue (like bone). Polylactic-polyglycolic acid copolymers and tricalcium phosphate [Ca3(PO4)2] are typical examples of bioresorbable materials [11]. Gypsum, calcium carbonate, and calcium oxide are some additional typical materials which were used throughout the past three decades. Various types of biomaterials present along with their advantages, disadvantages, and uses are given in Figure 20.2 and Table 20.2, respectively.

## 20.1.3 BIOMATERIAL SURFACE

The effectiveness of host reactions to biomaterials, referred to as biocompatibility, relies on chemical concepts, which are the basis of each of the body's cell signaling networks as well as the surface layout of the biomaterial. The in vivo biochemical signal-transduction systems must be appropriately matched to the surface physico-chemical properties of biomaterials, which remains a crucial gap [12]. With the goal to promote the creation of neo-tissue, the proposed biomaterials towards use in tissue engineering ought to possess a high affinity for the targeted cells. The chemistry of the substances and the intended biological framework are completely necessary for establishing significant beneficial relationships among the biomaterial interface and cells. Considering the perspective of materials engineering, any physicochemical characteristic of the outermost layer of the biomaterial (such as functional groups, topographical characteristics, stiffness, and interfacial free energy) may have a

**TABLE 20.2**

**Different Types of Biomaterials Currently used along with Advantages, Disadvantages, and Possible Usage are Shown**

| Type | Advantages | Disadvantages | Applications |
|---|---|---|---|
| Metal and alloys of metal E.g.,: gold, platinum, titanium, steel, chromium, cobalt | • Increased strength of material <br> • They are simple to make and sterilize | • Highly corrosive in nature <br> • Aseptic loosening is seen <br> • Contains high elastic modulus | • Orthopedic implants, screws, pins, and plates |
| Ceramics and compounds of carbon E.g.,: calcium phosphate salts (HA), glass, oxides of aluminum and titanium | • Biocompatibility is enhanced <br> • Material strength observed is greater <br> • Improved resistance to corrosion | • Molding is difficult <br> • High modulus of elasticity | • Bioactive orthopedic implants <br> • Dental implants <br> • Artificial hearing aids |
| Polymers E.g.,: PMMA—poly (methyl methacrylate), Polycaprolactone (PCL), PLA, polycarbonates, polyurethanes | • Biodegradability and Biocompatibility is observed <br> • Molding is easy and can be found easily <br> • Has adequate mechanical strength | • Can be leachable in body fluids <br> • Sterilization is hard and difficult | • Orthopedic and dental implants <br> • Prostheses <br> • Tissue engineering scaffolds <br> • Drug delivery systems |
| Composites E.g.,: Dental filling composites, carbon fiber reinforced methyl methacrylate bone cement + ultra-high molecular weight polyethylene | • Very good mechanical characteristics <br> • Produce very good resistance to corrosion | • Costly <br> • Extremely difficult fabrication method is used | • Porous orthopedic implants <br> • Dental fillings <br> • Rubber catheters and gloves |

significant impact on the biological processes [2,3]. Regarding a biomaterial to work at its best, the surface physico-chemistry must be altered depending on the intended location, as shown in Figure 20.3. In fact, physicochemical surface characteristics such as architecture, surface roughness, charge, energy, and functional groups are important factors to consider when choosing biomaterials for use in tissue engineering. Therefore, it is important to carefully research in vitro and in vivo the way every physico-chemical surface feature affects the biological activity of biomaterials [13]. The physico-chemical characteristics of the biomaterial surface, including biological moieties, functional groups, charges, and ion enrichment, are crucial in determining how biological processes react to the material. Diverse mechanotransduction, physiological, macromolecular adsorption, and biochemical signaling mechanisms that might vary from tissue to tissue are important in considering a biological standpoint. The exterior physico-chemistry of the biomaterial

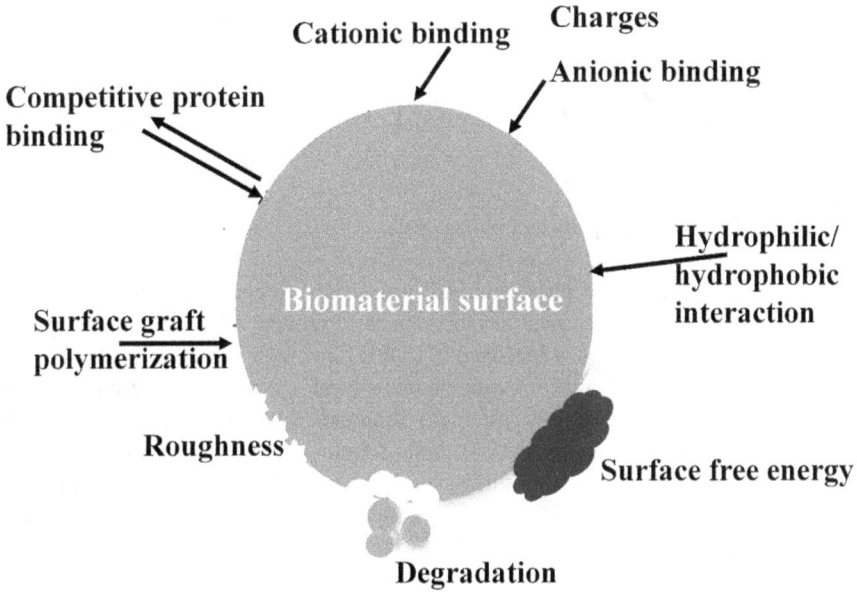

**FIGURE 20.3** Various processes that take place on a biomaterial surface.

will determine whether or not plasma components such as proteins, lipids, carbohydrates, and ions may attach to it. The circulatory exudate also contains platelets, that with aggregation and coagulation produce a fibrin-rich clots. The clot that forms serves as an interim framework that promotes cellular and molecular activities. The adsorbed proteins attach to the particular binding sites, which include proline-histidine-serine-arginine-asparagine (PHSRN), N-termini, C-termini, and arginine-glycine-aspartic acid (RGD), and engage with inherent immune system cells like neutrophils, fibroblasts, monocytes, and endothelial cells [14,15].

## 20.1.4 SURFACE PROPERTIES OF BIOMATERIALS

Whether biomaterials react to their biological surroundings is significantly influenced by their exterior qualities, which also have an important effect upon their biocompatibility and biofunctionality [16]. The major parameters that affect the surface properties of biomaterial are shown in Figure 20.4.

Roughness, appearance, and layout of the surface may have an impact on cell attachment, motility, and proliferation. It is actually possible to change the topography of the covering to regulate cell behavior and tissue assembly. The chemical composition of this interface controls its interactions with amino acids, cells, and several other substances. Cell adsorption, adhesion, and signaling mechanisms are all impacted by the surface makeup [17]. This may be possible to improve biological compatibility and some biological processes by altering the surface's chemical constituents. The wetting ability of surfaces as determined as the water contact angle ($\theta$) shown requires reliance on the protein adherence on the

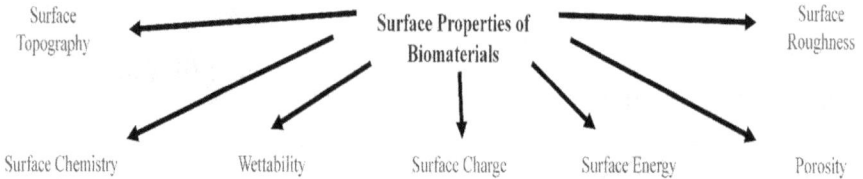

**FIGURE 20.4**   Surface properties of biomaterials.

lower density polyethylene platforms. Adsorption of protein, coagulation of blood, and cell and bacterial attachment are all impacted by wettability [16,18]. By glow-discharge plasma alterations, a spectrum of LDPE films with varying degrees of water wettability—from hydrophilic to hydrophobic—were created. Surface charges and electronically reactive dynamic microenvironments each have the potential to enhance the adhesion, proliferation, and differentiation of neuron-like cells. Piezoelectric materials have the ability to emit brief charges on their surfaces in response to mechanical stimulation. Concerning the impact upon surface charge, cells grown during static circumstances on PVDF films that had been charged (PVDF+ and PVDF-) showed noticeably stronger flattening/elongation than cells grown on PVDF NP. The deficiency of charges in the surface in cells grown on PVDF NP and the control medium varies little within static and dynamic circumstances, indicating that the mechanical stimulation by itself, without first being supplemented by surface charge deviation, has no immediate impact on adhesion of cells [19]. The energy from the surface is well known to regulate protein adhesion and cellular response downward in the realm of biomedical implant development. Surface energy of biomaterials might be seen while the physical labor put forth by intermolecular forces when they act to boost phase surface area. As a result, the electrical charge and polarization of the biomaterial's exterior functional units affect surface energy. Polar functional chains can enhance surface energy, while more energetic substrates containing more polar sections, which results in an increased hydrophilic interface. By placing greater number of porosity unit cells in the areas closest to the implant's limitations, in which bony ingrowth becomes most essential and slowly decreasing the level of porosity as one approaches the implant's interior parts, which have a solid center able to support the utilized musculoskeletal load, the situation is possible to satisfy both the requirements for extremely porous patterns that make it possible for bone-like development as well as sturdy concepts that offer adequate mechanical assistance [8,16]. When the lattice structure's porosity develops, permeable bioma-terials' mechanical features often deteriorate. A larger porosity gives a smaller amount for the actual stress that corresponds to the identical amount of repetitions to failure, much as the modulus of elastic force and stress yielding cases.

## 20.1.5   SURFACE DEGRADATION OF BIOMATERIALS

Whenever a substance comes in touch with a biological surroundings, such as simulating bodily fluid, serum, in vitro cell culture media, or in vivo post-implantation, it is referred to as having biodegradability. Decrease in structural durability, modifications to density that result in variations between micro- as well as macro-porosity, and shifts in the component's size and/or weight are only a few of the physico-chemical alterations that can happen in a substance undergoing biodegradation [20]. Because of an alteration in the amount of ions, these substance shifts could also be followed by an alteration in pH within the area just around the scaffolds. Throughout the course of creation, preservation, and usage, biomaterials can experience a variety of mechanical deterioration mechanisms [9]. Mechanical deterioration may occur in an actual setting as a result of shear stresses, tension, and/or compression. The mean molecular weight of any polymers decreases as a result of its mechanical deterioration. Additionally, primary chain breakage by stretching, grinding, milling, or any other sort of polymer shearing action results in the production of radicals that are free. When heated, these free radicals target the polymeric framework and trigger radical reorganization processes that trigger more disintegration reactions. The deterioration of biomaterials has been divided into three phases: quasi-stable, a reduction in weight and mechanical capabilities, and lastly the actual disintegration of the scaffold. In reality, when the scaffold breaks down, all of these stages may result in variations in cell movement (such as gene expression, proliferation, and metabolism of extracellular matrix components). Modifications in the biomaterial's surface features, which appear as variations in the energy of the surface, protein adsorption, and cell environment and a result of the discharge of byproducts, may be blamed for the impact left by deterioration on cells [9]. The persistence and degree of attachment of cells inside PLGA scaffolds are negatively correlated to the pace of deterioration, underscoring the value of comprehending the constantly changing relationship among extracellular matrix synthesis, proliferation of cells, and biomaterial breakdown [20]. It is crucial to meticulously create biomaterials with acceptable disintegration rates, stability, and biocompatibility in order to reduce potential hazards brought on by surface degradation, which can be seen in Figure 20.5. Optimizing the efficiency and security of biomaterials in medical devices may be achieved by comprehending the

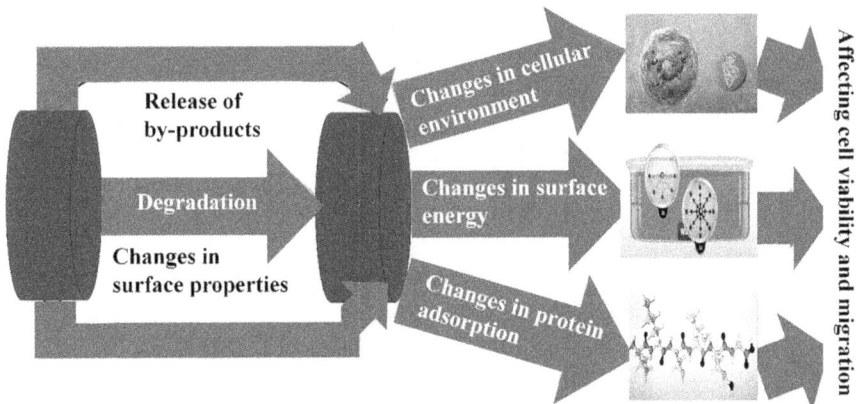

**FIGURE 20.5**   Illustration showing various degradation processes on biomaterial surface.

causes of degradation, choosing appropriate materials, and using surface enhancement methods. To be able to identify any symptoms of deterioration or related problems quickly, frequent monitoring and assessment of transplanted biomaterials is also essential.

### 20.1.6 DIFFERENT ROUTES FOR SURFACE MODIFICATION IN BIOMATERIALS

The surface qualities of biomaterials can be changed in a variety of ways to improve their biological compatibility, usefulness, or specific relationships with biological systems.

As shown in the orthopedic industry, certain individuals' anatomical intricacy or allergies may prevent conventional implants from fitting them properly. Customized implants with a certain level of intricacy or structures that are porous and appropriate for hard-tissue engineering can be created using the technique of additive manufacturing (AM) [21]. With AM of metallic biomaterials, the words "shape complexity" primarily refer to the exact manipulation of the interior form and microarchitecture of the produced objects to obtain a specified set of attributes, that can't be achieved using standard production techniques. The outermost layers of an implant must be correctly constructed to optimize its biological performance, such as protein adsorption, surface thrombogenicity, adhesion of cells, development, and differentiation, control of fibrous capsule development, etc. [22]. This is because the initial cell-surface relationships take place at the interface. The pathogen responsible adhesion to implant surfaces is typically influenced by three factors: (i) implant material features like surface topography, roughness of the surface, chemical composition of the surface, and surface energy; (ii) bacterial traits like the charge on the surface and surface wettability (hydrophilicity/hydrobicity); and (iii) host surroundings features like antibiotic type, temperature, pH, contact time, and concentration of bacteria. Protein adsorption as well as adhesion of cells, for example, are greatly influenced by the physico-chemical interface characteristics of metallic prostheses at the material-tissue contact [5]. Based on in vitro and in vivo study, a biomaterial's degree of roughness, size of pores, and amount of porosity play crucial roles in the development of bones. The best choice of these features leads to improved bone ingrowth [21]. Although extremely roughened surfaces might encourage bacterial adherence, excessive porosity, on the contrary, decreases mechanical characteristics. The development of innovative materials or the modification of the interfaces of surgical instruments so that potential biofilm pathogens aren't drawn to them was a useful technique for lowering the risk of bacterial adhesion. Important characteristics of antimicrobial interfaces for biomaterials include microbe-repelling surfaces, microbe-killing surfaces, anti-adhesive surfaces, including surfaces that release biocides. It is necessary to modify the surface's topography in order to stop microbial development. Various routes for surface modification are given in Figure 20.6.

## 20.2  HEAT TREATMENT OF BIOMATERIALS

The layer of oxide is stabilized through heat treatment. It's crucial to pick the right temperature and time for the heat treatment. In order to provide the opportunity for

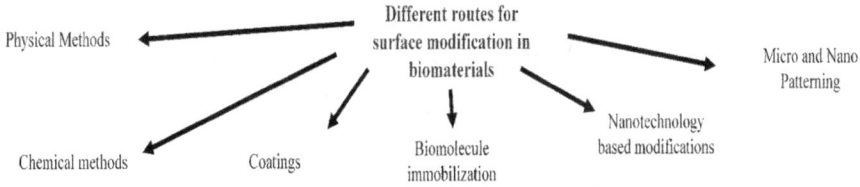

**FIGURE 20.6**   Different routes for surface modification in biomaterials.

the stabilization of the oxide layer, the heat-treatment temperature must be sufficient as well as the time prolonged enough. Let's use the anodizing procedure as an example for creating a uniform hierarchical oxide layer over the selectively laser-melted porous titanium structures [23]. Following that, their anodized porous alloys underwent heat treatment for one or two hours at 400, 500, or 600°C, depending upon the temperature. According to the heat treatment process's settings, the nanotubes' post-heat treatment architectures varied greatly. There's no significant breakdown of the nanotube levels occurred at 400°C for one or two hours. Once samples were subjected to heating at 500°C for one hour, the texture of the nanotube strata remained mostly unchanged. As the heat treatment lasted for two hours, the frequency of disruptive nanoparticles (white dots) rose [24]. The heat treatment at 600°C had a significant impact on the nanotube layers' internal structure. After one hour of heat treatment at 600°C, the oxide layer's morphology underwent substantial modifications, including the incomplete closure of nanotube and the emergence of novel crystal forms. Once the heat treatment went on for a further two hours, the nanotubes were virtually completely closed. After being heated at 600°C for two hours, anodized specimens showed significant micro-scale sized phases on their surface. Both the crystal structure as well as the shape of the sample are significantly influenced by the heat treatment procedure's variables [25]. Thus, it's crucial to maximize the heat treatment process's variables in order to maximize the bioactivity of the substances that arise. Different types of heat treatment processes are mentioned below.

## 20.2.1   SOLUTION HEAT TREATMENT

In order to gain an understanding of this heat treatment procedure, it is possible to study the impact of solution heat treatment temperatures (500, 750, and 1000°C) upon the phase transitions, microscopic structure, microhardness, and Young's modulus of Ti-25Ta-xZr alloys intended for biomedical usage [26]. For the purpose to create specimens with the $\alpha''$, $\beta + \alpha''$, and phases, the Ti-25Ta-xZr alloy bars were heated up in an arc furnace having five distinct percentages (x = 0, 10, 20, 30, and 40 wt%). The findings demonstrated that the examined alloys' microstructure and mechanical characteristics may be modified in accordance with the temperatures employed for solution within the Ti-25Ta-xZr setup. Hardness typically increases with greater solution heat treatment temperatures for single-phase alloys because to increased phase stabilization, but in mainly predominantly $\beta$ or $\alpha'' + \beta$ alloys, hardness falls because of inhibition of phase $\alpha''$. Yet, performing solution heat treatment at 1000°C leads the alloys' elastic modulus to decline. Generally speaking, solution heat treatment carried out at elevated temperatures greater effectively stabilises the phase, optimizing the alloys' reduced modulus. The procedure involved melting the billets in an arc furnace and hot rolling the specimens in a temperature of 1000°C to treat it [27]. Following this process, the cooling by water was then followed by six-hour thermal solution treatments at 500, 750, and 1000°C. Using a microdurometer, hardness was calculated. Five indentations were created in the specimens using a force of 0.245 N with a weight of 25 g for an interval of 60 s. An approach centred around impulse excitation is utilized for the calculation of elastic modulus.

## 20.2.2 ANNEALING

Thermal processes that go by the general label of "annealing" are often used to produce several properties. These are used on semi-finished goods made by the casting process, either cold or hot plastic deformation (free forging or deliberately in a mold, laminating, extruding, drawing), or welding (metal structures, machine components, and complicated tools). The IVA Schmetz GmbH, Menden, Germany's IU 72/1 F 2RV 60×60×40×10 bar CP type I vacuum furnace was used to carry out the heat treatment on the Ti-Mo-Zr-Ta alloys [23]. The outer shell houses the hot area, heat exchanger, high-capacity rotary fan featuring an electric motor, and gas pipes. A rail-guided loading vehicle is used to accurately put the goods in the area of greatest heat while the furnace is loaded and unloaded from the front. To avoid overpressure, the swiveling furnace lid will be hydraulically secured to the furnace case after loading. The treatment process then continues entirely on its own. The combustion chamber features a control panel that enables the person using it to schedule each phase of the treatment process while taking into account variables like temperature, cooling rate, gas pressure, ventilation, etc. Following securing the furnace's entrance, a vacuum was generated and kept for 16 minutes to ensure that no traces of residue or additional chemicals had been on the substances. A high-temperature quenching process was subsequently carried out in the following order: in a vacuum for 25 minutes at 650°C, 20 minutes at 850°C, and 20 minutes at 950°C to equalize the temperature within the samples' middle and core. N2 gas was used to carry out the process of cooling. This initially entered into the container at a 9-bar pressure for 37 minutes; following that, it was heated until it reached a tempering temperature of 550°C at 1.5 bar pressure and maintained for 2 hours and 10 minutes at a 2-bar pressure; finally, it was allowed to cool to ambient temperature at 1.5-bar pressure [28]. This heat treatment may be employed to complete excessive heated structures, lower the degree of residual tension created in the metallic weight of the products (stress relief annealing), recrystallize the cross grain after cold plastic deformation, or just alter the framework created by the casting process. It may also be employed to lower the degree of hardness of the metallic component in order to increase workability.

## 20.2.3 AGING

By adopting the right heat treatment procedures to convert metastable phases into the desired α phase, one may strengthen β-Ti alloys. Fast quenched β Ti alloys may be subjected to ageing heat treatment in order to avoid formation of α phase, which is more durable and rigid compared with orthorhombic martensite and β phase. But ageing of β Ti alloys may trigger the undesired ω phase to precipitate. The mechanical durability of an alloy is greatly increased by ω phase precipitation, yet at the cost of extreme brittleness [29]. Furthermore, the existence of ω phase in a β phase matrix could be crucial in functioning to serve as a nucleation substrate to facilitate the genesis of α phase, resulting in an even and homogeneous distribution of α phase in the β phase matrix. Alloy specimens Ti-5Mo, Ti-7.5Mo, Ti-10Mo, and Ti-15Mo had been aged for up to 100 hours at 523 K, 573 K, 623 K, and 723 K, respectively. After the heat treatments, the component phases were determined using the Vickers hardness and X-ray diffraction examinations. Considering the rise in ageing temperature, Sample 5 showed hardness peaks. Sample 5 aged at 523 K exhibited a barely perceptible rise in hardness until 40 min, that subsequently leveled out, after which the rise in hardness were far more evident after 16 h. Comparable behavior was observed in the specimen aged at 573 K, yet the initial hardness rise appeared following just 10 minutes and became more prominent after 2 hours. Forty hours of aging led to a reduction in hardness. The levels of hardness achieved was always greater with the aging times lower compared to those produced with ageing at 523 K, as predicted. The hardness vs. aging time graphs for the aging heat treatments at 623 K and 723 K show that hardness increases more quickly than predicted by the C-curve, which is supported by the data. Additionally, the maximum levels of hardness were achieved at greater ageing temperatures within a shorter amount of time. Following aging at 623 K, the reduced and better coherent hardness was reached, that, based on the X-ray diffraction data, was caused by the precipitation of the α phase. The hardness vs. aging time graph shows how the hardness started increasing after 40 min of aging at 623 K while overaging happened after 30 hours, although hardness levels must be carefully examined. On the contrary, at 573 K, the degree of hardness started to rise around 1 hour of aging, and overaging happened around 40 hours [30]. Once more, this supports C-curve behaviour. The maximum hardness level was produced by more severe precipitation that followed ageing at 723 K, and overaging starting 4 hours later. The initial temperature, on the contrary, shows greater consistency in values during 16 and 56 hours of aging and a modest drop in hardness beyond 80 hours, while the latter temperature clearly shows a propensity towards overaging beyond 40 hours. Yet, after 16 hours of heat treatment, a more severe precipitation that took place above an elevated temperature (623 K) resulted in greater hardness readings and overaging. The intensive ω phase precipitation within the first 5 minutes of aging was shown by the rapid hardening that occurred due to a consequence of ageing at 723 K. The results support the hypothesis because the main cause of hardness was ω phase precipitation, that was accompanied by α phase precipitation and began to significantly coarsen after 12 hours of aging. After 12 hours of heat treatment, the ω phase precipitates that served as the nucleation sites for the α phase appeared to be totally gone. Additionally, after 12 hours of aging, a noticeable loss in hardness was noted.

### 20.2.4 ALKALI HEAT TREATMENT

Alkali heat treatment (AHT) is presented to be a workable and affordable method to improve bioactivity and biocompatibility. This process can result in distinctive, interconnecting nano-pores upon the outermost layer of the material, which may serve as the sites of nucleation for Ca-P precipitation that eventually ensure the material's adhesion to bone tissues. To put it another way, the permeable nanostructured surface promotes the interchange of nutrients and oxygen up the cell metabolism while offering the proper environment for growth of bones and osseointegration [31]. The process of protein attachment, adhesion of cells, and proliferation can all be enhanced by the porous nanostructured material surface. In order to control adhesion of cells and development, it is important to improve surface qualities by AHT, like boosting surface hydrophilicity and creating distinctive nano-topography. Ti-27Nb alloy was alkali-treated (soaked in 5 M NaOH solution for 24 hours at 60°C, then annealed for 1 hour at 600°C) to produce a nanostructured sodium titanate ceramic surface that aided in the nucleation and development of the Ca-P phase. AHT may be used to create nanostructured ceramic titanate on a Ti substrate, and in vitro research has shown that this process can promote surface bioactivity and cytocompatibility. The alkali heat-treated Ti implantation displayed the bioactive induction, according to in vivo experiments. Alkali treatment was employed to fabricate the Ti-6Al-4V alloy with a nanostructure at ideal temperature circumstances. Their findings show that alkali treatment at 30°C leads in a thin, non-crackable nano-porous layer. They discovered that the use of alkali heat treatment significantly improved the adherence of bone marrow mesenchymal stem cells (BMMSCs) to the outermost layer of the Ti-6Al-4V alloy [32]. Despite being effectively demonstrated both in vitro and in vivo, bioactive stimulation of alkali heat treatment requires a lengthy response time (approximately three days). Thus, the creation of simple, inexpensive methods that need fewer computations time is required. Lately, an innovative fast AHT method was developed by accelerating the nucleation and development of nanostructured titanate in alkali solution by employing a fairly high reaction pressure as an accelerator. Their findings demonstrated that the surface that was treated exhibited a nanostructure resembling flaking with considerably increased biocompatibility.

### 20.2.5 STRESS-RELIEVING HEAT TREATMENT

Ti6Al4V(ELI)-3Cu (SR)'s SEM-EDX examinations reveal that Cu is effectively dispersed in the matrix, despite the finding of some Cu-enriched regions that were linked to the melt pool border. Following HT, $\alpha'$ changes into $\alpha + \beta$, and V and Cu atoms are ejected to borders. In stress-relieving materials, CuTi2 intermetallic phase precipitation showed up and this was improved following heat treatment (HT) [33]. Following SR, it was possible to see thin martensite laths (less than 1.5 μm), and also submicrometric precipitates that appeared both along and within $\alpha'$ boundaries. The round and oblong precipitates had a uniform dispersion inside the matrix. This circular CuTi2 intermetallic precipitates was recognized, whereas the elongated V-enriched precipitates belonged to the $\beta$ phase. They measured nearly 200 nm in length and less than 100 nm in diameter. The internal structure changed as a result of the HT. The formation of fresh $\alpha$, CuTi2 intermetallic, and sustained $\beta$ occurred following cooling post-heating to the $\beta$ area (over 820°C). As a consequence, the microstructure of the $\alpha$ phase laths reached a dimension of above 4–5 µm, and Cu- and V-rich precipitates formed between the $\alpha$ laths. Following HT, $\beta$ as well as CuTi2 intermetallic precipitates became entirely connected in comparison to the SR circumstance, while the morphology of the Cu precipitates altered into an extended form. $\beta$ and CuTi2 intermetallic precipitates grew from a diameter of less than 100 nm up to a length of 1.2 µm along with a length increase of 200 nm up to 1.5 µm, correspondingly [34]. Precipitates were found mainly among $\alpha$ laths following the HT. To reduce tensions along to start phase changes in the substance, the SR and HT post-treatments were applied. When Ti6Al4V is cooled rapidly (such as when it is as-printed), martensitic transformation occurs, while the $\alpha$ phase is the equilibrium phase that develops following HT or SR and gradual cooling.

## 20.3 SUMMARY

The chapter starts with a brief introduction of the biomaterials. An overall idea about what are biomaterials and their potential application is studied. Biocompatibility is one of the important parameters while working with biomaterials. It refers to their ability to interact harmoniously with biological systems, minimizing adverse reactions and promoting successful integration. Some international standards are also mentioned in the chapter. Different types of biomaterials present and their uses are listed. Further, what a biomaterial surface has and the surface properties of the biomaterials are discussed. As we know if a surface is present, then its degradation may happen due to various reasons. Some of the reasons are studied. To overcome the degradation, we apply different surface modification techniques. A few of the surface modification routes are also mentioned here. Finally, in the latter part, the heat treatment done in biomaterials is discussed. We saw that heat treatment is necessary and the different heat treatment processes employed are discussed in a detailed manner. All the processes are studied one by one by using various examples. Some of the major heat treatment processes used in biomaterials are solution heat treatment, annealing, aging, alkali heat treatment, and stress-relieving heat treatment. Therefore, we can conclude that heat treatment can be used to modify the surface properties of biomaterials, such as their biocompatibility, bioactivity, and mechanical properties.

## REFERENCES

1. V. Migonney, "History of Biomaterials," in *Biomaterials*, Hoboken, NJ, USA: John Wiley & Sons, Inc., 2014, pp. 1–10. doi: 10.1002/9781119043553.ch1.
2. A. Pandey, R. K. Sharma, and K. Balani, "Introduction to Biomaterials," in *Biosurfaces*, Hoboken, NJ, USA: John Wiley & Sons, Inc, 2015, pp. 1–64. doi: 10.1002/9781118950623.ch1.
3. G. Khang, S. J. Lee, M. S. Kim, and H. B. Lee, "Biomaterials: Tissue Engineering and Scaffolds," in *Encyclopedia of Medical Devices and Instrumentation*, Hoboken, NJ, USA: John Wiley & Sons, Inc., 2006. doi: 10.1002/0471732877.emd029.

4. S. Sankaran, S. Zhao, C. Muth, J. Paez, and A. del Campo, "Toward light-regulated living biomaterials," *Adv. Sci.*, vol. 5, no. 8, p. 1800383, Aug. 2018, doi: 10.1002/advs.201800383.

5. P. X. Ma, T. W. Eyster, and Y. Doleyres, "Tissue Engineering Biomaterials," in *Encyclopedia of Polymer Science and Technology*, Wiley, 2018, pp. 1–47. doi: 10.1002/0471440264.pst471.pub2.

6. B. D. Ratner, "New ideas in biomaterials science—a path to engineered biomaterials," *J. Biomed. Mater. Res.*, vol. 27, no. 7, pp. 837–850, Jul. 1993, doi: 10.1002/jbm.820270702.

7. H. Chen, X. Li, and Y. Du, "1D~3D Nano-engineered Biomaterials for Biomedical Applications," in *Integrated Biomaterials for Biomedical Technology*, Hoboken, NJ, USA: John Wiley & Sons, Inc., 2012, pp. 1–33. doi: 10.1002/9781118482513.ch1.

8. C. A. Custódio, R. L. Reis, and J. F. Mano, "Engineering biomolecular microenvironments for cell instructive biomaterials," *Adv. Healthc. Mater.*, vol. 3, no. 6, pp. 797–810, Jun. 2014, doi: 10.1002/adhm.201300603.

9. M. Niinomi *et al.*, "Development of low rigidity β-type titanium alloy for biomedical applications," *Mater. Trans.*, vol. 43, no. 12, pp. 2970–2977, 2002, doi: 10.2320/matertrans.43.2970.

10. O. Modi, D. Mondal, B. Prasad, M. Singh, and H. Khaira, "Abrasive wear behaviour of a high carbon steel: Effects of microstructure and experimental parameters and correlation with mechanical properties," *Mater. Sci. Eng. A*, vol. 343, no. 1–2, pp. 235–242, Feb. 2003, doi: 10.1016/S0921-5093(02)00384-2.

11. H.-I. Kim, M.-W. Han, S.-H. Song, and S.-H. Ahn, "Soft morphing hand driven by SMA tendon wire," *Compos. Part B Eng.*, vol. 105, pp. 138–148, Nov. 2016, doi: 10.1016/j.compositesb.2016.09.004.

12. L. Le Guéhennec, A. Soueidan, P. Layrolle, and Y. Amouriq, "Surface treatments of titanium dental implants for rapid osseointegration," *Dent. Mater.*, vol. 23, no. 7, pp. 844–854, Jul. 2007, doi: 10.1016/j.dental.2006.06.025.

13. M. Honda *et al.*, "In vitro and in vivo antimicrobial properties of silver-containing hydroxyapatite prepared via ultrasonic spray pyrolysis route," *Mater. Sci. Eng. C*, vol. 33, no. 8, pp. 5008–5018, Dec. 2013, doi: 10.1016/j.msec.2013.08.026.

14. Y. Sato and Y. Guo, "Shape-memory-alloys enabled actuatable fiber sensors via the preform-to-fiber fabrication," *ACS Appl. Eng. Mater.*, vol. 1, no. 2, pp. 822–831, Feb. 2023, doi: 10.1021/acsaenm.2c00226.

15. S. Ramesh, C. Y. Tan, I. Sopyan, M. Hamdi, and W. D. Teng, "Consolidation of nanocrystalline hydroxyapatite powder," *Sci. Technol. Adv. Mater.*, vol. 8, no. 1–2, pp. 124–130, Jan. 2007, doi: 10.1016/j.stam.2006.11.002.

16. H. Amani *et al.*, "Controlling cell behavior through the design of biomaterial surfaces: A focus on surface modification techniques," *Adv. Mater. Interfaces*, vol. 6, no. 13, p. 1900572, Jul. 2019, doi: 10.1002/admi.201900572.

17. M. J. Webber, "Engineering responsive supramolecular biomaterials: Toward smart therapeutics," *Bioeng. Transl. Med.*, vol. 1, no. 3, pp. 252–266, Sep. 2016, doi: 10.1002/btm2.10031.

18. K. D. Jandt, "Evolutions, revolutions and trends in biomaterials science – A Perspective," *Adv. Eng. Mater.*, vol. 9, no. 12, pp. 1035–1050, Dec. 2007, doi: 10.1002/adem.200700284.

19. A. Afshar, M. Ghorbani, N. Ehsani, M. Saeri, and C. Sorrell, "Some important factors in the wet precipitation process of hydroxyapatite," *Mater. Des.*, vol. 24, no. 3, pp. 197–202, May 2003, doi: 10.1016/S0261-3069(03)00003-7.

20. H. Denis *et al.*, "Antibacterial properties and abrasion-stability: Development of a novel silver-compound material for orthodontic bracket application," *J. Orofac. Orthop. / Fortschr. Kieferorthop.*, pp. 1–13, Jul. 2022, doi: 10.1007/s00056-022-00405-7.

21. S. Jin and K. Ye, "Nanoparticle-mediated drug delivery and gene therapy," *Biotechnol. Prog.*, vol. 23, no. 1, pp. 32–41, Feb. 2007, doi: 10.1021/bp060348j.

22. I. B., S. K., Soyoung Yang, and Sujeong Lee, "Hydrothermal treatment of Ti surface to enhance the formation of low crystalline hydroxyl carbonate apatite," *Biomater. Res.*, 2015.

23. M. S. Baltatu, C. Chiriac-Moruzzi, P. Vizureanu, L. Tóth, and J. Novák, "Effect of heat treatment on some titanium alloys used as biomaterials," *Appl. Sci.*, vol. 12, no. 21, p. 11241, Nov. 2022, doi: 10.3390/app122111241.

24. M. Tirrell, E. Kokkoli, and M. Biesalski, "The role of surface science in bioengineered materials," *Surf. Sci.*, vol. 500, no. 1–3, pp. 61–83, Mar. 2002, doi: 10.1016/S0039-6028(01)01548-5.

25. B.-H. Lee, Y. Do Kim, J. H. Shin, and K. Hwan Lee, "Surface modification by alkali and heat treatments in titanium alloys," *J. Biomed. Mater. Res.*, vol. 61, no. 3, pp. 466–473, Sep. 2002, doi: 10.1002/jbm.10190.

26. A. R. Luz, L. S. Santos, C. M. Lepienski, P. B. Kuroda, and N. K. Kuromoto, "Characterization of the morphology, structure and wettability of phase dependent lamellar and nanotube oxides on anodized Ti-10Nb alloy," *Appl. Surf. Sci.*, vol. 448, pp. 30–40, Aug. 2018, doi: 10.1016/j.apsusc.2018.04.079.

27. P. A. B. Kuroda, F. de Freitas Quadros, K. dos S. J. Sousa, T. A. G. Donato, R. O. de Araújo, and C. R. Grandini, "Preparation, structural, microstructural, mechanical and cytotoxic characterization of as-cast Ti-25Ta-Zr alloys," *J. Mater. Sci. Mater. Med.*, vol. 31, no. 2, p. 19, Feb. 2020, doi: 10.1007/s10856-019-6350-7.

28. G. K. Padhy, C. S. Wu, and S. Gao, "Friction stir based welding and processing technologies - processes, parameters, microstructures and applications: A review," *J. Mater. Sci. Technol.*, vol. 34, no. 1, pp. 1–38, Jan. 2018, doi: 10.1016/j.jmst.201 7.11.029.

29. D. R. Unune, G. R. Brown, and G. C. Reilly, "Thermal based surface modification techniques for enhancing the corrosion and wear resistance of metallic implants: A review," *Vacuum*, vol. 203, p. 111298, Sep. 2022, doi: 10.1016/j.vacuum.2022 .111298.

30. R. A. Gittens, R. Olivares-Navarrete, R. Tannenbaum, B. D. Boyan, and Z. Schwartz, "Electrical implications of corrosion for osseointegration of titanium implants," *J. Dent. Res.*, vol. 90, no. 12, pp. 1389–1397, Dec. 2011, doi: 10.1177/00220345114 08428.

31. F. Shahriyari, A. Razaghian, R. Taghiabadi, A. Peirovi, and A. Amini, "Effect of friction hardening pre-treatment on increasing cytocompatibility of alkali heat-treated Ti-6Al-4V alloy," *Surf. Coatings Technol.*, vol. 353, pp. 148–157, Nov. 2018, doi: 10.1016/j.surfcoat.2018.08.051.

32. Y. Zhou *et al.*, "Alkali-heat treatment of a low modulus biomedical Ti–27Nb alloy," *Biomed. Mater.*, vol. 4, no. 4, p. 044108, Aug. 2009, doi: 10.1088/1748-6041/4/4/ 044108.

33. A. Martín Vilardell *et al.*, "Effect of heat treatment on osteoblast performance and bactericidal behavior of Ti6Al4V(ELI)-3at.%Cu fabricated by laser powder bed fusion," *J. Funct. Biomater.*, vol. 14, no. 2, p. 63, Jan. 2023, doi: 10.3390/jfb14020063.

34. Y. Zhuang, L. Ren, S. Zhang, X. Wei, K. Yang, and K. Dai, "Antibacterial effect of a copper-containing titanium alloy against implant-associated infection induced by methicillin-resistant Staphylococcus aureus," *Acta Biomater.*, vol. 119, pp. 472–484, Jan. 2021, doi: 10.1016/j.actbio.2020.10.026.

# 21 Surface Modification of Magnetic Nanoparticles for Biomaterials

*Malihe Pooresmaeil*
Polymer Research Laboratory, Department of Organic and Biochemistry, Faculty of Chemistry, University of Tabriz, Tabriz, Iran

*Hassan Namazi*
Polymer Research Laboratory, Department of Organic and Biochemistry, Faculty of Chemistry, University of Tabriz, Tabriz, Iran

Research Center for Pharmaceutical Nanotechnology, Biomedicine Institute, Tabriz University of Medical Science, Tabriz, Iran

## 21.1 INTRODUCTION

Today, the development of nanotechnology for the reduction of human life problems has become an important subject. In this context, magnetic NPs with distinct properties have received enormous attention. Magnetic NPs commonly are a combination of magnetically responsive materials with magnetic features in the presence of an external magnetic field. This type of NPs commonly consists of metallic (Co, Fe, Mn, Ni,), bimetallic (FePt, CoPt$_3$, FeZn), and their oxides. Due to the unwanted toxicity of the magnetic NPs containing nickel, cobalt, and manganese, in recent decades high attention has been devoted to iron oxide NPs like MNPs (Fe$_3$O$_4$) [1]. So, among the numerous magnetic nanomaterials, magnetite displays good paramagnetic behavior and magnetic susceptibility [2,3]. On the other hand, MNPs are favorable agents for a variety of applications because of their superior features like low toxicity, easy recovery after usage, high level of accumulation in tissues, suitable function, easy modification, high specific surface area, antibacterial activity, near-infrared (NIR) absorbance, and controlled transporting that distinguishes them from other nanomaterials [4–11]. So, these considerably widened their uses in numerous areas for instance catalysts, pigments, recording devices, environmental and pharmaceutical biology, and materials science [12,13]. Despite these advantages, the bare MNPs display low colloidal stability with a high tendency to aggregation due to dipole-dipole interactions, upscalability, low stability in acid media, and

DOI: 10.1201/9781003429920-24

oxidization in the air atmosphere, which leads to the restraining of their functionality [14–16]. To overcome these, modifications are essential to avoid the variation in their physico-chemical features and achieve improved features. Different materials have been used for MNPs modification, in which modifications with polymers are ideal candidates. Up to now, different types of polymers have been utilized for this aimpoly(vinyl pyrrolidone) (PVP), poly(vinyl alcohol) (PVA), poly(ethylene glycol) (PEG), and biopolymers are some of them [6,17–20]. In comparison, synthetic polymers suffer from environmentally polluting because of their stability. For these reasons, the researchers turn their attention to using natural polymers. The natural polymers family is composed of proteins (soy, caffeine, etc.), polysaccharides (cellulose, CS, St, chitin, Alg, κ-Car, etc.), and hydrocarbon elastomers (natural rubber) sub-families [21,22]. In between, polysaccharides, cellulose, CS, St, Pec, κ-Car, and Alg are hopeful materials to improve the biological features of MNPs because of upgraded safety with higher stability, easy availability from natural products, reduced toxicity, non-immunogenicity, eco-friendly, and good physico-chemical properties, as well as they have numerous adsorption sites for strong and proper interaction with metal ions [23]. In some cases, MNPs are modified with two or more polysaccharides to obtain a new system with positive features of all components. Therefore the modification of MNPs with biopolymeric material is one of the main areas of intense research that leads to stabilization by electrostatic and steric repulsions and helps to construct a new product that can be tailored for certain applications [24]. It should be noted that the modification of MNPs with polysaccharides can be performed in three different ways: (I) in situ, (II) blending, and (III) grafting-onto. In the in situ way, the synthesis of MNPs occurs in the existence of the biopolymer. While in the blending and grafting onto techniques, respectively, the mixing of pre-synthesized biopolymer with MNPs and chemical grafting of biopolymer on MNPs were done. Considering all of these, the current chapter reports the important and basic principles of $Fe_3O_4$ modified systems along with their usage in different areas. It is required to note that due to the wide range of research in this field, only some cases as the examples have been selected and studied as samples. Therefore, we expect that because of the ongoing capability of $Fe_3O_4$ NPs in several areas, this chapter is promising for a researcher that works in this field to develop new $Fe_3O_4$ NP-based systems with improved properties.

## 21.2   $Fe_3O_4$NPs MODIFICATION WITH POLYSACCHARIDES

### 21.2.1   $Fe_3O_4$NPs MODIFIED WITH CELLULOSE OR ITS DERIVATIVES

Cellulose is one of the more abundant biomasses on the planet that can be extracted from plants, trees, algae, and animals [25]. This biopolymer is organized at a macromolecular level within fibrils having glucose units in a crystalline and linear arrangement, together with lignin and hemicellulose. The features of the extracted cellulose fibrils can be affected by the isolation process type, typically physical or chemical methods [26,27]. Owing to its biodegradability, low cost, mechanical strength, hydrophilicity, eco-friendly, renewable, modifiable, and nontoxic nature, as well as abundant resources, this polysaccharide is a good candidate for various applications. Cellulose has two distinct zones: the crystalline part and the amorphous part in which the chemical reactions typically do not occur on the crystalline area [3,28]. This polysaccharide consists of hundreds of linear chains of β-(1,4) linked D-glucose units as a $(C_6H_{10}O_5)_n$ formula [29]. The molecular chains of this biopolymer are composed of inter- and intra-molecular hydrogen bond networks, where the higher hydrogen basicity of the solvents couldfail the hydrogen bonds of the cellulose, leading to the cellulose dissolution [30]. Furthermore, both free hydroxyl groups and a large surface area of cellulose can act as ligands and also reducing agents for the synthesis and stabilization of NPs. This biopolymer also simplifies the metal NPs dispersion and avoids their aggregating [29,31]. The extracted cellulose from different sources could be converted to nanocellulose (NC) by employing several techniques like acidic hydrolyzingor mechanical treatment [28]. In comparison to cellulose, NC has received more attention because of its higher surface area and unique properties. This polymer also can be used for MNP modification so the encapsulation to them within the polymer matrix can improve the biocompatibility, colloidal stability, surface-to-volume ratio, etc., and produce the system with the combination of uniqueness and peculiarities of both materials [3,32].

### 21.2.1.1 MNPs/Cellulose Systems Prepared via In Situ Way

In recent years, a widespread range of active compounds was designed and fabricated through the in situ synthesis of $Fe_3O_4$ NPs in the presence of cellulose to obtain a new system having special features. For example, in one systematic effort, the microbial cellulose/$Fe_3O_4$ (MMC/$Fe_3O_4$) composite was designed as a three-dimensional electro-Fenton system (3D-EF) and then its catalytic activity was studied for tetracycline (TC) degradation and mineralization. The findings of this research displayed that the complete degradation and 65% mineralization of TC occurred within 20 min in the 3D-EF process (0.5 g $L^{-1}$ MMC/$Fe_3O_4$, neutral pH, and 10 mM NaCl electrolyte). The high degradation efficiency of TC was attributed to notable superoxide ($O_2^{\bullet-}$), single oxygen ($^1O_2$) participation, and less hydroxyl radical (OH•). Based on these, the authors proposed that this hopeful technology can open a new viewpoint on the nonradical paths of degradation and mineralization of different pollutants [33]. A hybrid of cellulose nanofibrils and magnetite was also prepared using in situ way for the degradation of methylene blue (MB) dye. Structure analysis results displayed the good dispersion of the magnetite in the mat of cellulosic nanofibrils. In the constructed Cel:Mag, the present Fe ions in magnetite catalyzed the generation of hydroxyl radicals from $H_2O_2$ and degraded the MB dye. So, degradation kinetics of $H_2O_2$ and MB display 90% discoloration and complete (100%) within 180 min with the magnetite and Cel:Mag hybrid, respectively [26]. In another research, the in situ way was used for the construction of the core-shell system by employing the superparamagnetic MNPs with pristine nano-cellulose and also its modified form. After detailed characterization, the systems were evaluated as an adsorbent for radioactive ions removal. As a consequence, the outcome showed that the nano magnetite coating with modified nano-cellulose especially nano-cellulose citrate improved the adsorption process and achieved the best results [28]. Considering the favorability of the new functional group's insertion in the composites' structure for enhancing their applicability, MNPs were in-situ synthesized in the cellulose vicinity and thereafter used as a platform for grafting of itaconic acid and the P(MB-IA)-g-MNCC was obtained using potassium persulfate ($K_2S_2O_8$) as free radical initiator and ethylene glycol dimethacrylate (EGDMA) as a cross-linker. Then the system surface was functionalized with 2-mercaptobenzamide and the sulfhydryl and carboxyl functional groups were decorated on the system. Figure 21.1a displays the adopted procedure for the P(MB-IA)-g-MNCC preparation. The change of pH pzc from 6.9 for MNCC to 6.1 for P(MB-IA)-g-MNCC (Figure 21.1b) indicated the success in the occurred reaction, which led to a more negative surface and can favor the adsorption of Cd(II) onto the P(MB-IA)-g-MNCC via the electrostatic interaction. Overall, based on the findings, it was detected that the P(MB-IA)-gMNCC with regeneration capability is an effective adsorbent for the Cd(II) removal from wastewater [3]. It is approved that the magnetically actuated liquid crystals are probable alternatives for the usual liquid crystal systems owing to their benefits of electrodeless operation, low cost, and remote control. Despite these, their real use faces a major challenge, specifically the strong optical absorption of the magnetic components. To remove this, in one interesting and beneficial research, the in situ way was utilized for the fabrication of magnetic cellulose microcrystals with tunable magneto-optical responses. It was approved that the introduction of the

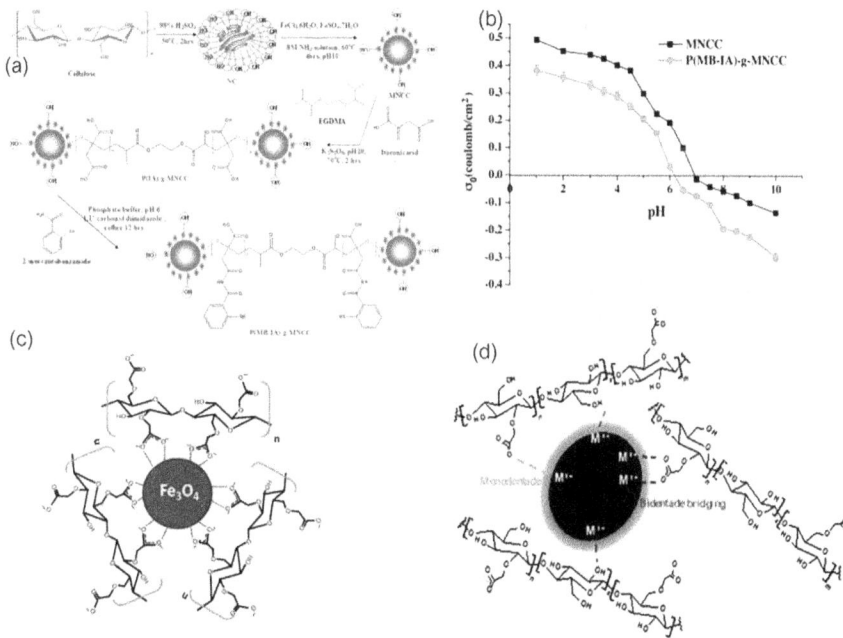

**FIGURE 21.1** The fabrication pathway and surface charge density of MNCC and P(MB-IA)-g-MNCC (a and b), schematic of CMC adsorption on magnetite interface (c), and the NPs/CMC interactions (d). Reproduced with permission from [2,35].

rod-shaped cellulose microcrystals in the structure offered the advantage of the birefringence and optical transparency of the cellulose microcrystals, also their anisotropic shape which allows their effective orientational alignment once their surfaces are functionalized by MNPs. Altogether, the results of current work expose the vast capability of the magnetic assembly approach and make it a hopeful applicant for anti-counterfeiting usages [34]. Cellulose derivatives are the other samples in which MNPs can be in situ synthesized in their vicinity. Up to now, several derivatives of cellulose were prepared and used in different areas. Carboxymethyl cellulose (CMC) is a water-soluble derivative of cellulose that in its structure some of the hydroxyl groups on glucopyranose unit are substituted by carboxymethyl groups. Therefore with the aim of evaluating the role of CMC polymer on the particle size of synthesized MNPs, CMC capped magnetite NPs were fabricated via oxidation-precipitation technique in different reaction temperatures and then utilized for MB removal. A schematic diagram of CMC adsorbed on a magnetite interface is presented in Figure 21.1c. The interaction of carboxylate of CMC with magnetite was specified as the bidentate chelating via Fourier-transform infrared (FT-IR) spectroscopy. It was detected that the addition of the CMC at the nucleation stage led to the control of the particle size through the controlling of the steric hindrance. Sorption studies displayed that the smaller particles presented higher adsorption of MB because of the high surface area [2]. Magnetic iron oxide NPs (MIONPs) were also in situ synthesized in the presence of CMC in core-shell nanohybrids. Additionally, cobalt doped ($Co_xFe_{3-x}O4$, Co-MION) NPs were also in situ synthesized. In the constructed structure, simultaneously, the CMC acted as a functional biocompatible organic coating and also stabilizing ligand. It was proposed that two different types of coordination, bidentate and monodentate bridging, are possible interactions between the CMC and $Fe_3O_4$ or $CoxFe_{3-x}O_4$ NPs, as illustrated in Figure 21.1d. The findings verified the significant influence of carboxylate groups concentration and CMC molecular weight on the zeta potential, hydrodynamic dimension, and produced heat via the magnetic hyperthermia of magnetic iron oxide nanoconjugates. Moreover, the in vitro hyperthermia outcomes demonstrated the outstanding killing activity of the system toward brain cancer cells (U87 cells) as a result of incubation with MION-CMC and the use of an alternating magnetic field [35].

### 21.2.1.2  MNPs/Cellulose Systems Prepared via Blending

Taking advantage of the cellulose in the preparation of polymeric nanocomposites, R. Correa et al. mixed the dispersion of $Fe_3O_4$ NPs in water with viscose (9%, w/w of cellulose in the viscose). The collection was then dispersed in kerosene and oleic acid while stirring at room temperature. A study of the acquired SEM images approved the composite formation via the cross-linking of the cellulose with MNPs [12]. A nanocomposite of $Fe_3O_4$ and bacterial cellulose (BC) was also developed in 2023 via the ex-situ technique. The change of the polymer color from white to dark brown approved the incorporation of the magnetic particles into the BC structure (Figure 21.2a). Next to verification of its successful production via the analytical characterization, its catalytic potential was explored in different textile matrices treatment. The findings confirmed that the use of the BC/$Fe_3O_4$ composite as an iron

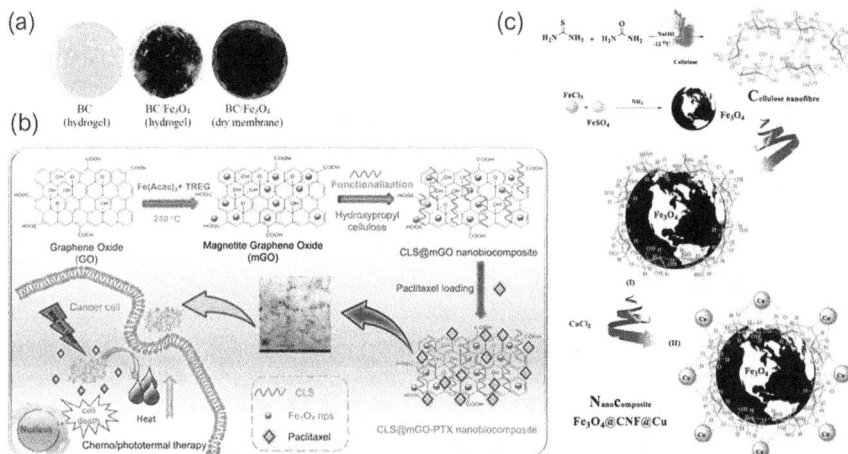

**FIGURE 21.2** Digital photo of the BC membrane and the BC/Fe$_3$O$_4$ nanocomposite (a), Fe$_3$O$_4$@CNF@Cu magnetic nanocomposite preparation procedure (b), and schematic of the CLS@mGO-PTX nanobiocomposite fabrication and application (c). Reproduced with permission from [29,32,37].

catalyst has a high efficiency for photocatalytic degradation by the advanced oxidation process of textile dyes, as well as because of the power of reuse and high catalytic activity will reduce treatment costs [32]. Cellulose also can be used for the modification of MNPs to obtain a new system with more applicability in the catalyst area. Taking this, the Fe$_3$O$_4$@CNF@Cu magnetic nanocomposite was developed via a step-wise immobilization of Cu NPs on cellulose-modified magnetic NPs. The general process for the synthesis of Fe$_3$O$_4$@CNF@Cu magnetic nanocomposite is presented in Figure 21.2b. Evaluation of the system's performance as a catalyst for the synthesis of Ullmann and Sonogashira cross-coupling reactions in a water mixture as environmentally benign and green reaction media approve the functionality of the system as a catalyst [29].

Published works evaluations show that the composites of cellulose and MNPs prepared via blending were also abundantly explored from the water treatment viewpoint. Therefore, with the aim of exploring the opportunity of cellulose-based adsorbents for effective removal of Pb$^{2+}$ from water, Luo et al. designed and prepared the millimeter-scale magnetic cellulose-based nanocomposite beads via an optimal extrusion dropping way by blending carboxyl-functionalized magnetite NPs with cellulose and acid-activated bentonite in a NaOH/urea aqueous solution. The performed thermodynamic study after the optimization specified that the Pb$^{2+}$ adsorption procedures were spontaneous, feasible, endothermic, and mostly controlled by chemical mechanisms [36]. In another similar example, the high-speed mechanical mixing of pre-synthesized nanomagnetite, oleic acid (OA), and NC aqueous solutions combined with a freeze-drying technique obtained a three-dimensional hierarchical aerogels network structure with a low-density NCA/OA/Fe$_3$O$_4$. The fabricated aerogel presented the capability of oil and organic solvents like cyclohexane, vacuum pump oil, and ethyl acetate adsorption removal [25].

Functionalization of cellulose with other small molecules or polymers obtains a new structure with exceptional features which can be used for MNPs modification. In this regard, in 2019, the pre-synthesized MNPs with thermal decomposition were modified with hydrophilic cellulose nanocrystals-poly citric acid (CNC-PCA/$Fe_3O_4$) and afterward evaluated as MRI contrast agents. The achievement verified the important improvement influence of nanocomposite as a dual negative and positive contrast agent. Furthermore, the applicability of CNC-PCA/$Fe_3O_4$ in biomedicine was approved through in vitro cellular uptake study, in vitro cytotoxicity, and colloidal stability investigation results [16]. Hydroxypropyl cellulose (CLS) as another water-soluble derivative of cellulose was used for the modification of the magnetite graphene oxide via the mixing method (CLS@mGO). Typically the mGO nanostructure was constructed through the in situ synthesis of the $Fe_3O_4$ NPs onto GO and then the modification of mGO with CLS occurred. In the designed structure, the CLS could interact with mGO through the hydrogen bonding interactions to stabilize the nanomaterial. Schematic of the CLS@mGO-PTX nanobiocomposite fabrication and its application for anticancer paclitaxel (PTX) delivery are shown in Figure 21.2c. The findings showed that the functionalization of the mGO surface with CLS not only improved colloidal stability, biocompatibility, and PTX entrapment efficiency, but also enhanced the PTX release rate [37].

### 21.2.1.3 MNPs/Cellulose Systems Prepared via Grafting-onto

Polymer-functionalized MNPs in their pristine or hybrid form with other NPs extensively explored in several fields. For example, the MNPs can be combined with graphene oxide (GO) to obtain new materials with high surface area and easy magnetic separation; in this regard, Niakan *et al.* fabricated and developed the cellulose-modified magnetite-graphene oxide nanocomposite through the click reaction. Afterward, the uniform and high stable dispersion of Pd NPs was obtained on nanocomposites through the abundant -OH functional groups of cellulose. The schematic of the performed synthesis reactions is presented in Figure 21.3a. The study of the chemical composition of the prepared catalyst via energy dispersive X-ray (EDX) spectroscopy revealed the important peaks consistent with C, N, O, Si, Fe, and Pd elements in the EDX spectrum of GO-$Fe_3O_4$-Cellulose-Pd. Furthermore, EDX elemental mapping images of GO-$Fe_3O_4$-Cellulose-Pd exposed that all of the elements were homogeneously distributed on the structure. The constructed nano-composite was explored as a heterogeneous catalyst for the Sonogashira and Heck coupling reactions [31]. For avoiding prolonging this section, the outcomes of the other published research works in this area are listed in Table 21.1.

### 21.2.2 $Fe_3O_4$NPs MODIFIED WITH CS OR ITS DERIVATIVES

Next to cellulose, CS is the most extensively existing biopolymer in the world. This polysaccharide is usually attained from the deacetylation of chitin as the major component of crustacean shells, mainly shells of crabs, krill, prawns, and shrimp [38,39]. CS is a linear polyamino saccharide consisting of randomly distributed β-(1-4)-linked D-glucosamine and N-acetyl-D-glucosamine. Pristine CS has two hydroxyl groups (–OH) and also one amino group (–$NH_2$) per glucosamine unit [7].

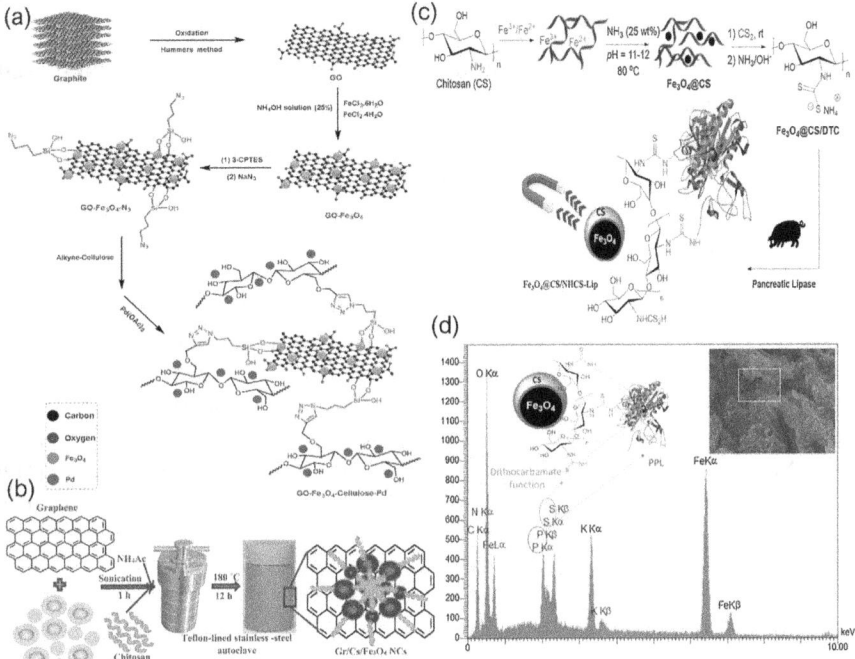

**FIGURE 21.3** GO-Fe$_3$O$_4$-Cellulose-Pd (a) CS/Fe$_3$O$_4$ and Gr/Fe$_3$O$_4$/Cs NCs (b), Fe$_3$O$_4$@CS/NHCSLip(c) synthesis procedure and EDX spectrum of Fe$_3$O$_4$@CS/NHCSLip (d). Reproduced with permission from [31,43,48].

The amino and hydroxyl groups in the backbone of CS can act as active sites to qualify it as a hopeful candidate for various applications. As well, these abundant functional groups can serve as reaction sites for useful modification or functionalization to obtain a new high-performance construct [40]. So, this polymer is one of the most extensively used natural polymers for the stabilization of MNPs because of its excellent biocompatibility, non-toxicity, biodegradability, and antibacterial activity. Because of these features, CS is one of the most significant bio-polymeric materials used in textiles, biomedicine, paper, food industries, etc. [7,41,42].

### 21.2.2.1  MNPs/CS Systems Prepared via In Situ

The multifunctional systems combined with IONPs and biopolymers like CS have been interesting, owing to their potential to be used in several areas. By considering these, in one interesting research work, an easy hydrothermal way was utilized for the construction of multifunctional CS/Fe$_3$O$_4$ and Gr/CS/Fe$_3$O$_4$ nanocomposites in which the Fe$_3$O$_4$ NPs were synthesized in the polymer platform. A schematic illustration of the one-step synthesis is provided in Figure 21.3b. The uniform appearance of the sphere-shaped Fe$_3$O$_4$ on CS and Gr surfaces approved its successful preparation. The fabricated composite exhibited acceptable photocatalytic efficiency in a short time toward rhodamine B, bromothymol blue, MB, and

**TABLE 21.1**

**Summary of Studies on a Collection of Polysaccharide-modified MNP-based Systems for Different Applications**

| Short Description of the Construct | Type of Polysaccharide | Studied Application Area | Ref. |
|---|---|---|---|
| CEC/Fe$_3$O$_4$ | Cellulose | Antimicrobial agent | [113] |
| (N-CNC)-Fe$_3$O$_4$ and (S-CNC)-Fe$_3$O$_4$ films | Cellulose | Biosensors | [114] |
| Fe$_3$O$_4$NP-INS-(DABC-EDA-Bzl) | Cellulose | Antimicrobial and cytotoxic activities and chemotherapy | [115] |
| M-CMC | Cellulose | – | [116] |
| BC-Fe$_3$O$_4$ | Cellulose | Removal of chromium(VI) | [117] |
| Fe$_3$O$_4$/PDA/CA | Cellulose | Removal of MB | [118] |
| PLA/M-NC | Cellulose | Co-delivery of 5-Fu and CUR | [119] |
| MCL | Cellulose | Anion exchange platform | [120] |
| MGN | Cellulose | Removal of bacterial from water | [121] |
| CMC-g-AM with embedded modified magnetite and PC | Cellulose | Removal of Pb-ions and MB | [122] |
| NMPap, ID-MPap, and D-MPap | Cellulose | Made of papers | [123] |
| DA/Fe$_3$O$_4$NPs@CNCs | Cellulose | Nano-biocatalyst for biomass conversion | [124] |
| M-JEPs | Cellulose | Removal of reactive Blue from a dye-contaminated ethanol-water mixture | [125] |
| MC | Cellulose | 5-Fu delivery | [126] |
| MCNC | Cellulose | Adsorption of doxycycline | [127] |
| GCM | Cellulose | Au adsorbent | [128] |
| MCBs | Cellulose | Dopamine delivery | [129] |
| MGO@Cellulose@Lipase | Cellulose | Removal of MB and MG | [130] |
| Fe$_3$O$_4$/MPC-[IL] | Cellulose | Chemical reaction catalyst | [131] |
| (Gr)/BC/Fe$_3$O$_4$ | Cellulose | Antibacterial agent and visible-light-driven photocatalyst | [132] |
| BC-Fe$_3$O$_4$-HA | Cellulose | Bone tissue engineering | [133] |
| CMC coated LSMO | Cellulose | – | [134] |
| CMC-Fe$_3$O$_4$/BC | Cellulose | Removal of pyrene and industrial field soil contaminated with PAHs and γ-HCH | [135] |
| CMC/magnetiteNPs | Cellulose | – | [136] |
| MCNC | Cellulose | Cr(VI) adsorption | [137] |
| Fe$_3$O$_4$/CNC-In(III) | Cellulose | Chemical reaction catalyst | [138] |
| MNP-PANI-BC | Cellulose | – | [139] |
| CCM | Cellulose | St degradation | [140] |
| MSNC | Cellulose | Biodiesel production | [141] |

**TABLE 21.1** *(Continued)*
## Summary of Studies on a Collection of Polysaccharide-modified MNP-based Systems for Different Applications

| Short Description of the Construct | Type of Polysaccharide | Studied Application Area | Ref. |
|---|---|---|---|
| MNP-C | Cellulose | – | [142] |
| NC-MA/L-MG and MC-O/ L-MG | Cellulose | Removal of arsenic | [143] |
| $Fe_3O_4$-CS | Cellulose | Removal of arsenic(V) | [144] |
| FF/$Fe_3O_4$ | Cellulose | Removal of color from indigo carmine solutions | [145] |
| GO/MNPs/HAP@CA | Cellulose | Removal of Cr(VI), Se(IV), and MB | [146] |
| MMTCB | Cellulose | Fluoride removal | [147] |
| CNT-g-Cellulose/$Fe_3O_4$ | Cellulose | Sorption of various oil/organic solvents | [148] |
| CS-mag | CS | Magnetic Cell Separation | [149] |
| $Fe_3O_4$/GO/CS | CS | Biofuel application | [150] |
| CSM | CS | α-amylase activity study | [151] |
| NMag-CS | CS | Removal of copper, lead, cadmium, chromium, and nickel metal ions | [152] |
| Nd-Ce doped $Fe_3O_4$-CS | CS | Fenton degradation of Direct Red 81 | [153] |
| Cs/CPL/$Fe_3O_4$-NC | CS | Removal of Pb(II) | [154] |
| CS@$Fe_3O_4$/CPE | CS | Working electrode | [155] |
| CH-MNP | CS | Pb(II) ions adsorption | [156] |
| Gr/CS/$Fe_3O_4$ | CS | Antibacterial | [157] |
| MNs | CS | – | [158] |
| $Fe_3O_4$-CHI | CS | DOX delivery | [159] |
| CS-PAA | CS | Application in MRI | [160] |
| MC-PYO and MC-PYS | CS | Silver sorption and application to metal recovery from waste X-ray photographic films | [161] |
| M-PPy/CS/GO | CS | Removal of Ponceau 4 R (P4R) dye and antimicrobial and antifungal efficacy | [162] |
| $Fe_3O_4$@$SiO_2$/SiTMC | CS | Glyphosate removal | [163] |
| CS-FeNPs | CS | Eradication of bacterial biofilms | [164] |
| CTS-$Fe_3O_4$ | CS | Anaerobic digestion processes under acid stress | [165] |
| CCM | CS | Adsorption of Cu(II) and As(V) | [166] |
| PC-$Fe_3O_4$ | CS | BSA delivery | [167] |
| CH/CMC-$Fe_3O_4$ | CS | Purification of peroxidase | [168] |
| MPyTMChi | CS | Degradation of MB | [169] |
| CS-$Fe_3O_4$-NPs | CS | Methotrexate delivery | [170] |
| CTM@$Fe_3O_4$ | CS | Adsorption of tetracycline | [171] |

*(Continued)*

**TABLE 21.1** *(Continued)*

## Summary of Studies on a Collection of Polysaccharide-modified MNP-based Systems for Different Applications

| Short Description of the Construct | Type of Polysaccharide | Studied Application Area | Ref. |
|---|---|---|---|
| MGCH | CS | Removal of 2-naphthol | [172] |
| Cur-CS-Fe$_3$O$_4$-RGO | CS | CUR delivery | [173] |
| LUT-CS/Alg-Fe$_3$O$_4$-NPs | CS | Lutein delivery | [174] |
| CSMNP | CS | Thrombolytic drug delivery | [175] |
| ChM | CS | Modified electrode sensor | [176] |
| MAG-CH | CS | Identification of blood plasma proteins | [177] |
| MagOPIC | CS | Pb$^{2+}$ biosorbent | [178] |
| Ch-Fe | CS | Pb(II) adsorption | [179] |
| MGC | CS | Adsorption of Hg(II) | [180] |
| MGC | CS | Bimodal magnetic resonance/ fluorescence imaging and 5-Fu delivery | [181] |
| MP@[Chi-SiO$_2$] and MP@ SiO$_2$ | CS | Adsorbent of gold(III) ion | [182] |
| NiO/Fe$_3$O$_4$@CS | CS | Catalyst for urea electro-oxidation | [183] |
| Pmc/CS/f-Fe$_3$O$_4$ x-Fe$_3$þ | CS | Strain sensors | [184] |
| Fe$_3$O$_4$@Chi | CS | Catalyst in multicomponent reaction | [185] |
| Magnetite/CS | CS | – | [186] |
| Magnetite-CS nanostructures | CS | – | [187] |
| Chit-MNPs | CS | Focus on hyperthermic and anti-amyloid activities | [188] |
| MCMNCs | CS | Meloxicam delivery | [189] |
| CS@Fe$_3$O$_4$ | CS | – | [190] |
| SPION-loaded CS coated bilosomes | CS | Resveratrol delivery | [191] |
| Magnetite CS microparticles | CS | Nitrate adsorption | [192] |
| Magnetic CS NPs | CS | Tacrine delivery | [193] |
| CS-magnetite films | CS | – | [194] |
| Fe$_3$O$_4$-CSNPs | CS | – | [195] |
| CFO | CS | Hyperthermia application | [196] |
| CMNs | CS | Removal of Fe(III) | [197] |
| Magnetic NPs coated with CHI | CS | – | [198] |
| ChMNs | CS | DNA and rhEGF separation | [199] |
| FMCH | CS | Adsorption and detection of Cr (VI) | [200] |
| CS-coated Fe$_3$O$_4$ | CS | Removal of Pb(II) and Ni(II) | [201] |
| Fe$_3$O$_4$/CS | CS | – | [202] |
| Fe$_3$O$_4$-aminated CS | CS | Study of its interactions with a cell model membrane | [203] |

**TABLE 21.1** *(Continued)*

## Summary of Studies on a Collection of Polysaccharide-modified MNP-based Systems for Different Applications

| Short Description of the Construct | Type of Polysaccharide | Studied Application Area | Ref. |
|---|---|---|---|
| MNPs@CS-SO$_3$H | CS | Catalyst of chemical reaction | [204] |
| PTh.M.Cs.NC. and PTh.M.Cs.NPs.NC. | CS | Removal of heavy metals and selective mercury | [38] |
| Fe$_3$O$_4$@Cs-PEG | CS | 5-Fu delivery | [205] |
| CS-SPIONs | CS | – | [206] |
| Ch-Fe$_3$O$_4$ MNPs | CS | Application in the synthesis of γ-D-glutamyl-L-tryptophan (SCV-07) | [207] |
| cMNPs | CS | Support for insolubilization of inulinase | [208] |
| CANF | CS | MO adsorption | [209] |
| HA-MG-CH | CS | Ni(II) and Pb(II) recovery | [210] |
| CS-g-poly(NIPAAm-co-DMAAm) | CS | DOX delivery | [211] |
| CHF1 | CS | Adsorption of Rodamine 6 G | [212] |
| CHF2 | CS | Adsorption of Rodamine 6 G | [212] |
| Fe$_3$O$_4$@CS@NaOL | CS | Treat emulsified oil wastewater | [213] |
| RC-Fe$_3$O$_4$-CS | CS | Removal of lead ions | [214] |
| C-Fe$_3$O$_4$ | CS | Hyperthermia application | [215] |
| CTN | CS | Adsorption of As(V) | [216] |
| Fe$_3$O$_4$-SiO$_2$-CS MNPs | CS | Adsorption of Pb$^{2+}$ and Cd$^{2+}$ | [217] |
| Fe$_3$O$_4$@CS@Ag | CS | Antimicrobial performance | [218] |
| Ser and Ala-functionalized magnetic-CS | CS | Adsorption of uranyl ions | [219] |
| AS-Fe$_3$O$_4$ | St | Human serum albumin immobilization | [220] |
| Fe$_3$O$_4$@apple seed starch-In(III) | St | Catalyst for chemical reactions | [221] |
| s-Fe$_3$O$_4$ | St | Catalyst for chemical reactions | [222] |
| St@MNPs | St | – | [223] |
| St-coated Fe$_3$O$_4$ | St | – | [224] |
| Aminated St coated magnetic NPs | St | CUR delivery | [225] |
| CMCS@Fe$_3$O$_4$ | St | DOX adsorption | [226] |
| HES MNCs | St | Oncocalyxone A delivery | [227] |
| St/Fe$_3$O$_4$-NPs | St | – | [228] |
| Fe$_3$O$_4$/silica-St/PNVCL NC | St | Acetazolamide delivery and adsorbents in cephalexin removal | [229] |
| St-magnetite | St | Hyperthermia | [230] |
| CS-magnetite | St | Hyperthermia | [230] |

*(Continued)*

**TABLE 21.1** *(Continued)*

**Summary of Studies on a Collection of Polysaccharide-modified MNP-based Systems for Different Applications**

| Short Description of the Construct | Type of Polysaccharide | Studied Application Area | Ref. |
|---|---|---|---|
| ST-coated magnetite | St | Removal of Optilan Blue | [231] |
| St/Fe₃O₄/zeolite-bionanocomposite | St | Catalyst for the oxygen reduction reaction | [232] |
| SIONPs | St | Removal of Cr(VI) | [233] |
| Magnetic St nanocomposite | St | Adsorption of tetracycline | [234] |
| ST-coated Fe₃O₄ NPs | St | Removal of Optilan Blue | [235] |
| CM-ABRS SPIONs | St | DOX delivery | [236] |
| Magnetic NPs coated with ALG | Alg | – | [198] |
| Fe₃O₄@-SiO₂@al/CQDs | Alg | DOX delivery | [237] |
| SPION-alginate | Alg | – | [238] |
| Fe₃O₄-SA-PVA-BSA | Alg | DOX delivery | [239] |
| Fe₃O₄/Alg-Ag NPs | Alg | Treat the human lung carcinoma | [240] |
| Fe₃O₄@AMALG @Ag | Alg | Catalytic activity toward p-NP removal and enzymatic-mimic activity for solid-colorimetric H₂O₂ detection | [241] |
| PBMA-gft-Alg/Fe₃O₄ | Alg | Corrosion | [242] |
| Alg-Fe₃O₄ | Alg | Phenol removal | [243] |
| PTPMNs | Alg | PTX delivery &MRI | [244] |
| Fe₃O₄@SiO₂/SiTBA-ALG | Algand Car | Removal of ciprofloxacin | [245] |
| MCA | CS and Alg | Phenol biodegradation | [246] |
| Magnetic NPs double coated with dextran and CS | Dextran and CS | – | [247] |
| MCCs-I and MCCs-I-II | Cellulose and CS | Enzymatic catalyst activity | [248] |
| MCNCs/St-g-(AMPS-co-AA) | Cellulose and St | Removal of CV and MB | [249] |

methyl orange (MO) [43]. In the following research for evaluation of the systems prepared from the MNPs in situ synthesized in the presence of CS, the ternary MCG composite based on Fe₃O₄, CS, and GO was facilely constructed via a solvothermal way. The synthesized MCG displayed impartial adsorption on divalent heavy metals in six consecutive cycles [40]. Similarly, Bulinin 2023 prepared a ternary nanohybrid based on magnetite (M)-CS-GO (MCSGO) in a simple way through the simultaneous reduction-precipitation at room temperature. Exploring the MCSGO potential approved its adsorption efficiency for Hg(II) and Zn(II) even after six consecutive cycles, signifying its promising removal efficacy [44]. It was reported that the poor dispersibility as well as the low solubility of CS in aqueous solutions with neutral pH commonly confine its usage. Therefore, carboxylation of CS, which produces carboxylated chitosan (CMCS), is one of the used methods to improve its moisturizing and solubility properties [41]. In this regard, Lian *et al.* in situ prepared MNPs in the presence of CMCS as a main water-soluble derivative of CS and CS oligosaccharide (COS) powder (CMC/COS). In the following, ethylenediaminetetraacetic acid (EDTA) was grafted on the system employing a large number of carboxyl, hydroxyl, and amino groups of CMC/COS and $Fe_3O_4$@CMCCOS-EDTA was obtained. In the prepared structure the CMCS and also EDTA offered the active sites to encapsulate the heavy metal ions via strong binding and remove them from aqueous solutions. The batch adsorption experiment results indicated the maximum adsorption capacity of pb(II) at about 432.34 mg/g at a pH of 5 and temperature of 308 K [45]. All findings of these works provide references for tailoring effectively $Fe_3O_4$/ CS-based architecture in the water treatment area. The capability of the prepared $Fe_3O_4$/CS-based architecture via in situ was also studied for bone tissue engineering. For instance, Heidari *et al.* designed and fabricated the CS/HA/magnetite through the in situ synthesis of magnetite in a mixture of pre-extracted natural HA and CS. Thereafter, the system was assessed from the elastic modulus, bending strength, compressive strength, and hardness viewpoints. Additionally, the cyto-compatibilities of these composites were assessed and discussed through the use of human mesenchymal stem cells (hMSCs) [46]. CS was also can be functionalized with aminic reagents to obtain a new system with more amine functional groups. According to these, Ziegler-Borowska and co-workers obtained the $Fe_3O_4$-CS via in situ magnetite synthesis in the presence of CS and after reaction with epichlorohydrin, oxidized it with sodium periodate. The formed aldehyde moieties and the free amino group of CS reacted with glutaraldehyde and, finally, functionalization with ethylenediamine gained a new magnetic $Fe_3O_4$-CS-$(NH_2)_3$ with abundant surface free amino functional groups. The performed ninhydrin test determined the amount of available free primary amino groups on the $Fe_3O_4$-CS-$(NH_2)_3$ surface at about 8.34 mM/g. In the end, the applicability of the fabricated NPs were assessed for bioligand immobilization using free amino and carboxylic groups of ligands, lipase from *Candida rugosa* and human serum albumin as a model protein. It was detected that owing to the more effective immobilization of both studied bioligands, the fabricated magnetic NPs are appropriate for numerous applications [47]. Similarly, with the aim of enzyme stabilization via their immobilizingon green support with more surface area the in situ fabricated magnetic CS nanocomposite was used to bring dithiocarbamate moieties on their surface. After functionalization with $CS_2$ as a result of the post-modification reaction, the porcine pancreas lipase (PPL) was immobilized on the surface. The schematic for the covalently bonded pancreatic lipase onto the $Fe_3O_4$@CS/NHCS$_2$H is reported in Figure 21.3c. The detection of the Fe element in the energy-dispersive X-ray spectroscopy (EDX) spectrum of the final system can be introduced as the main sign of $Fe_3O_4$ in situ synthesis in the presence of CS (Figure 21.3d). The outcome showed that compared with free lipase, the activity of immobilized lipase did not have a notable decrease. Additionally, it was confirmed that $Fe_3O_4$-Chit-CS-Lip was able to recycle at least six cycles without important loss in the activity [48]. Besides the organic reagents, the CS-coated MNPs also can offer the sites for the growth of inorganic NPs to attain a new system with more functionality. For example, the in situ synthesized MNPs in the presence of CS was used as a platform for silver NPs (Ag NPs) growth on it. The fabricated synthesized was tested for cationic and anionic dyes removal from the aqueous solutions. The proposed mechanism of MB and MO adsorption with $CS/Fe_3O_4/$ AgNPs demonstrated the main role of CS as a modifying agent in dye removal. Additionally, the antibacterial test results specified that the $CS/Fe_3O_4/AgNPs$ can efficiently prevent the growth of microbial cells [7].

### 21.2.2.2 MNPs/CS Systems Prepared via in Blending

MNPs also can be coated with CS through a facile mixing way to achieve a new system with improved and unique properties. Therefore, for the evaluation of CS-coated MNPs, Li *et al.* in 2022 synthesized the magnetic ferroferric oxide NPs in a simple hydrothermal way and then coated them with CS via a cross-linking reaction (MCS). Afterward, MCS was chemically grafted with four different amino acids, for instance, alanine, glycine, serine, or L-cysteine, and respectively, the Ala-MCS, Gly-MCS, Ser-MCS, and Cys-MCS were produced. Figure 21.4a displays the

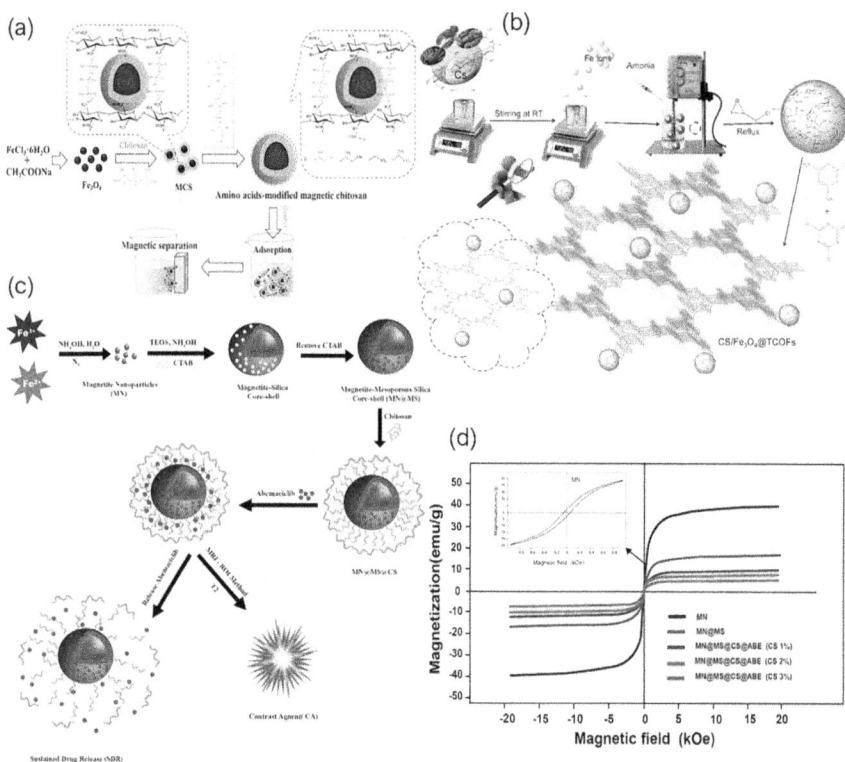

**FIGURE 21.4** Construction of amino acids modified MCS and its evaluation for the uranium adsorption process (a), $CS/Fe_3O_4$@TCOFs synthesis procedure (b), illustration of MN@MS@CS@ABE formula synthesis for use as a theranostic nanocarrier (c), and VSM curves of fabricated samples during MN@MS@CS@ABE synthesis (d). Reproduced with permission from [49,51,54].

pathway of all these synthesis ways. As reported in this paper, the changes in the value of the magnetization and also the zeta potential can be a powerful sign for the success in the performed reactions. The systems assessing showed the potential of the amino acids modified MCS for uranium removal from waste water [49]. In a similar viewpoint, the CS-coated MNPs were also functionalized with aminothiazole (ATA@MC) and also imidazole carboxamide (AIC@MC) to improve Cr(VI) removal. Before the valuation of the system for chromate sorption, they were chemically and physically characterized by a wide variety of analytical techniques. It was found that the AIC@MC is more selective for Cr(VI) removal between equimolar Ca(II), Cu(II), Cd(II), Zn(II), Ni(II), and Cr(VI), whereas ATA@MC has a wider reactivity for a broader family of metal ions [50]. In similar research in the water treatment area, the Namazi research team used the CS-coated $Fe_3O_4$ NPs as a platform for triazine-based covalent organic frameworks growth is presented in Figure 21.4b. The achieved $CS/Fe_3O_4@TCOFs$ presented the suitable performance for diclofenac sodium removal that arose from some special feature of the final nanosystem in which the ami groups of CS biopolymer was one of them [51]. This research group also used the CS-coated $Fe_3O_4@Cd$-MOF for amoxicillin removal. The output of the adsorption study displayed that the efficiency of removal is influenced by CS [52]. CS-coated magnetite NPs were also evaluated for drug delivery applications after modification with a secondary polymer. Typically, the magnetic IONPs were prepared in a co-precipitation way, and then were coated with CS in a simple mixing. In the next step, the [2-(methacryloyloxy) ethyl] trimethyl ammonium chloride was introduced to the structure via the radical polymerization way using the diethylene glycol dimethacrylate (DEGDMA) as a cross-linker, and $K_2S_2O_8$ as an initiator. The CS-IONPs-METAC polymeric material system was explored as a PTX drug carrier to improve the in vivo efficacy of this hydrophobic drug by avoiding solubility concerns and providing targeted delivery. In the apoptosis test, the PTX alone and CS-IONPs-METAC-PTX, respectively, presented an MCF-7 cells viability equal to 56.55 and 56.75%, respectively. All reported findings indicated the potency of the synthesized material for PTX delivery [53]. In another try, in biomedicine to obtain a theranostics platform, a new nanoformula comprising MNPs, mesoporous silica (MS), and CS coating was also prepared and further loaded with a chemotherapeutic agent, abemaciclib (ABE). A schematic of fabrication and the changes in the Ms value of MNPs as a result of modification with organic non-magnetic materials are provided respectively in Figure 21.6b and Figure 21.6c. Results of formula examination as a contrast agent of MRI stimulated a decrease in signal intensity in the MR T2-weighted imaging method. In conclusion, the results of current research work developed a new nanoformula that can be a talented theranostic agent in diagnosis and cancer therapy [54]. Besides the neat CS, its derivatives were also evaluated from their potential as a MNPs modifying agent. In this regard, the $Fe_3O_4$ NPs were fabricated in a facile chemical co-precipitation way and then encapsulated with CMCS and $Fe_3O_4@CYCTS$ obtained. The fabricated hybrid with a large specific surface area was used for snailase immobilization through physical adsorption for ginsenoside transformation and displayed a maximum value of 67.67 mg enzyme/g of support, principally driven through electrostatic interactions. It is reported that the immobilization of enzyme led to improved storage stability. Additionally, it was detected that the immobilized enzyme can be reused in nine long cycles (48 h/ transformation cycle) while retaining about 56.0% of its initial catalytic activity [41].

### 21.2.2.3 MNPs/CS Systems Prepared via Grafting-onto

CS also can be inserted onMNPs surface via chemical grafting in a pristine or derivative form. For example, with the aim of evaluating the water treatment capability of CS functionalized magnetic NPs, the $Fe_3O_4$ NPs were prepared in the presence of 1,2,3,4-butanetetracarboxylic acid (TCA) to obtain the magnetic NPs having carboxylic acid groups. In the following, the CS was grafted on the surface via an amidation reaction after activation by using 1-ethyl-3-(3-dimethylaminopropyl) carbodiimide (EDAC), and the product named $Fe_3O_4$@TCA@CS. The system presented a narrow particle size distribution in DLS analysis. Exploring the system capability for the Pb(II) adsorption process revealed that the pseudo-first-order and Freundlich models were more fitted models to define the mechanism of Pb(II) adsorption onto $Fe_3O_4$@TCA@CS and the maximum adsorption capacity was obtained at about 204.92 mg/g [42]. Magnetic-cored dendrimers (MDs) were also covalently modified with CMCS. Figure 21.5a,b,c displays the involved steps in the CMCS-modified magnetic-cored dendrimers (CCMDs) preparation procedure. The outcome of the adsorption tests showed that with the substitution of amino groups of MD with CMCS moieties, the adsorption sites for cationic materials were significantly increased, mainly via the electrostatic interaction [39]. For avoiding

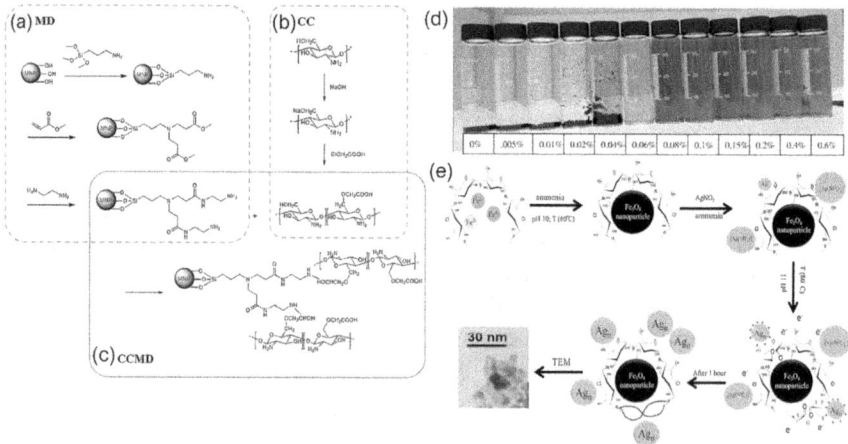

**FIGURE 21.5** Schematic of MD (b) CC (b), and CCMD (c) synthesis, the influence of St concentration on the physical stability of MNPs for the samples prepared at 0.1 g/L as Fe with different concentrations of St, taken photo 1 h after synthesis (d), Ag/$Fe_3O_4$ nanocomposites synthesis in the presence of St (e). Reproduced with permission from [39,60,61].

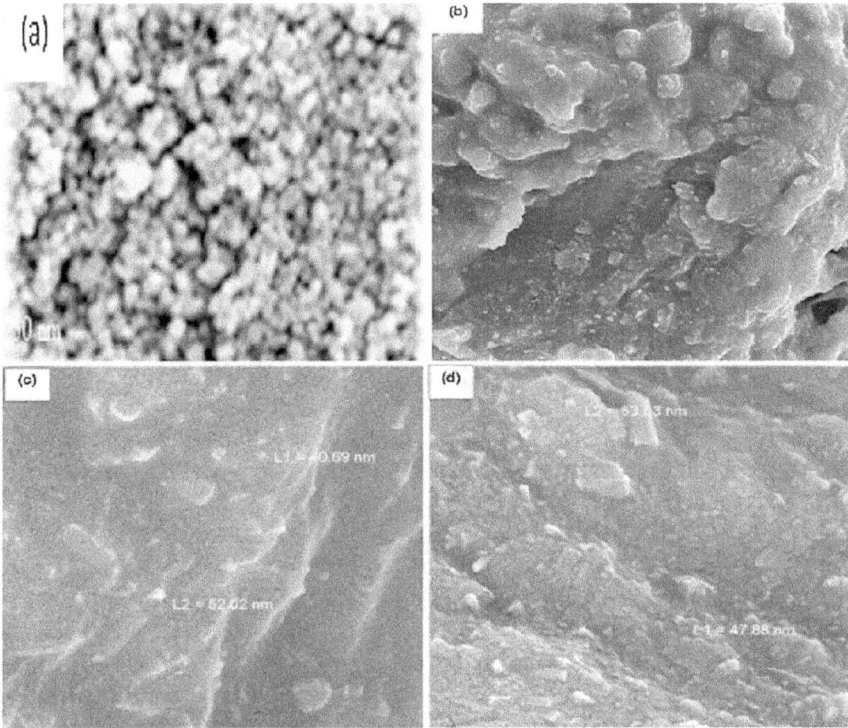

**FIGURE 21.6** SEM images of $Fe_3O_4$ NPs (a) poly(ethylene phthalate) (b), $Fe_3O_4$/starch (c) $Fe_3O_4$/starch-g-poly(ethylene phthalate) (d). Reproduced with permission from [65].

prolonging this section, the outcomes of the other published research works in this area are listed in Table 21.1.

### 21.2.3 $Fe_3O_4$NPs Modified with St or its Derivatives

St is a biopolymer derived from plants like potato, corn, rice, etc. The structure of this polysaccharide is composed of linear amylase and branched amylopectin with the identical backbone structure as amylose but with several $\alpha$-1, 6'- linked branch points [55,56]. Particularly, St is recognized as a semi-crystalline polymer displaying a complex composition of carbohydrates [57]. Because of the presence of numerous hydroxyl groups in the St structure, it can act as a capping agent. Additionally, owing to its biocompatible, biodegradable, inexpensive, edible, and renewable features, it received a substantial amount of academic and industrial efforts [55,58].

#### 21.2.3.1 MNPs/St Systems Prepared via In Situ Way

In regards to the special structure and features of the St biopolymer and with the aim of reach to an enhancement in the MNPs performances, reducing their toxicity, and increasing their stability, the $Fe_3O_4$ NPs can be in situ synthesized in the presence of St biopolymer. Considering these, in one research paper, a common co-precipitation procedure was used for the in situ synthesis of $Fe_3O_4$ NPs in the existence of St polysaccharide. The authors then evaluated the effect of the used coat on the toxicity of $Fe_3O_4$ NPs via a transcriptome sequencing (RNA-seq) procedure to illustrate the liver and gill transcriptomes from adult zebrafish when exposed to bare and St-stabilized $Fe_3O_4$ NPs for seven days. They detected that the neat $Fe_3O_4$ NPs exerted greater toxicity with respect to the St-coated $Fe_3O_4$ NPs in gills. In contrast, St-$Fe_3O_4$ NPs caused more severe harm to the liver, via both bare and stabilized NPs appeared to share analogous regulatory mechanisms. Summation of the obtained results displayed that the surface coatings play a main role in determining the NPs' toxicity, which in turn influences the cell uptake and biological responses, accordingly affecting the potential efficacy and safety of nanomaterials [59]. The capability of in situ fabricated $Fe_3O_4$ NPs in the presence of a water-soluble St solution as a stabilizer was also tested for perfluorooctanoic acid (PFOA) removal from the aqueous solution after full characterization. The results of the experimental tests showed that the presence of St at ≥0.2 wt% can fully stabilize 0.1 g/L as Fe of the $Fe_3O_4$ NPs (Figure 21.5d). Overall findings promise the St-stabilized $Fe_3O_4$ NPs utilization as a "green" adsorbent for the efficient removal of PFOA from groundwater and soil [60]. St-coated $Fe_3O_4$ NPs also can be further modified with other organic or inorganic materials to give the new multifunctional system with desired features. In this way, Ghaseminezhad and coworkers fabricated and used St-coated $Fe_3O_4$ NPs as a reducing agent for the reduction of Ag ions in an alkaline condition. All involved green synthesis steps for the construction of the Ag/$Fe_3O_4$ nanocomposites in the presence of soluble St are presented in Figure 21.8b. All of the results of antibacterial activity evaluation toward *Escherichia coli* (*E. coli*) via the minimum inhibitory concentration (MIC) technique are a sign that the produced nanocomposite has more capability as targeted antibacterial healing of infection sites under an external magnetic field [61]. In another research, the St-coated magnetite NPs were used as a platform for the MIL-88(Fe) growth on it, an example of the organometallic construct. Tetracycline (TC) was loaded into the system. The in vitro release profile assessment showed about 26% of TC release after 12 h, which was followed by a sustained and controlled release in the simulated physiological environment at pH 7.4. Furthermore, the TC-loaded St/$Fe_3O_4$/MIL-88(Fe) presented higher antibacterial potential toward both *E. coli* and *S. aureus* with the MIC value of 128 and 64 $\mu g \cdot mL^{-1}$, respectively [62]. Carboxymethylcellulose/polyacrylic acid (CMC/PAA) is the other organic agent which was introduced on the structure of the St coated magnetite for the targeted cancer treatment through co-delivery. After full characterization of the prepared materials employing different techniques, the anticancer doxorubicin hydrochloride (DOX) and 5-fluorouracil (5-Fu) were individually loaded on the final system. In-vitro drug release studies in the medium mimicking the gastrointestinal tract (GIT) displayed that CMC/PAA/St-$Fe_3O_4$ nanocomposites can avoid 5-Fu and DOX release in the stomach acidic environment because of their pH-sensitive swelling nature, and also improve the drug dosing stability in the colorectal, which are useful for each oral drug carrier. All of these results, along with the observed important cytotoxicity against human colon cancer cell lines (SW480) for the drug-loaded nanocomposites, could demonstrate the applicability of the system as an efficient carrier for colon cancer treatment [63].

Besides the neat St, the carboxymethyl Assam Bora rice starch as one of the main water-soluble derivatives of St was used as a backbone for in situ preparation of superparamagnetic IONPs, which were used as a targeted drug delivery system for cancer treatment. The results of in vitro studies showed that the prepared platform could be used for the design and optimization of actively targeted drug carriers [1]. $Fe_3O_4$ NPs were also in situ prepared in the medium containing St and a secondary polymer. Taking this, M. Badawy *et al.* in 2023 synthesized the magnetic $Fe_3O_4$ NPs in the presence of muscovite (Mus) and St carbohydrate polymer. The developed multifunctional Mus/St/MNPs nanocomposite presented brilliant adsorption properties for MO and hexavalent chromium. The constructed adsorbent was simply reactivated and reused numerous times so the regenerated Mus/St/MNPs composite presented nearly 79% of Cr(VI) and 85% of MO adsorption capacities even after the fourth adsorption-desorption cycle. Mechanism studying showed that the Cr(VI) ions were removed from the solutions through the adsorption process with a combination of non-horizontal and horizontal orientations on Mus/St/MNPs active sites, while the MO dye adsorption was accomplished via a vertical orientation with a multi-interactions mechanism [57].

### 21.2.3.2 MNPs/St Systems Prepared via Blending

St biopolymercan also be coated on $Fe_3O_4$ NPs after their synthesis. In this regard, the boiling rice starch extract (BRE) was utilized for spherical IONP coating, which was then used for cancer treatment via photoacoustic imaging (PAI)-guided chemophototermal therapy. Typically, at first, the IONPs were synthesized by the solvothermal method. Then they were coated with St, employing the physical immobilization technique. DOX molecules were loaded at about 78% and DOX-loaded BRE-IONPs were achieved. The drug was released in a pH-sensitive manner with a higher percentage at acidic pH. Biocompatible properties, excellent photothermal stability, and high NIR absorption for PAI-guided PTT treatment BRE-IONPs were the sign of this fact that the rice St is an appropriate bioactive coating for probable theranostic usages [64]. The hybrid of $Fe_3O_4$ with other NPs are another systems that were assessed for the modification with St to get the St advantages in nanohybrid modification. For example, constructed $Fe_3O_4$/Ag NPs impregnated on LDH were coated with St for the stabilization of the $Fe_3O_4$/Ag@Ca-Al LDH hybrid (FAL hybrid) and to improve the surface area for nitrophenols reduction (FALS-BNC). The reduction reaction of 2-nitrophenol and 4-nitrophenol employing FALS-BNC specified more catalytic activity because of their large numbers of hydroxyl groups as a capping agent. But the greater kinetic performance was observed for 4-NP with respect to 2-NP [55]. In the continuing upgrade of the potential of $Fe_3O_4$/St via modification of both organic and inorganic components, Hamidian and coworkers modified the pre-synthesized $Fe_3O_4$ NPs with St and then functionalized the achieved hybrid with polyester through the graft copolymerization and $Fe_3O_4$/St-g-poly (ethylene phthalate) hydrogel nanocomposite was obtained. The structural characterization of the $Fe_3O_4$/St-g-poly(ethylene phthalate) hydrogel nanocomposite employing the FT-IR verified the presence of all feed components in the nanocomposite. The recorded scanning electron microscope (SEM) images presented that the $Fe_3O_4$ NPs were homogeneously dispersed in a polymeric matrix (Figure 21.6). H3PW12O40 (HPA) as a model of the drug was loaded and then its release was considered [65]. On the other hand, Ravichandran et al. designed and fabricated the St-decorated $Fe_3O_4$@Ag nanocomposite through the one-step grinding of the $Fe_3O_4$, St, and $AgNO_3$ mixture. The transmission electron microscopy (TEM) technique obtained the average particle size of the nanocomposite at about 9–11 nm. The results of the ultraviolet-visible diffuse reflectance spectroscopy (UV-Vis DRS) and X-ray diffraction (XRD) analysis presented greater stability even for more than a month for the synthesized nanocomposite. The schematic for the preparation of

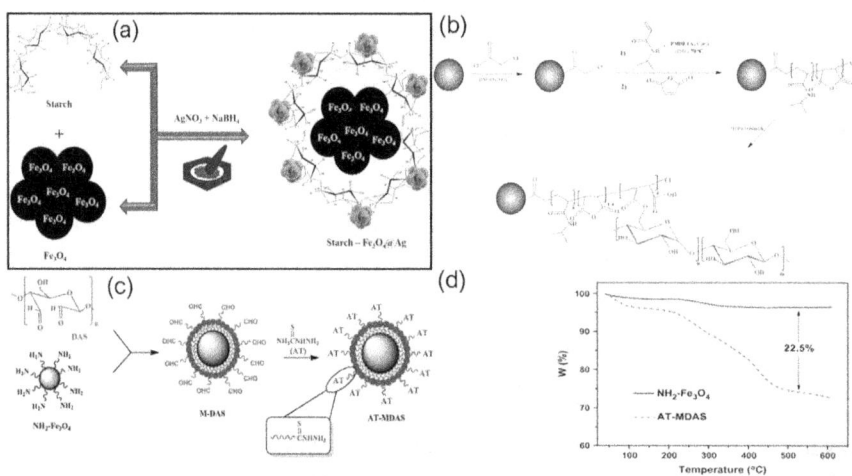

FIGURE 21.7   St-$Fe_3O_4$@Ag synthesis scheme (a), $Fe_3O_4$-g-(PNIPAAm-co-PMA)@St MNHGM fabrication strategy (b), schematic route for the synthesis of AT-MDAS nano-composite (c), and thermal behaviors of $NH_2$-$Fe_3O_4$ and the AT-MDAS nanocomposite (d). Reproduced with permission from [56,58,67].

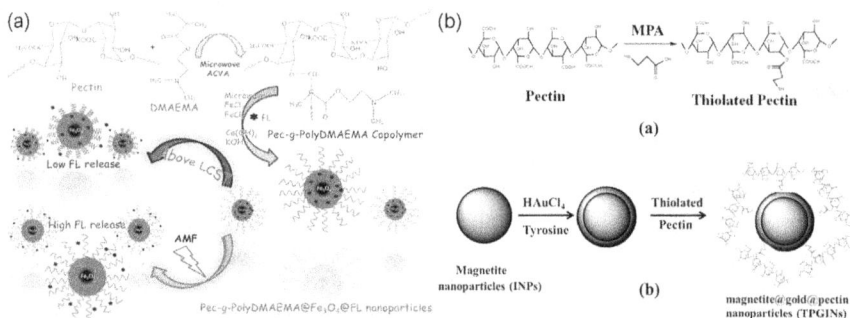

FIGURE 21.8   Schematic of the Pec-g-PolyDMAEMA@$Fe_3O_4$ (a) and TPGINs synthesis (b). Reproduced with permission from [15,68].

nanocomposite is displayed in Figure 21.7a. From the applicability viewpoint, a higher yield was detected for the conversion of benzaldehyde to benzoic acid at ambient conditions. Additionally, the nanocomposite revealed the brilliant catalytic activity for reductive degradation of different carcinogenic water pollutants for example methyl red (MR) and MB, respectively, within 18 min and 14 min [56].

In the area of materials application, it is very valuable to reach a new product with the desired structure. Cationic St is the other derivative of St with amin functional groups that can obtain a new system with cationic nature. By considering this goal, the cationic tapioca St (CTS) consisting of quaternary ammonium groups was reported for MNPs' functionalization (CTS@$Fe_3O_4$). The constructed hybrid presented equilibrium adsorption capacities of micromolecular pigments i.e., caffeic acid (CA), gallic acid (GA), and melanoidin (ME) at about 185, 160, and 580 mg $g^{-1}$ at the optimal conditions. These are owing to the presence of the quaternary ammonium groups on the St surface, which provided a stable positive charge and confirmed excellent adsorption performance for micromolecular and macromolecular pigments in sugarcane juice [66].

### 21.2.3.3  MNPs/St Systems Prepared via Grafting-onto

With the aim of developing a nanohybrid with complementary or synergetic behavior, in one interesting research work for surface modification of MNPs with St biopolymer via covalent bonding at first, MNPs were functionalized with chloroacetyl chloride moiety and then N-isopropylacrylamide (NIPAAm) and maleic anhydride (MA) monomers grafted on it via atom transfer radical polymerization (ATRP) technique. The achieved $Fe_3O_4$-g-(PNIPAAm-co-PMA) was cross-linked via the reaction between the hydroxyl groups of St and anhydride group of MA and $Fe_3O_4$-g-(PNIPAAm-co-PMA)@St MNHG was produced. The overall approach for the $Fe_3O_4$-g-(PNIPAAm-co-PMA)@St MNHG synthesis is presented in Figure 21.7b. The detected reduction in magnetization value of the $Fe_3O_4$-g-(PNIPAAm-co-PMA)@St MNHG in comparison with $Fe_3O_4$ NPs can be a sign of its successful surface modification. DOX loading, as well as temperature- and pH-responsive drug release performance of MNHG were also assessed and the highest drug release rate was obtained at 42°C and pH 5.3 that shows the suitability of developed MNHG as a de novo drug delivery system [58]. Dialdehyde St is another St derivative that was used for magnetite NPs modification via covalent bonding. In this way and in one exciting effort for the preparation of the appropriate heavy metal adsorbent, at first, amine-functionalized $Fe_3O_4$ NPs were prepared as a result of modification with 2-aminoethanol. Thereafter, the dialdehyde St was grafted on them via covalent linking by a Schiff base. Finally, more modification with aminothiourea occurred. The schematic route for the synthesis of AT-MDAS nano-composite is presented in Figure 21.7c. Further, focus on the thermal behaviors of $NH_2$-$Fe_3O_4$ and AT-MDAS nanocomposites obtained the content of DAS in AT-MDAS at about 22.5% (Figure 21.7d). The accomplished batch experiments presented the exceptional selectivity for Hg(II) among the mixed metal ions solution; also, good regeneration performance of nano-composite has been established [67]. For avoiding prolonging this section, the outcomes of the other published research works in this area are listed in Table 21.1.

## 21.2.4 Fe$_3$O$_4$NPs MODIFIED WITH PEC OR ITS DERIVATIVES

Pec is formed from (1–4) linked α-D-galacturonic acid backbone with some methyl esters and also many carboxyl groups that are partially in methyl ester form. This polysaccharide is commercially obtained from the citrus fruit peel and vegetable extract [68,69]. This naturally occurring anionic biopolymer can be used in several areas. Today, it specially attracted significant attention in biomedicine owing to its appealing features; for example, controllable biologic activity, low cost, non-toxicity, pH sensitivity, highly biodegradable nature, as well as the flexible chains that let the modulation of this polysaccharide to a desired shape [5,8]. Additionally, because of its special structure, this polysaccharide commonly has been considered to act as a polymer biodevice for the treatment of some kinds of GI diseases that influence specific areas of the GI tract, e.g., the colon, so it could be only digested with the pectinase present in the colon [5,68].

### 21.2.4.1 MNPs/Pec Systems Prepared via In Situ

Owing to the presence of polar functional groups in the structure of Pec, this polymer can be used as a platform for in situ synthesis of MNPs to obtain well-dispersed and biocompatible MNPs in a green way. Taking into account, Pec was used to stabilize the magnetic NPS which were then used for the cesium removal from wastewater. So, at first, the Pec was complexed with graphene oxide. In the next step, the collection was utilized for the magnetic NP formation via the reduction precipitation method. Finally, Prussian blue was introduced in the structure (PSMGPB). The enhanced adsorption capacity of PSMGPB nanocomposites was ascribed to the Pec-stabilized separation of graphene oxide sheets, as well as the improved distribution of magnetites on the graphene oxide surface [70]. In situ Fe$_3$O$_4$ synthesis was also used for the preparation of ultrasmall Fe$_3$O$_4$@C NPs with an average size of 7 nm employing citrus Pec as a carbon precursor. The brilliant outcome of this research was the MB removal from aqueous solution with superior recyclability up to 20 cycles and a maximum adsorption capacity of 141.3 mg g$^{-1}$ [71]. In another interesting research in the water treatment area, in 2020, Kulal and coworkers highlighted the potential of Pec/secondary polymer for in situ MNP fabrication. In this way, Pec-based gel was prepared via grafting of N-hydroxyethylacrylamide (HEAA) on Pec employing N,N-methylenebisacrylamide (MBA) as a cross-linker and K$_2$S$_2$O$_8$ as an initiator under microwave irradiation. To continue, the Fe$_3$O$_4$ NPs were formed inside this gel through in situ diffusion of Fe$^{2+}$ and Fe$^{3+}$ followed by a reaction with ammonia solution. The magnetic property measurement via a vibrating-sample magnetometer (VSM) test specified the ferromagnetic nature of the fabricated Pec-g-PHEAA/Fe$_3$O$_4$, which is a powerful sign for the successful MNPs synthesis. The Pec-g-PHEAA and Pec-g-PHEAA/Fe$_3$O$_4$ systems approved their efficiency for Rhodamine 6 G (R6G), a cationic dye; whereas, Cu(II) and Hg(II) ion removal from aqueous solution with more adsorption capacity can be observed in case of Pecg-PHEAA/Fe$_3$O$_4$. In conclusion, the results showed that the nanocomposite hydrogel could be introduced as a prospective adsorbent for the efficient removal of metal ions and cationic dyes from wastewater [69]. Similarly, the synthesized Pec-graft-poly(dimethylaminoethyl methacrylate) in a green and rapid approach was utilized as a platform for in situ formation of MNPs (Pec-gPolyDMAEMA@Fe$_3$O$_4$) (Figure 21.8a). After detailed characterization, the vitro release studies of 5-Fluorouracil (FL) were accomplished via altering the pH (5.5 and 7.4), temperature (37 and 44°C), and presence of an AMF. It was detected that the drug release was pH- and thermo-sensitive, as well as it was notably improved (100%) with the presence of a 50 mT magnetic field. Overall, it can be concluded that modified Fe$_3$O$_4$ NPs with Pec-g-PolyDMAEMA rendered a new system that can be considered appropriate applicants for drug delivery [68]. Iron NPs coated Pec was also prepared via in situ way and thereafter was analyzed from the feasibility of its use as probiotic strain *Lactobacillus plantarum* CIDCA 83114 delivery systems. Concerning bacterial viability, no decays were detected when Pec-decorated NPs were exposed to simulated fluids nor when stored at 4°C for 60 days. Together, the composites engineered in this work seem as suitable for probiotic bacteria delivery systems, whose target is the gut [72].

### 21.2.4.2   MNPs/Pec Systems Prepared via Blending

Coating of MNPs with a Pec biopolymer achieves a new nanosystem with abundant polar functional groups. In this regard, Sahu and coworkers constructed the super-paramagnetic MNPs coated with Pec as a result of mixture cross-linking with Ca$^{2+}$ ions in the form of spherical calcium pectinate nanostructures (MCPs). A reduction in Ms' value with increasing precursor Pec concentrations could be associated with the quenching of magnetic moments by the development of a magnetic dead layer on the MNPs. The uniform size distribution was detected by dynamic light scattering (DLS) measurement and the average size reached 400 nm in an aqueous medium. Overall based on findings the authors proposed the prepared system can be used in a wide range of bio-medicinal applications [73]. With some changes, in the other research, Pec was coated on Fe$_3$O$_4$ magnetic nanospheres (PCMNs) employing the sonochemical technique. The obtained saturation magnetization of 32.69 emu/g for the prepared nanoconstruct specified that the pre-synthesized magnetic NPs have been coated by Pec successfully, magnetite content of the system was determined up to 63% [74]. MNPs/ Pec samples also can be further modified with a second portion to reach a new multifunctional system. Therefore, in the following research works for the use of MNPs/polysaccharides as a platform for metal NPs synthesis, Zhang *et al.* used the Pec-coated Fe$_3$O$_4$ NPs as a platform for Pd decoration on it to obtain a new system with cardiovascular protective effects. In the mentioned research the functionalization with Pec not only offered the sites for the biogenic reduction of Pd$^{2+}$ ions but also stabilizes the ferrite NPs from agglomeration. The prepared catalyst presented outstanding reactivity and good to excellent yields in C-N and C-C cross-coupling reactions, respectively. Moreover, the nanocatalyst-treated cell cutlers pointedly (p $\leq$ 0.01) decreased the DNA fragmentation and caspase-3 activity in the 3-[4,5-dimethylthiazol-2-yl]-2,5 diphenyl tetrazolium bromide (MTT) assay, which displays its cardiovascular protective activities. It also enhanced the cell viability and mitochondrial membrane potential in the more concentration of Mitoxantrone-treated HAEC, HPAEC, and HCAEC cells and therefore could be directed as a cardiovascular protective drug for cardiovascular diseases treatment next to approval in the clinical trial studies in humans [8]. Besides pristine Pec, it was reported that thiolated Pec can be used for $Fe_3O_4$ NP coating in their hybrid form with Au NPs. In this contribution, IONPs were synthesized by coprecipitation and then coated with a gold shell. In the final step, the developed NPs were coated with thiolated Pec (TPGINs) as a result of the formation of coordinate bonds with thiol groups. Figure 21.8b presents the schematic of TPGINs synthesis. The fabricated system combines the magnetic feature of $Fe_3O_4$ NPs, surface plasmon resonance of gold, as well as stability because of Pec coating. The outcomes of curcumin (CUR) delivery, MTT, and annexin V-PI assay, furthermore TPGIN evaluation for imparting contrast in MRI specified the potential of TPGINs for MR imaging [15]. The Pec was modified with GMA in another reported Pec-based material for the MNPs modification. P. da Silva et al. used an ultrasound-induced cross-linking/ polymerization reaction between the glycidyl methacrylate modified Pec and vinyl groups of modified $TiO_2$ in the presence of magnetite and Pec to develop the polymer-coated system based on a water-in-oil emulsion. The prepared microspheres were evaluated for amoxicillin delivery and a sustained release was found in the acid medium. The authors proposed that the observed behavior is because of the NP disposition inside the polymer device. The cytotoxicity assay presented adequate biocompatibility for Pec microspheres, even after the introduction of $Fe_3O_4$ and $TiO_2$ [5]. For avoiding prolonging this section, the outcomes of the other published research works in this area are listed in Table 21.1.

## 21.2.5  $Fe_3O_4$NPs MODIFIED WITH κ-CAR OR ITS DERIVATIVES

κ-Car is an anionic water-soluble sulfated linear polysaccharide that commonly is extracted from red seaweeds in an alkaline medium. This biopolymer is the most important commercial form of carrageenans that is extensively used as a gelling agent in the cosmetic and food industry with respect to two other generic carrageenan families (iota or lambda) [75]. Besides these, κ-Car with the more polar functional groups, eco-friendly properties, low cost, biodegradable nature, and natural availability, as well as bioactive qualities like antibacterial, antioxidant, antiviral, anticoagulant, antihyperlipidemic, immunomodulating, and anticancer features, is one of the very attractive sources for MNPs modification [24,76]. The structure of κ-Car consists of repeating disaccharide units; 4-linked 3, 6-anhydro-d-galactose and 3-linked-d-galactose-4-sulfate. The presented structure contains one negative charge for each disaccharide unit that corresponds to ca. 20 wt% sulfate content. Owing to the special structure, κ-Car can form hydrogels through alkali-metal cations, such as potassium ions to offer gelation properties [77,78]. In the structure of this polymer, the sulfonate anions of κ-Car are responsible groups for the interaction with cations.

### 21.2.5.1  MNPs/κ-Car Systems Prepared via In Situ

The employment of the in situ way has shown promise in $Fe_3O_4$ NPs synthesis in the presence of κ-Car, so the products of this way were tested for possible applications in different areas. For example, Duman et al. utilized the in situ way for the preparation of magnetic oxidized multiwalled carbon nanotube (OMWCNT)-$Fe_3O_4$ and OMWCNT-κ-Car-$Fe_3O_4$ nanocomposites that were then used as an adsorbent for the MB removal from aqueous solution. A recorded SEM image clearly approved the in situ synthesis of MNPs. The findings of the adsorption studies presented a better adsorption performance for magnetic OMWCNT-$\kappa$-Car-$Fe_3O_4$ nanocomposite respect with to magnetic OMWCNT-$Fe_3O_4$ nanocomposite for the removal of MB from aqueous solution. This can be related because the $\kappa$-Car is a highly negatively charged natural polysaccharide; therefore, it is a potential candidate for the removal of cationic dyes like MB from aqueous solution [75]. In the other published research work, this research team also evaluated the performance of magnetic OMWCNT-$\kappa$-Car-$Fe_3O_4$ and OMWCNT-$Fe_3O_4$ for the removal of anionic RB5 and cationic CV dyes from aqueous solution. Adsorption studies verified that the affinity of OMWCNT-$\kappa$-Car-$Fe_3O_4$ was higher than OMWCNT-$Fe_3O_4$ for the cationic crystal violet (CV) dye adsorption, while for the anionic RB5 dye adsorption, the modification of OMWCNT with negatively charged $\kappa$-carrageenan led to a decrease in the adsorption capacity [79]. The existence of anionic functional groups in the $\kappa$-Car structure leads to its sensitivity to pH changes; considering this, in 2022, Geyik *et al.* used the in situ way to synthesize the MNPs in the presence of dimethylaminoethyl methacrylate (DMA) grafted onto $\kappa$-Car copolymer platform through a microwave-induced coprecipitation technique. A schematic illustration of the $Fe_3O_4@\kappa$-CG-g-PDMA@FL NPs preparation process is presented in Figure 21.9a. As presented in the occurred mechanism, the sulfate groups of $\kappa$-CG-g-PDMA copolymer can electrostatically interact with $Fe^{3+}/Fe^{2+}$ ions in iron salts. The XRD pattern of the $Fe_3O_4@\kappa$-CG-g-PDMA@FL NPs presented clearly and sharply defined peaks matching with a $Fe_3O_4$ magnetite phase (Figure 21.9b). This observation can be a main sign for the $Fe_3O_4@\kappa$-CG-g-PDMA@FL NPs successful synthesis. Consistent

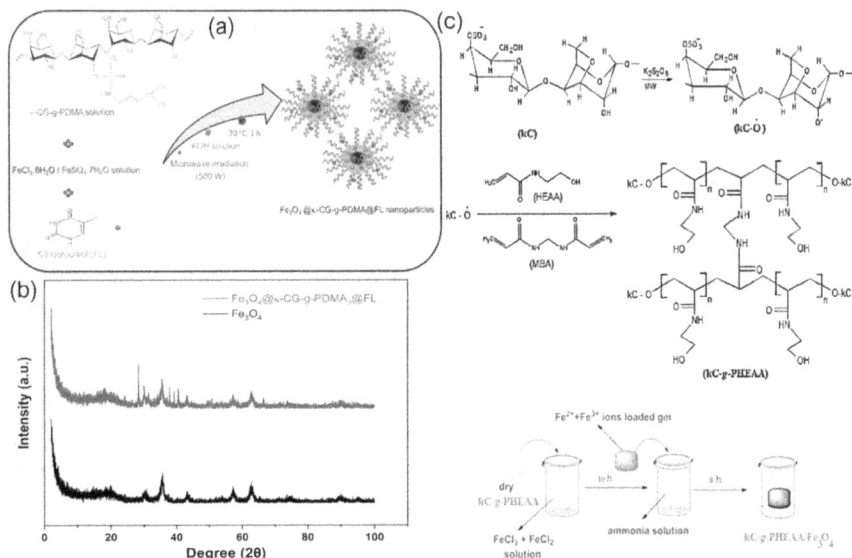

**FIGURE 21.9** $Fe_3O_4@\kappa$-CG-g-PDMA@FL NPs synthesis pathway (a), XRD patterns of $Fe_3O_4$ and $Fe_3O_4@\kappa$-CG-g-PDMA3@FL (b) and $\kappa$C-g-PHEAA/$Fe_3O_4$ nanocomposite fabrication (c). Reproduced with permission from [24,77].

with the collected results from a variety of experiments, the $Fe_3O_4$@κ-CG-g-PDMA NPs proposed a hopeful potential for cancer therapy [24]. In the other similar research, first, the κ-Car grafted with N-hydroxyethylacrylamide (κC-gPHEAA) was synthesized through microwave-assisted free radical polymerization method. Then the $Fe_3O_4$ NPs were grown into the κC-g-PHEAA network. Figure 21.9c reports the used way for its fabrication. As reported in this figure, the detection of the superparamagnetic behavior in the VSM analysis for the nanocomposite is a good sign for its preparation via in situ way. The system presented the MB and rhodamine 6 G (R6G), as well as Cu(II) and Hg(II) removal from an aqueous solution. Therefore, this could be used in water pollution management [77].

### 21.2.5.2  MNPs/κ-Car Systems Prepared via Blending

κ-Caralso can entrap the $Fe_3O_4$ NPs to obtain a new system with new surface functional groups. In this regard, the κ-Car coated magnetic iron oxide NPs were constructed and then explored as an adsorbent for the magnetically assisted removal of MB. Typically, at first, the magnetic iron oxide NPs were synthesized via the co-precipitation technique. Thereafter, acid-treated MNPs were obtained by washing with an aqueous solution of $HNO_3$ to increase the amount of the adsorbed κ-Car onto the MNPs surface. The FT-IR bands at 1229 $cm^{-1}$ and 842 $cm^{-1}$, respectively, are associated with the S-O asymmetric stretching of the sulfonate group of κ-Car and the α(1–3)-D-galactose C-O-S stretch and can confirm the successful coating procedure. Based on the outcome the authors proposed that the magnetic NPs coated with κ-Car was a highly efficient environmentally friendly dye adsorbent, with a rapid and facile separation process [78]. With some changes, in the other interesting research work, Alipour *et al.* in 2023 designed and fabricated the magnetite embedded κ-Car-based double network nanocomposite hydrogel. In the prepared structure, the κ-Car induced temperature-responsive shape memory features. Additionally, the use of amine-modified $Fe_3O_4$ NPs leads to an important role in adjustable mechanical strength, outstanding elasticity, and shape memory behavior, which allows the system to be a hopeful candidate for use as flexible electrical devices. Overall, these hydrogels are well guided in a zigzag path employing the external magnet which displays their potential for application as magnetic catheters and actuators [80]. Recently, considering the enhanced performance for the hybrid systems, the polydopamine-coated microporous composite of metal-organic frameworks and MNPs wrapped with carrageenan as a hydrogel bead form was constructed and then evaluated for the dispersive magnetic solid-phase extraction of parabens. It was reported that the use of metal organic frameworks and polydopamine improved the adsorption of parabens by π-π, hydrogen bonding, and hydrophobic interactions. The high-performance liquid chromatography (HPLC) chromatograms implied that the fabricated composite PDA@MIL101@$Fe_3O_4$@Car hydrogel sorbent was effective and can be utilized for the parabens extraction in various matrix interferences before analysis by HPLC [81].

### 21.2.5.3  MNPs/κ-Car Systems Prepared via Grafting-onto

MNPs were also can be covalently functionalized with κ-Car to obtain a new hybrid system having the functionality of both used reagents. Therefore, employing this way, F. Soares *et al.* fabricated the $Fe_3O_4@SiO_2/SiCRG$ in two distinct stages. Firstly, the $Fe_3O_4NPs$ were prepared by alkaline hydrolysis of $FeSO_4 \cdot 7H_2O$ under the $N_2$ stream. In the next step, the $Fe_3O_4NPs$ were chemically modified with the amorphous silica shells having κ-Car biopolymer. As transmission electron microscopy (TEM) analysis revealed the outcome of performed reactions was the MNPs with κ-Car hybrid siliceous thin shells. The resultant magnetic bio-hybrid eliminated the metoprolol tartrate (MTP) from water with a maximum MTP adsorption capacity of ca 300% higher than other reported sorbents in literature. The high affinity of sulfonate groups from κ-Car was used as a surface modifier to metoprolol molecules can be some reason for these results [76]. Similarly considering a more request for the development of low-price techniques to attain highly pure immunoglobulin G from the original complex matrices without any loss of its biological activity and stability, magnetic IONPs were also coated with hybrid shells of a siliceous material modified with the anionic κ-Car to obtain a system with high functionality. In the constructed structure, the polymeric shells can increase the nanoparticle's colloidal and chemical stability and inhibit the unwanted damage due to ionic and pH strength conditions which might be vital for the stability of the biological molecules [82]. For avoiding prolonging this section, the outcomes of the other published research works in this area are listed in Table 21.1.

## 21.2.6    $Fe_3O_4NPs$ MODIFIED WITH ALG OR ITS DERIVATIVES

Sodium alginate (SA), also termed *alginic acid sodium salt*, is an anionic biopolymer that commonly is extracted from seaweeds and brown algae cell walls [6]. As presented in the structure of this polymer contains α-L-glucuronic acid and β-D-mannuronic acid that are linked by (1–4). In fact, in the structure of this polymer, the monomers are covalently bound together in different sequential blocks like homopolymeric blocks with consecutive M-residues, G-residues, or alternative M- and G-residues. The carboxylate and hydroxyl functional groups in the Algmoiety render the oxygen sites to bind with the present metal cations in the acidic solution [13,23,83]. Additionally, the presence of $-COO^-$ functional group in the structure promotes the use of it for the fabrication of COOH functionalized $Fe_3O_4$ NPs. So, then the COOH-functionalized $Fe_3O_4$ NPs could be more connected with the other compounds to produce a new bi-functional system. Excellent biocompatibility, renewability mild encapsulation process at room temperature, and biodegradability are some advantages of this polysaccharide [6,84].

### 21.2.6.1    MNPs/Alg Systems Prepared via In Situ

It is proved that the use of polysaccharides as stabilizing molecules can prevent their agglomeration [4]. Considering this advantage, the $Fe_3O_4/Alg$ nanocomposites were fabricated based on the reported procedure for in situ way in which a mixture of $FeCl_2$ and ethanol were added into the aqueous solutions of Alg. Afterward, petroleum ether was also added under strong magnetic stirring. The addition of an alkaline solution and then exposure to air at room temperature achieved the final product. Magnetization measurements approved the magnetite wrapping in polymer via detection of superparamagnetic nature at room temperature [13]. The insitu way assisted with the ultrasound procedure was also utilized for the fabrication of stable water-dispersible superparamagnetic $Fe_3O_4$@Alg nanocomposite. In this procedure, the formed radical hydroxyl (HO·) during the ultrasound cavitation procedure was employed to oxidize ferrous ($Fe^{2+}$) to ferric ($Fe^{3+}$) cations in the existence of NaOH, and the $Fe_3O_4$ NPs were obtained. Interaction between coating Algand IONPs was identified by FT-IR. The dispersible composite exhibited superparamagnetic behavior with a Ms value of 10.96 emu $g^{-1}$ and negligible coercive field (Hc) and remanent magnetization (Mr). All findings offered $Fe_3O_4$@Alg nanocomposite as a potential vehicle to be employed as an MRI contrast agent [85]. In the other research, taking the magnetic mats with nanofibrous structure capability in promotion of cell adhesion and tumor-killing within an alternating magnetic field, the Alg-based electrospun mats were designed, fabricated, and thereafter were treated via a covalent or ionic cross-linking technique. Eventually, it was followed by chelation with $Fe^{2+}/Fe^{3+}$ for chemical co-precipitation of $Fe_3O_4$ nanoparticles. Both prepared magnetic mats exposed no important cytotoxicity and the $Fe_3O_4$-SA/PVA mat presented greater influence on tumor cells than the $Fe_3O_4$ NPs in vitro

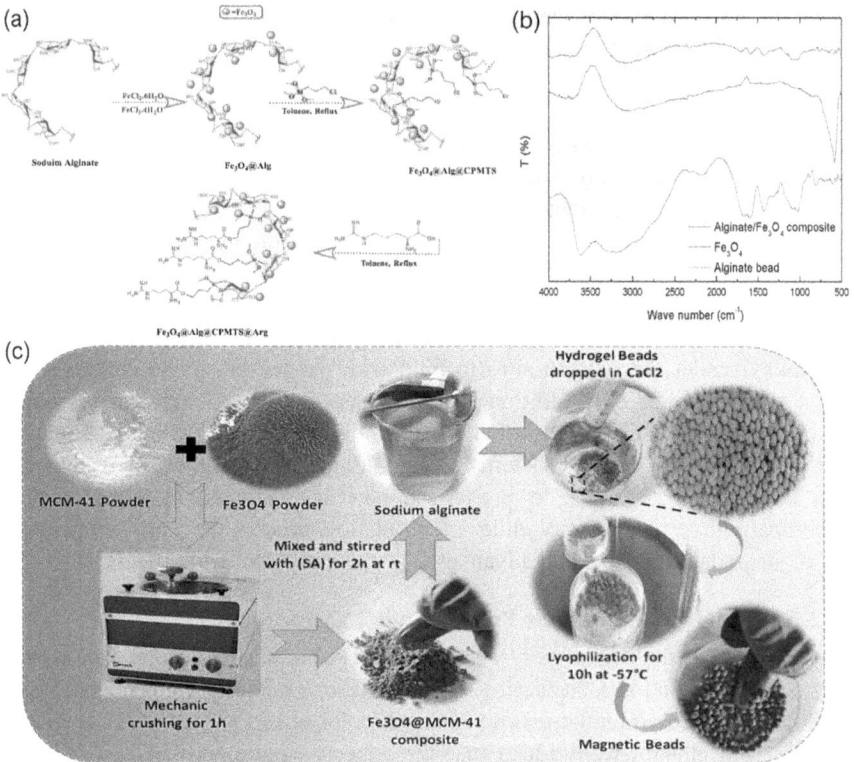

**FIGURE 21.10** Preparation pathway of $Fe_3O_4$@Alg@CPTMS@Arg magnetic nanocomposite (a), FT-IR spectrum of the Alg/$Fe_3O_4$ composite (b), a pathway of $Fe_3O_4$@MCM-41 encapsulated in calcium alginate in the beads (c). Reproduced with permission from [6,89,93].

hyperthermia [83]. The synthesized superparamagnetic IONPs/Alg by in situ protocol can also be coated with secondary reagents like L-arginine (Arg) to obtain a new system with more efficiency. A schematic of the $Fe_3O_4$@Alg@CPTMS@Arg preparation is reported in Figure 21.10a. The synthesized system was explored for the fabrication of 2,4,5-triarylimidazoles derivatives through a one-pot three-component reaction between aldehyde derivatives, ammonium acetate, and benzil under reflux. Apart from this activity, an antibacterial study presented as more inhibitory for the amino acid-modified system, compared with the bare Alg. The combined effect of L-Arg and Algis assumed the reason for this observation [6].

Besides the pristine Alg, the combination of it with other polymers was also studied as a platform for in situ $Fe_3O_4$NPs preparation. So, in one interesting research, $K_2S_2O_8$ as an initiator was used to form the free radical centers on the SA backbone which were then reacted with the present double bonds (vinylic) in the methyl methacrylate (MMA) structure to construct the graft copolymer. In the following, the $Fe_3O_4$ NPs were in-situ synthesized from the iron precursors ($FeSO_4$ and $FeCl_3$) in the alkaline medium of $NH_4OH$, and SA-g-PMMA/$Fe_3O_4$ was developed. Finding approved that the synthesized grafted co-polymeric nanocomposites have better performance than the unmodified SA itself in the electrochemical and gravimetric analysis [23].

### 21.2.6.2  MNPs/Alg Systems Prepared via Blending

Pre-synthesized $Fe_3O_4$ NPs can be coated with Alg biopolymer in the presence or absence of any cross-linker to obtain a new hybrid structure with advanced functionality. In this regard, Xu et al. designed and fabricated the Alg-coated $Fe_3O_4$ NPs with a core-shell structure through the so-called two-step technique. Generally, at first, the $Fe_3O_4$ NPs were prepared in a co-precipitation procedure and next they were coated with Alg with different amounts in a chemisorption way to optimize the colloidal stability of $Fe_3O_4$ NPs modified with Alg through the zeta potential recording [86]. In the other research, with some differences, the SA solution was mixed with the MNPs, and then dripped into a calcium chloride solution and the beads were formed. The appearance of major peaks related to magnetite in the XRD pattern of the final system indicated the successful fabrication of magnetic sorbent without any change in the chemical structure of magnetite [87]. In another interesting research, $Fe_3O_4$ NPs were added to the SA solution to elaborate magnetic alginate beads after ionic cross-linking with $Ca^{2+}$ ions (MABs). Thereafter the bioabsorption of copper ions was evaluated. The FT-IR results made on the material after adsorption identified the presence of copper carboxylate complex in the unidentate form and confirm its adsorption by the system [88]. Similarly, in one other interesting effort, the Alg/$Fe_3O_4$ composite synthesized via this technique employing different amounts of $Fe_3O_4$ (varied from 0.5 to 10 wt% (w/w)). FT-IR spectroscopy approved the success of Alg/$Fe_3O_4$ composite synthesis via the appearance of both absorption bands for neat Alg bead and $Fe_3O_4$ (Figure 21.10b). The achieved system was then evaluated for Sr adsorption behavior after detailed characterization. It was observed that in a system containing the mixed cation (Na, Mg, Ca, K, and Sr), Sr was the only cation adsorbed on the Alg/$Fe_3O_4$ composite, while the Sr adsorption capacity was reduced because of the presence of the coexisting ions. Considering the Sr adsorption capacity as well as magnetic separation the 1% of $Fe_3O_4$ was introduced as the optimal amount for $Alg/Fe_3O_4$ composite [89]. Norfloxacin (NOF) is the other pollutant its removal was studied by Niu*et al.* using the constructed $Alg/Fe@Fe_3O_4$ core/shell structured NPs as heterogeneous Fenton nanocatalysts. It is anticipated that the ALG/Fe coat can protect the Fe(II)/Fe(III) species on the inner $Fe_3O_4$ core from precipitation at pH 4.5–6.5, so that the OH generation could happen simply. Together the observations can propose that the NOF degradation could be performed in an extensive range of pH for $Fe_3O_4@ALG/Fe$-$H_2O_2$ system [90]. Magnetite/coir pith supported SA beads was also prepared and then employed for the removal of malachite under a parametric optimized condition of the adsorption procedure. Briefly, a solution of SA, $Fe_3O_4$ NPs, and coir pith were mixed and then converted as Fe/CP/NaAl beads using the ionic gelation method by freshly prepared $CaCl_2$ at room temperature. Additionally, the eco-toxicology test displayed that the fabricated Fe/CP/NaAl bead is an effective water treatment system without harmful influence on non-specific targets [91]. In a relatively similar way, the $Fe_3O_4$ and semiconductor zinc oxide (ZnO) powder were either separately or together wrapped by calcium alginate (Ca-Alg) beads to form a hybrid photocatalyst. ZnO/$Fe_3O_4$/Ca-Alg beads were evaluated to remove RR180 dye and photoreduction of Cr(VI) and acceptable findings were obtained. Repeated use of hybrid $ZnO/Fe_3O_4$/Ca-Alg beads also showed very good performance after ten cycles [92]. In some instances, the $Fe_3O_4$ NPs were also composited with other NPs, and the obtained nanohybrid then was coated with Alg. For example, magnetic biochar (BC) was also combined with SA employing glutaraldehyde as a cross-linker to prepare BC-G-SA nanocomposite with a surface area of 5.6709 for Pb (II) adsorption under several conditions studied [84]. Hachemaoui*et al.* combined the $Fe_3O_4$ NPs were with mesoporous silica (MCM-41) and then the fabricated $Fe_3O_4@MCM$-41 was encapsulated in calcium alginate in the beads form (hydrogel and aerogel), as reported in Figure 21.10c. The systems presented an acceptable catalytic activity towards the reduction reaction of two dyes: Orange G (OG) and MB [93]. The $Fe_3O_4$ coated with Alg can further improve via the insertion of new reagents in the final structure. Employing this way the magnetic alginate microparticles (MAMs) were prepared through the embedding of IONPs in calcium alginate gels. The magnetically templated hydrogel was processed with ethylenediaminetetraacetic acid (EDTA) and then the MAM was removed. The in vitro studies on cultured rat Schwann cells on templated hydrogels to model peripheral nerve injury repair reveal their tendency for providing cell guidance along the length of the channel which is a good sign in tissue repair applications [94]. Ghorbani-Vaghei *et al.* used the Alg covered magnetite NPs for the gold NPs decoration on them ($Fe_3O_4@Alg$-AuNPs). In the structure of the fabricated system, the anionic SA shell on the $Fe_3O_4$ core acted as a stabilizing agent for Au NPs immobilization. Thereafter, the catalytic performance of the final system was studied for the reduction of 4-nitrophenol to 4-aminophenol and outstanding catalytic activity was obtained in the 4-NP reduction with a high rate constant [95].

### 21.2.6.3 MNPs/Alg Systems Prepared via Grafting-onto

IONPs can be covalently modified with SA or its derivatives to exhibit new performance in comparison with both constituents. In this regard, Peng and coworkers fabricated a novel onco-theranostic vehicle by combining the redox/pH dual responsibility of Alg-coated SPION, in which the reducible Alg coating can answer to the more concentration of glutathione (GSH) and the acidic medium of cancer cells. A schematic of the SPIONs modification by introducing a layer of polymer (functionalized Alg with cystaminedihydrochloride for the introduction of disulfide bonds) onto the surface of SPIONs is presented in Figure 21.11a. The observed increase in the particle size from 19.6 nm for SPION-2 to 89.7 nm for SPIONAlgSS and 91.3 nm for SPIONAlg, respectively, can verify that the pH/redox dual responsive nanogels were effectively obtained owing to the introduction of polymer on the surface of SPIONs. DOX was loaded respectively at about 48.98 wt% and 42.70 wt% on SPIONAlg and SPIONAlgSS. Figure 21.11b and Figure 21.11c show the profiles of drug release under different conditions. As it is shown, the acidic release medium which simulates the acidic tumor extracellular pH led to a higher release rate compared with PBS with a pH 7.4. Additionally, the amount of released DOX for SPIONAlg was not dependent on the GSH concentration, but DOX-loaded SPIONAlgSS was evidently accelerated by adding GSH. The burst release of DOX can be ascribed to the disulfide bonds induced by the structure deformation of SPIONAlgSS. Findings can support the co-triggered nano-theranostic role of synthesized systems for both diagnostic and antitumor chemotherapy applications [96]. Khazaie *et al.* in 2021 designed and developed the

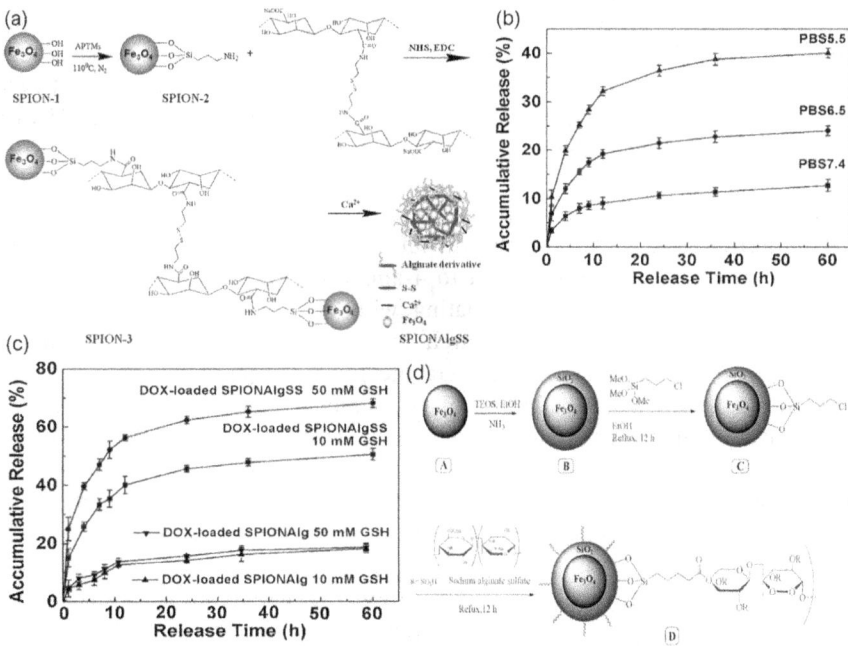

**FIGURE 21.11**  The procedure of SPIONAlgSS fabrication (a), pH-triggered drug release from DOX-loaded SPIONAlgSS in PBS (b), redox-triggered drug release from DOX-loaded SPIONAlg and SPIONAlgSS inPBS7.4 with GSH (c), and the involved steps in Fe$_3$O$_4$@SiO$_2$-SAS synthesis (d).Reproduced with permission from [96,97].

magnetic core-hydrophilic shell NPs as a well-organized draw solute for forward osmosis (FO) procedures to provide better surface specificity. As it is presented in Figure 21.11d, initially the pre-synthesized $Fe_3O_4$ NPs were coated with a silica shell. Thereafter, the functionalization was accomplished by the nucleophilic substitution of chlorine groups in the $Fe_3O_4@SiO_2(CH_2)_3Cl$ with carboxylate groups of SAS and $Fe_3O_4@SiO_2$-SAS was achieved. The finding approved that binding of hydrophilic SAS onto the magnetic $Fe_3O_4$ surface achieved the hybrid NPs with high solubility, which might increase the osmotic pressure and can desalinate and real wastewater saline feed solutions up to 0.2 M NaCl (261). For avoiding prolonging this section, the outcomes of the other published research works in this area are listed in Table 21.1.

## 21.2.7   $Fe_3O_4$NPs Modified with Two Polysaccharides

A survey of the published literature shows that in some cases the mixture of two different was used for the modification of MNPs to obtain a new construct having the positive feature of both of them.

### 21.2.7.1   MNPs/Binary Polysaccharides Systems Prepared via In Situ

In situ $Fe_3O_4$ NPs can perform in the presence of both different polysaccharides to achieve a new system with more stability. For example, a cationic polysaccharide simply can form the polyelectrolyte complex (PEC) with the anionic polysaccharides via the reaction of opposite charges. Taking advantage of the PEC, Yang and coworkers first prepared CS-conjugated magnetic NPs via the co-precipitation method. In the next step, the CS-magnetite captured HA on its surface at pH 6 through the formation of PEC [98]. Alg in a combined form with dextran was also used for the encapsulation of magnetite, silica gold, or titanium dioxide NPs to protect their stability and microstructures. The as-prepared composite gels have shown outstanding glucose sensing as well as catalytic reduction of 4-nitrophenol to 4-aminophenol with good recyclability and reusability [99]. $Fe_3O_4$ NP coating with dual polysaccharides was also evaluated by the Mahdavinia research group, so, they first synthesized the magnetite in the presence of κ-Ca via the in situ technique. Thereafter, the attained magnetic-κ-Car was cross-linked, employing the CS as the polycation biopolymer via the electrostatic interactions between negatively charged sulfate groups on κ-Car and positively charged amine groups on CS. Swelling capacity studies showed that all of the fabricated hydrogels have a high swelling amount in basic solutions. A pathway representative of the fabrication process of magnetic bio-nanocomposites is exposed in Figure 21.12a. Methotrexate was loaded on hydrogels and the in vitro release results showed that the release is pH-dependent, with a high percentage at pH 7.4 [100]. With some changes, this research group also fabricated magnetic CS/κ-Car and explored it as the bioadsorbents for anionic eriochrome black-T (EBT) removal. So in this work, initially magnetic CS was fabricated in the presence of CS with different molecular weights and then the hybrid was cross-linked with κ-Car to enhance its stability under acidic conditions. A scheme displaying the involved steps for preparing a magnetic

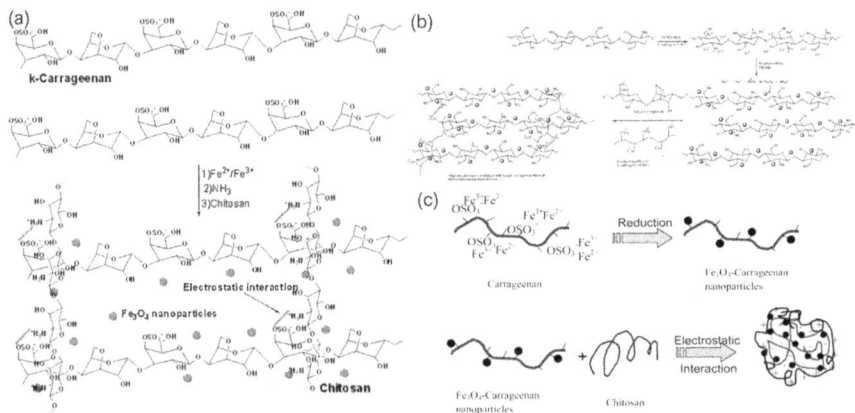

**FIGURE 21.12** Utilized procedure for the magnetic κ-Car/CS complex (a), magnetic CS-κ-Car complexes (b), and CS/κ-Car $Fe_3O_4$ nanocomposites (c) fabrication. Reproduced with permission from [100–102].

chitosan-carrageenan system is shown in Figure 21.12b. Studies of the EBT behavior obtained the maximum adsorption capacities of about 199, 235, and 280 mg/g respectively for bioadsorbents prepared with high, medium, and low molecular weights of CS [101]. In another similar research MNPs in situ synthesized in the κ-Car then were coated with CS in varying mass ratios of CS to $Fe_3O_4$-κ-Car nanocomposite. The carried-out typical procedure was reported in Figure 21.12c. Afterward, the nanocomposite was explored for bovine serum albumin as a model of protein delivery applications. The observed controlled release in an intestinal medium with respect to acidic medium, sustained release up to 85% in 30 min suggested the capability of CS/κ-Car nanocomposites as magnetically targeted therapeutic macromolecules delivery in a controlled manner [102]. MNPs were also in situ fabricated in the coated form with CMCS and κ-Car pH-responsive beads form to reach a controlled drug release behavior. The appearance of the XRD peaks of neat MNPs in the recorded spectra for the hydrogels was the main reason for the successful formation of these NPs within the hydrogel matrix. The diclofenac sodium release studies displayed smart behaviors on the subject of external alternating magnetic fields and physiological simulated pHs, observing a maximum releasing about 82% at pH 7.4 [103]. In similar research, the IONPs modified by a binary mixture of κ-Car/CMCS effectively adsorbed the bovine serum albumin from the aqueous solution. Firstly, the biopolymers solution was slowly added into $Fe^{2+}/Fe^{+3}$, and the beads were formed via ionic cross-linking. Then beads were kept in an ammonia solution to obtain $Fe_3O_4$ NPs in a biopolymer matrix. Eventually, the fabricated magnetic nanocomposites were suspended in a saline solution containing $CaCl_2$ and KCl salts for cross-linking via electrostatic interaction of $Ca^{2+}$ cations with a carboxylate group ($-CO_2^-$) on CMCS and also $K^+$ cations with sulfate groups ($-OSO_3^-$) on κ-Car [104].

### 21.2.7.2  MNPs/Binary Polysaccharides Systems Prepared via Blending

$Fe_3O_4$ NPs were coated with CS/agarose by mixing all three portions and the product used for CUR delivery. The observed sustained and pH-responsive release pattern could be related to the employed polymers as a coat along with the magnetic nature. The possible pathway of the nanocarrier fabrication, CUR loading, and its releases are given in Figure 21.13a [105]. CS also can be combined with Alg in the PEC form to coat the $Fe_3O_4$ surface as the consecutive layers over the MNP. Han *et al.* used this platform for Ag NPs and constructed $Fe_3O_4$/CS-Alg/Ag NPs as a catalyst of multicomponent reaction. A schematic of the $Fe_3O_4$/CS-Alg/Ag nanocomposite preparation and its use for the one-pot four-component synthesis of some 2H-indazolo[2,1-b]phthalazine-triones is presented in Figure 21.13b. Additionally, the performed cellular tests displayed that the catalyst could be used as a lung protective drug for the lung disease treatment after verifying the clinical trial studies in humans relay its biocompatibility, antioxidant nature [106].

Another report of magnetite NPs modification was reported by Zeng *et al.* in 2022, when they fabricated the magnetic Alg-CS porous beads based on iron (Fe)

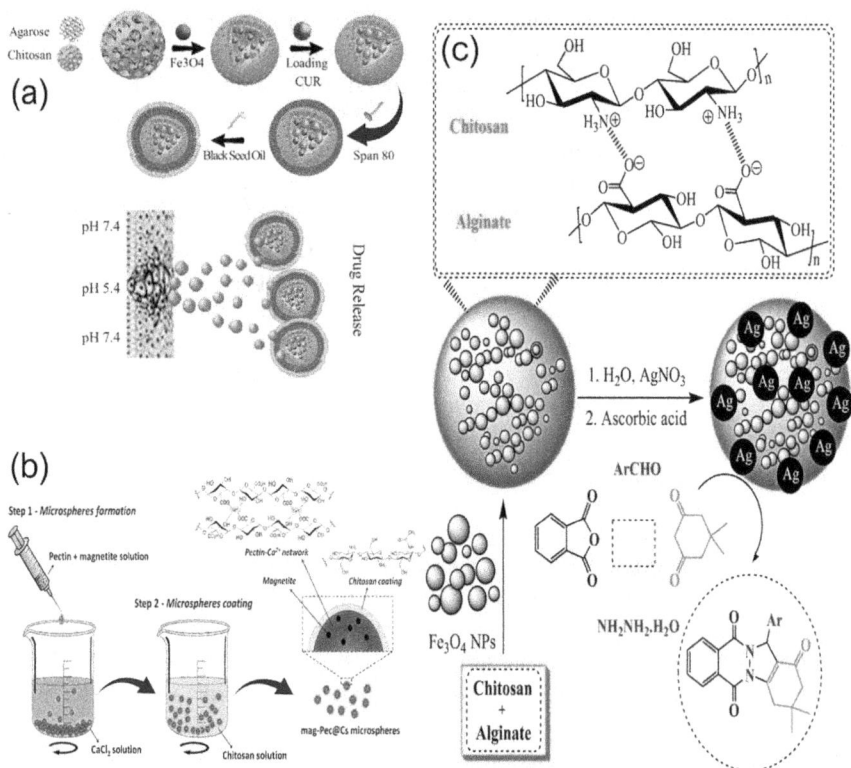

**FIGURE 21.13**  Schematic of $Fe_3O_4$/CS/agarose fabrication, CUR loading, adding surfactants, and drug release (a) $Fe_3O_4$/CS-Alg/Ag synthesis and application (b) and mag-Pec@Cs microspheres synthesis (c). Reproduced with permission from [105, 106, 110].

sludge (M-ACFBs) in a two-step procedure. They recorded strong saturation magnetization; ~15.0 emu/g was good evidence for the magnetic nature of the final construct. The optimum condition was determined for As(V) adsorption after assessing the effects of contact time, pH, temperature, and coexisting ions as well as the adsorption isotherms and kinetics were thoroughly explored. Eventually, the $4.2 \pm 0.4$ mg/g was determined as the maximum adsorption capacity [107]. In similar research, the CS-St functionalized $Fe_3O_4$ nanocomposite was used as a platform for Pd NPs synthesis on it ($Fe_3O_4$@CS-Starch/Pd), which then was used as a catalyst of Suzuki-Miyaura coupling. The used CS-St composite reduced the leaching possibility of metal NPs during the catalysis [14]. CS/St binary polysaccharides modified magnetite NPs were also used as a platform for AgNPs synthesis. For this, a mixture of CS-St was prepared by dissolving a mixture of them. Thereafter, $Fe_3O_4$@CS-St nanocomposite acted as a platform for in situ Ag NPs preparation. All outcomes of the MTT assay approved the anticancer feature of $Fe_3O_4$@CS-Starch/Ag. Therefore, the authors mentioned that $Fe_3O_4$@CS-Starch/Ag could be applied as an effective drug in human breast cancer treatment after the clinical study [108]. A completely green route was utilized for the fabrication of multi-stimuli-responsive magnetic CS-cross-linked κ-Car/MMt as a carrier for sunitinib-controlled delivery. So, initially, $Fe_3O_4$ NPs in situ were synthesized in the presence of MMt nanoclay. Then mMMt was blended with κ-Car. In the end, the mixture of magnetic κ-Car/MMt was poured into an acidic CS solution. The electrostatic interactions between the protonated amine groups on CS with anionic sulfate groups of κ-Car produced the magnetic CS-cross-linked κ-Cs/MMt carrier. Furthermore, the hydrogen bonding between the used ingredients is probable [109]. Pec is another polysaccharide that was used in the combined form with CS for magnetite coating. Typically, S. A. Lemos et al. first prepared the magnetic-Pec microspheres by ionotropic gelation of mixture with $CaCl_2$; afterward, polyelectrolyte complexation occurred with s CS addition. The used approach for the magnetic-Pec microspheres preparation and its coating by CS is presented in Figure 21.13c. After complete characterization, metamizole (Mtz) was a drug model encapsulated at about 85%. It accomplished release experiments in pH-simulated gastric and intestinal fluids recommended that the release is pH-dependent and the release value increased after the application of an external magnetic [110].

In the other interesting research, the hydrogel nanocomposite of CAG-NaAlg-cl-polyAA/$Fe_3O_4$ was constructed via a free radical cross-linking way using MBA and acrylic acid, respectively, as a cross-linker and monomer by the addition of $Fe_3O_4$ NPs before gelation. From the dye adsorption test results, it was observed that the system is suitable for auramine O, crystal violet, and malachite green dye removal [111]. Yang et al. fabricated the magnetic Alg microparticles (MAM) and CS-coated magnetic Alg (CMAM) via a microfluidic approach with uniform size distribution for controlled amoxicillin delivery under mild conditions. Drug release behavior was pH sensitive for all studied samples and detected that the CS coating on the magnetic Alg surface gives CMAM time extension in drug release by two times, attaining controlled and sustained drug release [112]. To avoid prolonging this section, the outcomes of the other published research works in this area are listed in Table 21.1.

## ACKNOWLEDGMENTS

This research was supported with a research grant of the University of Tabriz (number SAD/3930-14001225).

## REFERENCES

1. Mohapatra S, Asfer M, Anwar M, Ahmed S, Ahmad FJ, Siddiqui AA. Carboxymethyl Assam Bora rice starch coated SPIONs: Synthesis, characterization and in vitro localization in a micro capillary for simulating a targeted drug delivery system. International Journal of Biological Macromolecules. 2018;115:920–932.
2. Anushree C, Philip J. Efficient removal of methylene blue dye using cellulose capped Fe3O4 nanofluids prepared using oxidation-precipitation method. Colloids and Surfaces A: Physicochemical and Engineering Aspects. 2019;567:193–204.
3. Anirudhan T, Shainy F. Adsorption behaviour of 2-mercaptobenzamide modified itaconic acid-grafted-magnetite nanocellulose composite for cadmium (II) from aqueous solutions. Journal of Industrial and Engineering Chemistry. 2015;32:157–166.
4. Alexandrovskaya YM, Pavley YR, Grigoriev YV, Grebenev V, Shatalova T, Obrezkova M. Thermal behavior of magnetite nanoparticles with various coatings in the range 30–1000° C. Thermochimica Acta. 2022;708:179120.
5. da Silva EP, Sitta DL, Fragal VH, Cellet TS, Mauricio MR, Garcia FP, et al. Covalent TiO2/pectin microspheres with Fe3O4 nanoparticles for magnetic field-modulated drug delivery. International Journal of Biological Macromolecules. 2014;67:43–52.
6. Amirnejat S, Nosrati A, Javanshir S, Naimi-Jamal MR. Superparamagnetic alginate-based nanocomposite modified by L-arginine: An eco-friendly bifunctional catalysts and an efficient antibacterial agent. International Journal of Biological Macromolecules. 2020;152:834–845.
7. Abdelaziz MA, Owda ME, Abouzeid RE, Alaysuy O, Mohamed EI. Kinetics, isotherms, and mechanism of removing cationic and anionic dyes from aqueous solutions using chitosan/magnetite/silver nanoparticles. International Journal of Biological Macromolecules. 2023;225:1462–1475.
8. Zhang W, Veisi H, Sharifi R, Salamat D, Karmakar B, Hekmati M, et al. Fabrication of Pd NPs on pectin-modified Fe3O4 NPs: A magnetically retrievable nanocatalyst for efficient C–C and C–N cross coupling reactions and an investigation of its cardiovascular protective effects. International Journal of Biological Macromolecules. 2020;160:1252–1262.
9. Barkhordari S, Alizadeh A, Yadollahi M, Namazi H. One-pot synthesis of magnetic chitosan/iron oxide bio-nanocomposite hydrogel beads as drug delivery systems. Soft Materials. 2021;19(4):373–381.
10. Pooresmaeil M, Namazi H. Fabrication of a smart and biocompatible brush copolymer decorated on magnetic graphene oxide hybrid nanostructure for drug delivery application. European Polymer Journal. 2021;142:110126.
11. Namvari M, Namazi H. Magnetic sweet graphene nanosheets: preparation, characterization and application in removal of methylene blue. International Journal of Environmental Science and Technology. 2016;13:599–606.

12. Correa JR, Bordallo E, Canetti D, León V, Otero-Díaz LC, Negro C, et al. Structure and superparamagnetic behaviour of magnetite nanoparticles in cellulose beads. Materials Research Bulletin. 2010;45(8):946–953.

13. Srivastava M, Singh J, Yashpal M, Gupta DK, Mishra R, Tripathi S, et al. Synthesis of superparamagnetic bare $Fe_3O_4$ nanostructures and core/shell ($Fe_3O_4$/alginate) nanocomposites. Carbohydrate Polymers. 2012;89(3):821–829.

14. Veisi H, Joshani Z, Karmakar B, Tamoradi T, Heravi MM, Gholami J. Ultrasound assisted synthesis of Pd NPs decorated chitosan-starch functionalized $Fe_3O_4$ nanocomposite catalyst towards Suzuki-Miyaura coupling and reduction of 4-nitrophenol. International Journal of Biological Macromolecules. 2021;172:104–113.

15. Sood A, Arora V, Kumari S, Sarkar A, Kumaran SS, Chaturvedi S, et al. Imaging application and radiosensitivity enhancement of pectin decorated multifunctional magnetic nanoparticles in cancer therapy. International Journal of Biological Macromolecules. 2021;189:443–454.

16. Torkashvand N, Sarlak N. Fabrication of a dual T1 and T2 contrast agent for magnetic resonance imaging using cellulose nanocrystals/$Fe_3O_4$ nanocomposite. European Polymer Journal. 2019;118:128–136.

17. Pooresmaeil M, Javanbakht S, Nia SB, Namazi H. Carboxymethyl cellulose/ mesoporous magnetic graphene oxide as a safe and sustained ibuprofen delivery bio-system: Synthesis, characterization, and study of drug release kinetic. Colloids and Surfaces A: Physicochemical and Engineering Aspects. 2020;594:124662.

18. Pooresmaeil M, Namazi H. β-Cyclodextrin grafted magnetic graphene oxide applicable as cancer drug delivery agent: Synthesis and characterization. Materials Chemistry and Physics. 2018;218:62–69.

19. Namvari M, Namazi H. Preparation of efficient magnetic biosorbents by clicking carbohydrates onto graphene oxide. Journal of Materials Science. 2015;50:5348–5361.

20. Namvari M, Namazi H. Synthesis of magnetic citric-acid-functionalized graphene oxide and its application in the removal of methylene blue from contaminated water. Polymer international. 2014;63(10):1881–1888.

21. Mikhaylova M, Kim DK, Bobrysheva N, Osmolowsky M, Semenov V, Tsakalakos T, et al. Superparamagnetism of magnetite nanoparticles: Dependence on surface modification. Langmuir. 2004;20(6):2472–2477.

22. Zhang Y, Kohler N, Zhang M. Surface modification of superparamagnetic magnetite nanoparticles and their intracellular uptake. Biomaterials. 2002; 23(7):1553–1561.

23. Kesari P, Udayabhanu G, Roy A. Biopolymer sodium alginate based titania and magnetite nanocomposites as natural corrosion inhibitors for mild steel in acidic medium. Journal of Industrial and Engineering Chemistry. 2023;122:303–325.

24. Geyik G, Işıklan N. Multi-stimuli-sensitive superparamagnetic κ-carrageenan based nanoparticles for controlled 5-fluorouracil delivery. Colloids and Surfaces A: Physicochemical and Engineering Aspects. 2022;634:127960.

25. Gu H, Zhou X, Lyu S, Pan D, Dong M, Wu S, et al. Magnetic nanocellulose-magnetite aerogel for easy oil adsorption. Journal of Colloid and Interface Science. 2020;560:849–856.

26. Arantes ACC, das Graças Almeida C, Dauzacker LCL, Bianchi ML, Wood DF, Williams TG, et al. Renewable hybrid nanocatalyst from magnetite and cellulose for treatment of textile effluents. Carbohydrate Polymers. 2017;163:101–107.

27. Namazi H, Jafarirad S. Preparation of the new derivatives of cellulose and oligomeric species of cellulose containing magneson II chromophore. Journal of Applied Polymer Science. 2008;110(6):4034–4039.

28. Abd Elrhman H. Synthesis and characterization of core-shell magnetite nanoparticles with modified nano-cellulose for removal of radioactive ions from aqueous solutions. Results in Materials. 2020;8:100138.

29. Kargar PG, Len C, Luque R. Cu/cellulose-modified magnetite nanocomposites as a highly active and selective catalyst for ultrasound-promoted aqueous O-arylation Ullmann and sp-sp2 Sonogashira cross-coupling reactions. Sustainable Chemistry and Pharmacy. 2022;27:100672.

30. Baghaei B, Skrifvars M. All-cellulose composites: A review of recent studies on structure, properties and applications. Molecules. 2020;25(12):2836.

31. Niakan M, Masteri-Farahani M, Shekaari H, Karimi S. Pd supported on clicked cellulose-modified magnetite-graphene oxide nanocomposite for CC coupling reactions in deep eutectic solvent. Carbohydrate Polymers. 2021;251:117109.

32. da Rocha Santana RM, Napoleão DC, Rodriguez-Diaz JM, de Mendonça Gomes RK, Silva MG, de Lima VME, et al. Efficient microbial cellulose/Fe$_3$O$_4$ nanocomposite for photocatalytic degradation by advanced oxidation process of textile dyes. Chemosphere. 2023;326:138453.

33. Alizadeh Z, Jonoush ZA, Rezaee A. Three-dimensional electro-Fenton system supplied with a nanocomposite of microbial cellulose/Fe$_3$O$_4$ for effective degradation of tetracycline. Chemosphere. 2023:137890.

34. Chen X, Ye Z, Yang F, Feng J, Li Z, Huang C, et al. Magnetic cellulose microcrystals with tunable magneto-optical responses. Applied Materials Today. 2020;20:100749.

35. Leonel AG, Mansur HS, Mansur AA, Caires A, Carvalho SM, Krambrock K, et al. Synthesis and characterization of iron oxide nanoparticles/carboxymethyl cellulose core-shell nanohybrids for killing cancer cells in vitro. International Journal of Biological Macromolecules. 2019;132:677–691.

36. Luo X, Lei X, Xie X, Yu B, Cai N, Yu F. Adsorptive removal of Lead from water by the effective and reusable magnetic cellulose nanocomposite beads entrapping activated bentonite. Carbohydrate Polymers. 2016;151:640–648.

37. Işıklan N, Hussien NA, Türk M. Hydroxypropyl cellulose functionalized magnetite graphene oxide nanobiocomposite for chemo/photothermal therapy. Colloids and Surfaces A: Physicochemical and Engineering Aspects. 2023;656:130322.

38. Morsi RE, Al-Sabagh AM, Moustafa YM, ElKholy SG, Sayed MS. Polythiophene modified chitosan/magnetite nanocomposites for heavy metals and selective mercury removal. Egyptian Journal of Petroleum. 2018;27(4):1077–1085.

39. Kim H-R, Jang J-W, Park J-W. Carboxymethyl chitosan-modified magnetic-cored dendrimer as an amphoteric adsorbent. Journal of Hazardous Materials. 2016;317:608–616.

40. Bulin C. Combination mechanism of the ternary composite based on Fe$_3$O$_4$-chitosan-graphene oxide prepared by solvothermal method. International Journal of Biological Macromolecules. 2023:123337.

41. Li W, Zhang X, Xue Z, Mi Y, Ma P, Fan D. Ginsenoside CK production by commercial snailase immobilized onto carboxylated chitosan-coated magnetic nanoparticles. Biochemical Engineering Journal. 2021;174:108119.

42. Algamdi M, Alshahrani A, Alsuhybani M. Chitosan grafted tetracarboxylic functionalized magnetic nanoparticles for removal of Pb (II) from an aqueous environment. International Journal of Biological Macromolecules. 2023;225:1517–1528.

43. Maruthupandy M, Muneeswaran T, Anand M, Quero F. Highly efficient multifunctional graphene/chitosan/magnetite nanocomposites for photocatalytic degradation of important dye molecules. International Journal of Biological Macromolecules. 2020;153:736–746.

44. Bulin C. Formation mechanism of a ternary nanohybrid based on magnetite-chitosan-graphene oxide according to HSAB theory. Journal of Physics and Chemistry of Solids. 2023;176:111260.

45. Lian Z, Li Y, Xian H, Ouyang X-k, Lu Y, Peng X, et al. EDTA-functionalized magnetic chitosan oligosaccharide and carboxymethyl cellulose nanocomposite: Synthesis, characterization, and Pb (II) adsorption performance. International Journal of Biological Macromolecules. 2020;165:591–600.

46. Heidari F, Razavi M, Bahrololoom ME, Bazargan-Lari R, Vashaee D, Kotturi H, et al. Mechanical properties of natural chitosan/hydroxyapatite/magnetite nanocomposites for tissue engineering applications. Materials Science and Engineering: C. 2016;65:338–344.

47. Ziegler-Borowska M, Chełminiak D, Siódmiak T, Sikora A, Marszałł MP, Kaczmarek H. Synthesis of new chitosan coated magnetic nanoparticles with surface modified with long-distanced amino groups as a support for bioligands binding. Materials Letters. 2014;132:63–65.

48. Baghban A, Heidarizadeh M, Doustkhah E, Rostamnia S, Rezaei PF. Covalently bonded pancreatic lipase onto the dithiocarbamate/chitosan-based magnetite: Stepwise fabrication of $Fe_3O_4$@ CS/NHCS-Lip as a novel and promising nanobiocatalyst. International Journal of Biological Macromolecules. 2017;103:1194–1200.

49. Li Y, Dai Y, Tao Q, Gao Z, Xu L. Ultrahigh efficient and selective adsorption of U (VI) with amino acids-modified magnetic chitosan biosorbents: Performance and mechanism. International Journal of Biological Macromolecules. 2022;214:54–66.

50. Hamza MF, Hamad DM, Hamad NA, Adel A-H, Fouda A, Wei Y, et al. Functionalization of magnetic chitosan microparticles for high-performance removal of chromate from aqueous solutions and tannery effluent. Chemical Engineering Journal. 2022;428:131775.

51. Pooresmaeil M, Namazi H. Positively charged covalent organic framework modified magnetic chitosan as a smart device for efficient diclofenac sodium removal from water. Chemical Engineering Journal. 2023;452:139557.

52. Pooresmaeil M, Namazi H. Chitosan coated $Fe_3O_4$@Cd-MOF microspheres as an effective adsorbent for the removal of the amoxicillin from aqueous solution. International Journal of Biological Macromolecules. 2021;191:108–117.

53. Manjusha V, Rajeev M, Anirudhan T. Magnetic nanoparticle embedded chitosan-based polymeric network for the hydrophobic drug delivery of paclitaxel. International Journal of Biological Macromolecules. 2023;235:123900.

54. El-Shahawy AA, Zohery M, El-Dek S. Theranostics platform of Abemaciclib using magnetite@ silica@ chitosan nanocomposite. International Journal of Biological Macromolecules. 2022;221:634–643.

55. Dinari M, Dadkhah F. Swift reduction of 4-nitrophenol by easy recoverable magnetite-Ag/layered double hydroxide/starch bionanocomposite. Carbohydrate Polymers. 2020;228:115392.

56. Ravichandran R, Annamalai K, Annamalai A, Elumalai S. Solid state–Green construction of starch-beaded $Fe_3O_4$@ Ag nanocomposite as superior redox catalyst. Colloids and Surfaces A: Physicochemical and Engineering Aspects. 2023;664:131117.

57. Badawy AM, Farghali AA, Bonilla-Petriciolet A, Selim AQ, Seliem MK. Effective removal of Cr (VI) and methyl orange by nano magnetite loaded starch/muscovite biocomposite: Characterization, experiments, advanced modeling, and physico-chemical parameters interpretation. International Journal of Biological Macromolecules. 2023;224:1052–1064.

58. Massoumi B, Mozaffari Z, Jaymand M. A starch-based stimuli-responsive magnetite nanohydrogel as de novo drug delivery system. International Journal of Biological Macromolecules. 2018;117:418–426.

59. Zheng M, Lu J, Zhao D. Effects of starch-coating of magnetite nanoparticles on cellular uptake, toxicity and gene expression profiles in adult zebrafish. Science of the Total Environment. 2018;622:930–941.

60. Gong Y, Wang L, Liu J, Tang J, Zhao D. Removal of aqueous perfluorooctanoic acid (PFOA) using starch-stabilized magnetite nanoparticles. Science of the Total Environment. 2016;562:191–200.

61. Ghaseminezhad SM, Shojaosadati SA. Evaluation of the antibacterial activity of Ag/Fe$_3$O$_4$ nanocomposites synthesized using starch. Carbohydrate Polymers. 2016;144:454–463.

62. Abbasian M, Khayyatalimohammadi M. Ultrasound-assisted synthesis of MIL-88 (Fe) conjugated starch-Fe$_3$O$_4$ nanocomposite: A safe antibacterial carrier for controlled release of tetracycline. International Journal of Biological Macromolecules. 2023;234:123665.

63. Mohammadi R, Saboury A, Javanbakht S, Foroutan R, Shaabani A. Carboxymethylcellulose/polyacrylic acid/starch-modified Fe$_3$O$_4$ interpenetrating magnetic nanocomposite hydrogel beads as pH-sensitive carrier for oral anticancer drug delivery system. European Polymer Journal. 2021;153:110500.

64. Vo TMT, Mondal S, Park S, Choi J, Bui NT, Oh J. Rice starch coated iron oxide nanoparticles: A theranostic probe for photoacoustic imaging-guided photothermal cancer therapy. International Journal of Biological Macromolecules. 2021;183:55–67.

65. Hamidian H, Tavakoli T. Preparation of a new Fe$_3$O$_4$/starch-g-polyester nanocomposite hydrogel and a study on swelling and drug delivery properties. Carbohydrate Polymers. 2016;144:140–148.

66. Yin J, Fang K, Li J, Du N, Hu D, Cao D, et al. Competitive adsorption mechanisms of pigments in sugarcane juice on starch-based magnetic nanocomposites. International Journal of Biological Macromolecules. 2023:123134.

67. Wang Y, Zhang Y, Hou C, Qi Z, He X, Li Y. Facile synthesis of monodisperse functional magnetic dialdehyde starch nano-composite and used for highly effective recovery of Hg (II). Chemosphere. 2015;141:26–33.

68. Işıklan N, Polat S. Synthesis and characterization of thermo/pH-sensitive pectin-graft-poly (dimethylaminoethyl methacrylate) coated magnetic nanoparticles. International Journal of Biological Macromolecules. 2020;164:4499–4515.

69. Kulal P, Badalamoole V. Magnetite nanoparticle embedded Pectin-graft-poly (N-hydroxyethylacrylamide) hydrogel: Evaluation as adsorbent for dyes and heavy metal ions from waste water. International Journal of Biological Macromolecules. 2020;156:1408–1417.

70. Kadam AA, Jang J, Lee DS. Facile synthesis of pectin-stabilized magnetic graphene oxide Prussian blue nanocomposites for selective cesium removal from aqueous solution. Bioresource Technology. 2016;216:391–398.

71. Zhang W, Zhang LY, Zhao XJ, Zhou Z. Citrus pectin derived ultrasmall Fe$_3$O$_4$@ C nanoparticles as a high-performance adsorbent toward removal of methylene blue. Journal of Molecular Liquids. 2016;222:995–1002.

72. Ghibaudo F, Gerbino E, Copello GJ, Dall'Orto VC, Gómez-Zavaglia A. Pectin-decorated magnetite nanoparticles as both iron delivery systems and protective matrices for probiotic bacteria. Colloids and Surfaces B: Biointerfaces. 2019;180:193–201.

73. Sahu S, Dutta RK. Novel hybrid nanostructured materials of magnetite nanoparticles and pectin. Journal of Magnetism and Magnetic Materials. 2011;323(7):980–987.

74. Dai J, Wu S, Jiang W, Li P, Chen X, Liu L, et al. Facile synthesis of pectin coated Fe$_3$O$_4$ nanospheres by the sonochemical method. Journal of Magnetism and Magnetic Materials. 2013;331:62–66.

75. Duman O, Tunç S, Polat TG, Bozoğlan BK. Synthesis of magnetic oxidized multiwalled carbon nanotube-κ-carrageenan-Fe$_3$O$_4$ nanocomposite adsorbent and its

application in cationic Methylene Blue dye adsorption. Carbohydrate Polymers. 2016;147:79–88.

76. Soares SF, Simoes TR, Antonio M, Trindade T, Daniel-da-Silva AL. Hybrid nanoadsorbents for the magnetically assisted removal of metoprolol from water. Chemical Engineering Journal. 2016;302:560–569.

77. Kulal P, Badalamoole V. Hybrid nanocomposite of kappa-carrageenan and magnetite as adsorbent material for water purification. International Journal of Biological Macromolecules. 2020;165:542–553.

78. Salgueiro AM, Daniel-da-Silva AL, Girão AV, Pinheiro PC, Trindade T. Unusual dye adsorption behavior of κ-carrageenan coated superparamagnetic nanoparticles. Chemical Engineering Journal. 2013;229:276–284.

79. Duman O, Tunç S, Bozoğlan BK, Polat TG. Removal of triphenylmethane and reactive azo dyes from aqueous solution by magnetic carbon nanotube-κ-carrageenan-$Fe_3O_4$ nanocomposite. Journal of Alloys and Compounds. 2016;687:370–383.

80. Alipour S, Pourjavadi A, Hosseini SH. Magnetite embedded κ-carrageenan-based double network nanocomposite hydrogel with two-way shape memory properties for flexible electronics and magnetic actuators. Carbohydrate Polymers. 2023:120610.

81. Klongklaew P, Bunkoed O. The enrichment and extraction of parabens with polydopamine-coated microporous carrageenan hydrogel beads incorporating a hierarchical composite of metal-organic frameworks and magnetite nanoparticles. Microchemical Journal. 2021;165:106103.

82. Magalhães FF, Almeida MR, Soares SF, Trindade T, Freire MG, Daniel-da-Silva AL, et al. Recovery of immunoglobulin G from rabbit serum using κ-carrageenan-modified hybrid magnetic nanoparticles. International Journal of Biological Macromolecules. 2020;150:914–921.

83. Chen Y-H, Cheng C-H, Chang W-J, Lin Y-C, Lin F-H, Lin J-C. Studies of magnetic alginate-based electrospun matrices cross-linked with different methods for potential hyperthermia treatment. Materials Science and Engineering: C. 2016;62:338–349.

84. Abdelwahab MS, El Halfawy NM, El-Naggar MY. Lead adsorption and antibacterial activity using modified magnetic biochar/sodium alginate nanocomposite. International Journal of Biological Macromolecules. 2022;206:730–739.

85. Ferreira A, Campello S, de Araújo A, Rodrigues A, Pereira G, Azevedo W. One-pot ultrasound synthesis of water dispersible superparamagnetic iron oxide@ alginate nanocomposite. Solid State Sciences. 2022;128:106870.

86. Xu X, Shen H, Xu J, Xie M, Li X. The colloidal stability and core-shell structure of magnetite nanoparticles coated with alginate. Applied Surface Science. 2006;253(4):2158–2164.

87. Lim S-F, Zheng Y-M, Zou S-W, Chen JP. Uptake of arsenate by an alginate-encapsulated magnetic sorbent: Process performance and characterization of adsorption chemistry. Journal of Colloid and Interface Science. 2009;333(1):33–39.

88. Germanos G, Youssef S, Abboud M, Farah W, Lescop B, Rioual S. Diffusion and agglomeration of iron oxide nanoparticles in magnetic calcium alginate beads initiated by copper sorption. Journal of Environmental Chemical Engineering. 2017;5(4):3727–3733.

89. Hong H-J, Jeong HS, Kim B-G, Hong J, Park I-S, Ryu T, et al. Highly stable and magnetically separable alginate/$Fe_3O_4$ composite for the removal of strontium (Sr) from seawater. Chemosphere. 2016;165:231–238.

90. Niu H, Meng Z, Cai Y. Fast defluorination and removal of norfloxacin by alginate/ Fe@ $Fe_3O_4$ core/shell structured nanoparticles. Journal of Hazardous Materials. 2012;227:195–203.

91. Sarkar S, Tiwari N, Basu A, Behera M, Das B, Chakrabortty S, et al. Sorptive removal of malachite green from aqueous solution by magnetite/coir pith supported sodium alginate beads: Kinetics, isotherms, thermodynamics and parametric optimization. Environmental Technology & Innovation. 2021;24:101818.

92. Bilici Z, Işık Z, Aktaş Y, Yatmaz HC, Dizge N. Photocatalytic effect of zinc oxide and magnetite entrapped calcium alginate beads for azo dye and hexavalent chromium removal from solutions. Journal of Water Process Engineering. 2019;31:100826.

93. Hachemaoui M, Mokhtar A, Mekki A, Zaoui F, Abdelkrim S, Hacini S, et al. Composites beads based on $Fe_3O_4$@ MCM-41 and calcium alginate for enhanced catalytic reduction of organic dyes. International Journal of Biological Macromolecules. 2020;164:468–479.

94. Singh I, Lacko CS, Zhao Z, Schmidt CE, Rinaldi C. Preparation and evaluation of microfluidic magnetic alginate microparticles for magnetically templated hydrogels. Journal of Colloid and Interface Science. 2020;561:647–658.

95. Ghorbani-Vaghei R, Veisi H, Aliani MH, Mohammadi P, Karmakar B. Alginate modified magnetic nanoparticles to immobilization of gold nanoparticles as an efficient magnetic nanocatalyst for reduction of 4-nitrophenol in water. Journal of Molecular Liquids. 2021;327:114868.

96. Peng N, Ding X, Wang Z, Cheng Y, Gong Z, Xu X, et al. Novel dual responsive alginate-based magnetic nanogels for onco-theranostics. Carbohydrate Polymers. 2019;204:32–41.

97. Khazaie F, Shokrollahzadeh S, Bide Y, Sheshmani S, Shahvelayati AS. Forward osmosis using highly water dispersible sodium alginate sulfate coated-$Fe_3O_4$ nanoparticles as innovative draw solution for water desalination. Process Safety and Environmental Protection. 2021;146:789–799.

98. Yang P-F, Lee C-K. Hyaluronic acid interaction with chitosan-conjugated magnetite particles and its purification. Biochemical Engineering Journal. 2007;33(3):284–289.

99. Thomas M, Naikoo GA, Sheikh MUD, Bano M, Khan F. Architecture of Ba/ alginate/dextran stabilized Au, $Fe_3O_4$, TiO2 & silica nanoparticles gels and their applications for reduction of 4-nitrophenol and glucose sensing. Reactive and Functional Polymers. 2016;105:78–88.

100. Mahdavinia GR, Mosallanezhad A, Soleymani M, Sabzi M. Magnetic-and pH-responsive κ-carrageenan/chitosan complexes for controlled release of methotrexate anticancer drug. International Journal of Biological Macromolecules. 2017;97:209–217.

101. Karimi MH, Mahdavinia GR, Massoumi B, Baghban A, Saraei M. Ionically cross-linked magnetic chitosan/κ-carrageenan bioadsorbents for removal of anionic eriochrome black-T. International Journal of Biological Macromolecules. 2018;113:361–375.

102. Long J, Yu X, Xu E, Wu Z, Xu X, Jin Z, et al. In situ synthesis of new magnetite chitosan/carrageenan nanocomposites by electrostatic interactions for protein delivery applications. Carbohydrate Polymers. 2015;131:98–107.

103. Mahdavinia GR, Etemadi H, Soleymani F. Magnetic/pH-responsive beads based on caboxymethyl chitosan and κ-carrageenan and controlled drug release. Carbohydrate Polymers. 2015;128:112–121.

104. Mahdavinia GR, Etemadi H. Surface modification of iron oxide nanoparticles with κ-carrageenan/carboxymethyl chitosan for effective adsorption of bovine serum albumin. Arabian Journal of Chemistry. 2019;12(8):3692–3703.

105. Pourmadadi M, Ahmadi M, Yazdian F. Synthesis of a novel pH-responsive $Fe_3O_4$/ chitosan/agarose double nanoemulsion as a promising Nanocarrier with sustained release of curcumin to treat MCF-7 cell line. International Journal of Biological Macromolecules. 2023;235:123786.

106. Han Y, Gao Y, Cao X, Zangeneh MM, Liu S, Li J. Ag NPs on chitosan-alginate coated magnetite for synthesis of indazolo [2, 1-b] phthalazines and human lung protective effects against α-Guttiferin. International Journal of Biological Macromolecules. 2020;164:2974–2986.

107. Zeng H, Sun S, Xu K, Zhao W, Hao R, Zhang J, et al. Adsorption of As (V) by magnetic alginate-chitosan porous beads based on iron sludge. Journal of Cleaner Production. 2022;359:132117.

108. He C, Guo Y, Karmakar B, El-kott A, Ahmed AE, Khames A. Decorated silver nanoparticles on biodegradable magnetic chitosan/starch composite: Investigation of its cytotoxicity, antioxidant and anti-human breast cancer properties. Journal of Environmental Chemical Engineering. 2021;9(6):106393.

109. Jafari H, Atlasi Z, Mahdavinia GR, Hadifar S, Sabzi M. Magnetic κ-carrageenan/chitosan/montmorillonite nanocomposite hydrogels with controlled sunitinib release. Materials Science and Engineering: C. 2021;124:112042.

110. Lemos TS, de Souza JF, Fajardo AR. Magnetic microspheres based on pectin coated by chitosan towards smart drug release. Carbohydrate Polymers. 2021;265:118013.

111. Sharma AK, Gupta A, Dhiman A, Garg M, Mishra R, Agrawal G. $Fe_3O_4$ embedded κ-carrageenan/sodium alginate hydrogels for the removal of basic dyes. Colloids and Surfaces A: Physicochemical and Engineering Aspects. 2022;654:130155.

112. Yang D, Gao K, Bai Y, Lei L, Jia T, Yang K, et al. Microfluidic synthesis of chitosan-coated magnetic alginate microparticles for controlled and sustained drug delivery. International Journal of Biological Macromolecules. 2021;182:639–647.

113. Dacrory S, Moussa M, Turky G, Kamel S. In situ synthesis of $Fe_3O_4$@ cyanoethyl cellulose composite as antimicrobial and semiconducting film. Carbohydrate Polymers. 2020;236:116032.

114. Tracey CT, Torlopov MA, Martakov IS, Vdovichenko EA, Zhukov M, Krivoshapkin PV, et al. Hybrid cellulose nanocrystal/magnetite glucose biosensors. Carbohydrate Polymers. 2020;247:116704.

115. Chaabane L, Chahdoura H, Mehdaoui R, Snoussi M, Beyou E, Lahcini M, et al. Functionalization of developed bacterial cellulose with magnetite nanoparticles for nanobiotechnology and nanomedicine applications. Carbohydrate Polymers. 2020;247:116707.

116. Habibi N. Preparation of biocompatible magnetite-carboxymethyl cellulose nanocomposite: Characterization of nanocomposite by FTIR, XRD, FESEM and TEM. Spectrochimica Acta Part A: Molecular and Biomolecular Spectroscopy. 2014;131:55–58.

117. Stoica-Guzun A, Stroescu M, Jinga SI, Mihalache N, Botez A, Matei C, et al. Box-Behnken experimental design for chromium (VI) ions removal by bacterial cellulose-magnetite composites. International Journal of Biological Macromolecules. 2016;91:1062–1072.

118. Mahmoud ME, Abdelwahab MS. Fabricated and functionalized magnetite/phenylenediamine/cellulose acetate nanocomposite for adsorptive removal of methylene blue. International Journal of Biological Macromolecules. 2019;128:196–203.

119. Bakr EA, Gaber M, Saad DR, Salahuddin N. Comparative study between two different morphological structures based on polylactic acid, nanocellulose and magnetite for co-delivery of flurouracil and curcumin. International Journal of Biological Macromolecules. 2023;230:123315.

120. Beyki MH, Mohammadirad M, Shemirani F, Saboury AA. Magnetic cellulose ionomer/layered double hydroxide: An efficient anion exchange platform with enhanced diclofenac adsorption property. Carbohydrate polymers. 2017;157:438–446.

121. Malakootikhah J, Rezayan AH, Negahdari B, Nasseri S, Rastegar H. Glucose reinforced $Fe_3O_4$@ cellulose mediated amino acid: Reusable magnetic glyconanoparticles with enhanced bacteria capture efficiency. Carbohydrate Polymers. 2017;170:190–197.

122. Kamel S, El-Gendy AA, Hassan MA, El-Sakhawy M, Kelnar I. Carboxymethyl cellulose-hydrogel embedded with modified magnetite nanoparticles and porous carbon: Effective environmental adsorbent. Carbohydrate Polymers. 2020;242:116402.

123. Mashkour M, Mashkour M. A Simple and scalable approach for fabricating high-performance superparamagnetic natural cellulose fibers and Papers. Carbohydrate Polymers. 2021;256:117425.

124. Ariaeenejad S, Motamedi E, Salekdeh GH. Immobilization of enzyme cocktails on dopamine functionalized magnetic cellulose nanocrystals to enhance sugar bio-conversion: A biomass reusing loop. Carbohydrate Polymers. 2021;256:117511.

125. Zhou S, Xia L, Fu Z, Zhang C, Duan X, Zhang S, et al. Purification of dye-contaminated ethanol-water mixture using magnetic cellulose powders derived from agricultural waste biomass. Carbohydrate Polymers. 2021;258:117690.

126. Yusefi M, Lee-Kiun MS, Shameli K, Teow S-Y, Ali RR, Siew K-K, et al. 5-Fluorouracil loaded magnetic cellulose bionanocomposites for potential colorectal cancer treatment. Carbohydrate Polymers. 2021;273:118523.

127. Soliman AI, Baca JAD, Fatehi P. One-pot synthesis of magnetic cellulose nanocrystal and its post-functionalization for doxycycline adsorption. Carbohydrate Polymers. 2023;308:120619.

128. Ghazitabar A, Naderi M, Haghshenas DF, Ashna DA. Graphene aerogel/cellulose fibers/magnetite nanoparticles (GCM) composite as an effective Au adsorbent from cyanide solution with favorable electrochemical property. Journal of Molecular Liquids. 2020;314:113792.

129. Shah A, Kuddushi M, Mondal K, Jain M, Malek N. Magnetically driven release of dopamine from magnetic-non-magnetic cellulose beads. Journal of Molecular Liquids. 2020;320:114290.

130. Mahmoud ME, El-Sharkawy RM, Ibrahim GA. A novel bionanocomposite from doped lipase enzyme into magnetic graphene oxide-immobilized-cellulose for efficient removal of methylene blue and malachite green dyes. Journal of Molecular Liquids. 2022;368:120676.

131. Jahanbakhshi A, Farahi M. Immobilized sulfonic acid functionalized ionic liquid on magnetic cellulose as a novel catalyst for the synthesis of triazolo [4, 3-a] pyrimidines. Arabian Journal of Chemistry. 2022;15(12):104311.

132. Maruthupandy M, Riquelme D, Rajivgandhi G, Muneeswaran T, Cho W-S, Anand M, et al. Dual-role of graphene/bacterial cellulose/magnetite nanocomposites as highly effective antibacterial agent and visible-light-driven photocatalyst. Journal of Environmental Chemical Engineering. 2021;9(5):106014.

133. Torgbo S, Sukyai P. Fabrication of microporous bacterial cellulose embedded with magnetite and hydroxyapatite nanocomposite scaffold for bone tissue engineering. Materials Chemistry and Physics. 2019;237:121868.

134. Khan AS, Nasir MF, Murtaza A. Study of carboxymethyl cellulose (CMC) coated manganite as potential candidate for magnetic hyperthermia applications. Materials Chemistry and Physics. 2022;286:126198.

135. Gao Y, Xue Y, Zhen K, Guo J, Tang X, Zhang P, et al. Remediation of soil contaminated with PAHs and γ-HCH using Fenton oxidation activated by carboxymethyl cellulose-modified iron oxide-biochar. Journal of Hazardous Materials. 2023;453:131450.

136. Maccarini M, Atrei A, Innocenti C, Barbucci R. Interactions at the CMC/magnetite interface: Implications for the stability of aqueous dispersions and the magnetic

properties of magnetite nanoparticles. Colloids and Surfaces A: Physicochemical and Engineering Aspects. 2014;462:107–114.

137. Mikhaylov VI, Torlopov MA, Vaseneva IN, Sitnikov PA. Magnetically controlled liquid paraffin oil-in-water Pickering emulsion stabilized by magnetite/cellulose nanocrystals: Formation and Cr (VI) adsorption. Colloids and Surfaces A: Physicochemical and Engineering Aspects. 2021;622:126634.

138. Kolvari E, Marandi A, Kheyroddin N. Magnetic cellulose nanocrystals as efficient support for indium (III) in the synthesis of tetrazoles and phthalazines. Colloids and Surfaces A: Physicochemical and Engineering Aspects. 2022;655:130154.

139. Park M, Cheng J, Choi J, Kim J, Hyun J. Electromagnetic nanocomposite of bacterial cellulose using magnetite nanoclusters and polyaniline. Colloids and Surfaces B: Biointerfaces. 2013;102:238–242.

140. Namdeo M, Bajpai S. Immobilization of α-amylase onto cellulose-coated magnetite (CCM) nanoparticles and preliminary starch degradation study. Journal of Molecular Catalysis B: Enzymatic. 2009;59(1-3):134–139.

141. El-Nahas AM, Salaheldin TA, Zaki T, El-Maghrabi HH, Marie AM, Morsy SM, et al. Functionalized cellulose-magnetite nanocomposite catalysts for efficient biodiesel production. Chemical Engineering Journal. 2017;322:167–180.

142. Habibi N. Functional biocompatible magnetite–cellulose nanocomposite fibrous networks: Characterization by fourier transformed infrared spectroscopy, X-ray powder diffraction and field emission scanning electron microscopy analysis. Spectrochimica Acta Part A: Molecular and Biomolecular Spectroscopy. 2015;136:1450–1453.

143. Taleb K, Markovski J, Veličković Z, Rusmirović J, Rančić M, Pavlović V, et al. Arsenic removal by magnetite-loaded amino modified nano/microcellulose adsorbents: Effect of functionalization and media size. Arabian Journal of Chemistry. 2019;12(8):4675–4693.

144. Nagarajan D, Venkatanarasimhan S. Magnetite microparticles decorated cellulose sponge as an efficacious filter for improved arsenic (V) removal. Journal of Environmental Chemical Engineering. 2019;7(5):103386.

145. Ravelo-Nieto E, Ovalle-Serrano SA, Gutiérrez-Pineda EA, Blanco-Tirado C, Combariza MY. Textile wastewater depuration using a green cellulose based $Fe_3O_4$ bionanocomposite. Journal of Environmental Chemical Engineering. 2023;11(2):109516.

146. Al-Wafi R, Ahmed M, Mansour S. Tuning the synthetic conditions of graphene oxide/magnetite/hydroxyapatite/cellulose acetate nanofibrous membranes for removing Cr (VI), Se (IV) and methylene blue from aqueous solutions. Journal of Water Process Engineering. 2020;38:101543.

147. Nehra S, Dhillon A, Sharma R, Nair M, Kumar D. Water–dispersible magnetite montmorillonite encapsulated cellulose beads for fluoride removal and their kinetics and mechanism. Environmental Nanotechnology, Monitoring & Management. 2022;18:100690.

148. Neelamegan H, Yang D-K, Lee G-J, Anandan S, Wu JJ. Synthesis of magnetite nanoparticles anchored cellulose and lignin-based carbon nanotube composites for rapid oil spill cleanup. Materials Today Communications. 2020;22:100746.

149. Honda H, Kawabe A, Shinkai M, Kobayashi T. Development of chitosan-conjugated magnetite for magnetic cell separation. Journal of fermentation and bioengineering. 1998;86(2):191–196.

150. John JA, Samuel MS, Selvarajan E. Immobilized cellulase on $Fe_3O_4$/GO/CS nanocomposite as a magnetically recyclable catalyst for biofuel application. Fuel. 2023;333:126364.

151. Bindu V, Mohanan P. Thermal deactivation of α-amylase immobilized magnetic chitosan and its modified forms: A kinetic and thermodynamic study. Carbohydrate Research. 2020;498:108185.

152. Lasheen M, El-Sherif IY, Tawfik ME, El-Wakeel S, El-Shahat M. Preparation and adsorption properties of nano magnetite chitosan films for heavy metal ions from aqueous solution. Materials Research Bulletin. 2016;80:344–350.

153. Alimard P. Fabrication and kinetic study of Nd-Ce doped $Fe_3O_4$-chitosan nanocomposite as catalyst in Fenton dye degradation. Polyhedron. 2019;171:98–107.

154. Javanbakht V, Ghoreishi SM, Habibi N, Javanbakht M. A novel magnetic chitosan/clinoptilolite/magnetite nanocomposite for highly efficient removal of Pb (II) ions from aqueous solution. Powder Technology. 2016;302:372–383.

155. Sanchayanukun P, Muncharoen S. Chitosan coated magnetite nanoparticle as a working electrode for determination of Cr (VI) using square wave adsorptive cathodic stripping voltammetry. Talanta. 2020;217:121027.

156. Cheraghipour E, Pakshir M. Process optimization and modeling of Pb (II) ions adsorption on chitosan-conjugated magnetite nano-biocomposite using response surface methodology. Chemosphere. 2020;260:127560.

157. Maruthupandy M, Rajivgandhi G, Muneeswaran T, Anand M, Quero F. Highly efficient antibacterial activity of graphene/chitosan/magnetite nanocomposites against ESBL-producing Pseudomonas aeruginosa and Klebsiella pneumoniae. Colloids and Surfaces B: Biointerfaces. 2021;202:111690.

158. Malar CG, Seenuvasan M, Kumar KS. Improvisation of diffusion coefficient in surface modified magnetite nanoparticles: A novel perspective. Materials Science and Engineering: C. 2019;103:109832.

159. Adimoolam MG, Amreddy N, Nalam MR, Sunkara MV. A simple approach to design chitosan functionalized $Fe_3O_4$ nanoparticles for pH responsive delivery of doxorubicin for cancer therapy. Journal of Magnetism and Magnetic Materials. 2018;448:199–207.

160. Feng B, Hong R, Wu Y, Liu G, Zhong L, Zheng Y, et al. Synthesis of monodisperse magnetite nanoparticles via chitosan–poly (acrylic acid) template and their application in MRI. Journal of Alloys and Compounds. 2009;473(1-2):356–362.

161. Hamza MF, Adel A-H, Hawata MA, El Araby R, Guibal E, Fouda A, et al. Functionalization of magnetic chitosan microparticles–Comparison of trione and trithione grafting for enhanced silver sorption and application to metal recovery from waste X-ray photographic films. Journal of Environmental Chemical Engineering. 2022;10(3):107939.

162. Salahuddin NA, EL-Daly HA, El Sharkawy RG, Nasr BT. Nano-hybrid based on polypyrrole/chitosan/grapheneoxide magnetite decoration for dual function in water remediation and its application to form fashionable colored product. Advanced Powder Technology. 2020;31(4):1587–1596.

163. Soares SF, Amorim CO, Amaral JS, Trindade T, Daniel-da-Silva AL. On the efficient removal, regeneration and reuse of quaternary chitosan magnetite nanosorbents for glyphosate herbicide in water. Journal of Environmental Chemical Engineering. 2021;9(3):105189.

164. Saravanakumar K, Sathiyaseelan A, Manivasagan P, Jeong MS, Choi M, Jang E-S, et al. Photothermally responsive chitosan-coated iron oxide nanoparticles for enhanced eradication of bacterial biofilms. Biomaterials Advances. 2022;141:213129.

165. Nie W, Lin Y, Wu X, Wu S, Li X, Cheng JJ, et al. Chitosan-$Fe_3O_4$ composites enhance anaerobic digestion of liquor wastewater under acidic stress. Bioresource Technology. 2023;377:128927.

166. Cho D-W, Jeon B-H, Chon C-M, Kim Y, Schwartz FW, Lee E-S, et al. A novel chitosan/clay/magnetite composite for adsorption of Cu (II) and As (V). Chemical Engineering Journal. 2012;200:654–662.

167. Shagholani H, Ghoreishi SM, Mousazadeh M. Improvement of interaction between PVA and chitosan via magnetite nanoparticles for drug delivery application. International Journal of Biological Macromolecules. 2015;78:130–136.

168. Kurt BZ, Uckaya F, Durmus Z. Chitosan and carboxymethyl cellulose based magnetic nanocomposites for application of peroxidase purification. International Journal of Biological Macromolecules. 2017;96:149–160.

169. Abdelwahab N, Morsy E. Synthesis and characterization of methyl pyrazolone functionalized magnetic chitosan composite for visible light photocatalytic degradation of methylene blue. International Journal of Biological Macromolecules. 2018;108:1035–1044.

170. Ali EM, Elashkar AA, El-Kassas HY, Salim EI. Methotrexate loaded on magnetite iron nanoparticles coated with chitosan: Biosynthesis, characterization, and impact on human breast cancer MCF-7 cell line. International Journal of Biological Macromolecules. 2018;120:1170–1180.

171. Ahamad T, Naushad M, Al-Shahrani T, Al-Hokbany N, Alshehri SM. Preparation of chitosan based magnetic nanocomposite for tetracycline adsorption: Kinetic and thermodynamic studies. International Journal of Biological Macromolecules. 2020;147:258–267.

172. Rebekah A, Bharath G, Naushad M, Viswanathan C, Ponpandian N. Magnetic graphene/chitosan nanocomposite: A promising nano-adsorbent for the removal of 2-naphthol from aqueous solution and their kinetic studies. International Journal of Biological Macromolecules. 2020;159:530–538.

173. Kazemi S, Pourmadadi M, Yazdian F, Ghadami A. The synthesis and characterization of targeted delivery curcumin using chitosan-magnetite-reduced graphene oxide as nano-carrier. International Journal of Biological Macromolecules. 2021;186:554–562.

174. Bulatao BP, Nalinratana N, Jantaratana P, Vajragupta O, Rojsitthisak P, Rojsitthisak P. Lutein-loaded chitosan/alginate-coated $Fe_3O_4$ nanoparticles as effective targeted carriers for breast cancer treatment. International Journal of Biological Macromolecules. 2023;242:124673.

175. Chen J-P, Yang P-C, Ma Y-H, Wu T. Characterization of chitosan magnetic nanoparticles for in situ delivery of tissue plasminogen activator. Carbohydrate Polymers. 2011;84(1):364–372.

176. Freire T, Dutra LM, Queiroz D, Ricardo N, Barreto K, Denardin J, et al. Fast ultrasound assisted synthesis of chitosan-based magnetite nanocomposites as a modified electrode sensor. Carbohydrate Polymers. 2016;151:760–769.

177. das Merces AAD, da Silva Ferreira R, Silva KJS, Salu BR, da Costa Maciel J, Aguiar JAO, et al. Identification of blood plasma proteins using heparin-coated magnetic chitosan particles. Carbohydrate Polymers. 2020;247:116671.

178. Sayin F, Akar ST, Akar T, Celik S, Gedikbey T. Chitosan immobilization and $Fe_3O_4$ functionalization of olive pomace: An eco-friendly and recyclable Pb2+ biosorbent. Carbohydrate Polymers. 2021;269:118266.

179. Rasoulzadeh H, Dehghani MH, Mohammadi AS, Karri RR, Nabizadeh R, Nazmara S, et al. Parametric modelling of Pb (II) adsorption onto chitosan-coated $Fe_3O_4$ particles through RSM and DE hybrid evolutionary optimization framework. Journal of Molecular Liquids. 2020;297:111893.

180. Bulin C, Zheng R, Song J, Bao J, Xin G, Zhang B. Magnetic graphene oxide-chitosan nanohybrid for efficient removal of aqueous Hg (II) and the interaction mechanism. Journal of Molecular Liquids. 2023;370:121050.

181. Hassani S, Gharehaghaji N, Divband B. Chitosan-coated iron oxide/graphene quantum dots as a potential multifunctional nanohybrid for bimodal magnetic

resonance/fluorescence imaging and 5-fluorouracil delivery. Materials Today Communications. 2022;31:103589.

182. Nuryono N, Miswanda D, Sakti SCW, Rusdiarso B, Krisbiantoro PA, Utami N, et al. Chitosan-functionalized natural magnetic particle@ silica modified with (3-chloropropyl) trimethoxysilane as a highly stable magnetic adsorbent for gold (III) ion. Materials Chemistry and Physics. 2020;255:123507.

183. Hefnawy MA, Medany SS, El-Sherif RM, Fadlallah SA. Green synthesis of NiO/ $Fe_3O_4$@ chitosan composite catalyst based on graphite for urea electro-oxidation. Materials Chemistry and Physics. 2022;290:126603.

184. Li S-N, Li B, Yu Z-R, Gong L-X, Xia Q-Q, Feng Y, et al. Chitosan in-situ grafted magnetite nanoparticles toward mechanically robust and electrically conductive ionic-covalent nanocomposite hydrogels with sensitive strain-responsive resistance. Composites Science and Technology. 2020;195:108173.

185. Liandi AR, Cahyana AH, Yunarti RT, Wendari TP. Facile synthesis of magnetic $Fe_3O_4$@ Chitosan nanocomposite as environmentally green catalyst in multi-component Knoevenagel-Michael domino reaction. Ceramics International. 2022;48(14):20266–20274.

186. Li B, Jia D, Zhou Y, Hu Q, Cai W. In situ hybridization to chitosan/magnetite nanocomposite induced by the magnetic field. Journal of Magnetism and Magnetic Materials. 2006;306(2):223–227.

187. Bezdorozhev O, Kolodiazhnyi T, Vasylkiv O. Precipitation synthesis and magnetic properties of self-assembled magnetite-chitosan nanostructures. Journal of Magnetism and Magnetic Materials. 2017;428:406–411.

188. Khmara I, Molcan M, Antosova A, Bednarikova Z, Zavisova V, Kubovcikova M, et al. Bioactive properties of chitosan stabilized magnetic nanoparticles–Focus on hyperthermic and anti-amyloid activities. Journal of Magnetism and Magnetic Materials. 2020;513:167056.

189. Subbiah L, Palanisamy S, Thamizhmurasu S, Joseph ABM, Thangavelu P, Ganeshan M, et al. Development of Meloxicam-chitosan magnetic nanoconjugates for targeting rheumatoid arthritis joints: Pharmaceutical characterization and preclinical assessment on murine models. Journal of Magnetism and Magnetic Materials. 2021;523:167571.

190. Arévalo-Cid P, Isasi J, Caballero AC, Martin-Hernandez F, González-Rubio R. Effects of shell-thickness on the powder morphology, magnetic behavior and stability of the chitosan-coated $Fe_3O_4$ nanoparticles. Boletín de la Sociedad Española de Cerámica y Vidrio. 2022;61(4):300–312.

191. Abbas H, Refai H, El Sayed N, Rashed LA, Mousa MR, Zewail M. Superparamagnetic iron oxide loaded chitosan coated bilosomes for magnetic nose to brain targeting of resveratrol. International Journal of Pharmaceutics. 2021;610:121244.

192. Oh DW, Kang JH, Kim YJ, Na S-B, Kwon TK, Kim S, et al. Preparation of inhalable N-acetylcysteine-loaded magnetite chitosan microparticles for nitrate adsorption in particulate matter. International Journal of Pharmaceutics. 2023;630:122454.

193. Elmizadeh H, Khanmohammadi M, Ghasemi K, Hassanzadeh G, Nassiri-Asl M, Garmarudi AB. Preparation and optimization of chitosan nanoparticles and magnetic chitosan nanoparticles as delivery systems using Box–Behnken statistical design. Journal of pharmaceutical and biomedical analysis. 2013;80:141–146.

194. Bhatt AS, Bhat DK, Santosh M. Electrical and magnetic properties of chitosan-magnetite nanocomposites. Physica B: Condensed Matter. 2010;405(8):2078–2082.

195. Li G-y, Jiang Y-r, Huang K-l, Ding P, Chen J. Preparation and properties of magnetic $Fe_3O_4$–chitosan nanoparticles. Journal of Alloys and Compounds. 2008;466(1-2):451–456.

196. Jamir M, Islam R, Pandey LM, Borah J. Effect of surface functionalization on the heating efficiency of magnetite nanoclusters for hyperthermia application. Journal of Alloys and Compounds. 2021;854:157248.

197. Namdeo M, Bajpai S. Chitosan–magnetite nanocomposites (CMNs) as magnetic carrier particles for removal of Fe (III) from aqueous solutions. Colloids and Surfaces A: Physicochemical and Engineering Aspects. 2008;320(1-3):161–168.

198. Castelló J, Gallardo M, Busquets MA, Estelrich J. Chitosan (or alginate)-coated iron oxide nanoparticles: A comparative study. Colloids and Surfaces A: Physicochemical and Engineering Aspects. 2015;468:151–158.

199. Pérez AG, González-Martínez E, Águila CRD, González-Martínez DA, Ruiz GG, Artalejo AG, et al. Chitosan-coated magnetic iron oxide nanoparticles for DNA and rhEGF separation. Colloids and Surfaces A: Physicochemical and Engineering Aspects. 2020;591:124500.

200. Luo Y, Hu Z, Lei X, Wang Y, Guo X. Fluorescent magnetic chitosan-based hydrogel incorporating Amino-Functionalized $Fe_3O_4$ and cellulose nanofibers modified with carbon dots for adsorption and detection of Cr (VI). Colloids and Surfaces A: Physicochemical and Engineering Aspects. 2023;658:130673.

201. Tran HV, Dai Tran L, Nguyen TN. Preparation of chitosan/magnetite composite beads and their application for removal of Pb (II) and Ni (II) from aqueous solution. Materials Science and Engineering: C. 2010;30(2):304–310.

202. Zhang W, Jia S, Wu Q, Wu S, Ran J, Liu Y, et al. Studies of the magnetic field intensity on the synthesis of chitosan-coated magnetite nanocomposites by co-precipitation method. Materials Science and Engineering: C. 2012;32(2):381–384.

203. Piosik E, Klimczak P, Ziegler-Borowska M, Chełminiak-Dudkiewicz D, Martyński T. A detailed investigation on interactions between magnetite nanoparticles functionalized with aminated chitosan and a cell model membrane. Materials Science and Engineering: C. 2020;109:110616.

204. Masteri-Farahani M, Shahsavarifar S. Chemical functionalization of chitosan biopolymer and chitosan-magnetite nanocomposite with sulfonic acid for acid-catalyzed reactions. Chinese Journal of Chemical Engineering. 2021;39:154–161.

205. Khoee S, Saadatinia A, Bafkary R. Ultrasound-assisted synthesis of pH-responsive nanovector based on PEG/chitosan coated magnetite nanoparticles for 5-FU delivery. Ultrasonics Sonochemistry. 2017;39:144–152.

206. Braim FS, Ab Razak NNAN, Aziz AA, Ismael LQ, Sodipo BK. Ultrasound assisted chitosan coated iron oxide nanoparticles: Influence of ultrasonic irradiation on the crystallinity, stability, toxicity and magnetization of the functionalized nanoparticles. Ultrasonics Sonochemistry. 2022;88:106072.

207. Saini M, Gupta R. Fabrication of chitosan-coated magnetite nanobiocatalyst with Bacillus atrophaeus γ-glutamyl transpeptidase and its application to the synthesis of a bioactive peptide SCV-07. Process Biochemistry. 2022;122:238–249.

208. Paripoorani KS, Ashwin G, Vengatapriya P, Ranjitha V, Rupasree S, Kumar VV, et al. Insolubilization of inulinase on magnetite chitosan microparticles, an easily recoverable and reusable support. Journal of molecular catalysis B: Enzymatic. 2015;113:47–55.

209. Tanhaei B, Ayati A, Lahtinen M, Sillanpää M. Preparation and characterization of a novel chitosan/Al2O3/magnetite nanoparticles composite adsorbent for kinetic, thermodynamic and isotherm studies of Methyl Orange adsorption. Chemical Engineering Journal. 2015;259:1–10.

210. Hamza MF, Wei Y, Mira H, Adel A-H, Guibal E. Synthesis and adsorption characteristics of grafted hydrazinyl amine magnetite-chitosan for Ni (II) and Pb (II) recovery. Chemical Engineering Journal. 2019;362:310–324.

211. Yuan Q, Venkatasubramanian R, Hein S, Misra R. A stimulus-responsive magnetic nanoparticle drug carrier: Magnetite encapsulated by chitosan-grafted-copolymer. Acta Biomaterialia. 2008;4(4):1024–1037.

212. Al-Hussainawy MK, Mehdi ZS, Jasim KK, Alshamsi HA, Saud HR, Kyhoiesh HAK. A single rapid route synthesis of magnetite/chitosan nanocomposite: Competitive study. Results in Chemistry. 2022;4:100567.

213. Xu Z, Zhu Q, Bian J. Preparation of a recyclable demulsifier for the treatment of emulsified oil wastewater by chitosan modification and sodium oleate grafting $Fe_3O_4$. Journal of Environmental Chemical Engineering. 2021;9(4):105663.

214. Usman UL, Singh NB, Allam BK, Banerjee S. Plant extract mediated synthesis of $Fe_3O_4$-chitosan composite for the removal of lead ions from aqueous solution. Materials Today: Proceedings. 2022;60:1140–1149.

215. Jamir M, Borgohain C, Borah J. Chitosan modified $Fe_3O_4$ nanoparticles for hyperthermia application. Materials Today: Proceedings. 2022;65:2484–2489.

216. Gogoi P, Thakur AJ, Devi RR, Das B, Maji TK. Adsorption of As (V) from contaminated water over chitosan coated magnetite nanoparticle: Equilibrium and kinetics study. Environmental Nanotechnology, Monitoring & Management. 2017;8:297–305.

217. Amin KF, Gulshan F, Asrafuzzaman F, Das H, Rashid R, Hoque SM. Synthesis of mesoporous silica and chitosan-coated magnetite nanoparticles for heavy metal adsorption from wastewater. Environmental Nanotechnology, Monitoring & Management. 2023;20:100801.

218. Yeamsuksawat T, Zhao H, Liang J. Characterization and antimicrobial performance of magnetic $Fe_3O_4@$ Chitosan@ Ag nanoparticles synthesized via suspension technique. Materials Today Communications. 2021;28:102481.

219. Al-Ghamdi AA, Galhoum AA, Alshahrie A, Al-Turki YA, Al-Amri AM, Wageh S. Mechanistic studies of uranyl interaction with functionalized mesoporous chitosan-superparamagnetic nanocomposites for selective sorption: Characterization and sorption performance. Materials Today Communications. 2022;33:104536.

220. Ziegler-Borowska M. Magnetic nanoparticles coated with aminated starch for HSA immobilization-simple and fast polymer surface functionalization. International Journal of Biological Macromolecules. 2019;136:106–114.

221. Marandi A, Nasiri E, Koukabi N, Seidi F. The $Fe_3O_4@$ apple seed starch core-shell structure decorated In (III): A green biocatalyst for the one-pot multicomponent synthesis of pyrazole-fused isocoumarins derivatives under solvent-free conditions. International Journal of Biological Macromolecules. 2021;190:61–71.

222. Verma P, Pal S, Chauhan S, Mishra A, Sinha I, Singh S, et al. Starch functionalized magnetite nanoparticles: A green, biocatalyst for one-pot multicomponent synthesis of imidazopyrimidine derivatives in aqueous medium under ultrasound irradiation. Journal of Molecular Structure. 2020;1203:127410.

223. Robinson MR, Abdelmoula M, Mallet M, Coustel R. Starch functionalized magnetite nanoparticles: New insight into the structural and magnetic properties. Journal of Solid State Chemistry. 2019;277:587–593.

224. Tancredi P, Botasini S, Moscoso-Londoño O, Méndez E, Socolovsky L. Polymer-assisted size control of water-dispersible iron oxide nanoparticles in range between 15 and 100 nm. Colloids and Surfaces A: Physicochemical and Engineering Aspects. 2015;464:46–51.

225. Saikia C, Das MK, Ramteke A, Maji TK. Effect of cross-linker on drug delivery properties of curcumin loaded starch coated iron oxide nanoparticles. International Journal of Biological Macromolecules. 2016;93:1121–1132.

226. Fang K, Li K, Yang T, Li J, He W. Starch-based magnetic nanocomposite as an efficient absorbent for anticancer drug removal from aqueous solution. International Journal of Biological Macromolecules. 2021;184:509–521.

227. Sousa AC, Romo AI, Almeida RR, Silva AC, Fechine LM, Brito DH, et al. Starch-based magnetic nanocomposite for targeted delivery of hydrophilic bioactives as anticancer strategy. Carbohydrate Polymers. 2021;264:118017.

228. Abdullah NH, Shameli K, Abdullah EC, Abdullah LC. A facile and green synthetic approach toward fabrication of starch-stabilized magnetite nanoparticles. Chinese Chemical Letters. 2017;28(7):1590–1596.

229. Rami MR, Meskini M, Qarebaghi LM, Salami M, Forouzandehdel S, Cheraghali M. Synthesis of magnetic bio-nanocomposites for drug release and adsorption applications. South African Journal of Chemical Engineering. 2022;42:115–126.

230. Kim D, Lee S, Im K, Kim K, Kim K, Shim I, et al. Surface-modified magnetite nanoparticles for hyperthermia: Preparation, characterization, and cytotoxicity studies. Current Applied Physics. 2006;6:e242–e246.

231. Stan M, Lung I, Soran M-L, Opris O, Leostean C, Popa A, et al. Starch-coated green synthesized magnetite nanoparticles for removal of textile dye Optilan Blue from aqueous media. Journal of the Taiwan Institute of Chemical Engineers. 2019;100:65–73.

232. Abdullah NH, Shameli K, Nia PM, Etesami M, Abdullah EC, Abdullah LC. Electrocatalytic activity of starch/$Fe_3O_4$/zeolite bionanocomposite for oxygen reduction reaction. Arabian Journal of Chemistry. 2020;13(1):1297–1308.

233. Singh P, Tiwary D, Sinha I. Improved removal of Cr (VI) by starch functionalized iron oxide nanoparticles. Journal of Environmental Chemical Engineering. 2014;2(4):2252–2258.

234. Okoli CP, Naidoo EB, Ofomaja AE. Role of synthesis process variables on magnetic functionality, thermal stability, and tetracycline adsorption by magnetic starch nanocomposite. Environmental Nanotechnology, Monitoring & Management. 2018;9:141–153.

235. Stan M, Lung I, Soran M-L, Opris O, Leostean C, Popa A, et al. Data on the removal of Optilan Blue dye from aqueous media using starch-coated green synthesized magnetite nanoparticles. Data in Brief. 2019;25:104165.

236. Mohapatra S, Asfer M, Anwar M, Sharma K, Akhter M, Ahmad FJ, et al. Doxorubicin loaded carboxymethyl Assam bora rice starch coated superparamagnetic iron oxide nanoparticles as potential antitumor cargo. Heliyon. 2019;5(6):e01955.

237. Molaei MJ, Salimi E. Magneto-fluorescent superparamagnetic $Fe_3O_4$@ $SiO2$@ alginate/carbon quantum dots nanohybrid for drug delivery. Materials Chemistry and Physics. 2022;288:126361.

238. Ma H-l, Qi X-r, Maitani Y, Nagai T. Preparation and characterization of superparamagnetic iron oxide nanoparticles stabilized by alginate. International Journal of Pharmaceutics. 2007;333(1-2):177–186.

239. Prabha G, Raj V. Sodium alginate–polyvinyl alcohol–bovin serum albumin coated $Fe_3O_4$ nanoparticles as anticancer drug delivery vehicle: Doxorubicin loading and in vitro release study and cytotoxicity to HepG2 and L02 cells. Materials Science and Engineering: C. 2017;79:410–422.

240. Huang J, Guo J, Zhu J, Zou X. Supported silver nanoparticles over alginate-modified magnetic nanoparticles: Synthesis, characterization and treat the human lung carcinoma. Journal of Saudi Chemical Society. 2022;26(1):101393.

241. Ismail SM, Abd-Elaal AA, Abd El-salam FH, Taher FA, Aiad I, Shaban SM. Synthesis of silver decorated magnetic $Fe_3O_4$/alginate polymeric surfactant with controllable catalytic activity toward p-NP removal and enzymatic-mimic activity for solid-colorimetric H2O2 detection. Chemical Engineering Journal. 2023;453:139593.

242. Alzahrani E, Abo-Dief HM, Algethami F. Electrochemical investigations of hydrochloric acid corrosion for carbon steel and coating effect by Poly (butyl Methacrylate)-grafted alginate/$Fe_3O_4$. Arabian Journal of Chemistry. 2021;14(5):103100.

243. Marjani A, Zare MH, Sadeghi MH, Shirazian S, Ghadiri M. Synthesis of alginate-coated magnetic nanocatalyst containing high-performance integrated enzyme for phenol removal. Journal of Environmental Chemical Engineering. 2021;9(1):104884.

244. Arora V, Sood A, Kumari S, Kumaran SS, Jain TK. Hydrophobically modified sodium alginate conjugated plasmonic magnetic nanocomposites for drug delivery & magnetic resonance imaging. Materials Today Communications. 2020;25:101470.

245. Soares SF, Rocha MJ, Ferro M, Amorim CO, Amaral JS, Trindade T, et al. Magnetic nanosorbents with siliceous hybrid shells of alginic acid and carrageenan for removal of ciprofloxacin. International Journal of Biological Macromolecules. 2019;139:827–841.

246. Shahabivand S, Mortazavi SS, Mahdavinia GR, Darvishi F. Phenol biodegradation by immobilized Rhodococcus qingshengii isolated from coking effluent on Na-alginate and magnetic chitosan-alginate nanocomposite. Journal of Environmental Management. 2022;307:114586.

247. Barbosa-Barros L, García-Jimeno S, Estelrich J. Formation and characterization of biobased magnetic nanoparticles double coated with dextran and chitosan by layer-by-layer deposition. Colloids and Surfaces A: Physicochemical and Engineering Aspects. 2014;450:121–129.

248. Rodriguez LC, Restrepo-Sánchez N, Pelaez C, Bernal C. Enhancement of the catalytic activity of carbonic anhydrase by covalent immobilization on magnetic cellulose crystals. Bioresource Technology Reports. 2023;21:101380.

249. Moharrami P, Motamedi E. Application of cellulose nanocrystals prepared from agricultural wastes for synthesis of starch-based hydrogel nanocomposites: Efficient and selective nanoadsorbent for removal of cationic dyes from water. Bioresource Technology. 2020;313:123661.

# 22 Phenomenology of Treatment of Biomaterial Surfaces

*Bijayinee Mohapatra*
PG Department of Physics, Government Autonomous College, Angul, Odisha, India

*Tapash R. Rautray*
Biomaterials and Tissue Regeneration Lab, CETMS, Institute of Technical Education and Research, Siksha 'O'Anusandhan (Deemed to be University), Bhubaneswar, Odisha, India

## 22.1 INTRODUCTION

Biomaterial-assisted bone scaffold is one of the most emerging approaches of recent research as critical bone defects that previously were treated by autografts, allografts, or xenografts have severe limitations, such as donor site scarcity, pathogen transfer, immuno rejection, etc. Present-day researchers are thus trying to focus on the improvement of synthetic biomaterials that can be the best match for bone defect sites and can reduce the gap between bone graft substitutes and overall clinical need of patients. Multiple factors are to be addressed simultaneously, starting from physical, mechanical, and biochemical compatibility to surface functionalization of synthetic biomaterials for immobilization of proteins, growth factors, and stem cells, which may better mimic the bone tissue [1–4]. Besides providing 3D space and support for formation of new tissues, biomaterials guide tissue growth, help in delivering cells at specific sites of the body that may not be that much successful with direct injection of cell suspension [5]. Depending on the age, sex of the patient, the sites at which the biomaterial is to be incorporated in the human body, the load and stress to withstand by the target site, the selection, fabrication, and development of biomaterials vary [6]. Since ancient times, biomaterials have been in use as implant materials; however, a lot of progress has occurred in the development of these materials, keeping in view the compatibility to structural and biochemical properties [7–11]. The biomaterials must be biocompatible, bioactive, biodegradable, osteoinductive, and osteoconductive. In addition to this, the materials must be capable of enhancing bone regeneration by the action of progenitor cells, increasing vascularization, and have suitable degradation kinetics *in vivo* [7,10].

Basing on the orthopedic application, implant site, and patient condition, the chosen biomaterials vary. Most commonly used biomaterials for bone wound healing applications are metals (Ti, Co-Cr alloys, Mg alloys), ceramics (calcium phosphate (CaP) based, bioglass, etc.), polymers (natural and synthetic), and composites of two or more of these, each having their own advantages and disadvantages [7]. Mechanical stability of the metallic biomaterials makes them suitable for BTE applications. However, surface inertness and poor osseointegration are the major limitations associated with these types of materials. CaP-based ceramics are worthy candidates for bone tissue construct because of their chemical resemblance with the inorganic component of bone. Furthermore, easy functionalization of surface of ceramic materials and tunable degradation properties fetch the interest of material scientists in these materials. Though tailorable surface properties of polymers are advantageous for growth factor, stem cell induction, and drug immobilization, lack of mechanical strength and low degradation kinetics constrain their applications. To overcome the lacunae associated with the individual materials, composite materials of polymer-ceramic, metal-ceramic, ceramic-ceramic, or two different polymers are the focus of advanced research [7,12,13].

Keeping in view the current trends, challenges, and demands of the present biomaterial research, we tried to explore different aspects of biomaterials in BTE applications, their interaction with biological systems, clinical needs and availability, etc. Moreover, in this book chapter, future prospects of hybrid, composite, multifunctional biomaterials, and their ability for bone regeneration are also described.

## 22.2   HISTORY OF BIOMATERIALS

Biomaterials are in use since ancient times, when natural plant or animal materials like corals, shells, wood, animal bones and teeth, metals such as gold and silver were in use for replacement of lost bones of humans, to heal bone wounds and to restore missing bone function. Starting from the use of biomaterials, this field is continually evolving and growing towards sophistication with technological revolution and involvement of research from several multidimensional approaches. Previously used biomaterials were mainly biologically inert and hence interaction of these materials with physiological environment is less. Some degradation products of these materials proved toxic to the human body. Considering the need for biomaterials, a series of natural and synthetic biomaterials are being tried. Selection and fabrication of biomaterials, however, is greatly affected since 1970 with the molecular biology revolution as specific proteins required for osteogenesis can be incorporated into bioinert biomaterials by enhancing their bioactivity [7,8,14,15].

## 22.3   CLINICAL NEED FOR SELECTION AND DESIGN OF BIOMATERIALS

Tissue regeneration or new tissue formation demands a 3D biodegradable support matrix that can provide a base for cell attachment and must create a micro environment for guided bone formation. Therefore, the prime requisite for a biomaterial is tailorable shape and structural integrity of the biomaterial scaffold to be maintained until new bone cell formation and maturation, even under high mechanical stress conditions [15]. Biomaterial-based tissue constructs must be capable of controlling varied cellular and molecular phenomena and resorption rate, besides being osteoinductive, osteoconductive, non-cytotoxic, bioactive, and printable [2,8,15]. Moreover, the biomaterials should serve as a storehouse of different bioactive agents, growth factors, and must have controlled release kinetics [5,16,17].

Considering the design of biomaterials, several factors such as porosity, pore size, pore structure, interconnectivity, and scaffold shape are to be taken care of as implant materials are site specific and most commonly greater surface-area-to-volume ratio is required without comprising the mechanical stability [5,8]. There must be sufficiency in nutrient and oxygen transfer through the implant materials [6]. Besides, modifiable structures are often desired for producing a biomimetic environment [7,18].

## 22.4  CLASSIFICATION OF BIOMATERIALS

As discussed in the previous section, orthopedic applications require biomaterials involving specific mechanical and biological properties. Depending on the origin, the materials may be classified into natural and synthetic materials. In spite of having many superior features, such as biocompatibility, low antigenicity, low cytotoxicity, and biodegradability, natural biomaterials suffer limitations because of their substantial purification methods. On the other hand, the synthetic materials have tunable mechanical properties that act as promising candidates for BTE applications [3,4].

A. Natural materials [metals (Au, Ag, etc.), ceramics (hydroxyapatite (HA) based), polymers (chitosan, alginate, gelatin, etc.)
B. Synthetic materials [metals (Ti and alloys, stainless steel (SS), Mg alloys, etc.), ceramics (CaP based materials, bioglass, etc.), polymers (PLA, PGA, PLGA, etc.)

In the subsequent sections, we will mainly focus on the application of different synthetic biomaterials as implant materials for BTE.

## 22.5  METALLIC BIOMATERIALS FOR BONE WOUND HEALING

### 22.5.1  General Characteristics and Basic Requirements

Metallic materials, because of their high mechanical stability and excellent physical properties, are often used as implant materials in the case of load-bearing applications [19–21]. Besides stability, these materials have high ductility, tensile strength, fatigue resistance, elastic modules, and better electrical and magnetic properties that make them a suitable choice for orthopedic surgery, dentistry, and spinal injuries [19]. However, the metallic materials very often encounter problems like corrosion resistance and wear resistance, which can be overcome by combining more than one material coating of a suitable material, decreasing the oxygen content, passive film formation, or by different surface treatment techniques. Surface processing, property enhancing, and shaping processes are generally followed for improving the mechanical and physical properties of metallic materials. Some of them to be listed are heat treatments, work hardening, solution hardening, particle dispersion hardening, precipitation hardening, grain refining, composites strengthening, etc. Similarly, for enhancing biological properties of the elements, coating (dry coating, spray coating), cold spraying, sputtering, physical vapor deposition (PVD), chemical vapor deposition (CVD), and surface functionalization techniques are applied [19,22–24]. In some cases, metals have mechanical strength greater than that of bone such that stress shielding occurs, leading to bone resorption and loosening of human bone [19,21]. The physical properties alone cannot decide the fate of new bone, biological characteristics depending on the shape, design orientation, and structure are to be taken care of with equal importance [19]. Hence, porous metal scaffolds are usually preferred where biocompatibility and mechanical strength go hand in hand. Besides, for a suitable metallic implant, the degradation product must be non-cytotoxic and non-allergenic [19,21,22,25].

### 22.5.2 Physical Approach to Metallic Biomaterial Design

An ideal porous scaffold must have with adequate number of interconnected pores and pore size suitable for nutrient and oxygen transport. In addition to this, the scaffold should provide sufficient space for cell adhesion and growth. The shape of the scaffold is also an important factor determining cell growth as, for example, smooth and continuous pore structures allow better cell proliferation. Literature reports that an increase in porosity surface area thereby increases the permeability of the scaffold. It is found that bone cell growth in the scaffold remarkably increases for a particular pore size and tensile strength and elastic modulus usually decreases with an increase in pore size and porosity. Thus, pore geometry is one of the major factors determining the strength of the scaffold [20,21]. Pore structure can be controlled by regulating fabrication method of metallic implants. Generally, computer-aided designs (CAD) and additive manufacturing (AM) techniques have great control over porosity and these advanced methods can prepare structures that closely resemble the bone matrix. Besides, as the computer designs are single-step processes, unlike that of conventional techniques where individual parts are assembled following different steps, the above methods are cost- and time-saving procedures for preparation of porous metallic implants [20,23]. Frequently used computer-aided methods are direct metal laser sintering (DMLS), selective laser sintering (SLS), fused deposition modeling (FDM), electron beam melting (EBM), ink jet printing, selective laser melting (SLM), etc. [20,21,26].

### 22.5.3 Metals Used as Implants

Frequently used metals as implant materials in clinical cases may be classified into non-biodegradable metals (Ti, alloys of Ti, stainless steel, Co-Cr alloy, tantalum

(Ta), Ni Ti alloy, etc.) and biodegradable implants (Mg, Zn, Fe, alloys, etc.). Along with conventionally used metal alloys, some high entropy alloys (Ti Zr Ta Hf Nb, Ti Zr Hf Cr Mo, Ti Nb Ta Zr Mo, Co-Cr, Mo, etc.) comprising a variety of metals, are also used for obtaining higher strength, corrosion resistance, and better biocompatibility [20]. We will discuss some of the metal alloys in the following section.

### 22.5.3.1  Ti and its Alloys

For load-bearing applications, Ti and its alloys are the most suitable candidates of choice. High strength, high corrosion resistance, lightweight biocompatibility, and non-toxicity of Ti make them suitable for bone formation and ossification processes. The strength of Ti and its alloys can further be increased by quenching, annealing, and various thermal treatment methods [20,21,27,28]. The ease of formation of thin oxide film over Ti metal raises its corrosion resistance and biocompatibility and makes these promising candidates for artificial hip joints, knee caps, bone plates, bone screws, etc. [20,21,23,29]. Ti metals with comparatively low densities have mechanical properties compared to that of natural cancellous bone and with high densities are close to that of cortical bone. But stress shielding or mechanical mismatch between Ti and its alloys and cancellous bones occur because of higher elastic modules of Ti and very poor vascularization, leading to bone resorption and osteolysis [30–32]. Though Ti is highly durable, in the case of osteolysis or local inflammation, a second surgery is required [30,31]. Also, Ti and its alloys very often face wear and corrosion, which can be counteracted by oxide layer formation over the whole metal. Surface coating techniques such as 'diamond like carbon' coating, PVD, etc. play an efficient role in enhancing the biocompatibility of Ti [21,23,33]. Another fact is that because of weak osseointigration of Ti, there is every chance of connective tissue formation between the bone implant site impacting bone regeneration [2,8]. Recent studies focus on development of Ti alloys for orthopedic applications. Ti6Al4V ($\alpha+\beta$ type), Ti- Nb ($\beta$ type), Ti-15Mo-5Zr-3Al, Ti-15Zr-4Nb-2Ta-0.2 Pd, and Ti-12 Mo-6Zr-2 Fe are some of the commonly used biocompatible alloys [30,31,34,35] whose biocompatibility can further be improved by following different surface treatment procedures. With more $\alpha$-phase (pure Ti), the mechanical stability decreases while $\beta$-type Ti alloys are favorable because of possessing lower Young's modulus and hence Ti alloys comprising of group 4 and group 5 elements in the periodic table are current research topics. Since the last two decades, twinning induced plasticity (TWIP) and transformation induced plasticity (TRIP) were tried on Ti alloys to produce metastable $\beta$-alloys, which are the materials of future generations. Ti-Ni and Ti-Mo are also used in dentistry [19,34]. Ti6Al4V is one of the most important alloys used in hard TE applications, preferably for long bone repair. These scaffolds, when fabricated with 3D-printing method, have controllable pore structure (diamond or dodecahedron shaped pores with 60 to 70% porosity and 500 to 700 μm pore size) and hence promote bone regeneration and reduce bone resorption [36–38].

### 22.5.3.2  Stainless-steel Alloys

Stainless-steel (SS) implants have been used as biomedical implants since 1920, as these materials have the desired mechanical strength, greater corrosion resistance than that of pure metals, and hence there is very low risk of complications post-operation. Commonly used SS for biomedical applications are Ni free as Ni reduces biocompatibility of the material, though it increases corrosion resistance. By the addition of Mo, corrosion resistance can further be improved [19,21]. These materials are used as bone fracture plates, screws, nails, low cost copies of actual implants by medical practitioners, dental implants, femoral heads and stems, etc. [21]. The lower cost of SS materials in comparison to other metallic biomaterials is one of the greatest advantages [23]. A popularly known SS biomaterial is 316 L stainless steel, where 316 represents the metal grade and L signifies low carbon content. A study reported that by coating graphene oxide (GO) and polyprolactone (PCL)/gelatin, the overall performance of 316LSS increases [39,40]. Wire and spindle-shaped SS can be prepared by using forged SS. Among all the metallic biomaterials, SS is cost effective and easy to design. Surface passivation can also be done easily, employing nitric acid, which leads to the removal of unwanted ions from the surface [40]. Poly L-lactic acid (PLLA) coating on SS improves biocompatibility and osteogenesis [41]. Having higher elastic modules than that of Ti and Ta alloys, SS often causes stress shielding and hence by adjusting pore size, porosity, and several of the factors, the strength and modulus of the steel scaffolds is brought to required levels [20].

### 22.5.3.3  Co-Cr-based Alloys

Co-Cr alloys are promising candidates for hip replacement and long bone repair because of superior corrosion and wear resistance and ease of fabrication methods. Casting and forging power processing are the frequently used processes for fabrication of Co-Cr alloys and also the Co-Cr-Mo system [42]. The behavior of Co-Cr-Mo and Co-Ni-Cr-Mo alloys were studied in physiological solutions and it was found that both systems passivate to a better extent by the formation of a $Cr_2O_3$ layer [21–23,43,44].

### 22.5.3.4  Mg-based Alloys

Magnesium (Mg) and its alloys grouped under third-generation biomaterials are lightweight, have a high strength to weight ratio, have antibacterial properties, and hence can be used for bone tissue regeneration. It is a biodegradable material with a comparatively lower elastic modulus than other metallic components, but closer to that of bone and hence decreases the chance of stress shielding [20,45–48]. These can be used in the form of plates, screws, rods, wires, pins, etc. [45]. The major concern in the case of Mg-based materials is their higher degradation rate and lower corrosion resistance. But one advantage is that the degradation product of Mg further enhances new bone formation [45]. Coatings of HA and polymers can be applied to improve the corrosion resistance and biocompatibility [46]. To enhance the mechanical strength, alloying, mechanical processing, topology optimization, etc. is carried out [47]. Fabrication routes play a great role in determining the behavior of Mg-based scaffolds. The Mg scaffolds fabricated by AM technology, solid free form fabrication (SFF) technique, powder metallurgy, and liquid Mg casting have controllable porosity and pore size, resulting in greater surface area for cell interaction [21,45]. Mg can be alloyed with Sr, Zn (decreasing corrosion), Mn (increasing ductility and mechanical stability), Ca (improving creep resistance of precipitates), Zn (enhancing mechanical strength and as grain refining agent), Y (increasing solubility), Nd, Ce(for mechanical strength), etc. [45].

### 22.5.3.5 Other Metal Alloys

Tantalum (Ta), a non-degradable biocompatible corrosion-resistant material is very often used as artificial joints because of having superior bone cell fixation ability. But a higher modulus of Ta limits its applications and hence alloying is done, usually with Ti [21].

Fe, a prime element required by the human body, is biocompatible and can be used as a scaffold for bone repair, along with its alloys (Fe-Mn, etc.), but with limitations, like a slower degradation rate [20].

Zn having a closer degradation rate with that of natural bone makes itself promising for orthopedic applications when prepared with taliorable pore size and porosity by AM technologies [20,49].

Ti-Ni alloy (Ninitol), a shape memory alloy, is emerging in the biomedical field [21]. Besides, some high entropy alloys (Ti-Zr-Ta-Hf, Ti-Zr-Hf-Cr-Mo, etc.) have gained research attention because of their enhanced biocompatibility, along with high wear and corrosion resistance [20,50].

### 22.5.4 Challenges Faced by Metallic Implants

Metallic biomaterials possess mechanical strength adequate with that of bone or sometimes even greater than that, but biocompatibility is a major issue [19]. CADs for biodegradable scaffolds generally avoid a second surgery option; however, permanent metallic implants are a choice of interest. Design methods play a significant role, as mimicking bone structure is a great challenge. Surface functionalization techniques are equally important for improving biocompatibility of most metallic biomaterials [20,21].

## 22.6 CERAMIC MATERIALS IN BTE

### 22.6.1 General Properties and Basic Needs

Bioceramics are identified by their outstanding biocompatibility, superior osteoconductivity, excellent corrosion resistance, tailorable surface properties but with brittleness, low elasticity, and low fracture toughness. These materials found a lot of application in different TE. Bone consists of collagen fibrils and HA as inorganic parts and thus bioceramics comprising of Ca, P, and ions incorporated into Ca and P (Sr, Zn, Mg, Si, Mn, etc.) are the best match for bone wound healing [51–54]. However, HA- and CaP-based ceramics very often requires biomineralization, which can be possible by coating the surface with different polymers [55–57]. Generally scaffolds made up of ceramic materials can easily be combined with bone morphogenic proteins (BMPs) and human bone marrow stromal cells (hMSCs), proving themselves beneficial for bone growth [51]. Moreover, the degradation products of ceramic biomaterials create an environment conducive for new bone formation [53]. Cancellous bone can suitably be repaired by ceramic scaffolds while repair of long bone defects with these biomaterials is quite unusual [55]. For orthopedic applications, the ceramics used must have sufficient mechanical strength with adequate porosity i.e., there must be a complete trade-off between porosity, pore size, and mechanical strength [58,59].

### 22.6.2 DESIGNING OF CERAMIC BIOMATERIALS AS BTE SCAFFOLDS

An ideal bone scaffold requires adequate porosity, pore interconnectivity, pores of different sizes for transport of nutrients, oxygen, and ease of vascularization without jeopardizing the mechanical strength. Here comes the impact of the fabrication method and designing parameters for obtaining a suitable porous ceramic structure for a bone defect site [59,60]. At the micro level, bone is a very complex structure and mimicking the exact bone structure by conventional fabrication methods such as polymers sponge, sol-gel, gas foaming, solvent casting, freeze drying, and electrospinning is hard to achieve. Besides, the scaffold prepared by the above methods may have remnants of toxic organic solvents, additional phases, inadequate mechanical strength, higher production cost, and higher energy consumption for preparation. To overcome the limitations associated with traditional preparation processes, several AM technologies and CADs, such as bioprinting, SLS,SLM, powder bed selective laser processing (PBSLP), and 3D printing, are followed [51,55,58,59]. AM technologies have the capability to replicate the complex geometry, control overpore size, and pore connectivity and thus are in the forefront of research [51,58,59].

### 22.6.3 TYPES OF CERAMIC IMPLANTS

Taking into consideration structure, composition, and properties of bioceramic materials, these may be classified into

A. Bioinert ceramics ($Al_2O_3$, zirconia, etc.)
B. Bioactive ceramics (bioglass and other glass ceramics, etc.)
C. Bioresorpable ceramics (CaP-based biomaterials).

The bioinert ceramics provide structural support (example used is a femoral head), whereas bioactive and bioresorbable ceramics act as bone fillers and in short bone defect repair [51].

We shall discuss some of the commonly used ceramic biomaterials in this section.

### 22.6.3.1 Alumina-based Implants

Alumina-based biomaterials are generally used for hip replacement and dental implantation, etc. These materials possess high hardness and high wear resistance. But because of low fracture toughness, usually alumina is combined with other materials [51].

### 22.6.3.2 Zirconia Implants

Zirconia-based bioceramics have found their application in TE since 1969. These materials are represented by their large elastic modulus, high wear resistance, high fracture toughness, better biocompatibility, low temperature conductance, tooth-like shade, etc. As zirconia can be colored to different shades of teeth, it is preferable to Ti in the field of dentistry. It is one of the highest strength ceramics known as 'steel ceramics' and hence is one of the best matches for orthopadic applications. It can be doped with $CaO, Y_2O_3$, $CeO_2$, MgO, etc., becoming more favorable for orthopedic and dentistry applications. Many studies report that taking zirconia as basement material and coating with bioactive materials will prove it as a promising candidate for biomimetic scaffold. Besides, zirconia itself is used as a coating material to improve the biocompatibility of metallic biomaterials [51,61,62].

### 22.6.3.3 Bioglass Implants

Bioglass (comprising of $SiO_2$, CaO, $P_2O_5$, $Na_2O$) has superb bioactive properties. The ions degraded from bioglass (Ca, P, Si, Na, etc.), thus enhances osteogenesis. In addition to this, Sr, Ag, B, Mg, Zn, and K are incorporated into bioglass to enhance the biological response [51]. Mainly polymer bioglass composites are used to obtain mechanical strength and biocompatibility as well [63,64]. These are capable of bond formation with hard and soft tissues. Normally sol-gel process, melt quenching, polymer foam, and SFF methods are followed for bioglass scaffold fabrication [51]. The reaction of bioglass with body fluid produces a hydroxyl carbonate layer, and can promote angiogenesis and osteogenesis, and help in drug delivery [65]. However, the fracture toughness required for long bone defect and load-bearing applications cannot be met by bioactive glass and hence strong bioglass scaffolds are the clinical need. Moreover, the properties can be tailored by controlling composition and processing methods [66]. Nano porous bioglass scaffolds were tried in an animal model and it was found that bone regeneration occurs with angiogenesis and without any inflammatory reaction [66].

### 22.6.3.4 CaP-based Implants

Hydroxyapatite (HA), $\alpha$-TCP, $\beta$-TCP, and BCP are some of the CaP-based ceramics often used in BTE. CaP implants are osteoconductive, osteoinductive, bioresorbable, easy to design, and have low production cost. These are used in knee, tibia, vertebral cage, dentistry, etc. However, their use in load-bearing applications is limited. Rather, some of them are used as bone cements that solidify inside the body [51]. Biphasic CaP (BCP), one of the CaP compounds, is similar in composition to natural bone. It has the ability to promote cell growth, proliferation, and differentiation [67,68]. Composites of CaP with other ceramics and polymers are formed for hard tissue engineering applications. Calcium phosphate cement (CPC), when incorporated with stem cells, induces osteogenesis and angiogenesis to a great extent [68,69]. CaP scaffolds loaded with growth factor stem cells fabricated via the PBSLP method provide a great opportunity for wound healing [55]. Bone consists of HA as its inorganic part. So, synthetic HA composites consisting of various biopolymers are the promising candidates for hard tissue engineering. However, load-bearing applications need higher mechanical strength materials that cannot be fulfilled by HA alone [69].

### 22.6.4 FATE OF BIOCERAMICS

Bioceramics possess almost all the required qualities for bone wound healing, such as biocompatibility, bioresorbability, osteoconductivity, osteoinductivity, tailorable geometry and degradation kinetics, varied forms, and porous structure, but with the limitations of lower elastic modulus and fracture toughness. Hence, for load-bearing applications, mechanical stability of the bioceramics is to be improved [51]. The above problem is approached from scaffold fabrication methods and many bioceramic composite scaffolds are to be produced with 3D printing and AM technologies. Customized 3D porous ceramic implants are therefore trends of current research [51,55].

## 22.7 POLYMERIC BIOMATERIALS IN BTE

### 22.7.1 GENERAL CHARACTERISTICS

Various natural (collagen, hyaluronic acid, fibrin, keratin, fibronectin, alginate, gelatin, etc.) and synthetic polymers (PLA, PCL, PGA, PLGA, etc.) are used in BTE because of their closely similar structure with bone ECM, biodegradability, biocompatibility, bioresorbability, moldable forms (films, sponge, scaffolds, etc.). Synthetic polymers can easily be modified according to the requirement as compared to natural polymers and their chemical properties can also be manipulated. Moreover, the degradation products of synthetic polymers cause no harm to surrounding tissues in most cases. But polymers alone lack all the desired properties of long bone defect repair in terms of bioactivity and mechanical integrity and hence compositing with bioceramics or loading or grafting of polymers is a basic requirement. For enhanced performance, polymers can also be loaded with bioactive molecules and stem cells. The cross-linking ability of polymers and swelling index are also important factors to be considered [70,71]. Cross-linking of polymers can be done by adding $Ca^{2+}$, $Mg^{2+}$, and $Zn^{2+}$ ions. Besides, shape memory polymers are also being tried, but fewer cases are reported in BTE [71].

### 22.7.2 FABRICATION METHOD OF POLYMERS

Polymers can be processed into scaffolds by solvent casting, polymer sponge, gas foaming (preparation of porous polymer matrix), electrospinning (preparation of nano fibrous polymer matrix), thermal induced phase separation (TIPS) (interconnected post structure formation), rapid prototyping technique (replicating complex geometry), CADs, bioprinting, and AM technologies like SLA, SLS, etc. [72].

### 22.7.3 SYNTHETIC POLYMERS AS IMPLANTS FOR HARD TE

Among the synthetic polymers, PLA plays a significant role in BTE applications because it possesses mechanical stability, biocompatibility, non-cytotoxic degradation products, and degradation rate matching that of bone. PLA nanofibers loaded with BMP prove to be a suitable choice for formation of bone-inducing ostogenesis [73,74]. A PLA scaffold produced by AM technology has a good, porous, interconnected network structure providing many advantages for bone scaffolds [74]. PLGA-based scaffolds prepared by 3D printing and osteogenic properties were evaluated. It was found that different ion-loaded PLGA scaffolds can significantly help in improving biocompatibility [75]. One of the studies reported 3D-printed PLGA scaffolds loaded with BMP-9 are found to be very promising for treatment of large bone defects [76]. Besides being biocompatible and biodegradable, PCL is easily available, cost effective, and modifiable, which makes them a better choice for hard tissue applications [77]. PCL scaffolds were fabricated by a precision extruding deposition (PED) technique and mechanical and biological properties are analyzed using various characterization techniques. The fabricated scaffolds have controlled pore size and pore interconnectivity, as required for bone wound healing [78].

Long bone defect repair is quite difficult with pure polymers and very few cases are reported. So in the next section, we will discuss in detail polymer composites for hard TE.

### 22.7.4 FUTURE ASPECTS OF POLYMERIC BIOMATERIALS

For two to three decades, tremendous improvement has been made in polymeric scaffolds. Combining natural and synthetic polymers, compositing with ceramics, surface modification to improve biological ability, and encapsulation of biomolecules are carried out mainly for BTE applications. Still, there is a lot more to be explored and developing multifunctional immunogenic smart polymeric biomaterials are the topics of research.

## 22.8 SIGNIFICANCE OF COMPOSITES

Mismatch of one or more properties of pure metals, ceramics, or polymers with that of natural bone is quite common. Metals have excellent physical properties; wear product of metals and corrosion properties associated with them are the major limitations. Stress shielding is another factor linked with most metal materials. Ceramics and their degradation products have very good osteoinductive and osteoconductive properties, but lack physical properties as desired. Similarly, polymers have tailorable surface properties and modifiable biophysical characteristics but without adequate mechanical integrity. A combination of different ceramics, polymer, and ceramics enhances mechanical properties along with bone regeneration and bone cell interaction. Thus, combining two or more materials to get the advantages of individual components is logical [79,80].

## 22.8.1 COMPOSITE BIOMATERIALS AS BONE IMPLANTS

Extensive research has been carried out on the role of composite biomaterials in orthopedic applications and the number of works is vast. Here we will report a few works aimed at designing multifunctional composite biomaterials.

HA nanoparticles and Mg particles are incorporated in polyurethane (PU) and coated over Ti, following a freeze-drying procedure in a very cost-effective manner. This is a very efficient outcome relating to bone regeneration [81]. In another study, $TiO_2$ nanoparticles in varied concentrations were made into composites with poly(D,L lactic acid) (PDLLA) and MG 63 cell proliferation were investigated. It is found that a polymer $TiO_2$ composite significantly enhances the bioactivity compared to $TiO_2$ nanoscaffolds alone [82–84]. One of the study reports an increase in comprehensive modulus, wettability with addition of $TiO_2$ to PLGA, and thereby increasing ALP activity in comparison with a pure PLGA scaffold [85–87]. Ta nanoparticles were added to PCL and there was a great enhancement of mechanical strength and osteogenic differentiation, which could be beneficial for hard TE [88]. Steel, when added with Mo, corrosion resistance is greatly enhanced [19]. In one of the studies, steel was coated with GO and PCL/gelatin/fosterite solution over the first layer. The bilayered scaffold thus prepared showed outstanding cell-binding abilities [39]. PLA coating on SS induces osteogenic properties and proves the capability of the scaffold for bone defect repair [41].

Nano hydroxyappetite (n-HA) particles were uniformly distributed on a PLGA scaffold by the method of fused deposition modeling. The composite scaffolds were analyzed as highly biocompatible and create a very low immunogenic reaction [89]. Collagen-HA, chitosan-HA, collagen-hyaluronic acid, HA/β-TCP/hyaluronic acid, gelatin-silk fibroin (SF), collagen-graphene-HA, and PCL-chitosan are some of the polymer composites used in BTE applications [90,91]. A PLA scaffold with HA/β-TCP as filler material was prepared by a gas-forming method and evaluated for mechanical and biological behavior. These scaffolds have satisfactory pore geometry and interconnectivity of pores [92–94]. A composite of CaP-based ceramics with different polymers are very popular in bone wound healing. In a work, 3D-printed composite scaffolds were prepared, comprising CaP and collagen and inserted into the animal model. The implant material showed a marked increase in material, physical, and biological properties [95]. CaP and calcium silicates (CS) fabricated by the BPSLP method resulted in implant materials suitably fit in long bone defects [55,96]. In a study, composite of chitosan, HA, and sodium alginate were prepared and the scaffold demonstrated very good biological, chemical, material antibacterial, and bone-inducing characteristics [97–100]. Polydopamine (PDA) was used to bio-mineralize an HA coating on several implant materials and the PDA-HA coating thus serves in bone wound healing, antibacterial activity, bone regeneration, etc. [101]. Zirconia is coated with HA, CaP, BG, etc. for enhancement of biological and physical activities [63]. A PDLLA bioglass composite showed promising results in a MG-63 cell attachment and can thus support bone formation on implant material [102,103].

## 22.8.2 Future Directions of Composite Materials

In the field of TE, composite biomaterials can be treated as superb alternatives of conventional bone defect repair methods and prove to be more useful than individual components. However, much has to be investigated regarding improvement of required characteristics as bone implants and compatibility of different components of composites is also to be taken care of.

## 22.9 SUMMARY

Biomaterials (metal, ceramics, polymers, and composites) are the research hotspot in the field of TE and this field is continuously emerging with the development of smart multifunctional materials and the advancement of fabrication technologies. In this chapter, we tried to discuss different biomaterials used as bone implants and the advantages and limitations associated with each type of biomaterial. Composites of metal matrix, ceramic-polymers, ceramic reinforcement, polymer-polymer matrix, and ion incorporation into biomaterials are described to some extent. It is observed that composite materials are found to be the best fit for bone wound healing and several aspects are to be addressed for successful bone regeneration.

## 22.10 CHALLENGES AND FUTURE OUTLOOK

BTE strongly requires mechanical integrity for load-bearing cases. Fabricating efficient bioengineered materials needs thorough understanding of bone regeneration and the desired biological cues at cellular and molecular levels. Thus, mechanical and biological factors of the tissue construct are to be taken care of simultaneously. Mimicking complex geometry of boned defects is one of the greatest challenges faced by material scientists. A deep knowledge of bone implant interaction is the demand of correct research. Success of bone defect treatments thus requires several factors, such as selection of suitable biomaterial; knowing the physical, chemical, biological, and material aspects of the biomaterial; understanding the architecture of the defect site; knowledge of fabrication technologies; bone implant interaction; etc. Reduction of production cost, reproducible scaffolds, and ease of fabrication are also the challenges to be met.

A lot has been explored, but there is far more to go in this field. It is probable that, in the future, cost-effective advanced composite biomaterials will be produced that can replace the traditional bone graft substitutes and will closely mimic natural bone.

## REFERENCES

1. Mohapatra, Bijayinee, and Tapash R. Rautray. "Strontium-substituted biphasic calcium phosphate scaffold for orthopedic applications." *Journal of the Korean Ceramic Society* 57 (2020): 392–400.

2. Stevens, Molly M. "Biomaterials for bone tissue engineering." *Materials Today* 11, no. 5 (2008): 18–25.

3. Han, Xuejiao, Aqu Alu, Hongmei Liu, Yi Shi, Xiawei Wei, Lulu Cai, and Yuquan Wei. "Biomaterial-assisted biotherapy: A brief review of biomaterials used in drug delivery, vaccine development, gene therapy, and stem cell therapy." *Bioactive Materials* (2022).

4. Sayed Shithima, Faruq Omar, and Uzer Gunes. "A review of biomaterials for bone tissue regeneration." *Orthoplastic Surgery & Orthopedic Care International Journal* 2, no. 5 (2022)

5. Lee, Esther J., F. Kurtis Kasper, and Antonios G. Mikos. "Biomaterials for tissue engineering." *Annals of Biomedical Engineering* 42 (2014): 323–337.

6. Sharma, Krati, Mubarak A. Mujawar, and Ajeet Kaushik. "State-of-art functional biomaterials for tissue engineering." *Frontiers in Materials* 6 (2019): 172.

7. Dolcimascolo, Anna, Giovanna Calabrese, Sabrina Conoci, and Rosalba Parenti. "Innovative biomaterials for tissue engineering." In *Biomaterial-supported tissue reconstruction or regeneration*. IntechOpen, 2019.

8. Qu, Huawei, Hongya Fu, Zhenyu Han, and Yang Sun. "Biomaterials for bone tissue engineering scaffolds: A review." *RSC Advances* 9, no. 45 (2019): 26252–26262.

9. Swain, S., and T. R. Rautray. "Silver doped hydroxyapatite coatings by sacrificial anode deposition under magnetic field." *Journal of Materials Science: Materials in Medicine* 28 (2017): 1–5.

10. Rautray, Tapash R., Bijayinee Mohapatra, and Kyo-Han Kim. "Fabrication of strontium–hydroxyapatite scaffolds for biomedical applications." *Advanced Science Letters* 20, no. 3–4 (2014): 879–881.

11. Lyons, Joseph G., Mark A. Plantz, Wellington K. Hsu, Erin L. Hsu, and Silvia Minardi. "Nanostructured biomaterials for bone regeneration." *Frontiers in Bioengineering and Biotechnology* 8 (2020): 922.

12. Lee, Ki-Won, Cheol-Min Bae, Jae-Young Jung, Gi-Bong Sim, Tapash Ranjan Rautray, Hyo-Jin Lee, Tae-Yub Kwon, and Kyo-Han Kim. "Surface characteristics and biological studies of hydroxyapatite coating by a new method." *Journal of Biomedical Materials Research Part B: Applied Biomaterials* 98, no. 2 (2011): 395–407.

13. Swain, Subhasmita, and Tapash R. Rautray. "Estimation of trace elements, antioxidants, and antibacterial agents of regularly consumed Indian medicinal plants." *Biological Trace Element Research* 199, no. 3 (2021): 1185–1193.

14. Huebsch, Nathaniel, and David J. Mooney. "Inspiration and application in the evolution of biomaterials." *Nature* 462, no. 7272 (2009): 426–432.

15. Chen, Fa-Ming, and Xiaohua Liu. "Advancing biomaterials of human origin for tissue engineering." *Progress in Polymer Science* 53 (2016): 86–168.

16. Mohapatra, Bijayinee, and Tapash R. Rautray. "Facile fabrication of Luffa cylindrica-assisted 3D hydroxyapatite scaffolds." *Bioinspired, Biomimetic and Nanobiomaterials* 10, no. 2 (2021): 37–44.

17. Mishra, Saswati, and Tapash R. Rautray. "Silver-incorporated hydroxyapatite–albumin microspheres with bactericidal effects." *Journal of the Korean Ceramic Society* 57 (2020): 175–183.

18. Swain, Subhasmita, Rabindra Nath Padhy, and Tapash Ranjan Rautray. "Electrically stimulated hydroxyapatite–barium titanate composites demonstrate immunocompatibility in vitro." *Journal of the Korean Ceramic Society* 57, no. 5 (2020): 495–502.

19. Nakano, T. "Mechanical properties of metallic biomaterials." In *Metals for biomedical devices*, pp. 71–98. Woodhead Publishing, 2010.

20. Lv, Yuting, Binghao Wang, Guohao Liu, Yujin Tang, Eryi Lu, Kegong Xie, Changgong Lan, Jia Liu, Zhenbo Qin, and Liqiang Wang. "Metal material,

properties and design methods of porous biomedical scaffolds for additive manufacturing: A review." *Frontiers in Bioengineering and Biotechnology* 9 (2021): 641130.

21. Metallic biomaterials: Current challenges and opportunities

22. Nayak, Gargi Shankar, Adele Carradò, Patrick Masson, Geneviève Pourroy, Flavien Mouillard, Véronique Migonney, Céline Falentin-Daudre, Caroline Pereira, and Heinz Palkowski. "Trends in metal-based composite biomaterials for hard tissue applications." *JOM* 74, no. 1 (2022): 102–125.

23. Santos, G. "The importance of metallic materials as biomaterials." *Adv Tissue Eng Regen Med Open Access* 3, no. 1 (2017): 300–302.

24. Rautray, T. R., and K-H. Kim. "Nanoelectrochemical coatings on titanium for bioimplant applications." *Materials Technology* 25, no. 3-4 (2010): 143–148.

25. Swain, Subhasmita, Chris Bowen, and Tapash Rautray. "Dual response of osteoblast activity and antibacterial properties of polarized strontium substituted hydroxyapatite—Barium strontium titanate composites with controlled strontium substitution." *Journal of Biomedical Materials Research Part A* 109, no. 10 (2021): 2027–2035.

26. Rautray, Tapash R., and Kyo Han Kim. "Synthesis of Mg2+ incorporated hydroxyapatite by ion implantation." In *Key engineering materials*, vol. 529, pp. 114–118. Trans Tech Publications Ltd, 2013.

27. Swain, Subhasmita, and Tapash Ranjan Rautray. "Effect of surface roughness on titanium medical implants." *Nanostructured Materials and their Applications* (2021): 55–80.

28. Rautray, Tapash R., R. Narayanan, and Kyo-Han Kim. "Ion implantation of titanium based biomaterials." *Progress in Materials Science* 56, no. 8 (2011): 1137–1177.

29. Kim, Kyo-Han, Tapash R. Rautray, and R. Narayanan. *Surface modification of titanium for biomaterial applications.* Nova Science Publ., 2010.

30. Wen, C. E., Y. Yamada, K. Shimojima, Y. Chino, H. Hosokawa, and M. Mabuchi. "Novel titanium foam for bone tissue engineering." *Journal of Materials Research* 17 (2002): 2633–2639.

31. Zuo, Weiyang, Lingjia Yu, Jisheng Lin, Yong Yang, and Qi Fei. "Properties improvement of titanium alloys scaffolds in bone tissue engineering: A literature review." *Annals of Translational Medicine* 9, no. 15 (2021).

32. Behera, Dipti Rani, Pratibindhya Nayak, and Tapash Ranjan Rautray. "Phosphatidylethanolamine impregnated Zn-HA coated on titanium for enhanced bone growth with antibacterial properties." *Journal of King Saud University-Science* 32, no. 1 (2020): 848–852.

33. de Viteri, V. Sáenz, and Elena Fuentes. "Titanium and titanium alloys as biomaterials." *Tribology-Fundamentals and Advancements* 1, no. 5 (2013): 154–181.

34. Hanawa, Takao. "Titanium–tissue interface reaction and its control with surface treatment." *Frontiers in Bioengineering and Biotechnology* 7 (2019): 170.

35. Swain, Subhasmita, R. D. K. Misra, C. K. You, and Tapash R. Rautray. "TiO2 nanotubes synthesised on Ti-6Al-4V ELI exhibits enhanced osteogenic activity: A potential next-generation material to be used as medical implants." *Materials Technology* 36, no. 7 (2021): 393–399.

36. Gu, Yifei, Yi Sun, Sohaib Shujaat, Annabel Braem, Constantinus Politis, and Reinhilde Jacobs. "3D-printed porous Ti6Al4V scaffolds for long bone repair in animal models: A systematic review." *Journal of Orthopaedic Surgery and Research* 17, no. 1 (2022): 68.

37. Swain, Subhasmita, Joo L. Ong, Ramaswamy Narayanan, and Tapash R. Rautray. "Ti-9Mn β-type alloy exhibits better osteogenicity than Ti-15Mn alloy in vitro."

*Journal of Biomedical Materials Research Part B: Applied Biomaterials* 109, no. 12 (2021): 2154–2161.

38. Rautray, Tapash R., R. Narayanan, Tae-Yub Kwon, and Kyo-Han Kim. "Surface modification of titanium and titanium alloys by ion implantation." *Journal of Biomedical Materials Research Part B: Applied Biomaterials* 93, no. 2 (2010): 581–591.

39. Khosravi, Fatemeh, Saied Nouri Khorasani, Shahla Khalili, Rasoul Esmaeely Neisiany, Erfan Rezvani Ghomi, Fatemeh Ejeian, Oisik Das, and Mohammad Hossein Nasr-Esfahani. "Development of a highly proliferated bilayer coating on 316L stainless steel implants." *Polymers* 12, no. 5 (2020): 1022.

40. Walley, Kempland C., Mergim Bajraliu, Tyler Gonzalez, Ara Nazarian, and James A. Goulet. "The chronicle of a stainless steel orthopaedic implant." *The Orthopaedic Journal at Harvard Medical School* 17 (2016): 68–74.

41. Branquinho, Mariana V., Sheila O. Ferreira, Rui D. Alvites, Adriana F. Magueta, Maxim Ivanov, Ana Catarina Sousa, Irina Amorim et al. "In vitro and in vivo characterization of plla-316l stainless steel electromechanical devices for bone tissue engineering—A preliminary study." *International Journal of Molecular Sciences* 22, no. 14 (2021): 7655.

42. Patel, Bhairav, Gregory Favaro, Fawad Inam, Michael J. Reece, Arash Angadji, William Bonfield, Jie Huang, and Mohan Edirisinghe. "Cobalt-based orthopaedic alloys: Relationship between forming route, microstructure and tribological performance." *Materials Science and Engineering: C* 32, no. 5 (2012): 1222–1229.

43. Kocijan, A., I. Milošev, D. Kek Merl, and B. Pihlar. "Electrochemical study of Co-based alloys in simulated physiological solution." *Journal of Applied Electrochemistry* 34 (2004): 517–524.

44. Aherwar, Amit, Amit Kumar Singh, and Amar Patnaik. "Cobalt based alloy: A better choice biomaterial for hip implants." *Trends in Biomaterials & Artificial Organs* 30, no. 1 (2016).

45. Antoniac, Iulian, Marian Miculescu, Veronica Mănescu, Alexandru Stere, Pham Hong Quan, Gheorghe Păltânea, Alina Robu, and Kamel Earar. "Magnesium-based alloys used in orthopedic surgery." *Materials* 15, no. 3 (2022): 1148.

46. Rahman, Mostafizur, Naba K. Dutta, and Namita Roy Choudhury. "Magnesium alloys with tunable interfaces as bone implant materials." *Frontiers in Bioengineering and Biotechnology* 8 (2020): 564.

47. Uppal, Gavish, Amit Thakur, Amit Chauhan, and Saroj Bala. "Magnesium based implants for functional bone tissue regeneration–A review." *Journal of Magnesium and Alloys* 10, no. 2 (2022): 356–386.

48. Praharaj, Rinmayee, Snigdha Mishra, R. D. K. Misra, and Tapash R. Rautray. "Biocompatibility and adhesion response of magnesium-hydroxyapatite/strontium-titania (Mg-HAp/Sr-TiO2) bilayer coating on titanium." *Materials Technology* 37, no. 4 (2022): 230–239.

49. Swain, Subhasmita, Sapna Mishra, Abhishek Patra, Rinmayee Praharaj, and Tapash Rautray. "Dual action of polarised zinc hydroxyapatite-guar gum composite as a next generation bone filler material." *Materials Today: Proceedings* 62 (2022): 6125–6130.

50. Swain, Subhasmita, and Tapash Ranjan Rautray. "Assessment of polarized piezoelectric SrBi4Ti4O15 nanoparticles as an alternative antibacterial agent." *bioRxiv* (2021): 2021-01.

51. Pina, Sandra Cristina Almeida, Rui L. Reis, and Joaquim M. Oliveira. "Ceramic biomaterials for tissue engineering." (2018).

52. Kaushik, Neha, Linh Nhat Nguyen, June Hyun Kim, Eun Ha Choi, and Nagendra Kumar Kaushik. "Strategies for using polydopamine to induce biomineralization of

hydroxyapatite on implant materials for bone tissue engineering." *International Journal of Molecular Sciences* 21, no. 18 (2020): 6544.

53. Sabree, Israa, Julie E. Gough, and Brian Derby. "Mechanical properties of porous ceramic scaffolds: Influence of internal dimensions." *Ceramics International* 41, no. 7 (2015): 8425–8432.

54. Rautray, T. R., V. Vijayan, and S. Panigrahi. "Synthesis of hydroxyapatite at low temperature." *Indian Journal of Physics* 81 (2007): 95–98.

55. Kamboj, Nikhil, Antonia Ressler, and Irina Hussainova. "Bioactive ceramic scaffolds for bone tissue engineering by powder bed selective laser processing: A review." *Materials* 14, no. 18 (2021): 5338.

56. Rautray, Tapash R., and Kyo Han Kim. "Synthesis of silver incorporated hydroxyapatite under magnetic field." In *Key Engineering Materials*, vol. 493, pp. 181–185. Trans Tech Publications Ltd, 2012.

57. Mishra, Saswati, and Tapash R. Rautray. "Fabrication of Xanthan gum-assisted hydroxyapatite microspheres for bone regeneration." *Materials Technology* 35, no. 6 (2020): 364–371.

58. Kamboj, N. Bioactive Ceramic Scaffolds. Encyclopedia. Available online: https://encyclopedia.pub/entry/14464 (accessed on 29 April 2023).

59. Gabor, Alin Gabriel, Virgil-Florin Duma, Mihai MC Fabricky, Liviu Marsavina, Anca Tudor, Cosmin Vancea, Petru Negrea, and Cosmin Sinescu. "Ceramic scaffolds for bone augmentation: Design and characterization with SEM and confocal microscopy." *Materials* 15, no. 14 (2022): 4899.

60. Hamdy, Tamer M. "Polymers and ceramics biomaterials in orthopedics and dentistry: A review article." *Egyptian Journal of Chemistry* 61, no. 4 (2018): 723–730.

61. Sakthiabirami, Kumaresan, Vaiyapuri Soundharrajan, Jin-Ho Kang, Yunzhi Peter Yang, and Sang-Won Park. "Three-dimensional zirconia-based scaffolds for load-bearing bone-regeneration applications: Prospects and challenges." *Materials* 14, no. 12 (2021): 3207.

62. Weng, Wenxian, Weiwei Wu, Mengdie Hou, Taotao Liu, Tianlin Wang, and Huazhe Yang. "Review of zirconia-based biomimetic scaffolds for bone tissue engineering." *Journal of Materials Science* 56 (2021): 8309–8333.

63. Daskalakis, Evangelos, Boyang Huang, Cian Vyas, Anil Ahmet Acar, Ali Fallah, Glen Cooper, Andrew Weightman, Bahattin Koc, Gordon Blunn, and Paulo Bartolo. "Novel 3D bioglass scaffolds for bone tissue regeneration." *Polymers* 14, no. 3 (2022): 445.

64. Will, Julia, Lutz-Christian Gerhardt, and Aldo R. Boccaccini. "Bioactive glass-based scaffolds for bone tissue engineering." *Tissue Engineering III: Cell-Surface Interactions for Tissue Culture* (2012): 195–226.

65. Fu, Qiang, Eduardo Saiz, Mohamed N. Rahaman, and Antoni P. Tomsia. "Bioactive glass scaffolds for bone tissue engineering: State of the art and future perspectives." *Materials Science and Engineering: C* 31, no. 7 (2011): 1245–1256.

66. Wang, S., M. M. Falk, A. Rashad, M. M. Saad, A. C. Marques, R. M. Almeida, M. K. Marei, and H. Jain. "Evaluation of 3D nano–macro porous bioactive glass scaffold for hard tissue engineering." *Journal of Materials Science: Materials in Medicine* 22 (2011): 1195–1203.

67. Beheshtizadeh, Nima, Mahmoud Azami, Hossein Abbasi, and Ali Farzin. "Applying extrusion-based 3D printing technique accelerates fabricating complex biphasic calcium phosphate-based scaffolds for bone tissue regeneration." *Journal of Advanced Research* 40 (2022): 69–94.

68. Wang, P., L. Zhao, W. Chen, X. Liu, M. D. Weir, and H. H. K. Xu. "Stem cells and calcium phosphate cement scaffolds for bone regeneration." *Journal of Dental Research* 93, no. 7 (2014): 618–625.

69. Islam, Mohammad Shariful, Mohammad Abdulla-Al-Mamun, Alam Khan, and Mitsugu Todo. "Excellency of hydroxyapatite composite scaffolds for bone tissue engineering." *Biomaterials* 10 (2020): 1–22.

70. Shi, Chen, Zhangqin Yuan, Fengxuan Han, Caihong Zhu, and Bin Li. "Polymeric biomaterials for bone regeneration." *Ann. Jt* 1 (2016): 27.

71. Kalirajan, Cheirmadurai, Amey Dukle, Arputharaj Joseph Nathanael, Tae-Hwan Oh, and Geetha Manivasagam. "A critical review on polymeric biomaterials for biomedical applications." *Polymers* 13, no. 17 (2021): 3015.

72. Devi, GV Yashaswini, Sukumaran Anil, and Jayachandran Venkatesan. "Biomaterials and scaffold fabrication techniques for tissue engineering applications." *Engineering Materials for Stem Cell Regeneration* (2021): 691–706.

73. Fattahi, Farnaz-sadat, Akbar Khoddami, and Ozan Avinc. "Poly (lactic acid)(PLA) nanofibers for bone tissue engineering." *Journal of Textiles and Polymers* 7, no. 2 (2019): 47–64.

74. Donate, Ricardo, Mario Monzón, and María Elena Alemán-Domínguez. "Additive manufacturing of PLA-based scaffolds intended for bone regeneration and strategies to improve their biological properties." *e-Polymers* 20, no. 1 (2020): 571–599.

75. Sun, Fengbo, Xiaodan Sun, Hetong Wang, Chunxu Li, Yu Zhao, Jingjing Tian, and Yuanhua Lin. "Application of 3D-printed, PLGA-based scaffolds in bone tissue engineering." *International Journal of Molecular Sciences* 23, no. 10 (2022): 5831.

76. Song, Xiaoliang, Xianxian Li, Fengyu Wang, Li Wang, Li Lv, Qing Xie, Xu Zhang, and Xinzhong Shao. "Bioinspired protein/peptide loaded 3D printed PLGA scaffold promotes bone regeneration." *Frontiers in Bioengineering and Biotechnology* 10 (2022): 832727.

77. Dwivedi, Ruby, Sumit Kumar, Rahul Pandey, Aman Mahajan, Deepti Nandana, Dhirendra S. Katti, and Divya Mehrotra. "Polycaprolactone as biomaterial for bone scaffolds: Review of literature." *Journal of Oral Biology and Craniofacial Research* 10, no. 1 (2020): 381–388.

78. Shor, Lauren, Selçuk Güçeri, Robert Chang, Jennifer Gordon, Qian Kang, Langdon Hartsock, Yuehuei An, and Wei Sun. "Precision extruding deposition (PED) fabrication of polycaprolactone (PCL) scaffolds for bone tissue engineering." *Biofabrication* 1, no. 1 (2009): 015003.

79. Davis, H. E., and J. K. Leach. "Hybrid and composite biomaterials in tissue engineering." *Topics in Multifunctional Biomaterials and Devices* 10 (2008): 1–26.

80. Lei, Bo, Baolin Guo, Kunal J. Rambhia, and Peter X. Ma. "Hybrid polymer biomaterials for bone tissue regeneration." *Frontiers of Medicine* 13 (2019): 189–201.

81. Agour, Mahmoud, Abdalla Abdal-Hay, Mohamed K. Hassan, Michal Bartnikowski, and Sašo Ivanovski. "Alkali-treated titanium coated with a polyurethane, magnesium and hydroxyapatite composite for bone tissue engineering." *Nanomaterials* 11, no. 5 (2021): 1129.

82. Boccaccini, A. R., L-C. Gerhardt, S. Rebeling, and J. J. Blaker. "Fabrication, characterisation and assessment of bioactivity of poly (d, l lactid acid)(PDLLA)/TiO2 nanocomposite films." *Composites Part A: Applied Science and Manufacturing* 36, no. 6 (2005): 721–727.

83. Rautray, Tapash R., Subhasmita Swain, and Kyo-Han Kim. "Formation of anodic TiO2 nanotubes under magnetic field." *Advanced Science Letters* 20, no. 3-4 (2014): 801–803.

84. Praharaj, Rinmayee, Snigdha Mishra, and Tapash R. Rautray. "Growth mechanism of aligned porous oxide layers on titanium by anodization in electrolyte containing Cl." *Materials Today: Proceedings* 62 (2022): 6216–6220.

85. Eslami, Hossein, Hamidreza Azimi Lisar, Tahereh Sadat Jafarzadeh Kashi, Mohammadreza Tahriri, Mojtaba Ansari, Tohid Rafiei, Farshid Bastami, Alireza Shahin-Shamsabadi, Fatemeh Mashhadi Abbas, and Lobat Tayebi. "Poly (lactic-co-glycolic acid)(PLGA)/TiO2 nanotube bioactive composite as a novel scaffold for bone tissue engineering: In vitro and in vivo studies." *Biologicals* 53 (2018): 51–62.

86. Swain, Subhasmita, Tapash Ranjan Rautray, and Ramaswamy Narayanan. "Sr, Mg, and Co substituted hydroxyapatite coating on TiO2 nanotubes formed by electrochemical methods." *Advanced Science Letters* 22, no. 2 (2016): 482–487.

87. Praharaj, Rinmayee, Snigdha Mishra, and Tapash R. Rautray. "The structural and bioactive behaviour of strontium-doped titanium dioxide nanorods." *Journal of the Korean Ceramic Society* 57 (2020): 271–280.

88. Xiong, Zixuan, Wenbin Liu, Hu Qian, Ting Lei, Xi He, Yihe Hu, and Pengfei Lei. "Tantalum nanoparticles reinforced PCL scaffolds using direct 3D printing for bone tissue engineering." *Frontiers in Materials* 8 (2021): 609779.

89. Babilotte, Joanna, Benoit Martin, Vera Guduric, Reine Bareille, Rémy Agniel, Samantha Roques, Valérie Héroguez et al. "Development and characterization of a PLGA-HA composite material to fabricate 3D-printed scaffolds for bone tissue engineering." *Materials Science and Engineering: C* 118 (2021): 111334.

90. Fraile-Martínez, Oscar, Cielo García-Montero, Alejandro Coca, Miguel Angel Álvarez-Mon, Jorge Monserrat, Ana M. Gómez-Lahoz, Santiago Coca et al. "Applications of polymeric composites in bone tissue engineering and jawbone regeneration." *Polymers* 13, no. 19 (2021): 3429.

91. Satpathy, Anurag, Rinkee Mohanty, and Tapash R. Rautray. "Bio-mimicked g uided tissue regeneration/guided bone regeneration membranes with hierarchical structured surfaces replicated from teak leaf exhibits enhanced bioactivity." *Journal of Biomedical Materials Research Part B: Applied Biomaterials* 110, no. 1 (2022): 144–156.

92. Mathieu, Laurence Marcelle, Thomas L. Mueller, Pierre-Etienne Bourban, Dominique P. Pioletti, Ralph Müller, and Jan-Anders E. Månson. "Architecture and properties of anisotropic polymer composite scaffolds for bone tissue engineering." *Biomaterials* 27, no. 6 (2006): 905–916.

93. Swain, Subhasmita, Janardhan Reddy Koduru, and Tapash Ranjan Rautray. "Mangiferin-enriched Mn–hydroxyapatite coupled with β-TCP scaffolds simultaneously exhibit osteogenicity and anti-bacterial efficacy." *Materials* 16, no. 6 (2023): 2206.

94. Swain, Subhasmita, Abhishek Patra, Shubha Kumari, Rinmayee Praharaj, Satrujit Mishra, and Tapash Rautray. "Corona poled gelatin-Magnesium hydroxyapatite composite demonstrates osteogenicity." *Materials Today: Proceedings* 62 (2022): 6131–6135.

95. Inzana, Jason A., Diana Olvera, Seth M. Fuller, James P. Kelly, Olivia A. Graeve, Edward M. Schwarz, Stephen L. Kates, and Hani A. Awad. "3D printing of composite calcium phosphate and collagen scaffolds for bone regeneration." *Biomaterials* 35, no. 13 (2014): 4026–4034.

96. Moshiri, Ali, Neda Tekyieh Maroof, and Ali Mohammad Sharifi. "Role of organic and ceramic biomaterials on bone healing and regeneration: An experimental study with significant value in translational tissue engineering and regenerative medicine." *Iranian Journal of Basic Medical Sciences* 23, no. 11 (2020): 1426.

97. Liu, Dingkun, Zhihui Liu, Jundong Zou, Lingfeng Li, Xin Sui, Bizhou Wang, Nan Yang, and Bowei Wang. "Synthesis and characterization of a hydroxyapatite-sodium

alginate-chitosan scaffold for bone regeneration." *Frontiers in Materials* 8 (2021): 648980.

98. Priyadarshini, Itishree, Subhasmita Swain, Janardhan Reddy Koduru, and Tapash Ranjan Rautray. "Electrically polarized withaferin a and alginate-incorporated biphasic calcium phosphate microspheres exhibit osteogenicity and antibacterial activity in vitro." *Molecules* 28, no. 1 (2022): 86.

99. Swain, Subhasmita, Tae Yub Kwon, and Tapash R. Rautray. "Fabrication of silver doped nano hydroxyapatite-carrageenan hydrogels for articular cartilage applications." *bioRxiv* (2021): 2020-12.

100. Rautray, Tapash Ranjan, and Kyo Han Kim. "Synthesis of controlled release Sr-hydroxyapatite microspheres." In *Bioceramics-24*. 2012.

101. Kaushik, Neha, Linh Nhat Nguyen, June Hyun Kim, Eun Ha Choi, and Nagendra Kumar Kaushik. "Strategies for using polydopamine to induce biomineralization of hydroxyapatite on implant materials for bone tissue engineering." *International Journal of Molecular Sciences* 21, no. 18 (2020): 6544.

102. Swain, Subhasmita, Shubha Kumari, Priyabrata Swain, and Tapash Rautray. "Polarised strontium hydroxyapatite–xanthan gum composite exhibits osteogenicity in vitro." *Materials Today: Proceedings* 62 (2022): 6143–6147

103. Verrier, Sophie, Jonny J. Blaker, Veronique Maquet, Larry L. Hench, and Aldo R. Boccaccini. "PDLLA/Bioglass® composites for soft-tissue and hard-tissue engineering: An in vitro cell biology assessment." *Biomaterials* 25, no. 15 (2004): 3013–3021.

# 23 LASER-based Surface Modification Techniques for Fatigue Life Improvement of Biomaterials

*T. Rajesh Kumar Dora*
Gitam School of Technology (GST), GITAM, Visakhapatnam, AP, India

*Karthik Dhandapani*
Department of Physics, School of Engineering, Presidency University, Bangalore, India

*Sarada Prasanna Mallick*
Department of Biotechnology, Koneru Lakshmaiah Education Foundation, Guntur, AP, India

*Pratik Shukla*
The Manufacturing Technology Centre (MTC), Coventry, UK

## 23.1 INTRODUCTION-SURFACE ENGINEERING TECHNIQUES

Surface engineering techniques became an integral part of devices that are continuously being exposed to a cyclic loading condition or in continuous contact with an abrasion surface [1]. Out of these conditions mentioned, the former becomes more predominant to achieve a prolonged fatigue life for the products that are popularly used in both automobile and aerospace applications. The expected fatigue life of a shaft depending on its application is supposed to have at least 100 a million number of cycles before its failure; however, it is quite difficult to achieve such surface property with both conventional (casting, forming, and machining) and unconventional (electric discharge machining, chemical machining, and laser cutting) manufacturing technology [2]. Therefore, it is important to introduce a secondary process that can improve/enhance the surface property of an as-processed metal/alloy

DOI: 10.1201/9781003429920-26

that can prolong their service life. The sole purpose of adopting a secondary process is to impart compressive residual stresses on the surface that can act as a barrier for fatigue crack propagation [3].

**What Is the Significance of Surface Engineering?**
Nowadays, there is a huge scope of surface engineering techniques; those are used to counter the issues related to degradation of material due to friction, corrosion, and wear [4]. Apart from these, this technique is also used to induce surface compressive residual stress that can enhance the fatigue life of devices subjected to cyclic loading during the service life of the engineering devices [5]. Surface compressive residual stress helps in delaying the initial crack propagation stage that can add extra life to the devices used in the medical industry because these devices are difficult to retract once deployed in the human body [6]. Surface coatings are applied on the medical devices to improve their biocompatibility of these devices.

There are many surface engineering techniques that are regularly being practiced by industries over many decades, and out of these methods, some of the techniques are given in Table 23.1.

**Advantages and Disadvantages of Surface Engineering Techniques**
Mechanical processes such as, shot peening/blasting and burnishing techniques are known to be most successful commercially adopted method so far; however, such a contact-type method for imparting surface residual stress has limitations, namely, (i) difficulty in controlling the surface finish; (ii) non-uniform coverage; (iii) risk of developing surface crack; and (iv) difficulty in handling unsymmetrical geometry [7]. As far as thermal processes are concerned, except laser hardening, there is limited scope of surface engineering for miniature and unsymmetrical objects, and for thermo-chemical processes, the surface engineering is not much effective due to very low cased depth [8]. Therefore, the requirement of a non-contact-type surface peening method becomes necessary to overcome the capability of contact-type surface peening process. However, for residual stress generation, shot peening/shot blasting is considered the best among all surface engineering techniques [9].

### 23.1.1 IMPROVEMENT OF FATIGUE LIFE THROUGH RESIDUAL STRESS GENERATION

a. **Fatigue in Metals/Alloys:** Out of all mechanical failures, 90% of the machine components made up of metallic alloys fail prematurely below their UTS due to alternating/fluctuating stress during their service life [10]. And this is mainly due to fast propagation of pre-existing cracks developed during their functioning in respective applications. Fatigue failures are broadly classified into two types, namely, (i) high cycle fatigue; (ii) low cycle fatigue. The most important factor such as crack initiation process is greatly dependent on the *Geometrical Stress Concentration* of the machine component, the stress concentration may vary depending on the design that includes notches or contours [11]. Besides this, the surface finish of the machine component can also affect the crack initiation process rather than crack propagation. And sometimes environmental factors such as corrosion

**TABLE 23.1**
**Methods of Various Surface Engineering Techniques**

Surface Engineering Techniques

| Without Changing Surface Chemistry | | Other Process | By Altering the Surface Chemistry | By Applying Coating |
|---|---|---|---|---|
| Thermal process | Mechanical process | 1. Chemical/ Electro-etching | 1. Electro-chemical process | 1. Thermal spraying |
| 1. Laser hardening | 1. Shot peening | 2. Laser Shock Peening | 2. Thermo-chemical Diffusion | 2. Electroplating/ electro-less plating |
| 2. Induction hardening | 2. Shot blasting | | 3. Chemical conversion coating | 3. Galvanizing |
| 3. Flame hardening | 3. Explosive hardening | | 4. Ion implantation process | 4. PVD/CVD |
| | 4. Burnishing | | | 5. Powder coating |
| | 5. Machine hammer peening | | | 6. Cladding |

and atmospheric/surrounding temperature may enhance/accelerate fatigue failure in the case of metals and alloys [12]. Various techniques are being adopted worldwide to improve fatigue life of machine components and they are briefly discussed below.

b. **Techniques for improving fatigue life:** The fatigue behavior of a metallic alloy can be influenced by modifying their material property; however, it can also be influenced by applying localized modification of the machine component surface to pin the surface crack initiation process [13]. Several surface treatments, such as mechanical (shot peening, cold rolling, grinding), thermal (flame hardening, Induction hardening, laser hardening), and thermo-chemical (case hardening, nitriding, plating) methods, are used to modify the surface. Out of these, mechanically modified surface such as shot peening/sand blasting are most popularly used method to induce surface compressive residual stress [14]. The residual stress distribution profile from the shot peened surface is demonstrated in Figure 23.1, where the compressive stress first increases to a maximum value ($\sigma_{max}$) and then decreases to a depth ($\delta$ = effective plastically affected depth) [15]. For this process, spherical shots/balls made of cast iron or ceramic are made to strike the surface, and as a result of localized plastic deformation below the surface, compressive residual stress will be developed [16].

The maximum compressive stress ($\sigma_{max}$) and plastically affected depth ($\delta$) can be controlled by choosing suitable shot type, shot diameter, pressure, and velocity of the particle. The maximum compressive residual stress that can be achievable by using shot peening is half of the material's yield strength ($\sigma_0$) [17]. Moreover, generating surface compressive residual stress using such method is more suitable for higher strength alloys compared to lower strength material [18]. There are certain limitations to this method such as unable to respond high alternating stress cycle and it is more effective to high cycle fatigue compared to low cycle fatigue [19]. As far as thermal treatments such as flame hardening and induction hardening are concerned, the treated surface/case tend to have more hardness and tensile strength compared to the core [20]. Due to the variation in stress at the case and core

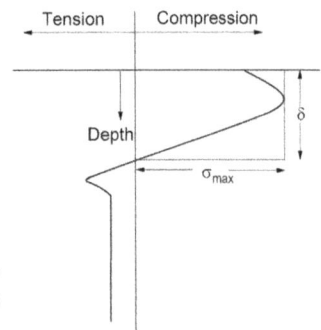

**FIGURE 23.1** Residual stress distribution profile due to shot peening, and effective depth ($\delta$) of compressive residual stress.

of the machine component. a compressive residual stress will be developed along the layers of the material and the fatigue life is expected to improve. However, in the case of thermo-chemical treatment such as carburizing and nitriding, the surface is hardened by supplying carbon and nitrogen externally from the external source for the respective process [5]. The sole idea of these processes is to improve wear resistance of the alloy, but these processes also improve the fatigue resistance. When it comes to nickel and chrome plating, they are more prone to induce unfavorable tensile residual stress compared to other coatings such as zinc, cadmium, lead, and tin [13].

## 23.2   BASICS OF LASER SHOCK PEENING (LSP)

There are studies available where shock waves generated by confining laser-produced plasma within the boundaries of the target material and transparent overlay were used to induce compressive surface residual stresses [21]. Such processes are known as laser shock peening (LSP), and it was found that the residual stress generation is due to the transmission of shock waves into the bulk material which causes the plastic deformation on the surface. The shock wave generation can be explained by the high pressure of the plasma state at the interface of the target material and transparent overlay which is a function of laser power ($I_0$). The magnitude of such pressure (GPa) is given by equation (23.1), that is directly proportional to the square root of the laser power. In addition to this, there are other operating parameters such as absorptivity level of laser pulses, reduced shock impedance of target material, and confinement media (equation 23.2). A schematic of laser shock peening process is shown in Figure 23.9, where a guided and focused laser beam falls on the metallic target to generate shock waves responsible for plastic deformation. There are other factors such as laser spot diameter (d), overlap ratio (%), the direction of LSP (Zig-zig/spiral) that affects the surface property of the target material. Besides these instrumental parameters, other factors affect the severity of surface deformation such as LSP with a coating (Al/Fe foil or black paint) and LSP without coating. Generally, the coating is provided to protect the surface for degradation/ablation/burning.

$$P(\text{GPa}) = A\left(\frac{\alpha}{\alpha + 3}\right)^{1/2} Z^{1/2} I_0^{1/2} \tag{23.1}$$

$$\frac{1}{Z} = \frac{1}{Z_1} + \frac{1}{Z_2} \tag{23.2}$$

where P = Plasma pressure (GPa), A = Constant, $Z_1$ = Shock impedance of target material, $Z_2$ = Shock impedance of confinement, $\alpha$ = Absorptivity of laser pulses, and $I_0$ = Laser power (GW.cm$^{-2}$) (Figure 23.2).

$$P = BI^{1/2} \tag{23.3}$$

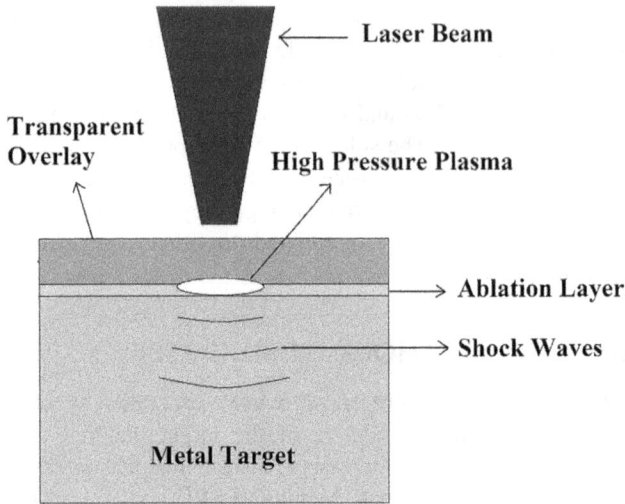

**FIGURE 23.2** Schematic of laser shock peening process (LSP).

The shock pressure that is directly proportional to the square root of the laser power ($I_0$) also depends on the type of confinement layer. During the LSP process, water and BK 7 glass are commonly used as confinement layer and the proportionality constant (B) for water and borosilicate glass is given by 10.1 and 21, respectively (equation 23.3) [21]. In addition to this, researchers have used numerous pressure ~ time profiles for numerical analysis (FEM), given in Table 23.2 [22]. Out of these pressure profiles, most of the studies have followed Gaussian [23] and a triangular-type pressure profile [24].

### 23.2.1 BRIEF DETAILS OF LASER SOURCE

It is also important to discuss about the LASER source used for shock peening study. The difference in LASER light over an ordinary light is due to its monochromatic, directional, and coherent nature that makes it an obvious choice to use it for the shock peening process [25]. The enormous amount of kinetic energy required to induce plastic deformation is not achievable by using a normal LASER light, therefore the energy needs to be amplified and it can only be achieved by opting for a Q-switching, Nd: YAG (neodymium-doped glass or yttrium aluminum garnet) laser source. Q-Switching is a technology by which LASER rays with very short (ns) and powerful (MW) pulse can be achieved, and typically for LSP process the laser wavelength of size 532–1064 nm and pulse duration of 10–100 ns are popularly used. The key phenomena of Q-switching are based on mode locking in which the interference of laser rays produces a train of pulses [26]. There are two methods to improve the frequency of the laser rays due to interference: (1) Second Harmonic Generation (SHG); and (2) Third Harmonic Generation (THG). Q-Switching is categorized into two types, such as *Active* (when switching is controlled by the operator) and *Passive* (when switching is controlled by the laser

**TABLE 23.2**
**Pressure Profile with Time**

| Sl no | Type of Pressure Profile | Pressure vs Time | Ref. |
|---|---|---|---|
| 1 | Spatial | | [22] |
| 2 | Temporal | | [22] |
| 3 | Gaussian | | [23] |
| 4 | Triangular | | [24] |

process itself), and it can be obtained by using both Mechanical Switch and Optical Switch by using *Rotating mirror* and *Pockel's cell,* respectively.

## 23.2.2 RESIDUAL STRESS GENERATION THROUGH LASER SHOCK PEENING

The compressive residual stress generation in the case of the metallic alloy sample is only possible in the direction of transmission of laser shock wave deep due to plastic deformation. The depth of the plastic deformation will only be confined to a region where the peak pressure generated by the plasma limits itself to the materials Hugonoit Elastic Limit (HEL), which depends on both dynamic yield point and Poisson's ratio of the material. The equation (23.4) that is derived from the Johnson–Rhode theory explains the relationship between the yield strength of the material under the uniaxial shock loading and dynamic yield point of the material [27].

$$\text{Hugonoit Elstic Limit(Ph)} = \frac{1 - v}{1 - 2v}\sigma_y^{dyn} \tag{23.4}$$

where, $\sigma_y^{dyn}$ = Dynamic yield point of material (GPa), Ph = Yield strength under a uniaxial shock loading condition, and $v$ = Poissons's ratio.

For a better understanding of the deformation mechanism during shock loading, an analytical model was proposed by *Ballard et. al., 1991* for an elastic-plastic material [28]. They tried to establish a relation between the plastic strain and residual stress field generated for impact pressure (P) and pressure pulse duration ($\tau$). The proposed model was developed by making some assumptions, such as (i) The shock wave must propagate longitudinally in a perfectly elastic-plastic half-space of the material; (ii) the plastic deformation of the material must follow *Von – Mises* plasticity criterion; and (iii) the pressure applied to the half-space must be uniform. The stress and strain tensor for the induced plastic deformation are given by equation (23.5).

$$\varepsilon = \begin{bmatrix} 0 & 0 & 0 \\ 0 & 0 & 0 \\ 0 & 0 & \varepsilon \end{bmatrix}; \ \sigma = \begin{bmatrix} \sigma_r & 0 & 0 \\ 0 & \sigma_r & 0 \\ 0 & 0 & \sigma_x \end{bmatrix}; \ \epsilon_p = \begin{bmatrix} -\varepsilon_p/2 & 0 & 0 \\ 0 & -\varepsilon_p/2 & 0 \\ 0 & 0 & \varepsilon_p \end{bmatrix} \tag{23.5}$$

The stresses generated ($\sigma_x, \sigma_r$) in the longitudinal and radial direction for the half-space of the material were shown in Figure 23.10 is due to a uniform pressure field $P(t)$ which is a function of time. Hooke's law with Lame's constant $\lambda$ and $\mu$, which are given by equation (23.6), can be applied to the above physical condition.

$$\sigma = \lambda \, tr(\varepsilon) + 2\mu(\varepsilon - \varepsilon_p) \tag{23.6}$$

where $\sigma$ and $\varepsilon$ are the stress and strain tensors. The equation (23.6) can be explained depending on the assumption made for the analytical model.

For completely elastic deformation: $\sigma = (\lambda + 2\mu)(\varepsilon); \sigma_r = \lambda\varepsilon$          (23.7)

For elastic–plastic deformation: $\sigma = (\lambda + 2\mu)\varepsilon - 2\mu\varepsilon_p; \sigma_r = \lambda\varepsilon + \mu\varepsilon_p$ (23.8)

By applying the Von-Misses criterion of plastic deformation ($\sigma_r - \sigma_x = YS$), a relationship between Hugonoit Elastic Limit and compressive static yield strength of the material was obtained, given by equation (23.9).

$$HEL = \left(\frac{1 + \lambda}{2\mu}\right)Y \qquad (23.9)$$

The obtained relationship was further modified by considering the existence of an initial surface residual stress ($\sigma_0$) and given by equation (23.10).

$$HEL = \left(\frac{1 + \lambda}{2\mu}\right)(\sigma_0 + YS) \qquad (23.10)$$

The deformation mechanism due to shock loading can be presented by Figure 23.10, where the entire deformation was explained in various stages. From point 0–1, the material is deformed elastically to 1 × HEL, but from point 1–2 that lies between 1 × HEL and 2 × HEL, plastic deformation takes place to 2 × HEL. Beyond 2 × HEL, there is no further deformation allowed and elastic unloading will take place from points 2–3 followed by plastic unloading from points 3–4. The dotted line represents the Von-Misses criterion ($\sigma_r - \sigma_x = YS$). (Figure 23.3)

The plastically affected depth ($L_p$) is a function of elastic shock velocity, plastic shock velocity, and pressure pulse duration and that is given by equation (23.11). The $C_{el}$ and $C_{pl}$ values can be calculated both experimentally and analytically (equation 23.12, 23.13). Similarly, the surface residual stress can also be calculated by using impact pressure (P), laser spot radius (r), Poisson's ratio ($v$), and plastically affected depth ($L_p$) given by equation (23.14). It was further modified due to the presence of initial surface residual stress ($\sigma_0$) and presented by equation (23.15). The validity of these equations is subjected to the condition that the impact pressure must be greater than the Hugonoit Elastic Limit (HEL) of the material.

$$L_p = \frac{C_{el}C_{pl}\,\tau}{C_{el} - C_{pl}} \qquad (23.11)$$

where, $L_p$ = Plastically affected depth (mm), $C_{el}$ = Elastic shock wave velocity, $C_{pl}$ = Plastic shock wave velocity, and $\tau$ = Pressure pulse duration (ns).

$$C_{el} = \sqrt{\frac{\lambda + 2\mu}{\rho}} \qquad (23.12)$$

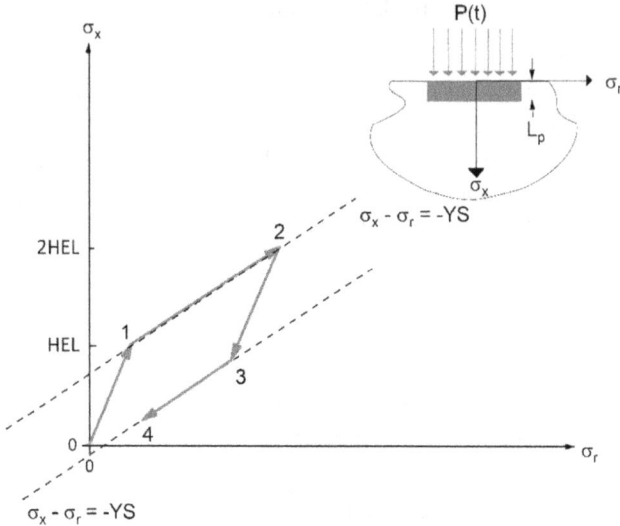

**FIGURE 23.3**  Plastic deformation mechanism during the LSP process.

$$C_{pl} = \sqrt{\frac{\lambda + \frac{\mu}{3}}{\rho}} \qquad (23.13)$$

$$\sigma_{Surf} = -\frac{P}{2\left(1 + \frac{\lambda}{2\mu}\right)}\left[1 - \frac{4\sqrt{2}}{\pi r}(1 + \vartheta)L_p\right] \qquad (23.14)$$

$\sigma_{Surf}$ = Surface compressive residual stress (MPa), $\lambda$ and $\mu$ = Lame's constant, r = Radius of impact (mm), $v$ = Poissons's ratio, $L_p$ = Plastically affected depth (mm), P = Shock wave pressure (for P > HEL).

$$\sigma_{Surf} = \sigma_0 - (YS + \sigma_0)\left[1 - \frac{4\sqrt{2}}{\pi r}(1 + \vartheta)L_p\right] \qquad (23.15)$$

$\sigma_0$ = Initial residual stress of the material, and YS = Yield strength of the material.

Few experimental results of surface residual stress were induced due to laser shock peening on numerous metallic alloys, shown in Figure 23.11, that can be correlated to the analytical model. In the proposed model where the residual stress ($\sigma_0$) is proportional to the plastically affected depth ($L_p$) by keeping the impact pressure (P) constant. For the below-mentioned results, the LSP test was conducted at 6 GPa pressure using a glass confined media. The surface residual stress increases with the increase in plastically affected depth (Figure 23.4).

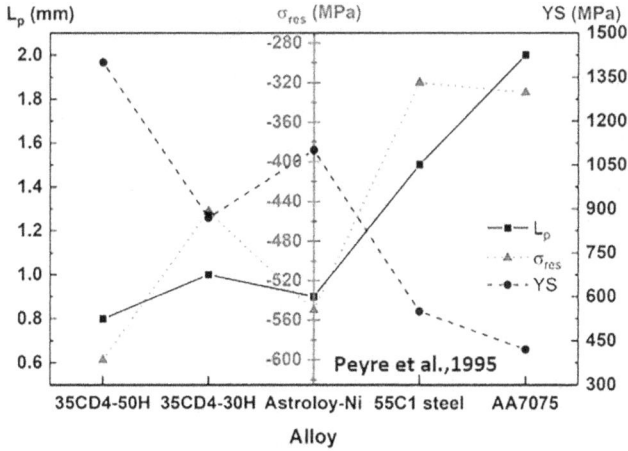

**FIGURE 23.4** Plastically affected depth ($L_p$) and surface residual stress ($\sigma_{res}$) of alloys determined by the analytical model [29].

## 23.3 PREVIEW OF MATERIALS USED IN MEDICAL DEVICES

Any implant or medical device's success is highly dependent on the biomaterial employed. Metals, polymers, ceramics, and composites are examples of synthetic materials that have made important contributions to many established medical equipment. The goal of this publication is to provide a basic understanding of the effect of non-contact-type surface treatment on implant fatigue life. The most critical parameters for any material used in biomedical implants are non-carcinogenic, biocompatibility, host-tissue reactivity to implants, and cytotoxicity. The density, strength, and hardness are all secondary factors. Biomaterial requirements are categorized into four categories: (a) biocompatibility, (b) sterilizability, (c) functionability, and (d) manufacturability. Figure 23.5 depicts a broader classification of biomaterials based on the criteria listed above [30]. After meeting the basic criteria, the material must additionally meet other criteria such as mechanical qualities and surface features. Mechanical parameters identified to be most relevant for

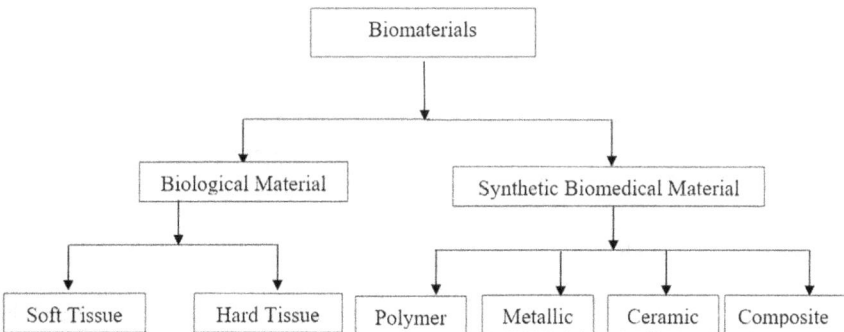

**FIGURE 23.5** Classification of biomaterials.

biomaterials include modulus of elasticity, ultimate tensile strength, elongation to failure, and fracture toughness. All these properties can be modified by fine-tuning the processing route and thermal treatment methods. However, the surface properties of the biomaterial can be modified using a variety of surface engineering techniques, such as shot peening, nitriding, or coating. These surface engineering approaches are crucial in changing the surface properties of biomaterials; because it was believed that the cell adhesion probability tends to increase. But the prime objective of surface engineering is to impart surface residual stresses that increase the fatigue life of the biomaterials.

### 23.3.1 CLASSIFICATION OF METALLIC ALLOYS FOR BIOMEDICAL APPLICATIONS

The medical community has started to pay attention to metallic alloys because of the high modulus and improved corrosion resistance of several recently discovered alloys. The greatest biomedical implants are made of titanium, Co-Cr, and stainless-steel alloys. Some SMART materials, such as shape memory alloys, can also successfully serve the dental and medical industries because of their superelastic nature. The newly developed Ni-Ti shape memory alloy for medical purposes is best known for its use in stents and arch wires for cardiovascular and orthodontic applications, respectively. The upper plateau stress that comes from the Austenite to Martensite phase transformation is decreased or alleviated during the Martensite to Austenite phase transformation [31]. Instead of using metals with excellent corrosion resistance, efforts were made in the latter part to manufacture biodegradable or corrodible metallic implants. Due to the failures of non-biodegradable implants, the demand for developing biodegradable implants has emerged. Mg-based alloys with certain compositions, such as Mg-Ca and Mg-Zn binary alloys, have been successfully produced as biodegradable implants [32]. The complexity in the design of biomedical devices utilized in the human body led to initiatives to build these components additively rather than traditionally [33]. A broad classification of metallic implants is given in Figure 23.6.

### 23.4 EFFECT OF LASER SHOCK PEENING ON METALLIC ALLOYS AND COMPOUNDS

#### 23.4.1 TITANIUM ALLOYS

After any surface engineering, the residual stress distribution provides a true evaluation of the procedure. During laser shock peening, it is anticipated that the metallic surface will experience significant plastic deformation and generate compressive residual stress. However, to maintain a proper force balance, the compressive residual stress tends to be more tensile as we move deeper into the cross-section or thickness direction. Relative residual stress varies from alloy to alloy and is dependent on the elastic modulus of the alloy. Figure 23.7a shows the variation in residual stress following LSP for different titanium alloys. The comparison was made while keeping the energy source (5 J) constant in order to evaluate the variation in residual stresses for titanium alloys. When peened with

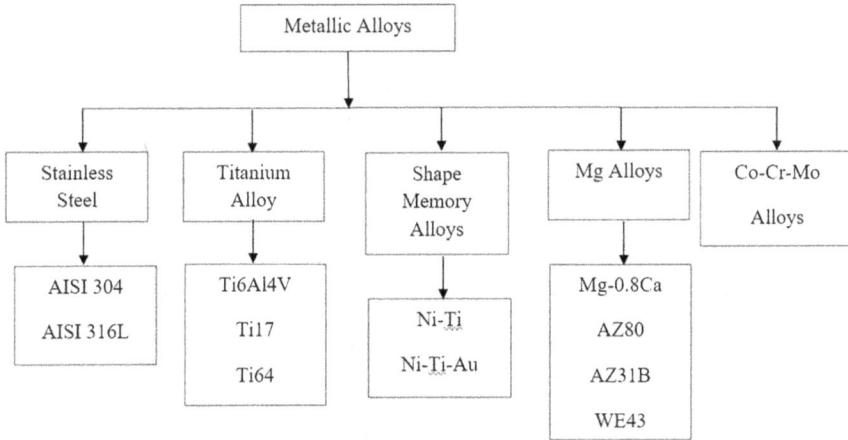

**FIGURE 23.6**  Broad classification of metallic alloys for biomaterials.

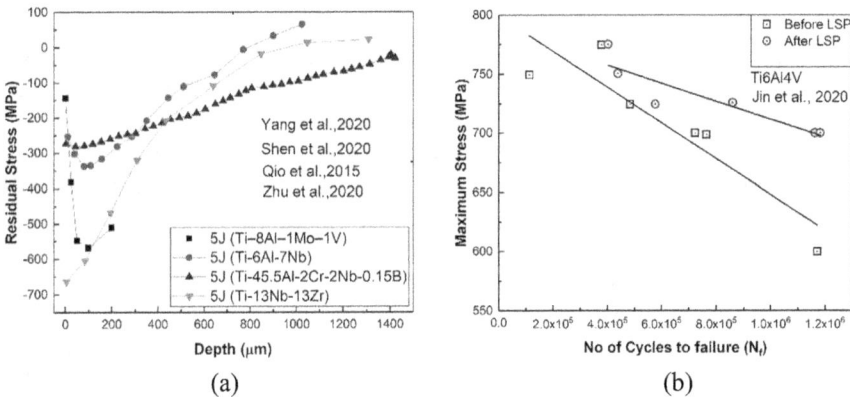

**FIGURE 23.7**  (a) Residual stress distribution along the depth of various titanium alloys peened with 5 J Laser energy; (b) comparison of fatigue life after and before LSP of Ti6Al4V alloy.

similar laser intensity, Ti45.5Al2Cr2Nb0.15B has the least residual stress on the surface, but Ti13Nb13Zr tends to have the most residual stress at the surface compared to all other Ti-based alloys [34–37]. The effect of compressive residual stress tries to decrease after 500 microns of depth, reaches the base value, and then stays flat. More time is needed for fatigue fracture formation the higher the compressive residual stress at the surface, which extends the material's fatigue life. Figure 23.7b, which plots the maximum stress against the number of cycles to failure for the untreated sample and the LSPed sample, displays the fatigue findings for the Ti6Al4V alloy. It was discovered that, compared to the untreated sample, the fatigue life had improved after laser therapy [38].

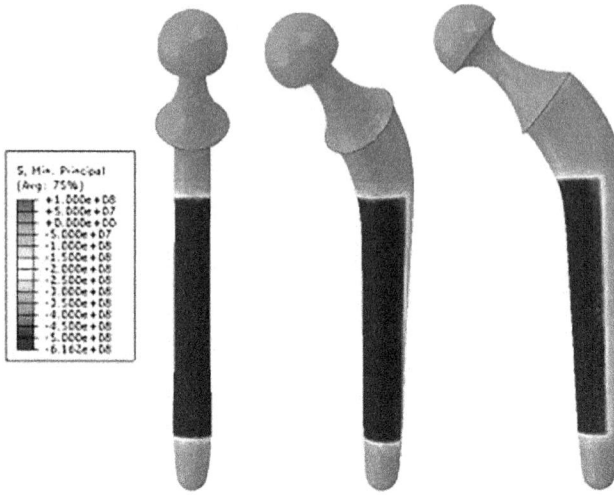

**FIGURE 23.8**  Representation of residual stress drveloped on a hip implant made up of Ti6Al4V.

A 2.4 J laser source was used to laser peen a standard Charnley stem-head form (hip replacement geometry) in order to extend its fatigue life. Figure 23.8 depicts the residual stress (von Mises stress) that occurred on the surface following laser shock peening. Figure 23.9a presents the residual stress variation for two distinct paths (1 and 2) with respect to depth, whereas Figure 23.9b depicts the internal residual stress distribution of the geometry [39].

## 23.4.2  STAINLESS STEEL

The most often used metallic biomaterial for biomedical implants is austenitic stainless steel because it is less expensive, easier to produce, stiffer, and has better chemical stability than Co-Cr alloys, pure Ti, and Ti alloys. Comparing austenitic SS to Ti-based and Co-Cr-Mo-based biomaterials, however, raises questions about its biocompatibility. However, many efforts have been made to increase the corrosion resistance of these stainless steels by utilizing cutting-edge surface engineering techniques, one of which is laser shock peening. Evaluating the compressive residual stress distribution in the alloy created following laser treatment is equally significant to corrosion resistance. In the discussion that follows, Figure 23.10a shows the residual stress distribution of laser-peened 304 and 316 L stainless steel, both of which have attained almost comparable compressive stresses (300 to 600 MPa). However, there is one case where 316 L steel is preferred over 304 steels [40,41]. In contrast to 304, where the compressive stress persisted up to a depth of 1000 microns, 316 L's compressive stress becomes tensile after a depth of 100 microns. The fatigue life of 316 L improved after the laser treatment, as illustrated in Figures 23.10b before and after LSP [40].

(a)  (b)

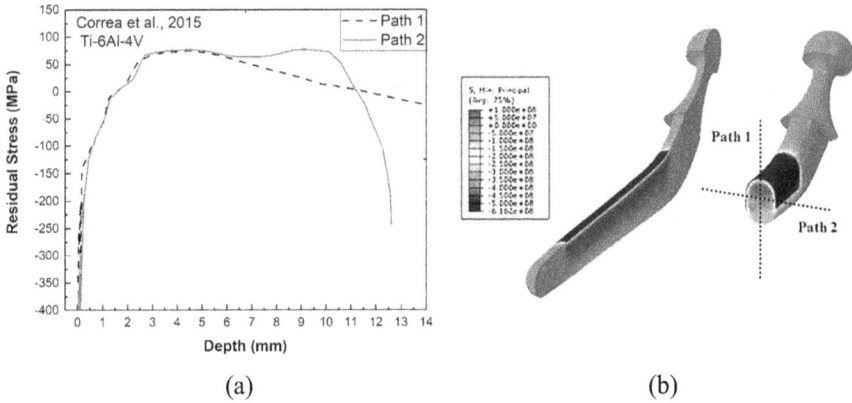

**FIGURE 23.9** (a) Residual stress profile obtained along path 1 and path 2 with respech to depth; (b) internal stress distribution in the hip replacement.

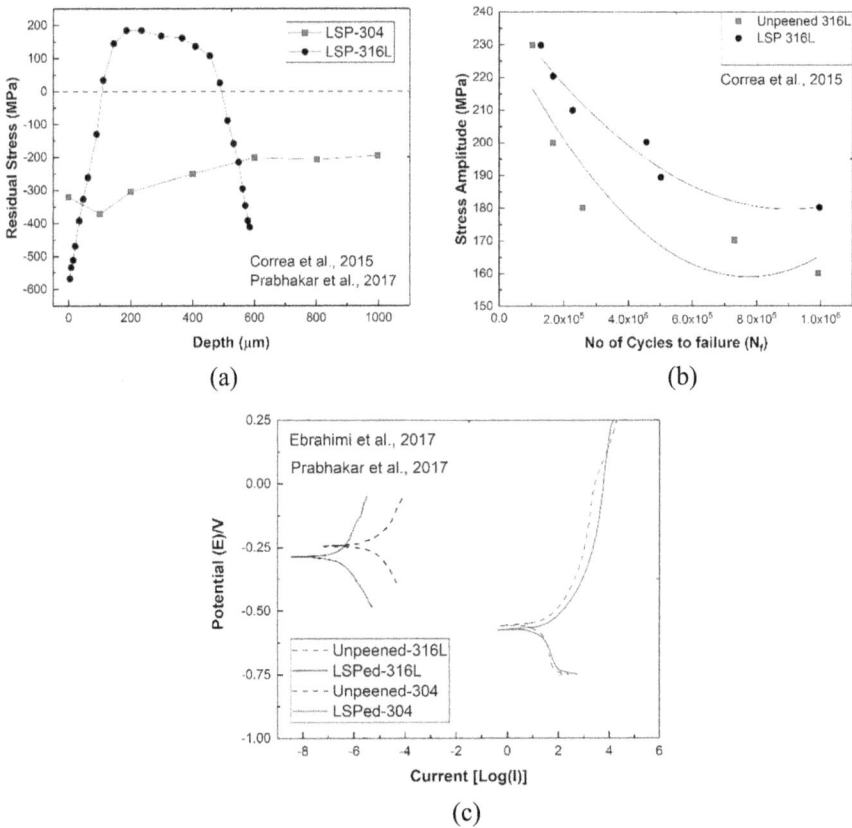

(a)  (b)

(c)

**FIGURE 23.10** (a) Residual stress distribution across thickness of 304 and 316 L steel; (b) comparison of fatigue life after and before LSP of 316 L stainless steel;(c) comparison of tafel plot of 304 and 316 L steel before LSP and after LSP.

The cathodic and anodic curves between unpeened and laser-peened steel show little difference in the corrosion study of 304 and 316 L stainless steel shown in Figure 23.10c. Although the corrosion potential of the LSPed sample is slightly shifted in the direction of the negative potential, this does not imply that the samples' corrosion resistance has decreased. The intriguing fact, however, is that for 304 steels, the corrosion current density for laser peened sample is a little lower than the unpeened sample. This suggests that the laser-peened sample will have a lower corrosion rate (mm/y) than the untreated sample [42]. In both the untreated and LSP conditions, the 304 SS has greater corrosion resistance than the 316 L alloy.

### 23.4.3   Ni-Ti Shape Memory Alloys

Shape memory alloys were accidentally discovered, but their thermoelastic nature, which was intended to be employed as an energy material for the use of power generation, was initially the main emphasis. But later, the biomedical industry became interested in this material because of its superb superelasticity (strain recovery behavior close to 11%). It has been discovered that the elastic modulus of Ni-Ti shape memory alloy is extremely similar to the elastic modulus of bone tissues. Initial worries about this alloy's corrosion resistance, particularly the presence of the cancer-causing Ni ion, were raised. An allergic reaction, skin blistering, and inflammation may result from the excessive leaking of Ni-ion into the human body. However, it was later discovered that the titanium in the alloy prevents the leaching of nickel ions by creating a barrier known as titanium dioxide. Even so, it's crucial to monitor the Ni-ion release both before and after applying a surface engineering technique.

The corrosion resistance of the Ni-Ti binary alloy has been greatly increased by laser shock peening. The tafel plot in Figure 23.11a shows that by decreasing the corrosion current, the corrosion potential has shifted towards the positive side [43]. Spectroscopy analysis was used to evaluate the metal ion release from the binary alloy following its immersion in SBF solution for a predetermined amount of time. Figures 23.11b and 23.11c, which detail the release of Ti and Ni metal ions, respectively. When compared to untreated samples, laser shock peening was found to reduce the concentration (in ppb) of both Ni-ion and Ti-ion release. In addition, a cell culture research utilizing a LIVE/DEAD staining kit was carried out on both untreated and LSPed Ni-Ti samples. After 24 hours of incubation, the vitality of the ADS cells depicted in Figure 23.11d was evaluated. Under a fluorescent light source, live cells appear green while dead cells appear red. When compared to an untreated sample, it was discovered that laser shock peening increased the number of live cells. By measuring the cell length-width ratio, it is possible to evaluate the cell adherence and growth on the implant surface. The monochrome images of the cells (shown in Figure 23.11e) are displayed in pseudo color to demonstrate the cell adhesion. The better the cell development, the smaller the ratio, although in this case, it was discovered that the untreated sample's (357 $\mu m^2$) cell area was smaller than that of the LSP-treated sample (357 $\mu m^2$).

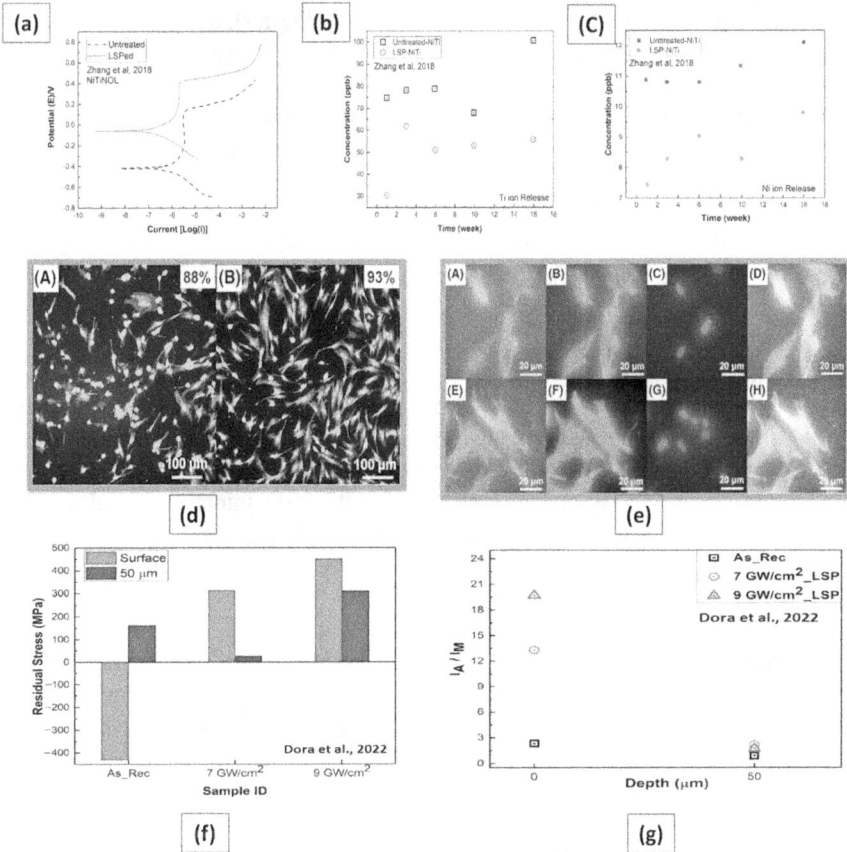

**FIGURE 23.11** (a) Comparison of tafel plot of 304 and 316 L steel before LSP and after LSP; (b) Concentration of Ti-ion release in PPB of Ni-Ti sample; (c) Concentration of Ni-ion release in PPB of Ni-Ti sample; (d) Live/Dead staining of cells before LSP – A and after LSP – B; (e) Cell adhesion pattern to untreated samples – A to D and LSP samples – E to H; (f) Residual stress at surface and 50 μm below surface of untreated and LSP sample; (g) XRD Intensity ratios of Austenite to Martensite phase at surface and 50 μm below surface of untreated and LSP sample.

It's crucial to talk about residual stress generation during laser shock peening in Ni-Ti shape memory alloys in addition to studies on cell viability and corrosion. Due to some complexity in the laser-material interaction and dynamic behavior, not enough research has been done on the fatigue life and residual stress behavior of laser peened Ni-Ti SMAs. It is frequently discovered that shape memory alloys are either entirely Austenite or totally Martensite. There is a chance that Austenite and Martensite phase mixtures can be discovered in Ni-Ti shape memory alloys at room temperature, depending on the transition temperatures, a study that was done on Ni-Ti alloys with both phases present at ambient temperature. The alloy was laser peened with 7 GW/cm$^2$ and 9 GW/cm$^2$ power while being contained by black tape and water, respectively. After laser shock peening, the residual stress, which was compressive in the untreated condition, has changed to tensile (shown in Figure 23.11f). In an effort to determine the cause of this peculiar behavior, phase analysis employing XRD was carried out at both the surface and at the sub-surface below 50 µm. The Austenite-Martnesite ratios were calculated and compared (shown in Figure 23.11g).

In comparison to the 7 GW/cm$^2$ sample and the received material, the value of this ratio for the 9 GW/cm$^2$ sample was higher near the surface. This suggests that during peening, martensite was changing into austenite, with the effect being more pronounced at higher laser power densities. When compared to their surface counterparts, the ($I_A(110)/I_M(111)$) for 7 and 9 GW/cm$^2$ was significantly lower at a depth of 50 µm, suggesting that the martensite fraction was higher for these samples.It was found that the peened material saw a more dramatic shift in ($I_A(110)/I_M(111)$) with depth, whereas the as-received material was only suffering a slight decrease in this value. This suggests that the austenite and martensite phase distribution was being significantly affected by the laser interaction, and that this had an impact on the material's mechanical characteristics [44].

### 23.4.4   BIODEGRADABLE MG-ALLOYS

A secondary material that can be taken out of the human body after surgery is always needed in addition to the commonly accessible metallic biomaterial. Mg-based alloys are chosen above all other materials for biodegradable implants. Although Mg-Ca based alloys are claimed to have excellent biocompatibility, they have poor corrosion resistance and degrade quickly. Low modulus and tensile strength of Mg-based alloys present another issue, hence any non-contact peening process, such as laser shock peening, is favored over a contact type peening method. Figure 23.12a, where Mg-0.8 Ca and AZ91D alloys are peened with 5.1 GW/cm$^2$ and 5.4 GW/cm$^2$ laser power, respectively, shows the residual stress distribution with respect to depth [45,46]. For nearly identical laser power, the AZ91D alloy tends to have higher compressive residual stress than the Mg-0.8Ca alloy. Additionally, the fatigue life of the AZ91D alloy (shown in Figure 23.12b) has improved compared to the untreated sample after laser shock treatment [47]. The tafel plot of both the WE43 alloy and the AZ80 alloy is shown in Figure 23.12c. The corrosion resistance of both alloys was investigated both before and after laser treatment [48,49]. In the case of WE43 alloy the corrosion resistance of untreated alloy was found to be slightly better compared to the laser peened alloy, whereas in case of AZ80 alloy the scenario is opposite. The corrosion resistance has improved after laser shock peening. The Mg and Al ion leakage (shown in Figure 23.12d and 23.12e) was calculated in an α-MEM solution. When compared to laser shock peened samples, the untreated sample was found to reduce the concentration (in ppm) of both Mg-ion and Al-ion release. The only concern is that as compared to magnesium, aluminum ion release is considered more toxic or carcinogenic.

In addition, cell culture research utilizing a LIVE/DEAD staining kit was carried out on both untreated and LSPed AZ31B samples. More live cells were found in

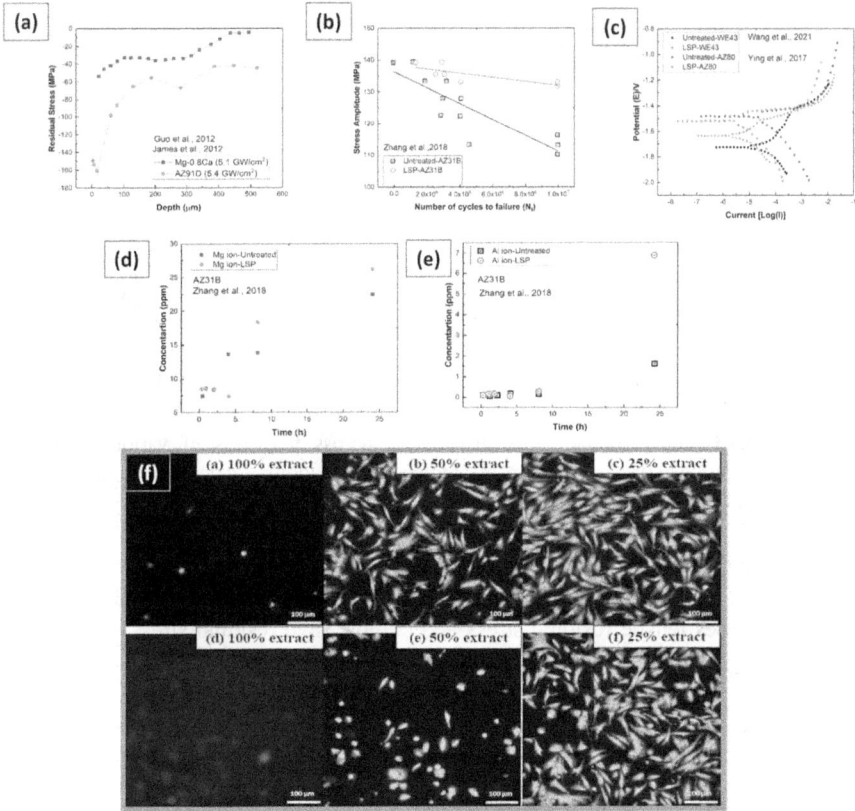

**FIGURE 23.12** (a) Residual stress distribution across thickness of WE43 and AZ91D Mg-alloy; (b) Comparison of fatigue life after and before LSP of AZ31B Mg-alloy; (c) Comparison of tafel plot of WE43 and AZ80 Mg-alloy before LSP and after LSP;(d) Concentration of Mg-ion release in PPM of AZ31B Mg-alloy; (e) Concentration of Al-ion release in PPM of AZ31B Mg-alloy; (f)Live/Dead staining of cells before LSP – 'a' to 'c' and after LSP – 'd' to 'f'.

25% extract media for untreated samples compared to the laser peened sample. This may be due to the Al-ion leaking from the magnesium alloy.

## 23.4.5 Additively Manufactured Alloys

Historically, casting, or forging techniques have been used to create biomedical implants made of titanium. Small implants can be made using the traditional method of manufacturing, but when design complexity increases, new production techniques must be used. One of the innovative processes that will be useful in making jawbone implants and hip replacements is additive manufacturing. There are a few instances of biomaterials made from titanium and nickel-titanium that were laser peened and additively fabricated to increase their fatigue life. Figure 23.13a shows a comparison

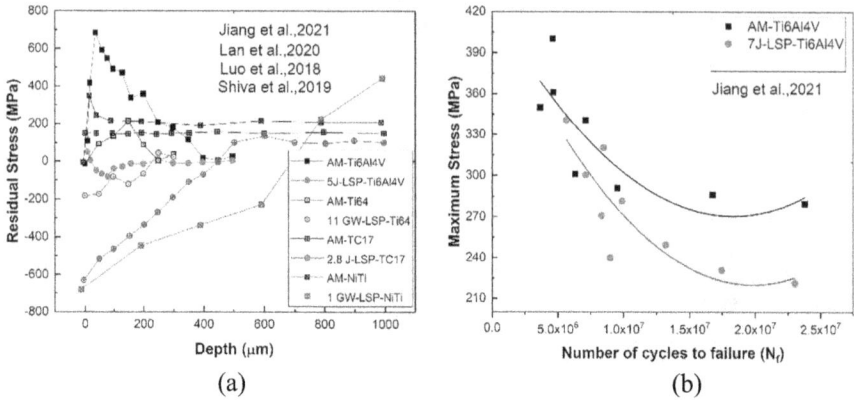

(a)                                                                  (b)

**FIGURE 23.13** (a) Residual stress distribution across thickness of various additively manufactured alloys; (b) comparison of fatigue life of additively manufactured Ti6Al4V alloy after and before LSP.

of the residual stress distribution profile of alloys produced additively before and after laser peening at different laser powers. The residual stress of an untreated NiTi alloy, which possesses tensile residual stress and compressive residual stress created after laser peening, was one intriguing discovery. Figure 23.13b illustrates the fatigue life of Ti6Al4V alloy produced additively before and after laser peening. After laser peening, the fatigue life appears to be reducing [50–53].

### 23.4.6 EFFECT ON SURFACE ROUGHNESS AND CONTACT ANGLE

When it comes to surface engineering procedures, surface roughness and contact angle are two aspects that frequently complement one another. Because the contact type peening technique leaves the surface rough, electropolishing of medical implants is occasionally done after the surface treatments are finished. A somewhat high surface roughness could result in some significant cracks, which would shorten the implants' fatigue life. Therefore, it is crucial to evaluate the surface characteristics of the treated surface, including its contact angle and surface roughness. Surface contact angle and surface energy are connected, and both have an impact on how quickly cells proliferate on implant surfaces. In the following discussion, Figures 23.14a and 23.14d show, respectively, the surface roughness and contact angle of biomaterials produced conventionally and additively before and after laser treatment [35,37,41,43,48,50]. As a result of the debate, it was discovered that all of the alloys' surfaces are rougher than they are before laser treatment. A variation was seen in the instance of contact angle, where additively generated Ti6Al7Nb and conventionally prepared alloys like AISI 304 and AZ31B have contact angles that increase after laser treatment. However, the unpeend sample exhibits a higher contact angle in the case of Ti-22Nb and NiTi shape memory alloy than the LSPed sample.

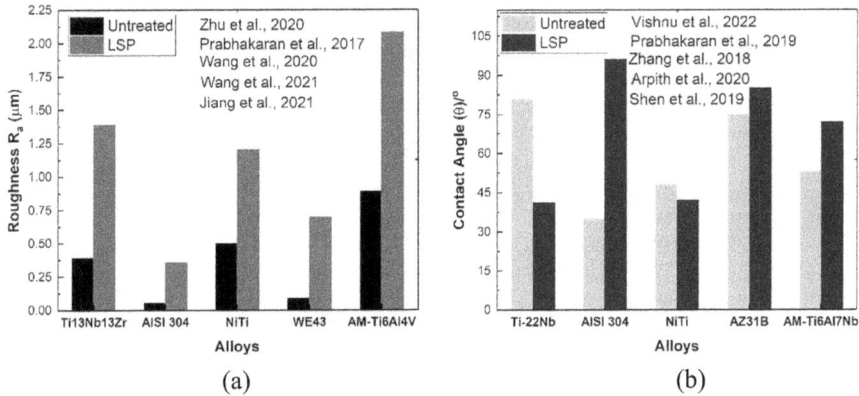

**FIGURE 23.14** (a) Surface roughness of various metallic alloys before and after LSP treatment; (b) contact angle of various metallic alloys before and after LSP treatment.

## 23.5 CONCLUSION

This chapter goes into great detail about the potential for laser shock peening various biomaterials used as implants. In-depth coverage was given to how laser peening affected the residual stress distribution profiles of stainless steel, shape memory alloys, titanium alloys, and biodegradable magnesium alloys. The improvement of fatigue life for the majority of the alloys used as implants was also given special consideration before and after laser treatment. Before and after laser treatment, the impact on a few key parameters, including the alloy's corrosion resistance and the concentration of elemental degradation (leakage of metal ions into the treating media), was briefly examined. The unique draw of this chapter is how laser peening affects cell viability and proliferation on laser-peened Ni-Ti shape memory alloys and biodegradable Mg-based alloys. In a separate section, the impact of laser shock peening on residual stress distribution profile and fatigue life was studied with special focus on additively produced alloys used in biomedical devices. To acquire a bird's eye view of the surface effect caused by non-contact type surface engineering techniques, the effect of laser treatment on surface roughness and contact angle of all implant alloys is examined at the end. approaches for surface engineering.

## ACKNOWLEDGMENT

The authors would like to thank the editor and associate editor for giving an opportunity to include this chapter in the present edition of the book titled, "Surface Engineering for Biomaterials". We are also thankful to the publishers who gave their copyright consent for a few images related to cell culture.

## REFERENCES

1. T. Burakowski and T. S. Wierzchon, *Surface engineering of metals: principles, equipment, technologies*, CRC Press, 1998.

2. K. Holmberg and A. Matthews, *Coatings tribology: properties, mechanisms, techniques and applications in surface engineering*, Elsevier, 2009.

3. J. F. Throop, *Residual stress effects in fatigue*, vol. 776, ASTM International, 1982.

4. A. Matthews and A. Leyland, "Hybrid techniques in surface engineering.," *Surface and Coatings Technology*, vol. 71, no. 2, pp. 88–92, 1995.

5. M. E. Fitzpatrick and A. Lodini, *Analysis of residual stress by diffraction using neutron and synchrotron radiation*, CRC Press, 2003.

6. B. Tomkins, "Fatigue crack propagation—an analysis," *Philosophical Magazine*, vol. 18, no. 155, pp. 1041–1066, 1968.

7. V. Schulze, F. Bleicher, P. Groche, Y. B. Guo, and Y. S. Pyun, "Surface modification by machine hammer peening and burnishing," *CIRP Ann.*, vol. 65, no. 2, pp. 809–832, 2016, doi: https://doi.org/10.1016/j.cirp.2016.05.005.

8. J. L. Dossett and H. E. Boyer, *Practical heat treating*, ASM International, 2006.

9. L. Wagner, *Shot peening*, vol. Vol. 8, John Wiley & Sons., 2003.

10. J. Schijve, *Fatigue of structures and materials*, Springer Science & Business Media, 2001.

11. M. P. Savruk and A. Kazberuk, *Stress concentration at notches*, Cham (Switzerland): Springer, 2017, p. p. 516.

12. F. C. Campbell, *Elements of metallurgy and engineering alloys*, ASM International, 2008.

13. R. W. Hertzberg, R. P. Vinci and J. L. Hertzberg, *Deformation and fracture mechanics of engineering materials*, John Wiley & Sons., 2020.

14. F. Czerwinski, "Thermochemical treatment of metals," *Heat Treatment–Conventional and Novel Applications*, vol. 5, pp. 73–112, 2012.

15. I. C. Noyan and J. B. Cohen, *Residual stress: measurement by diffraction and interpretation*, Springer, 2013.

16. K. J. Marsh, *Shot peening: techniques and applications*, vol. 320, United Kingdom: Engineering Materials Advisory Service Ltd., 1993.

17. M. Kobayashi, T. Matsui and Y. Murakami, "Mechanism of creation of compressive residual stress by shot peening.," *International Journal of Fatigue*, vol. 20, no. 5, pp. 351–357, 1998.

18. G. E. Totten, Handbook of residual stress and deformation of steel, ASM International, 2002.

19. K. Chaudhary, Shot peening: process, equipment and applications, Educreation Publishing, 2017.

20. S. Brill and D. M. Schibisch, "Induction hardening versus case hardening: a comparison," *Induc. Technol. Rep*, vol. 1, pp. 76–83, 2015.

21. P. Peyre and R. Fabbro, *Opt Quant Electron. J*, vol. 27, p. 1213, 1995.

22. H. K. Amarchinta, R. V. Grandhi, A. H. Clauer, K. Langer and D. S. Stargel, "Simulation of residual stress induced by a laser peening process through inverse optimization of material models," *Journal of Materials Processing Technology*, vol. 210, no. 14, 2010.

23. P. Peyre, I. Chaieb and C. Braham, "FEM calculation of residual stresses induced by laser shock processing in stainless steels," *Modelling and Simulation in Materials Science and Engineering*, vol. 15, no. 3, p. 205, 2007.

24. X. Wang, W. Xia, X. Wu and C. Huang, "Scaling law in laser-induced shock effects of NiTi shape memory alloy," *Metals*, vol. 8, no. 3, p. 174, 2018.

25. S. W. T, *Laser fundamentals*, Cambridge University Press, 2004.

26. M. P. W and E. J. H, *Lasers*, Chichester: Wiley, 1988.

27. C. S. Montross, T. Wei, L. Ye, G. Clark and Y. W. Mai, "Laser shock processing and its effects on microstructure and properties of metal alloys: a review," *International Journal of Fatigue*, vol. 24, no. 10, pp. 1021–1036, 2002.

28. P. Ballard, J. Fournier, R. Fabbro and J. Frelat, "Residual stresses induced by laser-shocks," *Le Journal de Physique IV*, vol. 1, no. C3, pp. C3–487, 1991.

29. P. Peyre and R. Fabbro, "Laser shock processing: a review of the physics and applications," *Optical and Quantum Electronics*, vol. 27, pp. 1213–1229, 1995.

30. S. H. Teoh, *Engineering materials for biomedical applications*, vol. 1, World Scientific, 2004.

31. M. Niinomi, "Recent metallic materials for biomedical applications," *Metallurgical and Materials Transactions A*, vol. 33, pp. 477–486, 2002.

32. Y. F. Zheng, X. N. Gu and F. Witte, "Biodegradable metals," *Materials Science and Engineering: R: Reports*, vol. 77, pp. 1–34, 2014.

33. S. K. D. Bose, H. Sahasrabudhe and A. Bandyopadhyay, "Additive manufacturing of biomaterials," *Progress in Materials Science*, vol. 93, pp. 45–111, 2018.

34. K. Yang, Q. Huang, B. Zhong, Q. Wang, Q. Chen, Y. Chen, N. Su and H. Liu, "Enhanced extra-long life fatigue resistance of a bimodal titanium alloy by laser shock peening," *International Journal of Fatigue*, vol. 141, p. 105868, 2020.

35. X. Shen, P. Shukla, A. Subramaniyan, A. Zammit, P. Swanson, J. Lawrence and M. Fitzpatrick, "Optics & Laser Technology," *Residual Stresses Induced by Laser Shock Peening in Orthopaedic Ti-6Al-7Nb Alloy*, vol. 131, p. 106446, 2020.

36. H. Qiao, J. Zhao and Y. Gao, "Experimental investigation of laser peening on TiAl alloy microstructure and properties," *Chinese Journal of Aeronautics*, vol. 28, no. 2, pp. 609–616, 2015.

37. R. Zhu, Y. Zhang, C. Lin and Y. Chen, "Residual stress distribution and surface geometry of medical Ti13Nb13Zr alloy treated by laser shock peening with flat-top laser beam," *Surface Topography: Metrology and Properties*, vol. 8, no. 4, p. 045026, 2020.

38. X. Jin, L. Lan, S. Gao, B. He and Y. Rong, "Effects of laser shock peening on microstructure and fatigue behavior of Ti–6Al–4 V alloy fabricated via electron beam melting," *Materials Science and Engineering: A*, vol. 780, p. 139199, 2020.

39. C. Correa, A. Gil-Santos, J. Porro, M. Díaz and J. Ocaña, "Eigenstrain simulation of residual stresses induced by laser shock processing in a Ti6Al4V hip replacement," *Materials & Design*, vol. 79, pp. 106–114, 2015.

40. C. Correa, L. De Lara, M. Díaz, A. Gil-Santos, J. Porro and J. Ocaña, "Effect of advancing direction on fatigue life of 316 L stainless steel specimens treated by double-sided laser shock peening,". *International Journal of Fatigue*, vol. 79, pp. 1–9, 2015.

41. S. Prabhakaran, A. Kulkarni, G. Vasanth, S. Kalainathan, P. Shukla and V. Vasudevan, "Laser shock peening without coating induced residual stress distribution, wettability characteristics and enhanced pitting corrosion resistance of austenitic stainless steel.," *Applied Surface Science*, pp. 17–30, 2018.

42. M. Ebrahimi, S. Amini and S. Mahdavi, "The investigation of laser shock peening effects on corrosion and hardness properties of ANSI 316 L stainless steel," *The International Journal of Advanced Manufacturing Technology*, vol. 88, pp. 1557–1565, 2017.

43. R. Zhang, S. Mankoci, N. Walters, H. Gao, H. Zhang, X. Hou, H. Qin, Z. Ren, X. Zhou, G. Doll and A. Martini, "Effects of laser shock peening on the corrosion behavior and biocompatibility of a nickel–titanium alloy," *Journal of Biomedical Materials Research Part B: Applied Biomaterials*, vol. 107, no. 6, pp. 1854–1863, 2019.

44. R. Tamiridi, R. Goud, P. Subramaniyan, K. Sivaperuman, A. Subramaniyan, I. Charit and S. Gollapudi, "Contrasting Effects of Laser Shock Peening on Austenite and Martensite Phase Distribution and Hardness of Nitinol," *Crystals*, vol. 12, no. 9, p. 1319, 2022.

45. Y. Guo, M. Sealy and C. Guo, "Significant improvement of corrosion resistance of biodegradable metallic implants processed by laser shock peening," *CIRP Annals*, vol. 61, no. 1, pp. 583–586, 2012.

46. J. A. Russo, *The Effects of Laser Shock Peening on the Residual Stress*, Cincinnati, 2012.

47. R. Zhang, X. Zhou, H. Gao, S. Mankoci, Y. Liu, X. Sang, H. Qin, X. Hou, Z. Ren, G. Doll and A. Martini, "The effects of laser shock peening on the mechanical properties and biomedical behavior of AZ31B magnesium alloy," *Surface and Coatings Technology*, vol. 339, pp. 48–56, 2018.

48. W. Wang, C. Hung, L. Howe, J. Chen, K. Wang, V. Ho, S. Lenahan, M. Murayama, N. Vinh and W. Cai, "Enabling High-Performance Surfaces of Biodegradable Magnesium Alloys via Femtosecond Laser Shock Peening with Ultralow Pulse Energy," *ACS Applied Bio Materials*, vol. 4, no. 11, pp. 7903–7912, 2021.

49. Y. Xiong, Q. Hu, R. Song and X. Hu, "LSP/MAO composite bio-coating on AZ80 magnesium alloy for biomedical application," *Materials Science and Engineering: C*, vol. 75, pp. 1299–1304, 2017.

50. Q. L. S. Z. C. Z. B. Z. Y. Jiang, "Effects of laser shock peening on the ultra-high cycle fatigue performance of additively manufactured Ti6Al4V alloy," *Optics & Laser Technology*, vol. 144, p. 107391, 2021.

51. L. Lan, R. Xin, X. Jin, S. Gao, B. He, Y. Rong and N. Min, "Materials," *Effects of laser shock peening on microstructure and properties of Ti–6Al–4V titanium alloy fabricated via selective laser melting*, vol. 13, no. 15, p. 3261, 2020.

52. S. Luo, W. He, K. Chen, X. Nie, L. Zhou and Y. Li, "Regain the fatigue strength of laser additive manufactured Ti alloy via laser shock peening," *Journal of Alloys and Compounds*, vol. 750, pp. 626–635, 2018.

53. S. Shiva, I. Palani, C. Paul and K. Bindra, "Laser shock peening of Ni-Ti bulk structures developed by laser additive manufacturing.," *Journal of Materials Engineering and Performance*, vol. 30, no. 8, pp. 5603–5613, 2021.

# 24 Coatings of High Entropy Alloys in Biomedical Applications

*Mojtaba Najafizadeh*
Faculty of Chemical and Materials Engineering, Shahrood University of Technology, Shahrood, Iran

Department of Innovation Engineering, University of Salento, Lecce, Italy

*Payam Sarir*
College of Civil Engineering, Tongji University, Shanghai, China

*Ehsane Marzban Shirkharkolaei*
Departement of Mechanical Engineering, Isfahan University of Technology, Isfahan, Iran

*Mansoor Bozorg*
Faculty of Chemical and Materials Engineering, Shahrood University of Technology, Shahrood, Iran

*Morteza Hosseinzadeh*
Department of Engineering, Ayatollah Amoli branch, Islamic Azad University, Amol, Iran

*Negar Sarrafan*
School of Dentistry, Urmia University of Medical Sciences, Urmia, Iran

*Pasquale Cavaliere*
Department of Innovation Engineering, University of Salento, Lecce, Italy

## 24.1 INTRODUCTION

The specific classifications of the materials are biomaterials. These materials have been developed for treating and diagnosing of different diseases. The biomaterials are significantly used in medical applications such as controlled systems of drug

DOI: 10.1201/9781003429920-27

527

delivery, surgical devices, support and replacement of body tissue, and scaffolds for tissue engineering. The general classification of biomaterials divided into five categories such as natural derived materials, polymers, ceramics, composites, and metals [1]. The preliminary classification of biomaterials is metallic materials for using those wildly in over many years, due to this reason the market of metallic materials is increasing as compared to the other categories [2]. The antiquity of the biomaterials started as early as 500 BC for using gold as a dental material, and from the 15th century, the dentistries used noble metals in dentures, crowns, bridges, and fillings [3]. The design and manufacturing of dental and orthopedic implants using engineered metals started in the 1940s of modern times [4]. The most common metals and alloys are used in medical applications such as titanium and its alloys, stainless steel, noble metal alloys, and cobalt-based alloys for having individual characterization [5]. The important characterizations for leading metallic materials widely used for synthesizing biomaterials are high wear resistivity, toughness, fatigue resistivity, ductility, elastic modulus, strength, magnetic property, electrical conductivity, minimum total cost, and ability to tailor the shape [4,6]. The essential disadvantages of using metallic materials as biomaterials are [7].

Therefore, the necessary solutions for having suitable bulk mechanical specifications for the metallic biomaterials with defeating of the aforesaid disadvantages are solved by coating metals and modification of surfaces [8].

The biocompatibility, wear, and corrosion properties of the metallic biomaterials are improved by the formation of the modified layer on the surface of the metal [9]. Moreover, the metallic biomaterial with the metal coating surrounded by the tissue simplifies integration, increase antibacterial property and protect the mechanical properties of the bulk metal [10]. Different techniques are used for surface modification such as pattering, roughening, multilayer films, grafting, and coating. Coatings such as plasma spray, chemical vapor deposition (CVD), solvent evaporation, and physical vapor deposition (PVD) can be chosen as per the dominant parameter requirement in various application [11]. The dignity of spallation of film under the mechanical loads is reduced by coating the surface of the metal with the metallic materials (metals and alloys) caused the distribution of the internal stress along the thickness of the film with the chemical homogeneity between the modified layer and substrate [12]. The alloying process can be used to better understand about coating surfaces with metallic alloys.

The alloy production process consists of adding other elements to the base metal and using them to increase the properties of the material in the past until now [13]. Nevertheless, traditional alloying guidelines in important amounts result in the formation of unfavorable intermetallic phases that make the alloy brittle [14]. The philosophy of the classic design of the alloy was broken and shifted to the other paradigm in 2004 and the researchers in the same year introduced a new classification of alloys [15]. Moreover, increasing elements in an alloy claimed that the total free energy of the entropic contribution dominates the enthalpic contribution. Therefore, an irregular solid solution of crystal shapes (body-centered cubic (BCC), hexagonal closed-pack (HCP), and face centered cubic (FCC)) is stabilized and the constitution of the mixed intermetallic compounds is repressed [16].

The Gibbs free energy of the system should be at the minimum for the formation of the stable solid solution. The value mixing of the entropy in HEA system is higher than in the simple alloying system. The effect of the high entropy is stabilized in the system by approaching Gibbs free energy to zero [17]. This class of materials is named by Yeh et al.; high entropy alloys (HEAs) due to the mixing of high entropy can reduce the number of phases in an alloy system [18]. The high entropy alloys (HEAs) made by Cantor et al. with another explanation (the alloy should hold at least five constitutive elements in an equal or near equal ratio and the difference between the solvent and solute are not noticeable) and named equiatomic multicomponent alloys [19]. Moreover, the HEAs contain the main elements and are below the 5% ratio of the minor elements [20]. The name of the HEAs is a term because the structure and characterization of the HEAs are affected by entropy and other factors [21]. Figure 24.1 shows the different terms of the element number in multiprincipal alloy and conventional alloy.

The properties such as oxidation, wear, and corrosion resistivity by multiple elements arranged randomly in the solid solution lead to the chemical reactivity, especially local instability[22]. Moreover, the HEAs with suitable alloying elements and individual multiprincipal element composition can achieve phenomenal diversity properties such as high toughness, strength, hardness, thermal stability, and ultrahigh fracture toughness at cryogenic temperatures [23]. Replacing one or several elements in composition can strongly improve the quality of obtaining from HEAs as compared to the original alloys. In addition, the metallographic structure can be changed by adding different ratio elements to the alloy, and the characteristics of the alloy can be impacted by the structure [24]. The new classifications of HEAs used as biomaterial are produced by combining innovative elements and can achieve special properties in biocompatibility and corrosion resistance [25]. The researchers are focused on the HEAs as metallic biomaterials because of having preferable surface properties and aforesaid properties [26].

Usually, the cost of the HEAs is higher than the conventional alloys because the HEAs contain expensive elements and this reason reduces the usability of the HEAs in different industrial applications. Therefore, the coatings and film of the HEAs have been investigated and developed because of the border application fields and low cost [27]. The HEA coatings and films as compared to the traditional coatings

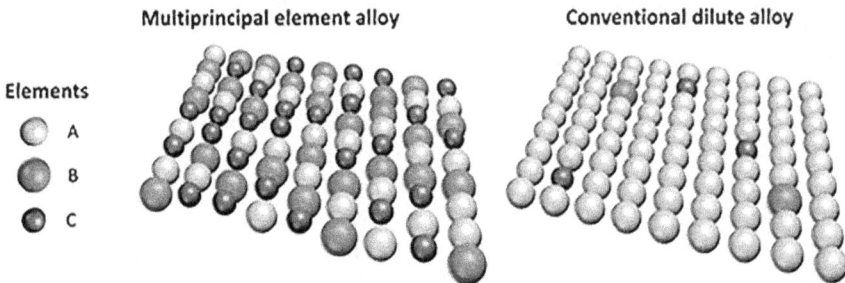

**FIGURE 24.1** The philosophy of the multiprincipal alloy and conventional alloy.

have different advantages such as high hardness and strength, excellent corrosion and wear resistance, high toughness for wide temperature applications, thermal stability, irradiation resistance, etc. Moreover, the HEA coatings are developed based on different technologies by the researchers [28]. The different thicknesses and various surface modification techniques recently used by researchers for producing HEAs coating and films with the excellent advantage of resistance for the bio application [22]. In addition, the HEA coatings and films mostly have a homogeneous structure with high wear resistivity and thermal stability. The HEA coating protects and minimizes the substrate from the electrochemical reaction by the formation of stable passive films versus the diffusion of corrodes and high ionic resistivity [29].

## 24.2 THE COATINGS OF THE HIGH ENTROPY ALLOYS (HEAS)

Metallic materials in biological applications have appropriate mechanical characterization but the surface of the metals have problem with biocompatibility and for solving these problems different techniques for surface modification are used. On the other hand, the surface of the metals can be modified by surface coating for improving significantly the required biocompatibility characterisation (such as integration with adjacent tissue and antibacterial properties), wear resistivity, corrosion resistivity, and the mechanical properties of the biomaterial have been protected by the surface coating [30]. The principles and methods for using HEA biomaterials as a coating surface have been given in the following sections. The physical and chemical surface modification of the metallic biomaterials is separated from the coating process. The HEA coating materials have the potential for use in different environmental conditions because of the unique properties and high performance of the HEA coating. Thus, the development and research on the coating of the HEAs started in 2004 [31]. The following sections discuss the fabrication and classification of different HEA-based coatings.

### 24.2.1 TECHNIQUES FOR FABRICATING HEA-BASED COATING

The properties of the surface component have a correlation with the performance of the coating in applications of practical. The quality of the surface has a significant effect on the performance and lifetime of the material and cannot be ignored in the design [32]. The practical and economical method for having specific surface characteristics and more enhanced with the coating (thin film or thicker layer) of the incompatible material is possible because of the wide availability of surface modification technologies [19]. A nearly unlimited number of films or coatings and numerous processes on the surface of the materials are used for applying films or coatings. Generally, selecting the right coating materials and the combination of the coating techniques for the special application needs specific knowledge [33]. The coating is described as a surface on the substrate covered by a thick layer and film is described as a surface layer on the substrate with the fraction thickness of the nanometers to the several micro-meters. Nevertheless, the prospects of the films and coatings overlap in application and science [34].

The HEAs have significant functional properties such as thermal stability, anti-oxidation, wear resistivity, and corrosion resistance, these days the concentration of research studies on HEA coating of surface modification such as thermal insulation, corrosive environments, intense stress applications, and tribological applications are very valuable [35]. The HEA coatings have more excellent performance than the bulk HEAs [36]. The HEA coatings are used for increasing the physical and mechanical properties of different types of materials such as corrosion resistivity, hardness, temperature resistivity and wear resistivity [37]. The fabrication methods of the HEA coatings have been developed in the last few years. Different techniques are described in this section. The HEA-based coatings are fabricated by the following main classification [38]:

- Vapor deposition: vacuum arc deposition and magnetron sputtering
- Thermal spraying: cold spraying, plasma spraying, and high-velocity oxygen fuel spraying
- Laser deposition: plasma cladding, laser surface cladding, laser surface alloying

### 24.2.1.1 Vapor Deposition (VD)

The most common depositing techniques for producing hard coatings are chemical vapor deposition (CVD) and physical vapor deposition (PVD). The surface modification of these techniques is modified by chemical and physical specifications. The vapor is a specific feature of CVD and PVD. The vapor in the CVD technique is fabricated by the chemical reaction on the surface of the substrate and formed thin coatings and in the PVD technique is fabricated from molecules and atoms that simply condense on the substrate [39].

The HEA-based coating materials with demanded surface properties are fabricated by CVD and PVD techniques such as vacuum arc deposition and magnetron sputtering. It was found that the HEA coatings fabricated by the VD methods frequently contained metallic HEA coatings, nitrides, and carbides. However, the Hf, Zr, Ti, V, and Nb in HEA-based coatings fabricated nitride on the coating because of the strong reaction between those elements and the nitrogen. In addition, the composition of the HEA-based coating can significantly develop the nitride phases for fabricating the HEA-based coating with the nitride [40]. Table 24.1 shows the description of the different HEA coatings fabricated by the VD processes on the different biomedical substrates [41]. The thickness of the coating fabricated by the VD methods has limitations according to Table 24.1.

### 24.2.1.1.1 Vacuum Arc Deposition

The surface of a substrate is coated by the target materials evaporated by the arc heating energy in a vacuum environment in the vacuum arc deposition method. The performance of the surface and microstructure of the fabricated nitride HEA coatings are significantly related to the condition of the coating deposition [42]. Thus, the vacuum arc deposition method according to the previous research has been used rarely for fabricating the HEAs coating on the biomedical substrate. Figure 24.2 shows the schematic of the vacuum arc deposition.

**TABLE 24.1**

**The Description of the HEA-based Coating Fabricated by the VD Methods**

| Process | Base-metal | Coating | Crystal Structure | Thickness (µm) |
|---|---|---|---|---|
| Vacuum arc deposition | Ti-6Al-4V | Ti-Ag-W-Nb-Ta-Zr | Non-crystalline | 1.1 |
| | Stainless steel (SS) | (Ti-Cr-Al-Y-Nb-Zr)N | BCC+FCC | 7 |
| | C 45 | (Ti-V-Hf-Nb-Zr)N | FCC | 4.78 |
| Magnetron sputtering | Ti-6Al-4V | Ti-Zr-Hf-Ta-Nb | Non-crystalline | 0.8 |
| | SS201 | Fe-Co-Mn-Cr-Al-Cu | FCC | 1.8 |
| | SS 304 | Cu-V-Ta-Mo-W | BCC+FCC | 0.9 |
| | SS 304 | Fe-Ni-Cr-Co-$Mo_{0.1}$ | FCC | 0.85 |
| | SS 304 | Cr-Zr-Nb-Mo-Ti-Mo | Non-crystalline | 1.2 |

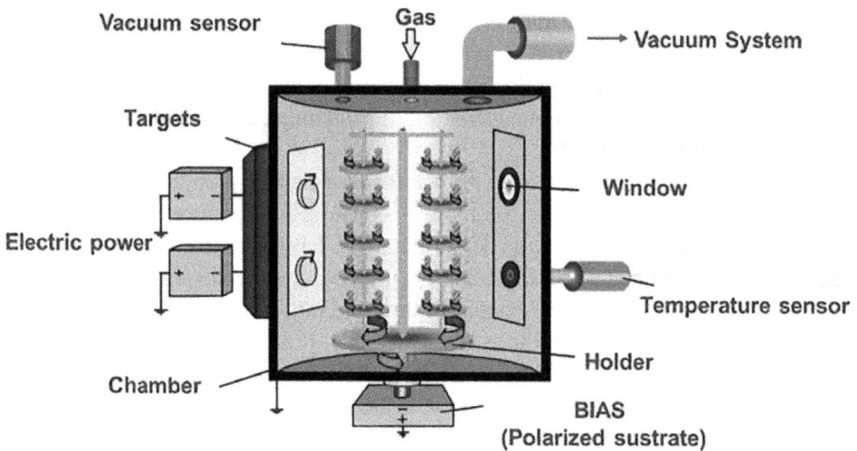

**FIGURE 24.2**  Schematic of the vacuum arc deposition.

### 24.2.1.1.2 Magnetron Sputtering

The significant technique in the last decade for coating the different substrates in different applications of the industry is magnetron sputtering in this method. The applied coating of the magnetron sputtering method has high quality and characterization such as ornamental coating, corrosion-resistant coating, low friction coatings, hard coating, wear-resistant coating, and electrical and optical properties. The performance of the magnetron sputtering coating is higher than the other PVD deposition techniques and compared to the other techniques for applying coating on the surface of the substrate with the thicker coatings have the same performance [43]. The advantages of using the magnetron sputtering technique for applying coating are high adjustability and low deposition temperature, control of target powers by changing the stoichiometry, creation of a concentration gradient for one or more elements, and control of the coating thickness [44]. Figure 24.3 shows the schematic of the magnetron sputtering.

The most distinguished technique for fabricating HEA coatings is deposited by sputtering. The process parameters and chemical composition of a given target can be changed during sputtering to control stoichiometry effectively. In addition, the effective method for fabricating the HEA coatings with this method has carbide, nitride, or oxide because depositing used the active gases the same as $C_2H_2$, $N_2$, or $O_2$ and this method developed the effective method for producing the new HEA coating systems [34], whereas researchers have a lot of attention on the magnetron sputtering process that can deposit HEA-based ceramic coatings including carbide and nitride for protecting the surface of the substrate.

For dominating the natural negative properties of the ceramic coatings, magnetron sputtering is used for developing the multilayered composite or metallic HEA-based coating alongside the ceramic HEA-based coating [42]. The wear properties and microstructure of the Ti-6Al-4V biomedical alloy as a substrate were investigated by applying the Ti-Zr-Hf-Ta-Nb HEA coating [45]. The results show the multi-component equimolar Ti-Zr-Hf-Ta-Nb HEA coating deposited by magnetron sputtering on Ti-6Al-4V substrate improved the mechanical properties because of an amorphous structure with the small grain and the surface topography is homogeneous. The wear properties of the alloy are improved by increasing the hardness and elastic modulus of the surface coating. The Ti-Zr-Hf-Ta-Nb HEA coating is formed homogeneous and dense on the surface of the Ti-6Al-4V substrate

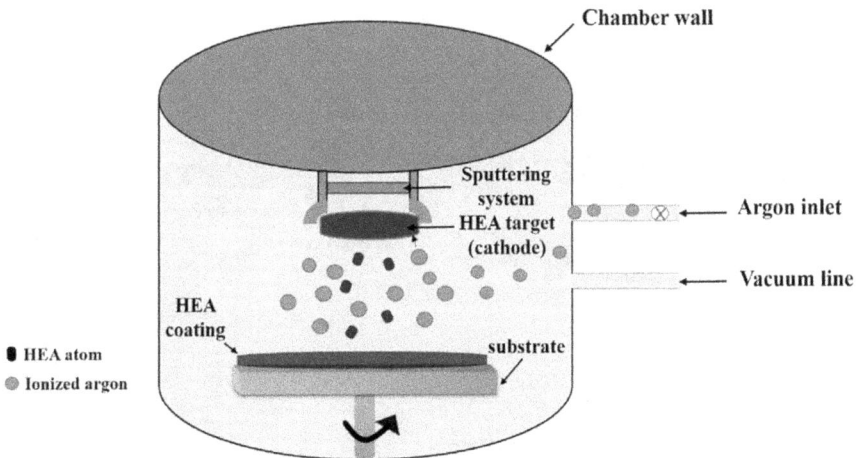

**FIGURE 24.3**   Schematic of the magnetron sputtering.

improving the mechanical properties and the surface of the substrate protected from crack and wear. The important characteristic of biomedical implants used in long-term orthopedics is having the impact dynamic contact load the same as replacements of the knee or hip [45].

Generally, the ultra-thin coatings of HEA with thicknesses of less than a micron, especially the ceramic coatings of the HEA are fabricated by magnetron sputtering.

### 24.2.1.2  Thermal Spraying (TS)

The TS method is the coating process for applying on the surface of the materials and the surface of the engineered components is recoated with this method. The schematic of the TS method is shown in Figure 24.4.

The energy of thermal in the TS method is produced by the electrical or chemical functions [46,47]. The coating feedstock materials (composite materials, ceramics, metals, and alloys) in the form of wire, rod, or powder are applied on the surface of the substrate by melting or semi melting the coating materials with the energy sources [19]. The acceleration of the droplets (semi molten or molten) material is dependent on the kinetic energy of the diffusion particle speed [48]. When the particles are impacted on the surface of a substrate deform significantly because of the high speed and temperature of the particles. The splats are lamellar or thin layers produced by the deforming particles after spraying on the substrate with excellent adhesion. The shape of that layer is like a pancake and a single impact particle [49]. Eventually, the mechanisms of the coating by the TS on the substrate are as follows [50]:

- The spraying particles on the substrate manufactured the mechanical bonding between the particles and the substrate, and by interlocking the particles manufactured the roughened on the surface of the substrate.
- The diffusion of the coating material into the substrate is locally
- The van Der Waals forces are described as the bonding mechanism

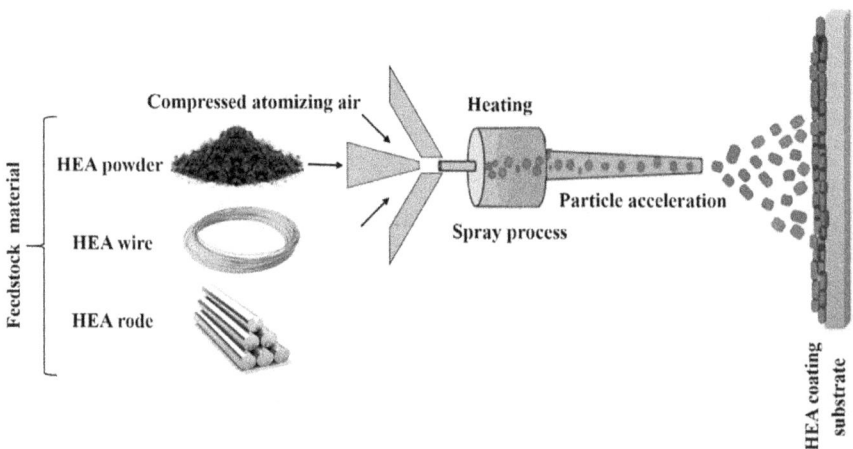

**FIGURE 24.4**   Schematic of the TS process.

The TS methods have the important advantage of using a wide range of materials for fabricating the coating on the surface of the substrate. All the materials approximately can be melted by using the TS method provided the feedstock is not decomposed at the melting temperature. The other significant advantage of the TS methods is about having the ability to fabricate a coating on a substrate without intensive heating. Nevertheless, the deposition of the coating with the TS methods has a significant disadvantage without ignoring that is about the sight line. On the other hand, all the TS methods have followed the techniques of the sight line (the components in the way of the spray can be directly coated on the substrate) [51]. The usability of the TS methods is significantly increased in new applications such as biomedical engineering, dielectrics, and electronic coatings because of the new equipment developed and the quality of the materials improved for fabricating the coating [51]. Currently, researchers are focused on TS methods for depositing HEAs on the substrate to protect the base metal from corrosion, wear, and oxidation at high temperatures and aggressive environments. Based on the energy resources, the TS methods are classified into three main categories as follows:

- Using gas expansion energy for having processes at low temperatures such as cold spray (CS)
- Using electrical energy for fabricating arc or plasma such as plasma spray (PS)
- Using combustion heat sources such as high-velocity oxygen fuel spraying (HVOF)

Recently, researchers have reported that TS methods such as CS, PS, and HVOF are used for fabricating HEA-based coating materials on the substrate to improve functional and mechanical properties. Table 24.2 shows the description of the different HEA coatings fabricated by the TS processes on the different biomedical substrates [41].

**TABLE 24.2**
**The Description of the HEA-based Coating Fabricated by the TS Methods**

| Process | Base-metal | Coating | Crystal Structure | Thickness (µm) |
|---------|-----------|---------|-------------------|----------------|
| CS | Low alloy steel | Co-Ni-Fe-Cr-Mn | FCC | 1000 |
| PS | Mild steel | Al-Ni-Cr-Co-Fe | FCC+BCC | – |
| | S235 | $(Co-Fe-Ni-Cr)_{95}Nb_5$ | FCC | 500 |
| | SS 304 | Fe-Mn-Cr-Ni-Co | FCC | 195 |
| | SS316 | Al-Ti-Fe-Cr-Ni-Co/Ni60 | BCC+FCC | – |
| | SS 304 | Fe-Si-Cr-Ni-Co-Al$_x$ | FCC+BCC | 160 |
| HVOF | SS 304 L | Ti-Fe-Mo-Nb-Mn | BCC | – |
| | A572 | Al$_{0.6}$-Ni-Co-Ti-Cr-Fe | BCC | 300 |
| | SS Austenitic | Al-Fe-Co-Ni-Cr-Ti$_{0.5}$ | BCC | 260 |

*24.2.1.2.1  Cold Spray (CS)*

The feedstock material with the shape of wire or powder in traditional methods of TS coatings needs to become melted or semi-melted and the molten droplets need to be cooled down on the surface of the substrate for fabricating a coating with the lamellae structure [52]. The traditional methods in TS used a high temperature for fabricating coatings on the surface of the substrate and this can affect the mechanical, physical, and electrochemical properties of the coatings because of the high levels of oxidation, high residual stresses, phase transformation, and crack formation [53]. The commercial applications of the coating fabricated with the traditional methods of the TS despite those limitations, the materials are sensitive to temperature. The cold spray (CS) coating method in TS can eliminate or minimize all those problems [54].

The traditional TS methods have many similarities with the CS method. However, the fundamental difference between the CS and traditional TS methods makes the CS method and materials highly exclusive [55]. The CS method compared to the traditional method of TS has a specific property about the feedstock not melted in the CS process and the major driving force for feedstock material consolidation and adhesion to the substrate is provided by the kinetic energy between the metallurgical bonding. On the other hand, the adherence of the sprayed particles to the substrate is not increased by thermal energy, and this is increased by high kinetic energy [56]. The gas flow temperature is much lower than the melting point of the material of the feedstock particles during the CS process [42]. The temperature of the expanded gas stream exhausted from the nozzle is low for the CS method making the operation of the CS method safer [53].

The coating produced by the CS process has several unique specifications, such as high bond stress, high efficiency, high rate of deposition, selection of the substrate is flexible, and the ultrathick coatings are formed due to the CS process is a solid state process [57]. It can be mentioned that the coatings fabricated by the CS method have low porosity and wrought microstructure because of the low gas energy and high kinetic energy [53]. The CS process has another advantage of coating the surface of the substrates with a thickness of lower than 1000 μm without any damage to the substrate. The thick coatings and multilayer coatings can be produced by the compressive pressures with the well-bond. The ductility of the coating fabricated by the CS process is reduced because of the plastic deformation process and this is the significant disadvantage of the CS method. Applying the coating by the CS method on the ceramic substrates has a limitation for binding strength, so the ductility of the substrate has an effect on the formation of the bonded coatings [57].

The HEAs tend to form a simple structure, and the solid-state methods are used for fabricating HEA coatings same as the CS method is significantly widely used. Different types of coatings are successfully deposited on various materials such as composites, polymers, ceramics, and metals by using the CS method; this method can be demonstrated to have the capability to apply HEA coatings [58]. The important advantage of the CS method is that it keeps the natural alloy feedstock quality incomparable and the CS method has a strong limitation for fabricating the HEA coating layers in research [59].

## 24.2.1.2.2   Plasma Spray (PS)

One of the TS processes for fabricating coating on the substrate is plasma spraying (PS); this method can deposit the molten or semi-molten different materials such as metallic and non-metallic on the prepared substrate for increasing the usability of the different materials into the thermal barrier, corrosion resistivity, and wear resistivity. The particles of the materials in the PS method before solidifying on the substrate are precipitated by feeding into the plasma jet or plasma for transforming the particles to molten or semi-molten [60]. The source of the heating in the PS method is produced by a direct current plasma arc and the energy of the plasma is high enough for spraying and depositing the vast range of the coating materials such as metallic and non-metallic materials with high melting points [42]. The best choice without failure instead of the other coating methods for fabricating coatings with high density and minimum porosity is the PS method [50]. The most important advantage of the PS method is the vast range of spraying temperatures can be used for melting high-temperature materials such as refractory metals can be used for depositing the coating on the substrate. The significant disadvantage of the PS method is the complexity of the parameters and without having the interaction parameters causing the peeling and cracks on the coating [60]. The PS techniques can be classified based on the fabrication environment as follows:

- Atmospheric plasma spray (APS)
- Vacuum plasma spray (VPS)
- Low pressure plasma spray (LPPS)

Recently, researchers reported that the PS method is an excellent method for applying HEA coatings because of the interfacial bonding high strength, plasma source heat energy, high temperature of the flame, low dilution, and high efficiency in deposition [42]. The characterization of the HEA coatings applied by the PS method reduced the wear rate and increased wear resistivity [61] and also improved the mechanical and physical properties same as increased hardness, saturated magnetism, and low porosity [62]. The Al-Ti-Fe-Cr-Ni-Co/Ni60 HEA coating fabricated by the PS has an effect on the wear resistivity, microstructure, and microhardness properties of the 316 stainless steel alloy substrate. The addition of the large particles of Ni60 to the Al-Ti-Fe-Cr-Ni-Co HEA coating shows that the strength of the coating increased significantly. The composition of the Ni60 is a Ni-Cr-Fe-based alloy and this has excellent wear, corrosion, and heat resistivity with high hardness. The cohesion strength of the Al-Ti-Fe-Cr-Ni-Co is increased by adding the large particles of the Ni60 because the mechanical bonding is strong. Moreover, the Al-Ti-Fe-Cr-Ni-Co HEA coating with the large particle of the Ni60 has fantastic wear properties. According to the results of the researchers, it is found that adding Ni60 to the HEA coatings improves wear resistivity [63].

## 24.2.1.2.3   High Velocity Oxygen Fuel Spray (HVOF)

The performance of the substrate surface is improved or protected by using HVOF for fabricating coating with the thermal spraying procedure [51]. The surface of the cermet, ceramic, and metallic materials can be coated by the HVOF method and the coating on the surface is dense with having high corrosion and wear resistivity and excellent mechanical properties the same as high bonding strength, low oxide content, and low porosity [42,64]. The HOVF method for melting feedstock used fuel gases such as kerosene, methane, and propylene with oxygen and hydrogen. However, the efficiency of the HVOF method is controlled by using high kinetic energy with the high velocity of the particle in the jet and also can be used at a low processing temperature by controlling the thermal energy [50]. The HVOF method with these advantages can be considered a competitive candidate to replace the PS method [42]. The HVOF method has other advantages such as having a smoother surface after applying a coating, the capability of increasing the thickness, and the impact on the environment is less (decreased oxidation, decarbonization, and loss of the key elements by vaporing) [65]. The important disadvantage of the HVOF is the complexity of the method for applying a coating on the surface of the substrate [66].

The HEA coating materials can be produced by using TS methods. The mechanical and functional properties of the HEA coatings produced by the TS methods are significantly high. The average thickness of the coatings produced by the CS, PS, and HVOF methods started from 160 to 1000 μm. Moreover, the HEA coatings can be applied on several biomedical substrates using the TS methods by the researchers.

### 24.2.1.3  Laser Deposition (LD)

Laser beams have been widely used in the modification of metal surfaces in recent decades for having a frequency at the narrow bandwidth, high coherence, high energy density, and directionality [67]. The common technique for fabricating a coating with the laser beam is LD; in this method, the laser beam focuses on heating a volatile precursor of the local place on the substrate to form a thin deposition layer [68]. The variable thicknesses of the coating from several micro-meters to macro-meters are fabricated on the different types of materials by using an appropriate laser in the LD method [69]. The popular and attractive methods for fabricating HEA coatings are LD and the other related methods to the LD because the coating fabricated with these methods has a quick process, high energy density, eco-friendly process and the wastage of the material is low. According to the ultrahigh heating and cooling rates involved in the process, it can be used with the high power of the laser for melting substrate and achieving the strengthening phase at very high temperatures. The metallurgical bonding between substrate and coating provided by the LD method is strong and the forms of the microstructure are fine and uniform and the substrate can be a little damaged by the thermal [70].

The main categories for the LD processing are classified into the three methods as follows:

- Plasma cladding (PC)
- Laser surface cladding (LSC)
- Laser surface alloying (LSA)

**TABLE 24.3**

**The Description of the HEA-based Coating Fabricated by the LD Methods**

| Process | Base-metal | Coating | Crystal Structure | Thickness (μm) |
|---------|------------|---------|-------------------|----------------|
| PC | CP Ti | $Fe-Cr-Ni-Co-Nb_x$ | BCC+FCC | 1500 |
| LSC | SS | $Al_2-Fe-Ni-Cr-Mo_x$ | BCC | – |
| | S235 | $Mg-Nb-Fe-Mo-Ti_2-Y_x$ | BCC+FCC | 1300 |
| | Ti-6Al-4V | Al-Mo-Nb-V-Ti | BCC | – |
| | CP Ti | Co-Nb-Fe-Ni-Cr | $Cr_2Ti+BCC+Cr_2Nb$ | 500-600 |
| | Ti-6Al-4V | $Co-Fe-Cr-V_{0.5}-Ni_2-Ti_{0.75}$ | $Ti-rich+BCC+(Co, Ni)Ti_2$ | 800 |
| LSA | SS 304 | Fe-Al-Co-Ni-Cr | BCC | 600 |
| | Ti-6Al-4V | Ni-Ti-Cr-Al-V-Co | $HCP+(Co, Ni)Ti_2+BCC$ | – |
| | Ti-6Al-4V | Ni-Al-Cr-Ti-Co-V | $(Co, Ni)Ti_2+BCC$ | – |

Table 24.3 shows the description of the different HEA coatings fabricated by the LD processes on the different biomedical substrates [41].

### 24.2.1.3.1 *Plasma Cladding (PC)*

The beams with high energy are mostly used as cladding heating sources. Those can be used to produce alloy coating of some materials with high melting points whenever the steel substrate is less diluting. The advantages of the plasma beams are higher than the other high-energy beams such as material compatibility, high-speed heating, heating concentration, and during cladding can be melted metallic powders with high melting points. Moreover, the blowing force of the laser during cladding is lower than the plasma beams because the plasma beams can form a uniform multi-component alloy [71]. The PC method can produce a large scale of coating with high gradient wear resistivity and has different advantages such as low thermal distortion, substrate materials dilution, high energy density, low-cost operation, low-cost equipment, and excellent interfacial bonding strength between coating and substrate [72].

The heating and cooling rates of the PC as compared with the other processes of the LD are lower and can be held in a melting pool for a long term; this can help to fabricate homogenized microstructure and minimize the defects significantly [73]. Moreover, the coating produced by the TS methods as compared with the PS has displayed a high degree of porosity and a weak bond to a substrate. In addition, the thin coatings fabricated by the TS methods with the structure for heavy loading same as gears and bearings are appropriate. Furthermore, the fully metallurgical bonding between coating and substrate is fabricated by the PC method. The characterization of the coating for the PC method is reduced porosity, high bonding strength, and larger thickness [74]. Thus, the PC method with these characterizations has excellent benefits in producing and depositing HEA coatings [75]. Table 24.3 shows the characterization of the HEA coating fabricated by the PC method. The schematic of the PC method is in Figure 24.5.

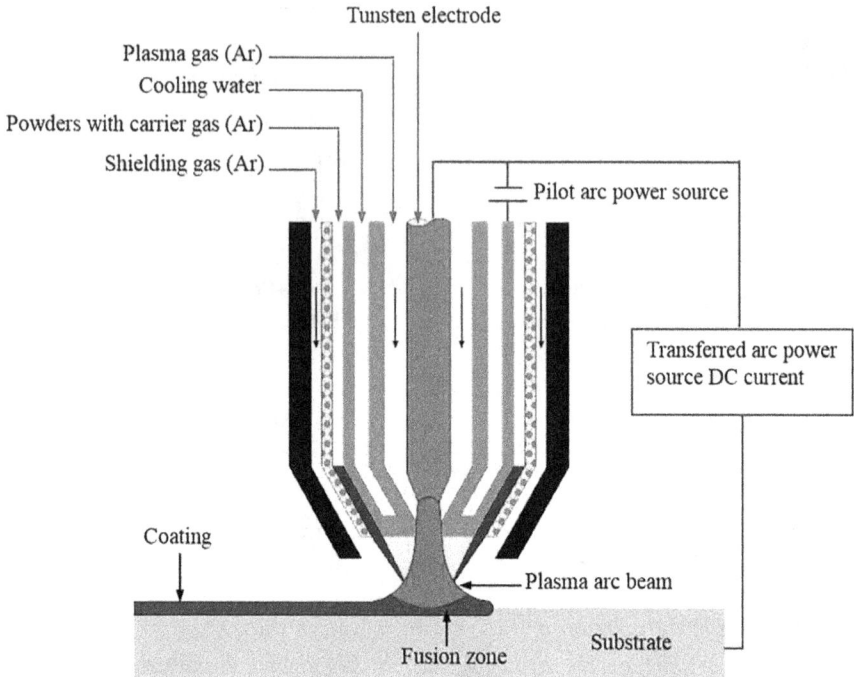

**FIGURE 24.5**   Schematic of the PC method.

### 24.2.1.3.2  Laser Surface Cladding (LSC)

The cladding method can be use to deposite a thin metallic or ceramic layer with a thickness range of several micrometers to the micrometers on a substrate for fabricating bulk coatings [76]. The laser beam used in LSC has simultaneously melted the powders or wires for fabricating a solidified thin surface layer on the substrate; the coating fabricated by this method has strong metallurgical bonding with the substrate [77]. Figure 24.6 shows the schematic of the LSC method.

The LSC method as compared with the other methods of surface coatings has a high density in microstructure and metallurgical bonding with a low porosity between coating and substrate [67]. One of the important features of the LSC is the diluted rate of the cladding in a single track, the coating can be mixed with a significant amount of the material substrate by increasing the dilution rate and the characterization of the coating can be changed. Also, the strong metallurgical bonding between the substrate and the coating layer deposited by the LSC with the low dilution rate is difficult to achieve and the optimized dilution rate for applying the coating layer with the laser cladding layer is between 10% to 25%, respectively [78]. Moreover, the LSC method has a tight bonding and dilution rate and as compared to other surface modifications, the LSC due to having ultrahigh heating temperature and ultrafast cooling rate has other advantages as the microstructure is homogeneous, uniform and fine, with small deformation, superficial heat-affected zone [79]. In addition, the unmelted strengthening phases and intermediate phases

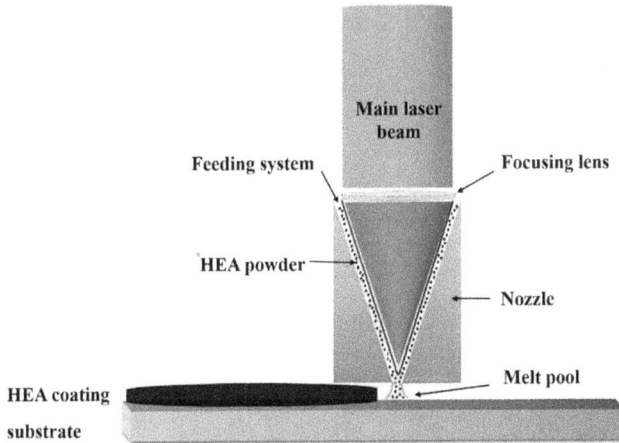

**FIGURE 24.6**  Schematic of the LSC method.

are fabricated by the LSC; these phases are desirable for improving wear resistivity, magnetic characterization, thermal fatigue resistivity, and corrosion resistivity [80]. The LSC by having those advantages is a primary producing method for the HEA-based coatings.

The HEA coating with a composition of Fe-Cr-Ni-Co-Nb$_x$ was applied on the surface of a pure Ti by laser pulse cladding. The Fe-Cr-Ni-Co-Nb$_x$ HEA coating was successfully applied on the surface of the substrate with excellent metallurgical bonding. Comparing the hardness of the HEA coating with the pure Ti, the HEA coating areas have higher hardness than the pure Ti base metal. The hardness of the Fe-Cr-Ni-Co HEA coating with the 1 wt% Nb is significantly higher than the pure Ti substrate and similar composition of the Fe-Cr-Ni-Co HEA alloy. As illustrated in Table 24.3, the different substrates have HEA coatings by the laser cladding method.

### 24.2.1.3.3  *Laser Surface Alloying (LSA)*

The laser beam sources in the LSA method have high power density for melting the metallic materials coatings and a small part of the base metal [17]. The nozzle in the LSA method is used for melting the powders into the melting pool by the laser generating and then the melted substrate and the additives have chemical interactions. Eventually, during the solidification of the melt pool created by the laser source, a surface coating layer is developed on the substrate with a thickness of approximately 0.5 to 1 mm [81].

The melting happens in the LSA method on the surface of the substrate for a short period of time, which causes the substrate to stay cold. This method can act like an unending heat sink. Whereas the temperature of the solid substrate has a lot of difference from the melting surface, the results show that the process has fast self-quenching and resolidification. The LSA method for having different varieties of microstructure and chemical composition is interesting and attractive because the liquid phases have rapid quenching in this method [82]. With all these advantages,

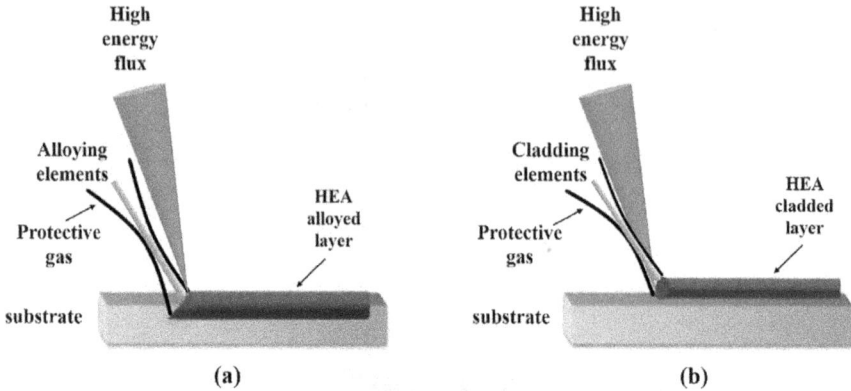

**FIGURE 24.7** The difference between a) LSA and b) LSC methods.

the main disadvantages of the LSA method are high residual stresses, heat-affected zone formation, and the HEA coating and substrate have elemental dilution [32].

The mixing of the substrate with the coating materials has a main difference between LSA and LSC, as shown in Figure 24.7. The coating material in LSA is completely blended with the substrate to create the new alloy composition with the new phases on the surface of the substrate but in the LSA method; the mixture of the coating is on the interface or above the interface of the substrate and the characterization and composition of the cladding materials is preserved. The thickness of the LSA method is lower than the LSC method [70]. The dilution rate of the coatings fabricated by the LSA is significantly greater than the LSC [83].

The most useful method for fabricating HEA coatings on the biomedical substrate is the LSA method. The LSA method can be used for applying the HEA coatings with the base of Ni, Co, and Cr on the different substrates, the same as Ti-6Al-4V and Stainless steel, as mentioned in Table 24.3.

The LD methods due to having a high energy density in a laser are suitable for applying the HEA coatings on important biomedical substrates, especially by LSC and LSA because the coating by these methods has dense structures and strong metallurgical bonding. The most common and important disadvantage of using laser surface treatment is porosities and cracks because those are inherent properties of used from laser methods. According to Table 24.3, the thickness of the coatings fabricated by Pc, LSA, and LSC methods is between 500 to 1500 μm.

## 24.2.2  HEA Coatings Classification

The classification of HEA coatings on the substrate is divided into three categories. The classification of the HEA coatings is as follows:

- HEA composites coating
- HEA metallic coating
- HEA ceramics coating

The development of HEA surface modification and coating methods can be forecasted with the development or subject to changes in those classifications [32].

### 24.2.2.1  HEA Coatings on Composites Substrate

HEA-based coatings on composites offer several advantages, such as high wear resistivity and hardness at elevated temperatures, along with excellent adherence and chemical stability [84]. The synthesized HEA-based coatings are formed on hard ceramic or lightweight metallic reinforcements that can reinforce the HEA matrix [85]. As an example, by applying Cr-Al-Fe-Ni-Cu HEA coating on a reinforcing particle of the Ti(N, C) ceramic can be fabricated the composite coating and exhibit the wear resistance and microhardness of the coated substrate is improved significantly as compared to the uncoated substrate [86]. Moreover, the biocompatibility of the substrate with the HEA-based composite coatings improved significantly. The biocompatibility of the implants is improved significantly by applying the refractory HEA coating with the composition of the $(Nb-Ti-Zr)_{14}$-Mo-Sn [87]. Moreover, the implants with the $(Nb-Ti-Zr)_{14}$-Mo-Sn coating have excellent corrosion resistivity and wear behavior.

### 24.2.2.2  HEA Coatings on Metallic Substrate

Applying a coating based on high-entropy alloys (HEA) onto metal substrates has been shown to prevent the wear and corrosion of the underlying metal [88]. The transitional elements in HEA-based coatings on the metallic substrate mostly consist of Fe, Co, Ti, Ni, Cr, Al, Co, V, Ni, and Mn. By revealing the transition elements into the HEA-based coatings the wear and high-temperature oxidation of the metallic substrates are improved [89]. Recently, researchers have been interested in the effect of refractory elements on HEA coating structure on the metallic substrate instead of the transition elements [55,56]. The mechanical properties of the metallic substrate can be improved by adding the refractory elements Ti, Zr, Hf, Ta, Nb, V, Cr, W, and Mo into the HEA-based coating [90]. The researchers by applying the HEA coating with the refractory elements on the S 45 substrate recognized that the softening resistivity and microhardness of the material increased [91].

### 24.2.2.3  HEA Coatings on Ceramics Substrate

The ceramic substrates with the HEA coating are typically practical due to having low diffusion coefficient, high hardness, and thermal stability [92]. By applying the HEA coating on the ceramic substrate it can exhibit exceptional surface properties for protecting the substrate such as low diffusivity, thermal stability, high hardness, and anti-corrosive performance [93]. The main components of the HEA coatings are the elements Zr, Ti, Si, Cr, Al, and Nb because they form strong oxides, nitrides, and carbides [94]. Recently, researchers investigated the benefits of introducing impurities such as N and C into the HEA coatings in the solid solution state, which has demonstrated improvements in the wear and mechanical properties of the coatings while still protecting the effect of the high entropy [95]. In addition, in the subsequent sections will be discussed about the properties of the HEA coatings.

## 24.3    CHARACTERIZATION AND PROPERTIES OF THE HEAS COATINGS

The unique structure and exceptional properties of the HEAs are special candidates for the coating materials [96] and the other variety applications [46]. The main characteristics and properties of HEAs are presented as follows:

- High corrosion resistivity
- Great biocompatibility
- High wear resistivity
- Outstanding mechanical property
- Diverse surface topography

The characteristics and functional properties of the HEA coatings are reviewed in this section.

### 24.3.1    HEAs Corrosion Resistivity

The corrosion resistance of HEAs is the second property discussed in this section. Excellent corrosion resistivity is achieved by arranging the multiple unique elements in the structure of HEAs leads to the chemical environment with local disordering [29,47]. This feature makes HEAs ideal for use as coating materials, and several studies have been conducted in the past decade to investigate this property [97]. The HEAs coating has been successfully applied on different base metals such as Si, steels, and aluminum alloys by using laser cladding, magnetron spattering, and electro-spark deposition [98].

The formation of a compact amorphous passive layer by HEA films acts as a barrier to the diffusion of oxygen, protecting against corrosion. This passive layer prevents corrodents from penetrating the substrate and minimizes electrochemical reactions under the HEA coating [29,99]. The appropriate selection elements for the HEA are crucial to achieving high corrosion resistance. By adding copper and aluminum to the HEA, it can transform the microstructure and segregate the elements; those are cases of reducing the resistivity of the corrosion by the formation of the ununiformed films of passive. Moreover, the probability formation of the pitting reduced by fabricating a steady protective oxide film can be presented by passivation elements same as chromium. The protection of the passive HEA films can be significantly increased by the anodic and heat treatment [29]. In addition, the corrosion resistivities of the HEAs coating can be improved by the heat treatment because of minimizing the segregation elements and equalizing microstructure [100].

### 24.3.2    HEAs Wear Resistivity

The wear resistivity of the materials with the high hardness is low and the HEAs have high hardness [101]. Recent studies have investigated the wear resistance of HEAs and found that their microstructure is a crucial factor [102]. The wear resistance of the HEA coating improved by reducing the friction of the surface by forming the β phase on the surface with the BCC lattice structure [103]. The wear resistivity of the HEAs subsurface is related to the temperature and load. The HEA coating can achieve high wear resistivity at high temperatures by designing the HEA elemental composition. The wear resistivity of the HEA coating is improved by the formation of the different phases; for example, by adding Al the B2 and $Cr_3Si$ phases are fabricated [104]. For applying the HEA coating on the different substrates, the method is important for improving the wear properties; the laser re-melting method can improve microhardness and strengthen the grain boundary for high wear resistance.

### 24.3.3   HEAs Mechanical Properties

Moreover, HEAs exhibit exceptional mechanical properties such as high hardness, tensile strength, and fracture toughness [105]. The high mixing entropy of HEAs creates a high degree of lattice distortion and provides an increased resistance to deformation, which leads to their excellent mechanical properties [106]. The mechanical properties of HEAs can also be enhanced by introducing nanoscale precipitates, which can improve their strength, ductility, and fracture toughness [107]. Furthermore, the mechanical behavior of HEAs is strongly dependent on the testing temperature, strain rate, and loading mode [108]. For example, HEAs show enhanced ductility and strain hardening rate at elevated temperatures due to the activation of additional deformation mechanisms such as dynamic recrystallization, dynamic recovery, and dynamic precipitation [109].

#### 24.3.3.1   HEAs Tensile and Compressive Strengths

The deformation mechanisms of HEAs have been studied extensively, revealing that HEAs possess good ductility due to mechanisms such as deformation-induced twinning and transformation-induced plasticity [110]. These deformation mechanisms are closely related to the chemical composition and crystal structure of HEAs, which can be optimized to achieve the desired mechanical properties.

In addition to their mechanical properties, HEAs also exhibit excellent thermal properties, such as high melting points, good thermal stability, and low thermal expansion coefficients [111]. These properties make HEAs suitable for high-temperature applications, such as in gas turbines, nuclear reactors, and aerospace components [112]. The thermal properties of HEAs can be further improved by designing and fabricating HEAs with a high density of defects, such as stacking faults, twin boundaries, and nanocrystalline structures, which can enhance phonon scattering and reduce thermal conductivity [113].

Furthermore, HEAs exhibit excellent corrosion and wear resistance, making them ideal candidates for coating materials in harsh environments [114]. The corrosion resistance of HEAs is due to the formation of a passive oxide film on their surface, which protects the underlying metal from further corrosion [115]. The wear resistance of HEAs is attributed to their high hardness and strength, as well as their ability to form protective oxide layers and reduce friction through self-lubrication [116]. HEAs can be further optimized for wear resistance by designing and fabricating composite coatings with hard ceramic reinforcements or metallic alloys [117].

### 24.3.3.2  HEAs Toughness

The high fracture toughness values can be achieved for the HEAs coating with having a single FCC phase structure same as Cr-Ni-Mn-Co-Fe due to the occurrence of multiple deformation mechanisms. These mechanisms include twinning, dislocation slip, and mechanical twinning-induced transformation. The deformation mechanisms in HEAs with dual-phase structures, such as BCC and FCC or BCC and martensite phases, result in intermediate fracture toughness values. However, the structure of the BCC phase for the HEAs revealed low fracture toughness due to the limited number of deformation mechanisms and the absence of dislocation nucleation sources at the crack tip [118].

### 24.3.3.3  HEAs Stiffness

Stiffness is a property that indicates how resistant an object is to deformation under an external force [119]. Multiple studies have shown that HEAs exhibit high stiffness [120]. One of the main reasons for this is the lattice distortion caused by the multi-component solid solution phases of HEAs [121]. The stiffness of most HEA coatings improved by increasing the adhesive energy in lattice structures between the required high shear stress and atoms for breaking the bonds of the metals [122]. The local lattice direction in HEAs can be predicted by using machine learning modeling; this method can serve as a guide for adjusting their stiffness in a quick and computationally efficient way [123].

### 24.3.3.4  HEAs Elasticity

Elasticity is a crucial property for studying the mechanical behavior of alloys and determining how crystals respond to external forces [124]. The external and internal factors such as fatigue, temperature, applied force, ordering, and alloy elements have affected the elasticity of the HEAs [125]. The researchers checked the effect of adding aluminum and hafnium on the elasticity properties of the (Zr-Ta-Nb-Ti)-$Al_x$-$Hf_{1-x}$ HEA [126]. The results show that, with adding aluminum, the elastic modulus of the HEA increased from 59500 MPa to 160000 MPa because the BCC phase is stabilized. Computational modeling is also used to predict the elastic properties of HEAs [114]. The researchers investigated the elasticity and thermodynamic properties of several HEAs by the computational modeling of the single phase of BCC structures [114]. According to the results of the study, the experiment is approximately consistent with the calculation of the elastic properties.

### 24.3.3.5  HEAs Ductility

According to the different studies HEAs can give superior ductility [116,127]. The valance electron concentration is one of the important key factors in the ductility of HEAs, which can be optimized to achieve the desired ductility level. The higher ductility in HEAs can be achieved by forming the FCC structure with a high VEC. Also, the strength of the HEAs is improved by forming BCC phase structure with a low VEC. Based on the design composition and specific application requirements of the HEAs can be achieved to the balance between ductility and strength [128]. HEAs with dual-phase structures comprising both FCC and BCC phases have been found to exhibit superior ductility and high strength [117,129]. The tensile ductility of the HEAs with the dual phase structure (FCC-BBC) is largely increased because the two crystal structures coexist to induce transformation [117].

### 24.3.3.6 HEAs Hardness

HEAs have high hardness, which can be further improved through various strategies [118]. HEAs with different crystal structures, such as BCC/B2 and HCP, exhibit ultra-high hardness values [130]. The values of the HEA hardness can be predicted by machine learning methods. The HEAs have excellent mechanical properties, same as high hardness, stiffness, ductility, compressive, and tensile strength, and for these properties, the HEAs are suitable for biomedical applications as a coating. HEAs can be designed and optimized to achieve specific properties by optimizing the structure parameter and designing the composition, enabling their low-cost production. The mechanical properties of the HEAs can be compared and replaced with the hard tissues. Nevertheless, the design of the composition and mechanical surface treatment in HEAs can be improved by reducing the strength of the compressive. Table 24.4 provides a comparison between the mechanical properties of the hard tissue and the biomedical HEAs [41].

### TABLE 24.4
### Compared the Mechanical Properties of the Hard Tissues with the HEA Biomedical

| Type | Sample | Tensile strength (MPa) | Compressive strength (MPa) | Hardness (GPa) | Elastic modulus (GPa) |
|------|--------|------------------------|----------------------------|----------------|------------------------|
| Hard Tissues | Enamel human tooth | 10 | 384 | 3.3 | 41.4 |
| | Dentin human tooth | 52 | 295 | 2.23–2.54 | 18.6 |
| | Cortical bone | 35–283 | 88–164 | 0.43 | 5–23 |
| | Cancellous bone | 1.5–38 | 2–12 | 0.46 | 0.01–1.57 |
| HEAs | Al-Fe-Co-Ni-Cr | 370 | – | 1.4 | 0.21 |
| | Al-Zn-Cu-Fe-Si | – | 1987 | 6.76–9.55 | 2.79 |
| | Al-Mn-Cr-Ni-Fe | – | 1091–1200 | 5.3 | – |
| | Ti-Nb-Zr-Ta-Hf | – | 800–985 | – | 0.08 |
| | Ti-Si-Mo-Ta-Zr | – | 800 | 3.8 | 0.07–0.09 |
| | Ti-Nb-Zr-Mo-Ta | – | 2600 | 4.9 | – |

### 24.3.4 Wettability and Surface Charge

The surface energy of a material influences its hydrophilicity or hydrophobicity, which in turn affects its interactions with liquids and its ability to adsorb proteins and promote cell adhesion [131]. For implant materials, surface wettability is a critical parameter that can impact their biocompatibility [132]. While metals are typically hydrophilic, there is a need to develop metal films with hydrophobic properties. The inertness and electron work function values are increased in HEAs' unique structure due to their hydrophobic characteristics. The hydrophobicity of the HEAs is increased by reducing the tendency for interfacial exchange of electrons between the atoms of the surface and molecules of the water [133]. Researchers have deposited HEA films on various substrates and observed their hydrophobicity; with results showing that HEA-coated surfaces have better hydrophobicity compared to uncoated samples. The important characteristic of the HEAs is the hydrophobic surface and the advantage of this potential has antifouling and antibacterial abilities, which warrant further investigation [134].

The HEAs may be beneficial for use as platforms of the catalyst because of their distinct surface charge density [135,136]. The surface microstructure and composition are related to the different elemental metals on the surface and this can be repatriated of surface charges for arranging the accumulated alternative elements. The metallic atom on the surface has a different charge density from the neighboring atom, and this can lead to a more active center for chemical transformation. This behavior enables HEAs to transfer their charge to adsorbates and polarize inert molecules. The critical point for controlling the signals of the cell and the adsorption of the protein is the surface charge of the biomaterials. Moreover, the effect of the unique HEA surface on the charge has not been studied yet for biomedical applications [135,137].

### 24.3.5 Biocompatibility and Surface Chemistry

In biomedical applications, surface characteristics play a crucial role in material preparation. The surface of biomaterials can be physically and chemically modified to improve biocompatibility by reducing protein adsorption and cell interactions [138]. Coating the surface of biomaterials with chemically suitable materials is a primary approach to optimize the interactions between the surface of synthetic materials and the biological environment [139]. Compared to classical alloys, the unique surface chemistry and tannable biocompatibility of HEAs arise from the complex distribution of a wide range of various chemical elements within their crystalline network. The ability to select the combination of chemical elements allows for the simultaneous improvement of both mechanical properties and biocompatibility, making HEAs promising materials for medical applications [140].

The cost of producing HEAs can be high due to the presence of expensive elements [141]. However, using HEAs as surface coatings instead of bulk materials can reduce the amount required and lower the overall cost while still providing excellent surface properties [29]. The process parameters of the coating method have a significant impact on the resulting films' stoichiometry and microstructure.

Khan et al. [142] conducted a study using a radio frequency magnetron sputtering technique to deposit thin films of Al-Co-Fe-Cr-Ni-Cu HEA and investigated the effect of coating process parameters on microstructure, surface chemistry, and composition. The results showed that all HEA thin films deposited had protective surface oxides of $Al_2O_3$ and $Cr_2O_3$, but at higher radio frequency powers, there was an increase in re-sputtering of lighter elements, leading to a reduction in Al content. In another study, the same group examined the effect of deposition pressure on the surface chemistry and microstructure of coated HEAs. The results indicated that films deposited at lower pressures were amorphous and had higher concentrations of Al atoms, while those deposited at higher pressures had protective surface oxide layers of $Al_2O_3$ and $Cr_2O_3$ and were composed of a mixture of FCC and BCC crystal structure phases [143]. Therefore, it is essential to consider the deposition process parameters to tune the surface chemistry and achieve the desired properties of HEA thin films as coatings [142,143].

HEAs containing refractory elements like W, Nb, Mo, Ta, and V, have been investigated for their potential in biomedical applications. These HEAs consist of transitional metals such as V-Ta-Nb-W-Mo, Ta-Zr-Nb-Ti-Hf, Hf-Ti-Nb-Zr-Ta, Mo-V-Nb-W-Ta, and Hf-Ti-Mo-Nb-Zr-Ta. Apart from vanadium, these elements are biocompatible. Studies have shown that Ti-Mo-Nb-Zr-Ta, Ti-Zr-Nb-Fe-Ta, Ti–W-Nb–Zr-Ta, Ti-Cr-Zr-Nb-Ta, and Ti-Hf-Nb-Zr-Ta HEAs exhibit superior biocompatibility compared to pure titanium [144–146]. The excellent biocompatibility, mechanical properties, and corrosion resistance of HEAs make them promising candidates for biomedical applications [145].

### 24.3.6 Topography, Porosity, and Surface Texture

Surface texture is a crucial factor affecting the tribological properties of coated surfaces, comprising roughness, waviness, lay, and flaw. Properly designed surface texture can decrease friction and wear, increase load-carrying capacity, and enhance fluid film stiffness. The influence of vacuum-arc coating parameters on the surface texture of Fe-V-Co-Al-Cu-Ni-Cr HEA coatings, observing a change in the preferential orientation of crystallites from <111> to <100> as the negative bias potential increased from −40 to −200 V [147]. HEA films exhibit minimal plastic deformation on their surfaces after applying normal loads, unlike uncoated substrates [45]. The Ti–Ta-Hf-Nb-Zr HEA coating displayed cauliflower-like structures, with a roughness of 2.78 nm, while the uncoated Ti–6Al–4 V substrate was relatively smooth, with a roughness of 0.85 nm.

The surface texture is determined by surface roughness and orientation and is characterized by peaks and valleys. In the biomedical field, surface topography plays a critical role in the effectiveness of coating materials. The patterning of substrate surfaces is a useful tool for regulating cell functions, inducing different cell functions depending on the pattern. However, despite the significant effect of surface topography on the functional properties of HEAs, there have been relatively few studies on this topic. One study examined the effect of as-cast and heat-treated conditions on the pattern formation and surface topography of Ti-Cu-Zr-Ni-Hf-Co HEA, finding that the shapes of nano-precipitates were dependent on the degree of homogeneity and morphological state of the microstructure [148]. Additionally, surface porosity is a crucial factor for controlling cell behavior and tissue integration in biomaterial applications, but few studies have investigated this issue in HEAs.

## 24.4  SUMMARY

HEAs, due to their high mixing entropy, can form stable single-phase microstructures depending on the atomic ratio, structural parameters, and number of elements. These result in a unique structure with excellent mechanical properties and wear resistance, making HEAs a promising coating material. Additionally, the random arrangement of elements in HEAs leads to high corrosion resistance and tuneable biocompatibility. However, the higher cost of HEAs compared to conventional alloys limits their application as bulk materials, and using them as coatings provides advantages such as coating delamination prevention, wear and corrosion resistance, and biocompatibility. However, issues such as pores, cracks, peelings, and dilutions can occur during HEA coating processes, which must be addressed through appropriate fabrication methods and parameters. Despite advancements in manufacturing techniques, these issues persist. Additive manufacturing technology offers a potential solution for producing complex HEA products with desirable performance and reducing common coating problems. Further research is necessary to develop HEAs coating for clinical applications by designing composition, tailoring properties, and microstructure.

## REFERENCES

1. Kulinets, I. "Biomaterials and their applications in medicine." *Regulatory affairs for biomaterials and medical devices*. Woodhead Publishing, 2015. 1–10.
2. Rokaya, Dinesh, et al. "Metallic biomaterials for medical and dental prosthetic applications." *Functional biomaterials: drug delivery and biomedical applications*. Singapore: Springer Singapore, 2022. 503–522.
3. Manappallil, John J. *Basic dental materials*. JP Medical Ltd, 2015.
4. Ghosh, Sougata, Sahil Sanghavi, and Parag Sancheti. "Metallic biomaterial for bone support and replacement." *Fundamental biomaterials: metals*. Woodhead Publishing, 2018. 139–165.
5. Nouri, Alireza, and Cuie Wen. "Noble metal alloys for load-bearing implant applications." *Structural Biomaterials*. Woodhead Publishing, 2021. 127–156.
6. Gong, Pan, et al. "Research on nano-scratching behavior of TiZrHfBeCu (Ni) high entropy bulk metallic glasses." *Journal of Alloys and Compounds* 817 (2020): 153240.
7. Bahraminasab, Marjan, et al. "Aseptic loosening of femoral components–Materials engineering and design considerations." *Materials & Design* 44 (2013): 155–163.
8. Yuan, Wei, et al. "A review on current research status of the surface modification of Zn-based biodegradable metals." *Bioactive Materials* 7 (2022): 192–216.
9. Zhenhuan, Wu, et al. "Physiochemical and biological evaluation of SLM-manufactured Ti-10Ta-2Nb-2Zr alloy for biomedical implant applications." *Biomedical Materials* 15.4 (2020): 045017.
10. Karaji, Z. Gorgin, et al. "A multifunctional silk coating on additively manufactured porous titanium to prevent implant-associated infection and stimulate bone regeneration." *Biomedical Materials* 15.6 (2020): 065016.

11. Qiu, Zhi-Ye, et al. "Advances in the surface modification techniques of bone-related implants for last 10 years." *Regenerative Biomaterials* 1.1 (2014): 67–79.

12. Motallebzadeh, A., et al. "Mechanical properties of TiTaHfNbZr high-entropy alloy coatings deposited on NiTi shape memory alloy substrates." *Metallurgical and Materials Transactions A* 49 (2018): 1992–1997.

13. Vaidya, Mayur, Garlapati Mohan Muralikrishna, and Budaraju Srinivasa Murty. "High-entropy alloys by mechanical alloying: A review." *Journal of Materials Research* 34.5 (2019): 664–686.

14. Katakam, Shravana, et al. "Laser assisted high entropy alloy coating on aluminum: Microstructural evolution." *Journal of Applied Physics* 116.10 (2014).

15. Yeh, J-W., et al. "Nanostructured high-entropy alloys with multiple principal elements: Novel alloy design concepts and outcomes." *Advanced Engineering Materials* 6.5 (2004): 299–303.

16. Castro, D., et al. "An overview of high-entropy alloys as biomaterials. Metals. 2021; 11: 648." *High Entropy Materials* (2021): 3.

17. Siddiqui, Anas Ahmad, A. K. Dubey, and C. P. Paul. "Geometrical characteristics in laser surface alloying of a high-entropy alloy." *Lasers in Engineering* 43 (2019): 237–259.

18. Lyu, Zongyang, et al. "Effects of constituent elements and fabrication methods on mechanical behavior of high-entropy alloys: A review." *Metallurgical and Materials Transactions A* 50 (2019): 1–28.

19. Meghwal, Ashok, et al. "Thermal spray high-entropy alloy coatings: a review." *Journal of Thermal Spray Technology* 29 (2020): 857–893.

20. Tsai, Ming-Hung, and Jien-Wei Yeh. "High-entropy alloys: A critical review." *Materials Research Letters* 2.3 (2014): 107–123.

21. George, Easo P., Dierk Raabe, and Robert O. Ritchie. "High-entropy alloys." *Nature Reviews Materials* 4.8 (2019): 515–534.

22. Zhang, Weiran, Peter K. Liaw, and Yong Zhang. "Science and technology in high-entropy alloys." *Sci. China Mater* 61.1 (2018): 2–22.

23. Kasem, MdRiad, et al. "Synthesis of high-entropy-alloy-type superconductors (Fe, Co, Ni, Rh, Ir) Zr 2 with tunable transition temperature." *Journal of Materials Science* 56 (2021): 9499–9505.

24. Geanta, Victor, et al. "High entropy alloys for medical applications." *Eng. Steels High Entropy Alloys* (2020).

25. Perumal, Gopinath, et al. "Enhanced biocorrosion resistance and cellular response of a dual-phase high entropy alloy through reduced elemental heterogeneity." *ACS Applied Bio Materials* 3.2 (2020): 1233–1244.

26. Nagase, Takeshi, et al. "Design and fabrication of Ti–Zr-Hf-Cr-Mo and Ti–Zr-Hf-Co-Cr-Mo high-entropy alloys as metallic biomaterials." *Materials Science and Engineering: C* 107 (2020): 110322.

27. Wang, Mingliang, et al. "A novel high-entropy alloy composite coating with core-shell structures prepared by plasma cladding." *Vacuum* 184 (2021): 109905.

28. Fang, Qihong, et al. "Microstructure and mechanical properties of FeCoCrNiNbX high-entropy alloy coatings." *Physica B: Condensed Matter* 550 (2018): 112–116.

29. Shi, Yunzhu, Bin Yang, and Peter K. Liaw. "Corrosion-resistant high-entropy alloys: A review." *Metals* 7.2 (2017): 43.

30. Nouri, A., and Cuie Wen. "Introduction to surface coating and modification for metallic biomaterials." *Surface Coating and Modification of Metallic Biomaterials* (2015): 3–60.

31. Duchaniya, Rajendra Kumar, Upender Pandel, and Premlata Rao. "Coatings based on high entropy alloys: An overview." *Materials Today: Proceedings* 44 (2021): 4467–4473.

32. Sharma, Ashutosh. "High entropy alloy coatings and technology." *Coatings* 11.4 (2021): 372.

33. Najafizadeh, Mojtaba, et al. "Thermal barrier ceramic coatings." *Advanced Ceramic Coatings*. Elsevier, 2023. 335–356.

34. Li, Wei, Ping Liu, and Peter K. Liaw. "Microstructures and properties of high-entropy alloy films and coatings: a review." *Materials Research Letters* 6.4 (2018): 199–229.

35. Liang, Chaojie, et al. "Mechanical and tribological properties of (FeCoNi) 88-x (AlTi) 12Mox high-entropy alloys." *International Journal of Refractory Metals and Hard Materials* 105 (2022): 105845.

36. Yan, Xue Hui, et al. "A brief review of high-entropy films." *Materials Chemistry and Physics* 210 (2018): 12–19.

37. Gao, Libo, et al. "Microstructure, mechanical and corrosion behaviors of CoCrFeNiAl0. 3 high entropy alloy (HEA) films." *Coatings* 7.10 (2017): 156.

38. Zhou, Jia-li, et al. "Composition design and preparation process of refractory high-entropy alloys: A review." *International Journal of Refractory Metals and Hard Materials* 105 (2022): 105836.

39. Dobrzański, L. A., et al. "Corrosion resistance of multilayer and gradient coatings deposited by PVD and CVD techniques." *Archives of Materials Science and Engineering* 28.1 (2007): 12–18.

40. Braic, V., et al. "Nanostructured multi-element (TiZrNbHfTa) N and (TiZrNbHfTa) C hard coatings." *Surface and Coatings Technology* 211 (2012): 117–121.

41. Ahmady, Azin Rashidy, et al. "High entropy alloy coatings for biomedical applications: A review." *Smart Materials in Manufacturing* 1 (2023): 100009.

42. Li, Junchen, et al. "A review on high entropy alloys coatings: Fabrication processes and property assessment." *Advanced Engineering Materials* 21.8 (2019): 1900343.

43. Kelly, Peter J., and R. Derek Arnell. "Magnetron sputtering: A review of recent developments and applications." *Vacuum* 56.3 (2000): 159–172.

44. Padamata, Sai Krishna, et al. "Magnetron sputtering high-entropy alloy coatings: A mini-review." *Metals* 12.2 (2022): 319.

45. Tüten, N., et al. "Microstructure and tribological properties of TiTaHfNbZr high entropy alloy coatings deposited on Ti6Al4V substrates." *Intermetallics* 105 (2019): 99–106.

46. Yan, Xuehui, and Yong Zhang. "Functional properties and promising applications of high entropy alloys." *Scripta Materialia* 187 (2020): 188–193.

47. Sim, Rui Ken, et al. "Microstructure, mechanical properties, corrosion and wear behavior of high-entropy alloy AlCoCrFeNix (x> 0 the) and medium-entropy alloy (x= 0)." *Journal of Materials Science* 57.25 (2022): 11949–11968.

48. Habib, K. A., et al. "Effects of thermal spraying technique on the remelting behavior of NiCrBSi coatings." *Surface and Coatings Technology* 444 (2022): 128669.

49. Crawmer, D. E., Cold Spray Process, 2004.

50. Talib, R. J., et al. "Thermal spray coating technology: A review." *Solid State SciTechnol* 11.1 (2003): 109–117.

51. Amin, Sagar, and Hemant Panchal. "A review on thermal spray coating processes." *Transfer* 2.4 (2016): 556–563.

52. Nouri, Alireza, and Antonella Sola. "Powder morphology in thermal spraying." *Journal of Advanced Manufacturing and Processing* 1.3 (2019): e10020.

53. Cavaliere, Pasquale, L. Cavaliere, and Lekhwani. *Cold-Spray Coatings*. Berlin/ Heidelberg, Germany: Springer, 2018.

54. Singh, Harminder, et al. "Development of cold spray from innovation to emerging future coating technology." *Journal of the Brazilian Society of Mechanical Sciences and Engineering* 35 (2013): 231–245.

55. Smith, M. F. "Comparing cold spray with thermal spray coating technologies." *The cold spray materials deposition process*. Woodhead Publishing, 2007. 43–61.

56. Moridi, Atieh, et al. "Cold spray coating: Review of material systems and future perspectives." *Surface Engineering* 30.6 (2014): 369–395.

57. Karthikeyan, Jeganathan. "The advantages and disadvantages of the cold spray coating process." *The cold spray materials deposition process*. Woodhead Publishing, 2007. 62–71.

58. Hushchyk, D. V., et al. "Nanostructured AlNiCoFeCrTi high-entropy coating performed by cold spray." *Applied Nanoscience* 10 (2020): 4879–4890.

59. Ahn, Ji-Eun, et al. "Tuning the microstructure and mechanical properties of cold sprayed equiatomic CoCrFeMnNi high-entropy alloy coating layer." *Metals and Materials International* 27 (2021): 2406–2415.

60. Gromov, Viktor Evgen'evich, et al. *Structure and properties of high-entropy alloys*. Cham, Switzerland: Springer International Publishing, 2021.

61. Xiao, Jin-Kun, et al. "Microstructure and tribological properties of plasma sprayed FeCoNiCrSiAlx high entropy alloy coatings." *Wear* 448 (2020): 203209.

62. Cheng, Kuei-Chung, et al. "Properties of atomized AlCoCrFeNi high-entropy alloy powders and their phase-adjustable coatings prepared via plasma spray process." *Applied Surface Science* 478 (2019): 478–486.

63. Tian, Lihui, Zongkang Feng, and Wei Xiong. "Microstructure, microhardness, and wear resistance of AlCoCrFeNiTi/Ni60 coating by plasma spraying." *Coatings* 8.3 (2018): 112.

64. Dongmo, E., M. Wenzelburger, and R. Gadow. "Analysis and optimization of the HVOF process by combined experimental and numerical approaches." *Surface and Coatings Technology* 202.18 (2008): 4470–4478.

65. Dorfman, Mitchell R. "Thermal spray coatings." *Handbook of environmental degradation of materials*. William Andrew Publishing, 2018. 469–488.

66. Ren, Jiangzhuo, and Yongsheng Ma. "A feature-based physical-geometric model for dynamic effect in HVOF thermal spray process." *Comput Aided Des Appl* 17 (2020): 561–574.

67. Weng, Fei, Chuanzhong Chen, and Huijun Yu. "Research status of laser cladding on titanium and its alloys: A review." *Materials & Design* 58 (2014): 412–425.

68. Abegunde, Olayinka Oluwatosin, et al. "Overview of thin film deposition techniques." *AIMS Materials Science* 6.2 (2019): 174–199.

69. Duta, Liviu, and Andrei C. Popescu. "Current research in pulsed laser deposition." *Coatings* 11.3 (2021): 274.

70. Chi, Yiming, et al. "Laser surface alloying on aluminum and its alloys: A review." *Optics and Lasers in Engineering* 100 (2018): 23–37.

71. Lu, Jinbin, et al. "Microstructure evolution and properties of CrCuFexNiTi high-entropy alloy coating by plasma cladding on Q235." *Surface and Coatings Technology* 328 (2017): 313–318.

72. Cai, Zhaobing, et al. "Design and microstructure characterization of FeCoNiAlCu high-entropy alloy coating by plasma cladding: In comparison with thermodynamic calculation." *Surface and Coatings Technology* 330 (2017): 163–169.

73. Peng, Yingbo, et al. "Effect of WC content on microstructures and mechanical properties of FeCoCrNi high-entropy alloy/WC composite coatings by plasma cladding." *Surface and Coatings Technology* 385 (2020): 125326.

74. Peng, Y. B., et al. "Microstructures and mechanical properties of FeCoCrNi high entropy alloy/WC reinforcing particles composite coatings prepared by laser cladding and plasma cladding." *International Journal of Refractory Metals and Hard Materials* 84 (2019): 105044.

75. Wang, Jiying, et al. "Study of high temperature friction and wear performance of (CoCrFeMnNi) 85Ti15 high-entropy alloy coating prepared by plasma cladding." *Surface and Coatings Technology* 384 (2020): 125337.

76. Tamanna, Nusrat, Roger Crouch, and Sumsun Naher. "Progress in numerical simulation of the laser cladding process." *Optics and Lasers in Engineering* 122 (2019): 151–163.

77. Bao, Yefeng, et al. "Effects of WC on the cavitation erosion resistance of FeCoCrNiB0. 2 high entropy alloy coating prepared by laser cladding." *Materials Today Communications* 26 (2021): 102154.

78. Norhafzan, B., C. M. Khairil, and S. N. Aqida. "Laser cladding process to enhanced surface properties of hot press forming die: A review." IOP Conference Series: Materials Science and Engineering. Vol. 1078. No. 1. IOP Publishing, 2021.

79. Han, Bin, et al. "Review and prospect of the influence of laser cladding process parameters on the properties of die cladding layer." *Materials Science Forum*. Vol. 990. Trans Tech Publications Ltd, 2020.

80. Liu, Jianli, et al. "Research and development status of laser cladding on magnesium alloys: A review." *Optics and Lasers in Engineering* 93 (2017): 195–210.

81. Tian, Y. S., et al. "Research progress on laser surface modification of titanium alloys." *Applied surface science* 242.1-2 (2005): 177–184.

82. Draper, C. W., and J. M. Poate. "Laser surface alloying." *International Metals Reviews* 30.1 (1985): 85–108.

83. Arif, Zia Ullah, et al. "Laser deposition of high-entropy alloys: A comprehensive review." *Optics & Laser Technology* 145 (2022): 107447.

84. Aliyu, Ahmed, and Chandan Srivastava. "Microstructure and corrosion properties of MnCrFeCoNi high entropy alloy-graphene oxide composite coatings." *Materialia* 5 (2019): 100249.

85. Pei, Zongrui. "An overview of modeling the stacking faults in lightweight and high-entropy alloys: Theory and application." *Materials Science and Engineering: A* 737 (2018): 132–150.

86. Wang, Mingliang, et al. "A novel high-entropy alloy composite coating with core-shell structures prepared by plasma cladding." *Vacuum* 184 (2021): 109905.

87. Guo, Yaxiong, Xingmao Li, and Qibin Liu. "A novel biomedical high-entropy alloy and its laser-clad coating designed by a cluster-plus-glue-atom model." *Materials & Design* 196 (2020): 109085.

88. Miracle, Daniel B., and Oleg N. Senkov. "A critical review of high entropy alloys and related concepts." *Acta Materialia* 122 (2017): 448–511.

89. Cui, Yan, et al. "Wear resistance of FeCoCrNiMnAlx high-entropy alloy coatings at high temperature." *Applied Surface Science* 512 (2020): 145736.

90. Chen, Lin, et al. "Lightweight refractory high entropy alloy coating by laser cladding on Ti–6Al–4V surface." *Vacuum* 183 (2021): 109823.

91. Zhang, Mina, et al. "Synthesis and characterization of refractory TiZrNbWMo high-entropy alloy coating by laser cladding." *Surface and Coatings Technology* 311 (2017): 321–329.

92. Lin, Shao-Yi, et al. "Mechanical performance and nanoindenting deformation of (AlCrTaTiZr) NCy multi-component coatings co-sputtered with bias." *Surface and Coatings Technology* 206.24 (2012): 5096–5102.

93. Wan, Hongxia, et al. "Corrosion behavior of Al0. 4CoCu0. 6NiSi0.2Ti0. 25 high-entropy alloy coating via 3D printing laser cladding in a sulphur environment." *Journal of Materials Science & Technology* 60 (2021): 197–205.

94. Tsai, Du-Cheng, et al. "Structural morphology and characterization of (AlCrMoTaTi) N coating deposited via magnetron sputtering." *Applied Surface Science* 282 (2013): 789–797.

95. Kuang, Shaofu, et al. "Improvement of the mechanical and the tribological properties of CrNbTiMoZr coatings through the incorporation of carbon and the adjustment of the substrate bias voltage." *Surface and Coatings Technology* 412 (2021): 127064.

96. Hua, Nengbin, et al. "Mechanical, corrosion, and wear properties of biomedical Ti–Zr–Nb–Ta–Mo high entropy alloys." *Journal of Alloys and Compounds* 861 (2021): 157997.

97. Vallimanalan, A., et al. "Corrosion behaviour of thermally sprayed Mo added AlCoCrNi high entropy alloy coating." *Materials Today: Proceedings* 27 (2020): 2398–2400.

98. Karlsdottir, Sigrun N., et al. "Phase evolution and microstructure analysis of CoCrFeNiMo high-entropy alloy for electro-spark-deposited coatings for geothermal environment." *Coatings* 9.6 (2019): 406.

99. Shuang, S., et al. "Corrosion resistant nanostructured eutectic high entropy alloy." *Corrosion Science* 164 (2020): 108315.

100. Lin, Chun-Ming, Hsien-Lung Tsai, and Hui-Yun Bor. "Effect of aging treatment on microstructure and properties of high-entropy Cu0.5CoCrFeNi alloy." *Intermetallics* 18.6 (2010): 1244–1250.

101. Huang, Xuejun, Jiashi Miao, and Alan A. Luo. "Lightweight AlCrTiV high-entropy alloys with dual-phase microstructure via microalloying." *Journal of Materials Science* 54.3 (2019): 2271–2277.

102. Cheng, Hu, et al. "Tribological properties of nano/ultrafine-grained FeCoCrNiMnAlx high-entropy alloys over a wide range of temperatures." *Journal of Alloys and Compounds* 817 (2020): 153305.

103. Firstov, S. A., et al. "Wear resistance of high-entropy alloys." *Powder Metallurgy and Metal Ceramics* 56 (2017): 158–164.

104. Joseph, Jithin, et al. "On the enhanced wear resistance of CoCrFeMnNi high entropy alloy at intermediate temperature." *Scripta Materialia* 186 (2020): 230–235.

105. Maresca, Francesco, and William A. Curtin. "Mechanistic origin of high strength in refractory BCC high entropy alloys up to 1900K." *Acta Materialia* 182 (2020): 235–249.

106. Fu, Zhiqiang, et al. "A high-entropy alloy with hierarchical nanoprecipitates and ultrahigh strength." *Science Advances* 4.10 (2018): eaat8712.

107. Yi, Jiaojiao, et al. "Excellent strength-ductility synergy in a novel single-phase equiatomic CoFeNiTiV high entropy alloy." *International Journal of Refractory Metals and Hard Materials* 95 (2021): 105416.

108. Chen, Liangbin, et al. "Gradient structure design to strengthen carbon interstitial Fe40Mn40Co10Cr10 high entropy alloys." *Materials Science and Engineering: A* 772 (2020): 138661.

109. Qin, Gang, et al. "Microstructures and mechanical properties of Nb-alloyed CoCrCuFeNi high-entropy alloys." *Journal of Materials Science & Technology* 34.2 (2018): 365–369.

110. Yang, Huijun, et al. "High strength and ductility in partially recrystallized Fe40Mn20Cr20Ni20 high-entropy alloys at cryogenic temperature."

111. Wang, Shao-Ping, and Jian Xu. "(TiZrNbTa)-Mo high-entropy alloys: dependence of microstructure and mechanical properties on Mo concentration and modeling of solid solution strengthening." *Intermetallics* 95 (2018): 59–72.

112. Huang, Shuo, et al. "Chemical ordering controlled thermo-elasticity of AlTiVCr1-xNbx high-entropy alloys." *Acta Materialia* 199 (2020): 53–62.

113. Roy, Ankit, et al. "Lattice distortion as an estimator of solid solution strengthening in high-entropy alloys." *Materials Characterization* 172 (2021): 110877.

114. Gao, Michael C., et al. "Computational modeling of high-entropy alloys: Structures, thermodynamics and elasticity." *Journal of Materials Research* 32.19 (2017): 3627–3641.

115. Li, Zhiming, and Dierk Raabe. "Strong and ductile non-equiatomic high-entropy alloys: design, processing, microstructure, and mechanical properties." *Jom* 69.11 (2017): 2099–2106.

116. Ding, Huaping, et al. "Enhancing strength-ductility synergy in an ex situ Zr-based metallic glass composite via nanocrystal formation within high-entropy alloy particles." *Materials & Design* 210 (2021): 110108.

117. Zhang, T., et al. "Transformation-enhanced strength and ductility in a FeCoCrNiMn dual phase high-entropy alloy." *Materials Science and Engineering: A* 780 (2020): 139182.

118. Li, Weidong, Peter K. Liaw, and Yanfei Gao. "Fracture resistance of high entropy alloys: A review." *Intermetallics* 99 (2018): 69–83.

119. Sun, Tao, and Binbin Lian. "Stiffness and mass optimization of parallel kinematic machine." *Mechanism and Machine Theory* 120 (2018): 73–88.

120. Huang, Shuo, et al. "Chemical ordering controlled thermo-elasticity of AlTiVCr1-xNbx high-entropy alloys." *Acta Materialia* 199 (2020): 53–62.

121. Owen, Lewis Robert, and Nicholas Gwilym Jones. "Lattice distortions in high-entropy alloys." *Journal of Materials Research* 33.19 (2018): 2954–2969.

122. Roy, Ankit, et al. "Lattice distortion as an estimator of solid solution strengthening in high-entropy alloys." *Materials Characterization* 172 (2021): 110877.

123. Kim, George, et al. "First-principles and machine learning predictions of elasticity in severely lattice-distorted high-entropy alloys with experimental validation." *Acta Materialia* 181 (2019): 124–138.

124. Zheng, Shu-min, Wen-qiang Feng, and Shao-qing Wang."Elastic properties of high entropy alloys by MaxEnt approach." *Computational Materials Science* 142 (2018): 332–337.

125. Huang, Shuo, Fuyang Tian, and Levente Vitos. "Elasticity of high-entropy alloys from ab initio theory." *Journal of Materials Research* 33.19 (2018): 2938–2953.

126. Li, Shaohui, Xiaodong Ni, and Fuyang Tian."Ab initio predicted alloying effects on the elastic properties of Al x Hf1− x NbTaTiZr high entropy alloys." *Coatings* 5.3 (2015): 366–377.

127. Lei, Zhifeng, et al. "Snoek-type damping performance in strong and ductile high-entropy alloys." *Science Advances* 6.25 (2020): eaba7802.

128. Chen, Ruirun, et al. "Composition design of high entropy alloys using the valence electron concentration to balance strength and ductility." *Acta Materialia* 144 (2018): 129–137.

129. Nene, Saurabh Sanjay, et al. "Enhanced strength and ductility in a friction stir processing engineered dual phase high entropy alloy." *Scientific Reports* 7.1 (2017): 16167.

130. Zhang, Mengdi, et al. "Novel Co-free CrFeNiNb0. 1Tix high-entropy alloys with ultra high hardness and strength." *Materials Science and Engineering: A* 764 (2019): 138212.

131. Li, Mei, et al. "Tuning the surface potential to reprogram immune microenvironment for bone regeneration." *Biomaterials* 282 (2022): 121408.

132. Thaik, Nyein, et al. "Bioactive surface-modified Ti with titania nanotube arrays to design endoprosthesis for maxillofacial surgery: structural formation, morphology, physical properties and osseointegration." *Biomedical Materials* 15.3 (2020): 035018.

133. Ng, Law Yong, et al. "Polymeric membranes incorporated with metal/metal oxide nanoparticles: A comprehensive review." *Desalination* 308 (2013): 15–33.

134. Wang, Ze, et al. "Wettability, electron work function and corrosion behavior of CoCrFeMnNi high entropy alloy films." *Surface and Coatings Technology* 400 (2020): 126222.
135. Katiyar, Nirmal Kumar, et al. "A perspective on the catalysis using the high entropy alloys." *Nano Energy* 88 (2021): 106261.
136. Glasscott, Matthew W. "Classifying and benchmarking high-entropy alloys and associated materials for electrocatalysis: A brief review of best practices." *Current Opinion in Electrochemistry* 34 (2022): 100976.
137. Xin, Yue, et al. "High-entropy alloys as a platform for catalysis: progress, challenges, and opportunities." *Acs Catalysis* 10.19 (2020): 11280–11306.
138. Tang, Liping, Paul Thevenot, and Wenjing Hu. "Surface chemistry influences implant biocompatibility." *Current Topics in Medicinal Chemistry* 8.4 (2008): 270–280.
139. Lifeng, Zhao, et al. "The underlying biological mechanisms of biocompatibility differences between bare and TiN-coated NiTi alloys." *Biomedical Materials* 6.2 (2011): 025012.
140. Mishra, Rajiv S., Ravi Sankar Haridas, and Priyanshi Agrawal. "High entropy alloys–Tunability of deformation mechanisms through integration of compositional and microstructural domains." *Materials Science and Engineering: A* 812 (2021): 141085.
141. Canter, Neil. "High-entropy alloys." *Tribology & Lubrication Technology* 71.3 (2015): 14.
142. Khan, Naveed A., et al. "RF magnetron sputtered AlCoCrCu0. 5FeNi high entropy alloy (HEA) thin films with tuned microstructure and chemical composition." *Journal of Alloys and Compounds* 836 (2020): 155348.
143. Khan, Naveed A., et al. "High entropy alloy thin films of AlCoCrCu0.5FeNi with controlled microstructure." *Applied Surface Science* 495 (2019): 143560.
144. Wang, Shao-Ping, and Jian Xu. "TiZrNbTaMo high-entropy alloy designed for orthopedic implants: As-cast microstructure and mechanical properties." *Materials Science and Engineering: C* 73 (2017): 80–89.
145. Geanta, Victor, et al. "High entropy alloys for medical applications." *Eng. Steels High Entropy Alloys* (2020).
146. Todai, Mitsuharu, et al. "Novel TiNbTaZrMo high-entropy alloys for metallic biomaterials." *Scripta Materialia* 129 (2017): 65–68.
147. Sobol, O. V., et al. "The use of negative bias potential for structural engineering of vacuum-arc nitride coatings based on high-entropy alloys." (2019).
148. Hinte, Christian, et al. "Pattern-forming nanoprecipitates in NiTi-related high entropy shape memory alloys." *Scripta Materialia* 186 (2020): 132–135.

# 25 Surface Strategies and Classification of Biomaterials

*Monireh Ganjali*
Biomaterials Group, Department of Nanotechnology &
Advanced Materials, Materials and Energy Research Center,
Karaj, Iran

*Arash Ghalandarzadeh*
School of Metallurgy and Materials Engineering, Iran
University of Science and Technology, Tehran, Iran

*Mansoureh Ganjali*
NourZoha Materials Engineering Research Group, Tehran,
Iran

## 25.1 INTRODUCTION

Alumina is a ceramic material that possesses remarkable attributes such as exceptional mechanical strength, fracture toughness, biocompatibility, as well as resistance to corrosion and abrasion. Nevertheless, in vivo studies conducted within the orthopedic community have revealed that the slow propagation of cracks in alumina implants can eventually result in the failure of the ceramic component over time. In contrast, zirconia, another ceramic biomaterial, is extensively utilized in medical applications, including implants, due to its favorable mechanical properties, improved biocompatibility, and excellent resistance to abrasion. Zirconia-based ceramics are frequently employed in hip replacement surgery, tooth restoration, and dental implants, and serve as an alternative to titanium implants in dental procedures. The utilization of zirconia implants has gained popularity, primarily due to their appealing aesthetic appearance in dental treatments. Research indicates that zirconia implants exhibit superior bone formation and enhanced compatibility compared to titanium implants. Furthermore, zirconia promotes the integrity of soft tissues by demonstrating low bacterial colonization and excellent biocompatibility. Additionally, zirconia possesses anti-cancer properties and does not induce any harmful carcinogenic effects. Zirconia is a suitable choice for various implant applications, encompassing dental and bone procedures. However, most ceramics are considered bioinert, meaning they lack interaction with the surrounding tissues,

DOI: 10.1201/9781003429920-28

potentially leading to the encapsulation of fibrous foreign bodies due to the absence of interaction with the local environment.

Surface engineering is an expansive field with the objective of enhancing the performance of materials for specific applications by optimizing their surface properties. This domain encompasses various techniques, including machining, chemical etching, sandblasting, coating, and laser treatment, among others. Laser treatment, in particular, has garnered considerable attention due to its capacity to create structured surface patterns that can influence cell adhesion. Another advantage of laser treatment is its non-contaminating nature during the process. Significant advancements have been observed in laser treatment for improving osseointegration, reducing bacterial formation on biomaterial surfaces, and enhancing soft tissue adhesion, especially in the context of dental implants. However, it should be noted that metals and ceramics exhibit distinct laser absorption mechanisms that differentiate them from each other. Metals absorb laser energy through free electrons proportional to the intensity of the laser, while ceramics lack free electrons. Consequently, ceramics necessitate much higher laser fluence, imposing limitations on the use of short- or ultra-high-pulse lasers as a viable energy source.

The objective of this chapter is to present a comprehensive review of the various applications involving bioceramics and surface texturing. It also aims to highlight the advancements made by researchers in this field, with a specific focus on laser-modified surfaces and microscopic observations. As far as we know, there is limited research available in this area, mostly comprising application-based studies published in recent years. In addition to comparing different texturing processes, this chapter also showcases recently published research on the applications of laser surface modification in bioceramics.

## 25.2  BIOCERAMICS

Bioceramics have been employed in the medical domain for more than four decades as implantable devices and substitutes for compromised or diseased hard tissues [1]. Their remarkable effectiveness has rendered them a valuable resource for addressing diverse medical conditions. The initial iteration of bioceramics, referred to as bioinert ceramics, saw successful implementation through the use of alumina during the early 1970s [2]. The subsequent generation of biomaterials comprises calcium phosphate ceramics, offering enhanced biological properties owing to their chemical composition resembling that of bone minerals.

Since the 1980s, various types of bioceramics, including hydroxyapatite, tricalcium phosphate, and biphasic calcium phosphate, have been utilized in surgical procedures [3]. Another category of bioceramics is represented by bioglasses, which are silicon-based materials that promote bioactivity through bone bonding and the development of bone-like apatite layers on their surface [4]. Depending on the specific type of bioceramics and their interactions with host tissue, they can be classified as bioinert [5], absorbable ($\alpha$-CSH) [6], or bioactive [1]. They can take the form of porous bulk material [7], granules [8], or coatings [9].

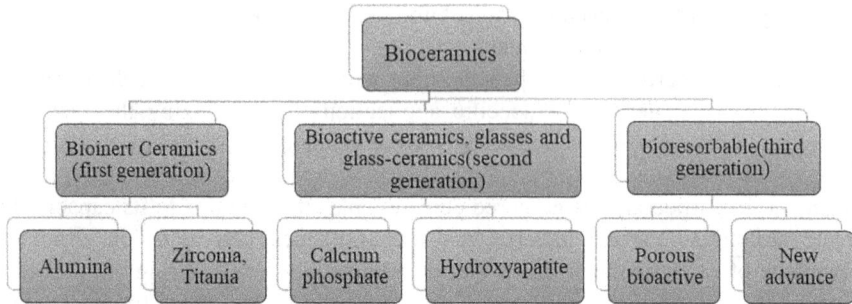

**FIGURE 25.1** Bioceramics classification [13].

Bioceramics, exemplified by $Al_2O_3$ and $ZrO_2$, are considered bioinert and showcase exceptional biocompatibility and mechanical properties such as excellent abrasion resistance, thermal and chemical stability, and an appealing aesthetic appearance. These qualities render them highly suitable for various biomedical applications [10–12]. In recent years, the field of medical equipment and tissue engineering has undergone a significant transformation due to the introduction of bioceramics like bioglass, hydroxyapatite, tricalcium phosphate, as well as biodegradable/bioresorbable ceramics (Figure 25.1).

## 25.2.1 INTERACTIONS OF TISSUES WITH DIFFERENT TYPES OF BIOCERAMICS

Following in vivo implantation, bioinert ceramics such as zirconia, alumina, and silicon nitride usually experience the formation of thick fibrous tissue around them. This encapsulation phenomenon, attributed to their chemical stability, can potentially result in the loosening of aseptic implants. On the other hand, bioactive ceramics necessitate a chemical reaction to establish bonds with the surrounding bone or soft tissue upon implantation in the human body [14]. Certain types of glass and glass ceramics fall into this category, and they have the ability to stimulate bone growth on their surfaces, thus enhancing the functional longevity of the implanted tissue. The classification of bioactive ceramics comprises various primary categories, such as calcium phosphate ceramics, bioactive glass, and glass ceramics [15–17]. In vivo implantation of bioresorbable materials leads to their gradual degradation, allowing them to be replaced by internal tissue. Extensive research has been conducted on bioceramics and glass ceramics based on bio-glass, calcium phosphate, and calcium silicate for the purpose of reconstructing skeletal bones.

An innovative advancement involves the replacement of second-generation bioactive ceramics with third-generation absorbable ceramics, achieved through the modification of existing materials or the development of new ones that incorporate biological agents. These ceramics can be utilized to create integrated structures. Both bioactive and absorbable ceramics can be produced as integrated structures or applied onto bioinert materials to enhance the bonding between the implant and surrounding tissue. It is crucial to acknowledge that the attachment mechanism of tissue to the implant varies depending on the type of tissue response. Consequently,

**TABLE 25.1**

**Tissue Response to Different Types of Materials**

| Type of Material | Tissue Response |
|---|---|
| Toxic | Death of the surrounding tissue |
| Non-toxic, biologically inactive | Formation of fibrous tissue |
| Non-toxic, biologically active | Formation of interfacial bond |
| Non-toxic, resorbable or degradable | Replacement of surrounding tissue |

**TABLE 25.2**

**Types of Bioceramic-Tissue Attachments**

| Types of Bioceramics | Types of Bioceramic-Tissue Attachment |
|---|---|
| Porous Hydroxyapatite | Attachment of Porous Implants by Bone Growth |
| Alumina, Zirconia | Attachment of Dense Bioinert Ceramics to the Surface of the Material with Bone Growth |
| Tricalcium Phosphate | Slow Replacement of Absorbable Bioceramics by Bone Growth |
| Bioactive Glasses and Glass-Ceramics | Attachment by Chemical Bonding with Bone |

no material can be categorized as completely inert since all materials interact with living tissue. Table 25.1 provides an overview of the four types of bioceramic reactions with tissues.

The propensity of bioactive ceramics to bond with surrounding tissue is influenced by factors such as the specific ceramic material, duration, resistance, and the mechanism of bonding. Table 25.2 provides a listing of different types of tissue attachments.

## 25.2.2 Bioceramic Properties

Bioceramics are ceramic materials used for biomedical applications because of their biocompatibility and bioactivity. They possess several properties that make them suitable for medical implants and devices, including the following key features:

1. Biocompatibility: One of the most important properties of bioceramics is their biocompatibility, which means they can be used in contact with biological tissues without causing adverse reactions.
2. Bioactivity: Bioceramics exhibit bioactivity and can bond directly with bone tissue, which is a useful property in dental and orthopedic implants.
3. Chemical stability: Bioceramics are chemically stable and do not react with body fluids or tissues, making them ideal for long-term implant applications.

4. Mechanical strength: Bioceramics are strong and durable, enabling them to withstand the loads and stresses placed on them within the body.
5. Wear resistance: Bioceramics are highly resistant to wear and friction, making them especially suitable for joint replacements.
6. Radiopacity: Bioceramics are radiopaque and visible on X-rays and other imaging techniques, which is an important property for monitoring the healing process after implantation.

Bioceramics present a distinctive combination of properties that make them a compelling choice for various medical applications.

### 25.2.3   ENHANCING BIOCERAMICS: SURFACE MODIFICATION STRATEGIES

The material surface exhibits a unique reactivity, characterized by properties that differ from its bulk. Surface modification aims to modify these properties in order to enhance the performance and functionality of materials in biomedical applications without altering their bulk properties. Bioceramics, commonly employed in bone and dental implants, require seamless integration with the surrounding tissue to promote healing and avoid rejection by the body's immune system. Surface modification can augment bioceramics' ability to bond with the surrounding tissue, elevate their bioactivity, and amplify their mechanical properties [18]. Furthermore, surface modification can optimize biocompatibility by mitigating adverse reactions and inflammation, promoting improved fusion with surrounding tissue, and guaranteeing the long-term success of the implant [19]. In summary, surface modification constitutes an essential factor in optimizing the performance of bioceramics for application across diverse biomedical domains [20].

Fig. 25.2 demonstrates the importance of the interaction between bone and implant. The figure illustrates the progression of the interface between bone and implant during the immediate post-surgery phase. In Fig. 25.2(a), the implant surface is surrounded by blood due to the bone cavity and both dead and living bone tissue. This phenomenon, known as the bone cohesion phenomenon, was originally defined by Brånemark in 1977. Additionally, remnants of bone may be observed around the implant surface. Following the surgical procedure, new bone formation takes place over a period of several weeks/months to bridge the gap between mature bone tissue and the implant surface (Figure 25.2b). Subsequently, the newly formed bone tissue gradually replaces the mature bone tissue around the implant's surface over several weeks or months after implant surgery, resulting in a bone-to-implant contact ratio, as depicted in Figure 25.2c.

Biomaterials researchers employ a range of techniques to promote favorable interactions between tissues and implants, including modifying physical and biological properties, establishing strong tissue bonds, and responding to changes in pH or temperature. Specifically, when considering that most tissue-implant interactions take place on the surface, two common approaches are employed: the creation of innovative bioactive materials or the modification of existing surfaces. While the mechanical connection between an implant and bone is crucial for stability, additional bioactive substances like hydroxyapatite, tricalcium phosphate,

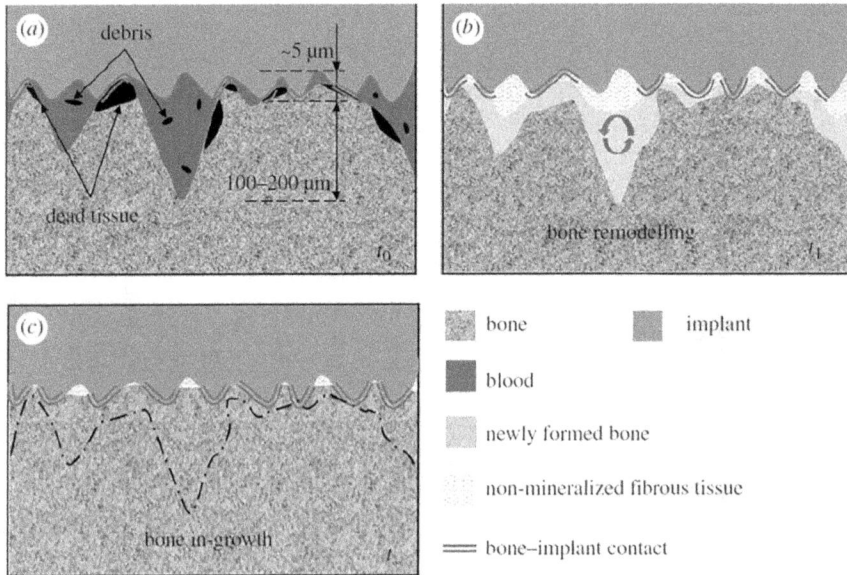

**FIGURE 25.2** Schematic of the bone and implant interface, showing (a) immediately after surgery (time t0), (b) during the period of bone remodeling (t1 - newly formed bone), and (c) after completion of osseointegration [21].

and bioactive glass can stimulate bone formation, encourage adhesion, proliferation, and facilitate the attachment of osteoblasts to both soft and hard tissues. However, due to the brittleness of bioactive materials, they are limited to use as coatings or in small bone grafts. Furthermore, both the composition of the material and the topography of the implant surface significantly influence their biological behavior and performance, even in bioinert ceramics. Several in vivo studies have provided evidence that surface-modified zirconia implants demonstrate superior osseointegration compared to non-modified ones. Importantly, Liu et al.'s recent study provides a comprehensive overview of various implantation techniques and their respective effects on osseointegration. A paramount challenge in implantology is achieving stable and robust adhesion to dentin; hence, surface modification has been implemented to enhance bond strength and ultimately prolong the clinical lifespan of zirconia implant restorations.

Controlled surface roughness has been shown to have a positive impact on cell proliferation and protein synthesis, exceeding the effects of a smooth surface, particularly in the context of bioactive ceramics. The biological properties of bioactive glass can be enhanced through the incorporation of nanoscale features, facilitating the formation of apatite, cell attachment, and proliferation. These surface modification techniques can be classified into three categories: physical, chemical, and biological (refer to Table 25.3). They can be utilized either independently or in combination. Each method possesses its own set of advantages and limitations, underscoring the importance of selecting an appropriate technique

**TABLE 25.3**
**Techniques for Surface Modification of Implants**

| Category | Techniques | Features | Ref. |
|---|---|---|---|
| Physical techniques | Grit blasting | Forcing particles against the surface: simple and cost-effective technique, enhances attachment of osteocytes and bacteria. | Jemat, Ghazali, Razali, and Otsuka [22] (2015) |
| | Additive manufacturing (AM) | Creating intricate 3D structures: straightforward process, conserves energy and materials. | Yuan, Ding, and Wen [23] (2019) |
| | Plasma spraying (PS) | Thermal technique: cost-effective and secure method. | Tang et al. [24] (2013) |
| | Physical vapor deposition (PVD) | Vacuum deposition; employs various inorganic and certain organic materials, ensuring robust adhesion. | Fedges, Wolke, Vredenberg, and Jansen [25] (2004) |
| | Machining | A straightforward approach to augment roughness, albeit with limited efficiency and speed. | Salou, Hoornaert, Louarn, andLayrolle [26] (2015) |
| | Laser treatment | Enables the attainment of intricate and accurate topography, characterized by its expeditious and pristine nature. | Hindy, Farahmand, and Tabatabaei [27] (2017) |
| Chemical techniques | Anodic oxidation (anodization) | A rapid electrochemical process that improves corrosion resistance and produces nanometer-scale features. | Hall et al. [28] (2017) |
| | Chemical vapor deposition (CVD) | Facilitates the formation of a compact and robust film, leading to the creation of both uniform and layered structures. | Li et al. [29] (2013) |
| | Sol-gel | A technique performed at low temperatures, suitable for drug delivery applications. | Adams et al. [30] (2009) |
| | Acid etching | A process that involves the removal of materials and the creation of surface roughness, influenced by factors such as acid concentration, temperature, and duration. | Jemat et al. [22] (2015) |
| | Alkali treatment | Uniformly extends the surface without causing any detrimental effects on mechanical properties. | Yao et al. [31] (2019) |
| Biological techniques | Cells | AMSCs, BMSCs, MSCs, embryonic stem cells | Heng et al. [32] (2011) |
| | Proteins | VEGF, ECM | Lewallen et al. [33] (2015) |

based on factors such as implant materials, application scenarios, and fabrication procedures.

Surface modification techniques, other than laser-based methods, have shown notable improvements in the biological properties of ceramic materials. These enhancements encompass the manipulation of surface roughness and wettability, resulting in heightened osseointegration. Research indicates that increasing the roughness of the implant surface contributes to both initial and long-term stability by facilitating bone anchoring. The primary objective of microtopography is to establish structures at the micro- and macro-cellular levels, aiming to enhance osseointegration. In contrast, nanotopography operates at molecular and atomic scales, promoting cell adhesion and mineralization.

Biocompatibility aside, the occurrence of microbial infection remains a crucial factor directly linked to implant success. Inadequate implant usage and insufficient environmental control can lead to bacterial colonization and biofilm formation, both during and after surgery. For instance, the absorption of salivary pellicle during dental implant healing can result in bacterial accumulation, potentially causing damage to the surrounding tissue and even implant failure. Several approaches exist to reduce adhesion and bacterial proliferation on the implant surface, including chemical modification through the application of antibacterial coatings and physical modification of surface morphology. However, it is important to note that chemical modification can introduce toxicity, exhibit reduced effectiveness over time, and promote the selection of chemical-resistant bacteria.

Conversely, studies have demonstrated the influence of implant surface topography on adhesion and bacterial proliferation. A randomly patterned surface can facilitate bacterial proliferation, while controlled tissue surfaces can minimize bacterial adhesion and the formation of biofilms. Furthermore, research has shown the efficacy of laser surface texturing in reducing bacterial adhesion and biofilm formation during tissue surface fabrication, while also stimulating the growth, proliferation, and metabolic activity of stem cells. Consequently, it can be concluded that laser surface texturing holds significant potential as a method to mitigate the risk of implant contamination.

## 25.2.4 Laser Surface Treatment: Enhancing Bioceramic Properties

Laser surface treatment demonstrates the potential to enhance the properties of bioceramics by modifying their surface topography, chemistry, and microstructure [34–36]. For example, lasers can selectively remove material from bioceramic surfaces, creating a roughened texture that facilitates integration with the surrounding bone tissue [19]. Moreover, lasers can introduce dopants to modify the chemical composition of bioceramics or apply a thin layer of a different material, thereby enhancing their mechanical strength and durability [37,38]. In summary, laser surface treatment holds promise in optimizing the performance and lifespan of biomedical implants, enabling the use of customized materials that result in improved patient outcomes [39].

### 25.2.4.1 Laser-Material Interactions and Mechanisms

Laser-material interactions encompass a wide array of physical and chemical processes that occur when a material is exposed to a high-intensity laser beam. The nature of this interaction is determined by various factors, including the material's properties and the specific characteristics of the laser beam, such as its wavelength, pulse duration, and intensity [40,41]. The interaction between laser beams and materials is governed by multiple mechanisms [42]:

1. Absorption: When a material absorbs a laser beam, it transfers its energy to the electrons present, causing them to transition to higher energy levels. As a result, this phenomenon leads to the heating and melting of the material.
2. Reflection: Specific materials possess the capacity to reflect laser light instead of absorbing it. This occurrence takes place when the laser beam encounters a highly reflective surface, such as a mirror or a polished metal surface.
3. Scattering: In certain cases, the laser beam disperses or scatters off the surface of the material. This scattered light can be utilized for the analysis of the material's properties.
4. Photochemical reactions: Laser light of high intensity has the ability to initiate chemical reactions within the material. For example, the laser beam can facilitate the breaking or formation of chemical bonds.
5. Plasma formation: Under extremely high intensities, the laser light ionizes the atoms of the material, resulting in the formation of a plasma. This process further increases the temperature and damages the material.

In conclusion, the interactions between lasers and materials are complex and depend on various factors. Understanding these mechanisms allows scientists and engineers to develop lasers customized for specific purposes, including cutting [43], welding [44], drilling [45], or materials analysis [46].

Laser material processing commonly utilizes two distinct pulse durations: long pulse durations, such as nanosecond pulses, and short pulse durations, including picosecond and femtosecond lasers. These pulse durations are suitable for achieving precise micro- and nanomachining [47–51]. In the case of nanosecond laser ablation, the process involves "melt expulsion" caused by the vapor pressure and recoil pressure induced by the laser. However, this process is often unstable, resulting in the re-solidification of the melted material. The dynamics of the fluid phase and the associated vapor conditions during this process are intricate, leading to imprecise and non-uniform ablated areas on the material's surface compared to those created by a femtosecond laser [52] (see Figure 25.3). Additionally, nanosecond laser ablation generates a heat-affected zone (HAZ).

### 25.2.4.2 Laser Surface Texturing: Versatile Patterns and Applications

Laser surface texturing (LST) utilizes a laser beam to create textures or patterns on material surfaces through the process of laser ablation [54]. This technique enables the removal of surface materials and the generation of diverse complex patterns with distinct geometries, such as linear grooves, crossed grooves, needle, and honeycomb shapes (as illustrated in Figure 25.4) [55,56]. Due to its exceptional controllability, high efficiency, accuracy, and favorable biocompatibility, the LST method has been extensively employed by researchers [57–59]. Its efficacy is evident in various engineering materials, including ceramics, metals, and polymers, as well as its applications in biomedical, coating, and tribology engineering [57,60].

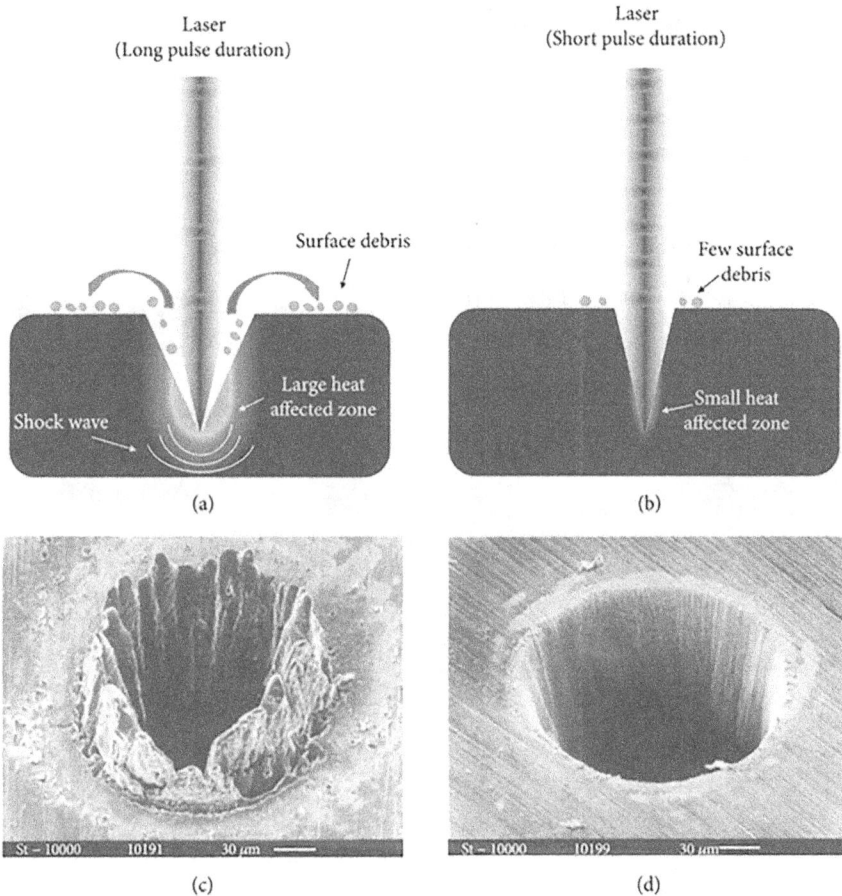

**FIGURE 25.3** Comparing the Effects of Long and Ultrashort Laser Pulses on Matter: (a) Long Pulse Duration vs. (b) Short Pulse Duration. SEM Images of Laser-Ablated Holes Fabricated on 100 μm Steel Foil by: (c) 780 nm Nanosecond Laser (3.3 ns, 0.5 J/cm2) (d) 780 nm Femtosecond Laser (200 fs, 0.5 J/cm2) [53].

### 25.2.4.2.1 Laser Ablation Techniques and Surface Roughness in Medical Applications

To create desired patterns using laser ablation, three primary techniques are employed: a) direct laser ablation, b) direct laser interference patterning (DLIP), and c) laser shock peening [61–63]. Recently, the impact of laser surface texturing (LST) on material surface roughness has been studied, particularly for medical purposes. Bioinert materials, such as zirconia, zirconia-alumina, and alumina composites, are considered ideal for medical and orthopedic applications. While their bioactivity may decrease bone formation efficiency, modifying the surface to increase roughness can enhance bone adhesion.

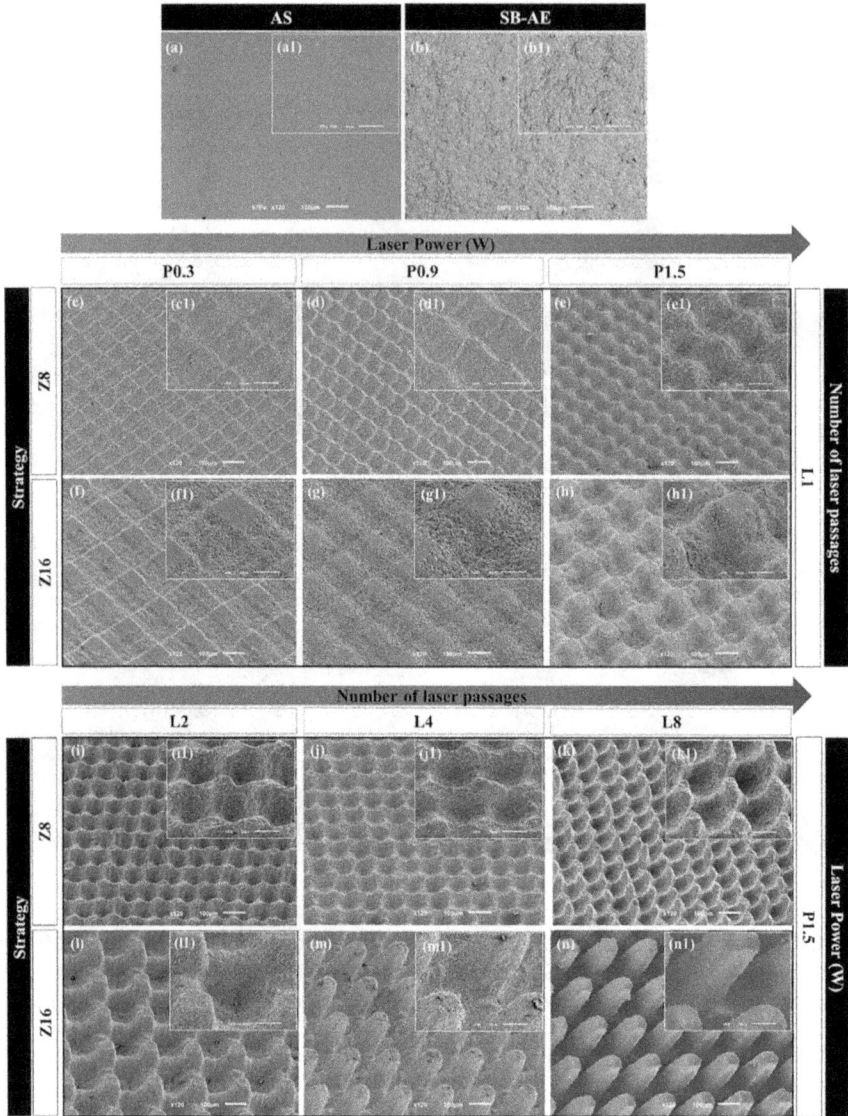

**FIGURE 25.4** SEM micrographs of laser-generated textures on zirconia surfaces as a function of laser power and number of laser passes [61].

*25.2.4.2.1.1 The LST Method: Laser Ablation, Surface Textures, and Applications* The LST method involves using a direct laser beam to remove materials from a solid surface. The laser energy is absorbed by the substrate, exciting its electrons and raising its thermal energy. The resulting temperature distribution depends on the heat balance determined by the heat flux. As the laser stimulates the surface, it heats up and can melt or evaporate. At high laser flux levels, an electromagnetic field causes ionization and plasma formation, generating heat [64].

Laser ablation is a process that uses a high-powered laser to remove material from a surface. The amount of material removed depends on several factors, including the intensity, pulse length, and wavelength of the laser, as well as the properties of the material being ablated [65]. In addition to removing material, laser ablation can create surface textures such as micro-dimples or grooves. The laser energy vaporizes the material, creating small pits or craters on the surface. These textures have various applications, such as improving friction or adhesion properties.

In recent studies, researchers have explored the application of laser surface texturing in various fields. For example, Francesco Baino et al. [62] implemented a consistent pattern of square-shaped "hills" and depressions on the outer surface of alumina/zirconia composite ceramics, specifically designed for hip joint prosthetics. Min Ji et al. [63] utilized a picosecond laser to generate micro grooves on zirconia surfaces, revealing the significant influence of these micro textures on the wettability and tribology of zirconia ceramics. Wassmann et al. [64] examined the impact of surface texture and wettability on initial bacterial adhesion to dental implants made of titanium and zirconium oxide. They observed that surface roughness (Ra) had no effect on the amount of *S. epidermidis* adhering to titanium or zirconia. However, S. epidermidis exhibited greater initial adhesion on hydrophobic surfaces compared to hydrophilic surfaces due to its hydrophobic properties. Carvalho et al. [65] conducted a study in which zirconia surfaces were functionalized with HAp using a hybrid laser technique that combined additive (laser sintering) and subtractive (laser machining) methods. The laser-generated line structures were textured through LST in both air and underwater, and a $CO_2$ laser was employed to sinter HAp powder onto the zirconia surfaces. Elia Marin et al. [66] utilized laser patterning to create cylindrical cavities on ZTA surfaces, which were subsequently filled with bioglass mixed with varying fractions of $Si_3N_4$ powder. Their findings illustrated that an increase in the fraction of $Si_3N_4$ resulted in enhanced cell proliferation on the ZTA surfaces.

*25.2.4.2.1.2 Direct Laser Interference Patterning (DLIP)*   Direct laser interference patterning (DLIP) is a technique employed in the field of laser surface texturing (LST), enabling the creation of periodic surface geometries such as lines, dots, and cross-like patterns that consist of micro- and nanoscale structures [67]. This method involves using a beam splitter to divide a laser beam into multiple beams, which are then merged and directed onto the surface of a sample using reflective mirrors. The size of the beam and the number of pulses can be controlled using a mechanical shutter and telescope, respectively. Figure 25.5 illustrates a schematic representation of the interference patterns generated during this process. DLIP offers a single-step solution for fabricating surface patterns on metals, polymers, and ceramics.

Claus Daniel and his team [69] utilized two beams with varying angles and low and high intensity to create a periodic line pattern on zirconia. The lines were spaced at a distance of 3.3 μm. XRD analysis in Figure 25.6 revealed that the laser processing did not induce any non-equilibrium phase transformations or alter the chemical composition, ensuring the preservation of yttria content that contributes to material stability. The internal structures of the material were affected by the number of laser pulses and the fluence [70,71].

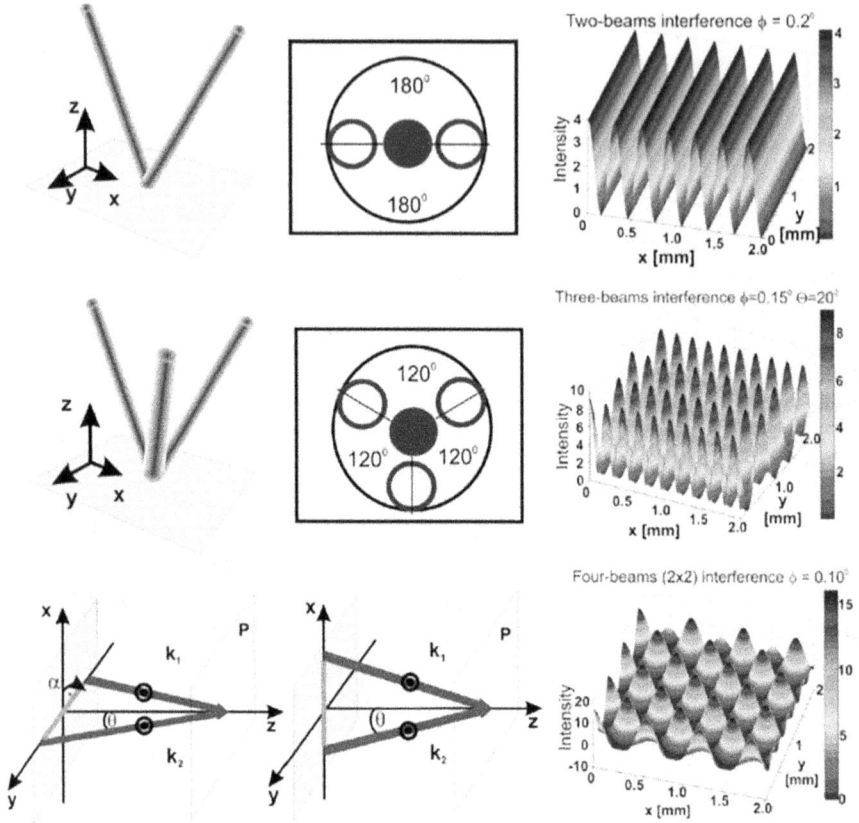

**FIGURE 25.5**    Interference patterns created by multiple coherent laser beams [68].

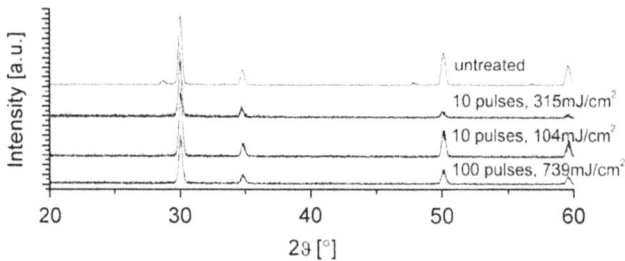

**FIGURE 25.6**    XRD patterns of laser DLIP tape-cast zirconia.

Claus et al. [72] conducted an investigation in which they utilized the DLIP technique to improve the mechanical properties of specific materials. The results showed a remarkable enhancement in fracture strength, with zirconia exhibiting an approximate 50% improvement and alumina showing a 40% improvement. In another study by Roitero et al. [73], it was demonstrated that the annealing treatment had a significant impact on laser-patterned Tetragonal Zirconia Polycrystal stabilized with

3 mol. % $Y_2O_3$ (3Y-TZP). The presence of a microcracks network on the surface of the laser-patterned zirconia caused a delayed transformation from the tetragonal to monoclinic phase.

*25.2.4.2.1.3 Laser Shock Peening (LSP)*   Laser shock peening (LSP) is a sophisticated technique widely used in various industries to enhance the mechanical properties of materials. This process involves directing precise laser pulses onto the surface of the material to induce controlled pre-stress, resulting in improved performance, quality, and durability [74,75]. Pre-stressing material is a well-recognized method for strengthening engineering components. It entails the impact of high-velocity particles or balls onto the material, compressing the surface layer and introducing residual stresses that contribute to enhanced performance, as illustrated in Figure 25.7.

Laser shock peening (LSP) represents an updated approach to conventional shot peening. Instead of using high-velocity balls, LSP employs pulsed lasers to generate shock waves, which introduce plastic deformation and residual stress into the material [77–79]. Compared to traditional shot peening, this method offers numerous advantages, including the ability to induce higher residual stresses, penetrate greater depths, and provide increased flexibility and control over the treatment process. However, it is important to note that LSP is associated with higher economic costs compared to traditional shot peening [80]. Laser shock peening is widely applied in the industry for treating metallic alloys, and multiple studies have investigated its effects on these conventional engineering materials. The capacity of LSP to modify the internal stresses of a specific surface is recognized for its potential to enhance various properties, such as fatigue resistance, fracture toughness, hardness, and corrosion behavior [81–83]. Nevertheless, the impact of LSP or LST on engineering ceramics remains uncertain, highlighting the need for further exploration in this field.

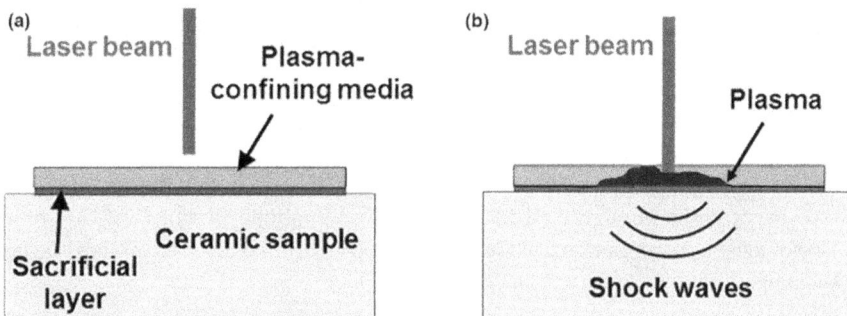

**FIGURE 25.7**   Illustration of Laser Shock Processing (LSP) in Ceramics: (a) A sacrificial layer and a plasma-confining medium are applied to the ceramic sample; (b) laser irradiation leads to plasma formation, and the resulting explosive expansion of the plasma generates shock waves that propagate into the bulk material [76].

Akita et al. [84]conducted a study applying laser shock peening (LSP) to polycrystalline $Si_3N_4$ ceramics at room temperature. X-ray diffraction results revealed that the application of compressive residual stresses ranging from 200–500 MPa enhanced the bending strength and Weibull modulus of $Si_3N_4$ ceramics. In a separate study, Shukla et al. [74] examined the effects of laser shock peening on alumina ceramics and found that adjusting LSP parameters could increase surface hardness by up to 100% and fracture toughness by 42%. Additionally, they performed $CO_2$ and fiber laser surface treatments on $ZrO_2$ and $Si_3N_4$ ceramics, respectively, and observed that different laser wavelengths and beam delivery systems influenced the modification of fracture toughness [85,86].

While these studies attributed the improved mechanical properties to the compressive residual stress induced by LSP, they did not investigate the microscale mechanisms underlying these effects or explore the structure-property relationship of LSP-treated ceramics [85]. LSP surpasses mechanical shot peening due to its ability to achieve deeper penetration, higher magnitudes of compressive residual stresses, faster processing speed, ease of control, and lack of contamination from shot peening materials [87,88]. Previous investigations on mechanical shot peening of structural ceramics such as $Si_3N_4$ and $Al_2O_3$ demonstrated that high compressive residual stresses of up to 2 GPa could be induced into the surface layer, resulting in a significant increase in near-surface strength and cyclic load capacity [89].

Various methods exist for introducing compressive surface stress in ceramics, including thermal quenching, surface chemical reactions, and stress-induced phase transformations [90–92]. However, laser shock peening (LSP) offers distinct advantages over these methods, such as deeper penetration of compressive residual stresses, faster processing speed, and easier control. The application of compressive surface stress can enhance the mechanical properties of ceramics, including fracture resistance, Weibull modulus, thermal shock resistance, wear resistance, and resistance to contact damage [93].

Experimental studies have shown that the benefits of LSP are linked to microstructural modifications near the material's surface during the process. In metallic systems, laser-generated shock waves induce significant plastic deformation, resulting in dislocations, deformation twins, and grain refinement [94–96]. Understanding the evolution of microstructure during laser-driven shock wave interactions in brittle ceramics is still limited. However, Wang et al. [97] have made significant progress in establishing a fundamental understanding of the relationship between processing, microstructure, and properties of ceramics such as polycrystalline alumina through the LSP process. Based on their findings, $Al_2O_3$ grains undergo elastic deformation, while plastic deformation may occur along grain boundaries.

## 25.3 CONCLUSION

The paper presents a comprehensive review of laser surface texturing (LST) in the context of biomedical applications. The author explores the utilization of different laser types, such as carbon-dioxide, excimer, and fiber lasers, to generate surface textures and evaluate their impact on crucial factors like proliferation, osseointegration, and cell adhesion. This extensive review primarily focuses on widely employed biomaterials like titanium, its alloys, and zirconia, effectively showcasing their significant contributions through well-documented micrographs and plots. Moreover, the paper delves into detailed studies of other materials utilized in biomedical applications, thereby highlighting the notable advancements achieved and emphasizing the essential role of research and development in advancing the field of LST.

a. Despite the extensive research conducted on the biomedical applications of titanium, its alloys, and zirconia, there still exists a substantial knowledge gap regarding the influence of laser parameters on the texturing process. It is imperative to conduct a thorough analysis of surface properties, including roughness, wettability, and hardness, to enhance the overall effectiveness of the process and enable its widespread adoption in this emerging field.

b. Further investigation is warranted to comprehend the underlying physics of laser surface texturing in other materials, accompanied by rigorous clinical trials to validate their suitability within the biomedical domain. It is worth noting that laser surface texturing has emerged as a promising technique for modifying biomaterial surfaces, demonstrating significant potential for successful integration into diverse biomedical applications.

## REFERENCES

1. S.M. Best, A.E. Porter, E.S. Thian, J. Huang, Bioceramics: Past, present and for the future, *Journal of the European Ceramic Society* 28(7) (2008) 1319–1327.
2. E. Champion, Sintering of calcium phosphate bioceramics, *Acta Biomaterialia* 9(4) (2013) 5855–5875.
3. A.J.J. Zhou, S.A.F. Peel, C.M.L. Clokie, An evaluation of hydroxyapatite and biphasic calcium phosphate in combination with Pluronic F127 and BMP on bone repair, *Journal of Craniofacial Surgery* 18(6) (2007) 1264–1275.
4. N. Jafari, M.S. Habashi, A. Hashemi, R. Shirazi, N. Tanideh, A. Tamadon, Application of bioactive glasses in various dental fields, *Biomaterials Research* 26(1) (2022) 31.
5. C. Piconi, A.A. Porporati, *Bioinert ceramics: Zirconia and alumina*, Handbook of Bioceramics and Biocomposites, Springer 2016, pp. 59–89.
6. S. Syam, Y.-C. Cho, C.-M. Liu, M.-S. Huang, W.-C. Lan, B.-H. Huang, T. Ueno, C.-H. Tsai, T. Saito, M.-S. Chen, An innovative bioceramic bone graft substitute for bone defect treatment: In vivo evaluation of bone healing, *Applied Sciences* 10(22) (2020) 8303.
7. X. Wang, J. Xue, B. Ma, J. Wu, J. Chang, M. Gelinsky, C. Wu, Black bioceramics: combining regeneration with therapy, *Advanced Materials* 32(48) (2020) 2005140.
8. X. Li, M. Wang, Y. Deng, X. Chen, Y. Xiao, X. Zhang, Fabrication and properties of Ca-P bioceramic spherical granules with interconnected porous structure, *Acs Biomaterials Science & Engineering* 3(8) (2017) 1557–1566.
9. M. Hosseini, J. Khalil-Allafi, M. Etminanfar, M.S. Safavi, N. Bloise, A. Ghalandarzadeh, Tackling the challenges facing the clinical applications of pure PEO hydroxyapatite layers: Co-deposition of YSZ nanoparticles, *Materials Chemistry and Physics* 293 (2023) 126899.

10. M. Giulio, I. Pierfrancesco Rossi, R. Luca, M. Paolo Francesco, Alumina and zirconia ceramic for orthopaedic and dental devices, in: P. Rosario (Ed.), *Biomaterials*, IntechOpen, Rijeka, 2011, p. Ch. 15.

11. A. Ghalandarzadeh, J. Javadpour, H. Majidian, M. Ganjali, The evaluation of prepared microstructure pattern by carbon-dioxide laser on zirconia-based ceramics for dental implant application: an in vitro study, *Odontology* (2022).

12. R.A. Shakir, R. Géber, Structure and properties of $ZrO_2$-$Al_2O_3$-MgO porous ceramic for biomedical applications, *Results in Engineering* 18 (2023) 101104.

13. K.K. Mallick, J. Winnett, 6 - 3D bioceramic foams for bone tissue engineering, in: K. Mallick (Ed.), *Bone Substitute Biomaterials*, Woodhead Publishing, 2014, pp. 118–141.

14. S. Davaie, T. Hooshmand, S. Ansarifard, Different types of bioceramics as dental pulp capping materials: A systematic review, *Ceramics International* 47(15) (2021) 20781–20792.

15. G. Kaur, V. Kumar, F. Baino, J.C. Mauro, G. Pickrell, I. Evans, O. Bretcanu, Mechanical properties of bioactive glasses, ceramics, glass-ceramics and composites: State-of-the-art review and future challenges, *Materials Science and Engineering: C* 104 (2019) 109895.

16. M. Tavoni, M. Dapporto, A. Tampieri, S. Sprio, Bioactive calcium phosphate-based composites for bone regeneration, *Journal of Composites Science*, 2021.

17. O. Peitl, E. Dutra Zanotto, L.L. Hench, Highly bioactive P2O5–Na2O–CaO–SiO2 glass-ceramics, *Journal of Non-Crystalline Solids* 292(1) (2001) 115–126.

18. S. Bose, S.F. Robertson, A. Bandyopadhyay, Surface modification of biomaterials and biomedical devices using additive manufacturing, *Acta Biomaterialia* 66 (2018) 6–22.

19. G. Soon, B. Pingguan-Murphy, K.W. Lai, S.A. Akbar, Review of zirconia-based bioceramic: Surface modification and cellular response, *Ceramics International* 42(11) (2016) 12543–12555.

20. B. Ben-Nissan, Nanoceramics in biomedical applications, *MRS Bulletin* 29(1) (2004) 28–32.

21. X. Gao, M. Fraulob, G. Haiat, Biomechanical behaviours of the bone-implant interface: A review, *Journal of The Royal Society Interface* 16 (2019) 20190259.

22. A. Jemat, M.J. Ghazali, M. Razali, Y. Otsuka, Surface modifications and their effects on titanium dental implants, *BioMed Research International* 2015 (2015).

23. L. Yuan, S. Ding, C. Wen, Additive manufacturing technology for porous metal implant applications and triple minimal surface structures: A review, *Bioactive Materials* 4 (2019) 56–70.

24. Z. Tang, Y. Xie, F. Yang, Y. Huang, C. Wang, K. Dai, X. Zheng, X. Zhang, Porous tantalum coatings prepared by vacuum plasma spraying enhance bmscs osteogenic differentiation and bone regeneration in vitro and in vivo, *PloS One* 8(6) (2013) e66263.

25. B. Feddes, J.G.C. Wolke, A.M. Vredenberg, J.A. Jansen, Initial deposition of calcium phosphate ceramic on polyethylene and polydimethylsiloxane by rf magnetron sputtering deposition: The interface chemistry, *Biomaterials* 25(4) (2004) 633–639.

26. L. Salou, A. Hoornaert, G. Louarn, P. Layrolle, Enhanced osseointegration of titanium implants with nanostructured surfaces: An experimental study in rabbits, *Acta biomaterialia* 11 (2015) 494–502.

27. A. Hindy, F. Farahmand, F.S. Tabatabaei, In vitro biological outcome of laser application for modification or processing of titanium dental implants, *Lasers in Medical Science* 32(5) (2017) 1197–1206.

28. D.J. Hall, R.M. Urban, R. Pourzal, T.M. Turner, A.K. Skipor, J.J. Jacobs, Nanoscale surface modification by anodic oxidation increased bone ingrowth and reduced fibrous tissue in the porous coating of titanium–alloy femoral hip arthroplasty

implants, *Journal of Biomedical Materials Research Part B: Applied Biomaterials* 105(2) (2017) 283–290.

29. X. Li, L. Wang, X. Yu, Y. Feng, C. Wang, K. Yang, D. Su, Tantalum coating on porous Ti6Al4V scaffold using chemical vapor deposition and preliminary biological evaluation, *Materials Science and Engineering: C* 33(5) (2013) 2987–2994.

30. C.S. Adams, V. Antoci Jr, G. Harrison, P. Patal, T.A. Freeman, I.M. Shapiro, J. Parvizi, N.J. Hickok, S. Radin, P. Ducheyne, Controlled release of vancomycin from thin sol-gel films on implant surfaces successfully controls osteomyelitis, *Journal of Orthopaedic Research* 27(6) (2009) 701–709.

31. Y.-t. Yao, S. Liu, M.V. Swain, X.-p. Zhang, K. Zhao, Y.-t. Jian, Effects of acid-alkali treatment on bioactivity and osteoinduction of porous titanium: An in vitro study, *Materials Science and Engineering: C* 94 (2019) 200–210.

32. B.C. Heng, P.P. Bezerra, P.R. Preiser, S.K. Alex Law, Y. Xia, F. Boey, S.S. Venkatraman, Effect of cell-seeding density on the proliferation and gene expression profile of human umbilical vein endothelial cells within ex vivo culture, *Cytotherapy* 13(5) (2011) 606–617.

33. E.A. Lewallen, S.M. Riester, C.A. Bonin, H.M. Kremers, A. Dudakovic, S. Kakar, R.C. Cohen, J.J. Westendorf, D.G. Lewallen, A.J. van Wijnen, Biological strategies for improved osseointegration and osteoinduction of porous metal orthopedic implants, *Tissue Engineering Part B: Reviews* 21(2) (2015) 218–230.

34. R. Vilar, A. Almeida, Laser surface treatment of biomedical alloys, Laser surface modification of biomaterials, *Elsevier* 2016, pp. 35–75.

35. A. Ghalandarzadeh, M. Ganjali, M. Hosseini, Effects of surface topography through laser texturing on the surface characteristics of zirconia-based dental materials: surface hydrophobicity, antibacterial behavior, and cellular response, *Surface Topography: Metrology and Properties* 11(2) (2023) 025007.

36. M. Lasgorceix, C. Ott, L. Boilet, S. Hocquet, A. Leriche, M. Asadian, N. De Geyter, H. Declercq, V. Lardot, F. Cambier, Micropatterning of beta tricalcium phosphate bioceramic surfaces, by femtosecond laser, for bone marrow stem cells behavior assessment, *Materials Science and Engineering: C* 95 (2019) 371–380.

37. X. Zhang, S. Pfeiffer, P. Rutkowski, M. Makowska, D. Kata, J. Yang, T. Graule, Laser cladding of manganese oxide doped aluminum oxide granules on titanium alloy for biomedical applications, *Applied Surface Science* 520 (2020) 146304.

38. L. Yin, X.F. Song, Y.L. Song, T. Huang, J. Li, An overview of in vitro abrasive finishing & CAD/CAM of bioceramics in restorative dentistry, *International Journal of Machine Tools and Manufacture* 46(9) (2006) 1013–1026.

39. M. Belwanshi, P. Jayaswal, A. Aherwar, A study on tribological effect and surface treatment methods of Bio-ceramics composites, *Materials Today: Proceedings* 44 (2021) 4131–4137.

40. M.V. Shugaev, C. Wu, O. Armbruster, A. Naghilou, N. Brouwer, D.S. Ivanov, T.J.Y. Derrien, N.M. Bulgakova, W. Kautek, B. Rethfeld, L.V. Zhigilei, Fundamentals of ultrafast laser–material interaction, *MRS Bulletin* 41(12) (2016) 960–968.

41. M.R.H. Knowles, G. Rutterford, D. Karnakis, A. Ferguson, Micro-machining of metals, ceramics and polymers using nanosecond lasers, The *International Journal of Advanced Manufacturing Technology* 33(1) (2007) 95–102.

42. A. Otto, M. Schmidt, Towards a universal numerical simulation model for laser material processing, *Physics Procedia* 5 (2010) 35–46.

43. N. Roy, A. Kuar, S. Mitra, Multi-objective optimization of nanosecond pulsed laser microgrooving of hydroxyapetite bioceramic, *Materials Today: Proceedings* 18 (2019) 5540–5549.

44. T. Morteza, S. Hedayat Mohammad, R. Ali, Laser Welding, in: C. Kavian Omar, C. Ronaldo Câmara (Eds.), *Engineering Principles*, IntechOpen, Rijeka, 2022, p. Ch. 1.

45. X. Jia, Y. Chen, L. Liu, C. Wang, J.a. Duan, Advances in Laser Drilling of Structural Ceramics, *Nanomaterials*, 2022.

46. J.S. Cowpe, R.D. Moorehead, D. Moser, J.S. Astin, S. Karthikeyan, S.H. Kilcoyne, G. Crofts, R.D. Pilkington, Hardness determination of bio-ceramics using Laser-Induced Breakdown Spectroscopy, *Spectrochimica Acta Part B: Atomic Spectroscopy* 66(3) (2011) 290–294.

47. E. Stratakis, A. Ranella, M. Farsari, C. Fotakis, Laser-based micro/nanoengineering for biological applications, *Progress in Quantum Electronics* 33(5) (2009) 127–163.

48. P. Shukla, D.G. Waugh, J. Lawrence, R. Vilar, 10 - Laser surface structuring of ceramics, metals and polymers for biomedical applications: A review, in: R. Vilar (Ed.), *Laser Surface Modification of Biomaterials*, Woodhead Publishing 2016, pp. 281–299.

49. H. Zhu, Z. Zhang, J. Xu, K. Xu, Y. Ren, An experimental study of micro-machining of hydroxyapatite using an ultrashort picosecond laser, *Precision Engineering* 54 (2018) 154–162.

50. M. Fikry, W. Tawfik, M.M. Omar, Investigation on the effects of laser parameters on the plasma profile of copper using picosecond laser induced plasma spectroscopy, *Optical and Quantum Electronics* 52(5) (2020) 249.

51. H. Zhou, C. Li, Z. Zhou, R. Cao, Y. Chen, S. Zhang, G. Wang, S. Xiao, Z. Li, P. Xiao, Femtosecond laser-induced periodic surface microstructure on dental zirconia ceramic, *Materials Letters* 229 (2018) 74–77.

52. R. Le Harzic, N. Huot, E. Audouard, C. Jonin, P. Laporte, S. Valette, A. Fraczkiewicz, R. Fortunier, Comparison of heat-affected zones due to nanosecond and femtosecond laser pulses using transmission electronic microscopy, *Applied Physics Letters* 80(21) (2002) 3886–3888.

53. Z. Lin, M. Hong, Femtosecond laser precision engineering: From micron, submicron, to nanoscale, *Ultrafast Science* 2021 (2021).

54. I. Etsion, State of the art in laser surface texturing, *Journal of Tribology* 127(1) (2005) 248–253.

55. A.G. Demir, P. Maressa, B. Previtali, Fibre laser texturing for surface functionalization, *Physics Procedia* 41 (2013) 759–768.

56. Y. Xing, J. Deng, X. Feng, S. Yu, Effect of laser surface texturing on $Si_3N_4$/TiC ceramic sliding against steel under dry friction, *Materials & Design (1980-2015)* 52 (2013) 234–245.

57. P. Šugár, J. Šugárová, M. Frnčík, Laser surface texturing of tool steel: textured surfaces quality evaluation, *Open Engineering* 6(1) (2016).

58. B. Mao, A. Siddaiah, Y. Liao, P.L. Menezes, Laser surface texturing and related techniques for enhancing tribological performance of engineering materials: A review, *Journal of Manufacturing Processes* 53 (2020) 153–173.

59. A. Riveiro, A.L.B. Maçon, J. del Val, R. Comesaña, J. Pou, Laser surface texturing of polymers for biomedical applications, *Frontiers in Physics* 6 (2018).

60. D. Li, X. Chen, C. Guo, J. Tao, C. Tian, Y. Deng, W. Zhang, Micro surface texturing of alumina ceramic with nanosecond laser, *Procedia Engineering* 174 (2017) 370–376.

61. D. Faria, S. Madeira, M. Buciumeanu, F.S. Silva, O. Carvalho, Novel laser textured surface designs for improved zirconia implants performance, *Materials Science and Engineering: C* 108 (2020) 110390.

62. F. Baino, M.A. Montealegre, J. Minguella-Canela, C. Vitale-Brovarone, Laser surface texturing of alumina/zirconia composite ceramics for potential use in hip joint prosthesis, *Coatings*, 2019.

63. M. Ji, J. Xu, M. Chen, M. El Mansori, Enhanced hydrophilicity and tribological behavior of dental zirconia ceramics based on picosecond laser surface texturing, *Ceramics International* 46(6) (2020) 7161–7169.

64. T. Wassmann, S. Kreis, M. Behr, R. Buergers, The influence of surface texture and wettability on initial bacterial adhesion on titanium and zirconium oxide dental implants, *International Journal of Implant Dentistry* 3(1) (2017) 32.

65. O. Carvalho, F. Sousa, S. Madeira, F.S. Silva, G. Miranda, HAp-functionalized zirconia surfaces via hybrid laser process for dental applications, *Optics & Laser Technology* 106 (2018) 157–167.

66. E. Marin, S. Horiguchi, M. Zanocco, F. Boschetto, A. Rondinella, W. Zhu, R.M. Bock, B.J. McEntire, T. Adachi, B.S. Bal, G. Pezzotti, Bioglass functionalization of laser-patterned bioceramic surfaces and their enhanced bioactivity, *Heliyon* 4(12) (2018) e01016.

67. A. Rosenkranz, M. Hans, C. Gachot, A. Thome, S. Bonk, F. Mücklich, Direct laser interference patterning: Tailoring of contact area for frictional and antibacterial properties, *Lubricants*, 2016.

68. J. Marczak, J. Kusinski, R. Major, A. Rycyk, A. Sarzynski, M. Strzelec, K. Czyż, Laser interference patterning of diamond-like carbon layers for directed migration and growth of smooth muscle cell depositions, *Optica Applicata* 44 (2014) 575–586.

69. C. Daniel, B. Armstrong, J. Howe, N. Dahotre, Controlled evolution of morphology and microstructure in laser interference-structured zirconia, *Journal of the American Ceramic Society* 91 (2008) 2138–2142.

70. E. Roitero, F. Lasserre, M. Anglada, F. Mücklich, E. Jiménez-Piqué, A parametric study of laser interference surface patterning of dental zirconia: Effects of laser parameters on topography and surface quality, *Dental Materials* 33 (2016).

71. J. Berger, M. Holthaus, N. Pistillo, T. Roch, K. Rezwan, A. Lasagni, Ultraviolet laser interference patterning of hydroxyapatite surfaces, *Applied Surface Science* 257 (2011) 3081.

72. C. Daniel, J. Drummond, R. Giordano, Improving flexural strength of dental restorative ceramics using laser interference direct structuring, *Journal of the American Ceramic Society* 91 (2008) 3455–3457.

73. E. Roitero, M. Ochoa, M. Anglada, F. Mücklich, E. Jimenez-Pique, Low temperature degradation of laser patterned 3Y-TZP: Enhancement of resistance after thermal treatment, *Journal of the European Ceramic Society* 38 (2017).

74. P. Shukla, P. Swanson, C. Page, Laser shock peening and mechanical shot peening processes applicable for the surface treatment of technical grade ceramics: A review, *Proceedings of the Institution of Mechanical Engineers Part B Journal of Engineering Manufacture* (2014).

75. A. Gujba, M. Medraj, Laser peening process and its impact on materials properties in comparison with shot peening and ultrasonic impact peening, *Materials* 7 (2014) 7925–7974.

76. F. Wang, C. Zhang, Y. Lu, M. Nastasi, B. Cui, Laser shock processing of polycrystalline alumina ceramics, *Journal of the American Ceramic Society* 100 (2016).

77. A. Clauer, Laser shock peening for fatigue resistance, 1997.

78. X. Hong, S. Wang, D. Guo, H. Wu, J. Wang, Y. Dai, X. Xia, Y. Xie, Confining medium and absorptive overlay: Their effects on a laser-induced shock wave, *Optics and Lasers in Engineering* 29 (1998) 447–455.

79. J. Ruschau, R. John, S. Thompson, T. Nicholas, Fatigue crack nucleation and growth rate behavior of laser shock peened titanium, *International Journal of Fatigue* 21 (1999).

80. C. Montross, T. Wei, L. Ye, G. Clark, Y.W. Mai, Laser shock processing and its effects on microstructure and properties of metal alloys: A review, *International Journal of Fatigue* 24 (2002) 1021–1036.

81. C. Rubio-González, J.L. Ocaña, G. Gomez-Rosas, C. Molpeceres, M. Paredes, A. Banderas, J. Porro, M. Morales, Effect of laser shock processing on fatigue crack

growth and fracture toughness of 6061-T6 aluminum alloy, *Materials Science and Engineering: A* 386 (2004) 291–295.

82. S. Srinivasan, D.B. Garcia, M.C. Gean, H. Murthy, T. Farris, Fretting fatigue of laser shock peened Ti–6Al–4V, *Tribology International - TRIBOL INT* 42 (2009) 1324–1329.

83. Y. Zhang, J. You, J. Lu, C. Cui, Y. Jiang, X. Ren, Effects of laser shock processing on stress corrosion cracking susceptibility of AZ31B magnesium alloy, *Surface and Coatings Technology* 204 (2010) 3947–3953.

84. K. Akita, Y. Sano, K. Takahashi, H. Tanaka, S.I. Ohya, Strengthening of $Si_3N_4$ ceramics by laser peening, *Materials Science Forum* 524 (2006) 141–146.

85. P. Shukla, J. Lawrence, Evaluation of fracture toughness of $ZrO_2$ and $Si_3N_4$ engineering ceramics following $CO_2$ and fibre laser surface treatment, *Optics and Lasers in Engineering* (2011).

86. P. Shukla, J. Lawrence, Fracture toughness modification by using a fibre laser surface treatment of a silicon nitride engineering ceramic, *Journal of Materials Science* 45 (2010) 6540–6555.

87. P. Shukla, *On the fracture toughness of a zirconia engineering ceramic and the effects thereon of surface processing with fibre laser radiation*, 2015.

88. O. Messé, S. Stekovic, M. Hardy, C. Rae, Characterization of plastic deformation induced by shot-peening in a Ni-base superalloy, *JOM: the Journal of the Minerals, Metals & Materials Society* 66 (2014).

89. W. Pfe, T. Frey, Advances in shot peening of silicon nitride ceramics, International Conference and Exhibition on Shot Peening, ICSP9 (2005).

90. J. Absi, J.C. Glandus, Improved method for severe thermal shocks testing of ceramics by water quenching, *Journal of the European Ceramic Society* 24 (2004) 2835–2838.

91. Y. Bao, F. Kuang, Y. Sun, Y.-M. Li, D. Wan, Z. Shen, D. Ma, L. He, A simple way to make pre-stressed ceramics with high strength, *Journal of Materiomics* 5 (2019).

92. E. Liu, G. Xiao, W. Jia, X. Shu, X. Yang, Y. Wang, Strain-induced phase transformation behavior of stabilized zirconia ceramics studied via nanoindentation, *Journal of the Mechanical Behavior of Biomedical Materials* 75 (2017).

93. X. Zhang, P. Zhou, P. Hu, W. Han, Toughening of laminated ZrB2–SiC ceramics with residual surface compression, *Journal of The European Ceramic Society - J EUR CERAM SOC* 31 (2011) 2415–2423.

94. J.Z. Lu, K. Luo, Y. Zhang, G. Sun, Y. Gu, J. Zhou, X.D. Ren, X.-C. Zhang, L.F. Zhang, K.M. Chen, C. Cui, Y. Jiang, Grain refinement mechanism of multiple laser shock processing impacts on ANSI 304 stainless steel, *Acta Materialia - ACTA MATER* 58 (2010) 5354–5362.

95. H. Ding, Y. Shin, Dislocation density-based modeling of subsurface grain refinement with laser-induced shock compression, *Computational Materials Science* 53 (2012) 79–88.

96. C. Ye, Y. Liao, S. Suslov, D. Lin, G. Cheng, Ultrahigh dense and gradient nano-precipitates generated by warm laser shock peening for combination of high strength and ductility, *Materials Science and Engineering: A* 609 (2014) 195–203.

97. F. Wang, Laser shock processing of ceramic materials, (2019).

# 26 Functional Surfaces for Biomaterials

*Saswati Mishra*

Departent of Biotechnology, School of Allied Health Sciences, Malla Reddy University, Hyderabad, Telangana, India

*Tapash R. Rautray*

Biomaterials and Tissue Regeneration Lab, CETMS, Institute of Technical Education and Research, Siksha 'O'Anusandhan (Deemed to be University), Bhubaneswar, Odisha, India

## 26.1 INTRODUCTION

Biomaterials science has proved to be a knife-edge field with sheer, strong growth over time of its existence. It is a multidisciplinary endeavor that encompasses several aspects of biology, medicine, chemistry, and more significantly material science [1]. This science is a culmination of both physical and biological aspects of materials along with its interaction with biological ambiance. Traditionally, the most significant and intense development focused on biomaterials development, testing, characterization, optimization, and most importantly the host material interaction [2]. Gradually, with intense study, most biomaterials showed stereotyped, non-specific biological reactions in the host [3]. So, with time and advanced studies, there was an emergence in the development and designing of engineered surfaces that could evoke highly precise reactions with biomolecules such as proteins and cells, customized for specific applications [4]. Indeed, the major goal turned to generating biocompatible material with pertinent host responses resisting all adverse effects. Such a case in point includes graft thrombosis due to incompatibility of synthetic implant surface that doctored flow dynamics at the site resulting in platelet adhesion and protein adsorption [5]. Bacterial colonization and biofilm formation is one of the leading issues in post-operative bone implant failures. As per reports, 2–14% of implant failure related to hip replacement, bone fracture surgeries, and knee replacement is due to bacterial infection [6]. So, a suitable crafted surface design pivotal for more productive and effective application of biomaterials is essential as surfaces directly interact with the biological medium for a considerable period [7]. Before initiation of any biological pathway in response to the implanted biomaterials, the surrounding cells locate the surface via filopodia of transmembrane-based cellular protein [8].

DOI: 10.1201/9781003429920-29

This is the primary natural way for the cell to investigate its affinity to the implant surface [9]. Several investigations report that both surface structure and chemistry play a role in influencing basic cellular behavior such as adhesion to the surface along with proliferation, migration, and finally differentiation on biomaterials [10]. Surface modification of biomaterials is an extensive subject with several aspects that include: properties of the material to be modified, essential surface features, stabilized modification, more significantly the utilization of the final by-product, and also the feasibility of the process adapted [11]. Both chemical and morphological/physical modification of the surfaces are suggested. In some cases, surface chemical modification is recommended, whereas in certain conditions surface morphology is modified [12]. Surface modification can also be induced by specific functional groups like amine groups, hydroxy, or methyl groups with an objective of promoting the non-specific protein at the cell surface interface. Further, the enhanced technique included biological macro molecules such as heparin or peptides, with a motive of fabricating bio mimetic active surfaces that showed better cell-surface interaction [13]. Polymers, metal, and ceramics are widely used materials in the implant industry and several different specific procedures are applied for the surface modifications. Metal is widely used as implant, and for such surface techniques designated to enhance their torpidity, like oxidation or increase in their bio activity via coating with ceramic or bioceramic with great potential in biomedical application [14–16]. As a whole, surface modification methodologies has evolved within a few decades such as PEG-induced modification, plasma polymerization, oxidation, peptide functionalization, heparinization, calcium phosphate deposition method, and certain surface topographical modification. Hence, this chapter will cover the methods of getting optimized surfaces in metals, polymers, and ceramics along with biological response with insights into future trends in biomaterials and biomedical sciences.

## 26.2　SURFACE FUNCTIONALIZATION BASED ON SURFACE TOPOGRAPHIES

Topographical features such as micro and macro texture influence and alter the cellular and host response, such as adhesion, proliferation, and alignment [17–19]. The surface topography significantly regulates *in vivo* effects that include phenotypic activities such as osteoblastic differentiation and neurite extension. Porosity also is an important parameter that regulates in growth of both hard and soft tissues [20–22]. The surface texture also affects epithelial inflammatory responses to percutaneous-based implants and fibrous encapsulation of subcutaneously fabricated materials [23]. Although the parameters eliciting a biological response to certain surface texture have been recognized, the exact mechanism behind such behavior is yet to be analyzed so several methods for producing surface texture can be classified based on the engineering of several topography and roughness of the target surface. Roughness of any surface is a complex and random pattern with dynamic spacing and amplitude, in a scale ranging from 10–20 um. Whereas surface topography has more organized, controlled, and well defined patterns on the surface, sand blasting, chemical etching, plasma spraying, and mechanical polishing are some of the non-specific methodologies used for modifying complex surface roughness topographies. The following section describes controlled topographies that can be generated by several micro and nano machining methodologies using glass, polymer, ceramics, and silicon as substrates [24].

## 26.2.1 PHOTO LITHOGRAPHY

Photo lithography is one of the pioneer technique, especially for micro topographic surfaces. The preliminary step involves spreading of thin layer photo resist substrate on a silicon wafer. A patterned mask is used to direct light on the resist, leading to formation of photosensitive pattern. The resist analogue to photography and both the exposed and non-exposed areas can be excluded, displaying a topography of resist surrounding exposed areas of silicon. In the final step, the resist is baked hard while the substrate can undergoes etching [25,26]. Etching can be done chemically i.e., wet or dry via the reactive ion bombardment method promoting the removal of hard resist and substrate etching. Etching is stopped in patterned material designed for biological application, to grow cells on their surfaces [27,28]. Currently, this technique can be used to fabricate biological features for 1–2 μm but not for nano features. In contrast, e-beam lithography is showing the potential for better nano-featured structures.

## 26.2.2 LASER HOLOGRAPHY

Laser holography is an interference technique developed to generate mask of ultra fine range for printing X-rays. Two mirror interferometer is used to produce linear patterns for a (period = $\lambda/2$ Sin θ); further samples are coated in photo resist. Depth patterns can be basically developed using dry etching. Patterns produced with a width of 260 nm and 100 nm depth have shown the proliferation of MDCK epithelial, BHK fibroblast, and neurite guidance [29]. However, this technique leads to a limited variety of patterns and it is not cost effective.

## 26.2.3 ELECTRON BEAM LITHOGRAPHY

Electron beam lithography has evolved to be one of the advanced computer-aided techniques to develop photo resist patterns with better resolution and precision. The electron beams are exposed in a controlled manner on the resist, to produce surface topography for nano fabrications. Yet use for biological application is restricted due to distorted beam deflected (>1 mm), restricting the pattern size. But, some recent researchers are using an interferometric control technique that prompts the movement of sample through e-beam writer and knitting the patterns [30]. Such large patterns promote cell testing, adhesion, and measurement of filopodial interaction with diminished focal-cell interaction on the surface [31]. Meng et al. reported a sub-micron fabrication method for polymer bioMEMS by electron beam lithography. They reported thin-film substrates, coated with Parylene C, Poly (chloral-p-xylene) to develop a Parylene-based neural probe as shown in Fig 26.1. The novelty of the work improves the challenges of using electrobeam lithography for polymers [32]. However, cheaper nano topography methodologies like colloidal topography are being preferred in biological sciences.

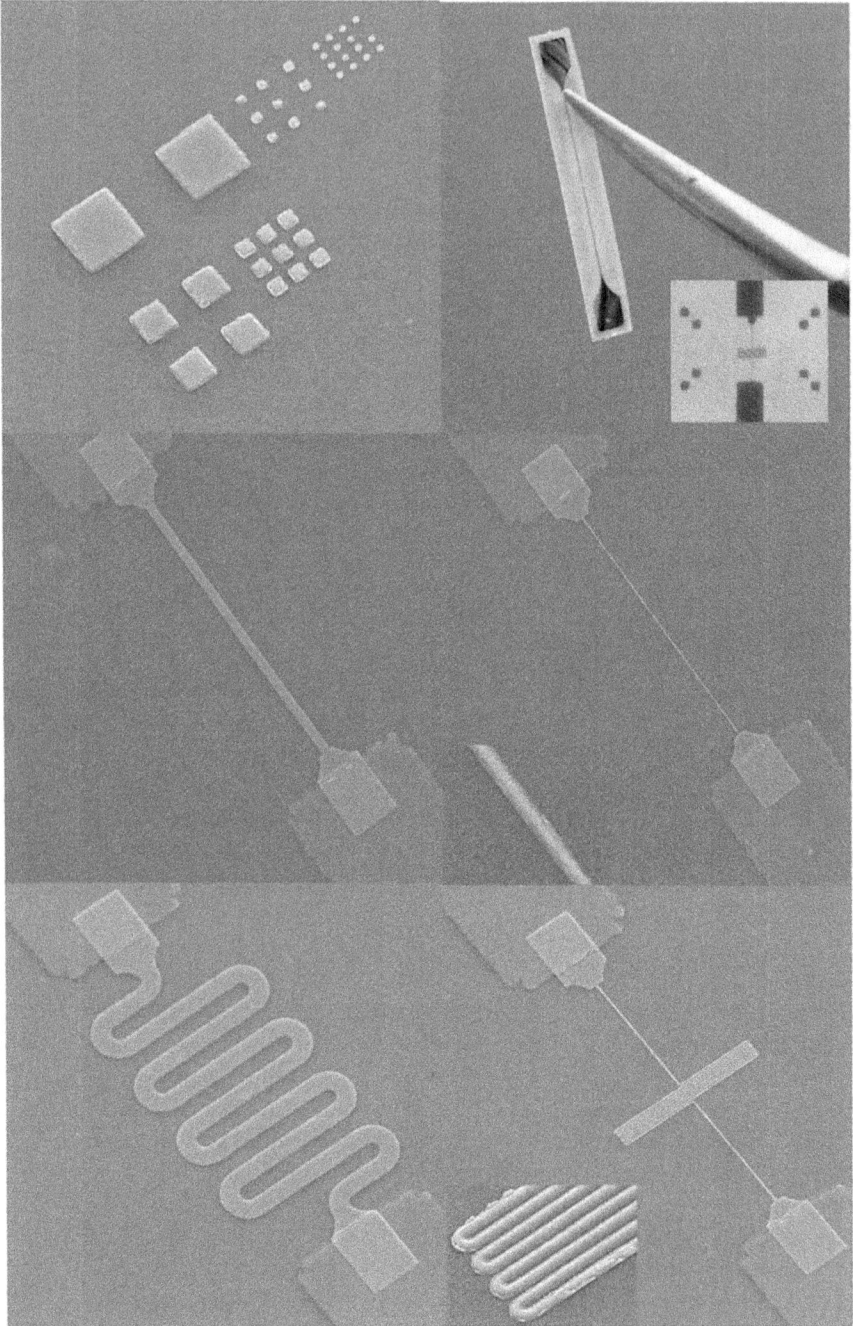

**FIGURE 26.1**    Micrographs of submicron metal structures pattern on Parylene C

*Source:* Scholten, K., Meng, E. Electron-beam lithography for polymer bioMEMS with submicron features. Microsyst Nanoeng 2, 16053 (2016). https://doi.org/10.1038/micronano.2016.53

## 26.2.4  Colloidal Lithography

Colloidal lithography is a technique to generate a large area of nano topographic surfaces in a short range. Nano colloids are utilized as an etching mask, spread over as a mono layer, and self-assembled electrostatically. The ion bombardment method is directed for etching colloid, surrounding the substrate material [33]. Colloidal-based techniques proposed roles in designing model of biomaterials with nano topographies. Such patterned substrate have been developed for biological applications, especially for protein adsorption [34]. This shows great promise for a controlled rate of cell adhesion and proliferation with respect to a give structure, inflammatory response, and altered gene regulation. Colloids used with high atomic numbers can be used to image individuals' nano features of any cyto-skeleton-based filamentous process.

## 26.2.5  Advanced Lithography Techniques

### 26.2.5.1  Soft Lithography

Photolithography is considered to be the primary technique in the field of electronics and biomedical sciences but with several loopholes with rigorous procedures, limiting its application in certain fields. Moreover, such techniques showed no significant control over the surface chemistry. So, unlike the conventional method, researchers are using elastomeric soft organic materials such as complex biochemicals and polymers for the microfabrication of specific patterns through moulding methodologies [35–37]. Soft lithography techniques basically fabricates and replicates structure through molding, embossing, and printing by using a mask, elastomeric stamps, and photo masks. Replica modeling, micro and nano transfer modeling, non-transfer molding, nano-skiving, and SAMIM i.e., solvent-assisted micro molding are some of the advanced patterning techniques [38–40].

## 26.3  SURFACE FUNCTIONALIZATION BASED ON WETTABILITY OF SURFACES

Wettability is an important and fundamental property of solid surfaces playing a significant part in our day-to-day life analyzing several natural surfaces, bio-mimics are being created through advanced technology to create material with apt wettability. Wettability of any surface can be determined by the contact angle (CA) of water with respect to a solid surface. The concept of CA and wettability correlation was drawn from Young's modulus. As per various theories and models, intrinsic wettability is classified into four types, i.e., hydrophobicity, hydrophilicity, oleophobicity, and oleophilicity. It can affect both topography and chemical content of the surface. In nature, we find endless paradigms of materials with unique wettability in insects (Spider, butterfly wings) and flowers (lotus leaf). Such inspirations are taken up by researchers to design surfaces with specific or desirable mutability for multidisciplinary applications. In general, several rational stimuli responsive-based surfaces have been designed. The component used for surfaces is light, electric field, pH, stress, ions, and magnetism. In particular, the fabrication methodologies have been divided into physical and chemical methods, as shown in Table 26.1. With advancement specific wettability surfaces are designed with a wide range of applications i.e., **SLIPS** (Super wetting, Slippery liquid-infused porous surfaces). So, the benefiting effect of wettability materials is gaining attention in biomedical materials [41–43]. The following section emphasizes some of the methods to obtain a surface with apt wettability [44,45].

## 26.3.1   3D PRINTING

3D printing is one of the advanced, revolutionary methodologies serving high efficiency, controlled procedure yet it is regarded as a simple solution for fabrication. They are capable of fabricating surface wettability at nano scale level. Laser polymerization or 2-photon laser writing are some of the progressive techniques of 3D printing. Different 3D-printed patterns such as spine structures, egg bitter structures, and pillar structures can be fabricated with super wettability performance. By adjusting the printing features, the liquid repelling ability of the desired surface can be modified accordingly. Hence, 3D printing is an ideal

---

**TABLE 26.1**
**Different Techniques to Improve Biomaterial Blood Interaction**

| Modifications | Description |
|---|---|
| Physical Immobilization | Polymer gelling (growth factor mixed with the material in the liquid state and change temp. pH or ion concentration to obtain a gel with nanopore |
| | Emulsion techniques (factors which are insoluble in aqueous solutions |
| | High-pressure gas foaming (incorporate GF into porous scaffold, without the use of solvent |
| Covalent modification | Surface distribution of ligands |
| | Distribution of ligands through the bulk of the material |
| Surface adsorption | Passive adsorption is driven by secondary interactions between the molecule and the proteins Self-assembled monolayers (SAMs) adsorption of the peptide(designed with hydrophobic and a spacer) from solution |
| | Micro-contact printing of alkanethiol SAMs, photo lithography (on hard materials), soft lithography on elastomeric materials |
| | Direct protein patterning, drop dispensing, microfluidic patterning |
| Cross linking | Photo/chemical cross-linking |
| Altering surface wettability | Ion bombardment |
| | UV irradiation |
| | Exposure to plasma discharge |
| Altering surface roughness | Deposition of polymer films islands, nanoparticles, metallographic paper or diamond paste Polishing, sandblasting, photolithpgraphy, and e-beam etching |

methodology to fabricate a perfect culmination of super wettability surface with high-resolution patterns of designing. However, 3D printing is time consuming and also an expensive method limiting its application in several areas [46–48].

## 26.3.2 TEMPLATE-BASED METHOD

Template-based methods have become skilled and suitable method to fabricate surfaces with desired wettability. In this method, surface roughness can be tuned to improve the wettability standard of the material by using different templates. Basically, several micro or nano based pattern designs are developed i.e., pillar shape, conical, textured, and honeycomb shape [49]. This method is a combination of three steps, i.e., development of the template with desired morphology, followed by moulding step, and finally removal of the templates. Both natural and artificial materials can be used as template. Artificial templates can be designed by using several lithographies and a microfluidic process [50,51]. Due to its simplicity, reproducibility, and effectiveness, this method is extensively being utilized and developed. However, there is always a risk to the surface, post-removal of the templates.

## 26.3.3 PLASMA TREATMENT

Plasma treatment is another effective way of functionalizing surfaces with super wettability. Constitutes both plasma etching and polymerization to fabricate materials with apt roughness and uniform coating. Plasma etching is suitable for fabricating micro and nanopatterns to enhance the surface roughness as depicted in Fig. 26.2 [52]. On the other hand, plasma polymerization relies on monomers with super wetting nature in the gaseous phase. Plasma treatment subjugates are rapid and selective, but again this involves expensive instrumentation with poor results. Apart from this technique, several conventional methods are used to fabricate materials desired wettability by exploiting the unstable mixture of multi-components. This methodology is termed as phase separation method. It has been instrumental in fabricating super waiting and super hydrophobic surfaces [53]. Spin spraying and spin coating add some of the traditional methods, involving thin films on the surface of substrates. Yet, this technique is found to be unconventional.

## 26.3.4 ELECTROSPINNING

Electrospinning is considered to be a high throughput technique for fabrication of super hydrophobic surface. The basic principle of fabrication involves polymer-based nano structured films deposited with arbitrary geometry on the required surface by controlling the deposition parameters such as applied voltage, given flow rate, and the distance of the needle from the surface and collector in culmination with the polymer composition. This technique results in highly porous structures [54]. So, several polymers such as PCL, polydimethylsiloxane (PDMS), poly-vinylidene fluoride (PVDF)-based nano fibers have been fabricated by electrospun method leading to the synthesis of hydrophobic surfaces with controlled surface roughness, antibacterial property, and adequate wettability [55,56].

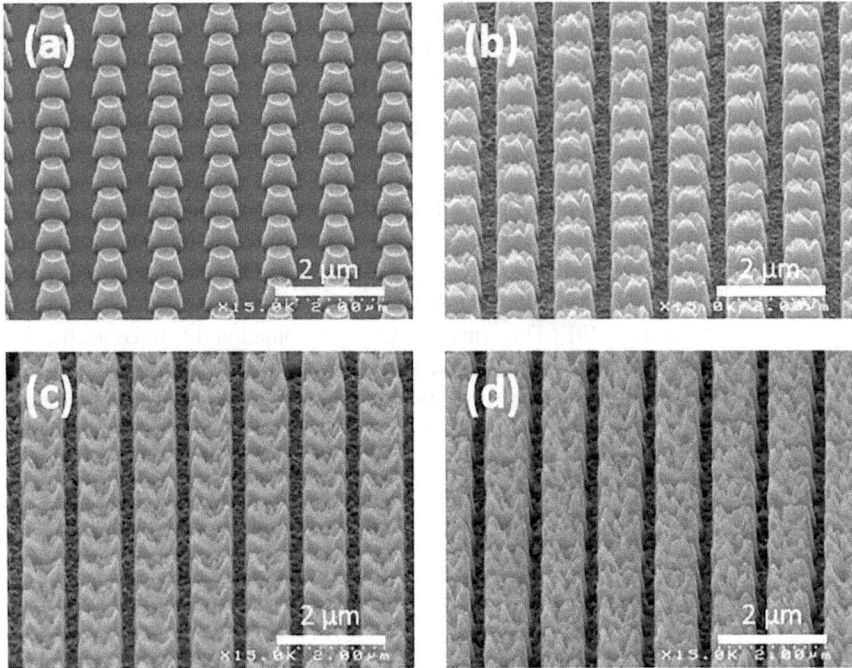

**FIGURE 26.2**    SEM analysis of nano structure patterns fabricated by Plasma Treatment.

*Source:* Durret, J., Szkutnik, P. D., Frolet, N., Labau, S., & Gourgon, C. (2018). Superhydrophobic polymeric films with hierarchical structures produced by nanoimprint (NIL) and plasma roughening. Applied Surface Science, 445, 97–106.

### 26.3.5  SOL GEL METHODS

Sol-gel is a chemical process of modification to alter the surface wettability of the surface. They are even considered as the primary methodology to fabricate super hydrophobic, hydrophilic and oleophobic surfaces [57,58]. This method is apt for manufacturing wettability surfaces based on composite materials. Layer by layer, one of such traditional techniques, is used to fabricate wet structures [59,60].

## 26.4  SURFACE MODIFICATION TO IMPROVE BLOOD MATERIAL INTERACTION

Biological-based surface modification have gained attention in the field of biomedical application will stop bio-molecule immobilized surfaces such as ligand, antibodies enzymes, pharmacological process, and diagnostic devices such as chips and sensors. The main principle of these materials includes combination of booth biological and synthetic components to enhance bio functionality and surface engineering that is suitable for a desirable biological response as shown in Table 26.2. Heparin immobilized onto polymer surfaces to improve the blood material compatibility. Polymer-based stents quoted with anti-hyper-plasic based

**TABLE 26.2**
**Biomolecules Used in the Biofunctionalization of Surfaces**

| Biomolecule | Application |
|---|---|
| Heparin | Blood-compatible surfaces: growth factor immobilization |
| Fibronectin, collagen Arginine-glycine-aspartate peptides | Cell adhesion and function in biosensors, arrays, devices, and tissue-engineered constructs |
| Antibodies | Biosensors;bioseparations;anticancer treatments |
| DNA plasmids Antisense oligonucleotides Small interfering RNA | Gene therapy for a multiple of diseases; DNA probes |
| Growth factor proteins and peptides | Anti cancer treatments; treatments for autoimmune and inflammatory conditions; enhanced wound repair |
| Enzymes | Biosensors;Bioreactors; anticancer treatment; anti thrombotic surfaces |
| Drugs and Antibiotics | Anti thrombotic agents; anticancer agents; a ntihyperplasia treatments; anti infection/inflammation treatments |
| Polysaccharides | Non-fouling supports for biosensors and bioseparations |

drugs, have been developed to decrease restenosis and achieve better patency. Apart from that adhesive ligand [adsorbed protein, laminin, and fibronectin] or synthetic tethered oligonucleotide RDG (arginine-glycine-aspartate) that promotes cell addition, which is applied in several tissue engineering and regeneration studies will stop important factors responsible for surface modification to improve blood material interaction are distribution, density, activity of the immobilized biomolecules. The following section elaborates on the method to immobilize important biomolecules on the surface by physical adsorption and physical entrapment.

### 26.4.1 PHYSICAL ADSORPTION

Physical adsorption is the facile efficient technique to immobilize biomacromolecules, i.e, protein, nucleic acid, and polysaccharide in order to fabricate biologically active surface. Conventional application of ECM best proteins collagen and fibronectin coating on the surface materials improvises cell adhesion properties. Protein addition is itself a complicated, dynamic procedure that includes electrostatic bonds, and hydrophobic bonds along with van der Waals forces with hydrogen bonding [60]. Structural parameters of proteins such as primary structure, structural stability, and size along with surface characteristics such as surface energy and its chemistry determine the biological activity of the biomolecules adsorbed to the surface. Such biological surfaces can be modified by displaying adsorbed biomolecules, cell mediated depositions, and matrix component remodeling in the biological mixture. This adsorption is achieved by crosslinking biological molecules to the surfaces. Physical immobilization of high affinity interaction of avidin and antigen-antibody strategies are used for both processing and diagnostic methods.

### 26.4.2  PHYSICAL ENTRAPMENT METHOD

It depends on matrix systems or diffusion barriers to regulate the bio-availability of biomolecules. Enzyme entrapment by sol-gel [nano scale level porosity] and protein therapeutics or drug encapsulation for improvised stability, segregation, and recovery of a biological medium in a regulated release kinetics [61]. The engineered encapsulated system can be utilized for biomolecule isolation or nonspecific degradation such as hydrolysis or specific enzyme-based degradation.

### 26.4.3  COVALENT IMMOBILIZATION

Covalent immobilization of soluble polymers combined with biomolecules can be Functionalized as the graft or network onto a biomaterial support involving carbodiimide CNBr and N-hydroxy sulfosuccinimide. Many biomolecules are immobilized via PEG leading to steric freedom and activities [62]. A strong relationship exists between PEG chain and resistance to protein adsorption leading to cell adhesion. Moreover, hydrophilic polymers such as polyacrylamide, phosphorylcholine, and poly carboxybetain, were resistant to protein adsorption [63–65]. Recent methodologies combine the active and passive features i.e., inflammation or bio-therapeutics regulate protein adsorption, self-cleaning biomaterials provide better reach to lessen protein adsorption successfully, enabling better presentation of bioactive particles.

## 26.5  SURFACE MODIFICATION WITH BIOACTIVE TO CONTROL INFECTION

Implant-based infection is one of the leading concerns in biomedical or regenerative science. Bacterial attachment, followed by colonization of bacteria leads to the formation of biofilms. Biofilms formed on a habitat, have strong colonies of both dead and live bacteria and a resistance to any stringent treatment. So, the most rational way of treating implant-associated infections, is to create biofilm resistance surfaces that prevent any growth rather than treating it with antibiotics. Advanced approaches concentrate on surface modification strategies basically anti-bacterial coatings onto the material surfaces. The general principle of such surfaces is to create anti-bacterial surfaces that prevent their attachment along with bacteriostatic and more significantly bactericidal effects [66,67]. Several such coating strategies include anti-adhesive polymer coating, contact killing coating, and anti-microbial agent cashes with controlled release kinetics.

### 26.5.1  COATINGS BASED ON HYDROPHOBIC POLYMER

Coatings based on hydrophobic polymer have been exclusively explored strategies for fabrication of antibacterial surfaces with protein resistant and non-fouling characteristics. But there are still challenges faced IN creating such coatings on polymer-based materials. Ran et al. translated zwitterionic coating i.e., PDA-associated deposition of branched copolymer of polyethylenimic-g- poly with sullfobetaine [68]. Such coatings showed high bacterial resistance i.e, 95% in short-term and 90% in long-term exposure [69,70]. Several such strategies involving copolymers were reported such as by Wang et al. who developed PEI-g-PMOXA i.e., poly(ethyleneimine) graft poly (2 methyl-2-oxazoline) showed enhanced resistance to deposition of protein. Xu et al reported the use of dopamine (polymerization initiator) assisting co-deposition of PDA via hydrogen bonding with a hydrophilic polymer as shown in Fig 26.3 [71–73]. It showed activity against 75% *P.aeruginosa* and 85% *E. coli*. Antibacterial-coated surfaces have shown promising applications in medical implants.

### 26.5.2 ANTI ADHESIVE HYDROGELS

Surface coated with hydrogel showed resistance to protein adsorption with antibiotic activity. Silicon catheters coated with antibacterial conformal gel coating [74–76]. This coating was prepared by a combination of shape-forming gel with oxygen inhibition effect of UV-based system, generating hydrogels with better mechanical stability. Hydrogel suppresses *E. coli* and *S. aureus* attachment for a short period. Another group reported robust Zwitterionic i.e., DURA-Z hydrogel antifouling coating applied especially on plain surfaces [77].

#### 26.5.2.1 Antimicrobial Peptide (AMP) Based Conjugates

Application of AMP is an alternative technology for the fabrication of antimicrobial based coatings on devices showing bio compatibility in several assays. Costa et al

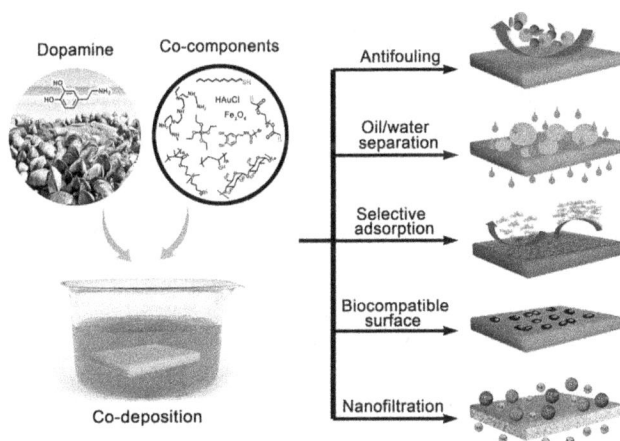

**FIGURE 26.3** Dopamine-assisted co-deposition: An emerging and promising strategy for surface modification.

*Source:* Wen-Ze Qiu, Hao-Cheng Yang, Zhi-Kang Xu, Dopamine-assisted co-deposition: An emerging and promising strategy for surface modification, Advances in Colloid and Interface Science, Volume 256,2018, PHey ages 111–125, https://doi.org/1 0.1016/j.cis.2018.04.011

showed immobilization of antimicrobial peptide sequence LLLFLLKKRKY i.e., Dhvar % with a cationic exposure, that improves antibacterial effect of chitosan coatings [78,79]. Lactoferin derived hLF1 conjugated with MPA-Ahx-Ahx-GRR-RRSVQWCA-NH2) used by Gallardo et al showed better anti-bacterial performance, using boot polymer polymer brushes or silanization [80,81]. AMP conjugated through polymer brushes shows better performance than silanization method. Thus polymeric surface interface shows modular characteristics, compliant with the functionalization of any thiol conjugated AMP showing diversified effect on multiple strains of bacteria.

### 26.5.3    NANOPILLAR ARRAY BASED COATINGS

Studies report that the wings of many insects like Dragonfly have enhanced antibacterial effects by virtue of nanopillar structures. Such a sharp structure results in a strong bactericidal effect. Many researchers report bio-mimicking of such nanopillars using silver, gold, graphene, and titanium material for the fabrication of antibacterial surfaces, as shown in Fig 26.4 [82]. Hasan et al fabricated silicon-based super hydrophobic coated nano-pillars [83–85]. Again another group of vertically aligned graphene sheets with sharp topographies when exposed to *E. coli* and *Staphylococcus stains* showed elevated bactericidal effect. ZnO-based nano-pillars along with ultra-thin layers of silver, and copper nanopillars are also fabricated and were also found to be very effective.

### 26.5.4    ANTIMICROBIAL COATINGS FOR BACTERICIDES

Another strategy included the introduction of nano particles-based coatings that lead to the effective killing of bacterial strains responsible for implant-based infection. Nanocomposite-based based coatings, coatings releasing AMPs, and coatings releasing antibiotics are some of the antimicrobial coating strategies.

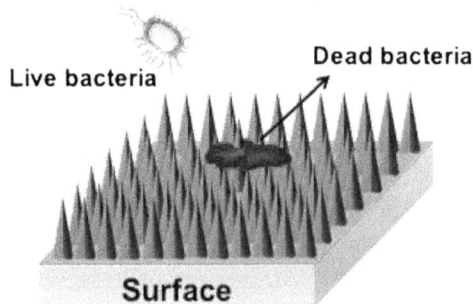

**FIGURE 26.4**   Sharp nano pillar structure showing bactericidal effect.

*Source:* Ahmadabadi, H. Y., Yu, K., & Kizhakkedathu, J. N. (2020). Surface modification approaches for prevention of implant-associated infections. Colloids and Surfaces B: Biointerfaces, 193, 111116. https://doi.org/10.1016/j.colsurfb.2020.111116

### 26.5.4.1 Nanocomposite Based Coatings

In this strategy, they use an array of antimicrobial metallic nano particles like silver, copper, bismuth, zinc, and **gold** in coatings to evade the bacteria from the desired surfaces. Such coatings with a cache of antibacterial nano-composite, release positively charged ions that are highly toxic even at very low concentrations. Silver coatings are exclusively used in implant or biomedical devices showing effects against both gram-positive and gram-negative [86–89]. Further several strategies reported the culmination of dopamine (PDA) with zwitterionic polymers and then finally modified to fabricated PDA-Ag binary layers, treated to catheters [90]. The binary layers are found to be effective for multiple bacterial strains i.e., *P. mirabilis*, *E. coli*, and *P. aeroginosa*. Schlaich et al conjugated dopamine with HPG (hyperbranched polyglycerol) to generate Ag-based spray coating prepared from mussel-inspired polymer (dendritic) [91,92]. Li et al used the same conjugates with copper nanoparticles (70 ppm) that showed 99.9% effectiveness for biofilms [93–96]. However, the fouling effect and its possible long-term anti-adhesive effect of the coating have not yet been studied.

### 26.5.4.2 Coating Releasing Antibiotics

Antibiotics-releasing coatings are being studied to fabricate anti-biofilm surfaces. The advantage of such modification is that it diminishes the side effects of antibiotics, also to an extent is cost effective. Bio-responsive coatings in culmination with tannic acid (TA) and antibiotics (gentamicin, polymixin B, tobramycin) with a self-assembly ability were fabricated by Zhuk et al [97–102]. Escobal et al. exclusively worked on multi-layer coated gentamicin (58%) and showed burst release with 6 hours evading 99.9% of *S, aureus* attachment in comparison to conventional gentamicin dipped surfaces. The same group also fabricated gentamicin-loaded Ti (titanium) mesoporous coatings with effective antibacterial activity and also promoting osseointegration.

### 26.5.4.3 Coating Releasing AMPs

Controlled release of AMPs were treated in nanotubes, hydrogel, and polymer coating. Here, $TiO_2$ (titanium oxide) and CaP (calcium phosphate) coatings were adsorbed to the surface through electrostatic and hydrophobic interaction. AMPs were combined with hydrogel leading to the controlled release of catalytic peptides [103–107]. Polymer lipid encapsulation is a technique developed for controlled release of AMPs like LL-37, these are otherwise coined as AMPs-PLEX i.e., polymer lipid encapsulation matrix. This strategy has shown promising effects on long-term microbial effects [108].

## 26.6 SUMMARY

Surface functionalization of biomaterials portrays encouraging paths to engineer and modify at the biomaterials-tissue interface to adapt biological reactions without affecting the bulk properties of materials. The chapter concentrates on several strategies, developed in recent times with respect to surface topography, wettability, the introduction of bioactive agents to functionalize surfaces, and the application of several anti-bacterial peptides, nano-composites, biological ligands, and great potential with releasing materials. Such approaches hold great potential with enhanced bioactivity, compatibility, and feasibility, escalating the biomaterials standards in regenerative medicine. Thus, it is anticipated that such technical quantum leaps in nano/microfabrication, bio-functionalization, synthetic chemistry, and surface characteristics will lead to the era of new-generation bioactive materials with unparalleled control on both cellular activities and healing response.

## REFERENCES

1. Clark AE, Hench LL, Paschall HA. The influence of surface chemistry on implant interface histology: a theoretical basis for implant materials selection. *J. Biomed. Mater. Res.* 1976;10:161–177.
2. Ratner BD, Bryant SJ. Biomaterials: where we have been and where we are going. *Annu Rev Biomed Eng* 2004;6:41–75.
3. Kim KH, Rautray TR, Narayanan R. *Surface modification of titanium for biomaterial applications*. Korea: Nova Sci Publ; 2010.
4. Stupp SI, Braun PV. Molecular manipulation of microstructures: biomaterials, ceramics, and semiconductors. *Science* 1997;277:1242–1248.
5. Gorbet MB, Sefton MV. Biomaterial-associated thrombosis: roles of coagulation factors, complement, platelets and leukocytes. *Biomaterials* 2004;25:5681–5703.
6. Zimmerli W, Trampuz A, Ochsner PE. Prosthetic-joint infections. *N Engl J Med* 2004;351:1645–1654.
7. Darouiche RO. Treatment of infections associated with surgical implants. *N Engl J Med* 2004;350:1422–1429.
8. Wu S, Liu X, Yeung KWK, Liu C, Yang X. Biomimetic porous scaffolds for bone tissue engineering. *Mater Sci Eng R* 2014;80:1–36.
9. Dalby MJ, Gadegaard N, Tare R, Andar A, Riehle MO, Herzyk P, et al. The control of human mesenchymal cell differentiation using nanoscale symmetry and disorder. *Nat Mater* 2007;6:997–1003.
10. Wu S, Liu X, Hu T, Chu PK, Ho JPY, Chan YL, et al. A bio mimetic hierarchical scaffold: natural growth of nanotitanates on three-dimensional microporous Ti-based metals. *Nano Lett* 2008;8:3803–3808.
11. Liu X, Chu PK, Ding C. Surface modification of titanium, titanium alloys, and related materials for biomedical applications. *Mater Sci Eng R* 2004;47:49–121.
12. Braceras I, Alava JI, Goikoetxea L, De Maeztu MA, Onate JI. Interaction of engineered surfaces with the living world: Ion implantation vs. Osseointegration. *Surface & Coatings Technology* 2007;201:8091–8098.
13. Yuan Y, Hays MP, Hardwidge PR, et al. Surface characteristics influencing bacterial adhesion to polymeric substrates. *Rsc Adv.* 2017;7(23):14254–14261.
14. Kim HY, et al. Martensitic transformation shape memory effect and superelasticity of Ti–Nb–binary alloys. *Acta Mater* 2006.
15. Lee K-W, Bae C-M, Jung J-Y, Sim G-B, Rautray TR, Lee H-J, Kwon T-Y, Kim K-H. Surface characteristics and biological studies of hydroxyapatite coating by a new method. *Journal of Biomedical Materials Research Part B: Applied Biomaterials* 2011;98:395–407.
16. Rautray TR, Narayanan R, Kwon T-Y, Kim K-H. Surface modification of titanium and titanium alloys by ion implantation. *Journal of Biomedical Materials Research Part B: Applied Biomaterials* 2010;93:581–591.

17. Curtis AS, Wilkinson CD. Reactions of cells to topography. *J Biomater Sci Polym Ed* 1998;9(12):1313e29.
18. Swain S, Rautray TR. Effect of surface roughness on titanium medical implants, In: Swain BP (Eds). *Nanostructured Materials and their Applications*, Springer; 2021. 55–80.
19. Flemming RG, Murphy CJ, Abrams GA, Goodman SL, Nealey PF. Effects of synthetic micro- and nano-structured surfaces on cell behavior. *Biomaterials* 1999;20(6):573e88.
20. Boyan BD, Hummert TW, Dean DD, Schwartz Z. Role of material surfaces in regulating bone and cartilage cell response. *Biomaterials* 1996;17(2):137e46.
21. Jansen JA, von Recum AF, Ratner BD, Hoffman AS, Schoen FJ, Lemons JE. Textured and porous materials. *Biomaterials Science: an Introduction to materials in medicine*. 2nd ed. San Diego: Academic Press; 2004. p. 218e25.
22. Rautray TR, Kim KH. Synthesis of Mg2+ incorporated hydroxyapatite by ion implantation. *Key Engineering Materials* 529, 114–118.
23. Chehroudi B, Brunette DM. Subcutaneous micro fabricated surfaces inhibit epithelial recession and promote long-term survival of percutaneous implants. *Biomaterials* 2002;23(1):229e37.
24. Flemming RG, Murphy CJ, Abrams GA, Goodman SL, Nealey PF. Effects of synthetic micro- and nano-structured surfaces on cell behavior. *Biomaterials* 1999;20(6):573e88.
25. Clark P, et al. Topographical control of cell behaviour. I. Simple step cues. *Development* 1987;99(3):439–448.
26. Clark P, et al. Topographical control of cell behaviour: II. Multiple grooved substrata. *Development* 1990;108(4):635–644.
27. Rautray TR, Kim KH. Nanoelectrochemical coatings on titanium for bioimplant applications. *Materials Technology* 2010;25(3-4):143–148.
28. Clark P, et al. Cell guidance by ultrafine topography in vitro. *J Cell Sci* 1991;99(Pt 1):73–77.
29. Dunn GA, Brown AF. Alignment of fibroblasts on grooved surfaces described by a simple geometric transformation. *J Cell Sci* 1986;83:313–340.
30. Wilkinson CDW, et al. The use of materials patterned on a nano and micro metric scale in cellular engineering. *Mat Sci Eng.* 2002;19:263–269.
31. Gallagher JO, et al. Interaction of animal cells with ordered nanotopography. *IEEE Transactions on Nanobioscience* 2002;1(1):24–28.
32. Scholten K, Meng E. Electron-beam lithography for polymer bioMEMS with submicron features. *Microsyst Nanoeng* 2016;2:16053. 10.1038/micronano.2016.53
33. Denis FA, et al. Fabrication of nanostructured polymer surfaces using colloidal lithography and spin coating. *Nanoletters* 2002;2:1419–1425.
34. Wood MA, Wilkinson CDW, Curtis ASG. The effects of colloidal nanotopography on initial fibroblast adhesion and morphology. *IEEE Trans. Nanobiosci* 2006;5(1) doi:10.1109/TNB.2005.864015.
35. Qin D, Xia Y, Whitesides GM. Soft lithography for micro-and nanoscale patterning. *Nat. Protoc.* 2010;5:491e502.
36. Xia Y, McClelland JJ, Gupta R, Qin D, Zhao XM, Sohn LL, Celotta RJ, Whitesides GM. Replica molding using polymeric materials: a practical step toward nanomanufacturing. *Adv. Mater.* 1997;9:147e149.
37. Rautray TR, Narayanan R, Kim KH. Ion implantation of titanium based biomaterials. *Prog Mater Sci* 2011;56:1137–1177.
38. King E, Xia Y, Zhao XM, Whitesides GM. Solvent-assisted microcontact molding: a convenient method for fabricating three-dimensional structures on surfaces of polymers. *Adv. Mater.* 1997;9:651e654.

39. Mohapatra B, Rautray TR. Strontium-substituted biphasic calcium phosphate scaffold for orthopedic applications. *Journal of the Korean Ceramic Society* 2020;57:392–400.

40. Xu Q, Rioux RM, Dickey MD, Whitesides GM. Nanoskiving: a new method to produce arrays of nanostructures. *Acc. Chem. Res.* 2008;41:1566e1577.

41. J, Chen F, Yang Q, Huo J, Hou X. Superoleophobic surfaces. *Chemical Society Reviews* 2017;46(14):4168–4217.

42. Chernozem RV, Guselnikova O, Surmeneva MA, Postnikov PS, Abalymov AA, Parakhonskiy BV, De Roo N, Depla D, Skirtach AG, Surmenev RA. *Appl. Mater. Today* 2020;20:100758.

43. Rautray TR, Vijayan V, Panigrahi S. Synthesis of hydroxyapatite at low temperature. *Indian Journal of Physics* 2007;81:95–98.

44. Sun L, Bian F, Wang Y, Wang Y, Zhang X, Zhao Y. Bioinspired programmable wettability arrays for droplets manipulation. *Proceedings of the National Academy of Sciences* 2020;117(9):4527–4532.

45. X, Luo H, Ma J, Wang P, Xu X, Jing G. A facile approach for fabrication of underwater superoleophobic alloy. *Applied Physics A* 2013;113:693–702.

46. Yang Y, Li X, Zheng X, Chen Z, Zhou Q, Chen Y. 3D-printed biomimetic super-hydrophobic structure for microdroplet manipulation and oil/water separation. *Advanced materials* 2018;30(9):1704912.

47. W, Zhang P, Zang R, Fan J, Wang S, Wang B, Meng J. Nacre-inspired mineralized films with high transparency and mechanically robust underwater superoleophobicity. *Advanced Materials* 2020;32(11):1907413.

48. Jafari R, Cloutier C, Allahdini A, Momen G. Recent progress and challenges with 3D printing of patterned hydrophobic and superhydrophobic surfaces. *The International Journal of Advanced Manufacturing Technology* 2019;103:1225–1238.

49. Dislaki E, Pokki J, Pané S, Sort J, Pellicer E. Fabrication of sustainable hydrophobic and oleophilic pseudo-ordered macroporous Fe–Cu films with tunable composition and pore size via electrodeposition through colloidal templates. *Applied Materials Today* 2018;12:1–8.

50. Arora JS, Cremaldi JC, Holleran MK, Ponnusamy T, He J, Pesika NS, John VT. Hydrogel inverse replicas of breath figures exhibit superoleophobicity due to patterned surface roughness. *Langmuir* 2016;32(4):1009–1017.

51. Honig F, Vermeulen S, Zadpoor AA, De Boer J, Fratila-Apachitei LE. Natural architectures for tissue engineering and regenerative medicine. *Journal of Functional Biomaterials* 2020;11(3):47.

52. Durret J, Szkutnik PD, Frolet N, Labau S, Gourgon C. Superhydrophobic polymeric films with hierarchical structures produced by nanoimprint (NIL) and plasma roughening. *Applied Surface Science* 2018;445:97–106.

53. Zhang YP, Li PP, Liu PF, Zhang WQ, Wang JC, Cui CX, … Qu LB. Fast and simple fabrication of superhydrophobic coating by polymer induced phase separation. *Nanomaterials* 2019;9(3):411.

54. Lim J-M, Yi G-R, Moon JH, Heo C-J, Yang S-M. Superhydrophobic films of electrospun fibers with multiple-scale surface morphology. *Langmuir.* 2007; 23(15):7981–7989.

55. Ma M, Mao Y, Gupta M, Gleason KK, Rutledge GC. Superhydrophobic fabrics produced by electrospinning and chemical vapor deposition. *Macromolecules.* 2005;38(23):9742–9748.

56. Radwan AB, Mohamed AMA, Abdullah AM, Al-Maadeed MA. Corrosion protection of electrospun PVDF-ZnO super hydrophobic coating. *Surface and Coatings Technology.* 2016;289:136–143.

57. Nishimoto S, Takiguchi T, Kameshima Y, Miyake M. Underwater superoleophobicity of Nb2O5 photocatalyst surface. *Chemical Physics Letters* 2019;726:34–38.
58. Swain S, Mishra S, Patra A, Praharaj R, Rautray T. Dual action of polarised zinc hydroxyapatite-guar gum composite as a next generation bone filler material. *Materials Today: Proceedings* 62:6125–6130.
59. Swain S, Patra A, Kumari S, Praharaj R, Mishra S, Rautray T. Corona poled gelatin-magnesium hydroxyapatite composite demonstrates osteogenicity. *Materials Today: Proceedings* 62:6131–6135.
60. Hoffman AS, Hubbell JA, Ratner BD, Schoen FJ, Lemons JE. *Surface-immobilized biomolecules. biomaterials science: an introduction to materials in medicine.* 2nd ed. San Diego: Academic Press; 2004. p. 225e33.
61. Behera DR, Nayak P, Rautray TR. Phosphatidylethanolamine impregnated Zn-HA coated on titanium for enhanced bone growth with antibacterial properties. *J King Saud Univ Sci* 2020;32:848–852.
62. McKay CS, Finn MG. Click chemistry in complex mixtures: bioorthogonal bioconjugation. *Chem Biol* 2014;21(9):1075e101.
63. Luk Y-Y, Kato M, Mrksich M. Self-assembled monolayers of alkanethiolates presenting mannitol groups are inert to protein adsorption and cell attachment. *Langmuir* 2000;16(24):9604e8.
64. Swain S, Padhy RN, Rautray TR. Electrically stimulated hydroxyapatite-barium titanate composites demonstrate immunocompatibility in vitro. *J Korean Ceram Soc* 2020;57:495–502.
65. Ruegsegger MA, Marchant RE. Reduced protein adsorption and platelet adhesion by controlled variation of oligomaltose surfactant polymer coatings. *J Biomed Mater Res* 2001;56(2):159e67.
66. Mishra S, Rautray TR. Silver-incorporated hydroxyapatite–albumin microspheres with bactericidal effects. *Journal of the Korean Ceramic Society* 2020;57:175–183.
67. Rautray TR, Kim KH. Synthesis of silver incorporated hydroxyapatite under magnetic field. *Key Engineering Materials* 2012;493:181–185.
68. Ran B, Jing C, Yang C, Li X, Li Y. Synthesis of efficient bacterial adhesion-resistant coatings by one-step polydopamine-assisted deposition of branched polyethylenimine-g-poly (sulfobetaine methacrylate) copolymers. *Applied Surface Science* 2018;450:77–84.
69. Swain S, Rautray TR. Assessment of Polarized Piezoelectric SrBi4Ti4O15 Nanoparticles as an alternative antibacterial agent. 10.1101/2021.01.02.425094
70. Zhang C, Liu S, Tan L, Zhu H, Wang Y. Star-shaped poly (2-methyl-2-oxazoline)-based films: rapid preparation and effects of polymer architecture on antifouling properties. *Journal of Materials Chemistry B* 2015;3(27):5615–5628.
71. Qiu W-Z, Yang H-C, Xu Z-K. Dopamine-assisted co-deposition: An emerging and promising strategy for surface modification. *Advances in Colloid and Interface Science* 2018;256:111–125.
72. Praharaj R, Mishra S, Rautray TR. Growth mechanism of aligned porous oxide layers on titanium by anodization in electrolyte containing Cl. *Materials Today: Proceedings* 2022;62:6216–6220.
73. Swain S, Kumari S, Swain P, Rautray T. Polarised strontium hydroxyapatite–xanthan gum composite exhibits osteogenicity in vitro. *Materials Today: Proceedings* 62:6143–6147.
74. Yong Y, Qiao M, Chiu A, Fuchs S, Liu Q, Pardo Y, Worobo R, Liu Z, Ma M. *Langmuir* 2019;35:1927.
75. Swain S, Bowen C, Rautray TR. Dual response of osteoblast activity and antibacterial properties of polarized strontium substituted hydroxyapatite-Barium strontioum titanate composites with controlled strontium substitution. *J Biomed Mater Res Part A* 2021;109:2027–2035.

76. Swain S, Padhy RN, Rautray TR. Polarized piezoelectric bioceramics composites exhibit antibacterial activity. *Mater. Chem. Phys.* 2020;239:122002.

77. Yuan K, Wang CY, Zhu LY, Cao Q, Yang JH, Li XX, ...Zhang DW. Fabrication of a micro-electromechanical system-based acetone gas sensor using CeO2 nanodot-decorated WO3 nanowires. *ACS applied materials & interfaces.* 2020; 12(12):14095–14104.

78. Costa FM, Maia SR, Gomes PA, Martins MCL. Dhvar5 antimicrobial peptide (AMP) chemoselective covalent immobilization results on higher antiadherence effect than simple physical adsorption. *Biomaterials* 2015;52:531–538.

79. Swain S, Ong JL, Narayanan R, Rautray TR. Ti-9Mn β-type alloy exhibits better osteogenicity than Ti-15Mn alloy in vitro. *J Biomed Mater Res B* 2021; 109:2154–2161.

80. Godoy-Gallardo M, Mas-Moruno C, Yu K, Manero JM, Gil FJ, Kizhakkedathu JN, Rodriguez D. Antibacterial properties of hLf1–11 peptide onto titanium surfaces: A comparison study between silanization and surface initiated polymerization. *Biomacromolecules* 2015;16(2):483–496.

81. Praharaj R, Mishra S, Misra RDK, Rautray TR. Biocompatibility and adhesion response of magnesium-hydroxyapatite/strontium-titania (Mg-HAp/ Sr-TiO2) bilayer coating on titanium *Materials Technology*, 1–10.

82. Ahmadabadi HY, Yu K, Kizhakkedathu JN. Surface modification approaches for prevention of implant associated infections. *Colloids and Surfaces B: Biointerfaces* 2020;193:111116.

83. Bhadra CM, Khanh Truong V, Pham VT, Al Kobaisi M, Seniutinas G, Wang JY, Juodkazis S, Crawford RJ, Ivanova EP, *Sci. Rep.* 2015;5:1.

84. Swain S, Rautray TR. Estimation of trace elements, antioxidants, and antibacterial agents of regularly consumed Indian medicinal plants. *Biological Trace Element Research*, 2021;199:1185–1193.

85. Hasan J, Raj S, Yadav L, Chatterjee K. Engineering a nanostructured "super surface" with superhydrophobic and superkilling properties. *RSC Adv* 2015; 5:44953–44959.

86. Marambio-Jones C, Hoek EM. A review of the antibacterial effects of silver nanomaterials and potential implications for human health and the environment. *Journal of nanoparticle research* 2010;12:1531–1551.

87. Swain S, Rautray TR. Silver doped hydroxyapatite coatings by sacrificial anode deposition under magnetic field. *Journal of Materials Science: Materials in Medicine* 2017;28:1–5.

88. Mohapatra B, Rautray TR. Facile fabrication of Luffa cylindrica-assisted 3D hydroxyapatite scaffolds. *Bioinspired, Biomimetic and Nanobiomaterials* 2021; 10:37–44.

89. Satpathy A, Mohanty R, Rautray TR. Bio-mimicked guided tissue regeneration/ guided bone regeneration membranes with hierarchical structured surfaces replicated from teak leaf exhibits enhanced bioactivity. *Journal of Biomedical Materials Research Part B: Applied Biomaterials* 110 (1).

90. Wang R, Neoh KG, Kang ET, Tambyah PA, Chiong E. Antifouling coating with controllable and sustained silver release for long-term inhibition of infection and encrustation in urinary catheters. *Journal of Biomedical Materials Research Part B: Applied Biomaterials* 2015;103(3):519–528.

91. Schlaich C, Li M, Cheng C, Donskyi IS, Yu L, Song G, Osorio E, Wei Q, Haag R. Advanced Mater. *Interfaces* 2018;5:1701254.

92. Mohanty S, et al. An investigation on the antibacterial, cytotoxic, and antibiofilm efficacy of starch-stabilized silver nanoparticles. *Nanomedicine: Nanotechnology, Biology, and Medicine* 2011;8(6):916–924.

93. Li M, Gao L, Schlaich C, Zhang J, Donskyi IS, Yu G, ...Ma N. Construction of functional coatings with durable and broad-spectrum antibacterial potential based on mussel-inspired dendritic polyglycerol and in situ-formed copper nanoparticles. *ACS applied materials & interfaces* 2017;9(40):35411–35418.
94. Priyadarshini I, Swain S, Koduru JR, Rautray TR. Electrically polarized withaferin A and alginate-incorporated biphasic calcium phosphate microspheres exhibit osteogenicity and antibacterial activity in vitro. *Molecules* 28 (1):86.
95. Rautray TR, Mohapatra B, Kim K-H. Fabrication of strontium–hydroxyapatite scaffolds for biomedical applications. *Advanced Science Letters* 2014;20: 879–881.
96. Praharaj R, Mishra S, Rautray TR. The structural and bioactive behaviour of strontium-doped titanium dioxide nanorods. *Journal of the Korean Ceramic Society* 2020;57:271–280.
97. Zhuk I, Jariwala F, Attygalle AB, Wu Y, Libera MR, Sukhishvili SA. Self-defensive layer-by-layer films with bacteria-triggered antibiotic release. *ACS nano* 2014; 8(8):7733–7745.
98. Swain S, Koduru JR, Rautray TR. Mangiferin-enriched mn–hydroxyapatite coupled with β-TCP scaffolds simultaneously exhibit osteogenicity and anti-bacterial efficacy. *Materials* 16(6), 2206.
99. Swain S, Kwon TY, Rautray TR. Fabrication of silver doped nano hydroxyapatite-carrageenan hydrogels for articular cartilage applications, 10.1101/2020.12.31.424664.
100. Rautray TR, Kim KH. Synthesis of controlled release sr-hydroxyapatite microspheres. *Bioceramics* 2012;24.
101. Mishra S, Rautray TR. Fabrication of Xanthan gum-assisted hydroxyapatite microspheres for bone regeneration. *Materials Technology* 2020;35:364–371.
102. Rautray TR, Kim KH. Synthesis of Mg2+ incorporated hydroxyapatite by ion implantation and their cell response. *Med Oral Patol Oral Cir Bucal* 2012;17:S276.
103. Escobar A, Muzzio N, Coy E, Liu H, Bindini E, Andreozzi P, Wang G, Angelomé P, Delcea M, Grzelczak M, Moya SE. Antibacterial layer-by-layer films of poly(acrylic acid)–gentamicin complexes with a combined burst and sustainable release of gentamicin. *Adv. Mater. Interfaces* 2019;6:1.
104. Swain S, Misra RDK, You CK, Rautray TR. TiO2 nanotubes synthesised on Ti-6Al-4V ELI exhibits enhanced osteogenic activity: A potential next-generation material to be used as medical implants. *Mater Tech* 2021;36:393–399.
105. Cheng H, Yue K, Kazemzadeh-Narbat M, Liu Y, Khalilpour A, Li B, Zhang YS, Annabi N, Khademhosseini A. Mussel-inspired multifunctional hydrogel coating for prevention of infections and enhanced osteogenesis. *ACS Appl. Mater. Interfaces* 2017;9:11428.
106. Swain S, Rautray TR, Narayanan R. Sr, Mg, and Co substituted Hydroxyapatite coating on TiO2 nanotubes by electrochemical methods. *Advanced Science Letters* 2016;22:482–487.
107. Rautray TR, Swain S, Kim K-H. Formation of Anodic TiO2 Nanotubes Under Magnetic Field. *Advanced Science Letters* 2014;20:801–803.
108. Riool M, de Breij A, de Boer L, Kwakman PH, Cordfunke RA, Cohen O, ...Zaat SA. Controlled Release of LL-37-Derived Synthetic Antimicrobial and Anti-Biofilm Peptides SAAP-145 and SAAP-276 Prevents Experimental Biomaterial-Associated Staphylococcus aureus Infection. *Advanced Functional Materials* 2017;27(20):1606623.

# Index

Pages in *italics* refer to figures and pages in **bold** refer to tables.